Semiconductor Nanocrystals and Metal Nanoparticles

Physical Properties and Device Applications

Advances in Materials Science and Engineering

Series Editor
Sam Zhang

Semiconductor Nanocrystals and Metal Nanoparticles

Physical Properties and Device Applications

Edited by
Tupei Chen • Yang Liu

CRC Press
Taylor & Francis Group
Boca Raton London New York

CRC Press is an imprint of the
Taylor & Francis Group, an **informa** business

MATLAB® is a trademark of The MathWorks, Inc. and is used with permission. The MathWorks does not warrant the accuracy of the text or exercises in this book. This book's use or discussion of MATLAB® software or related products does not constitute endorsement or sponsorship by The MathWorks of a particular pedagogical approach or particular use of the MATLAB® software.

CRC Press
Taylor & Francis Group
6000 Broken Sound Parkway NW, Suite 300
Boca Raton, FL 33487-2742

First issued in paperback 2019

© 2017 by Taylor & Francis Group, LLC
CRC Press is an imprint of Taylor & Francis Group, an Informa business

No claim to original U.S. Government works

ISBN-13: 978-1-4398-7830-9 (hbk)
ISBN-13: 978-0-367-86662-4 (pbk)

Library of Congress Cataloging-in-Publication Data

Names: Chen, Tupei, editor. | Liu, Yang, 1975 September 12- editor.
Title: Semiconductor nanocrystals and metal nanoparticles : physical properties and device applications / editors, Tupei Chen and Yang Liu.
Description: Boca Raton : Taylor & Francis, a CRC title, part of the Taylor & Francis imprint, a member of the Taylor & Francis Group, the academic division of T&F Informa, plc, [2017] | Series: Advances in materials science and engineering | Includes bibliographical references and index.
Identifiers: LCCN 2016009756 | ISBN 9781439878309 (alk. paper)
Subjects: LCSH: Nanoelectronics--Materials. | Semiconductor nanocrystals. | Semiconductor nanoparticles.
Classification: LCC TK7874.84 .S46 2017 | DDC 621.3815/2--dc23
LC record available at https://lccn.loc.gov/2016009756

Visit the Taylor & Francis Web site at
http://www.taylorandfrancis.com

and the CRC Press Web site at
http://www.crcpress.com

Contents

Preface

A material particle or cluster having a size in the range of one to several hundreds of nanometers is often referred to as "nanocrystal" or "nanoparticle." *Nanocrystal* is considered as crystalline clusters in either a single crystalline or a polycrystalline arrangement, while *nanoparticle* can represent both crystalline and noncrystalline clusters. Semiconductor nanocrystals with small dimensions are also described as quantum dots. Semiconductor nanocrystals and metal nanoparticles exhibit fascinating behaviors and unique physical properties as a result of quantum size effect. For example, for a semiconductor nanocrystal whose size is smaller than twice the size of its exciton Bohr radius, the excitons are squeezed, leading to quantum confinement. Quantum confinement will modify the electronic and optical properties of the semiconductor nanocrystal, for example, the bandgap of the nanocrystal increases with the decrease of the nanocrystal size. A direct experimental observation of the quantum confinement effect is that different sized quantum dots emit different color light. A good example of the fascinating behaviors of metal nanoparticles is the localized surface plasmon resonance (LSPR) in noble metal (for example, silver and gold) nanoparticles where the conduction electrons oscillate coherently under irradiation by light in the visible and near-infrared regions of the electromagnetic spectrum. The wavelength of excitation of the LSPR shows strong dependence on the size, shape, and dielectric environment of the metal nanoparticles.

Semiconductor nanocrystals and metal nanoparticles are the building blocks of the next generation of electronic, optoelectronic, and photonic devices. For instance, semiconductor quantum dots are particularly significant for optoelectronic/photonic device applications such as photovoltaic devices, light-emitting devices, and photodetector devices due to their tunable absorption spectrum and high extinction coefficient. In electronic applications, semiconductor nanocrystals or metal nanoparticles with sizes smaller than ~5 nm can be used to realize single-electron or few-electron devices (e.g., transistors and memory devices) as such tiny nanostructures exhibit the Coulomb blockade effect at room temperature.

This book examines in detail the physical properties and device applications of semiconductor nanocrystals and metal nanoparticles. It begins by giving a review on the synthesis and characterization of various semiconductor nanocrystals and metal nanoparticles and goes on to discuss in detail their electronic, optical, and electrical properties. Based on the knowledge of the physical properties, the book illustrates some exciting applications in nanoelectronic devices (e.g., memristors, single-electron devices), optoelectronic devices (e.g., UV detectors, quantum dot lasers, solar cells), and other applications (e.g., gas sensors, metallic nanopastes for power electronics packaging).

This book contains 13 chapters. It covers a rapidly developing, interdisciplinary field, and it is also a compilation of some research efforts in the field. It is intended as a reference book for senior undergraduate and graduate students and researchers in microelectronics/nanoelectronics, optoelectronics, photonics, nanoscience and nanotechnology, physics, and materials science.

Tupei Chen
Nanyang Technological University, Singapore
Yang Liu
University of Electronic Science and Technology of China, China

MATLAB® is a registered trademark of The MathWorks, Inc. For product information, please contact the following:

The MathWorks, Inc.
3 Apple Hill Drive
Natick, MA 01760-2098 USA
Tel: 508-647-7000
Fax: 508-647-7001
E-mail: info@mathworks.com
Web: www.mathworks.com

Editors

Tupei Chen received his PhD from the University of Hong Kong, Hong Kong, in 1994. He is currently an associate professor in the School of Electrical and Electronic Engineering, Nanyang Technological University, Singapore. His research areas include electronic and optoelectronic/photonic applications of nanoscale materials, novel memory devices, memristors and applications in artificial neural networks, Si photonics, and metal oxide thin films and applications in flexible/transparent devices. He is the author or coauthor of more than 250 peer-reviewed journal papers, more than 120 conference presentations, and 6 book chapters. He has filed several U.S. patents. He has supervised more than 20 PhD students and over 10 postdoctoral research staff.

Yang Liu received his BSc in microelectronics from Jilin University, China, in 1998 and his PhD from Nanyang Technological University, Singapore, in 2005. From May 2005 to July 2006, he was a research fellow at Nanyang Technological University. In 2006, he was awarded the prestigious Singapore Millennium Foundation Fellowship. In 2008, he joined the School of Microelectronics, University of Electronic Science and Technology, China, as a full professor. He is the author or coauthor of over 120 peer-reviewed journal papers and more than 80 conference papers. He has filed one U.S. patent and more than 30 Chinese patents. His current research includes advanced memory devices and their applications in artificial neural networks, radio frequency integrated circuits (RFICs), photonic/optoelectronic devices, and integrated circuits (ICs).

Contributors

M. Abdullah
Nassiriya Nanotechnology Research
  Laboratory
Thi-Qar University
Nassiriyah, Iraq

Amin H. Al-Khursan
Nassiriya Nanotechnology Research
  Laboratory
Thi-Qar University
Nassiriyah, Iraq

A.K. Arof
Centre for Ionics
Department of Physics
Faculty of Science
University of Malaya
Kuala Lumpur, Malaysia

Kit Au
Department of Applied Physics
The Hong Kong Polytechnic University
Kowloon, Hong Kong, People's Republic of
  China

Denzel Bridges
Department of Mechanical, Aerospace,
  and Biomedical Engineering
The University of Tennessee, Knoxville
Knoxville, Tennessee

M.H. Buraidah
Centre for Ionics
Department of Physics
Faculty of Science
University of Malaya
Kuala Lumpur, Malaysia

Ngai Yui Chan
Department of Applied Physics
The Hong Kong Polytechnic University
Kowloon, Hong Kong, People's Republic of
  China

Tupei Chen
School of Electrical and Electronic
  Engineering
Nanyang Technological University
Singapore, Singapore

Jiyan Dai
Department of Applied Physics
The Hong Kong Polytechnic University
Kowloon, Hong Kong, People's Republic of
  China

Jacky Even
FOTON Laboratory
National Institute for Applied Sciences
Rennes, France

Zhili Feng
Materials Science and Technology Division
Oak Ridge National Laboratory
Oak Ridge, Tennessee

P. Gangopadhyay
Materials Science Group
Indira Gandhi Centre for Atomic Research
Kalpakkam, Tamil Nadu, India

Zhiming Gao
National Transportation Research Center
Oak Ridge National Laboratory
Knoxville, Tennessee

Spiros Gardelis
Department of Solid State Physics
Faculty of Physics
National and Kapodistrian University of Athens
Athens, Greece

N. Gogurla
Department of Physics
Indian Institute of Technology
Kharagpur, West Bengal, India

Frédéric Grillot
Laboratory for Communication and Processing
  of Information
Telecom Paris Tech—Université Paris-Saclay
Paris, France

S. Hamad
School of Physics
University of Hyderabad
Hyderabad, Telangana, India

Anming Hu
Department of Mechanical, Aerospace, and
   Biomedical Engineering
The University of Tennessee, Knoxville
Knoxville, Tennessee

Ning-Cheng Lee
Indium Corporation
Clinton, New York

Helen La Wa Chan
Department of Applied Physics
The Hong Kong Polytechnic University
Kowloon, Hong Kong, People's Republic of
   China

Ruozhou Li
School of Electronic Science and
   Engineering
Southeast University
Nanjing, Jiangsu, People's Republic of China

and

Suzhou Key Laboratory of Metal Nano-
   Optoelectronic Technology
Suzhou Research Institute of Southeast
   University
Suzhou, Jiangsu, People's Republic of China

Yang Liu
State Key Laboratory of Electronic Thin Films
   and Integrated Devices
University of Electronic Science and
   Technology of China
Chengdu, Sichuan, People's Republic of
   China

Z. Liu
School of Materials and Energy
Guangdong University of Technology
Guangzhou, Guangdong, People's Republic of
   China

Wing Chong Lo
Department of Applied Physics
The Hong Kong Polytechnic University
Kowloon, Hong Kong, People's Republic of
   China

Farah T. Mohammed Noori
College of Science
Baghdad University
Baghdad, Iraq

I.M. Noor
Centre for Ionics
Department of Physics
Faculty of Science
University of Malaya
Kuala Lumpur, Malaysia

G. Krishna Podagatlapalli
Advanced Centre of Research in High Energy
   Materials
University of Hyderabad
Hyderabad, Telangana, India

T. Rakshit
Advanced Technology Development Centre
Indian Institute of Technology
Kharagpur, West Bengal, India

S.K. Ray
Department of Physics
Indian Institute of Technology
Kharagpur, West Bengal, India

Leon M. Tolbert
Department of Electrical Engineering and
   Computer Science
The University of Tennessee, Knoxville
Knoxville, Tennessee

S. Venugopal Rao
Advanced Centre of Research in High Energy
   Materials
University of Hyderabad
Hyderabad, Telangana, India

Cheng Wang
School of Information Science and Technology
ShanghaiTech University
Pudong, Shanghai, People's Republic of China

and

FOTON Laboratory
National Institute for Applied Sciences
Rennes, France

Zhiqiang Wang
Department of Electrical Engineering and
    Computer Science
The University of Tennessee, Knoxville
Knoxville, Tennessee

Zhiyong Wang
Institute of Laser Engineering
Beijing Institute of Technology
Chaoyang, Beijing, People's Republic of China

S.N.F. Yusuf
Centre for Ionics
Department of Physics
Faculty of Science
University of Malaya
Kuala Lumpur, Malaysia

Fan Zhang
Department of Applied Physics
The Hong Kong Polytechnic University
Kowloon, Hong Kong, People's Republic of
    China

1 Synthesis and Characterization of Nanocrystals and Nanoparticles

Yang Liu

CONTENTS

1.1 INTRODUCTION

Nanocrystals and nanoparticles are described as clusters having diameters in the range of one to several hundreds of nanometers. A "nanocrystal" is considered as a cluster of crystals, whereas a "nanoparticle" can consist of both crystalline and noncrystalline clusters. The synthesis of nanocrystals/nanoparticles has attracted extensive studies in the past several decades. These materials show size-dependent physical properties that are not shown by their bulk counterparts. For example, a Si nanocrystal shows photoluminescence (PL) and electroluminescence (EL), which cannot be achieved in bulk silicon due to its indirect bandgap. Therefore, precise control during the synthesis of nanocrystals/nanoparticles is of key importance for their applications.

Many techniques are used to synthesize nanocrystals/nanoparticles. Chemical vapor deposition (CVD), physical vapor deposition, ion implantation, sputtering, laser ablation, and spray pyrolysis are all gas-based techniques and have been used successfully for the synthesis of nanocrystals/nanoparticles. They provide the possibility of controlling the nanocrystal/nanoparticle size by adjusting the synthesis parameters such as temperature, pressure, power density, gas flow rate, and so on. Solution-based techniques are of intensive research interest, as they are highly effective in synthesizing nanocrystals/nanoparticles with good size control. Besides, they also have the advantages of low synthesis temperature, flexibility, and low cost. A large variety of methods have been developed to synthesize nanocrystals/nanoparticles, such as reduction, thermal decomposition, and processes with hydrolysis and alcoholysis. At the same time, various techniques have been utilized to characterize the structural and physical properties of nanocrystals/nanoparticles, such as transmission electron microscopy (TEM), scanning electron microscopy (SEM), atomic force microscopy (AFM), X-ray photoemission spectroscopy (XPS), X-ray diffraction (XRD), Raman scattering, and so on. In this chapter, we review the methods for synthesizing nanocrystals/nanoparticles. We also discuss the characterization of nanocrystals/nanoparticles using various techniques.

1.2 SEMICONDUCTOR NANOCRYSTALS

1.2.1 COMPOUND SEMICONDUCTOR NANOCRYSTALS

Compound semiconductor nanocrystals such as CdSe, CdS, GaAs, InAs, and CuCl have been widely investigated. In this section, we discuss the synthesis of compound semiconductor nanocrystals as well as their characterization using various techniques.

Synthesis of II–VI (A2B6) nanocrystals/nanoparticles such as CdS and CdSe is attractive because these materials have potential applications in photosensitive, photovoltaic, optical, and electronic devices. The shape, concentration, and size of the nanocrystals, as well as their size distribution, strongly depends on the synthesis process. CdSe nanoparticles showing quantum size effects was reported in 1993 [1]. Research on nanocrystalline cadmium selenide (CdSe) has been going on for years due to its excellent properties such as light emission, low bandgap (E_g = 1.75 eV), low toxicity, facile surface modification, and thermostability, with applications in light-emitting diodes, solar cells, hydrogen-producing catalysts, and biological imaging [2–15].

Chemical methods are simple, flexible, effective, and low cost to synthesize isolated CdSe nanocrystals, and include solvothermal methods and hydrothermal methods. The solvothermal methods always utilize TOP (tri-*n*-octylphosphine)/TOPO (tri-*n*-octylphosphine oxide), TOP/TOPO/HPA (hexylphosphonic), TOP/TOPO/HDA (hexadecylam), or other reagents to synthesize CdSe nanocrystals [16–23].

The TOP/TOPO method was reported in [16,20] and is illustrated in Figure 1.1. In [16], the synthesis of ZnS-capped CdSe nanocrystals was carried out by using TOP/TOPO with a size of 2.7–3.0 nm. In [20], CdSe nanocrystals were synthesized with a diameter of ~3 nm with TOP/TOPO. The nanocrystals were dried under N_2 flow and dissolved in toluene. Finally, the nanocrystal solution was filtered using a 0.45 mm syringe filter to remove other contaminations from the synthesis process.

The TOP/TOPO/HPA method was used in [17], where $Cd(CH_3)_2$ in heptane and Se/TBP (tri-*n*-butyphosphine) were mixed with TBP and the solution was injected into HPA (hexylphosphonic acid) in TOPO. The size of SeCd nanocrystals could be controlled in the range 5–20 nm by varying the injection volume. The TOP/TOPO/HDA method was reported in [19,21]. In [19], $Cd(CH_3)_2$ and Se/TOP were injected into TOPO and HDA. The size of prepared CdSe nanocrystals was in the range 4.5–5.0 nm. In [21], CdO/SA (stock solution of stearic acid with CdO) was injected into

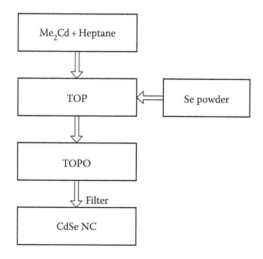

FIGURE 1.1 Process of the TOP/TOPO method.

TOPO and HDA, and the mixed solution was rapidly injected into Se/TOP. Finally, the mixture was purified by centrifugation. The size of CdSe nanocrystals was distributed in the range 2.6–5.9 nm.

In [18], the different solvent reactions for synthesizing CdSe nanocrystals were discussed. The Cd precursor was selected from $Cd(Ac)_2$, $CdCO_3$, or CdO, and the solvent from TOPO, SA/TOPO, SA, Tech TOPO, HPA/TOPO, or DA (dodecylamine)/TOPO. A Se solution (Se powder dissolved in TOP or TBP) was injected into the reaction vessel with the mixture of the above chemicals. CdSe nanocrystals of 4–25 nm were finally obtained. In [22], Se powder and cadmium myristate were added to a flask containing ODE (octadecene). Under a flow of Ar, the solution was stirred and heated to 240°C at the rate of 25 K min^{-1}. After the particle diameter reached ~3.0 nm, a solution of oleic acid in ODE was added to stabilize the growth of the nanocrystals.

There are several reports on hydrothermal methods for the synthesis of CdSe nanocrystals [9–13]. In [24], a solution of $Cd(ClO_4)_2 \cdot 6H_2O$ and the stabilizer in deionized (DI) water was adjusted to a pH of 11.2 using NaOH. The solution was mixed with the oxygen-free NaHSe solution. The size of CdSe nanocrystals obtained by this method ranged from 1.5 to 2.8 nm. In [27], a solution of $Cd(NO_3)_2$ was dissolved in distilled water containing trisodium citrate as the stabilizer. H_2Se gas was slowly passed through the above solution together with N_2 as buffer gas. The mean size of the nanocrystals was ~45 nm in the CdSe sol. In [26], an aqueous solutions was used for the synthesis of CdSe nanocrystals by reacting a solution $CdCl_2$ and Na_2SeSO_3 in gelatine solution in distilled water based on the following reaction:

$$Cd^{2+} + SeSO_3^{2+} + 2OH \rightarrow CdSe + SO_4^{2+} + H_2O$$

The size of the crystals could be controlled in the range 2.5–3.0 nm.

The synthesis of the MPA-capped CdSe nanocrystals was reported in [28]. $NaHSeO_3$ and $NaBH_4$ were injected into mixed solution of $CdCl_2 \cdot 2.5H_2O$, 3-MPA, and NaOH. The MPA-capped CdSe nanocrystals with an average size of 4–5 nm were finally obtained.

TEM, AFM, XRD, and XPS are often used to characterize nanocrystals/nanoparticles. TEM is widely used to characterize the size and nanostructure of CdSe nanocrystals/nanoparticles [16–19,21,22,24,25,28]. In [17], the TEM images of rod-, arrow-, teardrop-, and tetrapod-shaped CdSe Nanocrystals were presented. In [19], the TEM images of CdSe nanocrystals and CdSe/ZnS core–shell nanoparticles covered with monolayers of ZnS were shown. In [24], a TEM image of thioglycerol-stabilized CdSe nanocrystals was presented together with a single CdSe nanocrystal, and [25] provided TEM images of CdSe particles in different shapes. In [28], the TEM image showed the existence of MPA-CdSe nanocrystals obtained after 7 h of refluxing. In [18], by selecting an appropriate solvent system, CdSe nanocrystals in a very broad size range could be obtained from ~2 to >25 nm, as shown in Figure 1.2.

XRD is a very useful technique to obtain the material composition, internal information such as atomic or molecular structure, or form by the analysis of the X-ray diffraction pattern. The diameter can be determined by calculating the FWHM (full-width at half-maximum) of the XRD peaks with the Debye–Scherrer formula:

$$D_{hkl} = \frac{k\lambda}{\beta \cos\theta} \tag{1.1}$$

where

D_{hkl} is the crystallite diameter along the crystal direction perpendicular to the (*hkl*) plane
k is the Scherrer constant, which is usually 0.89
λ is the wavelength of X-rays
β is the Bragg diffraction angle
θ is the FWHM of the prominent XRD peak

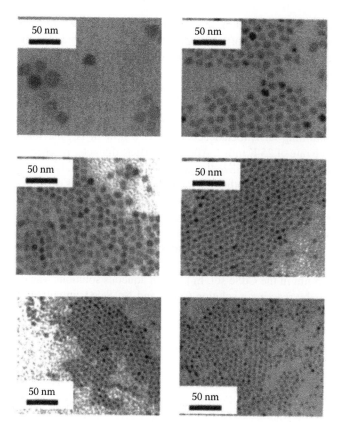

FIGURE 1.2   TEM images of wurtzite CdSe nanocrystals of different sizes synthesized in different solvent systems. (From Qu, L.H. et al., *Nano Lett.*, 1, 333, 2001.)

XRD has been widely used to characterize CdSe nanocrystals/nanoparticles size and crystal orientation [18,22,24,25,27,28]. In [18], CdSe nanocrystal of ~6 nm was demonstrated with high crystalline characteristics. In [22], the XRD patterns of zinc blende (ZB) CdSe nanocrystals with a diameter of 4.0 nm were provided. In [24], the authors presented the diffraction patterns of CdSe crystals of wurzite hexagonal and zinc blende cubic structures capped with different stabilizers. In [25], the relationship between CdSe nanocrystal structure and different reacting agents was demonstrated at different reaction temperatures through their XRD patterns. In [27], the XRD pattern of CdSe nanocrystals (zinc blende) grown by a reaction in an aqueous medium was shown. In [28], XRD patterns of CdSe nanocrystals after different refluxing times were presented. In [22], a typical XRD pattern of ZB CdSe nanocrystals with a diameter of 4.0 nm was illustrated, as shown in Figure 1.3, where the inset shows TEM image of identical size.

Many methods for preparing CdS nanocrystals have been reported, mainly including aqueous-phase synthesis [29–37], colloid technique [38–40], microemulsion by means of ultrasonic irradiation [41,42], ion implantation method [43], and precipitation [44–46]. Aqueous-phase synthesis was described in [29–37]. In [29], CdS nanocrystals were fabricated by in situ sulfuration of Cd^{2+} by using some organic and inorganic molecules containing sulfur with average size from 4 nm to several tens of nanometers. In [31], CdS nanocrystals of 3.2–6.0 nm were synthesized based on a mixture of CdM_2, oleic acid, toluene, and an aqueous solution of thiourea for different Cd/S molar ratios by the seeding-growth technique at temperatures of 180°C, 150°C, or 120°C. In [32], CdS nanocrystals of 20 nm were synthesized with a solution of thioacetamide (CH_3CSNH_2) and $CdCl_2$ mixed with sodium citrate in different concentrations. In [33], a $CdCl_2$ solution and a GSH

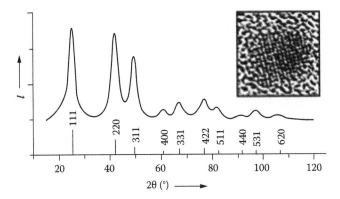

FIGURE 1.3 X-ray diffraction pattern of zinc blende CdSe nanocrystals with a diameter of 4.0 nm. The diffraction peak positions and relative intensities of bulk ZB CdSe are indicated. The inset shows a high-resolution TEM image of a CdSe nanocrystal with a size of ~4 nm. (From Yang, Y.A. et al., *Angew. Chem.*, 117, 6870, 2005.)

(L-glutathione) solution with an appropriate pH value were mixed with thiourea. CdS nanocrystals with an average diameter of 3.0 ± 0.3 nm were obtained by adding acetone followed by centrifuging and drying the solution. In [34], CdS nanoparticles with diameters below 30 nm were prepared by gradually adding $Cd(NO_3)_2 \cdot 4H_2O$ solution into the copolymer DADMAC/NVP (*N,N*-dimethyl *N,N*-diallyl ammonium chloride)/(*N*-vinyl pyrrolidinone) solution containing $Na_2S \cdot H_2O$. In [35], CdS nanocrystals with an average size of 3.1 nm were synthesized by using thioglycolic acid (TGA) as stabilizer in an aqueous solution containing $CdAc_2 \cdot 2H_2O$, Na_2S, and NaOH. Finally, ethanol was added and the solution was dried in vacuum. In [36], CdS nanocrystals of ~90 nm were synthesized from a mixture of $Cd(Ac)_2$ and Na_2S solutions.

A colloid synthesis technique was reported in [38–40]. In [38], CdS nanocrystals of 1–2 nm were prepared by a mixture of cadmium thioglicerate and sodium sulfide in dimethylformamide (DMF). In [39], CdS nanocrystals of 3–5 nm were prepared with the block copolymer PEG-*b*-poly (2-*N,N*-dimethylaminoerthyl methacrylate) (PAMA), Na_2S, and $CdCl_2$ solution. In [40], cadmium acetate hydrate, myristic acid, and octadecene (ODE) solutions with sulfur and the nucleation initiators (tetraethylthiuram disulfides [I1] and 2,2′-dithiobisbenzothiazole [I2]) were added into ODE. CdS nanocrystals of ~3.8 nm were obtained by heating the resulting solution at 240°C in an argon ambient.

In [42], a mixture of ethylendiamine and CS_2 solutions in oil phase was irradiated with ultrasound and was mixed with a solution of cadmium chloride under sonication. The resulting solution was centrifuged, washed with distilled water and with ethanol, and finally dried in air at room temperature. The CdS nanocrystal synthesized with this method was ~4 nm in size. In [45], CdS nanocrystals were prepared via a co-precipitation method with $Cd(NO_3)_2 \cdot 4H_2O$, cetyltrimethyl ammonium bromide (CTAB), and Na_2S solution. The size of nanocrystal could be varied, by controlling the heating temperature, from several nanometers to several hundred nanometers. In [46], CdS nanocrystals of 3–4 nm were synthesized by a wet chemical precipitation method by using cadmium sulfate ($CdSO_4$), sodium sulfide (Na_2S), and thio-glycerol.

TEM has been was extensively used to characterize CdS nanocrystals/nanoparticles [30–34,37, 40–42,45,46]. In [30], the TEM images of CdS:Mn nanocrystals without a ZnS shell and with the shell were presented. In [31], a TEM image of CdS nanocrystals synthesized by the seeding-growth technique was shown, while [32] showed images of CdS nanocrystals synthesized with different concentrations of sodium citrate. In [33], the TEM image of the as-prepared CdS monodisperse nanocrystals with an average diameter of 3.0 ± 0.3 nm was presented. In [34], the TEM images showed that CdS nanocrystals were in the range 50–70 nm obtained with different stabilizing copolymers. In [41], TEM images of the CdS nanocrystals synthesized with the ultrasound technique were presented. Figure 1.4 shows the TEM images of CdS nanocrystals of <10 nm.

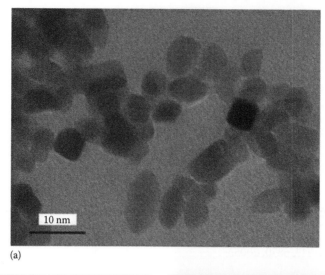

(a)

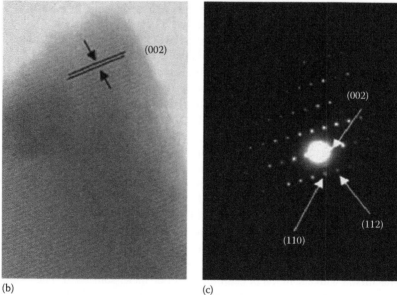

(b)   (c)

FIGURE 1.4 (a) TEM image of CdS nanocrystals, (b) HRTEM image, and (c) SAED of the CdS nanocrystals. (From Ghows, N. and Entezari, M.H., *Ultrasonics Sonochem.*, 18, 269, 2011.)

Figure 1.4b presents the TEM image of a single CdS nanocrystal in the direction (002) with a lattice spacing of 0.34 nm. The crystal structure and crystallinity of the nanocrystal are also shown in the selected area diffraction (SAED) pattern in Figure 1.4c.

XRD patterns of CdS nanocrystals were studied in [30–35,38,40–42,45,46]. In [30], the XRD patterns of CdS:Mn/ZnS core/shell nanocrystals synthesized by varying the water-to-surfactant molar ratios were presented, while in [32] the XRD patterns of CdS nanocrystals synthesized with 2.0 mM sodium citrate were shown. In [33], the XRD patterns exhibited a cubic zinc blende crystal phase of crystalline CdS. In [34], the XRD pattern of CdS nanocrystals stabilized with DADMAC-NVP (1/1) copolymer was shown. In [35], the XRD patterns of CdS nanocrystals showed the planes of cubic zinc blende structure with an average size of 3.1 nm. In [40], the authors presented the XRD pattern of CdS nanocrystals of the zinc blende phase. In [42], the XRD results of CdS nanocrystals synthesized from different precursors, different mole ratios of CdS, and different sonication temperatures and durations were shown. In [45], the authors provided the XRD patterns of samples for different synthesis durations and heating stages during thermal treatment. Figure 1.5 shows the wide-angle X-ray diffraction (WAXD) patterns of the CdS nanocrystals prepared at different temperatures and having the cubic zinc blende structure [31].

III–V compound semiconductor nanocrystals have attracted much attention because they exhibit strong, size-dependent optical and electrical properties, which provide new possible applications including large-gain low-threshold quantum lasers, light-emitting diodes, single-electron transistors, and so on. GaAs is one of the most important semiconductor materials with applications in integrated circuits, infrared detectors [47], quantum-well lasers [48], solar cells [49], and some other devices [50–53].

Electrochemical techniques were extensively used to synthesize GaAs nanocrystals with different sizes and densities [54–59]. Metallic Ga and As_2O_3 of high purity were separately dissolved in concentrated HCl and diluted with DI water. The solution of Ga was mixed with the As_2O_3 solution. The pH value was adjusted simultaneously by adding distilled water or acid [54–59]. A platinum anode and an indium–tin–oxide (ITO)-coated glass cathode were dipped in the solution and the whole system was kept in an ice bath at a constant current until it reached equilibrium. The GaAs nanocrystals were prepared at different current densities [54,56,58] or at different temperatures [55,59]. The sizes of GaAs nanocrystals were reported from 1.5 nm to several tens of nanometers [54–59].

Nanocrystalline GaAs was prepared by laser ablation [60–62], which is a promising way to synthesize nanocrystals with small size and high purity. In [60], GaAs nanocrystals were synthesized by laser-induced etching using a Nd:YAG ($\lambda = 1.06$ μm) laser. Figure 1.6 shows a schematic illustration of laser-induced etching. The sample is irradiated with a laser power density of 20 W cm^{-2} for a duration of 60 min to etch the GaAs wafer. After etching, GaAs nanocrystals of 3.5–5.4 nm were obtained after rinsing with ethanol and drying in nitrogen ambient. In [62], the deposition of GaAs onto Si was carried out using pulsed-laser deposition (PLD) with a Q-switched Nd:YAG laser. The size of GaAs nanocrystals could be controlled by the number of pulses of the laser from several to tens of nanometers. In [63,64], molecular beam epitaxy

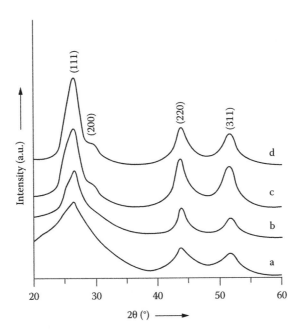

FIGURE 1.5 Wide-angle X-ray diffraction patterns of the CdS nanocrystals synthesized at (a) 180°C, (b) 120°C, and (c) 150°C, and (d) by the seeding-growth technique at 180°C. (From Wang, Q. et al., *Chem. Eur. J.*, 11, 3843, 2005.)

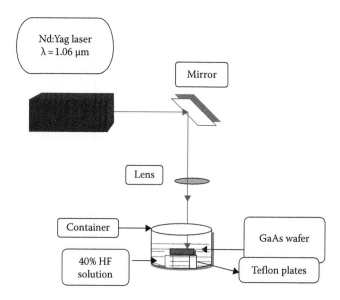

FIGURE 1.6 Schematic illustration of the setup for laser etching. (From Mavi, H.S. et al., *Mater. Sci. Eng. B*, 107, 148, 2004.)

(MBE) was used to prepare GaAs nanocrystals. GaAs nanocrystals were deposited on the Si substrate at the pressure of ~9 × 10⁻³ Pa for a size of ~20 nm [64].

A wet process was used to synthesize GaAs nanocrystals in [65]. Synthesis of GaAs nanocrystals was carried out in a vacuum chamber under nitrogen atmosphere. Triglyme containing 0.5 mmol of Ga(acac)$_3$ and 0.5 mmol of As(SiMe$_3$)$_3$ was heated at 216°C for 70 h, and then was filtered through a 0.2 mm PTFE filter. GaAs nanocrystals of size 1.5–9 nm were obtained finally.

TEM was used to study the microstructure of GaAs nanocrystal size, surface morphology, and lattice structure [55–59,61,65]. In [55–59], the researchers provided a series of TEM images and discussed the relationship between the nanocrystals and the different synthesis conditions by the electrochemical technique. In [61], the TEM characterization conducted after the sedimentation of larger particles showed the existence of GaAs nanocrystals, prepared by laser ablation, with size distribution ranging from 3 to 10 nm. In [65], TEM images of GaAs nanocrystals ranging from 1.5 to 9.0 nm were shown. A TEM image of 10 nm GaAs nanocrystals is shown in Figure 1.7 taken from [55]. The *d*-spacing measured from the lattice image was found to be 0.1673 nm close to the $d(211) = 0.1686$ nm of the orthorhombic GaAs phase.

In [54], the XRD patterns of GaAs nanocrystals prepared at different electrolysis current densities were presented. In [59], the authors discussed nanocrystalline GaAs deposited at different temperatures. In [55], the XRD patterns of orthorhombic nanocrystalline GaAs were shown. Figure 1.8 (taken from [59]) shows the XRD pattern recorded from nanocrystalline GaAs deposited on an ITO substrate. Three peaks were observed at 42.3191°, 44.307°, and 48.5141°, respectively. The *d*-values were calculated to be 2.1414 Å (*d*1), 2.0499 Å (*d*2), and 1.8815 Å (*d*3), respectively. *d*1 and *d*2 were close to the interplanar spacing of the (011) and (220) planes, respectively. *d*3 was close to the interplanar spacing of the (211) plane of cubic GaAs [55].

AFM was used to characterize GaAs nanocrystals [57,58,62]. In [57], the AFM image of PVA-capped GaAs nanocrystal film deposited on an ITO substrate having crystal size of 80–100 nm was presented. In [62], the authors showed the AFM images of PLD-synthesized GaAs for different number of pulses. Figure 1.9 (taken from [58]) shows the AFM images of GaAs nanocrystals synthesized at current densities of 6.0, 3.5, and 1.5 mA cm⁻², respectively, at 275 K. As shown in Figure 1.9a through c, with increase in the current density, the size of GaAs nanocrystals increases. At a higher temperature of 300 K, as shown in Figure 1.9d, GaAs nanocrystals have sizes of >250 nm.

1 nm

0.1673 nm

FIGURE 1.7 HRTEM image of a nanocrystalline GaAs. (From Nayak, J. and Sahu, S.N., *Appl. Surf. Sci.*, 229, 97, 2004.)

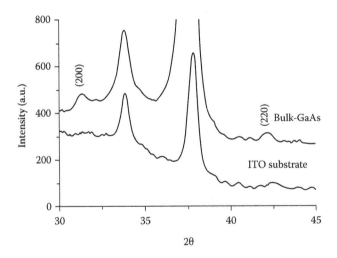

FIGURE 1.8 X-ray diffraction pattern recorded from nanocrystalline GaAs on an ITO substrate. (From Nayak, J. and Sahu, S.N., *Physica E*, 41, 92, 2008.)

Chemical synthesis, deposition, and epitaxy were used to synthesize InAs nanocrystals. Chemical synthesis for InAs nanocrystals were described in [66–69]. In [66], InAs nanocrystals of 2–6 nm were produced by the reaction between $InCl_3$ and $As[Si(CH_3)_3]_3$ with trioctylphosphine (TOP) as the solvent and capping agent. In [67], InAs nanocrystals of 2–8 nm were fabricated by solution-phase pyrolytic reaction of organometallic precursors. Surface passivation was carried out with the organic ligand TOP. In [68], the authors reported the synthesis of InAs nanocrystals with a size distribution of 7–60 nm by the solvothermal route at low temperature. In [70], InAs nanocrystals were grown by depositing InAs in a two-dimensional supperlattice formed by the GaAs matrix and AlAs barriers. In [71], the fabrication of InAs quantum dots of 60–65 nm on

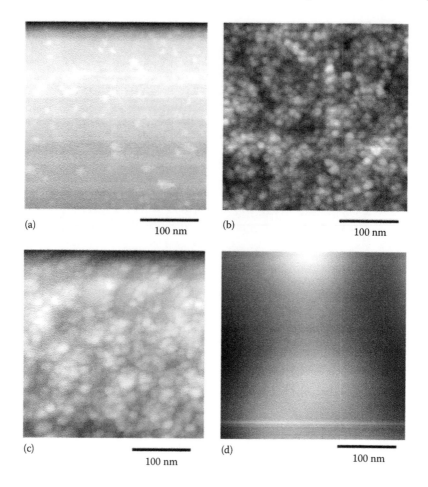

FIGURE 1.9 AFM image of GaAs thin films deposited on indium–tin–oxide (ITO) substrates at 275 K and current densities (a) 6.0 mA cm^{-2}, (b) 3.5 mA cm^{-2}, (c) 1.5 mA cm^{-2}. (d) Image with current density 2.5 mA cm^{-2} and temperature 300 K. (From Nayak, J. et al., *Appl. Surf. Sci.*, 252, 2867, 2006.)

SiO$_2$ patterned GaAs by metal organic chemical vapor deposition (MOCVD), followed by annealing in an arsine ambient, was described.

An epitaxial method was used to prepare InAs nanocrystals as reported in [75–79], such as organometallic/metalorganic vapor-phase epitaxy (OMVPE/MOVPE) and MBE. In [76], an InAs nanocrystal array of 75–280 nm was produced by MOVPE. A (100)-oriented InP wafer was exposed to phosphine (PH$_3$), with the introduction of arsine into the chamber. In [77], InAs nanocrystals smaller than 25 nm were grown using a solid-source MBE and finally with rapid cooling in arsenic ambient under certain pressure. In [78], InAs nanocrystals grown under a Volmer–Weber-like mode on Si nanomembranes that were synthesized by selective etching silicon-on-insulator substrate were described.

SEM images of InAs nanocrystals were presented in [71,78]. In [71], the authors showed SEM micrographs of InAs nanocrystals with a center-to-center spacing of 80 nm, as shown in Figure 1.10. The density of InAs nanocrystal could approach 1.8 × 10^{10} cm^{-2}. In [78], an SEM image of InAs nanocrystals deposited on a Si nanomembrane was presented.

TEM was extensively used to characterize InAs nanocrystals in [66,68,72–74]. In [73], TEM characterization was carried out to study Si/InAs nanocrystals before and after thermal annealing. Besides, [76] provided a cross-sectional view of InAs nanocrystals grown on InP crystal along

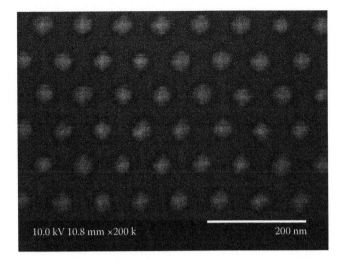

FIGURE 1.10   Array of InAs nanocrystals with a center-to-center spacing of 80 nm corresponding to a dot density of 1.8×10^{10} cm^{-2}. (From Dias, N.L. et al., *Appl. Phys. Lett.*, 98, 141112, 2011.)

different directions. In [77], the authors presented images of InAs nanocrystals on SiO$_2$ before capping and after capping with 20 nm thick SiO$_2$. Figure 1.11a shows InAs nanocrystals with an average diameter of 5 nm. A higher resolution image of nanocrystals with diameters ranging from 5 to 10 nm is shown in Figure 1.11b [66].

AFM studies on InAs nanocrystals were presented in [71,74,76–78]. In [71], the authors showed how single quantum dots and quantum dot clusters could be fabricated in pore diameters of 70 and 90 nm, respectively, as shown in Figure 1.12a and b. In [74], the topography of InAs nanocrystals/Si was characterized by AFM, as well as the elemental redistribution of Si and As. In [77], the AFM images of InAs nanocrystals grown at different temperatures by MBE were presented.

XRD patterns of InAs nanocrystals were shown in [66,68,73,78]. In [78], XRD patterns indicated a nonuniform strain state of InAs nanocrystals. In [66], the authors presented the XRD patterns of InAs nanocrystals of different sizes, as shown in Figure 1.13. The peak positions of the InAs nanocrystals showed good consistency with those of bulk InAs, indicating the same cubic zinc blende lattice structure. The width of peaks varied considerably, indicating a substantial change in the crystalline size.

CuCl was used in applications such as electrooptic modulators, optical filters [80], solid-state batteries [81], catalysts [82], adsorbent and air-purifying agents [83], and blue/UV light-emitting devices [84]. It may be possible to apply CuCl nanocrystals in high-density optical storage, laser

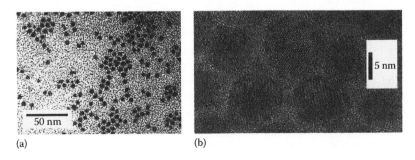

(a)   (b)

FIGURE 1.11   (a) TEM image of InAs nanocrystals of ~5 nm in diameter and (b) higher resolution TEM image of InAs nanocrystals of 5–15 nm in diameter. (From Guzelian, A.A. et al., *Appl. Phys. Lett.*, 69, 1432, 1996.)

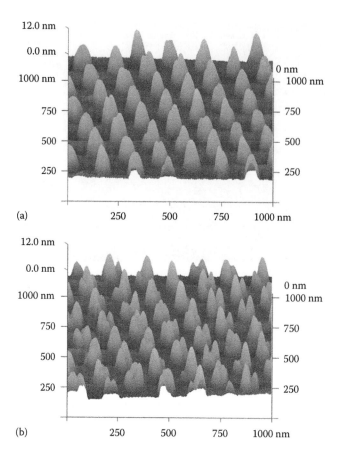

FIGURE 1.12 AFM profiles of quantum dots grown on 150 nm pitch pattern with (a) pore diameter of 70 nm and (b) pore diameter of 90 nm showing the transition from single QDs to QD clusters. (From Dias, N.L. et al., *Appl. Phys. Lett.*, 98, 141112, 2011.)

printing, projection displays, spectrophotometers, photocatalytic reactions, counterfeit detection, chemical detection, traffic signals, and medicine.

Chemical methods were used to obtain CuCl nanocrystals, including hydrothermal methods [85], ion implantation methods [86], double heat-treatment methods [87,88], and ionization in an alkali halide matrix [89–97].

The hydrothermal methods were discussed in [85]. A green hydrothermal route was used to prepare nanocrystalline CuCl powders via the reaction between $CuCl_2$ and α-D-glucose in distilled water. Double heat treatment methods were presented in [87,88]. The two-step glass melting was conducted with a 10 Hz, Q-switched Nd:YAG laser on silica glass with NaCl and CuO additives. CuCl microcrystallites were formed in the first heating step in silica glass. The second step of heating the glass containing CuCl microcrystallites induced nanometer-sized CuCl. The crystal size could be controlled by varying the temperature due to laser heating.

In [89–98], CuCl nanocrystals synthesized in an alkali halide matrix were discussed; there were two main approaches mentioned to fabricate single alkali halide crystals doped with CuCl, namely, the Czochralski method and the Bridgman method. There appeared to be no significant differences in the optical properties of crystals prepared by the two methods.

As can be seen in Figure 1.14, SEM characterization shows that the CuCl nanocrystals synthesized through hydrothermal methods exhibit lamellar crystallites [85]. XRD is widely used to

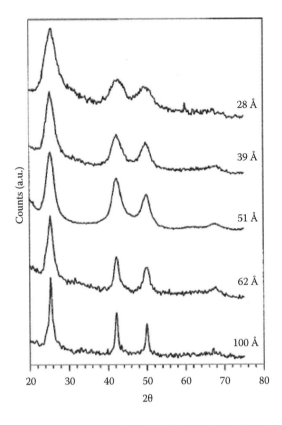

FIGURE 1.13 XRD patterns of InAs nanocrystals with different sizes. (From Guzelian, A.A. et al., *Appl. Phys. Lett.*, 69, 1432, 1996.)

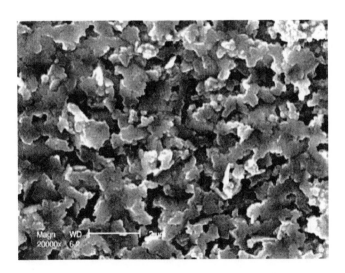

FIGURE 1.14 The SEM image of as-synthesized CuCl powders. (From Zhang, Y.C. and Tang, J.Y., *Mater. Lett.*, 61, 3708, 2007.)

characterize CuCl nanocrystals and crystal orientations [94,95]. A typical XRD pattern is presented in Figure 1.15. The XRD pattern of the CuCl nanocrystals synthesized using hydrothermal methods exhibits three diffraction peaks, namely, from the (111), (220), and (311) crystal planes of cubic phase. The lattice constant calculated from XRD data was 5.414 Å. The XRD characteristics of CuCl nanocrystals in NaCl matrix prepared with Czochralski method was discussed in [94,95]. Diffraction from the (111), (200), (220), (311), (400), and (331) planes of CuCl of zinc blende structure can be observed in the patterns.

1.2.2 ELEMENTAL SEMICONDUCTOR NANOCRYSTALS

Chemical etching, electrochemical etching, and dry etching have always been used to synthesize freestanding Si nanocrystals [100–103]. In [100], Si nanocrystals of several tens of nanometers were obtained from the pyrolysis of SiH_4 and etched subsequently in a two-phase solution of HF. The as-fabricated Si nanocrystals had a large size distribution from several nanometers to several tens nanometers. In [101], the synthesis of Si nanocrystals was carried out by the pyrolysis of SiH_4 and subsequent size reduction by chemical etching in hydrofluoric acid in a two-phase cyclohexane/propanol-2 solution. In [103], freestanding Si nanocrystals of 1.5–2 nm were obtained by etching off the oxide from the SiO_2/Si nanocomposite. The etching time was controlled for Si nanocrystal size reduction. Subsequently, organic ligands were attached to the Si nanocrystal surface via hydrosilylation.

   Electrochemical etching was used to synthesize Si nanocrystals [104–106]. In [104], the authors reported that Si nanocrystals with an average diameter of 3–5 nm were prepared by sonication of porous Si in an ultrasonic bath under Ar ambient. In [105], Si nanocrystals of 1.5–5 nm were synthesized by the electrochemical etching of boron and gallium-doped Si wafers in a mixture of hydrofluoric acid with pure ethanol. The Si nanocrystals were scratched off but aggregated in micrometer-sized grains, which were then subjected to a laser fragmentation process to achieve nanosized crystals. In [106], oxidized Si nanocrystals with the size of ~2.5–3 nm were prepared by electrochemical etching of monocrystalline Si wafers in a solution of HF, ethanol, and hydrogen peroxide and subsequent mechanical pulverization. In [107], the authors reported dry etching for obtaining Si nanocrystals smaller than 6 nm. Si nanocrystals of larger sizes were synthesized in a SiH_4-based plasma, which were then passed through a

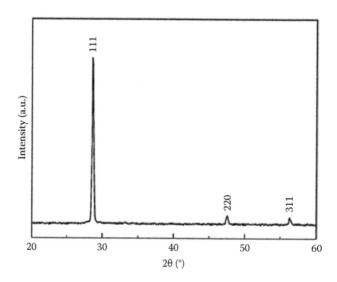

FIGURE 1.15 XRD pattern of as-synthesized CuCl particles. (From Zhang, Y.C. and Tang, J.Y., *Mater. Lett.*, 61, 3708, 2007.)

CF_4-based plasma for etching into a desirable size. In [108] a method to produce Si nanocrystals of about 2–3 nm capped with SiO_x by dehydration of SiH_4 was described.

Figure 1.16 shows SEM images of the as-prepared Si nanocrystals grains and after laser fragmentation for 55 min in ethanol [105]. Si nanocrystals were separated from the grains, and the average size was decreased. After undergoing the laser fragmentation process for 55 min, the mean size of the micrograins decreased from 25 μm to 5 μm [105]. In [103], the TEM image of large Si nanocrystals revealed the crystalline structure of the nanocrystals, as shown in Figure 1.17.

In [101], AFM was utilized to show the array growth of Si nanospikes, as presented in Figure 1.18. A monolayer of polystyrene balls having a diameter ~1.5 μm was prepared, as shown in Figure 1.18a and b. It was used as a mask to allow Si nanosize spike growth between three adjacent individual balls. After the deposition, the polystyrene balls were removed by some solvent, as shown in Figure 1.18c. An array of Si nanospikes was formed, as shown in Figure 1.18d.

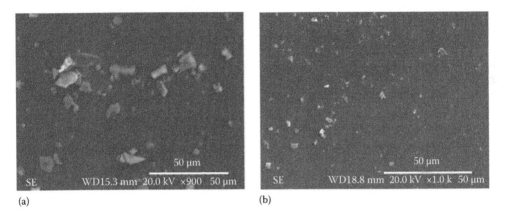

(a)   (b)

FIGURE 1.16   SEM images of Si micrograins (a) as prepared and (b) fragmented by nanosecond laser processing in ethanol for 55 min. (From Švrček, V. et al., *Acta Mater.*, 59, 764, 2011.)

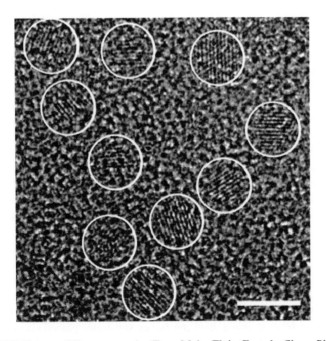

FIGURE 1.17   HRTEM image of Si nanocrystals. (From Maier-Flaig, F. et al., *Chem. Phys.*, 405, 175, 2012.)

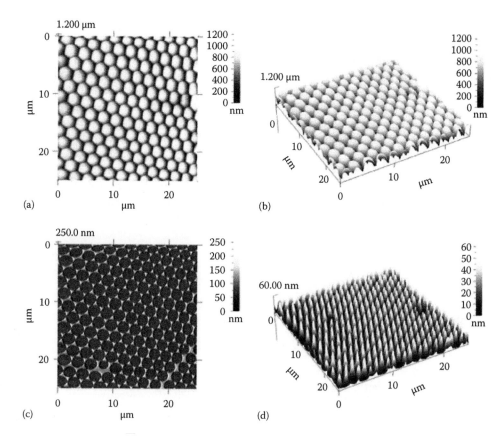

FIGURE 1.18 (a, b) A monolayer of polystyrene beads. (c) After removal of polystyrene. (d) Formation of array of Si nanospikes. (From Fojtik, A. et al., *Chin. Opt. Lett.*, 5, 250, 2007.)

Figure 1.19 shows the XRD pattern of Si nanocrystals synthesized at different pyrolysis temperatures [99]. As can be observed in the figure, the cubic crystalline structure was confirmed and the size of Si crystals was in the range 3–5 nm, which slightly increased with the pyrolysis temperature.

Methods of synthesizing freestanding Ge nanocrystals were reported, mainly using MBE [109] and chemical synthesis [110–112].

In [109], Ge nanocrystals were grown on Si (001) substrates pre-covered with carbon. The samples were prepared by MBE in the temperature range 350°C–750°C. The nonuniform distribution of carbon, together with the enhanced surface roughness, leads to the early onset of Ge nanocrystals having a size of 10–50 nm.

In [110], Ge nanocrystals with a broad size distribution of 2–70 nm were synthesized by the supercritical fluid method. Ge nanocrystals were synthesized by precipitation in hexane and octanol at 400°C–550°C under 20.7 MPa in a continuous-flow reactor. Diphenylgermane (DPG) and tetraethylgermane (TEG) were used as Ge precursors, which underwent thermolysis to crystalline Ge. Octanol was added to control the particle size and to serve as capping ligand.

In [111], Ge nanocrystals of 2.5–4.7 nm were obtained by the reaction of $GeCl_4$ and Mg in tetrahydrofuran (THF). The capping reaction of the remaining GeCl group was conducted by alkyl Grignard reagents in THF. The synthesis route to Ge nanocrystals is illustrated in Figure 1.20.

In [112], Ge nanocrystals were synthesized in a plasma reactor chamber via the decomposition of $GeCl_4$. Nanocrystal formation was realized in the plasma by the clustering of Ge atoms. Ge nanocrystals were collected from stainless steel meshes or substrates placed on meshes. The obtained crystals were characterized by SEM and TEM, as shown in Figure 1.21. The SEM images showed

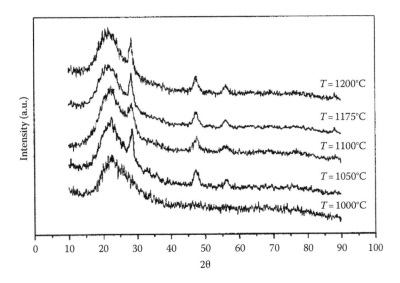

FIGURE 1.19 XRD spectra of the Si nanocrystals at different pyrolysis temperatures. (From Sorarù, G.D. et al., *Appl. Phys. Lett.*, 83, 749, 2003.)

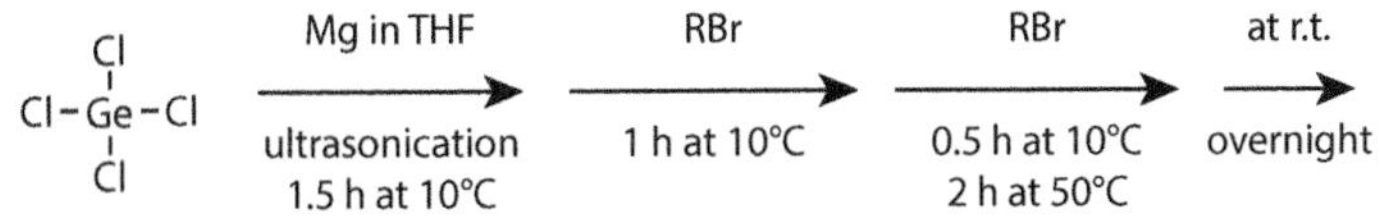

FIGURE 1.20 Synthesis route of Ge nanocrystals with $GeCl_4$ [p3]. R represents *n*-propyl or *tert*-butyl group. (From Watanabe, A. et al., *Appl. Organometal. Chem.*, 19, 530, 2005.)

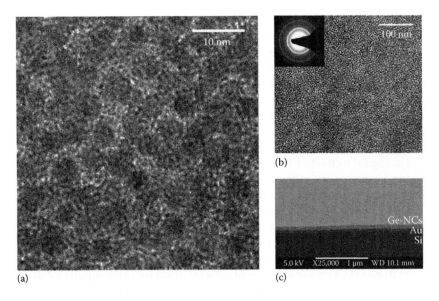

FIGURE 1.21 (a) Image for TEM characterization for Ge nanocrystals on Au/Si substrate. (b) Image of electron diffraction on a selected area. (c) Image of a cross-section of as-deposited sample by SEM. (From Holman, Z.C. and Kortshagen, U.R., *Langmuir*, 25, 11883, 2009.)

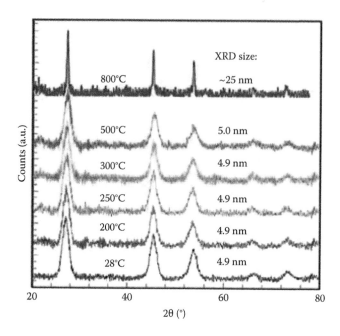

FIGURE 1.22 X-ray diffraction patterns of Ge nanocrystals of different sizes. (From Holman, Z.C. and Kortshagen, U.R., *Langmuir*, 25, 11883, 2009.)

that the Ge nanocrystals were formed on the Au/Si substrate, and TEM images showed that the nanocrystals had an average size of 4–5 nm. The XRD patterns in Figure 1.22 show Ge nanocrystals having different sizes synthesized by plasma decomposition of GeCl$_4$. The results indicated that the nanocrystal size could be controlled by varying the annealing temperature.

1.3 METAL NANOPARTICLES

Synthesis of metal nanoparticles for applications such as catalysis, electronics, optics, environmental, and biotechnology is of substantial interest. Gold, silver, and copper have been used mostly for the synthesis of stable nanoparticles, which are useful in areas such as photography, catalysis, biological labeling, photonics, optoelectronics, and surface-enhanced Raman scattering detection. Additionally, metal nanoparticles show surface-plasmon resonance absorption in the UV–visible region. Generally, metal nanoparticles can be prepared and stabilized by physical and chemical methods. Chemical approaches, such as chemical reduction, electrochemical techniques, and photochemical reduction, are widely used.

Many methods for preparing Au nanoparticles have been reported, mainly employing chemical, physical, and biological methods. Chemical methods have been more widely used to synthesize Au nanoparticles in aqueous media such as citrate or sodium borohydride [113–128].

In [113], the authors described Au nanoparticles with diameters of 1.5–20 nm synthesized by the reaction between HAuCl$_4$ · 3H$_2$O and N(C$_8$H$_{17}$)$_4$Br$_{(s)}$. Compounds C$_{12}$H$_{25}$NH$_{2(s)}$ and NaBH$_4$ were added to the above solution. Then black/brown precipitate in the organic solution was filtered and washed by ethanol. The gold nanoparticles that were filtered out were then dried in a vacuum chamber. Nanoparticles of 1–80 nm in diameter were prepared by mixing didodecyldimethylammonium bromide (DDAB) dissolved in toluene and gold chloride dissolved in the micelle solution by sonication [115,116]. Then NaBH$_4$ solution was added dropwise at room temperature. The color of the solution turned dark red after ~20 s. Au nanoparticles were obtained after stirring for another 15 min for completing the reaction. In [117], the synthesis of Au nanoparticles between 1.5 and 7 nm was reported. AuCl$_3$ was dissolved in toluene with ammonium surfactants, and then tetrabutylammonium borohydride (TBAB) was mixed with hydrazine in toluene to achieve Au nanoparticles. In [118], the authors reported Au nanoparticles of ~30 nm by the reaction of bis-(p-sulfonatophenyl)phenylphosphine dihydrate dipotassium (BSPP) and H_2O_2. Finally, sodium citrate solution and $HAuCl_4$ were added to obtain Au nanoparticles. In [119], dodecanethiol-capped Au nanoparticles of 2.1–8.3 nm were prepared by mixing $AuPPh_3Cl$ and dodecanethiol in benzene with the introduction of *tert*-butylamine-borane. Au nanoparticles were fabricated by $HAuCl_4$ aqueous solution in toluene with the aid of a phase-transfer agent (tetraoctylammonium bromide, TOAB) [121]. Phenylethylthiol was then added with $NaBH_4$ solution to complete the reaction. Au nanoparticles were dried by rotary evaporation and then collected. In [122], the authors described Au nanoparticles with diameters of 13 and 45 nm from a solution of gold chloride trihydrate with the addition of sodium citrate tribasic dehydrate ($C_6H_5Na_3O_7 \cdot 2H_2O$). In [123], it was reported that a piece of glassy carbon plate was placed into the seed solution of 4 nm Au colloid that was prepared by mixing $NaBH_4$ solution, $HAuCl_4$ solution, and trisodium citrate. Finally, Au nanoparticles on the glassy carbon surface was washed into a container with pure water, and then dried in nitrogen ambient. In [124], 15 ± 3 nm Au nanoparticles were prepared by adding sodium citrate to heated hydrogen tetrachloroaurate ($HAuCl_4$) solution, which was followed by a washing process. In [125], the borohydride reduction method was reported to synthesize Au nanoparticles of 8.1 ± 1.1 nm by injecting $HAuCl_4$ solution into sodium borohydride. In [126], gold (III) TOAB solution was synthesized by mixing a solution of $HAuCl_4 \cdot 3H_2O$ and another of tetraoctylammonium bromide (TOAB) in toluene and thiol derivative of calixarene [4]; and then $NaBH_4$ and MeOH solutions were added. Au nanoparticles of 2 ± 0.4 nm were obtained by dispersing and filtering the resulting solution. In [127], Au nanoparticles of 12–41 nm were achieved by the reaction of propanoic acid 2-(3-acetoxy-4,4,14-trimethylandrost-8-en-17-yl) (PAT) with an aqueous $HAuCl_4$ solution. In [128], Au nanoparticles were fabricated by mixing an ethanolic solution of tryptamine (TRA) and another ethanolic solution of $KAuCl_4$, followed with filtering, washing with pure cold water, and drying under vacuum.

Physical methods were also widely used to synthesize gold nanoparticles, such as RF sputtering [129,130], laser ablation techniques [131,132], proton beam irradiation [133], UV irradiation [134–136], atom beam co-sputtering [137], PLD [138], microwave (MW) irradiation [139], and multipulse methods (electrochemical synthesis) [140]. In [132], the authors reported the synthesis of Au nanoparticles by focusing a Nd:YAG laser beam on a gold target in pure DI water or in an aqueous solutions of PAMAM G5. In [133], proton beam irradiation was utilized to prepare 10 nm Au nanoparticles from a mixed solution of aqueous $HAuCl_4$, aqueous cetyltrimethylammonium bromide (CTAB), and aqueous NaOH in a tube. In [135], Au nanoparticles were synthesized by UV irradiation of $HAuCl_4$ dissolved in distilled water in the presence of poly(vinyl pyrrolidone) (PVP) under a nitrogen atmosphere. In [136], the authors reported the synthesis of Au nanoparticles of 10–40 nm by UV irradiation of an aqueous solution containing $AuCl_4^-$. In [139] , Au nanoparticles of <5 nm were fabricated using auric solutions placed inside a high-pressure microwave chamber. In [140], the synthesis of Au nanoparticles of ~27 nm by five times oxidation/reduction sequence in a $HAuCl_4$ solution was reported.

In [141–144], biological methods to synthesize Au nanoparticles were reported. In [141], a biological approach to fabricate Au nanoparticles of 6 nm was presented by adding trimethylphosphinchlorogold salt ($ClAuPMe_3$) mixed with a reducing agent. In [142], the authors reported the synthesis of Au nanocrystals with an average diameter of 5–13 nm by using $HAuCl_4 \cdot 3H_2O$ aqueous solution, biochemical β-D-glucose solution, and NaOH. Au nanoparticles of 5–50 nm were prepared by mixing the centrifuged supernatant and chloroauric acid ($HAuCl_4$) aqueous solution [143]. In [144], the synthesis of Au nanoparticles of 14–60 nm was reported using $HAuCl_4$ solution and aqueous *Terminalia arjuna* fruit extract, as shown in Figure 1.23.

TEM and HRTEM have been extensively used to characterize Au nanoparticles [114–122,124, 126–133,135,136,138,139,141–144]. In [114], TEM images of amine-capped Au nanoparticles were presented, while [116] presented the TEM images of ripening, segregation, and Au nanoparticle formation in colloids. In [117], the authors presented the TEM images of Au nanoparticles obtained using the single-phase approach. In [119], TEM images showed Au nanoparticles with different sizes

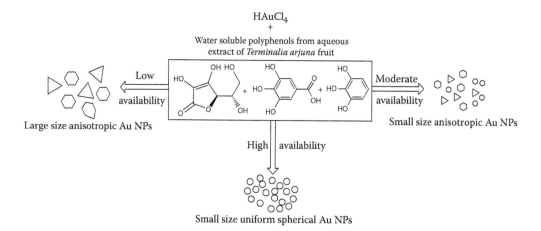

FIGURE 1.23   Illustration of the synthesis of Au nanoparticles with *Terminalia arjuna* fruit extract. (From Kumar, K.M. et al., *Spectrochim. Acta A: Mol. Biomol. Spectrosc.*, 116, 539, 2013.)

produced by varying the reaction solvent at different temperatures. In [122], TEM images of core-satellite structure containing Au nanoparticles were shown, whereas [126] showed the TEM images of Au nanoparticles before and after the addition of Co(II) in various concentrations. In [128], TEM studies showed a relatively narrow size distribution of Au nanoparticles in the range 36.6 ± 5.30 nm. In [132], TEM images of Au nanoparticles in water and in water plus PAMAM G5 as a stabilizing agent were shown. In [133], the authors presented TEM images of Au nanoparticles produced under irradiation of different fluences, durations, and solutions. In [144], TEM images of Au nanoparticles were shown with size and morphology variations depending on the quantities of extracts.

In [127], TEM images revealed the shape and size of Au nanoparticles synthesized using PAT having a size of 12–41 nm. In [135], the HRTEM images of Au nanoparticles fabricated at low or high concentration of $HAuCl_4$ as a precursor were shown. In [139], TEM images of Au nanoparticles obtained by a microwave-assisted method were presented. In [142], TEM images of β-D-glucose-stabilized Au nanoparticles (5–11 nm) were shown. In [143], a TEM study characterized polygonal Au nanoparticles of 5–50 nm. Figure 1.24 shows monodisperse Au nanoparticles of diameter of 2.2 nm (Figure 1.24a), and Au nanoparticles with lattice structure (Figure 1.24b), as given in [121].

XRD patterns of gold nanoparticles were shown in [113,114,121,127,128,134,140,143,144]. In [113], the authors reported the patterns of Au nanoparticles of four different sizes. In [114], the XRD patterns

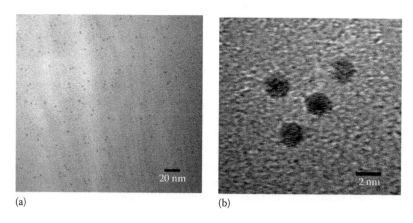

(a)   (b)

FIGURE 1.24   (a) TEM and (b) HRTEM images of gold nanoparticles. (From Qian, H.F. et al., *PNAS*, 109, 696, 2012.)

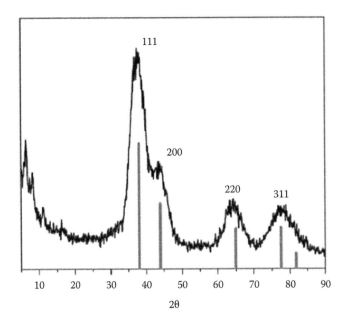

FIGURE 1.25 XRD patterns of gold nanoparticles exhibiting fcc structure. (From Qian, H.F. et al., *PNAS*, 109, 696, 2012.)

of amine-capped Au nanoparticles were presented. In [127], XRD patterns displayed PAT-stabilized Au nanoparticles of the face-centered cubic (fcc) structure. The XRD technique was used in [128] to investigate the fcc crystallinity of the Au-TRA nanoparticles. In [134], the XRD patterns of pure PMMA and Au–PMMA nanocomposites were shown before and after UV irradiation. In [142], the XRD patterns of β-ᴅ-glucose-stabilized Au nanoparticles were shown. In [143], XRD patterns of Au nanoparticles with main crystallographic planes were presented, and [144] showed discrete diffraction peaks in the XRD pattern that could be assigned to the planes of cubic Au nanoparticles. In [121], the XRD pattern showed the fcc structure of Au nanoparticles in Figure 1.25.

In [114,130,134,136,137], analyses of Au nanoparticles using the XPS technique were presented. In [114], the XPS spectra of amine-capped Au nanoparticles synthesized by different schemes were shown, based on which the stability of Au nanoparticles obtained from different synthesis schemes was discussed. In [130], the authors presented the XPS spectra of chemical state Au 4f from Au nanoparticles formed in the SiO_2 matrix. The binding energy difference between Au $4f_{7/2}$ and Au $4f_{5/2}$ was 3.62 eV, which was close to the standard value of 3.67 eV, indicating that metallic Au nanoparticles were formed in the SiO_2 matrix [130]. In [134], XPS core levels of Au–PMMA nanocomposites after UV irradiation were discussed. XPS spectra in [137] showed Au 4f and Si 3p core levels of the as-deposited and annealed Au–SiO_x nanocomposite films deposited on Si substrates.

Ag nanoparticles exhibit favorable properties for applications in optoelectronics, magnetic data storage, killing microbes, printable electronics, ultrasensitive chemicals, biological sensors, catalysts, and many others [145].

Many techniques were used to prepare Ag nanoparticles [146–158]. In [146], elemental Ag was evaporated at temperatures ranging from 1200 to 1500 K in a helium ambient. The flow stream was cooled over a short distance to 400 K to form Ag nanoparticles of 4–6 nm. Alkylthiol molecules were used for the etching and passivation of Ag nanoparticles. In [147], silver nanoparticles were produced and passivated by dodecanethiol monolayers. In [148], Ag nanoparticles were fabricated using a technique similar to that in [146,147] and were passivated by $SC_{12}H_{25}$ chain molecules.

The ion exchange/irradiation method was reported in [150,151]. The $Na^+ \rightarrow Ag^+$ ion-exchange technique was employed to form Ag ions in the surface layers of sodalime silicate glass by ion exchange in a AgNO$_3$/NaNO$_3$ solution at 300°C–380°C. Ions of 400–500 keV He, 1 MeV Ne, or 2 MeV Xe were used to induce the formation of Ag nanoparticles in sodalime silicate glass.

A chemical reduction method was reported for the synthesis of Ag nanoparticles [156,157], In [156], Ag nanoparticles were formed by adding aqueous AgNO$_3$ dropwise to a reducing solution of NaBH$_4$. Variations in particle size and distribution were obtained by adjusting the ratio between AgNO$_3$ and NaBH$_4$. In [157], a solution of AgNO$_3$ and sodium dodecyl sulfonate (SDS) was placed in an ice bath, to which was then added NaBH$_4$ with stirring for 3 h at room temperature. The pH value of the reaction solution was adjusted by an acetate buffer solution. Ag nanoparticles with an average size of 2 nm were then obtained.

TEM has been extensively used to characterize Ag nanoparticles [146–148,150,151,154–157]. In [147], the authors showed the TEM images of Ag nanoparticles. The atomic lattices and 3D morphology of the silver nanoparticles were observed. In [148], bright-field and dark-field TEM images of a monolayer of tetrahedral Ag nanoparticles were shown. In [151], the TEM images of Ag ion–exchanged BK7 glass after 1 MeV Xe irradiation were presented. In [154], the authors showed the images of Ag nanoparticle/CNT hybrid structure, and also revealed the morphologies of the connection region between Ag nanoparticles and CNTs. In [157], the TEM image indicated that the average size of Ag nanoparticles was about 2 nm and that the nanoparticles were spherical in shape. Figure 1.26 shows the Ag nanoparticles with typical sizes ranging from 50 to >100 nm [158].

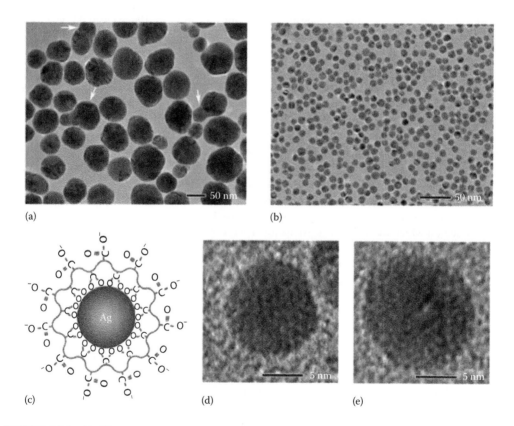

FIGURE 1.26 (a) Silver nanoparticles produced by traditional polyol process using polyvinylpyrrolidone (PVP) as the surfactant. The coagulation of growing particles results in polycrystalline products as indicated by the arrows. (b) A representative TEM image showing the uniform silver nanoparticles produced through the modified polyol process using poly(acrylic acid) (PAA) as the surfactant. (c) Schematic illustration of PAA-coated silver nanoparticles. HRTEM images of silver nanoparticles (d) without and (e) with defects. (From Hu, Y.X. et al., *J. Solid State Chem.*, 181, 1524, 2008.)

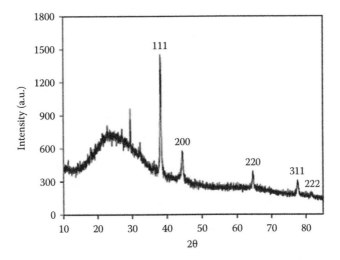

FIGURE 1.27 XRD pattern of Ag nanoparticles. (From Zheng, C.Z. et al., *J. Anal. Methods Chem.*, 2013, Article ID 261648, 2013.)

Figure 1.26b shows the images of uniform silver nanoparticles produced using the polyol process. Figure 1.26c illustrates the surface morphology of the silver nanoparticles synthesized with a surfactant. Figure 1.26d and e shows the lattice structures of Ag nanoparticles and the defects inside them (Figure 1.26e) [158].

XRD was used to characterize Ag nanoparticles [150,154,157]. In [150], the authors showed the XRD spectrum of an ion-exchanged sample irradiated with 500 keV He at a fluence of 1.1×10^{17} ions cm^{-2}. In [154], the XRD patterns of Ag nanoparticles/CNTs hybrid structures were shown. The angles $2\theta \approx 38.2°$ and $2\theta \approx 44.3°$ corresponded to Ag (111) and (200) planes, respectively. In [157], a typical XRD pattern of Ag nanoparticles was presented, as shown in Figure 1.27. The 2θ values of 38.12°, 44.28°, 64.43°, 77.48°, and 81.54° correspond to (111), (200), (220), (311), and (222) planes, respectively.

STM images of Ag nanoparticles were presented in [149,153]. In [149], the STM images of Ag nanoparticles and their lattice structure with different electronic transport properties were shown. In [153], the authors reported an STM study of Ag nanoparticles grown on a single crystal of SrTiO$_3$(001). The STM images in Figure 1.28 show that the top surface of Ag nanoparticles have two, three, or fivefold symmetry.

1.4 NANOCRYSTALS AND NANOPARTICLES EMBEDDED IN A DIELECTRIC FILM

The observation of visible photoluminescence from indirect-bandgap material systems, that is, nanocrystals in a dielectric matrix, presents potential applications in optoelectronic/optical devices [159]. The nanocrystals embedded in a dielectric film can be synthesized using techniques such as ion implantation, CVD, sputtering, and so on [159–199]. On the other hand, dielectric layers embedded with nanocrystals/nanoparticles have also been widely studied due to their possible applications in nonvolatile memory devices [160].

Si nanocrystals embedded in a dielectric matrix can be synthesized using techniques such as ion implantation, CVD, and sputtering. One promising technique to realize Si nanocrystals embedded in a dielectric matrix is by the implantation of Si ions into dielectric films [161–173]. First, dielectric films such as SiO$_2$, Si$_3$N$_4$, or Al$_2$O$_3$ were grown on a substrate. Then Si ions were accelerated in an arc chamber and implanted into the film at selected energies of 1–200 keV with dose ranging from

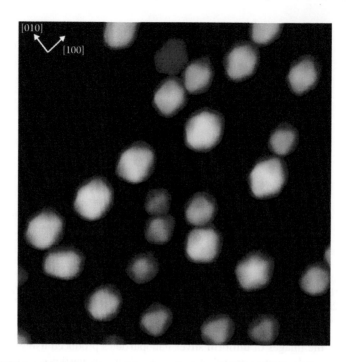

FIGURE 1.28 STM image of Ag nanoparticles deposited on a SrTiO₃ substrate. (From Silly, F. and Castell, M.R., *Appl. Phys. Lett.*, 87, 213107, 2005.)

10^{15} to 10^{17} cm^{-2}. A subsequent annealing at a high temperature (e.g., 1000°C) in N$_2$ or Ar ambient for a few minutes to a few hours was usually conducted to induce nanocrystal formation and remove defects. The size of Si nanocrystal synthesized with this technique was usually 1–10 nm.

Si nanocrystals can be grown by CVD in dielectrics such as SiO$_2$ and Si$_3$N$_4$. A Si-rich SiO$_2$ (SiO$_x$) film was formed using very-high frequency (VHF) PECVD from a N$_2$O:SiH$_4$ mixture. A subsequent high-temperature annealing (~1000°C–1300°C) of the SiO$_x$ films leads to the formation of nc-Si embedded in the SiO$_2$ matrix [184–187]. Low-pressure chemical vapor deposition (LPCVD) was used to grow Si dots on Si substrate. Then a thin oxide/nitride film was grown on the substrate, which also partially oxidized/nitridized the Si dots, resulting in Si nanocrystals embedded in the SiO$_2$/Si$_3$N$_4$ matrix [188,189]. Electron cyclotron resonance chemical vapor deposition (ECRCVD) was employed to grow Si-rich silicon nitride films by controlling the SiH$_4$ and PH$_3$ flow rates. A subsequent annealing was conducted to induce Si nanocrystal formation in Si$_3$N$_4$ [190].

Sputtering was used to synthesize Si nanocrystals embedded in a dielectric layer. In [176,179], SiO$_2$ films containing Si nanocrystals were prepared by co-sputtering Si and SiO$_2$ targets in an Ar gas atmosphere, which were annealed after deposition. Si nanocrystals embedded in SiC film prepared by magnetron co-sputtering were reported in [181–183]. The Si-rich amorphous silicon carbide films were deposited at room temperature by co-sputtering in an argon plasma. Subsequent high-temperature annealing was carried out in nitrogen ambient. Si nanocrystals embedded in Al$_2$O$_3$ film was synthesized by co-sputtering Si and Al$_2$O$_3$ targets [180].

PLD can be used to control Si nanocrystal size from 2 to 10 nm depending on the background gas species and pressure [192]. In [193–195], mechanical ball milling followed by spin-coating was used to prepare Si nanocrystals embedded in SiO$_2$ films. Si nanocrystals embedded in SiO$_2$ layer can also be fabricated by atomic layer deposition [191]. Such films can also be synthesized by thermal evaporation of SiO$_2$ powder onto rotating substrates, followed by thermal annealing [196].

TEM was extensively used to characterize Si nanocrystals embedded in a dielectric matrix [164,166,169,170,175,177,181–183,188–190]. In [164], the authors presented a cross-sectional view of Si nanocrystals prepared by Si ion implantation, as shown in Figure 1.29. Figure 1.29a shows the Si nanocrystals embedded SiO_2 film sandwiched between the capping layer and the Si substrate. Figure 1.29b shows the lattice fringes of Si nanocrystals. Figure 1.29c shows that the as-prepared Si nanocrystals had an average size of ~2 nm. In [169], the effect of Si implantation fluence on the microstructure of the Si nanocrystals and the SiO_2 matrix was examined using TEM; defects and faceting were observed in Si nanocrystals. In [170], a TEM image of Si nanocrystals of ~3 nm embedded in SiO_2 matrix was shown. In [175], the TEM micrograph showed a high density of Si nanocrystals with a mean radius of ~3 nm. In [177], TEM characterization showed Si nanocrystals of 3–5 nm embedded in a SiC matrix. TEM images in [181] illustrated the formation of Si nanocrystals/SiC films at different annealing temperatures. The size of Si nanocrystals in [182] characterized by TEM was in the range 3–5 nm. In [183], the TEM images of the Si-rich SiC films annealed at high temperature containing Si nanocrystals of 2–5 nm were shown, and in [190] those of a phosphorus-doped Si nanocrystals/SiNx film.

XRD studies of Si nanocrystals embedded in dielectric films were reported in [164,180,182,188,190]. In [180], the XRD patterns of Si nanocrystals in Al_2O_3 prepared with different magnetron powers and annealing temperatures were shown. The formation of Si nanocrystals embedded in SiC after annealing was confirmed by XRD measurements in [182]. And, [188] presented the XRD patterns of single and multiple Si nanocrystals layers. The XRD patterns in [190] showed the average Si nanocrystals/SiNx film fabricated by ECRCVD.

In [164], the small-angle XRD patterns for different ion implantation doses and annealing durations were given, which are shown in Figure 1.30. The Si atomic concentration of 15% and annealed for 1 h exhibited a size distribution of 2–10 nm, while the sample containing Si at 10% and annealed for 0.5 h showed a Gaussian distribution of nanocrystal sizes peaking at 6 nm.

XPS studied were carried out in [166,169,172,180,181,183,190]. In [166,197,198], the Si-implanted profile was obtained by the XPS analysis and secondary ion mass spectrometry (SIMS). In [169], the effect of Si implantation fluence on the microstructure of the Si nanocrystals and the SiO_2 matrix was examined by XPS. In [172], XPS was used to identify Si oxide and its electronic structure at different annealing temperatures for various durations. In [180], XPS depth profiling revealed the phase separation and formation of Si nanocrystals and SiO_2. In [181], the authors provided the chemical composition of the as-deposited layer containing C, Si, O, and N characterized by XPS.

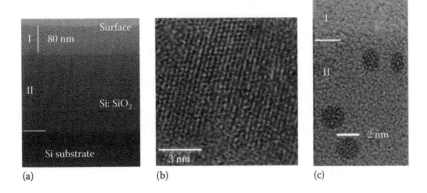

(a) (b) (c)

FIGURE 1.29 Cross-sectional TEM images of (a) the film structure. (b) Lattice structure of Si nanocrystals. (c) Average size of Si nanocrystals for about 2 nm. (From Guha, S. et al., *J. Appl. Phys.*, 88, 3954, 2000.)

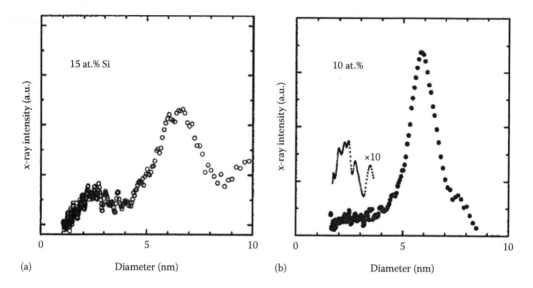

FIGURE 1.30 Small-angle X-ray diffraction intensity as a function of particle size from (a) 15 at.% (1 h annealing) and (b) 10 at.% (1/2 h annealing). (From Guha, S. et al., *J. Appl. Phys.*, 88, 3954, 2000.)

In [183], the as-deposited and annealed samples were characterized by XPS to study the chemical structure of Si-rich SiC co-sputtered films, and [190] presented the XPS analysis of the core levels of p-doped Si nanocrystals/SiNx films.

Figure 1.31 shows the deconvolution of the Si 2p peak for the as-implanted and annealed films. In [199], Si nanocrystal formation in Si-implanted SiO_2 films as a function of thermal annealing was investigated. There were five subpeaks corresponding to Si suboxidation states Si^{n+} ($n = 0, 1, 2, 3,$ and 4) in the SiO_2 films, as shown in Figure 1.31. The XPS results showed the evolution of the chemical structures and the formation of Si nanocrystals as functions of the annealing temperature and annealing time. In [197], depth profiles of the charging effect in Si nanocrystals were determined. The charging effect decreased with the increase of nc-Si concentration and vanished when a densely stacked nanocrystal layer was formed as a result of charge diffusion among the Si nanocrystals. In [198], the relative concentration of each oxidation state Si^{n+} ($n = 0, 1, 2, 3,$ and 4) at various depths was determined quantitatively by XPS analysis. The effects of annealing on both the oxidation states and their depth distributions were presented.

Ge nanocrystals embedded in a dielectric matrix show promising applications in optoelectronic, photovoltaic, and nonvolatile memory devices. Germanium has a larger dielectric constant and Bohr radius compared than Si [200,201] and hence the quantum size effects should be prominent in Ge nanocrystals even for a relatively large size.

Various techniques have been employed to synthesize Ge nanocrystals in the SiO_2 matrix. Ion implantation was found to be a promising technique to synthesize Ge nanocrystals in SiO_2 [202–209]. Ge ions were implanted into thermally grown SiO_2 films and followed by high-temperature annealing in inert ambient. A broad distribution of Ge nanocrystals embedded in relatively thick SiO_2 films had been achieved by the ion implantation at high energies from 75 to 450 keV [202,205–207,210]. On the other hand, an orderly two-dimensional array of Ge nanocrystals embedded in a thin SiO_2 film was realized with a low-energy (<10 keV) Ge ion implantation [211].

A sol–gel method for the synthesis of Ge nanocrystals in a SiO_2 matrix was reported in [212]. Silica glass containing 7 wt% Ge was prepared by the sol–gel method by using $Si(OC_2H_5)_4$ and $GeCl_4$ as starting materials. Ge nanocrystals of ~5 nm embedded in SiO_2 glass could be obtained after a heating treatment. A sol–gel method using TEOS and 3-trichlorogermanium propanoic acid (Cl_3–Ge–C_2H_4–COOH) as starting materials was purposed in [213,214]. The Ge nanocrystals were

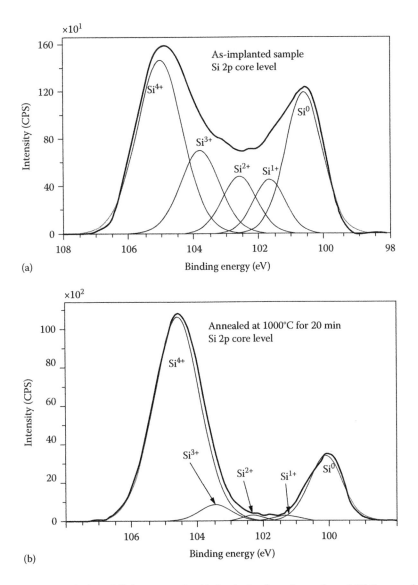

FIGURE 1.31 Deconvolution of Si 2p spectra for (a) the As-implanted sample and (b) the sample annealed at 1100°C for 20 min (b). (From Liu, Y. et al., *J. Phys. D: Appl. Phys.*, 36, L97, 2003.)

obtained in the range 1–13 nm by control of the reaction time and speed. Ge nanocrystals can be deposited on SiO_2 film surface by LPCVD and an additional capping SiO_2 was grown to make nanocrystals embedded in SiO_2 [215,216,237,256].

The synthesis of Ge nanocrystals embedded in SiO_2 has been demonstrated by the co-sputtering small pieces of Ge wafers attached on the pure SiO_2 target, followed by furnace-annealing at 800°C for 30 min [217,218]. Spherical Ge nanocrystals dispersed in the SiO_2 matrix with a mean size of ~6 nm could be achieved. To avoid the long process duration of the conventional furnace annealing, rapid thermal annealing (RTA) was employed for the formation of Ge nanocrystals [219,220–223].

Selective oxidation of a thin polycrystalline $Si_{1-x}Ge_x$ layer is another technique to synthesize Ge nanocrystals embedded in SiO_2 [224–233]. Deposition of a layer of $Si_{1-x}Ge_x$ film followed by dry or wet oxidation can realize Ge nanocrystals embedded in SiO_2. The deposition of the $Si_{1-x}Ge_x$ film can be done by many methods, including the co-sputtering of Si and Ge [224–226], LPCVD [227–230], ultrahigh vacuum CVD (UHVCVD) [231], ion implantation [232], and MBE [233]. The annealing ambient, that is, dry or wet oxidation, also affected the formation of Ge nanocrystals [227]. A proper control of the annealing conditions, for example, ambient, temperature, and duration, was crucial for the synthesis of Ge nanocrystals embedded in SiO_2 using the selective oxidation of $Si_{1-x}Ge_x$.

Besides the above-mentioned techniques, other techniques such as e-beam evaporation [234–236], PLD [238–240], and nano-patterning using a focused ion beam (FIB) [241,242] were demonstrated for the synthesis of Ge nanocrystals embedded in a SiO_2 matrix or on the SiO_2 surface.

TEM is an important technique to determine the crystallinity, external shape, size, and distribution of the nanocrystals [243–248]. Figure 1.32 shows typical cross-sectional TEM images of Ge nanocrystals embedded in SiO_2 synthesized by the co-sputtering technique. Figure 1.32a shows the electron diffraction pattern of the crystal structure of the Ge nanocrystals, and Figure 1.32b shows the structure of the Ge nanocrystal.

The coexistence of elemental Ge and Ge oxides in SiO_2 embedded with Ge has been studied by XPS [221,249,250]. Figure 1.33 shows a typical XPS analysis of the Ge nanocrystals embedded in a SiO_2 thin film deposited by CVD. The reduction of GeO_x and the formation of elemental Ge could be observed after thermal annealing. Moreover, the XPS analysis also provided useful information about the changes in the electronic structure of nanocrystals as a shift in the core-level binding energies. Shifts to the high values of the binding energy with the reduction of particles size have been reported in [251–255].

Al nanoparticles embedded in aluminum nitride (AlN) films or Al_2O_3 have attracted significant interest due to their applications in memory devices. Charge trapping/distrapping in the Al nanoparticles leads to a shift in the flat-band voltage (V_{FB}) of the metal–insulator–semiconductor (MIS) structure, which can be used in applications of memory devices [257–264].

The DC arc–discharge evaporation method was presented in [257]. The samples were prepared by evaporation using a DC arc discharge of metallic aluminum in an ambient of $N_2 + NH_3$. The volume ratio between NH_3 and ($N_2 + NH_3$) was adjusted to grow nanoparticles. AlN–Al nanocomposites were formed at a slightly lower ratio of <5%. RF magnetron sputtering synthesis was described

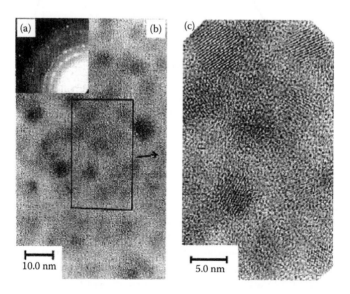

FIGURE 1.32 (a) Electron diffraction pattern and cross-sectional HRTEM image for a typical SiO_2 embedded with Ge nanocrystals synthesized by the co-sputtering technique. (b) Image of Ge nanocrystals in low resolution. (c) High resolution image of Ge nanocrystals. (From Fujii, M. et al., *Jpn. J. Appl. Phys.*, 30, 687, 2002.)

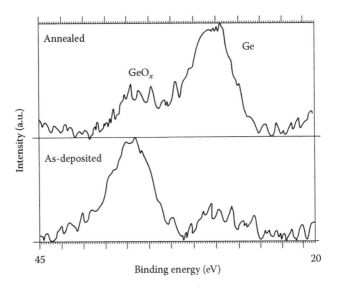

FIGURE 1.33 XPS spectra for the as-deposited and annealed Ge nanocrystals in SiO_2. (From Dutta, A.K., *Appl. Phys. Lett.*, 68, 1189, 1996.)

in [258–264]. The deposition was carried out by RF (13.56 MHz) magnetron sputtering of a pure Al target in a gas mixture of argon and nitrogen. The RF power was varied to achieve AlN films with different Al:N ratios.

TEM was extensively used to characterize Al nanocrystals [257–264]. In [257], the TEM micrograph and SAD pattern of AlN–Al nanocomposite were shown. In [258–264], the TEM images of Al nanoparticles/nanoclusters embedded in the AlN matrix were presented. A typical micrograph of Al nanoparticles is shown in Figure 1.34, where TEM image of Al nanocrystals (*nc*–Al) embedded in the AlN matrix can be observed with size of 4–8 nm [260].

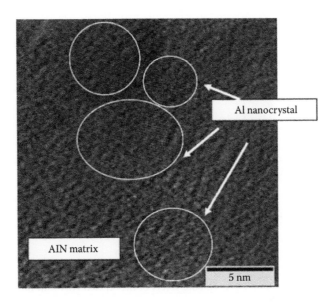

FIGURE 1.34 TEM image of Al nanoparticles embedded in AlN matrix. (From Liu, Y. et al., *Appl. Phys. A*, 93, 483, 2008.)

In [276], by varying the RF power between 100 W (sample 1), 150 W (sample 2), and 250 W (sample 3), three types of samples were obtained, which were characterized by XPS. The chemical composition of the deposited films could be changed by varying the RF power. The as-deposited films consisted of metallic Al embedded in the AlN and Al_2O_3 matrix, as shown in Figures 1.35 and 1.36, respectively. It was observed that at a higher RF power, the Al concentration was lower.

Al_2O_3 is considered a promising candidate for the gate dielectric of field-effect transistors due to its excellent electrical properties such as low leakage current, high dielectric constant (~8), and high breakdown voltage (~9 MV cm^{-1}). Recently, it was shown that Al nanoparticles embedded in Al_2O_3 thin film can be used to realize resistive memory devices [266,269,271,275] and floating-gate nonvolatile memory devices [273].

Two techniques were used for the synthesis of Al nanocrystals/nanoparticles embedded in Al_2O_3 matrix. The RF sputtering of Al target was carried out in O_2 ambient at a power of 250–310 W [265,267–270,272–275]. The thickness of the film was dependent on the sputtering duration. The Ar/O_2 ratio was set to different values to achieve different film properties. The films were annealed with a rapid thermal process for a duration varying from 30 s to 5 min. Aluminum anodization is another promising technique to fabricate Al nanoparticles [266,271]. A thin aluminum film was deposited on a silicon substrate by RF sputtering or other film growth techniques. Several hundreds of nanometers may be anodized. A platinum foil was used as the cathode. Anodization of the film was carried out in an oxalic acid solution at room temperature under a DC voltage of ~40 V for a duration of ~2 h. The thickness of the anodized aluminum oxide was nearly 30–50 nm [266,271].

XPS was used to characterize Al-rich Al_2O_3 thin films [267,270,272,273]. In [273], elemental Al was observed in Al 2p core levels, as shown in Figure 1.36a. In [267], XPS analysis indicated that thermal annealing could cause reactions at the interface between the Al-rich Al_2O_3 thin film and the Si substrate, leading to an increase in the Al concentration at the interface of Al_2O_3/Al. TEM characterizations were conducted in [270,273,274], where Al nanocrystals/nanoparticles were observed, as shown in Figure 1.36b. The size of Al nanocrystal/nanoparticle prepared by sputtering was in the range of several nanometers to tens of nanometers [270,273,274].

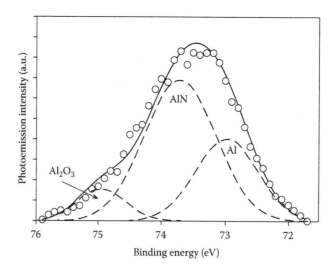

FIGURE 1.35 Peak decomposition of Al 2p core levels of the Al-rich AlN thin film. (From Liu, Y. et al., *Appl. Phys. Lett.*, 87, 033112, 2005.)

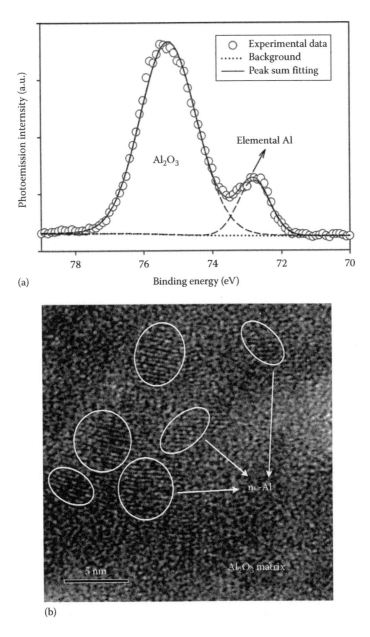

(a)

(b)

FIGURE 1.36 TEM image of AlN thin film embedded with Al. (a) XPS result of Al rich Al_2O_3. (b) TEM image of Al_2O_3 thin film embedded with Al nanocrystals. (From Liu, Z. et al., *Appl. Phys. Lett.*, 96, 173110, 2010.)

REFERENCES

1. C.B. Murray, D.J. Norris, and M.G. Bawendi, *J. Am. Chem. Soc.* 115 (1993) 8706.
2. A.P. Alivisatos, *Science* 271 (1996) 933.
3. W.W. Yu, L. Qu, W. Guo, and X. Peng, *Chem. Mater.* 15 (2003) 2854.
4. M. Achermann, M.A. Petruska, S. Kos, D.L. Smith, D.D. Koleske, and V.I. Klimov, *Nature* 429 (2004) 642.
5. A.H. Mueller, M.A. Petruska, M. Achermann, D.J. Werder, E.A. Akhadov, D.D. Koleske, M.A. Hoffbauer, and V.I. Klimov, *Nano Lett.* 5 (2005) 1039.

6. N. Pradhan, D. Goorskey, J. Thessing, and X. Peng, *J. Am. Chem. Soc.* 127 (2005) 17586.

7. M.J. Bruchez, M. Moronne, P. Gin, S. Weiss, and A.P. Alivisatos, *Science* 281 (1998) 2013.

8. X.G. Peng, L. Manna, W.D. Yang, J. Wickham, E. Scher, A. Kadavanish, and A.P. Alivisatos, *Nature* 404 (2000) 59.

9. E.D. Sone, and S.I. Stupp, *J. Am. Chem. Soc.* 126 (2004) 12756.

10. R. Baron, C.H. Huang, D.M. Bassani, A. Onopriyenko, M. Zayats, and I. Willner, *Angew. Chem., Int. Ed.* 44 (2005) 4010.

11. H. Tsuji and A.K. Kato, *Angew. Chem., Int. Ed.* 44 (2005) 3565.

12. C.B. Murray, D.J. Norris, and M.G. Bawendi, *J. Am. Chem. Soc.* 115 (1993) 8706.

13. R. Xie, X. Zhong, and T. Basche, *Adv. Mater.* 17 (2005) 2741.

14. M.J. Bowers II, J.R. McBride, and S.J. Rosenthal, *J. Am. Chem. Soc.* 127 (2005) 15378.

15. B.O. Dabbousi, J. Rodriguez-Viejo, F.V. Mikulec, J.R. Heine, H. Mattoussi, R. Ober, K.F. Jensen, and M.G. Bawendi, *J. Phys. Chem. B* 101 (1997) 9463.

16. M.A. Hines and P. Guyot-Sionnest, *J. Phys. Chem.* 100 (1996) 468–471.

17. L. Manna, E.C. Scher, and A. Paul Alivisatos, *J. Am. Chem. Soc.* 122 (2000) 12700–12706.

18. L.H. Qu, Z.A. Peng, and X.G. Peng, *Nano Lett.* 1 (2001) 333–337.

19. D.V. Talapin, A.L. Rogach, A. Kornowski, M. Haase, and H. Weller, *Nano Lett.* 1 (2001) 207–211.

20. K. Walzer, U.J. Quaade, D.S. Ginger, N.C. Greenham, and K. Stokbro, *J. Appl. Phys.* 92 (2002) 1434.

21. D.G. Wu, M.E. Kordesch, and P. Gregory Van Patten, *Chem. Mater.* 17 (2005) 6436–6441.

22. Y.A. Yang, H.M. Wu, K.R. Williams, and Y.C. Cao, *Angew. Chem.* 117 (2005) 6870–6873.

23. S.J. Wang, H. Kim, H.H. Park, Y.S. Lee, and H. Jeon, *J. Vac. Sci. Technol. A* 28 (2010) 559.

24. A.L. Rogach, A. Kornowski, M.Y. Gao, A. Eychmüller, and H. Weller, *J. Phys. Chem. B* 103 (1999) 3065–3069.

25. Q. Peng, Y.J. Dong, Z.X. Deng, and Y.D. Li, *Inorganic Chem.* 41 (2002) 5249.

26. A.E. Raevskaya, A.L. Stroyuk, S.Y. Kuchmii, Y.M. Azhniuk, V.M. Dzhagan, V.O. Yukhymchuk, and M.Y. Valakh, *Proceedings of 25th International Conference on Microelectronics (MIEL)*, Beirut, Lebanon, Vol. 1 (2006), p. 4244-0117.

27. J. Mu and X.B. Gao, *J. Dispers. Sci. Technol.* 26 (2005) 763–767.

28. X.M. Xu, Y.L. Wang, L.Y. Zhou, L.L. Wu, J.Z. Guo, Q. Niu, L.F. Zhang, and Q.F. Liu, *Micro Nano Lett.* 7 (2012) 589–591.

29. H. Mathieu, T. Richard, J. Allègre, P. Lefebvre, G. Arnaud, W. Granier, L. Boudes, J.L. Marc, A. Pradel, and M. Ribes, *J. Appl. Phys.* 77 (1995) 287.

30. H. Yang, P.H. Holloway, G. Cunningham, and K.S. Schanze, *J. Chem. Phys.* 121 (2004) 10233.

31. Q. Wang, D.C. Pan, S.C. Jiang, X.L. Ji, L.J. An, and B.Z. Jiang, *Chem. Eur. J.* 11 (2005) 3843–3848.

32. T. Serranoa, I. Gómeza, R. Colásb, and J. Cavazosb, *Colloids Surf. A: Physicochem. Eng. Aspects* 338 (2009) 20–24.

33. L. Zou, Z. Fang, Z.Y. Gu, and X.H. Zhong, *J. Lumin.* 129 (2009) 536–540.

34. B. Girginer, G. Galli, E. Chiellini, and N. Bicak, *Int. J. Hydrogen Energy* 34 (2009) 1176–1184.

35. Q. Xiao and C. Xiao, *Appl. Surf. Sci.* 255 (2009) 7111–7114.

36. J. Chen, Y.J. Ma, G.C. Fan, Y.F. Li, J.Y. Jiang, and Z.Y. Huang, *Mater. Lett.* 65 (2011) 1768–1771.

37. P.Q. Zhao, S.J. Xiong, X.L. Wu, and P.K. Chu, *Appl. Phys. Lett.* 100 (2012) 171911.

38. M.V. Artemyev, V. Sperling, and U. Woggon, *J. Appl. Phys.* 81 (1997) 6975.

39. I. Umezua, R. Koizumia, K. Mandaia, T. Aoki-Matsumotoa, K. Mizunoa, M. Inada, A. Sugimura, Y. Sunaga, T. Ishii, and Y. Nagasaki, *Microelectron. Eng.* 66 (2003) 53–58.

40. Y.C. Cao and J.H. Wang, *J. Am. Chem. Soc.* 126 (2004) 14336–14337.

41. N. Ghows and M.H. Entezari, *Ultrasonics Sonochem.* 18 (2011) 269–275.

42. N. Ghows and M.H. Entezari, *Ultrasonics Sonochem.* 18 (2011) 629–634.

43. D. Matsuura, Y. Kanemitsu, T. Kushida, C.W. White, J.D. Budai, and A. Meldrum, *Appl. Phys. Lett.* 77 (2000) 2289.

44. M.A. Zhukovskiy, A.L. Stroyuk, V.V. Shvalagin, N.P. Smirnova, O.S. Lytvyn, and A.M. Eremenko, *J. Photochem. Photobiol. A: Chem.* 203 (2009) 137–144.

45. M.F. Kotkata, A.E. Masoud, M.B. Mohamed, and E.A. Mahmoud, *Physica E* 41 (2009) 1457–1465.

46. K.S. Rathore, P.D. Deepika, N.S. Saxena, and K. Sharma, *AIP Conference Proceedings*, Malaysia, Vol. 1249 (2010), p. 145.

47. A.L. Korotkov, *J. Appl. Phys.* 93 (2003) 786.

48. W. Schrenk, N. Finger, S. Gianordoli, E. Gornik, and G. Strasser, *Appl. Phys. Lett.* 77 (2000) 3328.

49. N. Usami, Y. Azuma, T. Ujihara, G. Sazaki, and K. Nakajima, *Phys. Lett.* 77 (2000) 3565.

50. Y. Kanemitsu, H. Tanaka, S. Mimura, and T. Kushida, *J. Lumin.* 83–84 (1999) 301–304.

51. Y. Kanemitsu, H. Tanaka, T. Kushida, K.S. Min, and H.A. Atwater, *Physica E* 7 (2000) 322–325.
52. Y. Kanemitsu, H. Tanaka, T. Kushida, K.S. Min, and H.A. Atwater, *J. Lumin.* 87–89 (2000) 432–434.
53. K. Taniguchi, Y. Morishige, and Y. Kanemitsu, *Physica E* 17 (2003) 79–81.
54. J. Nayak and S.N. Sahu, *Appl. Surf. Sci.* 182 (2001) 407–412.
55. J. Nayak and S.N. Sahu, *Appl. Surf. Sci.* 229 (2004) 97–104.
56. J. Nayaka, R. Mythilib, M. Vijayalakshmib, and S.N. Sahu, *Physica E* 24 (2004) 227–233.
57. J. Nayak and S.N. Sahu, *Physica E* 30 (2005) 107–113.
58. J. Nayak, S.N. Sahu, and S. Nozaki, *Appl. Surf. Sci.* 252 (2006) 2867–2874.
59. J. Nayak and S.N. Sahu, *Physica E* 41 (2008) 92–95.
60. H.S. Mavi, A.K. Shukla, B.S. Chauhan, and S.S. Islam, *Mater. Sci. Eng. B* 107 (2004) 148–154.
61. R.A. Ganeev, A.I. Ryasnyanskiy, and T. Usmanov, *Opt. Commun.* 272 (2007) 242–246.
62. P. Dubček, B. Pivac, S. Milošević, N. Krstulović, Z. Kregar, and S. Bernstorff, *Appl. Surf. Sci.* 257 (2011) 5358–5361.
63. O.A. Neucheva, A.A. Evstrapov, Y.B. Samsonenko, and G.E. Cirlin, *Tech. Phys. Lett.* 33 (2007) 923–925.
64. H. Usui, S. Mukai, H. Asuda, and H. Mori, *J. Cryst. Growth* 311 (2009) 2269–2274.
65. H. Matsumoto, H. Uchida, H. Yoneyama, T. Sakata, and H. Mori, *Res. Chem. Intermed.* 20 (1994) 723–733.
66. A.A. Guzelian, U. Banin, A.V. Kadavanich, X. Peng, and A.P. Alivisatos, *Appl. Phys. Lett.* 69 (1996) 1432–1434.
67. O. Millo, D. Katz, Y. Levi, Y.W. Cao, and U. Banin, *J. Low Temp. Phys.* 118 (2000) 365–373.
68. X.X. Xu, K.H. Yu, W. Wei, B. Peng, S.H. Huang, Z.H. Chen, and X.S. Shen, *Appl. Phys. Lett.* 89 (2006) 253117.
69. S.A. Jewett and A. Ivanisevic, *Acc. Chem. Res.* 45 (2012) 1451–1459.
70. L. Rebohle, F.F. Schrey, S. Hofer, G. Strasser, and K. Unterrainer, *Appl. Phys. Lett.* 81 (2002) 2079.
71. N.L. Dias, A. Garg, U. Reddy, J.D. Young, and V.B. Verma, *Appl. Phys. Lett.* 98 (2011) 141112.
72. Y.L. Zi, K. Jung, D. Zakharov, and C. Yang, *Nano Lett.* 13 (2013) 2786–2791.
73. A. Tchebotareva, J.L. Brebner, S. Roorda, P. Desjardins, and C.W. White, *J. Appl. Phys.* 92 (2002) 4664.
74. S. Prucnal, S. Facsko, C. Baumgart, H. Schmidt, M.O. Liedke, L. Rebohle, A. Shalimov et al., *Nano Lett.* 11 (2011) 2814–2818.
75. S.A. Dayeh, E.T. Yu, and D. Wang, *J. Phys. Chem. C* 111 (2007) 13331–13336.
76. H.D. Fonseca-Filho, C.M. Almeida, R. Prioli, M.P. Pires, and P.L. Souza, *J. Appl. Phys.* 107 (2010) 054313.
77. M. Hocevar, P. Regreny, A. Descamps, D. Albertini, and G. Saint-Girons, *Appl. Phys. Lett.* 91 (2007) 133114.
78. C. Deneke, A. Malachias, A. Rastelli, L. Merces, M. Huang, F. Cavallo, O.G. Schmidt, and M.G. Lagally, *Am. Chem. Soc.* 6 (2012) 10287–10295.
79. M. Bennour, F. Saidi, L. Bouzaïene, L. Sfaxi, and H. Maaref, *J. Appl. Phys.* 111 (2012) 024310.
80. O. Ambacher, *J. Phys. D Appl. Phys.* 31 (1998) 2653.
81. A. Brune and S. Jiang, *Mater. Res. Bull.* 30 (1995) 573.
82. Z. Li, K.C. Xie, and R.C.T. Slade, *Appl. Catal. A: Gen.* 209 (2001) 107–115.
83. V. Recupero, L. Pino, M. Cordarö, A. Vita, F. Cipitĺ, and M. Laganà, *Fuel Process. Technol.* 85 (2004) 1445–1452.
84. F.O. Lucas, L. ÓReilly, G. Natarajan, P.J. McNally, S. Daniels, D.M. Taylor, S. William, D.C. Cameron, A.L. Bradley, and A. Miltra, *J. Cryst. Growth* 287 (2006) 112–117.
85. Y.C. Zhang and J.Y. Tang, *Mater. Lett.* 61 (2007) 3708–3710.
86. O.A. Podsvirov, A.I. Sidorov, V.A. Tsekhomskii, and A.V. Vostokov, *Phys. Solid State* 52 (2010) 1906–1909.
87. A. Nakamura and H. Ohmura, *J. Lumin.* 83–84 (1999) 97–103.
88. M.Y. Lee, *Bull. Korean Chem. Soc.* 16 (1995) 126–129.
89. M. Haselho, K. Reimann, and H.-J. Weber, *J. Cryst. Growth* 196 (1999) 135–140.
90. M. Haselhoff and H.-J. Weber, *Mater. Res. Bull.* 30 (1995) 607–612.
91. D. FröGhlich, H.F. Haselhoff, and K. Reimann, *Solid State Commun.* 94 (1995) 189–191.
92. M. Haselho, K. Reimann, and H.-J. Weber, *Eur. Phys. J. B* 12 (1999) 147–155.
93. M. Haselhoff and H.-J. Weber, *Phys. Rev. B* 58 (1998) 5052.
94. S. Mahtout, M.A. Belkhir, and M. Samah, *Acta Phys. Polonica A* 105 (2004) 279–286.
95. S. Mahtout, M.A. Belkhir, and M. Samah, *Semicond. Phys. Quant. Electron. Optoelectron.* 7 (2004) 185–189.

96. K. Yamanak, K. Edamatsu, and T. Itoh, *J. Lumin.* 76&77 (1998) 256–259.
97. P.M. Valov and V.I. Leiman, *Phys. Solid State* 51 (2009) 1703–1708.
98. T. Itoh, S. Yano, S. Iwai, K. Edamatsu, T. Goto, and A. Ekimov, *Mater. Sci. Eng. A* 217&218 (1996) 167–170.
99. G.D. Sorarù, S. Modena, P. Bettotti, G. Das, G. Mariotto, and L. Pavesi, *Appl. Phys. Lett.* 83 (2003) 749.
100. A. Fojtik, J. Valenta, The Ha Stuchlíková, J. Stuchlík, I. Pelant, and J. Kočka, *Thin Solid Films* 515 (2006) 775–777.
101. A. Fojtik, J. Valenta, I. Pelant, M. Kalal, and P. Fial, *Chin. Opt. Lett.* 5 (2007) 250–253.
102. T.L. Sudesh, L. Wijesinghe, E.J. Teo, and D.J. Blackwood, *Electrochim. Acta* 53 (2008) 4381–4386.
103. F. Maier-Flaig, E.J. Henderson, S. Valouch, S. Klinkhammer, C. Kubel, G.A. Ozin, and U. Lemmer, *Chem. Phys.* 405 (2012) 175–180.
104. E. Froner, E. D'Amato, R. Adamo, N. Prtljaga, S. Larcheri, L. Pavesi, A. Rigo, C. Potrich, and M. Scarpa, *J. Colloid Interface Sci.* 358 (2011) 86–92.
105. V. Švrček, T. Yamanari, Y. Shibata, and M. Kondo, *Acta Mater.* 59 (2011) 764–773.
106. K. Kůsová, *J. Non-Cryst. Solids* 358 (2012) 2130–2133.
107. X.D. Pi, R.W. Liptak, S.A. Campbell, and U. Kortshagen, *Appl. Phys. Lett.* 91 (2007) 083112.
108. J. Martin, F. Cichos, and C. von Borczyskowski, *J. Lumin.* 132 (2012) 2161–2165.
109. O. Leifeld, R. Hartmann, E. Müller, E. Kaxiras, K. Kern, and D. Gürtzmacher, *Nanotechnology* 10 (1999) 122–126.
110. X.M. Lu, K.J. Ziegler, A. Ghezelbash, K.P. Johnston, and B.A. Korgel, *Nano Lett.* 4 (2004) 969–974.
111. A. Watanabe, F. Hojo, and T. Miwa, *Appl. Organometal. Chem.* 19 (2005) 530–537.
112. Z.C. Holman and U.R. Kortshagen, *Langmuir* 25 (2009) 11883–11889.
113. D.V. Leff, P.C. Ohara, J.R. Heath, and W.M. Gelbart, *J. Phys. Chem.* 99 (1995) 7036–7041.
114. D.V. Leff, L. Brandt, and J.R. Heath, *Langmuir* 12 (1996) 4723–4730.
115. X.M. Lin and C.M. Sorensen, *Chem. Mater.* 11 (1999) 198–202.
116. X.M. Lin, C.M. Sorensen, and K.J. Klabunde, *J. Nanopart. Res.* 2 (2000) 157–164.
117. N.R. Jana and X.G. Peng, *J. Am. Chem. Soc.* 125 (2003) 14280–14281.
118. E. Hao, R.C. Bailey, G.C. Schatz, J.T. Hupp, and S.Y. Li, *Nano Lett.* 4 (2004) 327–330.
119. N.F. Zheng, J. Fan, and G.D. Stucky, *J. Am. Chem. Soc.* 128 (2006) 6550–6551.
120. C.-L. Lu, K.S. Prasad, H.-L. Wu, J. Annie Ho, and M.H. Huang, *J. Am. Chem. Soc.* 132 (2010) 14546–14553.
121. H.F. Qian, Y. Zhu, and R.C. Jin, *PNAS* 109 (2012) 696–700.
122. Z.Q. Weng, H.B. Wang, J. Vongsvivut, R.Q. Li, A.M. Glushenkov, J. He, Y. Chen, C.J. Barrow, and W.R. Yang, *Anal. Chim. Acta* 803 (2013) 128–134.
123. G. Chang, H.H. Shu, K. Ji, M. Oyama, X. Liu, and Y.B. He, *Appl. Surf. Sci.* 288 (2014) 524–529.
124. U. Bubniene, M. Oćwieja, B. Bugelyte, Z. Adamczyk, M. Nattich-Rak, J. Voronovic, A. Ramanaviciene, and A. Ramanavicius, *Colloids Surf. A: Physicochem. Eng. Aspects* 441 (2014) 204–210.
125. H.H. Deng, G.W. Li, L. Hong, A.L. Liu, W. Chen, X.H. Lin, and X.H. Xia, *Food Chem.* 147 (2014) 257–261.
126. D. Maity, R. Gupta, R. Gunupuru, D.N. Srivastava, and P. Paul, *Sensors Actuat. B* 191 (2014) 757–764.
127. M. Venkatachalam, K. Govindaraju, A. Mohamed Sadiq, S. Tamilselvan, V. Ganesh Kumar, and G. Singaravelu, *Spectrochim. Acta A Mol. Biomol. Spectrosc.* 116 (2013) 331–338.
128. M.A. de Carvalho, P.F. Andrade, F.C.A. Corbi, M. do C. Goncalves, A.L.B. Formiga, I.O. Mazali, J.A. Bonacin, and P.P. Corbi, *Synthet. Metals* 185–186 (2013) 61–65.
129. D. Barreca, A. Gasparotto, E. Tondello, G. Bruno, and M. Losurdo, *J. Appl. Phys.* 96 (2004) 1655.
130. C.C. Wang, J.Y. Tseng, T.B. Wu, L.J. Wu, C.S. Liang, and J.M. Wu, *J. Appl. Phys.* 99 (2006) 026102.
131. J.Y. Yang, J.H. Kim, W.J. Choi, Y.H. Do, C.O. Kim, and J.P. Hong, *J. Appl. Phys.* 100 (2006) 066102.
132. E. Giorgetti, A. Giusti, S.C. Laza, P. Marsili, and F. Giammanco, *Phys. Status Solidi (a)* 204 (2007) 1693–1698.
133. Y.-J. Kim, G. Cho, and J. Hee Song, *Nucl. Instrum. Methods Phys. Res. B* 246 (2006) 351–354.
134. M.K. Abyaneh, D. Paramanik, S. Varma, S.W. Gosavi, and S.K. Kulkarni, *J. Phys. D Appl. Phys.* 40 (2007) 3771–3779.
135. M.M. Chili, V.S.R. Rajasekhar Pullabhotla, and N. Revaprasadu, *Mater. Lett.* 65 (2011) 2844–2847.
136. S. Boufi, M.R. Vilar, A.M. Ferraria, and A.M. Botelho do Rego, *Colloids Surf. A: Physicochem. Eng. Aspects* 439 (2013) 151–158.
137. S. Mohapatra, Y.K. Mishra, D.K. Avasthi, D. Kabiraj, J. Ghatak, and S. Varma, *Appl. Phys. Lett.* 92 (2008) 103105.
138. K.C. Chan, P.F. Lee, and J.Y. Dai, *Appl. Phys. Lett.* 95 (2009) 113109.

139. C. Vargas-Hernandez, M.M. Mariscal, R. Esparza, and M.J. Yacaman, *Appl. Phys. Lett.* 96 (2010) 213115.
140. C. Fernández-Blanco, A. Colina, A. Heras, V. Ruiz, and J. López-Palacios, *Electrochem. Commun.* 18 (2012) 8–11.
141. R. Djalali, Y.F. Chen, and H. Matsui, *J. Am. Chem. Soc.* 125 (2003) 5873–5879.
142. J.C. Liu, G.W. Qin, P. Raveendran, and Y. Ikushima, *Chem. Eur. J.* 12 (2006) 2131–2138.
143. L. Karthik, G. Kumar, T. Keswani, A. Bhattacharyya, B. Palakshi Reddy, K.V. Bhaskara Rao, *Nanomed. Nanotechnol. Biol. Med.* 9 (2013) 951–960.
144. K.M. Kumar, B.K. Mandal, H.A. Kiran Kumar, and S.B. Maddinedi, *Spectrochim. Acta A Mol. Biomol. Spectrosc.* 116 (2013) 539–545.
145. V.I. Pokhmurs'kyi, V.M. Dovhunyk, M.M. Student, M.D. Klapkiv, V.M. Posuvailo, and A.R. Kytsya, *Mat. Sci.* 48 (2013) 636–641.
146. S.A. Harfenist, Z.L. Wang, M.M. Alvarez, I. Vezmar, and R.L. Whetten, *J. Phys. Chem.* 100 (1996) 13904–13910.
147. S.A. Harfenist, Z.L. Wang, R.L. Whetten, I. Vezmar, and M.M. Alvarez, *Adv. Mater.* 9 (1997) 817–822.
148. Z.L. Wang, S.A. Harfenist, I. Vezmar, R.L. Whetten, J. Bentley, N.D. Evans, and K.B. Alexander, *Adv. Mater.* 10 (1998) 808–812.
149. S.-H. Kim, G. Medeiros-Ribeiro, D.A.A. Ohlberg, R. Stanley Williams, and J.R. Heath, *J. Phys. Chem. B* 103 (1999) 10341–10347.
150. D.P. Peters, C. Strohhöfer, M.L. Brongersma, J. van der Elsken, and A. Polman, *Nucl. Instrum. Methods Phys. Res. B* 168 (2000) 237–244.
151. J.J. Penninkhof, A. Polman, L.A. Sweatlock, S.A. Maier, H.A. Atwater, A.M. Vredenberg, and B.J. Kooi, *Appl. Phys. Lett.* 83 (2003) 4137–4139.
152. J.M. Warrender and M.J. Aziz, *Appl. Phys. A* 79 (2004) 713–716.
153. F. Silly and M.R. Castell, *Appl. Phys. Lett.* 87 (2005) 213107.
154. Z.X. Wang, X.N. Li, C.L. Ren, Z.Z. Yong, J.K. Zhu, W.Y. Luo, and X.M. Fang, *Sci. China Ser. E Tech. Sci.* 52 (2009) 3215–3218.
155. P. Benzo, L. Cattaneo, C. Farcau, A. Andreozzi, M. Perego, G. Benassayag, B. Pécassou, R. Carles, and C. Bonafos, *J. Appl. Phys.* 109 (2011) 103524.
156. A. Thøgersen, J. Bonsak, C.H. Fosli, and G. Muntingh, *J. Appl. Phys.* 110 (2011) 044306.
157. C.Z. Zheng, H.P. Wang, L.Z. Liu, M.J. Zhang, J.G. Liang, and H.Y. Han, *J. Anal. Methods Chem.* 2013 (2013) Article ID 261648.
158. Y.X. Hu, J.P. Ge, D. Lim, T. Zhang, and Y.D. Yin, *J. Solid State Chem.* 181 (2008) 1524–1529.
159. T. Shimizu-Iwayama, K. Fujita, S. Nakao, K. Saitoh, T. Fujita, and N. Itoh, *J. Appl. Phys.* 75 (1994) 7779–7783.
160. S. Tiwari, F. Rana, H. Hanafi, A. Hartstein, and E.F. Crabbé, *Appl. Phys. Lett.* 68 (1996) 1377–1379.
161. C.E. Chryssou, A.J. Kenyon, T.S. Iwayama, C.W. Pitt, and D.E. Hole, *Appl. Phys. Lett.* 75 (1999) 2011.
162. S.-H. Choi and R.G. Elliman, *Appl. Phys. Lett.* 74 (1999) 3987.
163. P.G. Kik, M.L. Brongersma, and A. Polman, *Appl. Phys. Lett.* 76 (2000) 2325.
164. S. Guha, S.B. Qadri, R.G. Musket, M.A. Wall, and T. Shimizu-Iwayama, *J. Appl. Phys.* 88 (2000) 3954.
165. A.R. Wilkinson and R.G. Elliman, *Appl. Phys. Lett.* 83 (2003) 5512.
166. P. Pellegrino, B. Garrido, C. Garcia, J. Arbiol, J.R. Morante, M. Melchiorri, N. Daldosso, L. Pavesi, E. Scheid, and G. Sarrabayrouse, *J. Appl. Phys.* 97 (2005) 074312.
167. I. Sychugov, A. Galeckas, N. Elfström, A.R. Wilkinson, R.G. Elliman, and J. Linnros, *Appl. Phys. Lett.* 89 (2006) 111124.
168. Y. Liu, T.P. Chen, L. Ding, S. Zhang, Y.Q. Fu, and S. Fung, *J. Appl. Phys.* 100 (2006) 096111.
169. V. Levitcharsky, R.G. Saint-Jacques, Y.Q. Wang, L. Nikolova, R. Smirani, and G.G. Ross, *Surf. Coat. Technol.* 201 (2007) 8547–8551.
170. R.-J. Zhang, Y.-M. Chen, W.-J. Lu, Q.-Y. Cai, Y.-X. Zheng, and L.-Y. Chen, *Appl. Phys. Lett.* 95 (2009) 161109.
171. T.S. Iwayama, *Vacuum* 86 (2012) 1634–1637.
172. O. Korotchenkov, A. Podolian, B. Kuryliuk, B. Romanyuk, V. Melnik, and I. Khatsevich, *J. Appl. Phys.* 111 (2012) 063501.
173. R. Khelifi, D. Mathiot, R. Gupta, D. Muller, M. Roussel, and S. Duguay, *Appl. Phys. Lett.* 102 (2013) 013116.
174. G.S. Chang, J.H. Son, K.H. Chae, C.N. Whang, E.Z. Kurmaev, S.N. Shamin, V.R. Galakhov, A. Moewes, and D.L. Ederer, *Appl. Phys. A* 72 (2001) 303–306.
175. S. Kim, Y.M. Park, S.-H. Choi, and K.J. Kim, *J. Appl. Phys.* 101 (2007) 034306.

176. K. Watanabe, H. Tamaoka, M. Fujii, and S. Hayashi, *J. Appl. Phys.* 92 (2002) 4001.

177. D. Song, E.-C. Cho, G. Conibeer, Y. Huang, and M.A. Green, *Appl. Phys. Lett.* 91 (2007) 123510.

178. R. Limpens and T. Gregorkiewicz, *J. Appl. Phys.* 114 (2013) 074304.

179. J. Skov Jensen, D.A. Buttenschön, T.P. Leervad Pedersen, J. Chevallier, B. Bech Nielsen, and A. Nylandsted Larsen, *J. Appl. Phys.* 101 (2007) 056108.

180. I. Dogan, I. Yildiz, and R. Turan, *Physica E* 41 (2009) 976–981.

181. D. Song, E.-C. Cho, G. Conibeer, Y.-H. Cho, Y. Huang, S. Huang, C. Flynn, and M.A. Green, *J. Vac. Sci. Technol. B* 25 (2007) 1327.

182. D. Song, E.-C. Cho, G. Conibeer, Y. Huang, C. Flynn, and M.A. Green, *J. Appl. Phys.* 103 (2008) 083544.

183. A. Gencer Imer, I. Yildiz, and R. Turan, *Physica E* 42 (2010) 2358–2363.

184. T. Arguirov, T. Mchedlidze, M. Kittler, R. Rölver, B. Berghoff, M. Först, and B. Spangenberg, *Appl. Phys. Lett.* 89 (2006) 053111.

185. S. Hernández, P. Pellegrino, A. Martínez, Y. Lebour, B. Garrido, R. Spano, M. Cazzanelli et al., *J. Appl. Phys.* 103 (2008) 064309.

186. Y. Gong, J. Lu, S.-L. Cheng, Y. Nishi, and J. Vukovi, *Appl. Phys. Lett.* 94 (2009) 013106.

187. S. Prezioso, S.M. Hossain, A. Anopchenko, L. Pavesi, M. Wang, G. Pucker, and P. Bellutti, *Appl. Phys. Lett.* 94 (2009) 062108.

188. S. Gardelis, A.G. Nassiopoulou, P. Manousiadis, S. Milita, A. Gkanatsiou, N. Frangis, and C.B. Lioutas, *J. Appl. Phys.* 111 (2012) 083536.

189. P. Basa, Zs.J. Horváth, T. Jászi, A.E. Pap, L. Dobos, B. Pécz, L. Tóth, and P. Szöllösi, *Physica E* 38 (2007) 71–75.

190. P.-J. Wu, Y.-C. Wang and I.-C. Chen, *Nanoscale Res. Lett.* 8 (2013) 457.

191. T. Baron, A. Fernandes, J.F. Damlencourt, B. De Salvo, F. Martin, F. Mazen, and S. Haukka, *Appl. Phys. Lett.* 82 (2003) 4151.

192. L. Patrone, D. Nelson, V.I. Safarov, M. Sentis, W. Marine, and S. Giorgio, *J. Appl. Phys.* 87 (2000) 3829.

193. H.W. Lau, O.K. Tan, Y. Liu, C.Y. Ng, T.P. Chen, K. Pita, and D. Lu, *J. Appl. Phys.* 97 (2005) 104307.

194. H.W. Lau, O.K. Tan, Y. Liu, D.A. Trigg, and T.P. Chen, *Nanotechnology* 17 (2006) 4078.

195. H.W. Lau, O.K. Tan, B.C. Ooi, Y. Liu, T.P. Chen, and D. Lub, *J. Cryst. Growth* 288 (2006) 92.

196. U. Kahler and H. Hofmeister, *Appl. Phys. Lett.* 75 (1999) 5.

197. Y. Liu, T.P. Chen, C.Y. Ng, L. Ding, S. Zhang, Y.Q. Fu, and S. Fung, *J. Phys. Chem. B* 110 (2006) 16499–16502.

198. Y. Liu, Y.Q. Fu, T.P. Chen, M.S. Tse, S. Fung, J.-H. Hsieh, and X.H. Yang, *Jpn. J. Appl. Phys.* 42 (2003) L1394–L1396.

199. Y. Liu, T.P. Chen, Y.Q. Fu, M.S. Tse, J.H. Hsieh, and P.F. Ho, *J. Phys. D Appl. Phys.* 36 (2003) L97–L100.

200. J.D. Choi, J.H. Lee, W.H. Lee, K.S. Shin, Y.S. Yim, J.D. Lee, Y.C. Shin et al., *IEEE IEDM Tech.* (2000) 767–770.

201. J.D. Lee, J.H. Choi, D. Park, and K. Kim, *IEEE Electron Dev. Lett.* 24 (2003) 748.

202. K. Heinig, B. Schmidt, A. Markwitz, R. Grotzschel, M. Strobel, and S. Oswald, *Nucl. Instrum. Methods Phys. Res. B* 148 (1999) 969–974.

203. N. Arai, H. Tsuji, H. Nakatsuka, K. Kojima, K. Adachi, H. Kotaki, T. Ishibashi, Y. Gotoh, and J. Ishikawa, *Mater. Sci. Eng. B* 147 (2008) 230.

204. H. Fukuda, S. Sakuma, T. Yamada, S. Nomura, and M. Nishino, *J. Appl. Phys.* 90 (2001) 3524.

205. Y. Chen, G.Z. Ran, Y.K. Sun, Y.Q. Wang, J.S. Fu, and W. Chen, *Nucl. Instrum. Methods Phys. Res. B* 183 (2001) 305.

206. J. Zhang, X. Wu, and X. Bao, *Appl. Phys. Lett.* 71 (1997) 2505.

207. A. Markwitz, L. Rebohle, H. Hofmeister, and W. Skorupa, *Nucl. Instrum. Methods Phys. Res. B* 147 (1999) 361.

208. S. Duguay, S. Burignat, P. Kern, J.J. Grob, A. Souifi, and A. Slaoui, *Semicond. Sci. Technol.* 22 (2007) 837.

209. M. Klimenkov, J. Von Borany, W. Matz, R. Grotzschel, and F. Herrmann, *J. Appl. Phys.* 91 (2002) 10062.

210. L. Rebohle, J. Von Borany, H. Fröb, and W. Skorupa, *Appl. Phys. B* 71 (2000) 131.

211. J. Zhao, D. Huang, Z. Chen, W. Chu, B. Makarenkov, A. Jacobson, B. Bahrim, and J. Rabalais, *J. Appl. Phys.* 103 (2008) 124304.

212. M. Nogami and Y. Abe, *Appl. Phys. Lett.* 65 (1994) 2545.

213. H. Yang, X.S. Wang, H. Shi, F. Wang, X. Gu, and X. Yao, *J. Cryst. Growth* 236 (2002) 371.

214. H. Yang, R. Yang, X. Wan, and W. Wan, *J. Cryst. Growth* 261 (2004) 549.
215. Y. Shi, S.L. Gu, X.L. Yuan, Y.D. Zheng, K. Saito, H. Ishikuro, and T. Hiramoto, *Fifth International Conference on Solid-State and Integrated Circuit Technology*, 1998, Beijing, China, pp. 838–841.
216. T. Baron, P. Gentile, N. Magnea, and P. Mur, *Appl. Phys. Lett.* 79 (2001) 1175.
217. M. Fujii, S. Hayashi, and K. Yamamoto, *Jpn. J. Appl. Phys.* 30 (2002) 687.
218. Y. Maeda, N. Tsukamoto, Y. Yazawa, and Y. Kanemitsu, *Appl. Phys. Lett.* 59 (1991) 3168.
219. W. Choi, W. Chim, C. Heng, L. Teo, V. Ho, V. Ng, D. Antoniadis, and E. Fitzgerald, *Appl. Phys. Lett.* 80 (2002) 2014.
220. W. Choi, V. Ho, V. Ng, Y. Ho, S. Ng, and W. Chim, *Appl. Phys. Lett.* 86 (2005) 143114.
221. W. Choi, Y. Ho, S. Ng, and V. Ng, *J. Appl. Phys.* 89 (2001) 2168.
222. W. Choi, H. Thio, S. Ng, V. Ng, and B. Cheong, *Philos. Mag. B* 80 (2000) 729.
223. W.K. Choi, V. Ng, S.P. Ng, H.H. Thio, Z.X. Shen, and W.S. Li, *J. Appl. Phys.* 86 (1999) 1398.
224. E. Kan, W. Chim, C. Lee, W. Choi, and T. Ng, *Appl. Phys. Lett.* 85 (2004) 2349.
225. E.W.H. Kan, W. Choi, C.C. Leoy, W.K. Chim, D.A. Antoniadis, and E.A. Fitzgerald, *Appl. Phys. Lett.* 83 (2003) 2058.
226. E.W.H. Kan, W.K. Choi, W.K. Chim, E.A. Fitzgerald, and D.A. Antoniadis, *J. Appl. Phys.* 95 (2004) 3148.
227. A. Rodríguez, M. Ortiz, J. Sangrador, T. Rodríguez, M. Avella, A. Prieto, Á. Torres, J. Jiménez, A. Kling, and C. Ballesteros, *Nanotechnology* 18 (2007) 065702.
228. J. Wu and P. Li, *Semicond. Sci. Technol.* 22 (2006) S89.
229. T.C. Chang, S.T. Yan, P.T. Liu, C.W. Chen, S.H. Lin, and S.M. Sze, *Electrochem. Solid-State Lett.* 7 (2004) G17.
230. M. Avella, A. Prieto, J. Jimenez, A. Rodriguez, J. Sangrador, and T. Rodriguez, *Solid State Commun.* 136 (2005) 224.
231. P. Li, W. Liao, S. Lin, P. Chen, S. Lu, and M. Tsai, *Appl. Phys. Lett.* 83 (2003) 4628.
232. Y.C. King, T.J. King, and C. Hu, *IEEE Trans. Electron Dev.* 48 (2001) 696.
233. J. Zhang, Q. Fang, A. Kenyon, and I.W. Boyd, *Appl. Surf. Sci.* 208–209 (2003) 364.
234. J.G. Couillard and H.G. Craighead, *J. Mater. Sci.* 33 (1998) 5665.
235. C. Heng and T. Finstad, *Phys. E Low-Dimensional Syst. Nanostruct.* 26 (2005) 386.
236. T. Kobayashi, T. Endoh, H. Fukuda, S. Nomura, A. Sakai, and Y. Ueda, *Appl. Phys. Lett.* 71 (1997) 1195.
237. B. De Salvo, G. Ghibaudo, and G. Pananakakis, *IEEE Trans. Electron Dev.* 48 (2001) 1789.
238. J.K. Kim, H.J. Cheong, Y. Kim, J.Y. Yi, H.J. Bark, S.H. Bang, and J.H. Cho, *Appl. Phys. Lett.* 82 (2003) 2527.
239. Y. Kim, H.J. Cheong, K.H. Park, T.H. Chung, H.J. Bark, S.H. Bang, and J. Yi, *Semicond. Sci. Technol.* 17 (2002) 1039.
240. X. Ma, Z. Yan, B. Yuan, and B. Li, *Nanotechnology* 16 (2005) 832.
241. I. Berbezier, A. Karmous, A. Ronda, T. Stoica, L. Vescan, R. Geurt, A. Olzierski, E. Tsoi, and A. Nassiopoulou, *J. Phys. Conf. Ser.* 10 (2005) 73.
242. A. Karmous, I. Berbezier, and A. Ronda, *Phys. Rev. B* 73 (2006) 5.
243. M.L. Ostraat, J.W. De Blauwe, M.L. Green, L.D. Bell, M.L. Brongersma, J. Casperson, R.C. Flagan, and H.A. Atwater, *Appl. Phys. Lett.* 79 (2001) 433.
244. Y.Q. Wang, R. Smirani, and G.G. Ross, *Nano Lett.* 4 (2004) 2041.
245. P. Dimitrakis, E. Kapetanakis, D. Tsoukalas, D. Skarlatos, C. Bonafos, G.B. Asssayag, A. Claverie et al., *Solid-State Electron.* 48 (2004) 1511.
246. G.B. Assayag, C. Bonafos, M. Carrada, A. Claverie, P. Normand, and D. Tsoukalas, *Appl. Phys. Lett.* 82 (2003) 200.
247. Y.Q. Wang, R. Smirani, G.G. Ross, and F. Schiettekatte, *Phys. Rev. B* 71 (2005) 161310.
248. Y.H. Kwon, C.J. Park, W.C. Lee, D.J. Fu, Y. Shon, T.W. Kang, C.Y. Hong, H.Y. Cho, and K.L. Wang, *Appl. Phys. Lett.* 80 (2002) 2502.
249. A.K. Dutta, *Appl. Phys. Lett.* 68 (1996) 1189.
250. J.Y. Zhang, X.M. Bao, and Y.H. Ye, *Thin Solid Films* 323 (1998) 68.
251. V. Mulloni, P. Bellutti, and L. Vanzetti, *Surf. Sci.* 585 (2005) 137.
252. D. Schmeiber, O. Bohme, A. Yfantis, T. Heller, D.R. Batchelor, I. Lundstrom, and A.L. Spetz, *Phys. Rev. Lett.* 83 (1999) 380.
253. K. Borgohain, J.B. Singh, M.V. Rama Rao, T. Shripathi, and S. Mahamuni, *Phys. Rev. B* 61 (2000) 11093.
254. T.P. Chen, Y. Liu, C.Q. Sun, M.S. Tse, J.H. Hsieh, Y.Q. Fu, Y.C. Liu, and S. Fung, *J. Phys. Chem. B* 108 (2004) 16609.

255. C.Q. Sun, L.K. Pan, Y.Q. Fu, B.K. Tay, and S. Li, *J. Phys. Chem. B* 107 (2003) 5113.
256. T. Baron, F. Martin, P. Mur, C. Wyon, and M. Dupuy, *J. Cryst. Growth* 209 (2000) 1004.
257. S. Yu, D. Li, H. Sun, H. Li, H. Yang, and G. Zou, *J. Cryst. Growth* 183 (1998) 284–288.
258. Y. Liu, T.P. Chen, P. Zhao, S. Zhang, S. Fung, and Y.Q. Fu, *Appl. Phys. Lett.* 87 (2005) 033112.
259. Y. Liu, T.P. Chen, H.W. Lau, J.I. Wong, L. Ding, S. Zhang, and S. Fung, *Appl. Phys. Lett.* 89 (2006) 123101.
260. Y. Liu, T.P. Chen, H.W. Lau, L. Ding, M. Yang, J.I. Wong, S. Zhang, and Y.B. Li, *Appl. Phys. A* 93 (2008) 483–487.
261. M. Yang, T.P. Chen, Y. Liu, L. Ding, J.I. Wong, Z. Liu, S. Zhang, W. Zhang, and F. Zhu, *IEEE Trans. Electron Dev.* 55 (2008) 3605–3609.
262. Z. Liu, T.P. Chen, Y. Liu, L. Ding, M. Yang, J.I. Wong, Z.H. Cen, Y.B. Li, S. Zhang, and S. Fung, *Appl. Phys. Lett.* 92 (2008) 013102.
263. Y. Liu, T.P. Chen, M. Yang, Z.H. Cen, X.B. Chen, Y.B. Li, and S. Fung, *Appl. Phys. A* 95 (2009) 753–756.
264. Y. Liu, T.P. Chen, L. Ding, Y.B. Li, S. Zhang, and S. Fung, *J. Nanosci. Nanotechnol.* 10 (2010) 5796–5799.
265. W. Zhu, T.P. Chen, M. Yang, Y. Liu, and S. Fung, *IEEE Trans. Electron Dev.* 56 (2009) 2060–2064.
266. W. Zhu, T.P. Chen, Z. Liu, M. Yang, Y. Liu, and S. Fung, *J. Appl. Phys.* 106 (2009) 093706.
267. Z. Liu, T.P. Chen, Y. Liu, Z.H. Cen, S. Zhu, M. Yang, J.I. Wong, Y.B. Li, and S. Zhang, *IEEE Trans. Electron Dev.* 58 (2011) 33–38.
268. Z. Liu, T.P. Chen, Y. Liu, M. Yang, J.I. Wong, Z.H. Cen, S. Zhang, and Y.B. Li, *Appl. Phys. Lett.* 94 (2009) 243106.
269. W. Zhu, T.P. Chen, Y. Liu, and S. Fung, *J. Appl. Phys.* 112 (2012) 063706.
270. Y. Liu, T.P. Chen, W. Zhu, M. Yang, Z.H. Cen, J.I. Wong, Y.B. Li, S. Zhang, X.B. Chen, and S. Fung, *Appl. Phys. Lett.* 93 (2008) 142106.
271. W. Zhu, T.P. Chen, M. Yang, Y. Liu, and S. Fung, *IEEE Trans. Electron Dev.* 59 (2012) 2363–2367.
272. Z. Liu, T.P. Chen, Y. Liu, M. Yang, J.I. Wong, Z.H. Cen, and S. Zhang, *ECS Solid State Lett.* 1 (2012) Q4–Q7.
273. Z. Liu, T.P. Chen, Y. Liu, M. Yang, J.I. Wong, and Z.H. Cen, *Appl. Phys. Lett.* 96 (2010) 173110.
274. Y. Liu, T.P. Chen, L. Ding, M. Yang, Z. Liu, J.I. Wong, and S. Fung, *J. Appl. Phys.* 110 (2011) 096108.
275. W. Zhu, T.P. Chen, Y. Liu, M. Yang, and S. Fung, *IEEE Trans. Electron Dev.* 58 (2011) 960–965.
276. W. Zhu, Electrical properties and memory applications of Al-rich Al-based dielectric thin films, PhD thesis, Nanyang Technological University, Singapore, 2012.

2 Size- and Shape-Controlled ZnO Nanostructures for Multifunctional Devices

S.K. Ray, N. Gogurla, and T. Rakshit

CONTENTS

2.1 INTRODUCTION

Zinc oxide (ZnO), a II–VI compound semiconductor, has been in the focus of current research due to its size- and morphology-dependent electrical, optical, and chemical properties at the nanoscale. The thermodynamically stable phase of ZnO is the wurtzite hexagonal one with the lattice parameters $a = 0.3249$ and $c = 0.5207$ nm. This structure belongs to the point group $6mm$ and space group $P6_3mc$. Besides wurtzite ZnO, it can also exist in other crystalline phases, such as zinc blende (ZB) and rock salt. A few experimental and theoretical studies have addressed the growth and fundamental properties of metastable ZB ZnO [1]. However, ZnO is most stable in the wurtzite structure rather than the ZB form under ambient conditions due to its ionicity, which lies at the border line between those of covalent and ionic materials. Some important physical properties of the wurtzite and ZB ZnO structures are summarized in Table 2.1. Due to the noncentrosymmetric structure in the wurtzite form and large electromechanical coupling of ZnO, it also exhibits excellent piezo-electric and pyroelectric properties. The high breakdown strength and high saturation velocity of ZnO are attractive for electronic applications. With a wide direct bandgap of 3.34 eV at room temperature, ZnO is useful for high-temperature applications, transparent electrodes, and light emitting devices. It also has a large exciton binding energy of 60 meV, which is higher than the thermal energy at room temperature (26 meV). Therefore, it is expected to have an excitonic gain at room temperature for ZnO-based blue/UV light emitting diodes.

ZnO is intrinsically an n-type semiconductor due to the native defects, and it is difficult to make p–n homojunctions due to low solubility of p-type dopant. Much effort has been made to prepare p-type ZnO by doping with phosphorous, nitrogen, and arsenic [3–5]. However, achieving reproducible, reliable, high-quality p-type conductivity is still a challenge. Group III elements (Al, Ga, and In) have also been doped to enhance the n-type conductivity in ZnO [6,7]. Al- and In-doped ZnO thin films and nanostructures are widely used as transparent conductors for photonic devices. Various kinds of defects in ZnO such as zinc vacancy (V_{Zn}), oxygen vacancy (V_O), zinc interstitials (Zn_i), and oxygen antisites (O_{Zn}) play a vital role in controlling the optical, electrical, and magnetic properties. An excellent control over the surface defects is mandatory for realizing efficient optoelectronic devices. Under ambient condition, ZnO exhibits low conductivity due to the capturing of free electrons by oxygen molecules on its surface. This oxygen adsorption and desorption play significant roles on the photo- and gas-sensing properties of ZnO.

ZnO can be prepared in different morphologies using low-temperature processing techniques. The performance of ZnO devices depends on the structural quality, morphology, surface effects,

TABLE 2.1
Some Important Properties of Wurtzite and Zinc Blende ZnO Structures

Property	Wurtzite	Zinc Blende
Lattice	Hexagonal	Cubic
Point group	6mm	$\bar{4}3m$
Space group	$P6_3mc$	$F\bar{4}3m$
Parameters	$a = 0.3249$ nm and $c = 0.5207$ nm	$a = 0.447$ nm
Density	5.606 g/cm^3	
Melting point	1975°C	
Lattice expansion coefficients	A_0: 6.531026/°C	
	c_0: 3.031026/°C	
Thermal conductivity	69$_\parallel$, 60$_\perp$ W/m/K	
Refractive index	2.008, 2029	
Static dielectric constant	8.656	
Energy gap	3.34 eV	3.27 eV
Intrinsic carrier concentration	$<10^6$ cm^{-3}	
Electron Hall mobility at 300 K	200 cm^2/V/s	
Hole Hall mobility at 300 K	5–50 cm^2/V/s	
Electron and hole effective masses	0.24 and 0.59 m_0	

Sources: Pearton, S.J. et al., *Superlattices Microstruct.*, 34, 3, 2003; Ashrafi, A. and Jagadish, C., *J. Appl. Phys.*, 102, 071101(1–12), 2007.

and the rate of adsorption/desorption of oxygen molecules. Therefore, the structural properties and surface engineering can enhance the performance of devices. In this chapter, we mainly focus on the tunable properties of shape- and size-controlled ZnO nanostructures for multifunctional applications. The optical and electrical properties of these nanostructures have been studied. Enormous effort has been made to implement efficient electrical and optical devices using different morphologies of ZnO. However, there exist several challenges for the practical use of the grown nanostructures for commercial device applications. Recently, several sensitizers have been used to improve the performance of ZnO-based devices. Here, we review the progress of semiconductor- and metal-sensitized ZnO nanostructures for the realization of efficient optoelectronic and gas-sensing devices.

2.2 GROWTH OF ZnO NANOSTRUCTURES

Several techniques have been reported for growing ZnO nanostructures. The electrochemical method [8], template-based growth [9], and sol–gel processing [10,11] are some of the widely used methods for growing different types of ZnO nanostructures. Another effective solution-based method is the hydrothermal process [12], by which the nanostructures can be grown at relatively low temperatures. ZnO nanostructures have also been grown by other methods such as electrospinning [13], two-step mechanochemical–thermal process [14], chemical vapor deposition (CVD) [15], sonochemical synthesis [16], microwave-assisted combustion [17], anodization [18], precipitation [19], and so on. Another widely used method for growing ZnO nanostructures is the vapor transport process, in which Zn and oxygen or oxygen mixture vapors react with each other to form ZnO. Zn and oxygen vapors can be produced in several ways. The simplest one uses the decomposition of ZnO [20]; however, it requires a very high temperature (~1400°C). A relatively low temperature (500°C–800°C) can be used for growing ZnO nanostructures by heating Zn powder in an oxygen atmosphere [21,22]. The growth temperature, pressure, duration of evaporation, and gas flow rate need to be precisely controlled for achieving the desired morphology of ZnO nanostructures. The carbothermal method, which uses graphite and ZnO powder as the source materials [23], can be used for the growth at about 800°C–1100°C. Metalorganic vapor phase epitaxy using an organometallic zinc compound under appropriate flow of oxygen or nitrous oxide can also be used for the growth of ZnO nanostructures [24]. The widely used vapor transport method is usually classified into catalyst-assisted vapor–liquid–solid (VLS) and catalyst-free vapor–solid (VS) processes. Both the processes are capable of producing a wide variety of ZnO morphologies [25,26]. Among all the materials, ZnO nanostructures probably possess the widest range of morphologies, such as nanorods, nanowires, nanobelts, nanorings, nanosprings, nanopropellers, tripods, tetrapods, and so on [26,27]. Some of the reported morphologies of ZnO nanostructures are shown in Figure 2.1.

The morphology of ZnO nanostructures can be tuned by doping with appropriate elements. Doping ZnO tetrapods with a high concentration of Sn changes the morphology to flower-like multipods and nanowires [28]. ZnO nanocrystals exhibit different morphologies upon doping with different elements such as Na, Li, Pr, Cu, and Mg [29]. The morphology of S-doped ZnO nanostructures changes from nanonails to nanowires by varying the sulfur concentration [30]. The optical properties of ZnO nanostructures are largely influenced by this change in morphology. The near band–edge and defect state emissions depend significantly on the doping concentration of S and Sn [28,30]. The electrical conductivity and field emission properties of ZnO nanostructures can also be modulated upon doping with Sn and Mg [28,31].

2.3 OPTICAL PROPERTIES OF ZnO TRIPODS AND TETRAPODS

Tripods and tetrapods, consisting of three and four rod-shaped arms emerging from a central core, respectively, are the commonly exhibited morphologies of ZnO nanostructures [26,28,31–37]. They have a higher specific surface area than one-dimensional ZnO nanostructures such as nanorods,

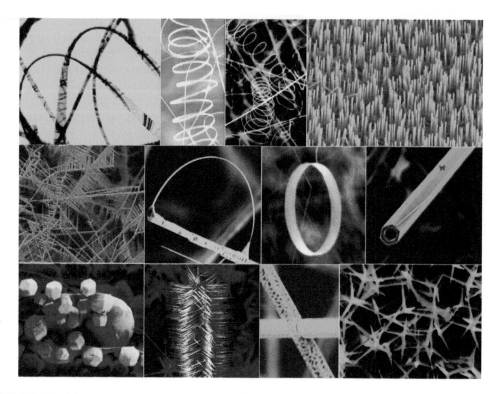

FIGURE 2.1 Micrographs of reported ZnO nanostructures. (Reprinted from *Mater. Today*, 7, Wang, Z.L., Nanostructures of zinc oxide, 26–33, © 2004, with permission from Elsevier.)

nanowires, and so on. Another advantage of tetrapods over other morphologies is their ability to spontaneously orient over the substrate with an arm directing perpendicular to the substrate plane, which can be useful for photovoltaic applications [38,39]. The fundamental study of the intrinsic optical properties of ZnO is important for their applications in photonic devices. For this, different types of experimental techniques such as optical transmission, absorption, reflection, photoreflection, cathodoluminescence, photoluminescence (PL), spectroscopic ellipsometry, and so on, have been used. In the following sections, we will discuss the optical properties of ZnO tripods and tetrapods studied using photoluminescence spectroscopy.

2.3.1 Undoped ZnO Tripods

Figure 2.2 shows the room-temperature PL characteristics of ZnO tripods and tetrapods, synthesized by the catalyst-free vapor–solid method [26]. The spectrum of bulk ZnO is also shown for comparison. A typical spectrum consists of a peak in the ultraviolet (UV) region, which is related to the free excitonic (FX) emission, and a peak in the visible region, which is due to defect-state transitions [40,41]. The defect-state emission intensity of ZnO tripods and tetrapods are much lower than that of bulk ZnO, with tripods having the lowest. The PL spectra indicate the improved structural ordering in ZnO tripods. Figure 2.3a shows the temperature-dependent PL spectra of ZnO tripods. The peak at 3.407 eV (bound exciton) is the most dominant up to 50 K, whereas the one at 3.425 eV (free exciton) becomes stronger with further increase of temperature and dominates up to room temperature. The increase of the FX peak intensity is due to the strong coupling between the excitons and phonons. The high intensity of the bound excitonic (BX) peak

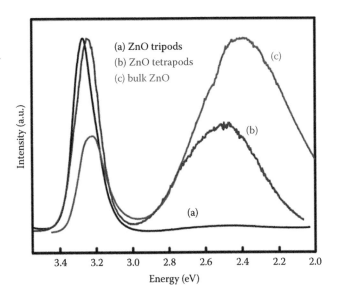

FIGURE 2.2 Room-temperature photoluminescence spectra of ZnO: (a) tripods, (b) tetrapods, and (c) bulk powder. (From Mandal, S. et al., *J. Appl. Phys.*, 105, 033513, 2009.)

at a temperature below 50 K suppresses the evolution of the FX peak. The quenching of the FX emission with temperature can be described by the equation [42]

$$I = \frac{I_0}{1 + A_1 \exp(-E_a/k_B T)} \tag{2.1}$$

where
I_0 is the emission intensity at $T = 0$ K
A_1 is the proportionality constant
k_B is the Boltzmann's constant
E_a is the activation energy

The activation energy has been found to be 66 meV, which is near the FX binding energy of bulk ZnO (60 meV). Figure 2.3b shows the PL emission characteristics of ZnO tripods at 10 K. Besides FX and BX emissions, the longitudinal optical (LO) phonon replicas of FX emission, which are designated as FX-1LO, FX-2LO, FX-3LO, FX-4LO, and FX-5LO, could be detected at the energies of 3.355, 3.281, 3.209, 3.135, and 3.067 eV, respectively. Similarly, the LO phonon replicas of BX transition, designated as BX-1LO, BX-2LO, and BX-3LO, are observed in Figure 2.3b. The presence of eight phonon replicas at 10 K indicates the growth of ZnO tripods with very low defect density. The variation of the FX peak energy with temperature can be described by a Bose–Einstein-type expression [43]:

$$E_g(T) = E_g(0) - \frac{K}{\exp(\theta_E/T) - 1} \tag{2.2}$$

where
$E_g(0)$ is the bandgap energy at $T = 0$ K
K is the electron–phonon coupling strength
θ_E is the Einstein temperature

The Einstein temperature obtained from the fitted plot is 175 K, compared to 240 K reported for bulk ZnO crystals [44].

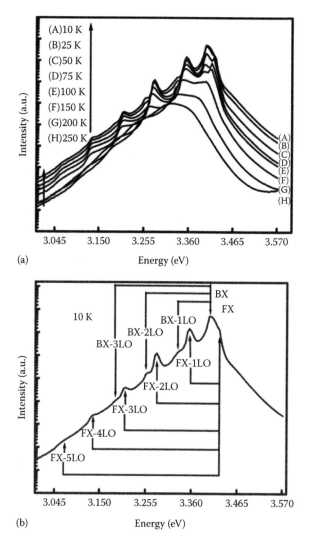

FIGURE 2.3 (a) Temperature-dependent PL spectra and (b) PL characteristics at 10 K of ZnO tripods. (From Mandal, S. et al., *J. Appl. Phys.*, 105, 033513, 2009.)

2.3.2 Undoped ZnO Tetrapods

Figure 2.4a shows the temperature-dependent PL spectra and Figure 2.4b displays the spectrum recorded at 20 K for ZnO tetrapods grown by the catalyst-free vapor–solid method [32]. The strongest peak at 20 K is found to be at the energy of 3.352 eV, which is related to the bound excitonic emission. The peak at 3.362 eV arises from the FX emission. The binding energy of the bound excitonic peak is found to be 10 meV, suggesting its origin as due to excitons bound to neutral donors (D^0X). Another strong peak (3.296 eV) at 20 K is related to the first-order transverse optical (TO) phonon replica of D^0X, and assigned as D^0X-TO. The high peak intensity indicates a strong coupling between D^0X and TO phonons. The peaks observed at 3.226 and 3.153 eV are the LO phonon replicas of D^0X-TO emission, which are designated as D^0X-TO-1LO and D^0X-TO-2LO transitions, respectively. The deformation potential mainly gives rise to TO phonon scattering, whereas the LO scattering occurs because of Fröhlich and deformation potentials [45]. The short-range interaction is

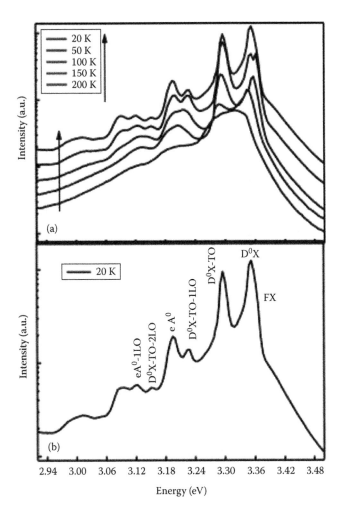

FIGURE 2.4 (a) Temperature-dependent PL spectra and (b) PL spectrum at 20 K of ZnO tetrapods. (Reprinted from *Ceram. Int.*, 41, Roy, N. and Roy, A., Growth and temperature dependent photoluminescence characteristics of ZnO tetrapods, 4154–4160. © 2015, with permission from Elsevier.)

usually influenced by the deformation potential in ZnO nanostructures [46]. Thus, the TO phonons are accompanied by a number of LO phonon replicas. The phonon peaks merge with other neighboring ones or become very weak at a high temperature, and therefore they can be distinguished only at temperatures below 150 K, as in Figure 2.4a. The temperature-dependent PL spectra in the range 20–150 K indicate the thermal ionization of donors at high temperature. The free electron–acceptor (eA^0) transition [40,47] results in the evolution of a new electron–acceptor recombination peak at 3.195 eV. With increase in temperature, the population of acceptor levels gradually increases, resulting in the decreased intensity of the eA^0 peak. Also the FX peak becomes stronger with increasing temperature and dominates over the eA^0 transition. The LO phonon replica of eA^0 peak is also observed at 20 K in Figure 2.4b.

2.3.3 Sn- AND P-Doped ZnO TETRAPODS

The effect of doping with Sn and P on the optical properties of ZnO tetrapods has been reported [28,33]. The introduction of Sn (up to 5 at.%) and P reduces the intensity of defect-state emission in ZnO tetrapods [28,33]. This may be due to the reduction in intrinsic defect density in ZnO

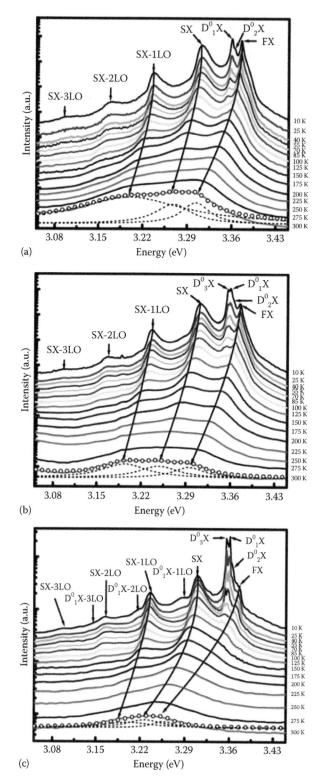

FIGURE 2.5　Temperature-dependent PL spectra of (a) pure ZnO, (b) 3 at.% Sn-doped, and (c) 5 at.% Sn-doped ZnO tetrapods. The spectrum at 300 K (solid line) has been fitted with Lorentzian curves (dash lines). The open circles denote the sum of the three Lorentzian curves. (From Rakshit, T. et al., *AIP Adv.*, 3, 112112, 2013.)

due to doping. The temperature-dependent excitonic PL emission bands of undoped ZnO, 3 at.% Sn-doped ZnO, and 5 at.% Sn-doped ZnO tetrapods are shown in Figure 2.5a through c, respectively. At 10 K, both undoped and Sn-doped ZnO samples have some common features [28]. The peak at 3.375 eV is related to the FX emission, and those at 3.360 and 3.364 eV are assigned to excitons bound to neutral donors, designated by D^0_1X and D^0_2X, respectively [41,48]. There is a transition at 3.311 eV in both pure and Sn-doped ZnO tetrapods. The binding energy of the peak (64 meV) indicates its excitonic nature. Therefore, excitons bound to defect states (SX) could be considered to be the cause of this emission [49,50]. The PL features observed at 3.238, 3.166, and 3.094 eV are the LO phonon replicas of SX, denoted by SX-1LO, SX-2LO, and SX-3LO, respectively. An additional peak is observed at 3.355 eV in Sn-doped ZnO tetrapods, which can be assigned to excitons bound to neutral donors, designated by D^0_3X. This emission may have originated from the deep-level energy states formed in ZnO due to Sn doping. Some additional peaks could be observed for 5 at.% Sn-doped ZnO samples. These peaks at 3.288, 3.216, and 3.144 eV are the LO phonon replicas of D^0_1X, and denoted by D^0_1X-1LO, D^0_1X-2LO, and D^0_1X-3LO, respectively. In all these samples, the D^0X peak disappears at a temperature above 125 K. As the temperature rises, the D^0X peak thermally dissociates into free excitons, resulting in a decrease in the intensity of D^0X emission. The spectrum at 300 K of all the samples, fitted with a Lorentzian function, exhibited deconvoluted peaks corresponding to FX, SX, and SX-1LO emissions. The variation of the FX peak energy with temperature of the pure and Sn-doped ZnO tetrapods could be fitted with Equation 2.2. The Einstein temperature from the plot is found to be 259 ± 3, 240 ± 2, and 179 ± 5 K, for pure ZnO, 3 at.% Sn-doped ZnO, and 5 at.% Sn-doped ZnO tetrapods, respectively. In comparison, the Einstein temperature is 175 K for pure ZnO tripods and 240 K for ZnO crystals.

The PL spectra of ZnO nanotetrapods grown by chemical vapor deposition revealed the presence of only neutral-donor-bound and FX emissions [33]. The PL spectra recorded with various excitation intensities at 10 K for P-doped ZnO nanotetrapods are shown in Figure 2.6a. The peaks at 3.313 and 3.355 eV are related to the free electrons to neutral-acceptor (FA) and neutral-acceptor-bound excitonic (A^0X) transitions, respectively. Both the peaks are invariant to the change in excitation intensity. With the increase of excitation intensity from 0.3 to 30 mW/cm^2, the peak at 3.287 eV exhibits a blue shift of ~10 meV. The close donor–acceptor pair (DAP) transitions undergo faster recombination, and therefore the Coulomb interaction between them is much stronger than that of the distant pairs. This results in a blue shift in DAP transitions with increase in excitation intensity. Hence, the peak at 3.287 eV is attributed to the DAP emission. The weak peak at 3.212 eV is related to the LO phonon replica of DAP transition, and denoted by DAP-1LO. The radiative recombination for DAP transitions results in a photon energy given by [51]

$$h\nu_{DAP} = E_g - E_D - E_A + \frac{e^2}{4\pi\varepsilon r} \qquad (2.3)$$

where
E_g is the bandgap energy
E_D and E_A are the donor and acceptor binding energy, respectively
e is the electronic charge
ε is the dielectric permittivity
r is the DAP distance

Using Equation 2.3, the acceptor binding energy was calculated to be 90–120 meV. Figure 2.6b shows the temperature-dependent PL spectra of P-doped ZnO nanotetrapods. The A^0X and DAP emission undergoes red and blue shifts, respectively, with the increase of temperature. The intensity of DAP emission shows a rapid reduction with temperature. There is no shift in the FA peak at temperatures below 30 K, but undergoes a red shift with further increase of temperature. Such a shift in

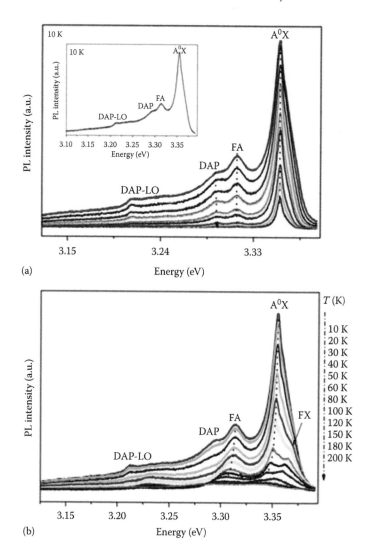

FIGURE 2.6 (a) Excitation intensity–dependent PL spectra recorded at 10 K and (b) temperature-dependent PL spectra of P-doped ZnO nanotetrapods. The inset in (a) shows the near-band edge emission of P-doped ZnO nanotetrapods. (Reprinted with permission from Yu, D. et al., Photoluminescence study of novel phosphorus-doped ZnO nanotetrapods synthesized by chemical vapour deposition, *J. Phys. D: Appl. Phys.*, 42, 055110, 2009. © 2009 by the IOP Publishing.)

the FA, DAP, and A^0X peaks implies that a fraction of the doped phosphorus atoms are acting like acceptors in ZnO nanotetrapods. The acceptors gradually ionize with the increase of temperature, and thus the peaks disappear at high temperatures. Using Equation 2.1, the binding energy between free excitons and acceptor was calculated to be 11.5 meV.

From the above discussion, it is apparent that doping plays a significant role in governing the optical properties of ZnO tetrapods. Doping with Sn in ZnO tetrapods resulted in the formation of a new peak due to excitons bound to neutral donors. Also, the LO phonon replicas of the emission related to excitons bound to neutral donors were observed only in 5 at.% Sn-doped ZnO tetrapods. When ZnO tetrapods were doped with P, the neutral-donor-bound excitonic peaks disappeared, and several new PL peaks evolved such as free electrons to neutral-acceptor, neutral-acceptor-bound excitonic transitions, as well as donor–acceptor pair emission and its LO phonon replica.

2.3.4 Mg- and **Mn-Doped ZnO** Tetrapods

The introduction of Mg and Mn dopants also significantly changes the optical properties of ZnO tetrapods [34,35]. The near-band edge (NBE) emission underwent a blue shift of about 5 nm upon adding Mg dopant in ZnO tetrapods [34]. However, the blue shift in the NBE peak was not monotonic. Since MgO has a higher bandgap (6.7 eV) than ZnO, the addition of Mg to ZnO is expected to increase the bandgap. But, at the same time, increasing the Mg fraction in ZnO changed the dimension of the tetrapods, resulting in a lower quantum confinement effect and decrease of the bandgap. Thus, no monotonic blue shift could be observed in the NBE transition with Mg addition. The emission related to the native shallow donors decreased with the increase in Mg concentration. Such an emission peak is due to the transition between the zinc interstitial energy level and the valence band [52]. With the increase of Mg mole fraction, the concentration of Zn interstitial decreased, resulting in the reduction of the emission due to the native shallow donors. However, Mn doping in ZnO tetrapods resulted in the reduction of the ratio of UV to defect-state emission, without altering the peak position [35].

2.4 ELECTRICAL PROPERTIES OF ZnO TETRAPODS

The electrical properties of ZnO have been widely investigated for applications in nanoelectronics. ZnO exhibits n-type semiconducting behavior due to the presence of native defects such as zinc interstitials and oxygen vacancies. In this section, we discuss the study of the electrical properties of pure and doped ZnO tetrapods using complex-plane impedance spectroscopy. Impedance spectroscopy provides a correlation between the microstructural and electrical transport properties. It is possible to separate the resistive and capacitive properties associated with different parts of a material by relating the individual components to various relaxation times (τ). The relaxation time, defined as $\tau = RC$, is a unique and important property of the material and does not change with the geometry of the sample. It is possible to obtain the time constant associated with the relaxation processes taking place within the material by analyzing the impedance spectrum.

2.4.1 Undoped ZnO Tetrapods

In order to find out the individual contribution of the arm and junction of a tetrapod to the overall electrical conduction, Huh et al. [36] carried out impedance measurements on a single ZnO tetrapod. Figure 2.7a and b shows the scanning electron microscopy (SEM) images of the fabricated devices (designated as devices A and B) used for electrical measurements. Terminals I, II, and III are the Ti/Au electrodes, which were deposited at different positions in devices A and B, such that the DC resistance measured between any pair of terminals in device A (R_{dc}^A) consists of the resistance due to the arm (R_A) and the junction (R_J) (i.e., $R_{dc}^A \cong R_A + R_J$). On the other hand, the measurement between terminals I and II in device B (R_{dc}^B) provides the resistance due to the arm (i.e., $R_{dc}^B \cong R_A$) only. Figure 2.7c and d shows the complex-plane plots measured at 380°C in the frequency range 10^{-1}–10^7 Hz in devices A and B, respectively, using the terminals I and II in both cases. The corresponding equivalent circuits are shown in the inset of the figures. The impedance spectrum obtained from device A (Figure 2.7c) consists of two partially overlapping semicircles appearing in the lower and higher frequency regions. These are due to the arm and junction of a single tetrapod, which are structurally and electrically two different regions. On the other hand, the impedance spectrum obtained from device B, shown in Figure 2.7d, comprises only a single semicircle, corresponding to the arm of a tetrapod. The activation energy of R_A (extracted from the impedance spectra of Figure 2.7d) was found to be 0.95 eV. On the other hand, the activation energy of the resistance corresponding to the semicircles in the higher and lower frequency regions (Figure 2.7c) was 0.9 and 0.73 eV, respectively. The activation energy values indicate that

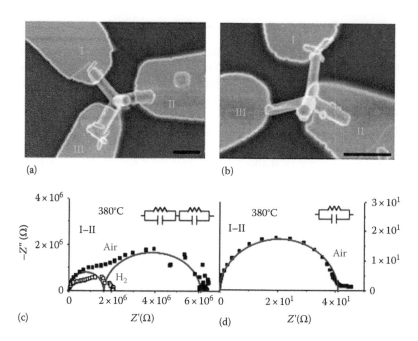

FIGURE 2.7 SEM micrographs of the fabricated single tetrapod devices for the measurement of dc resistance at (a) arm and junction (device *A*) and (b) arm (device *B*), and complex-plane impedance spectra measured at 380°C under air (solid squares) in (c) device *A* and (d) device *B*. The impedance spectrum measured at 380°C under hydrogen (open squares) is also shown in part (c). The solid curves in parts (c) and (d) are the corresponding fit. The corresponding equivalent circuits are shown in the insets (c) and (d). The scale bar shown in parts (a) and (b) corresponds to 1.0 μm. (Reprinted with permission from Huh, J., Kim, G.-T., Lee, J.S., and Kim, S., *Appl. Phys. Lett.*, 93, 042111, 2008. © 2008, American Institute of Physics.)

the semicircles in the higher and lower frequency regions of Figure 2.7c correspond to those of the arm and the junction of the tetrapod, respectively. Thus, the individual contribution of the arm and junction of a tetrapod to the overall electrical conductivity could be extracted, which may be useful for several electronic devices. Moreover, the resistance of the junction is higher than the arm, indicating the key role played by the junction in controlling the electrical property of a tetrapod, in spite of its smaller volume fraction compared to the arm.

2.4.2 Sn-Doped ZnO Tetrapods

Impedance measurements have also been carried on Sn-doped ZnO tetrapods [28]. Unlike the work reported on a single ZnO tetrapod [36], the study on Sn-doped ZnO [28] considered a high density of tetrapods grown on SiO_2/Si substrates. Al deposited over the tetrapods was used as the electrodes. The insulating SiO_2 layer between the tetrapods and the Si substrate eliminated the contribution of the substrate in the electrical measurement. In high-density tetrapods, there exists a connection between the arm of the tetrapod and the arm or junction of the neighboring one. Conduction occurs between the two Al electrodes through these interconnected tetrapods. Figure 2.8a through c shows the impedance spectra recorded for pure ZnO, 3 at.% Sn-doped ZnO, and 5 at.% Sn-doped ZnO tetrapods, respectively. The measurements were carried out at room temperature with the DC bias voltage varying from 0 to 1.5 V. In a typical plot, two overlapping semicircles are observed within the measured range: one in the low-frequency region (denoted by semicircle 1) and another relatively small semicircle in the high-frequency region (denoted by semicircle-2). The semicircle 2 is shown separately in the inset of the figures.

The semicircles could be attributed to the arm and junction of the tetrapods. The equivalent circuit, shown in Figure 2.8d, results in an equivalent impedance, given by

$$Z = R_S + \left(\frac{1}{R_A} + i\omega C_A \right)^{-1} + \left(\frac{1}{R_J} + i\omega C_J \right)^{-1} \tag{2.4}$$

where

R_S is the resistor accounting for the shift occurring along the real Z-axis in the high-frequency region

R_A, C_A are the resistance and capacitance due to the arms, respectively

R_J, C_J are the resistance and capacitance due to the junctions of the tetrapods, respectively

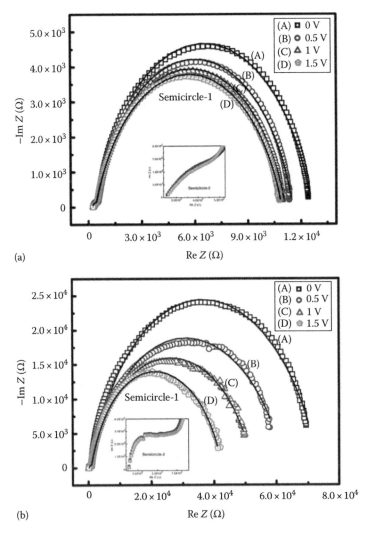

FIGURE 2.8 Complex-plane impedance spectra for (a) pure ZnO, (b) 3 at.% Sn-doped. (From Rakshit, T. et al., *AIP Adv.*, 3, 112112, 2013.)   (*Continued*)

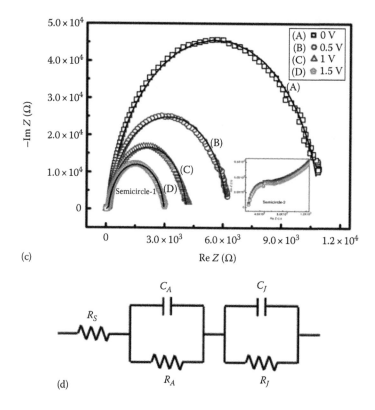

(c)

(d)

FIGURE 2.8 (Continued) Complex-plane impedance spectra for (c) 5 at.% Sn-doped ZnO tetrapods. The semicircle in the higher frequency region is shown in the inset of the figures. The experimental data are denoted by open symbols and corresponding fit by the solid curves. (d) Equivalent circuit of the device. (From Rakshit, T. et al., *AIP Adv.*, 3, 112112, 2013.)

The impedance spectra were fitted with Equation 2.4, and the best fit is shown by the solid curves in Figure 2.8a through c. The resistance extracted from semicircle 1 and semicircle 2, shown in Figure 2.9a and b, respectively, is higher in Sn-doped ZnO tetrapods as compared to undoped ZnO. As the DC bias increases, the resistance decreases for both semicircle 1 and 2. However, the change of resistance with DC bias of semicircle 2 is much less than for semicircle 1. For semicircle 1, the resistance of 5 at.% Sn-doped ZnO tetrapods is lower than that of 3 at.% Sn-doped ZnO samples from 0.5 V onward. However, the resistance value increases uniformly with increase in Sn doping for semicircle 2. When a bias is applied, carriers are injected into the tetrapods, which find a conducting path due to the defects (such as oxygen vacancies) present in the arms and junctions, thereby reducing the resistance. The defect density is generally more in junctions than in the solid arms. So, the change of resistance of the arms is expected to be less than that of the junction. Therefore, semicircle 1 and semicircle 2 correspond to the junction and arms of tetrapods, respectively. In 3 at.% Sn-doped ZnO tetrapods, the dominant carrier concentration is from defect states. When the doping concentration is increased to 5 at.%, and a high bias is applied, the injected carriers dominate over the defect-state contribution. This results in a decrease of the resistance of the junction of 5 at.% Sn-doped ZnO tetrapods compared to 3 at.% Sn-doped ZnO tetrapods. On the other hand, tetrapod arms have much lower defect density, so the conductivity is not much affected by the injected carriers. Therefore, an increase in the resistance of the arms with Sn doping is observed at all bias voltages. The relaxation time of Sn-doped ZnO tetrapods is higher than that of pure ZnO. Thus, by varying the concentration of Sn doping in ZnO tetrapods, both the conductivity and relaxation time could be tuned.

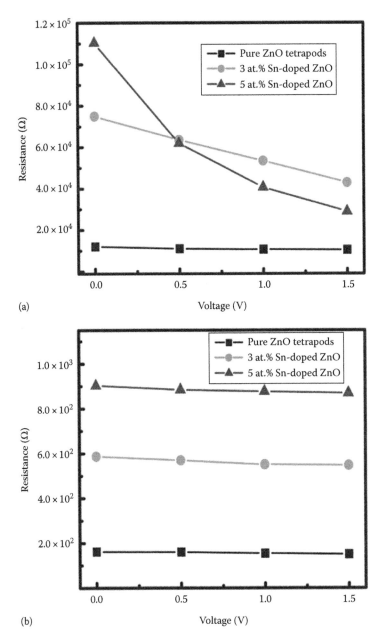

FIGURE 2.9 Variation of resistance extracted from semicircles (a) in the low frequency region (semicircle-1) and (b) in the high frequency region (semicircle-2) with dc bias voltage. (From Rakshit, T. et al., *AIP Adv.*, 3, 112112, 2013.)

2.5 FIELD-EMISSION PROPERTIES OF ZnO TETRAPODS

Field emission (FE) is strongly dependent on a number of factors of the emitter such as geometry, conductivity, work function, and so on. ZnO has a low work function, is highly stable both thermally and mechanically, and is extremely resistant to high-energy radiation. These properties make ZnO nanostructures a potential candidate for FE devices. High FE current density can be obtained at a relatively low electric field in ZnO nanotetrapods with high aspect ratio [53,54]. When ZnO tetrapods are used as field emitters, highly efficient electron emission can be obtained from the four arms, with an arm always directed perpendicular to the anode, resulting in a lower turn-on field [55]. Doping with selective elements such as Sn, Mg, and so on, serves as an effective way in tuning the electrical properties and hence to improve the FE performance [31,37].

2.5.1 Sn-Doped ZnO Tetrapods

The FE properties of Sn-doped ZnO tetrapods were evaluated by preparing cathodes with the tetrapods [37]. Such tetrapods were synthesized by thermal evaporation, with different concentrations of Sn dopant in the Zn and Sn mixture. The fraction of Sn was 0, 1, 3, 5, and 10 at.%, which are denoted as Sn0, Sn1, Sn3, Sn5, and Sn10 samples, respectively, hereafter. Tetrapods with low concentration of Sn doping (Sn1–Sn5) exhibited high aspect ratio with longer and thinner arms. The variation of current density as a function of applied electric field is shown in Figure 2.10a. The turn-on field is 3.91, 1.95, 1.96, 2.27, and 3.84 V/μm, for undoped ZnO, Sn1, Sn3, Sn5, and Sn10 sample, respectively. The threshold field is found to be ~4.58 and 5.35 V/μm for Sn1 and Sn3 samples, respectively. The highest current density of ~1.13 mA/cm^2 was obtained for 1 at.% Sn:ZnO at an applied electric field of 4.71 V/μm. Thus, low concentration of Sn doping in ZnO tetrapods shows significant improvement in the FE performance. The FE characteristics can be analyzed using the Fowler–Nordheim (F–N) equation, which is given by [53]

$$J = A\left(\frac{\beta^2 E^2}{\varphi}\right)\exp\left(\frac{-B\varphi^{3/2}}{\beta E}\right) \tag{2.5}$$

So,

$$\ln\left(\frac{J}{E^2}\right) = \ln\left(\frac{A\beta^2}{\varphi}\right) + \left(\frac{-B\varphi^{3/2}}{\beta}\right)\frac{1}{E} \tag{2.6}$$

where
  J and E are the current density (A/m^2) and the applied field, respectively
  coefficients $A = 1.54 \times 10^{-10}$ (A/V^2 eV) and $B = 6.83 \times 10^9$ (V/m/eV$^{3/2}$)
  φ is the work function of the emitter material (~5.4 eV for ZnO)
  β is the field-enhancement factor [37], which is a measure of the ability of the emitter to increase the local electric field

The field-enhancement factor strongly depends on the geometry, crystal structure, density of nanostructures, work function, and conductivity [31]. The F–N plots for Sn-doped ZnO devices are shown in Figure 2.10b, demonstrating that the characteristics are linearly dependent at high applied fields. The field-enhancement factor calculated from the F–N plots for Sn0, Sn1, Sn3, Sn5, and Sn10 was 1919, 9556, 9256, 8938, and 3614, respectively. Thus, this factor is much higher in lightly doped ZnO sample (Sn1) compared to undoped and heavily doped ones. Sample Sn1 also showed a stable emission current density over a period of 200 min under an applied field of ~4 V/μm (Figure 2.10c). The superior FE performance of the lightly Sn-doped ZnO sample can be attributed to the high electron density in the conduction band caused by the doping and high aspect ratio of the tetrapods.

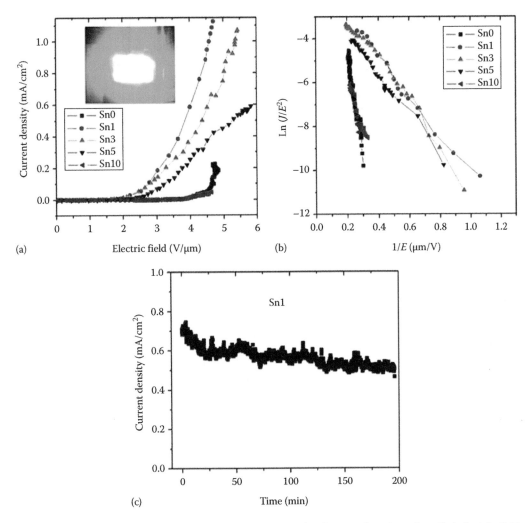

FIGURE 2.10 (a) Variation of the field emission current density as a function of applied electric field. The field emission micrograph of 1 at.% Sn:ZnO (Sn1) is shown in the inset. (b) Fowler–Nordheim (F–N) plot of the corresponding *J–E* curves. (c) Variation of field emission current density as a function of time for Sn1 sample under an applied field of ~4 V/μm. (Reprinted with permission from Zhou, X., Lin, T., Liu, Y., Wu, C., Zeng, X., Jiang, D., Zhang, Y.-A., and Guo, T., *ACS Appl. Mater. Interfaces*, 5, 10067–10073, 2013. © 2013, American Chemical Society.)

2.5.2 Mg-Doped ZnO Tetrapods

The FE performance of Mg-doped ZnO tetrapods, grown by thermal evaporation, has also been reported [31]. Two different types of samples were analyzed: one with higher Mg concentration but with thicker and shorter tetrapod arms compared to the other. The turn-on field for the higher and lower Mg atomic fraction was 2.6 and 2.8 V/μm, respectively. The corresponding current density for the samples reached 0.76 and 0.58 mA/cm^2, respectively, at an applied electric field of 6.0 V/μm. Using Equation 2.6, the field-enhancement factor was found to be 2327 and 1690 for higher and lower Mg fraction, respectively. The results indicate that the FE performance of the doped ZnO tetrapods with higher Mg content is better with higher current density. This could be attributed to the higher conductivity for higher concentration of dopants. The Mg-doped ZnO samples exhibited superior FE performance compared to the undoped ones due to the enhanced conductivity induced by Mg doping.

2.6 SHAPE- AND SIZE-DEPENDENT PROPERTIES OF ZnO NANOSTRUCTURES

The morphology of ZnO nanostructures largely controls their optical properties. Different excitonic emissions have been observed from PL measurements in ZnO nanostructures. The NBE emission of ZnO nanoneedles and nanonails at low temperatures consisted of several well-resolved peaks, whereas well-resolved phonon peaks were absent in nanorods having rounded tips [56]. The excitonic emissions together with their phonon replicas contributed differently to the room temperature PL of ZnO nanorods and nanopencils, resulting in a red shift of 52 meV in the NBE emission of nanopencils compared to the nanorods [57]. This was related to the different exciton–phonon interaction, which depends on the surface defects. ZnO nanopencils, with a large surface to volume ratio, exhibited strong exciton–phonon interactions leading to different excitonic emissions from ZnO nanorods. Similar behavior was observed in the excitonic emission from ZnO nanosheets and nanowires [58]. The enhanced exciton–phonon interactions in nanosheets resulted in a ~40 meV red shift of the NBE emission at room temperature compared to the nanowires. The optical properties are also influenced by the size of ZnO nanostructures. Room-temperature PL measurements on ultrathin ZnO nanobelts of 6 nm width exhibited a 120 meV blue shift in the NBE emission compared to 200 nm wide nanobelts [59]. A blue shift in the NBE emission was also observed with decrease in the size of ZnO quantum dots due to quantum confinement effect [60].

The shape and size of ZnO nanostructures also affect their electrical properties. ZnO nanowires and nanorods were utilized for fabricating field-effect transistors (FETs) [61,62]; the electrical transport properties such as carrier concentration and mobility of ZnO nanorods were almost independent of the size of the nanorods [62]. The electric FE performance of ZnO nanowires, nanoneedles, tetrapods, and so on. has been extensively studied [55,63,64]. At a current density of 0.1 μA/cm^2, the turn-on field was 6 V/μm for ZnO nanowires [63]. An improved FE performance was observed with ZnO tetrapods, in which a low turn-on field of 1.6 V/μm was found at a current density of 1 μA/cm^2 [55]. The higher aspect ratio of the tetrapods compared to nanowires could be the reason for the enhanced performance.

2.7 ZnO NANOSTRUCTURES FOR DEVICE APPLICATIONS

Over the years, an enormous amount of research has been carried out on the synthesis and application of ZnO nanostructures, motivated by their outstanding electrical and optical properties. ZnO nanostructures have great potential for a variety of functional applications including photovoltaics [65], photodetectors [66], light-emitting diodes (LEDs) [67], transparent conducting oxides [68], optical waveguides [69], field emitters [70], piezoelectric transducers [71], spintronic devices, gas sensors [72], and so on. Some of the device characteristics are discussed here.

2.7.1 ULTRAVIOLET (UV) EMITTERS

The wide and direct bandgap of ZnO makes it a promising candidate for UV light emitters. Due to the large exciton-binding energy of ZnO (60 meV), it has emerged as a potential candidate to replace GaN (25 meV). However, because of the lack of high-quality and reliable p-type ZnO, the homojunction ZnO UV emitters have rarely been investigated. Yang et al. have reported the electroluminescence (EL) from p–n junction ZnO nanowires fabricated by As$^+$ ion implantation [73]. The schematic of an As-doped ZnO nanowire diode p–n junction is shown in Figure 2.11a. The EL spectra of the ZnO device with different forward injection currents are shown in Figure 2.11b. An intense, narrow emission in the UV range is clearly observed. The inset of Figure 2.11b shows the injection current–dependent EL intensity. The ZnO homojunction device exhibited enhanced luminescence and lasing performance in the UV region at room temperature. A Fabry–Perot-type ZnO UV laser using n-type ZnO film and p-type Sb-doped ZnO nanowires has been reported [74].

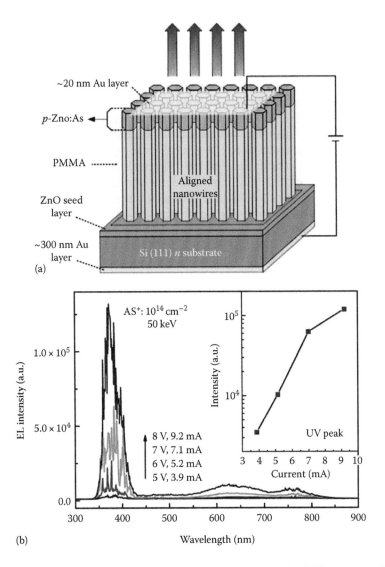

FIGURE 2.11 (a) Schematic illustration of the p–n junction ZnO nanowire LED structure. (b) EL spectra of the p–n ZnO device with different forward injecting currents. Inset shows the EL intensity as a function of injecting current. (Reprinted with permission from Yang, Y., Sun, X.W., Tay, B.K., You, G.F., Tan, S.T., and Teo, K.L., *Appl. Phys. Lett.*, 93, 253107(1–3). © 2008, American Institute of Physics.)

The study demonstrated the potential use of ZnO for diode lasers with lower cost and higher power as compared to GaN. However, efforts are still on to fabricate homojunction ZnO devices with improved performance. The formation of ZnO heterojunctions with several p-type semiconductors such as Si, SiC, GaN, AlGaN and NiO, and so on, has been investigated. Due to the well-matched lattice constant and thermal expansion coefficient of ZnO and GaN, ZnO-based heterojunction photonic devices have been fabricated on p-type GaN or AlGaN. Coaxial n-GaN/ZnO nanorods on p-GaN by metal-organic vapor phase epitaxy have been fabricated to achieve intense EL at near-UV wavelengths [75]. LED devices using ordered and aligned ZnO nanorods on p-GaN have shown strong UV emission at 390 nm [76]. The ZnO–MgO alloy resulted in UV emitters with broadened emission wavelength. Bandgap tuning of ZnO by doping with several transition metals such as Co, Mn, and Ni could be done to fabricate UV emitters over a broad spectral range.

2.7.2 UV Photodetectors

UV detectors based on ZnO are attractive owing to their simple fabrication process, high on/off current ratio, visible-light blindness, and potential for flexible electronics [77,78]. The first individual ZnO nanostructure-based detector was reported in 2002 [79]. The device showed a high sensitivity with a decreased resistance of four to six times in magnitude upon illumination with 365 nm [79]. A single ZnO nanowire UV photodetector with high internal photoconductive gain of ~10^8 has been reported [80]. Two main factors contribute to the high photoresponse of ZnO detectors: large surface to volume ratio, and reduced charge carrier transit time in low-dimensional structures. ZnO-based photodetectors can be implemented in different device configurations including metal–semiconductor–metal (MSM), Schottky barrier, and p–n or p–i–n junctions [81–83]. In particular, MSM photodetectors or photoconductors have been widely used because of their simple structure and ease of device fabrication. However, such photodetectors have slow response and poor gain due to charge recombination or charge trapping/detrapping at the defect levels of ZnO [84]. In order to improve the photoresponse of ZnO, the charge carrier density in ZnO needs to be increased. Recently, there have been considerable efforts in enhancing the UV photoresponse of ZnO using functionalized metal nanoparticles (NPs). Due to the localized surface plasmon resonances (LSPRs) of metal NPs, the absorption of light is enhanced, resulting in higher charge density in ZnO. The enhancement in photoresponse by metal nanoparticles can be explained by LSPR-induced absorption and scattering phenomena. The absorption (C_{abs}) and scattering cross-sections (C_{scat}) of metal NPs of average radius a are given by [85]

$$C_{abs} = 4\pi ka^3 \, \text{Im} \left[\frac{\varepsilon_s - \varepsilon_d}{\varepsilon_s + 2\varepsilon_d} \right] \tag{2.7}$$

$$C_{scat} = \frac{8\pi}{3} k^4 a^6 \left| \frac{\varepsilon_s - \varepsilon_d}{\varepsilon_s + 2\varepsilon_d} \right|^2 \tag{2.8}$$

where ε_s and ε_d are the dielectric constants of the metal sphere and surrounding dielectric medium, respectively. Therefore, the absorption and scattering processes vary with the radius of the metal spheres. Consequently, for smaller particles the absorption dominates, while for larger sized particles the scattering process dominates. Therefore, the plasmonic effect could be useful to tailor the photoresponse over a broad wavelength range.

2.7.2.1 Plasmonic ZnO Photodetector

Metal-functionalized ZnO nanostructures can be prepared by various methods. A simple photoreduction method is one of the new approaches to prepare metal-functionalized ZnO nanocomposites. In a typical process, the photogenerated electrons from ZnO under UV illumination react with metal salts in an aqueous solution to form metal NPs on the surface of ZnO [86]. Typical low- and high-resolution transmission electron microscopy (TEM) images and selected area electron diffraction pattern (SAED) of synthesized Au–ZnO plasmonic nanocomposites are shown in Figure 2.12. As shown in Figure 2.12a, the spherical Au NPs are well attached to ZnO nanosheets. The lattice fringes of Au–ZnO nanocomposites shown in Figure 2.12b reveal an interplanar spacing of 0.281 and 0.235 nm for ZnO($1\bar{1}00$) and Au(111) planes, respectively. The combined SAED pattern of Au–ZnO shows the hexagonal spot patterns with circular rings, as shown in Figure 2.12c. The hexagonal SAED pattern can be indexed to the ($\bar{1}100$), ($01\bar{1}0$), ($10\bar{1}0$), and ($\bar{1}010$) planes of ZnO nanosheets. The circular rings corresponding to (111), (200), and (220) crystallographic planes of Au NPs reveal their polycrystalline nature. The absorption spectra of ZnO and Au–ZnO plasmonic nanocomposites are presented in Figure 2.13. The Au–ZnO sample clearly exhibits two

(a)

(b)

(c)

FIGURE 2.12 Typical TEM images of plasmonic Au–ZnO nanocomposites at (a) low and (b) high resolution. (c) Combined SAED pattern of plasmonic Au–ZnO nanocomposites. (From Gogurla, N. et al., *Sci. Rep.*, 4, 6483(1–9), 2014.)

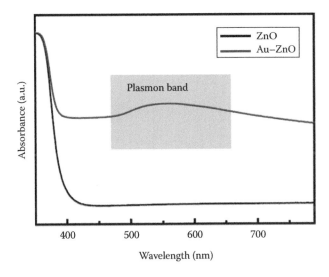

FIGURE 2.13 UV–visible absorption spectra of ZnO and plasmonic Au–ZnO nanocomposites. (From Gogurla, N. et al., *Sci. Rep.*, 4, 6483(1–9), 2014.)

peaks compared to one for pure ZnO. The peak in the UV region is due to the band-edge absorption of ZnO and the one in the visible region is attributed to the surface plasmon absorption of Au NPs. The surface plasmon peak of Au–ZnO nanocomposite is broad in the visible region due to the wide particle size distribution of Au NPs.

Photoconductive gain and spectral response are the key parameters to test the performance of photodetectors. ZnO and plasmonic Au–ZnO nanocomposite devices were tested in the lateral photoconductor configuration under UV illumination, the schematic of which is shown in Figure 2.14a. Typical *I–V* characteristics for ZnO and Au–ZnO devices in the dark and under UV illumination are shown in Figure 2.14b. For both the devices, a pronounced increase in the current under UV illumination is observed. The photocurrent of the Au–ZnO device is found to be enhanced compared to that of the control ZnO sample. The spectral photoresponse of ZnO

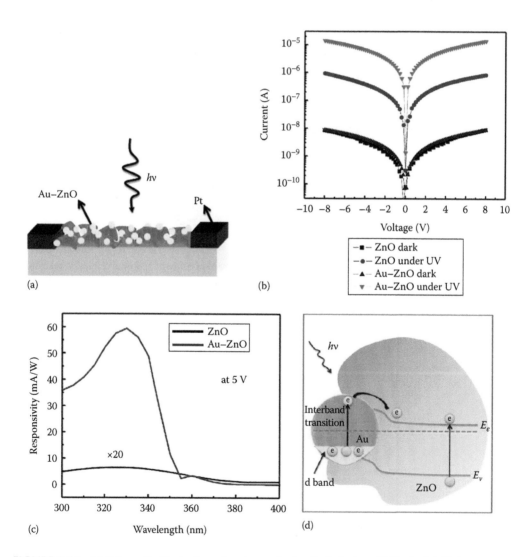

FIGURE 2.14 (a) Schematic illustration of a plasmonic Au–ZnO device. (b) Typical *I–V* characteristics of ZnO and plasmonic Au–ZnO devices under dark and UV illumination. (c) Spectral response of ZnO and Au–ZnO devices in the UV region at 5 V bias. (d) Schematic diagram representing the charge transfer mechanism involved in the Au–ZnO device. (From Gogurla, N. et al., *Sci. Rep.*, 4, 6483(1–9), 2014.)

and Au–ZnO devices at 5 V bias are shown in Figure 2.14c. The responsivity (R_λ) of the devices can be calculated from [87]

$$R_\lambda = \frac{\Delta I}{PA_d} \tag{2.9}$$

where
  P is the optical power density falling on the sample
  ΔI is the difference between the dark current and photocurrent
  A_d is the area of the lateral device.

The Au–ZnO nanocomposite device exhibited an enhanced photoresponse over a broad wavelength range compared to ZnO. The responsivity of Au–ZnO is notably higher for the wavelength range 300–350 nm and slowly decreases when the device is illuminated with a longer wavelength. As seen in Figure 2.14c, the peak responsivity of the plasmonic Au–ZnO device is enhanced by 80 times at 325 nm over the control ZnO sample.

The origin of the enhanced photoresponse in metal-functionalized semiconductor UV photodetectors can be explained by the LSPR effect of metal NPs [88,89]. However, in this case, Au nanoparticles show LSPR only in the visible region, as can be seen in Figure 2.13. Therefore, the enhanced UV photoresponse for the Au–ZnO device might be attributed to the interband transition in Au under UV illumination or to photogenerated hole trapping in Au nanoparticles [90,91]. Under UV illumination, the filled d-band electrons of Au NPs are excited to the conduction band and subsequently get transferred to ZnO due to the electric field at the junction formed between Au and ZnO, resulting in a higher photoconductivity. In case of ZnO, the responsivity in the UV region is attributed to the band-edge absorption and is much lower than that of the plasmonic nanocomposite. The absorption and transfer processes are schematically shown in Figure 2.14d. It is also possible that the photogenerated holes in ZnO can be trapped by Au nanoparticles, which may suppress the electron–hole recombination in the plasmonic device. This would result in an enhanced photoresponse in the UV region. Therefore, metal-sensitized ZnO nanostructures can be used as a novel platform for enhanced UV photodetectors.

Due to the negative real part and relatively low imaginary dielectric function of Al in the UV region, it is also useful to enhance the UV photoresponse of ZnO. Enhanced UV photodetection characteristics of ZnO with Al NPs have been reported [92]. ZnO nanorods prepared on quartz substrates by vapor-phase transport were decorated by Al nanoparticles using RF sputtering. A schematic MSM device diagram of the Al NP-decorated ZnO nanorods is shown in Figure 2.15a. Typical I–V characteristics of ZnO MSM photodetectors without and with Al NPs under dark and with UV illumination are shown in Figure 2.15b and c, respectively. The linear increment in the current with the voltage indicates the ohmic contact between the ZnO nanorods and Ag electrodes at the interfaces. Due to the lower work function of Al compared to ZnO [93], the electrons can move from Al to ZnO. These electrons contribute to the conduction under 5 V bias, resulting in a higher dark current for the Al–ZnO device. Under UV light illumination, the ZnO device with Al NPs has shown increased photocurrent as compared to that without Al NPs. This can be attributed to the LSPR coupling between the Al NPs and ZnO nanorods, leading to the enhanced electron–hole generation in ZnO.

Figure 2.16 shows the responsivity of ZnO UV photodetectors with and without Al NPs under UV–vis light illumination at a bias of 5 V. The responsivity spectra show that the devices are sensitive to UV light. It was also observed that the responsivity of the ZnO detector with Al NPs enhanced greatly in the UV region. The responsivity peak was observed at 375 nm for both the devices. UV radiation can be absorbed by Al NPs due to LSPR effect at the metal–dielectric interface. The LSPR of Al NPs can couple with the excitons of ZnO due to the matching of surface plasmon energy with ZnO interband transition. This significant energy coupling will lead to the pronounced excitation in ZnO, resulting in a faster response under UV illumination.

Several other metal nanoparticles have also been used to enhance the photoresponse of ZnO photodetectors. Enhanced photoresponse could be achieved in ZnO thin film MSM devices using Pt

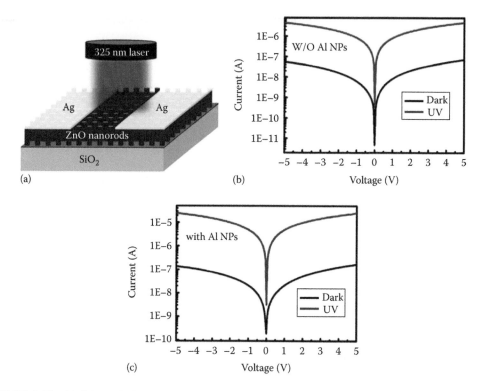

FIGURE 2.15 (a) Schematic illustration showing the device diagram of plasmonic Al–ZnO nanorods. *I–V* characteristics of (b) bare ZnO and (c) Al-decorated ZnO devices in the dark and 325 nm illumination. (From Lu, J., Xu, C., Dai, J. et al., Improved UV photoresponse of ZnO nanorod arrays by resonant coupling with surface plasmons of Al nanoparticles, *Nanoscale*, 7, 3396–3403, 2015. Reproduced by permission of The Royal Society of Chemistry.)

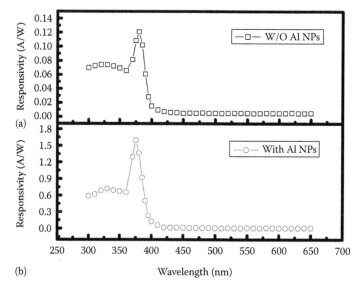

FIGURE 2.16 Photoresponsivity spectra (a) without and (b) with Al-decorated ZnO devices at a bias of 5 V. (From Lu, J., Xu, C., Dai, J. et al., Improved UV photoresponse of ZnO nanorod arrays by resonant coupling with surface plasmons of Al nanoparticles, *Nanoscale*, 7, 3396–3403, 2015. Reproduced by permission of The Royal Society of Chemistry.)

nanoparticles [94]. The photoresponse of ZnO devices with Pt NPs prepared by different sputtering times were recorded at 3 V bias. The responsivity of ZnO increased from 0.836 to 1.306 A/W using Pt nanoparticles [94]. The enhanced photoresponse in the Pt–ZnO devices could be well understood from the absorption and scattering phenomena of Pt nanoparticles.

2.7.3 ZnO Nanostructures for Photovoltaic Devices

Due to their high transparency in the visible region and their high electron mobility, ZnO nanostructures are attractive for dye- and quantum-dot-sensitized solar cells. Several studies have been carried out on ZnO nanostructure–based, dye-sensitized solar cells (DSSCs), which are summarized in different reviews [95,96]. Although the obtained maximum power conversion efficiency of 5.6% for ZnO-based DSSCs [96] is much lower than that of TiO_2 (efficiency of 11%) [97], it is still potentially attractive to replace TiO_2 due to the easy crystallization and anisotropic growth of ZnO. However, the degradation of dyes results in the lack of stability and long-term performance, which limits the applications of DSSCs. An effective approach to overcome the problems associated with DSSCs is to replace the dyes with polymers. Such hybrid solar cells comprise a polymer as the electron donor and metal oxide as the electron acceptor. Hybrid solar cells are expected to exhibit superior performance since they combine the merits of both polymers and metal oxides, for example, flexibility and low processing cost of polymers as well as the thermal and chemical stability, high electron mobility, high dielectric constant, and size-tunable optical properties of metal oxides. The ordering of donors and acceptors in a polymer–metal oxide hybrid solar cell is commonly done in two different ways. In one case, the polymer (donor) can be blended with nanostructured metal oxides (acceptor) to form polymer–metal oxide hybrid bulk heterojunction (BHJ) solar cells. Such solar cells have the advantage of low fabrication cost, since the blending of the polymer with ZnO nanostructures can be done by the solution-growth technique. The blending also leads to large donor–acceptor interfacial areas, which results in the formation of a large number of free charge carriers. However, the main challenge lies in controlling the morphology of the heterojunction so that definite transport pathways for the charge carriers can be obtained. The other approach is to grow ordered metal oxide nanostructures on the substrate and infiltrate the polymer in between them. Such solar cells offer a definite transport pathway for the charge carriers. But the main challenges to obtaining high performance in these solar cells are (1) the complete infiltration of polymer into the nanostructures, (2) growth of metal oxide nanostructures of smaller dimensions to obtain higher interfacial area, and (3) controlling the distance between the nanostructures to generate larger number of excitons within the excitonic diffusion length.

In particular, vertically aligned ZnO nanorods have been employed as electron acceptors along with the conductive organic polymer in hybrid solar cells due to their low reflectivity as well as efficient charge separation and collection [98]. In this regard, hybrid solar cells based on ZnO nanostructures and the regioregular poly(3-hexylthiophene) (P3HT) polymer have been studied by several research groups. Baeten et al. have demonstrated ZnO nanorods/P3HT solar cell with 0.76% efficiency [99]. The power conversion efficiency of ZnO/P3HT hybrid devices has been observed to be very low because of the incompatibility of the interface between ZnO and organic polymers. Some polymer processing techniques such as annealing at their melting point and subsequent cooling, spin coating from different organic solvents, and UV treatment on ZnO before polymer deposition can be employed to improve the performance [100,101]. An effective way to improve the photovoltaic performance of hybrid solar cells is to modify the surface of ZnO nanostructures with different inorganic semiconductors. Such modification improves the interfaces between the ZnO nanostructures and polymers to form an optimum morphology, which would ensure efficient charge transport with enhanced performance of the solar cells.

2.7.3.1 CdS-Sensitized ZnO Nanorods

Hybrid photovoltaic devices have been studied with an active layer consisting of a blend of CdS-nanoparticle-decorated ZnO nanorods and P3HT polymer [102]. CdS was deposited by pulsed-laser deposition on the surface of ZnO nanorods at varying deposition times of 10, 20, and 30 min, with the samples denoted as CSZO-1, CSZO-2, and CSZO-3, respectively. In CSZO-1, ZnO nanorods are only partially covered by CdS nanoparticles, whereas in the CSZO-2 sample a larger portion becomes covered. A continuous CdS layer of about 80 nm thickness was formed on the ZnO nanorod surface in the sample CSZO-3, as shown in Figure 2.17a. A photovoltaic device structure having Al/CdS-decorated ZnO nanorods:P3HT/PEDOT:PSS/ITO was fabricated and tested. The conductive PEDOT:PSS layer plays several roles: it assists in hole transport, blocks the excitons, and prevents the oxygen diffusion from ITO by keeping the anode material away from the active layer to prevent the formation of undesirable trap sites [103]. PL measurements showed that the intensity of P3HT quenched with the increase of ZnO concentration. The quenching of P3HT intensity further enhanced on modification of the surface of ZnO nanorods with CdS, indicating the higher rate of exciton dissociation.

The current density versus voltage characteristics of CdS-modified ZnO:P3HT devices (of weight ratio 5:1) with varying CdS surface coverage are shown in Figure 2.17b. The measurement

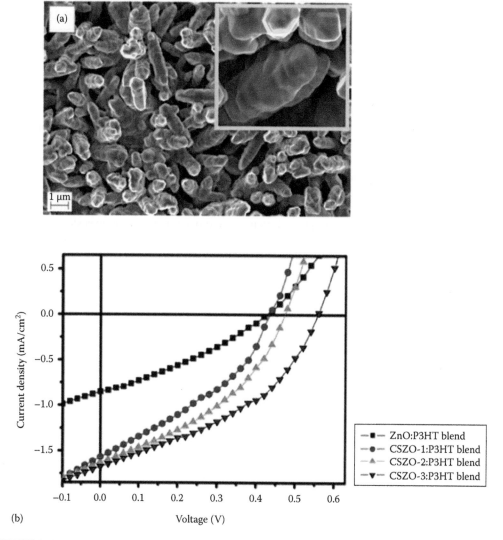

FIGURE 2.17   (a) FESEM image of CdS-decorated ZnO nanorods (CSZO-3 sample). Inset shows the magnified view of the micrograph. (b) Current density versus voltage characteristics of CdS-modified ZnO:P3HT and control ZnO:P3HT hybrid solar cells (blended in weight ratio of 5:1) with varying CdS surface coverage. (From Rakshit, T. et al., *ACS Appl. Mater. Interfaces*, 4, 6085, 2012.)

was performed under 100 mW/cm² AM 1.5 simulated solar irradiation. The surface modification of ZnO nanorods with CdS nanoparticles led to an increase in the open-circuit voltage (V_{OC}) and short-circuit current density (J_{SC}). These parameters were found to increase with the increase in CdS surface coverage. The highest power conversion efficiency (PCE) of 0.38% was obtained in the CSZO-3:P3HT devices, with J_{SC} of 1.67 mA/cm², V_{OC} of 557 mV, and fill factor (FF) of 40%. The efficiency of CdS-decorated ZnO (CSZO-3):P3HT device is higher by more than 300% compared to the control ZnO:P3HT device.

Figure 2.18a shows the external quantum efficiency (EQE) of ZnO:P3HT devices with different ZnO:P3HT weight ratios. The EQE initially increases with ZnO addition in the blend up to a weight

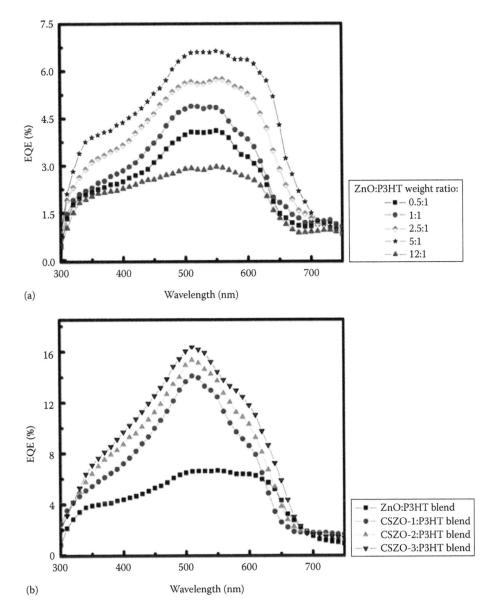

(a)

(b)

FIGURE 2.18 External quantum efficiency of (a) ZnO:P3HT hybrid solar cells, blended in different weight ratios, and (b) CdS-modified ZnO:P3HT and control ZnO:P3HT hybrid solar cells (blended in weight ratio of 5:1), with varying CdS surface coverage. (From Rakshit, T. et al., *ACS Appl. Mater. Interfaces*, 4, 6085, 2012.)

ratio of 5:1, but then decreases with the further increase of ZnO concentration. The highest EQE obtained for ZnO:P3HT devices is ~7% at around 510 and 550 nm. The EQE of CdS-decorated ZnO:P3HT devices (of weight ratio of 5:1) is shown in Figure 2.18b. The EQE of CdS-modified ZnO devices is higher than that of the unmodified one in the wavelength range 300–620 nm, which is enhanced with the increase of CdS surface coverage. The highest EQE for CdS-decorated ZnO:P3HT devices was found to be ~16% at around 510 nm, which is a factor of 2 higher than that of the control ZnO device.

The improved photovoltaic performance of CdS-decorated ZnO:P3HT devices can be explained from the energy band diagram shown schematically in Figure 2.19. The electrons flow efficiently from donors to acceptors if the energy difference between the lowest unoccupied molecular orbital (LUMO) of donors and the conduction band edge of acceptors is ~0.3–0.5 eV [104]. However, this difference is much higher (1.2 eV) in ZnO:P3HT devices. When the surface of ZnO nanorods is modified with CdS nanoparticles, a cascaded band structure is formed, since the conduction band of CdS has a higher energy than that of ZnO but lower than the LUMO of P3HT. This ensures efficient flow of electrons from P3HT to ZnO. Moreover, the hole transfer from P3HT to ZnO is blocked by CdS because of the alignment of the highest occupied molecular orbital (HOMO) of P3HT relative to the valence band energy of CdS. This reduces the recombination in ZnO, thereby increasing the lifetime of the carriers. Therefore a higher efficiency is observed in the CdS-modified ZnO devices. This efficiency can be further improved by enhancing the CdS coverage on the surface of ZnO nanorods. When the CdS nanoparticles partially cover the ZnO nanorod surface, a cascaded band structure is absent in the unmodified portion. The formation of the cascaded band structure occurs when the ZnO nanorods are fully covered with CdS nanoparticles, resulting in the smooth transfer of electrons, which enhances the photovoltaic performance of CdS-decorated ZnO devices.

2.7.3.2 PbS-Sensitized ZnO Nanorods

Hybrid photovoltaic devices have been studied with PbS quantum dot (QD)-sensitized ZnO nanorod arrays and poly[2-methoxy-5-(2-ethylhexyloxy-*p*-phenylenevinylene)] (MEH-PPV) polymer as the active layer [105]. The PbS QDs were deposited on the surface of ZnO nanorods by chemical bath deposition (CBD), with two, four, and six CBD cycles. For lower CBD cycles, PbS QDs partially

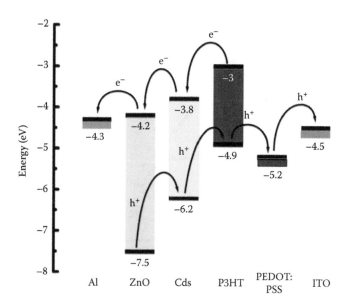

FIGURE 2.19 Schematic energy band diagram of CdS-decorated ZnO:P3HT hybrid solar cell. (From Rakshit, T. et al., *ACS Appl. Mater. Interfaces*, 4, 6085, 2012.)

covered the surface of ZnO nanorods, whereas ZnO nanorods were almost fully covered with six CBD cycles. The device structure used for the photovoltaic measurement was Au/PEDOT:PSS/ MEH-PPV/PbS/ZnO-nanorod/ZnO-seed layer/ITO. The current density versus voltage characteristics of PbS QD-sensitized ZnO devices, along with the control MEH-PPV/ZnO device, measured under 100 mW/cm^2 AM 1.5 simulated solar irradiation condition are shown in Figure 2.20a. The MEH-PPV/ZnO device exhibited J_{SC} of 1.06 mA/cm^2, V_{OC} of 0.25 V, FF of 30%, and PCE of 0.09%. The efficiency was found to be enhanced with the introduction of PbS QDs on the surface of ZnO nanorods. The highest efficiency of 0.42% was obtained for four-cycle CBD grown, PbS QD-sensitized device. Thus, the efficiency obtained in PbS QD-sensitized ZnO devices is higher by approximately five times that of the unmodified device.

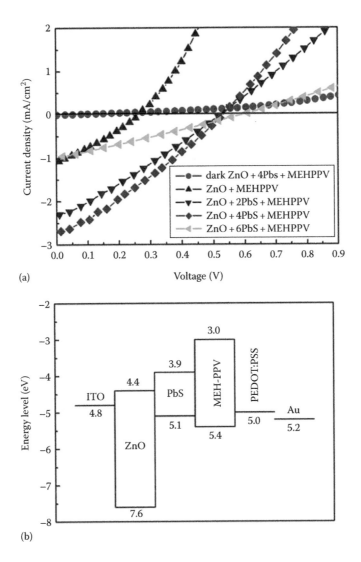

(a)

(b)

FIGURE 2.20 (a) Current density versus voltage characteristics of MEH-PPV/PbS QD-sensitized ZnO and control MEH-PPV/ZnO devices. ZnO + 2PbS, ZnO + 4PbS, and ZnO + 6PbS are PbS QD-sensitized ZnO nanorods, grown with two, four, and six CBD cycles, respectively. (b) Schematic energy band diagram of MEH-PPV/PbS QD-sensitized ZnO solar cell. (Reprinted with permission from *Nanoscale Res. Lett.*, Hybrid polymer/ZnO solar cells sensitized by PbS quantum dots, 7, 2012, 106-1, Wang, L., Zhao, D., Su, Z., and Shen, D. © 2012 by the SpringerOpen.)

Figure 2.20b shows the schematic energy band diagram of MEH-PPV/PbS QD-sensitized ZnO devices. The introduction of PbS QDs on the surface of ZnO nanorods results in the formation of a cascaded band structure, since the conduction band of PbS has a higher energy than that of ZnO. Such a cascaded structure aids in efficient transport of electrons and reduces the recombination of charge carriers. Also, the PbS coating passivates the surface states of ZnO nanorods, resulting in reduced recombination and charge trapping. This enhances the short-circuit current density and open-circuit voltage of the MEH-PPV/PbS QD-sensitized ZnO devices. The ZnO seed layer between the active layer and the ITO electrode prevents the leakage of holes into the ITO electrode. These factors result in an improved efficiency in the four-cycle CBD grown PbS QD-sensitized device. The decrease of efficiency at higher PbS coating (six CBD cycles) is attributed to the reduction in the exposed amount of ZnO. The higher PbS coating also reduces the spacing between the nanorods, resulting in poor infiltration of MEH-PPV into the ZnO nanorods. Thus, the efficiency of MEH-PPV/PbS QD-sensitized ZnO hybrid solar cells largely depends on the amount of surface coverage of ZnO nanorods by PbS QDs.

2.7.3.3 Cosensitized (CdS and CdSe) ZnO Nanowires

Hybrid solar cells have also been fabricated with ZnO nanowires cosensitized with two inorganic semiconductors, namely, CdS and CdSe [106]. A CdS layer of ~9 nm was uniformly deposited on the surface of ZnO nanowires by successive ionic layer adsorption and reaction (SILAR), over which CdSe was deposited by CBD. CdSe nanoparticles were deposited randomly on the surface of the CdS/ZnO nanowires. The structure of the fabricated devices used for photovoltaic measurements was Au/P3HT/CdSe/CdS/ZnO-nanowire/ZnO-film/ITO. The P3HT/CdSe/CdS/ZnO nanowires showed a high absorbance in the visible region of 400–700 nm. The spectral response was much broader than of pure ZnO, CdS/ZnO and CdSe/CdS/ZnO nanowires. Figure 2.21a shows the current density versus voltage characteristics of hybrid solar cells fabricated with P3HT/CdSe/CdS/ZnO, P3HT/CdS/ZnO, and P3HT/ZnO nanowires, measured under 100 mW/cm^2 AM 1.5G solar illumination. The introduction of CdS and CdSe quantum dots on the surface of ZnO nanowires greatly enhances the photovoltaic performance, yielding an efficiency of 1.5%, which is about 75 and 6 times higher than those of P3HT/ZnO and P3HT/CdS/ZnO devices, respectively. The corresponding short-circuit current density, open-circuit voltage, and fill factor of P3HT/CdSe/CdS/ZnO device are 4.2 mA/cm^2, 675.2 mV, and 51.8%, respectively. The performance was found to be highly stable even after 40 days of exposure in air.

The photovoltaic performance of P3HT/CdSe/CdS/ZnO nanowires was also studied as a function of the length of the ZnO nanowires varying from ~630 nm to 1.45 μm. The short-circuit current density increased with the increase of the length of the ZnO nanowires up to ~1 μm. This could be due to the higher loading of CdS and CdSe QDs in longer nanowires. However, further increase in the length of ZnO nanowires lowered the infiltration of P3HT through the nanowires, which reduced the short-circuit current density. On the other hand, the open-circuit voltage decreased with the increase of the length of the ZnO nanowires. This may be due to the enhanced density of recombination sites and inefficient transport of charges in longer ZnO nanowires. The highest efficiency of ~1.4% was obtained in devices with ~1 μm long ZnO nanowires. The cosensitized device exhibited a higher incident photon-to-current conversion efficiency in the visible region of 350–700 nm, with the maximum value of ~25% at 530 nm.

The energy band diagram and device structure of hybrid solar cells based on CdS- and CdSe-sensitized ZnO nanowires are shown in Figure 2.21b. When the surface of ZnO nanowires is sensitized with CdS and CdSe, a cascaded band structure is formed. The cascaded band structure is useful in the dissociation and efficient transport of charge carriers. This reduces the recombination probability and hence increases the carrier lifetime [107,108]. CdS and CdSe QDs also passivate the surface states of ZnO nanowires, thus reducing the density of charge trapping and recombination sites [107,108]. The shunt resistance of hybrid solar cells, which prevents the flow of photogenerated charge carriers through the internal defects, was also analyzed for cosensitized solar cells.

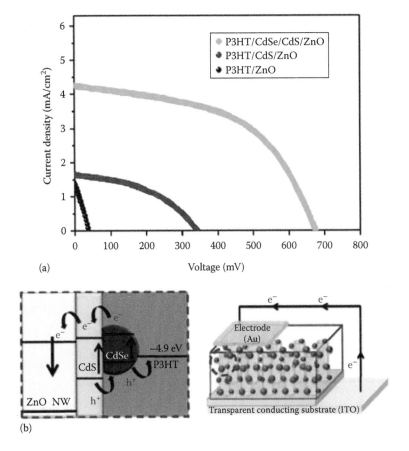

(a)

(b)

FIGURE 2.21 (a) Current density versus voltage characteristics of hybrid solar cells fabricated with P3HT/CdSe/CdS/ZnO, P3HT/CdS/ZnO and P3HT/ZnO nanowires. (b) Schematic energy band diagram and device structure of P3HT/CdSe/CdS/ZnO nanowire hybrid solar cell. (Reprinted with permission from Kim, H., Jeong, H., An, T.K., Park, C.E., and Yong, K., Hybrid-type quantum-dot cosensitized ZnO nanowire solar cell with enhanced visible-light harvesting, *ACS Appl. Mater. Interfaces*, 5, 268. © 2013, American Chemical Society.)

The shunt resistance of P3HT/ZnO solar cell was increased by about 17 times by modifying with CdS and by ~26 times after cosensitizing with CdS and CdSe QDs. All these factors resulted in an improved photovoltaic performance in P3HT/CdSe/CdS/ZnO nanowire solar cells.

Some of the reported results on hybrid solar cells based on semiconductor-sensitized ZnO nanostructures are summarized in Table 2.2. It is apparent that the modification of the surface of ZnO nanostructures with different inorganic semiconductors plays a significant role in controlling the photovoltaic performance of both ZnO/polymer BHJ and nanostructured ZnO/polymer hybrid solar cells.

2.7.4 GAS SENSORS USING ZnO NANOSTRUCTURES

The change in electrical conductivity of ZnO thin films in the presence of reactive gases was discovered in 1962 [115]. This property of ZnO has led many researchers to synthesize different kinds of ZnO nanostructures for novel gas sensing applications. ZnO is one of the most widely investigated sensor materials used for the detection of different gases and volatile organic compounds (VOCs) [116–120]. ZnO nanostructures exhibit pronounced surface effects owing to their high surface to volume ratio, leading to superior sensing performance. Gas sensors using a variety of ZnO nanostructures have been widely reported. Jing et al. have fabricated a porous ZnO nanoplate-based

TABLE 2.2

Photovoltaic Parameters for Hybrid Solar Cells Based on Semiconductor-Sensitized ZnO Nanostructures

Device Structure	Short-Circuit Current Density, J_{SC} (mA/cm²)	Open-Circuit Voltage, V_{OC} (V)	Fill Factor, FF (%)	Efficiency, η (%)	Irradiance (mW/cm²)	References
Al/CdS-decorated ZnO Nanorods:P3HT/PEDOT: PSS/ITO	1.674	0.557	40	0.38	100	[102]
Au/P3HT/CdS/ZnO Nanofiber/ ZnO thin film/ITO	2.25	0.619	46.5	0.65	100	[109]
Au/PEDOT:PSS/MEH-PPV/ CdS/ZnO nanorod/ZnO seed layer/ITO	2.87	0.78	29	0.65	100	[110]
Au/PEDOT:PSS/MEH-PPV/ CdS/ZnO nanorod/ITO	4.65	0.77	34.4	1.23	100	[111]
Al/MoO₃/P3HT/CdSe/ZnO Nanorod/ZnO seed layer/ITO	1.44	0.36	43	0.23	100	[112]
Au/P3HT/CdSe/CdS/ZnO Nanowire/ZnO thin film/ITO	4.2	0.6752	51.8	1.50	100	[106]
Au/PEDOT:PSS/MEH-PPV/ PbS/ZnO nanorod/ZnO seed layer/ITO	2.68	0.55	29	0.42	100	[105]
Pt/P3HT/CuS/ZnO Nanorod/ ZnO seed layer/ITO	4.9	0.391	53	1.02	100	[113]
Au/P3HT/TiO₂/ZnO nanorod/ ZnO seed layer/ITO	1.14	0.50	50	0.29	100	[114]

sensor that exhibited high response to chlorobenzene at low operating temperatures [121]. A strong response to ethanol was also observed for the sensor operated at high temperatures [121]. Park et al. have reported the gas sensing properties of nanograined ZnO nanowires, which showed an enhanced electrical response to NO_2 at 300°C [122]. Different morphologies of ZnO nanostructures have been utilized to sense ethanol and the results indicated that the nanofiber sensor shows desirable response at 270°C [123]. Some theoretical simulations have also been performed on ZnO gas sensors. The interaction of ZnO nanostructure $(10\bar{1}0)$ surface with NO_2, NO, O, and N was examined using density functional theory and ab initio molecular dynamics simulations [124]. A weak adsorption of nitrogen oxides and stable adsorption of O and N even at 700 K on the surface of ZnO were observed. The calculations also indicated that all the adsorbates act as electron acceptors on the surface. This potential sensing properties of ZnO nanostructures resulted in highly responsive gas sensors. However, the applications of these ZnO nanostructures are often limited by the high operating temperature and lack of stability, accuracy, and selectivity toward various gases and VOCs. VOCs have quite similar molecular structure and composition, which increases the difficulty of distinguishing one VOC from another.

2.7.4.1 Gas Sensing Parameters

The adsorption of oxygen on the surface of ZnO nanostructures and the rate of reaction between the target gas and oxygen depend strongly on the operating temperature. Other important parameters determining the performance of gas sensors are sensitivity, selectivity, response time, recovery time, and the limit of detection. These parameters are largely influenced by the surface area, donor density, porosity of the sensing material, and the use of catalysts [125]. The crystallite size of metal oxide also plays a significant role in determining the sensitivity of a gas sensor. When the size of the crystallite (D) is less than twice the thickness of the depletion layer (L), that is, $D < 2L$, the entire crystallite would become depleted of charge carriers. The energy bands become almost flat, leading to easy charge transport within the crystallites. This inter-crystallite conductivity largely controls the conductivity of the oxide. Then a small change in the thickness of the depletion layer due to the reaction of the oxygen ion with the target gas causes a large variation in the conductivity of the metal oxide. Thus, metal oxides having a small crystallite size are highly sensitive to the target gas.

One of the approaches of improving the selectivity and other sensing performances is to modify the surface of ZnO nanostructures with different metals or metal oxides. When the surface of ZnO nanostructures is sensitized by another metal oxide, the Fermi levels on both sides of the interface reach equilibrium through charge transfer, and a depletion layer is formed. This depletion layer plays a significant role in enhancing the performance of the gas sensor. To reduce the operating temperature and improve the sensing performance, several approaches such as doping or functionalizing with noble metals [126,127] and UV light stimulation [128] have been employed. UV light stimulation effectively improves the sensing performances at low operating temperatures.

The sensitivity of a gas sensor is defined in various ways. The most widely used one is defined as R_a/R_g for an n-type metal oxide in the presence of a reducing gas, where R_a and R_g are the resistance of the sensor in air/carrier gas and target gas, respectively. Similarly, in the presence of an oxidizing gas, the sensitivity is given by R_g/R_a for an n-type metal oxide. These become reversed in the case of a p-type material. The response and recovery time are usually defined as the time needed for the sensor to reach 90% of the variation in resistance upon exposure and removal of the target gas, respectively.

2.7.4.2 Surface-Modified ZnO Sensors

As discussed previously, several factors such as operating temperature, surface morphology, and growth technique control the performance of ZnO nanostructure–based gas sensors. The size, shape, and surface states of ZnO nanostructures largely depend on the growth techniques. Variation of any of these parameters can change the performance of sensors. To improve the gas sensing performance, the surface of ZnO nanostructures is often modified with different semiconductor metal oxides or metal nanoparticles. In this section, we discuss the sensing performance of gas sensors based on different semiconductor- and metal-sensitized ZnO nanostructures.

2.7.4.2.1 Semiconductor-Sensitized Gas Sensors

2.7.4.2.1.1 SnO₂ Nanowire/ZnO Nanorod Heterostructures
$2.7.4.2.1.1$ *SnO$_2$ Nanowire/ZnO Nanorod Heterostructures*   The sensing performance of brush-like SnO$_2$ nanowire/ZnO nanorod heterostructures, synthesized by a combination of pulsed laser deposition and hydrothermal methods, was tested with different VOCs such as acetone, ethanol, methanol, acetic acid, toluene, and triethylamine [129]. SnO$_2$ nanowire/ZnO nanorod heterostructures with varying length of the nanowires of ~25, 40, and 55 nm were utilized, which were designated as ZSO1, ZSO2, and ZSO3 samples, respectively. The density of SnO$_2$ nanowires was not uniform over the ZnO nanorod, so the surface of the heterostructures was porous and rough. The optimum operating temperature for sensing acetone, ethanol, acetic acid, toluene, and triethylamine was found to be 300°C, whereas that for methanol sensing was 250°C. Figure 2.22a through c shows the sensing transients of pure ZnO nanorods and SnO$_2$/ZnO heterostructures toward toluene, ethanol, and acetone, respectively, operated at a temperature of 300°C. The response (R_a/R_g) of ZSO1 sensor for toluene (14–115 ppm) is ~1.3 times higher than that of ZnO. However, the response of ZSO2 and ZSO3 sensors is lower than that of ZnO. The sensing characteristics for ethanol (29–230 ppm), shown in Figure 2.22b, reveals that the highest response is shown by ZSO3 sample, which is ~1.8–3 times higher than that of the control sample. On the other hand, the ZSO2-based sensor exhibits the highest response of 21.4–48.4 toward acetone (115–693 ppm), which is ~5.1–3 times higher than that of ZnO nanorods (Figure 2.22c). Sensing transients of SnO$_2$/ZnO heterostructures indicate that the performance depends strongly on the length of the SnO$_2$ nanowire brushes covering the ZnO surface.

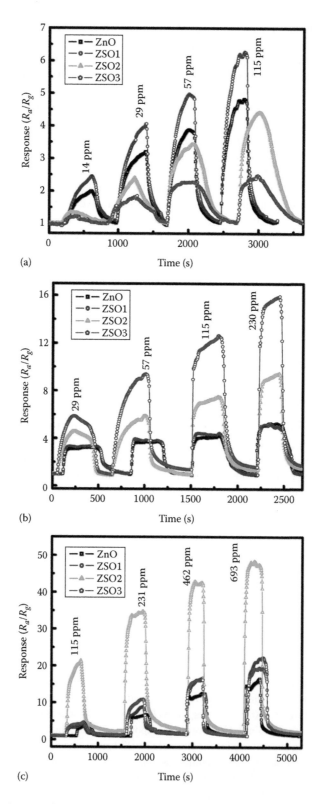

(a)

(b)

(c)

FIGURE 2.22 Sensing transients of ZnO nanorods and SnO$_2$/ZnO heterostructures toward (a) toluene, (b) ethanol, and (c) acetone, operated at 300°C. (From Rakshit, T. et al., *RSC Adv.*, 4, 36749, 2014.)

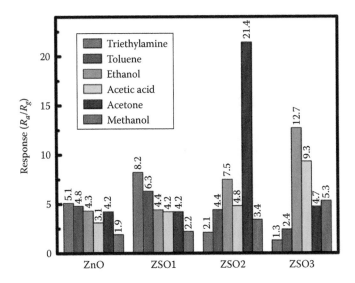

FIGURE 2.23 Sensing responses of ZnO nanorods and SnO₂/ZnO heterostructures to 115 ppm acetone, ethanol, methanol, acetic acid, toluene, and triethylamine. (From Rakshit, T. et al., *RSC Adv.*, 4, 36749, 2014.)

The selectivity of ZnO nanorods and SnO₂/ZnO heterostructures toward 115 ppm acetone, ethanol, methanol, acetic acid, toluene, and triethylamine has been investigated, and the results are shown in Figure 2.23. In pure ZnO nanorods and 25 nm long SnO₂ nanowire heterostructure, the response toward different VOCs is nearly same, and thus lacks any selectivity. The sensing response of 40 nm long SnO₂ nanowire heterostructure (ZSO2) is highest for acetone, which is more than three times higher than that for the other VOCs. On the other hand, the response toward ethanol and acetic acid is quite close to each other for ZSO3-based sensor, and thus lacks selectivity. Thus, SnO₂/ZnO heterostructures are capable of selectively detecting acetone by varying the length of SnO₂ nanowires on the surface of ZnO nanorods.

The response of a semiconducting oxide gas sensing element is usually represented as [130]

$$S = A_g P_g^\gamma \tag{2.10}$$

where

P_g is the partial pressure of the test gas
A_g is the prefactor
γ is the exponent

The variation of response of ZSO1, ZSO2, and ZSO3 with toluene, acetone, and ethanol vapors, respectively, when fitted with Equation 2.10, yields the γ value of 0.43 ± 0.06, 0.44 ± 0.07, and 0.48 ± 0.06. These values are slightly lower than that exhibited by an ideal microstructure (0.5). This may be attributed to the less sensitive zones or agglomeration of the microstructure [130]. The limit of detection for toluene, ethanol, and acetone sensing has been estimated to be ~2, 1, and <0.2 ppm, respectively, indicating the capability of SnO₂/ZnO nanorod heterostructures of detecting very low VOC concentrations.

2.7.4.2.1.2 Sensing Mechanism for SnO₂/ZnO Heterostructures The gas sensing mechanism of ZnO nanorods and SnO₂/ZnO heterostructures can be described by the adsorption and desorption processes occurring on the surface of nanostructures [131–133]. When exposed to air, the surface of nanostructures adsorbs oxygen molecules, which turn into oxygen ions (O^{2-}, O_2^-, and O^-) by accepting electrons from the conduction band, resulting in the formation of a depletion layer on the surface. This leads to narrowing of the conduction channel, which increases the resistance of ZnO. Figure 2.24a through d shows the schematic representation of the sensing mechanism of ZnO nanorods and SnO$_2$ nanowire/ZnO nanorod heterostructures. When the nanostructures are exposed to a VOC such as acetone, it reacts with the chemisorbed oxygen ions following the reaction [134]

$$C_3H_6O + 8O^- \rightarrow 3CO_2 + 3H_2O + 8e^- \tag{2.11}$$

The electrons are thus released back to the conduction band of the nanostructures, resulting in the reduction of the depletion layer thickness and widening of the conduction channel. The resistance of the oxide thus decreases.

The electron transport properties of nanostructures depend on the formation of junctions at the interfaces. In SnO$_2$/ZnO heterostructures, a homojunction potential barrier develops at the interface of SnO$_2$ nanowires, whereas a heterojunction potential barrier forms at the SnO$_2$ nanowire/ZnO nanorod interface. Since ZnO has a higher work function than SnO$_2$ [135], electron transfer will occur from SnO$_2$ to ZnO in the heterojunction. The process will continue till the Fermi levels of ZnO and SnO$_2$ equilibrate. Thus, a depletion layer is developed at their interface, resulting in bending of the energy band (shown in Figure 2.24e). The difference of the work function of ZnO and SnO$_2$ is ~0.3 eV, which is the magnitude of the potential formed at their interface. If the electron transport is modulated by an effective potential barrier (Φ_{eff}), which includes the contribution of both the homo- and heterojunction potential barriers, the conductivity can be expressed by [136]

$$\sigma = \sigma_c \exp\left(\frac{-\Phi_{eff}}{k_B T}\right) \tag{2.12}$$

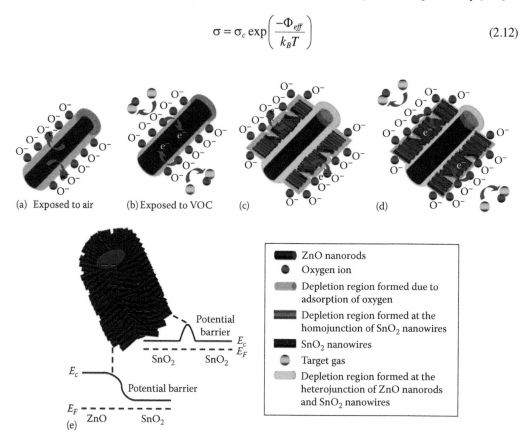

FIGURE 2.24 Schematic sensing mechanism of (a) ZnO nanorods exposed to air, (b) ZnO nanorods exposed to VOC, (c) SnO$_2$/ZnO heterostructures exposed to air, and (d) SnO$_2$/ZnO heterostructures exposed to VOC. (e) Band structure of SnO$_2$/ZnO heterostructures. (From Rakshit, T. et al., *RSC Adv.*, 4, 36749, 2014.)

where σ_c is a constant pre-exponential factor. When nanostructured samples are exposed to air, the adsorbed oxygen molecules extract electrons from the conduction band, increasing the height and width of the effective potential barrier. These trapped electrons are released back to the conduction band upon exposing the nanostructures to VOCs, which reduces the effective potential barrier [137,138]. Therefore, the electron transport in SnO_2/ZnO heterostructures is largely modulated by the homo- and heterojunction potential barriers formed at the interfaces, which results in a superior gas sensing performance as compared to control ZnO nanorods.

The strong dependence of the response toward different VOCs on the morphology of the SnO_2/ZnO heterostructures can be explained by different inter-nanorod spacing caused by the variation in the length of SnO_2 nanowires and difference in the molecular size of VOCs. The inter-nanorod spacing is quite large in short-length SnO_2 nanowire (ZSO1) sensor, allowing only a small fraction of the VOCs to be adsorbed on the surface of the heterostructures. This results in a low sensitivity toward VOCs of the ZSO1 sample. The SnO_2 nanowires in ZSO2 are longer than that of ZSO1, making the spacing between the nanorods smaller. As a result, the surface of the heterostructures adsorbs a relatively high fraction of different VOCs, leading to a higher response in ZSO2 compared to ZSO1. The size of an acetone molecule (having a kinetic diameter of 0.469 nm [139]) is larger than that of ethanol, methanol, and acetic acid molecule (having kinetic diameter of 0.45, 0.36, and 0.436 nm [139,140], respectively). So a higher fraction of acetone molecules get absorbed, leading to a higher acetone response in ZSO2. But at the same time, the narrow spacing between the nanorods makes the diffusion of toluene and triethylamine molecules more difficult, due to the relatively larger kinetic diameter of 0.585 and 0.78 nm [141,142], respectively. This results in a lower response toward toluene and triethylamine in the ZSO2 sensor than in ZSO1. In the longest SnO_2 nanowire (ZSO3) sensor, the spacing between the nanorods further decreases, leading to a higher response toward ethanol, methanol, and acetic acid than with ZSO2. But the diffusion of the larger acetone, toluene, and triethylamine molecules becomes more difficult, leading to a reduced sensitivity. Therefore, the selectivity observed in the SnO_2/ZnO heterostructures can be explained by the difference in the inter-nanorod spacing and molecular size of VOCs.

2.7.4.2.1.3 NiO-Sensitized ZnO Nanosheets To improve the sensing performance toward triethylamine, the surface of ZnO nanosheets was modified with a layer of NiO nanoparticles [143]. The interconnected nanosheets form a network-like structure, exhibiting a large surface area, which could enhance the density of adsorption sites. This leads to the adsorption of a large fraction of target gas molecules. The optimum operating temperature for sensing triethylamine with both ZnO nanosheets and NiO/ZnO nanosheet sensors was found to be 320°C. The NiO/ZnO nanosheet sensor exhibited a response (R_a/R_g) of 185.1–100 ppm for triethylamine at 320°C, whereas that for ZnO nanosheets was 78.4. Thus, the response of the NiO/ZnO nanosheets was ~2.4 times higher than that of the control ZnO. The fabricated sensors exhibited good repeatability, high stability, and low detection limit of 2 ppm. The response and recovery characteristics of the sensors to 10 ppm triethylamine at 320°C were also studied. The sensors exhibited fast response with a response time of 6–7 s, and the recovery time of ZnO nanosheet sensor was 22 s. However, the NiO/ZnO nanosheets required a slightly longer time of 33 s for recovery.

To examine the selectivity, the sensors were exposed to 100 ppm of different VOCs such as triethylamine, acetone, ethanol, *n*-hexane, 2-propanol, *p*-xylene, C_6H_{12}, C_6H_6, and CH_3OH at 320°C. The response of NiO/ZnO nanosheets for triethylamine was found to be much higher compared to that for the other interfering gases, indicating its excellent selectivity toward triethylamine. The effect of the variation of the thickness of NiO nanoparticle layer in NiO/ZnO nanosheets was also examined. The response initially increased with the increase of the thickness of the NiO nanoparticle layer, but decreased at very high thicknesses. The optimized NiO/ZnO nanosheet sensor exhibited a response of 42 toward 2 ppm triethylamine.

The basic sensing mechanism of ZnO nanosheets, as shown schematically in Figure 2.25a through d, is same as that of ZnO nanorods discussed earlier. However, the sensing mechanism of

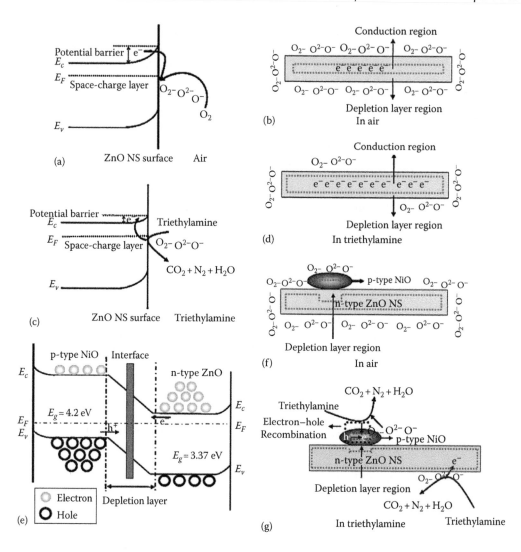

FIGURE 2.25 (a) Energy band diagram and (b) schematic sensing mechanism of ZnO nanosheets in air. (c) Energy band diagram and (d) schematic sensing mechanism of ZnO nanosheets in triethylamine. (e) Energy band diagram of p-type NiO/n-type ZnO nanosheets. Schematic sensing mechanism of NiO/ZnO nanosheets in (f) air and (g) triethylamine. (Reprinted from *Sens. Actuators B Chem.*, 200, Ju, D., Xu, H., Qiu, Z., Guo, J., Zhang, J., and Cao, B., Highly sensitive and selective triethylamine-sensing properties of nanosheets directly grown on ceramic tube by forming NiO/ZnO PN heterojunction, 288–296. © 2014, with permission from Elsevier.)

NiO/ZnO nanosheets is different, due to the formation of a p–n heterojunction at the interface of p-type NiO and n-type ZnO. When the surface of ZnO nanosheets is modified with a layer of NiO nanoparticles, the diffusion of electrons of ZnO and holes of NiO occurs in opposite directions due to the concentration gradient. This results in the formation of an internal electric field at the interface of NiO and ZnO, and the carrier diffusion becomes balanced. The depletion layer developed at the interface leads to the bending of energy bands till the Fermi levels of NiO and ZnO equilibrate, as shown in Figure 2.25e. Thus, the combined effect of the depletion layer formed as a result of the adsorbed oxygen molecules on the ZnO nanosheet surface and that formed at the interface of NiO and ZnO leads to a higher resistance in NiO/ZnO nanosheets than the control ZnO sensor, upon exposure to air (Figure 2.25f). When the NiO/ZnO nanosheets are exposed to triethylamine, the adsorbed oxygen ions react with the target gas following the reaction [129]

$$N(C_2H_5)_3 + O_2^- \rightarrow H_2O + CO_2 + N_2 + e^- \tag{2.13}$$

The electrons are thus released back to the conduction band of the nanosheets, thereby decreasing the resistance. Also, the electrons released by triethylamine into NiO recombine with the holes, leading to the reduction of the hole concentration of NiO. Following the law of mass action, the electron concentration of NiO increases, which decreases the concentration gradient at the interface. Thus, the depletion layer at the NiO/ZnO interfaces becomes narrow, resulting in the further reduction of the resistance, as shown in Figure 2.25g. In addition, the bond energy of triethylamine (C—N) is lower than that of acetone (C=O), hexane (C—C), ethanol/methanol (O—H), and benzene (C=C), making the reaction activity of triethylamine high. This also enhances the response of NiO/ZnO nanosheets to triethylamine. The low sensitivity of ZnO nanosheets covered by a very thick NiO nanoparticle layer could be attributed to the reduction of the adsorption sites of ZnO.

2.7.4.2.1.4   CuO/ZnO Nanorod Heterostructures   Gas sensors based on CuO/ZnO nanorod heterostructures for the detection of H_2S have been reported [144]. CuO nanoparticles were deposited by a photochemical process on the surface of hydrothermally grown ZnO nanorods to form CuO/ZnO nanorod heterostructures. The fabricated sensors exhibited a stable response to H_2S with rapid recovery in the temperature range 300°C–500°C. The sensitivity of CuO/ZnO nanorod heterostructures increased with the increase in operating temperature from 300°C to 500°C, and the response was found to be higher than that of control ZnO nanorods. The response and recovery characteristics of the sensors were examined as a function of the operating temperature. With the increase of the operating temperature, the recovery time of both CuO/ZnO heterostructures and ZnO nanorods decreased. However, the response time of CuO/ZnO heterostructures showed a different trend; it increased with the rise of operating temperature. The gas sensing performance was also studied by varying the surface coverage of ZnO nanorods by CuO nanoparticles. The response toward H_2S initially increased with the increase of the CuO surface coverage, but decreased at a very high surface coverage.

The energy band diagram of CuO/ZnO heterostructures upon exposure to air and H_2S gas is shown in Figure 2.26a and b, respectively. When the surface of n-type ZnO nanorods is sensitized

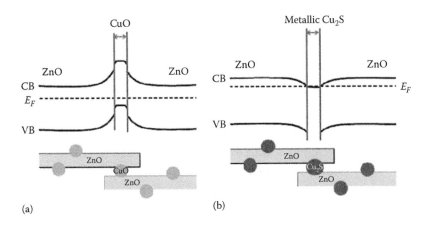

(a)   (b)

FIGURE 2.26   Schematic energy band diagram of CuO/ZnO heterostructures when exposed to (a) air and (b) H_2S gas. (Reprinted with permission from Kim, J., Kim, W., and Yong, K., *J. Phys. Chem. C*, 116, 15682–15691, 2012. © 2012, American Chemical Society.)

with p-type CuO nanoparticles, a p–n heterojunction is formed at the interface (Figure 2.26a). The surface of the heterostructures adsorbs oxygen molecules upon exposure to air, leading to the formation of a depletion layer and increase of resistance. When the CuO/ZnO heterostructures are exposed to H_2S gas, the gas reacts with CuO to form metallic Cu_2S following the reaction [144]

$$6CuO(s) + 4H_2S^-(ad) \rightarrow 3Cu_2S(s) + 4H_2O(g) + SO_2(g) \tag{2.14}$$

The formation of metallic Cu_2S changes the energy band diagram, as shown in Figure 2.26b. This leads to the reduction of the resistance of the heterostructures. When these heterostructures are exposed to air for recovery, Cu_2S reacts with the oxygen in air and converts back to CuO, following the reaction [144]

$$Cu_2S(s) + 2O_2 \rightarrow 2CuO(s) + SO_2(g) \tag{2.15}$$

As the operating temperature increases, the rate of conversion of CuO to Cu_2S becomes slower, resulting in an increase in the response time. When the surface coverage of ZnO nanorods by CuO nanoparticles is very high, the active sensing area decreases because of the high Cu_2S coverage. Also, only a small fraction of CuO is converted to metallic Cu_2S, and a large portion remains unconverted. Therefore, a low sensitivity at very high CuO coverage is observed. Thus, there exists an optimal surface coverage at which a portion of the ZnO nanorods remains exposed for H_2S sensing, and a considerable portion remains coated with CuO, for conversion to Cu_2S.

Some of the reported results on gas sensors based on semiconductor sensitized ZnO nanostructures are summarized in Table 2.3. It is clear that the gas sensing performance of ZnO nanostructures is strongly dependent on the surface modification of ZnO nanostructures with different inorganic semiconductors.

2.7.4.2.2   Metal-Sensitized ZnO Gas Sensors

2.7.4.2.2.1   Au-Functionalized ZnO Sensor   To improve the gas sensing performance, ZnO nanostructures have also been decorated with metal nanoparticles (Au, Pt, Pd, etc.). Metal nanoparticles on the surface of ZnO can enhance the interaction of the gas molecules with the adsorbed oxygen species. Gold-decorated ZnO nanowire CO gas sensors were fabricated by Chang et al. [153]. High-density ZnO nanowires were prepared on patterned ZnO:Ga finger electrodes on SiO_2/Si substrates. Gold nanoparticles were decorated by immersing ZnO nanowire samples in the ethanol/$HAuCl_4$ solution, and were kept under UV illumination for 4 min. The sensor measurements were performed by injecting different concentrations of CO gas into the sealed chamber at different operating temperatures. The measured sensitivities of Au/ZnO sensors operated at 350°C were around 30%, 37%, 46.5%, and 53% for 5, 20, 50, and 100 ppm concentrations of the CO gas, respectively [153]. It was also found that the Au/ZnO sensor response increased with temperature up to 250°C and then decreased with further increase in temperature. The higher response for Au-decorated ZnO nanowire device over ZnO might be attributed to the catalytic activity of Au nanoparticles at temperatures below 250°C [153]. The lower response above 250°C might be due to the difficulty in exothermic CO adsorption [153]. It is to be noted that the adsorbed oxygen species on the surface of ZnO play a significant role in understanding the gas sensing mechanism for different gases and VOCs. The reducing CO molecules react with oxygen and generate free electrons, which can be described as [153]

$$R + O^- \rightarrow RO + e^- \tag{2.16}$$

where R is the reducing gas. The free electrons contribute to the electrical conduction in ZnO, leading to a decrease in resistance. For the Au/ZnO sensor, this reaction rate may be increased because of the catalytic activity of Au nanoparticles. Thus, the Au/ZnO sensor showed higher response.

TABLE 2.3

Comparison of Sensing Performance of Semiconductor-Sensitized ZnO Nanostructures for Detection of Various Gases and VOCs

Gas/VOC	Sensitized Heterostructures	Optimum Operating Temperature (°C)	Response			Response Time (s)/ Recovery Time (s)	References
			Sensitivity, S	Conc. (ppm)	Temp. (°C)		
Ethanol	ZnO/α-Fe$_2$O$_3$ hierarchical nanostructures	370	53.6172[a]	100	370	6/7	[145]
Ethanol	Flower-like CuO/ ZnO nanorods	300	98.8[a]	100	300	7/9	[146]
Ethanol	PdO-decorated flower-like ZnO structures	320	35.4[a]	100	320	1/7	[147]
Ethanol	ZnO–SnO$_2$ core–shell nanowires	400	280[a]	200	400	—	[148]
Triethylamine	NiO/ZnO nanosheets	320	185.1[a]	100	320	7/33	[143]
Trimethylamine	Cr$_2$O$_3$-decorated ZnO nanowires	400	17.79[a]	5	400	~1–10/ 358–398	[149]
Acetone	SnO$_2$ nanowire/ ZnO nanorod heterostructures	300	21.4[a]	115	300	—	[129]
Acetone	NiO/ZnO heterostructures	330	~13[a,*]	100	330	—	[150]
Nitrogen dioxide	ZnO–SnO$_2$ core–shell nanowires	200	66.3[b]	10	200	~50/60*	[148]
Hydrogen	ZnO–In$_2$O$_3$ core–shell nanorods	—	15.5%[c]	100	Room temp.	—	[151]
Hydrogen sulfide	CuO/ZnO nanowires	200	~30[d,*]	5	200	360/1800	[152]

[a] $S = R_a/R_g$.

[b] $S = R_g/R_a$.

[c] $S = |(R_g - R_a)/R_a| \times 100\%$.

[d] $S = I_g/I_a$, where I_a and I_g are current values of the sensor in air and test gas, respectively.

* Value estimated from the reported graphical plot.

A chemical sensor based on Au-anchored ZnO hybrid material was studied by Liu et al. [154]. The dynamic response to different concentrations of ethanol at 310°C for ZnO and Au–ZnO sensors is presented in Figure 2.27a. As compared to ZnO, the Au–ZnO sensor exhibits significant enhancement in response toward each ethanol concentration, due to the presence of Au nanoparticles. The sensitivity of the Au–ZnO sensor shows a maximum value of 52.5 at 310°C, and then decreases with further increase in temperature [154]. It is known that the dynamic sensing reaction of a sensor is dramatically influenced by the operating temperature. The reaction between ethanol vapor and adsorbed oxygen species (O$^-$, O$_2^-$ and O^{2-}) on ZnO slowly increases with temperature and reaches

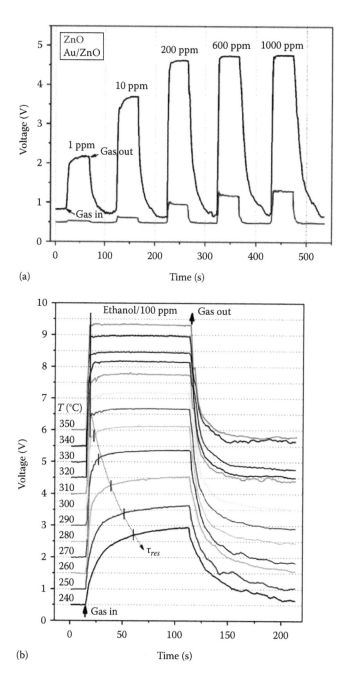

FIGURE 2.27 (a) Response of ZnO and Au/ZnO sensors to different ethanol concentrations operated at 310°C. (b) Sensing response and recovery curves of Au decorated ZnO device to 100 ppm ethanol at different operating temperatures. (From Liu, X., Zhang, J., Guo, X., Wu, S., and Wang, S., Amino acid-assisted one-pot assembly of Au, Pt nanoparticles onto one-dimensional ZnO microrods, *Nanoscale*, 2, 1178–1184, 2010. Reproduced by permission of The Royal Society of Chemistry.)

a maximum at 310°C. Thereafter, it gradually decreases at higher temperatures. Figure 2.27b shows the dynamic response and recovery of the Au–ZnO sensor for 100 ppm ethanol at different operating temperatures. As observed in the figure, the response time decreases with increasing temperature and there is a negligible effect on the recovery time. The response time is observed to be 45 and 7 s for operating temperatures of 240°C and 280°C, respectively. Therefore, a relatively fast response time of 3 s and recovery time of 2 min at an operating temperature exceeding 290°C suggests the potential use of the Au NP-sensitized ZnO surface for gas sensors.

The enhanced performance of Au–ZnO sensors over ZnO might be attributed to the spillover effect of Au nanoparticles or to the formation of an electron depletion layer at the interface of Au and ZnO. Guo et al. have also fabricated a high-performance gas sensor based on ZnO nanowires functionalized with Au nanoparticles and explained the sensing mechanism by the spillover effect [155]. Figure 2.28 shows the sensing mechanism involved in the improved performance of the Au–ZnO sensor. When the sensor is exposed to air, the oxygen molecules adsorb on the surface of ZnO by capturing the electrons and become negatively charged oxygen species, as shown in Figure 2.28a. These oxygen species react with the target gas molecules and provide the sensing response (Figure 2.28b). The spillover effect of Au nanoparticles in the Au–ZnO sensor may help adsorb more oxygen species on the surface of ZnO, which eventually enhances the interaction with ethanol (Figure 2.28c and d). Thus, the Au–ZnO sensors showed an enhanced performance compared to ZnO.

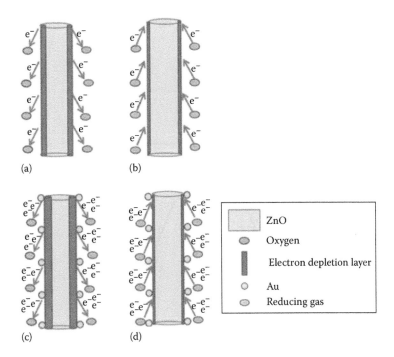

FIGURE 2.28 Schematic diagram showing the gas sensing mechanism for ZnO and Au/ZnO devices. (a) Formation of depletion region on the surface of ZnO due to adsorption of oxygen molecules by trapping free electrons. (b) Reaction of oxygen molecules with ethanol leads to thinner depletion layer. (c) Formation of thicker depletion layer due to Au nanoparticles. (d) Thinner depletion layer formed as a result of the interaction of oxygen molecules with ethanol. (Reprinted from *Sens. Actuators B*, 199, Guo, J., Zhang, J., Zhu, M., Ju, D., Xu, H., and Cao, B., High-performance gas sensor based on ZnO nanowires functionalized by Au nanoparticles, 339–345. © 2014, with permission from Elsevier.)

2.7.4.2.2.2 Photoinduced Au–ZnO Plasmonic Gas Sensors   Photoinduced gas sensors have attracted much interest to achieve lower operating temperatures and higher responsivity. Room-temperature sensors have the benefits for low-power applications including good compatibility to integrate with other devices on flexible platform, reduced energy consumption, and avoidance of major losses in passive electronic components. Although the UV-induced ZnO gas sensing property has been widely investigated, its responsivity toward gas still needs to be improved. Metal nanoparticles play a significant role in improving the sensing response for ZnO gas sensors due to their localized surface plasmon resonance (LSPR) characteristics. The LSPR effect mainly enhances the charge carrier density in ZnO surface via light scattering or charge transfer.

Wavelength-tunable gas sensors were fabricated using Au-functionalized ZnO nanosheets at room temperature [86]. The visible photoresponse of a Au–ZnO device is shown in Figure 2.29a. The Au–ZnO device is known to display a broad photoresponse, while the ZnO device does not show any photoresponse in the visible region. The visible photoresponse in Au–ZnO originates from the charge excitation in Au NPs due to the decay of LSPR followed by charge transfer to the conduction band of ZnO [156]. Therefore, the light illumination of Au–ZnO enhances the charge density for ZnO in both UV and visible ranges, which can tune the gas sensing over a broad wavelength range. Light-induced gas sensing study has been carried out in the presence of different gases using photoresistance measurements. The sensor response (S) can be calculated for a particular gas (NO) using the relation

$$S(\%) = \frac{R_{NO} - R_{air}}{R_{air}} \times 100 \qquad (2.17)$$

where
  R_{NO} is the resistance of the Au–ZnO film in the presence of NO gas
  R_{air} is the film resistance in presence of dry air

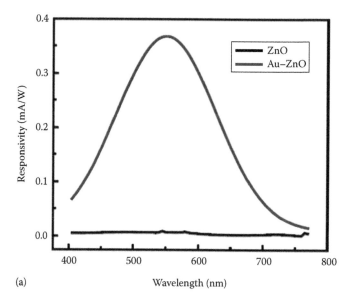

(a)   Wavelength (nm)

FIGURE 2.29   (a) Photoresponsivity of plasmonic Au–ZnO device in the visible region at 5 V. (*Continued*)

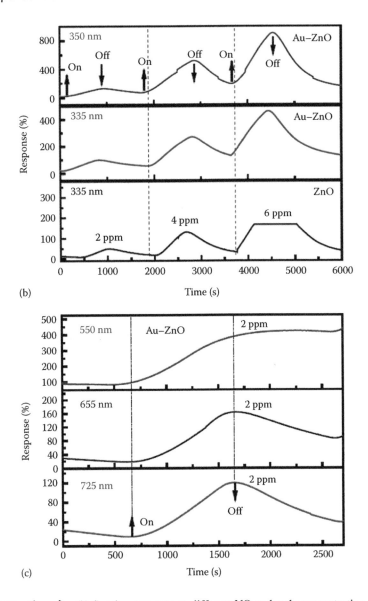

FIGURE 2.29 (*Continued*)   (b) Sensing response to different NO molecule concentration of Au–ZnO and ZnO devices on illumination with UV wavelengths at room temperature. (From Gogurla, N. et al., *Sci. Rep.*, 4, 6483(1–9), 2014.) (c) Response of Au–ZnO device to 2 ppm NO concentration at room temperature on illumination with visible wavelengths. (From Gogurla, N. et al., *Sci. Rep.*, 4, 6483(1–9), 2014.)

Figure 2.29b shows the room-temperature gas sensing response to different concentration of NO gas in dry air for both ZnO and Au–ZnO devices under UV illumination of wavelengths 335 and 350 nm. The time pulse sequence was maintained as "ON" for 15 min and "OFF" for 15 min for both the devices. As seen in the figure, both the devices show a prominent change in the response even at very low concentration levels. The response for Au–ZnO device upon exposure to 4 ppm NO is found to be enhanced three times higher than the ZnO device at 325 nm.

The sensing response of Au–ZnO device at different incident visible wavelengths at room temperature is also presented in Figure 2.29c. The response at a wavelength of 550 nm is found to be the maximum for the device.

The sensitivity as a function of wavelength of the Au–ZnO device at 2 ppm concentration of NO is presented in Figure 2.30a. The device exhibits a higher response at 550 nm illumination compared to other wavelengths. The sensitivity of the Au–ZnO device at 2 ppm is observed to be 194 and 50%/ppm for 550 and 335 nm illumination, respectively. The sensitivity at 550 nm is also higher as compared to the reported value (0.326%/ppm at 532 nm) of Au–ZnO sensor for 25 ppm C_2H_2 [157]. A histogram summarizing the selectivity of the Au–ZnO sensor to NO gas over CO and some selected VOCs is shown in Figure 2.30b. The Au–ZnO sensor exhibits much higher selectivity for NO compared to other gases measured at 335 nm. The limit of detection for the Au–ZnO sensor for NO is found to be 0.1 ppb under 335 nm illumination. The ability of very low concentration detection with high selectivity makes the photoconductive Au–ZnO nanocomposite attractive for breath sensing applications, where NO is a signature for asthma disease. Thus, the sensitivity and low detection limit of plasmonic devices operated at room temperature are comparable or even better in magnitude than the reported values for conventional sensors using metal oxides operating at higher temperatures [158,159].

2.7.4.2.2.3 Sensing Mechanism of Photoinduced ZnO Sensors The sensing mechanism of the light-induced Au–ZnO sensor can be explained by the photoconductive property of ZnO in the UV region and LSPR effect of Au nanoparticles in the visible region. Due to the large surface-to-volume ratio of ZnO nanostructures, oxygen molecules create a depletion layer with lower conductivity by capturing electrons: $O_2(gas) + e^- \rightarrow O_2^-(adsorption)$. Upon illumination, the photogenerated holes

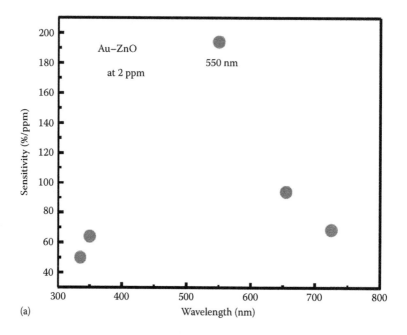

(a)

FIGURE 2.30 (a) Sensitivity as a function of wavelength of Au–ZnO device at 2 ppm NO concentration. (From Gogurla, N. et al., *Sci. Rep.*, 4, 6483(1–9), 2014.)   (*Continued*)

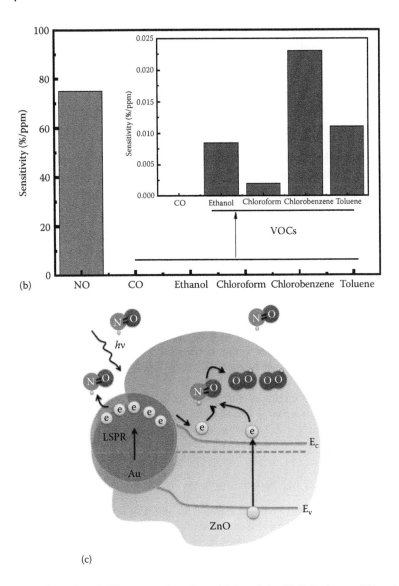

FIGURE 2.30 (*Continued*)   (b) Histogram plot of sensitivity of Au–ZnO device to different gases. Inset shows the sensitivity of the sensor upon exposure to CO and VOCs. (c) Schematic diagram representing the sensing mechanism of Au–ZnO device in the presence of light. (From Gogurla, N. et al., *Sci. Rep.*, 4, 6483(1–9), 2014.)

in ZnO migrate to the surface along the potential gradient and desorb oxygen from the surface $(O_2^-(\text{adsorption}) + h^+ \rightarrow O_2(\text{gas}))$. This results in an increase in the free carrier concentration in ZnO and a decrease in the width of the depletion layer. When the gas flow is switched "ON," because of the electron affinity of NO molecules, they can easily react with the photoelectrons of ZnO. The reaction can be expressed by [86]

$$2NO + e^- = N_2(g) + O_2^- \tag{2.18}$$

Here, the NO molecules interact with photoelectrons and generate O_2 molecules, which are again adsorbed on the surface of ZnO by capturing the photogenerated electrons, as shown in Figure 2.30c. As a result, there is an increase in the photoresistance of the ZnO sensor due to the broadening of surface depletion layer again. But for the Au–ZnO sensor, under the UV illumination, the excited electrons of Au NPs due to interband transition also participate in the reaction. These excited electrons in Au can be transferred to ZnO, or they can directly interact with the NO molecules. The enhanced interaction with NO molecules occurring as a result of the increased charge density in ZnO may lead to a higher photoresistance for Au–ZnO NO sensor.

The mechanism involved in the sensing in the visible region can be explained as due only to the LSPR effect of Au nanoparticles. When light interacts with Au nanoparticles, more electrons oscillate on the surface of Au due to plasmon resonance. The electron oscillation is more sensitive to the dielectric environment and charge density of Au [160]. The strong electron affinity of the NO molecule causes a strong interaction with the oscillating electrons of Au (Figure 2.30c). As a result, there is decrease in the charge density of Au, which finally increases the photoresistance of the film. It may be noted that the sensing response of the Au–ZnO sensor in the visible region due to surface plasmons is even higher than in the UV region. These results imply the potential of metal-functionalized ZnO for plasmonic gas sensors operated at room temperature and tunable over different wavelengths.

UV illumination–assisted gas sensors based on ZnO nanofibers decorated with Au nanoparticles were also fabricated for detecting ethanol at room temperature [161]. Under UV illumination (365 and 254 nm), the Au–ZnO sensor showed a higher response upon exposure to ethanol than that of pure ZnO nanofiber sensor operated at room temperature. The sensing mechanism involved in enhancing the performance of the Au–ZnO sensor could be explained by the photocatalytic reaction. Due to the Schottky junction between Au and ZnO, the photogenerated electron–hole pairs are separated and provide more charge carriers on the surface of ZnO for reaction. The photoxidation rate of organic vapor on the ZnO surface can also be improved by the catalytic action of Au nanoparticles.

2.7.5 Biosensors Using ZnO Nanostructures

ZnO nanostructures are also attractive for biosensing applications owing to their high surface-to-volume ratio, high surface reactivity and catalytic efficiency, fast electron transport, and the ability to adsorb and retain the activity of enzymes. The activity of the immobilized enzymes strongly depends on the temperature, toxic chemicals, pH, and humidity [162]. ZnO, having a high isoelectric point (IEP) of 9.5, presents a suitable electrode surface for immobilization of low-IEP enzymes such as glucose oxidase (GOx; IEP ~4.2). The positively charged ZnO nanostructures are capable of easily immobilizing negatively charged GOx at a physiological pH of 7.4, due to the electrostatic attractive force between them. This force promotes the efficient immobilization of GOx in mild chemical conditions while still retaining the enzymic bioactivity. Moreover, the high specific surface area of ZnO nanostructures enables high loading of enzyme, and also assists in efficient transfer of electrons between the electrode and GOx.

ZnO nanostructures of different morphologies have been used for fabricating glucose biosensors. Glucose biosensors based on inverse opals showed a broad, linear detection range of 0.01–18 mM, with a sensitivity of 22.5 $\mu A/cm^2/mM$ [163]. ZnO nanotube–based glucose biosensor exhibited a sensitivity of 30.4 $\mu A/cm^2/mM$, with a response time of less than 10 s [12]. The glucose biosensor fabricated using a single ZnO nanofiber exhibited a high sensitivity of 70.2 $\mu A/cm^2/mM$, with a fast response time of <4 s and a low detection limit of 1 μM [164]. Glucose sensors were also fabricated using ZnO tripods and nanorods [165]. ZnO tripods, synthesized by the vapor–solid method, were distributed uniformly over the Si substrate, as shown in Figure 2.31a. ZnO tripods and nanorods became interconnected with their neighboring ones after immobilization of GOx on their surfaces. Figure 2.31b shows the amperometric response of ZnO nanorod- and tripod-based glucose sensors.

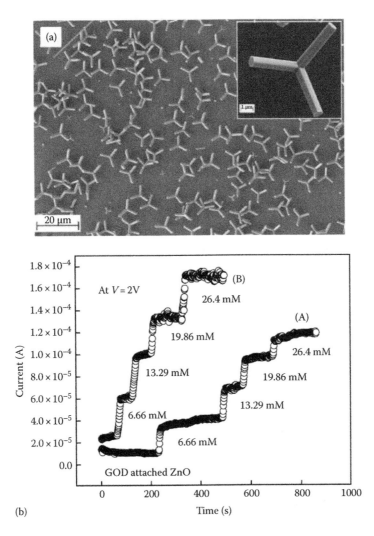

FIGURE 2.31 (a) Typical FESEM micrograph of ZnO tripods. Inset shows the magnified view of the micrograph. (b) Amperometric response of biosensors based on ZnO (A) nanorods and (B) tripods with successive addition of glucose. (From Mandal, S. et al., *Sens. Lett.*, 7, 635, 2009.)

Both sensors show a fast response on the addition of glucose, reaching 90% of the steady-state current in <10 s. This indicates that a rapid exchange of electrons occurs between the ZnO nanostructures and GOx. The sensitivity of the ZnO tripod- and nanorod-based glucose sensor is found to be 36.8 and 27.7 $\mu A/cm^2/mM$, respectively. The high specific surface area of ZnO tripods allows higher loading of GOx, resulting in an enhanced sensitivity compared to those of ZnO nanorods. Thus, the morphology of ZnO nanostructures plays an important role in controlling the performance of biosensors.

2.8 SUMMARY

In this chapter, we presented a review on ZnO-based nanostructures sensitized by semiconductors and metal nanoparticles for optoelectronic and gas sensor applications. The optical properties of ZnO tripods and tetrapods were discussed using low-temperature photoluminescence measurements. The optical and electrical properties of ZnO tetrapods can be tuned by doping with selective elements such as Sn, P, Mg, Mn, and so on, which may be useful for different optoelectronic applications. The photovoltaic and gas sensing performance can be improved by modifying the surface of ZnO nanostructures with selective inorganic semiconductors. The influence of metal nanoparticles on the performance of ZnO-based UV photodetectors was also discussed. The responsivity of ZnO devices could be enhanced 80 times with sensitized Au nanoparticles on the ZnO nanosheet surface. The interband transition and hole trapping in Au nanoparticles may be the possible reasons for enhanced photoresponse. Other metal nanoparticles (Al and Pt) are also used to enhance the photoresponse of ZnO UV photodetectors. The observed enhancement values for these detectors are found to be lower than those obtained for Au–ZnO nanocomposite devices. Metal nanoparticles can also be used to improve the performance of ZnO gas sensors. These sensors exhibit enhanced performance due to the LSPR effect of metal nanoparticles.

ACKNOWLEDGMENT

The academic contributions of Prof. I. Manna, Dr. S. Santra, and Dr. S. Mandal are gratefully acknowledged for several research results reported in this chapter.

REFERENCES

1. Pearton, S. J., Norton, D. P., Ip, K., Heo, Y. W., and Steiner, T. 2003. Recent progress in processing and properties of ZnO. *Superlattices Microstruct.* 34:3–32.
2. Ashrafi, A. and Jagadish, C. 2007. Review of zincblende ZnO: Stability of metastable ZnO phases. *J. Appl. Phys.* 102:071101(1–12).
3. Kim, K.-K., Kim, H.-S., Hwang, D.-K., Lim, J.-H., and Park, S.-J. 2003. Realization of p-type ZnO thin films via phosphorus doping and thermal activation of the dopant. *Appl. Phys. Lett.* 83:63–65.
4. Chavillon, B., Cario, L., Renaud, A. et al. 2012. P-type nitrogen-doped ZnO nanoparticles stable under ambient conditions. *J. Am. Chem. Soc.* 134:464–470.
5. Fan, J. C., Zhu, C. Y., Fung, S. et al. 2009. Arsenic doped *p*-type zinc oxide films grown by radio frequency magnetron sputtering. *J. Appl. Phys.* 106:073709(1–6).
6. Jun, M.-C., Park, S.-U., and Koh, J.-H. 2012. Comparative studies of Al-doped ZnO and Ga-doped ZnO transparent conducting oxide thin films. *Nanoscale Res. Lett.* 7:639(6pp).
7. Kim, D. H., Cho, N. G., Kim, H. G., and Choi, W.-Y. 2007. Structural and electrical properties of indium doped ZnO thin films fabricated by RF magnetron sputtering. *J. Electrochem. Soc.* 154:H939–H943.
8. Lee, J. and Tak, Y. 2001. Electrodeposition of ZnO on ITO electrode by potential modulation method. *Electrochem. Solid State Lett.* 4:C63–C65.
9. Li, Y., Meng, G. W., Zhang, L. D., and Phillipp, F. 2000. Ordered semiconductor ZnO nanowire arrays and their photoluminescence properties. *Appl. Phys. Lett.* 76:2011–2013.
10. Lakshmi, B. B., Dorhout, P. K., and Martin, C. R. 1997. Sol–gel template synthesis of semiconductor nanostructures. *Chem. Mater.* 9:857–862.
11. Chen, Y. W. and Liu, Y. C. 2005. Optical properties of ZnO and ZnO:In nanorods assembled by sol–gel method. *J. Chem. Phys.* 123:134701-1–134701-5.
12. Rakshit, T., Mandal, S., Mishra, P., Dhar, A., Manna, I., and Ray, S. K. 2012. Optical and bio-sensing characteristics of ZnO nanotubes grown by hydrothermal method. *J. Nanosci. Nanotechnol.* 12:308–315.
13. Park, J.-A., Moon, J., Lee, S.-J., Lim, S.-C., and Zyung, T. 2009. Fabrication and characterization of ZnO nanofibers by electrospinning. *Curr. Appl. Phys.* 9:S210–S212.
14. Rajesh, D., Lakshmi, B. V., and Sunandana, C. S. 2012. Two-step synthesis and characterization of ZnO nanoparticles. *Physica B* 407:4537–4539.
15. Wu, J.-J. and Liu, S.-C. 2002. Low-temperature growth of well-aligned ZnO nanorods by chemical vapor deposition. *Adv. Mater.* 14:215–218.
16. Zak, A. K., Majid, W. H. A., Wang, H. Z., Yousefi, R., Golsheikh, A. M., and Ren, Z. F. 2013. Sonochemical synthesis of hierarchical ZnO nanostructures. *Ultrason. Sonochem.* 20:395–400.
17. Kooti, M. and Sedeh, A. N. 2013. Microwave-assisted combustion synthesis of ZnO nanoparticles. *J. Chem.* 2013:1–4, Article ID: 562028.
18. Shetty, A. and Kar Nanda, K. 2012. Synthesis of zinc oxide porous structures by anodization with water as an electrolyte. *Appl. Phys. A* 109:151–157.

19. Kumar, S. S., Venkateswarlu, P., Rao, V. R., and Rao, G. N. 2013. Synthesis, characterization and optical properties of zinc oxide nanoparticles. *Int. Nano Lett.* 3:30-1–30-6.
20. Kong, X. Y. and Wang, Z. L. 2003. Spontaneous polarization-induced nanohelixes, nanosprings, and nanorings of piezoelectric nanobelts. *Nano Lett.* 3:1625–1631.
21. Chang, P.-C., Fan, Z., Wang, D. et al. 2004. ZnO nanowires synthesized by vapor trapping CVD method. *Chem. Mater.* 16:5133–5137.
22. Dang, H. Y., Wang, J., and Fan, S. S. 2003. The synthesis of metal oxide nanowires by directly heating metal samples in appropriate oxygen atmospheres. *Nanotechnology* 14:738–741.
23. Huang, M. H., Wu, Y., Feick, H., Tran, N., Weber, E., and Yang, P. 2001. Catalytic growth of zinc oxide nanowires by vapor transport. *Adv. Mater.* 13:113–116.
24. Park, W. I., Kim, D. H., Jung, S.-W., and Yi, G.-C. 2002. Metalorganic vapor-phase epitaxial growth of vertically well-aligned ZnO nanorods. *Appl. Phys. Lett.* 80:4232–4234.
25. Wang, Z. L. 2004. Zinc oxide nanostructures: Growth, properties and applications. *J. Phys.: Condens. Matter* 16:R829–R858.
26. Mandal, S., Dhar, A., and Ray, S. K. 2009. Growth and photoluminescence characteristics of ZnO tripods. *J. Appl. Phys.* 105:033513-1–033513-6.
27. Wang, Z. L. 2004. Nanostructures of zinc oxide. *Mater. Today* 7:26–33.
28. Rakshit, T., Manna, I., and Ray, S. K. 2013. Shape controlled Sn doped ZnO nanostructures for tunable optical emission and transport properties. *AIP Adv.* 3:112112-1–112112-12.
29. Jayanthi, K., Chawla, S., Sood, K. N., Chhibara, M., and Singh, S. 2009. Dopant induced morphology changes in ZnO nanocrystals. *Appl. Surf. Sci.* 255:5869–5875.
30. Shen, G., Cho, J. H., Yoo, J. K., Yi, G.-C., and Lee, C. J. 2005. Synthesis and optical properties of S-doped ZnO nanostructures: Nanonails and nanowires. *J. Phys. Chem. B* 109:5491–5496.
31. Pan, H., Zhu, Y., Sun, H., Feng, Y., Sow, C.-H., and Lin, J. 2006. Electroluminescence and field emission of Mg-doped ZnO tetrapods. *Nanotechnology* 17:5096–5100.
32. Roy, N. and Roy, A. 2015. Growth and temperature dependent photoluminescence characteristics of ZnO tetrapods. *Ceram. Int.* 41:4154–4160.
33. Yu, D., Hu, L., Qiao, S. et al. 2009. Photoluminescence study of novel phosphorus-doped ZnO nano-tetrapods synthesized by chemical vapour deposition. *J. Phys. D: Appl. Phys.* 42:055110-1–055110-6.
34. Rackauskas, S., Mustonen, K., Järvinen, T. et al. 2012. Synthesis of ZnO tetrapods for flexible and transparent UV sensors. *Nanotechnology* 23:095502-1–095502-7.
35. Roy, V. A. L., Djurišić, A. B., Liu, H. et al. 2004. Magnetic properties of Mn doped ZnO tetrapod structures. *Appl. Phys. Lett.* 84:756–758.
36. Huh, J., Kim, G.-T., Lee, J. S., and Kim, S. 2008. A direct measurement of the local resistances in a ZnO tetrapod by means of impedance spectroscopy: The role of the junction in the overall resistance. *Appl. Phys. Lett.* 93:042111-1–042111-3.
37. Zhou, X., Lin, T., Liu, Y. et al. 2013. Structural, optical, and improved field-emission properties of tetrapod-shaped Sn-doped ZnO nanostructures synthesized via thermal evaporation. *ACS Appl. Mater. Interfaces* 5:10067–10073.
38. Zhao, L. and Hu, L. 2012. Synthesis and applications of CdSe nano-tetrapods in hybrid photovoltaic devices. *Pure Appl. Chem.* 84:2549–2558.
39. Sun, B., Marx, E., and Greenham, N. C. 2003. Photovoltaic devices using blends of branched CdSe nanoparticles and conjugated polymers. *Nano Lett.* 3:961–963.
40. Reynolds, D. C., Look, D. C., Jogai, B. et al. 1998. Neutral-donor–bound-exciton complexes in ZnO crystals. *Phys. Rev. B* 57:12151–12155.
41. Teke, A., Özgür, Ü., Doğan, S. et al. 2004. Excitonic fine structure and recombination dynamics in single-crystalline ZnO. *Phys. Rev. B* 70:195207-1–195207-10.
42. Jiang, D. S., Jung, H., and Ploog, K. 1988. Temperature dependence of photoluminescence from GaAs single and multiple quantum-well heterostructures grown by molecular-beam epitaxy. *J. Appl. Phys.* 64:1371–1377.
43. Lautenschlager, P., Garriga, M., Logothetidis, S., and Cardona, M. 1987. Interband critical points of GaAs and their temperature dependence. *Phys. Rev. B* 35:9174–9189.
44. Wang, L. and Giles, N. C. 2003. Temperature dependence of the free-exciton transition energy in zinc oxide by photoluminescence excitation spectroscopy. *J. Appl. Phys.* 94:973–978.
45. Matsumoto, T., Kato, H., Miyamoto, K., Sano, M., Zhukov, E. A., and Yao, T. 2002. Correlation between grain size and optical properties in zinc oxide thin films. *Appl. Phys. Lett.* 81:1231–1233.
46. Liang, W. Y. and Yoffe, A. D. 1968. Transmission spectra of ZnO single crystals. *Phys. Rev. Lett.* 20:59–62.

47. Zhang, Y., Lin, B., Sun, X., and Fu, Z. 2005. Temperature-dependent photoluminescence of nano-crystalline ZnO thin films grown on Si(100) substrates by the sol–gel process. *Appl. Phys. Lett.* 86:131910-1–131910-3.
48. Meyer, B. K., Alves, H., Hofmann, D. M. et al. 2004. Bound exciton and donor–acceptor pair recombinations in ZnO. *Phys. Stat. Sol.* (*b*) 241:231–260.
49. Kurbanov, S. S. and Kang, T. W. 2010. Spectral behavior of the emission around 3.31 eV (A-line) from ZnO nanocrystals. *J. Lumin.* 130:767–770.
50. Kurbanov, S. S., Panin, G. N., and Kang, T. W. 2009. Spatially resolved investigations of the emission around 3.31 eV (A-line) from ZnO nanocrystals. *Appl. Phys. Lett.* 95:211902-1–211902-3.
51. Hwang, D.-K., Kim, H.-S., Lim, J.-H. et al. 2005. Study of the photoluminescence of phosphorus-doped p-type ZnO thin films grown by radio-frequency magnetron sputtering. *Appl. Phys. Lett.* 86:151917-1–151917-3.
52. Janotti, A. and Van de Walle, C. G. 2007. Native point defects in ZnO. *Phys. Rev. B* 76:165202-1–165202-22.
53. Ma, L. A. and Guo, T. L. 2013. Morphology control and improved field emission properties of ZnO tetrapod films deposited by electrophoretic deposition. *Ceram. Int.* 39:6923–6929.
54. Lee, G.-H. 2011. Synthesis and cathodoluminescence of ZnO tetrapods prepared by a simple oxidation of Zn powder in air atmosphere. *Ceram. Int.* 37:189–193.
55. Wan, Q., Yu, K., Wang, T. H., and Lin, C. L. 2003. Low-field electron emission from tetrapod-like ZnO nanostructures synthesized by rapid evaporation. *Appl. Phys. Lett.* 83:2253–2255.
56. Das, S. N., Kar, J. P., Choi, J.-H., Byeon, S., Jho, Y. D., and Myoung, J.-M. 2009. Influence of surface morphology on the optical property of vertically aligned ZnO nanorods. *Appl. Phys. Lett.* 95:111909-1–111909-3.
57. Ahn, C. H., Mohanta, S. K., Lee, N. E., and Cho, H. K. 2009. Enhanced exciton–phonon interactions in photoluminescence of ZnO nanopencils. *Appl. Phys. Lett.* 94:261904-1–261904-3.
58. Gu, X., Huo, K., Qian, G., Fu, J., and Chu, P. K. 2008. Temperature dependent photoluminescence from ZnO nanowires and nanosheets on brass substrate. *Appl. Phys. Lett.* 93:203117-1–203117-3.
59. Wang, X., Ding, Y., Summers, C. J., and Wang, Z. L. 2004. Large-scale synthesis of six-nanometer-wide ZnO nanobelts. *J. Phys. Chem. B* 108:8773–8777.
60. Lu, J. G., Ye, Z. Z., Zhang, Y. Z., Liang, Q. L., Fujita, S., and Wang, Z. L. 2006. Self-assembled ZnO quantum dots with tunable optical properties. *Appl. Phys. Lett.* 89:023122-1–023122-3.
61. Fan, Z., Wang, D., Chang, P.-C., Tseng, W.-Y., and Lu, J. G. 2004. ZnO nanowire field-effect transistor and oxygen sensing property. *Appl. Phys. Lett.* 85:5923–5925.
62. Yun, Y. S., Park, J. Y., Oh, H., Kim, J.-J., and Kim, S. S. 2006. Electrical transport properties of size-tuned ZnO nanorods. *J. Mater. Res.* 21:132–136.
63. Lee, C. J., Lee, T. J., Lyu, S. C., Zhang, Y., Ruh, H., and Lee, H. J. 2002. Field emission from well-aligned zinc oxide nanowires grown at low temperature. *Appl. Phys. Lett.* 81:3648–3650.
64. Zhu, Y. W., Zhang, H. Z., Sun, X. C. et al. 2003. Efficient field emission from ZnO nanoneedle arrays. *Appl. Phys. Lett.* 83:144–146.
65. Park, H., Chang, S., Jean, J. et al. 2012. Graphene cathode-based ZnO nanowire hybrid solar cells. *Nano Lett.* 13:233–239.
66. Tian, W., Zhang, C., Zhai, T. et al. 2014. Flexible ultraviolet photodetectors with broad photoresponse based on branched ZnS–ZnO heterostructure nanofilms. *Adv. Mater.* 26:3088–3093.
67. Zhu, G. Y., Xu, C. X., Lin, Y. et al. 2012. Ultraviolet electroluminescence from horizontal ZnO microrods/GaN heterojunction light-emitting diode array. *Appl. Phys. Lett.* 101:041110(1–4).
68. Lee, S.-H., Han, S.-H., Jung, H. S. et al. 2010. Al-Doped ZnO Thin film: A new transparent conducting layer for ZnO nanowire-based dye-sensitized solar cells. *J. Phys. Chem. C* 114:7185–7189.
69. Chen, B., Meng, C., Yang, Z. et al. 2014. Graphene coated ZnO nanowire optical waveguides. *Opt. Express* 22:24276–24285.
70. Li, C., Li, C., Di, Y., Lei, W., Chen, J., and Cui, Y. 2013. ZnO electron field emitters on three-dimensional patterned carbon nanotube framework. *ACS Appl. Mater. Interfaces* 5:9194–9198.
71. Wang, X., Zhou, J., Song, J., Liu, J., Xu, N., and Wang, Z. L. 2006. Piezoelectric field effect transistor and nanoforce sensor based on a single ZnO nanowire. *Nano Lett.* 6:2768–2772.
72. Hoffmann, M. W. G., Mayrhofer, L., Casals, O. et al. 2014. A highly selective and self-powered gas sensor via organic surface functionalization of p-Si/n-ZnO diodes. *Adv. Mater.* 26:8017–8022.
73. Yang, Y., Sun, X. W., Tay, B. K., You, G. F., Tan, S. T., and Teo, K. L., 2008. A p-n homojunction ZnO nanorod light-emitting diode formed by As ion implantation. *Appl. Phys. Lett.* 93:253107(1–3).
74. Chu, S., Wang, G., Zhou, W. et al. 2011. Electrically pumped waveguide lasing from ZnO nanowires. *Nat. Nanotechnol.* 6:506–510.

75. An, S. J. and Yi, G.-C. 2007. Near ultraviolet light emitting diode composed of *n*-GaN/ZnO coaxial nanorod heterostructures on a *p*-GaN layer. *Appl. Phys. Lett.* 91:123109(1–3).
76. Dong, J. J., Zhang, X. W., and Yin, Z. G. 2012. Ultraviolet electroluminescence from ordered ZnO nanorod array/p-GaN light emitting diodes. *Appl. Phys. Lett.* 100:171109(1–4).
77. Bai, S., Wu, W., Qin, Y., Cui, N., Bayerl, D. J., and Wang, X. 2011. High-performance integrated ZnO nanowire UV sensors on rigid and flexible substrates. *Adv. Funct. Mater.* 21:4464–4469.
78. Liu, N., Fang, G., Zeng, W. et al. 2010. Direct growth of lateral ZnO nanorod UV photodetectors with schottky contact by a single-step hydrothermal reaction. *ACS Appl. Mater. Interfaces* 7:1973–1979.
79. Kind, H., Yan, H., Messer, B., Law, M., and Yang, P. 2002. Nanowire ultraviolet photodetectors and optical switches. *Adv. Mater.* 14:158–160.
80. Soci, C., Zhang, A., Xiang, B. et al. 2007. ZnO nanowire UV photodetectors with high internal gain. *Nano Lett.* 7:1003–1009.
81. Dai, J., Xu, C., Xu, X. et al. 2013. Single ZnO microrod ultraviolet photodetector with high photocurrent gain. *ACS Appl. Mater. Interfaces* 5:9344–9348.
82. Cheng, G., Wu, X., Liu, B., Li, B., Zhang, X., and Du, Z. 2011. ZnO nanowire Schottky barrier ultraviolet photodetector with high sensitivity and fast recovery speed. *Appl. Phys. Lett.* 99:203105(1–3).
83. Wang, Z., Yu, R., and Wen, X., 2014. Optimizing performance of silicon-based p-n junction photodetectors by the piezo-phototronic effect. *ACS Nano* 8:12866–12873.
84. Afal, A., Coskun, S., and Unalan, H. E. 2013. All solution processed, nanowire enhanced ultraviolet photodetectors. *Appl. Phys. Lett.* 102:043503(1–5).
85. van de Hulst, H. C. 1981. *Light Scattering by Small Particles.* New York: Dover Publications.
86. Gogurla, N., Sinha, A. K., Santra, S., Manna, S., and Ray, S.K. 2014. Multifunctional Au–ZnO plasmonic nanostructures for enhanced UV photodetector and room temperature NO sensing devices. *Sci. Rep.* 4:6483(1–9).
87. Li, L., Wu, P., Fang, X. et al. 2010. Single-crystalline CdS nanobelts for excellent field-emitters and ultrahigh quantum-efficiency photodetectors. *Adv. Mater.* 22:3161–3165.
88. Bao, G., Li, D., Sun, X. et al. 2014. Enhanced spectral response of an AlGaN-based solar-blind ultraviolet photodetector with Al nanoparticles. *Opt. Express* 22:24286–24293.
89. Liu, Y., Zhang, X., Su, J., Li, H., Zhang, Q., and Gao, Y. 2014. Ag nanoparticles@ZnO nanowire composite arrays: An absorption enhanced UV photodetector. *Opt. Express* 22:30148–30155.
90. Xie, X. N., Zhong, Y. L., Dhoni, M. S. et al. 2010. UV-visible-near infrared photoabsorption and photodetection using close-packed metallic gold nanoparticle network. *J. Appl. Phys.* 107:053510(1–6).
91. Jin, Z., Gao, L., Zhou, Q., and Wang, J. 2014. High-performance flexible ultraviolet photoconductors based on solution-processed ultrathin ZnO/Au nanoparticle composite films. *Sci. Rep.* 4:4268(1–8).
92. Lu, J., Xu, C., Dai, J. et al. 2015. Improved UV photoresponse of ZnO nanorod arrays by resonant coupling with surface plasmons of Al nanoparticles. *Nanoscale* 7:3396–3403.
93. Aguilar, C. A., Haight, R., Mavrokefalos, A., Korgel, B. A., and Chen, S. 2009. Probing electronic properties of molecular engineered zinc oxide nanowires with photoelectron spectroscopy. *ACS Nano* 3:3057–3062.
94. Tian, C., Jiang, D., Li, B. et al. 2014. Performance enhancement of ZnO UV photodetectors by surface plasmons. *ACS Appl. Mater. Interfaces* 6:2162–2166.
95. Zhang, Q., Dandeneau, C. S., Zhou, X., and Cao, G. 2009. ZnO nanostructures for dye-sensitized solar cells. *Adv. Mater.* 21:4087–4108.
96. Xu, F. and Sun, L. 2011. Solution-derived ZnO nanostructures for photoanodes of dye-sensitized solar cells. *Energy Environ. Sci.* 4:818–841.
97. Chiba, Y., Islam, A., Watanabe, Y., Komiya, R., Koide, N., and Han, L. 2006. Dye-sensitized solar cells with conversion efficiency of 11.1%. *Jpn. J. Appl. Phys.* 45:L638–L640.
98. Wu, F., Shen, W., Cui, Q. et al. 2010. Dynamic characterization of hybrid solar cells based on polymer and aligned ZnO nanorods by intensity modulated photocurrent spectroscopy. *J. Phys. Chem. C* 114:20225–20235.
99. Baeten, L., Conings, B., Boyen, H.-G. et al. 2011. Towards efficient hybrid solar cells based on fully polymer infiltrated ZnO nanorod arrays. *Adv. Mater.* 23:2802–2805.
100. Olson, D. C., Lee, Y.-J., White, M. S. et al. 2007. Effect of polymer processing on the performance of poly(3-hexylthiophene)/ZnO nanorod photovoltaic devices. *J. Phys. Chem. C* 111:16640–16645.
101. Olson, D. C., Lee, Y.-J., White, M. S. et al. 2008. Effect of ZnO processing on the photovoltage of ZnO/poly(3-hexylthiophene) solar cells. *J. Phys. Chem. C* 112:9544–9547.
102. Rakshit, T., Mondal, S. P., Manna, I., and Ray, S. K. 2012. CdS-decorated ZnO nanorod heterostructures for improved hybrid photovoltaic devices. *ACS Appl. Mater. Interfaces* 4:6085–6095.

103. Saunders, B. R. 2012. Hybrid polymer/nanoparticle solar cells: Preparation, principles and challenges. *J. Colloid Interface Sci.* 369:1–15.

104. Sandberg, H. G. O., Frey, G. L., Shkunov, M. N. et al. 2002. Ultrathin regioregular poly(3-hexyl thiophene) field-effect transistors. *Langmuir* 18:10176–10182.

105. Wang, L., Zhao, D., Su, Z., and Shen, D. 2012. Hybrid polymer/ZnO solar cells sensitized by PbS quantum dots. *Nanoscale Res. Lett.* 7:106-1–106-6.

106. Kim, H., Jeong, H., An, T. K., Park, C. E., and Yong, K. 2013. Hybrid-type quantum-dot cosensitized ZnO nanowire solar cell with enhanced visible-light harvesting. *ACS Appl. Mater. Interfaces* 5:268–275.

107. Spoerke, E. D., Lloyd, M. T., McCready, E. M., Olson, D. C., Lee, Y.-J., and Hsu, J. W. P. 2009. Improved performance of poly(3-hexylthiophene)/zinc oxide hybrid photovoltaics modified with interfacial nanocrystalline cadmium sulfide. *Appl. Phys. Lett.* 95:213506-1–213506-3.

108. O'Regan, B. C., Scully, S., Mayer, A. C., Palomares, E., and Durrant, J. 2005. The effect of Al$_2$O$_3$ barrier layers in TiO$_2$/dye/CuSCN photovoltaic cells explored by recombination and DOS characterization using transient photovoltage measurements. *J. Phys. Chem. B* 109:4616–4623.

109. Wu, S., Li, J., Lo, S.-C., Tai, Q., and Yan, F. 2012. Enhanced performance of hybrid solar cells based on ordered electrospun ZnO nanofibers modified with CdS on the surface. *Org. Electron.* 13:1569–1575.

110. Wang, L., Zhao, D., Su, Z., Li, B., Zhang, Z., and Shen, D. 2011. Enhanced efficiency of polymer/ZnO nanorods hybrid solar cell sensitized by CdS quantum dots. *J. Electrochem. Soc.* 158:H804–H807.

111. Cui, Q., Liu, C., Wu, F. et al. 2013. Performance improvement in polymer/ZnO nanoarray hybrid solar cells by formation of ZnO/CdS-core/shell heterostructures. *J. Phys. Chem. C* 117:5626–5637.

112. Zhu, C., Pan, X., Ye, C., Wang, L., Ye, Z., and Huang, J. 2013. Effect of CdSe quantum dots on the performance of hybrid solar cells based on ZnO nanorod arrays. *Ceram. Int.* 39:2975–2980.

113. Liu, Z., Han, J., Han, L. et al. 2013. Fabrication of ZnO/CuS core/shell nanoarrays for inorganic-organic heterojunction solar cells. *Mater. Chem. Phys.* 141:804–809.

114. Greene, L. E., Law, M., Yuhas, B. D., and Yang, P. 2007. ZnO–TiO$_2$ core–shell nanorod/P3HT solar cells. *J. Phys. Chem. C* 111:18451–18456.

115. Seiyama, T., Kato, A., Fujiishi, K., and Nagatani, M. 1962. A new detector for gaseous components using semiconductive thin films. *Anal. Chem.* 34:1502–1503.

116. Müller, J. and Weißenrieder, S. 1994. ZnO-thin film chemical sensors. *Fresenius J. Anal. Chem.* 349:380–384.

117. Paraguay, D. F., Miki-Yoshida, M., Morales, J., Solis, J., and Estrada, L. W. 2000. Influence of Al, In, Cu, Fe and Sn dopants on the response of thin film ZnO gas sensor to ethanol vapour. *Thin Solid Films* 373:137–140.

118. Sarala Devi, G., Bala Subrahmanyam, V., Gadkari, S. C., and Gupta, S. K. 2006. NH$_3$ gas sensing properties of nanocrystalline ZnO based thick films. *Anal. Chim. Acta* 568:41–46.

119. Gaikwad, R. S., Patil, G. R., Pawar, B. N., Mane, R. S., and Han, S.-H. 2013. Liquefied petroleum gas sensing properties of sprayed nanocrystalline zinc oxide thin films. *Sens. Actuators B Phys.* 189:339–343.

120. Kalyamwar, V. S., Raghuwanshi, F. C., Jadhao, N. L., and Gadewar, A. J. 2013. Zinc oxide nanostructure thick films as H$_2$S gas sensors at room temperature. *J. Sens. Technol.* 3:31–35.

121. Jing, Z. and Zhan, J. 2008. Fabrication and gas-sensing properties of porous ZnO nanoplates. *Adv. Mater.* 20:4547–4551.

122. Park, S., An, S., Ko, H., Jin, C., and Lee, C. 2012. Synthesis of nanograined ZnO nanowires and their enhanced gas sensing properties. *ACS Appl. Mater. Interfaces* 4:3650–3656.

123. Wei, S., Wang, S., Zhang, Y., and Zhou, M. 2014. Different morphologies of ZnO and their ethanol sensing property. *Sens. Actuators B* 192:480–487.

124. Spencer, M. J. S. and Yarovsky, I. 2010. ZnO nanostructures for gas sensing: Interaction of NO$_2$, NO, O, and N with the ZnO (10$\bar{1}$0) surface. *J. Phys. Chem. C* 114:10881–10893.

125. Choi, J.-K., Hwang, I.-S., Kim, S.-J. et al. 2010. Design of selective gas sensors using electrospun Pd-doped SnO$_2$ hollow nanofibers. *Sens. Actuators B Chem.* 150:191–199.

126. Hongsith, N., Viriyaworasakul, C., Mangkorntong, P., Mangkorntong, N., and Choopun, S. 2008. Ethanol sensor based on ZnO and Au-doped ZnO nanowires. *Ceram. Int.* 34:823–826.

127. Chow, L., Lupan, O., Chai, G. et al. 2013. Synthesis and characterization of Cu-doped ZnO one-dimensional structures for miniaturized sensor applications with faster response. *Sens. Actuators A* 189:399–408.

128. Lu, G., Xu, J., Sun, J., Yu, Y., Zhang, Y., and Liu, F. 2012. UV-enhanced room temperature NO$_2$ sensor using ZnO nanorods modified with SnO$_2$ nanoparticles. *Sens. Actuators B Chem.* 162:82–88.

129. Rakshit, T., Santra, S., Manna, I., and Ray, S. K. 2014. Enhanced sensitivity and selectivity of brush-like SnO$_2$ nanowire/ZnO nanorod heterostructure based sensors for volatile organic compounds. *RSC Adv.* 4:36749–36756.

130. Scott, R. W. J., Yang, S. M., Chabanis, G., Coombs, N., Williams, D. E., and Ozin, G. A. 2001. Tin dioxide opals and inverted opals: Near-ideal microstructures for gas sensors. *Adv. Mater.* 13:1468–1472.

131. Tiemann, M. 2007. Porous metal oxides as gas sensors. *Chem. Eur. J.* 13:8376–8388.

132. Chiu, H.-C. and Yeh, C.-S. 2007. Hydrothermal synthesis of SnO$_2$ nanoparticles and their gas-sensing of alcohol. *J. Phys. Chem. C* 111:7256–7259.

133. Zhang, Y., Li, J., An, G., and He, X. 2010. Highly porous SnO$_2$ fibers by electrospinning and oxygen plasma etching and its ethanol-sensing properties. *Sens. Actuators B Chem.* 144:43–48.

134. Al-Hardan, N. H., Abdullah, M. J., and Abdul Aziz, A. 2013. Performance of Cr-doped ZnO for acetone sensing. *Appl. Surf. Sci.* 270:480–485.

135. Park, J. Y., Choi, S.-W., and Kim, S. S. 2011. A model for the enhancement of gas sensing properties in SnO$_2$–ZnO core–shell nanofibres. *J. Phys. D: Appl. Phys.* 44:205403-1–205403-4.

136. Weis, T., Lipperheide, R., Wille, U., and Brehme, S. 2002. Barrier-controlled carrier transport in micro-crystalline semiconducting materials: Description within a unified model. *J. Appl. Phys.* 92:1411–1418.

137. Sysoev, V. V., Goschnick, J., Schneider, T., Strelcov, E., and Kolmakov, A. 2007. A gradient microarray electronic nose based on percolating SnO$_2$ nanowire sensing elements. *Nano Lett.* 7:3182–3188.

138. Kolmakov, A., Klenov, D. O., Lilach, Y., Stemmer, S., and Moskovits, M. 2005. Enhanced gas sensing by individual SnO$_2$ nanowires and nanobelts functionalized with Pd catalyst particles. *Nano Lett.* 5:667–673.

139. Bowen, T. C., Noble, R. D., and Falconer, J. L. 2004. Fundamentals and applications of pervaporation through zeolite membranes. *J. Membr. Sci.* 245:1–33.

140. Borjigin, T., Sun, F., Zhang, J., Cai, K., Ren, H., and Zhu, G. 2012. A microporous metal-organic framework with high stability for GC separation of alcohols from water. *Chem. Commun.* 48:7613–7615.

141. Szostak, R. 1992. *Handbook of Molecular Sieves.* New York: Van Nostrand Reinhold.

142. Baertsch, C. D., Funke, H. H., Falconer, J. L., and Noble, R. D. 1996. Permeation of aromatic hydrocarbon vapors through silicalite–zeolite membranes. *J. Phys. Chem.* 100:7676–7679.

143. Ju, D., Xu, H., Qiu, Z., Guo, J., Zhang, J., and Cao, B. 2014. Highly sensitive and selective triethylamine-sensing properties of nanosheets directly grown on ceramic tube by forming NiO/ZnO PN heterojunction. *Sens. Actuators B Chem.* 200:288–296.

144. Kim, J., Kim, W., and Yong, K. 2012. CuO/ZnO heterostructured nanorods: Photochemical synthesis and the mechanism of H$_2$S gas sensing. *J. Phys. Chem. C* 116:15682–15691.

145. Huang, L. and Fan, H. 2012. Room-temperature solid state synthesis of ZnO/α-Fe$_2$O$_3$ hierarchical nano-structures and their enhanced gas-sensing properties. *Sens. Actuators B Chem.* 171–172:1257–1263.

146. Zhang, Y.-B., Yin, J., Li, L., Zhang, L.-X., and Bie, L.-J. 2014. Enhanced ethanol gas-sensing properties of flower-like p-CuO/n-ZnO heterojunction nanorods. *Sens. Actuators B Chem.* 202:500–507.

147. Lou, Z., Deng, J., Wang, L., Wang, L., Fei, T., and Zhang, T. 2013. Toluene and ethanol sensing performances of pristine and PdO-decorated flower-like ZnO structures. *Sens. Actuators B Chem.* 176:323–329.

148. Hwang, I.-S., Kim, S.-J., Choi, J.-K. et al. 2010 Synthesis and gas sensing characteristics of highly crystalline ZnO–SnO$_2$ core–shell nanowires. *Sens. Actuators B Chem.* 148:595–600.

149. Woo, H.-S., Na, C. W., Kim, I.-D., and Lee, J.-H. 2012. Highly sensitive and selective trimethylamine sensor using one-dimensional ZnO–Cr$_2$O$_3$ hetero-nanostructures. *Nanotechnology* 23:245501-1–245501-10.

150. Liu, Y., Li, G., Mi, R., Deng, C., and Gao, P. 2014. An environment-benign method for the synthesis of p-NiO/n-ZnO heterostructure with excellent performance for gas sensing and photocatalysis. *Sens. Actuators B Chem.* 191:537–544.

151. Huang, B.-R. and Lin, J.-C. 2012. Core–shell structure of zinc oxide/indium oxide nanorod based hydrogen sensors. *Sens. Actuators B Chem.* 174:389–393.

152. Datta, N., Ramgir, N., Kaur, M. et al. 2012. Selective H$_2$S sensing characteristics of hydrothermally grown ZnO-nanowires network tailored by ultrathin CuO layers. *Sens. Actuators B Chem.* 166–167:394–401.

153. Chang, S.-J., Hsueh, T.-J., Chen, I.-C., and Huang, B.-R. 2008. Highly sensitive ZnO nanowire CO sensors with the adsorption of Au nanoparticles. *Nanotechnology* 19:175502(5pp.).

154. Liu, X., Zhang, J., Guo, X., Wu, S., and Wang, S. 2010. Amino acid-assisted one-pot assembly of Au, Pt nanoparticles onto one-dimensional ZnO microrods. *Nanoscale* 2:1178–1184.

155. Guo, J., Zhang, J., Zhu, M., Ju, D., Xu, H., and Cao, B. 2014. High-performance gas sensor based on ZnO nanowires functionalized by Au nanoparticles. *Sens. Actuators B* 199:339–345.

156. Mubeen, S., Hernandez-Sosa, G., Moses, D., Lee, J., and Moskovits, M. 2011. Plasmonic photosensitization of a wide band gap semiconductor: Converting plasmons to charge carriers. *Nano Lett.* 11:5548–5552.

157. Zheng, Z. Q., Wang, B., Yao, J. D., and Yang, G. W. 2015. Light-controlled C_2H_2 gas sensing based on Au–ZnO nanowires with plasmon-enhanced sensitivity at room temperature. *J. Mater. Chem. C* 3:7067–7074.

158. Koshizaki, N. and Oyama, T. 2000. Sensing characteristics of ZnO-based NO sensor. *Sens. Actuators B* 66:119–122.

159. Shishiyanu, S. T., Shishiyanu, T. S., and Lupan, O. I. 2005. Sensing characteristics of tin doped ZnO thin films as NO_2 gas sensor. *Sens. Actuators B* 107:379–386.

160. Joy, N. A., Rogers, P. H., Nandasiri, M. I., Thevuthasan, S., and Carpenter, M. A. 2012. Plasmonic-based sensing using an array of Au-metal oxide thin films. *Anal. Chem.* 84:10437–10444.

161. Li, Y., Gong, J., He, G., and Deng, Y. 2012. Enhancement of photoresponse and UV-assisted gas sensing with Au decorated ZnO nanofibers. *Mater. Chem. Phys.* 134:1172–1178.

162. Hahn, Y.-B., Ahmad, R., and Tripathy, N 2012. Chemical and biological sensors based on metal oxide nanostructures. *Chem. Commun.* 48:10369–10385.

163. You, X., Pikul, J. H., King, W. P., and Pak, J. J. 2013. Zinc oxide inverse opal enzymatic biosensor. *Appl. Phys. Lett.* 102:253103-1–253103-5.

164. Ahmad, M., Pan, C., Luo, Z., and Zhu, J. 2010. A single ZnO nanofiber-based highly sensitive amperometric glucose biosensor. *J. Phys. Chem. C* 114:9308–9313.

165. Mandal, S., Sambasivarao, K., Mullick, H., Dhar, A., Maiti, T. K., and Ray, S. K. 2009. Amperometric detection of glucose biomolecules using ZnO tripods and nanorods: A comparative study. *Sens. Lett.* 7:635–639.

3 From Basic Physical Properties of InAs/InP Quantum Dots to State-of-the-Art Lasers for 1.55 μm Optical Communications

An Overview

Jacky Even, Cheng Wang, and Frédéric Grillot

CONTENTS

3.1 INTRODUCTION

3.1.1 TOWARD InP-BASED QUANTUM DOT LASERS FOR OPTICAL TELECOMMUNICATIONS

Semiconductor lasers play a crucial role in optical data communication and telecommunication applications. Nowadays, high-definition television, video on demand, broadband internet, and mobile phones are available all around the globe. The exponential rise in cost-effective information transmission would not have been possible without the introduction of optical transmission systems, which in turn are enabled by semiconductor lasers. A hundred million new semiconductor lasers are deployed in communication systems every year, generating several billion dollars of annual revenue at the component level [1]. Higher performance semiconductor laser sources are in strong demand because of the increasing data traffic in the WAN (wide area network), MAN (metropolitan area network), and LAN (local area network), which drives the development of novel semiconductor laser technologies.

Since the first demonstration of a semiconductor laser in 1962, the field has been witnessing the development from bulk-structure lasers and quantum-well (Qwell) to the advanced nanostructure quantum wire (Qwire), quantum dash (Qdash), and quantum dot (Qdot) lasers. In the 1960s, the bulk laser was developed on the basis of semiconductor heterostructures, which provided efficient confinement of charged carriers in the active region [2]. In particular, the double heterostructure (DH), which also yields optical confinement, has transformed semiconductor lasers from the laboratory to industry [3]. Quantum confinement occurs when one or more spatial dimensions of a nanocrystal approach the de Broglie wavelength of the carrier (on the order of 10 nm). The confinement of carriers leads to the quantization of the density of states, and splits the energy band of bulk semiconductors into energetic subbands [4]. In the 1970s, the first Qwell laser, in which carriers are confined in one dimension, was demonstrated [5]. Its recognized advantages over DH lasers were the reduced threshold current by decreasing the thickness of the active layer and the tunability of the wavelength via changing the Qwell thickness.

Increase of the confinement dimension leads to Qwire (2D confinement), Qdash (quasi-3D confinement), or Qdot structures (3D confinement). The 3D spatial confinement of Qdots results in an atomic-like density of states. The concept of Qdot semiconductor was proposed by Arakawa and Sakaki in 1982 [6], predicting temperature independence of the threshold current. Thereafter, reduction in threshold current density, high spectral purity, enhancement of differential gain, and chirp-free properties were theoretically discussed in the 1980s [7]. The most straightforward technique to produce an array of Qdots is to fabricate suitably sized mesa-etched quantum wells grown by metal-organic chemical vapor deposition (MOCVD) or molecular beam epitaxy (MBE). However, nonradiative defects produced during the etching procedure leads to a degradation of the material quality, which results in unsuitable structures for lasers. In the 1990s, both selective growth and self-assembled growth technique, which can avoid nonradiative defects, were well developed. Particularly, the Stranski–Krastanov growth mode turned out to be very successful for InGaAs/GaAs systems [8,9]. The strain-induced self-organization of InGaAs/GaAs quantum dots [10,11] yields threshold current densities as low as ~60 A cm^{-2} at room temperature [12]. Extensive work on the GaAs-based Qdot system has been carried out, which resulted in tremendous improvement in laser performance [13]. An ultralow threshold current density of 17 A cm^{-2} and a high output power of 7 W were achieved in InAs/GaAs lasers [14,15]. Nowadays, the InAs/GaAs Qdot products have already become commercially available in the market [8,16]. Meanwhile, several self-assembled growth techniques, such as solid-state MBE [17], gas-source MBE [18], MOCVD [19], and chemical-beam epitaxy (CBE) [20], have been improved and successfully used to grow Qdot materials.

Nevertheless, the GaAs-based Qdot laser devices emit usually in the O band (1260–1360 nm) of the telecommunication windows and are hardly able to reach the desirable long-haul communication window of ~1.55 μm. Instead, the InAs active region grown on InP substrates allows the realization of laser devices working in the C band window (1530–1565 nm) because of the smaller lattice mismatch. In the current market, InP-based 1.55 μm Qwell laser devices have shown substantial improvement in the optical characteristics in comparison with their DH counterparts. In order to improve the Qwell laser performance, InAs dots grown on InP substrates have attracted much attention in the research field. However, the formation of nanostructures on InP is much more challenging than on the GaAs substrate [21]. Although the InAs/InP and the InAs/GaAs systems have the same material in the dots, they differ in three aspects: (a) lattice mismatch in InAs/InP (~3%) is smaller than that in InAs/GaAs (~7%); (b) InAs/InP dots exhibit less confinement potential for electrons, but a stronger confinement for holes; (c) InAs/InP material shares the same cation (In), while the InAs/GaAs shares the same anion (As) at the interface [22]. The small lattice mismatch and the complex strain distribution can result in the formation of a new class of self-assembled Qdash nanostructures instead of Qdots. These are elongated, dot-like structures exhibiting interesting mixed characteristics in between those of the Qwell and the Qdot [23]. Strongly anisotropic QDash nanostructures are even closer to Qwire in their electronic properties [24]. Therefore, realization of real InAs/InP Qdots requires specific epitaxial growth procedures. This is done by employing conventional or miscut (100) InP substrates as well as (311) InP misoriented substrates along with various innovations in the growth process [25]. Nevertheless, reduction of the dot size and suppression of the size dispersion due to the self-assembly growth procedure are still challenges in achieving high-quality epitaxial material and hence better device performance. Currently, the size dispersions, characterized by the photoluminescence (PL) line-width in terms of the full-width at half-maximum (FWHM), are ~20 meV at 10 K for Qdots and ~50 meV at room temperature for Qdashes [26–29] and Qdots [30]. Further improvements in material quality are still required for them to be competitive with the mature InAs/GaAs Qdot systems. InAs dots formed on the (100)InP substrate usually have a low dot density (on the order of 10^9–10^{10} cm^{-2}) [31]. These laser structures usually require multiple stacked layers for sufficient material gain. In contrast, dots grown on high-index (311)B substrates can lead to a large increase of the Qdots density, commonly in the 5×10^{10}–10^{11} cm^{-2} range [32,33]. Figure 3.1 shows the atomic force microscopy (AFM) images of (a) InAs/InP Qdash, (b) Qdots on a (311)B substrate, and (c) Qdots on a (100) substrate. The first InAs/InP Qdot laser was reported by Ustinov et al. in 1998 [34,35], which emitted at ~1.9 μm at 77 K. Room-temperature operation was realized soon [36]. In the following, we review the development of InAs/InP Qdot lasers on both (311)B and (100)InP substrates employing the InAs/InGaAsP or InAs/InGaAlAs active material regions.

3.1.2 Lasers on (311)B InP Substrate

High-indexed (311)B InP substrates can offer a high density of nucleation points for the Qdot islands, which strongly reduces surface migration effects and leads to the formation of more symmetric Qdots in the planar direction. As a result, high density and uniform distribution of Qdots can be obtained. Many studies have been carried out on the InP (311)B substrate, and a relatively high density of 5–13×10^{10} cm^{-2} has been realized [36,38]. The inhomogeneous line-width broadening due to Qdot size dispersion can be constrained within 50 meV by using the double-capping-layer technique [30,39]. Figure 3.2a presents the AFM image of a typical InAs/InP dot on a (311)B substrate. The Qdot base resembles a circle rather than a square with a typical size of 30–50 nm in diameter, and the height for this kind of dot with cylindrical symmetry is usually several nanometers [33]. The cross-sectional scanning tunneling microscopy (X-STM) image of the Qdot structure in Figure 3.2b shows a truncated, faceted profile of the Qdot.

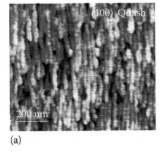
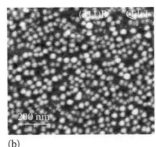
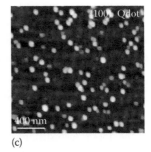

(a)   (b)   (c)

FIGURE 3.1 AFM image of InAs/InP (a) Qdashes on (100) orientation substrate, (b) Qdots on (311)B substrate, and (c) Qdots on (100) substrate. (a: From Zhou, D. et al., *Appl. Phys. Lett.*, 93, 161104, 2008; b: From Zhou, D. et al., *Appl. Phys. Lett.*, 93, 161104, 2008; c: From Bertru, N. et al., Two-dimensional ordering of self-assembled InAs quantum dots grown on (311)B InP substrate, *Proceedings of SPIE, QD laser on InP substrate for 1.55 um emission and beyond*, San Francisco, CA, Vol. 7608, p. 76081B, 2010.)

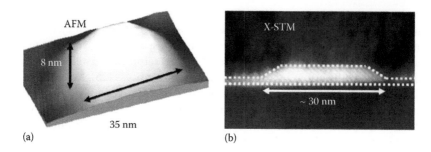

FIGURE 3.2 Structural investigations on InAs/InP Qdots on (311)B-oriented substrate. (a) AFM image: the bright areas represent the top of the Qdot; (b) X-STM image: the bright areas represent the rich InAs areas. (From Cornet, C. et al., *Phys. Rev. B*, 74, 035312, 2006.)

Using the InAs/InGaAsP active region, Nishi et al. demonstrated a Qdot laser with seven stacking layers grown by MBE [36]. The laser device had a dot density of 2×10^{10} cm^{-2} and a threshold current density of 4.8 kA cm^{-2}, and emitted at 1.4 μm at room temperature. By using a double-capping technique, Caroff et al. achieved a high dot density of 1.1×10^{11} cm^{-2}, emitting close to 1.55 μm at room temperature [40]. The laser had a high modal gain of 7 cm^{-1} per dot layer. The threshold current density and transparency current density were 190 A cm^{-2} (63 A cm^{-2} per layer) and 68 A cm^{-2} (23 A cm^{-2} per layer), respectively. These are the best threshold performances reported in this material system. Homeyer et al. reported a broad-area laser emitting at 1.54 μm at room temperature with a record internal quantum efficiency of 62% [41]. The laser consisted of only two stack layers with a modal gain of 8 cm^{-1} per layer. The dot density was also as high as $\sim 1.0 \times 10^{11}$ cm^{-2}. A single-active-layer laser device with a gain of 13 cm^{-1} was demonstrated in 2007, but it lased only up to 295 K [32]. Recently, a low-internal-loss laser of 6 cm^{-1} was reported, which included nine stack layers with a total modal gain of 25 cm^{-1} [42]. However, the InAs/InGaAsP material system usually has poor temperature stability with a characteristic temperature of only 25–50 K at room temperature due to the low conduction band offset [25].

By employing the InAs/InGaAlAs active region, Saito et al. first demonstrated a Qdot laser at the wavelength of 1.63 μm; the threshold current density was 660 A cm^{-2} (132 A cm^{-2} per layer) [43]. The extracted internal loss was 3.6 cm^{-1}, which is the lowest reported value for any InAs/InP system. Six years later, by exploiting the Al atoms in the spacer layers and employing the strain compensation technique, Akahane et al. demonstrated a 30-stack-layer laser with a threshold current density of 2.7 kA cm^{-2} (90 A cm^{-2} per layer) [44]. Subsequently, with improvement in the material growth quality, the threshold current density of the 30-stack laser was reduced to 1.72 kA cm^{-2} (57.4 A cm^{-2} per layer) with extremely high temperature stability. The characteristic temperature (T_0) was 114 K (20°C–75°C), which further improved to a record 148 K (25°C–80°C) in a 20-stack laser [45]. However, the laser had a high internal loss of ~26 cm^{-1} due to the imperfect coupling of the optical mode with the multistack gain medium.

3.1.3 LASERS ON (100)InP SUBSTRATE

The formation of self-assembled Qdot on the (100)InP substrate is more complicated. The formation of the dot or dash strongly depends on the growth conditions and the thickness of InAs layers. The major problem in obtaining good performance of the laser is the low dot density on this kind of substrate.

In the InAs/InGaAsP material system, the growth of dots is mostly based on CBE or MOCVD techniques. Allen et al. reported a Qdot laser using the CBE technique, which was operated in pulsed mode at room temperature [46]. The pulsed threshold current density was 3.56 kA cm^{-2} (713 A cm^{-2} per layer), where a dot height trimming procedure with growth interruptions was employed. The dot density was increased by using the higher energy barrier to $3–6 \times 10^{10}$ cm^{-2}. On the other hand, Lelarge et al. used a hybrid growth technique with an MBE-grown active region in conjunction with MOCVD grown p-doped cladding and contact layers [47]. The continuous wave (CW) threshold current density was 1.4 kA cm^{-2} (240 A cm^{-2} per layer) and T_0 was 56 K at room temperature. Particularly, the laser showed a very high modal gain of 64 cm^{-1} (10.7 cm^{-1} per layer). The laser grown by MOCVD showed a reduced threshold current density of 615 A cm^{-2} (123 A cm^{-2} per layer) [48]. The transparency current density was as low as 30 A cm^{-2} and the internal loss was only 4.2 cm^{-1}.

In the InAs/InGaAlAs material system, Kim et al. first demonstrated a Qdot laser using an assisted growth technique of a thin gas underlying layer before the growth of InAs dots in the InGaAlAs matrix [49,50]. The achieved dot density was 6.0×10^{10} cm^{-2}, and the laser exhibited a threshold current density of 2.8 kA cm^{-2} (400 A cm^{-2} per layer). The measured T_0 was 377 K for temperatures up to 200 K, and 138 K above 200 K. More recently, Gilfert et al. reported a high-gain Qdot laser using the MBE method [51]. A low internal loss of 4 cm^{-1} and a high gain of 15 cm^{-1} per layer were reported. The lasing threshold current density was 1.95 kA cm^{-2} (325 A cm^{-2} per layer).

Very recently, Mollet et al. reported a very high modal gain of 97 cm^{-1} for an InAs/InP(100) Qdash laser. However, the internal loss was also as high as 23 cm^{-1} [52]. Generally, the performance of InAs/InP lasers has improved significantly since its first demonstration. However, it still needs improvement to be comparable with InAs/GaAs laser devices.

3.1.4 Current Status of the Dynamic Performance of InP-Based Qdot Laser

In fiber-optic links, the laser transmitter may be either directly modulated, known as directly modulated (DM) laser, or externally modulated using a modulator. In the direct modulation scheme, the driving current carries the transmitted data and is directly applied to the laser. In the external modulation scheme, the laser, which is subjected to a constant bias current, emits a continuous wave, while an external modulator switches the optical power on or off according to the data stream. In the external modulation scheme, electrooptic (EO) or electroabsorption (EA) modulators are commonly used [53]. EO modulators, such as the Mach–Zehnder modulator, utilizes a signal-controlled crystalline material exhibiting the EO effect (Pockels effect) to modulate the CW laser light. The EO effect is the change in the refractive index of the material resulting from the application of a DC or low-frequency electric field. The EA modulator controls the intensity of a laser beam via an electric voltage. Its operation can be based on the Franz–Keldysh effect [54], that is, a change in the absorption spectrum caused by an applied electric field, which changes the bandgap energy but usually does not involve the excitation of carriers by the electric field. However, most EA modulators are made in the form of a waveguide with electrodes for applying an electric field in a direction perpendicular to the modulated light beam. For achieving a high extinction ratio, one usually exploits the quantum-confined Stark effect, which describes the effect of an external electric field upon the light absorption spectrum or emission spectrum of a quantum-well structure. Both EO and EA modulators are operated at a few volts (below 10 V). In comparison with EO modulators, EA modulators have the convenient feature that they can be integrated with the laser on a single chip to create a data transmitter in the form of a photonic integrated circuit [55,56].

In contrast, DM lasers are the most common, particularly for short-reach systems. They have the lowest cost and least energy consumption, but they usually suffer the chirp characteristics [68]. DM lasers generally produce more chirps for higher extinction ratios, leading to an optimum setting for trading off the signal-to-noise ratio and chirp penalty. Aiming to realize chirp-free DM laser devices, many attempts have been made to develop nanostructured semiconductor lasers. Table 3.1 summarizes the reported dynamic performance of InP-based Qdot and Qdash lasers in the literature, including the modulation bandwidth and the α-factor, which is linked to the frequency chirp under direct modulation.

TABLE 3.1

Dynamic Characteristics of Qdot and Qdash Lasers Grown on InP Substrate

References	Material	Bandwidth	α-Factor, $\leq I_{th}$	α-Factor, $> I_{th}$	Differential Gain, Gain Compression
[57]	Qdot, (311)B	4.8 GHz	~1.8	~6	7.3×10^{-15} cm^2
					6.4×10^{-16} cm^3
[58]	Qdot, (100)	5 GHz			
		15 Gbps			
[59]	Qdot, (100)	9 GHz			
		22 Gbps			
[60]	Qdot, (100) p-doped and tunnel injection	14.4 GHz	~0		0.8×10^{-15} cm^2
					5.4×10^{-17} cm^3
[61]	Qdash, (100)	7.6 GHz			
[62]	Qdash, (100)	6 GHz @ undoped	~1 @ p-doped		
		8 GHz @ p-doped	~0 @ p-doped and tunnel injection		
		12 GHz @ p-doped and tunnel injection			
[63]	Qdash, (100) p-doping	8 GHz			
[64]	Qdash, (100)	9.6 GHz		5–7	
		10 Gbps			
[65]	Qdash, (100) p-doped	10 GHz			1.1×10^{-15} cm^2
[66]	Qdash, (100)	10 GHz			
		20 Gbps			
[67]	Qdash, (100) p-doped	10 Gbps		2.2	
[52]	Qdash, (100)	~10 GHz		~5 @ undoped	$1–2 \times 10^{-15}$ cm^2
				~2.7 @ p-doped	

Martinez et al. reported a Qdot laser grown on the (311)B InP substrate with a modulation bandwidth of 4.8 GHz. The above-threshold α-factor was high (~6) and did not significantly depend on the bias current [57]. However, most reported dynamics of 1.55 μm lasers are for nanostructures grown on (100)InP substrates. Gready et al. demonstrated an InAs/InGaAlAs/InP Qdot laser with a 3 dB bandwidth of 5 GHz [58]. Interestingly, the laser showed a much larger signal modulation capability: 15 Gbps at 4 dB extinction ratio. The discrepancy between the small- and large-signal performances was attributed to the large nonlinear gain compression effect [69]. By optimizing the barrier width and the number of stack layers, the performance of the structure was improved by the same group. The small-signal modulation bandwidth was increased to 9 GHz, and the large signal modulation was operated up to 22 Gbps with an extinction ratio of 3 dB [59]. The best performance of InAs/InP Qdot laser was achieved by Bhowmick et al. recently [60]. The laser material system was InAs/InGaAlAs, and the active zone was grown on the (100)InP substrate. Tunnel injection and p-doping techniques enhanced the modulation bandwidth up to 14.4 GHz, and a near-zero α-factor was realized in this structure. For Qdash lasers on the (100)InP substrate, most works employed the p-doping technique to improve the modulation bandwidth and to reduce the α-factor [52,62,67,70]. Mi et al. showed that the modulation bandwidth could be increased from 6 GHz for the undoped laser to 8 GHz with the p-doped laser [62]. Mollet et al. demonstrated that the α-factor could be reduced from 5 down to 2.7 by p-doping [52]. On the other hand, the tunnel injection technique further increased the modulation bandwidth to more than 10 GHz while reducing the α-factor to near zero [60,62]. Consequently, a low chirp of 0.06 nm at a modulation frequency of 8 GHz was achieved in the Qdash laser and the same chirp level at 10 GHz in the Qdot laser.

3.2 BASIC PHYSICAL PROPERTIES OF BURIED InAs/InP QUANTUM DOTS

3.2.1 Quantum Dot Electronic Structure

Schematically, the active region of a Qdot laser often consists of a 3D separate confinement heterostructure (SCH, also known as barrier), a 2D carrier reservoir (RS, roughly corresponding to the wetting layer), and dots spatially confined in three dimensions. Figure 3.3 shows an illustration of the electronic structure in a Qdot laser.

Carriers in the barrier and the wetting layer can be treated as quasi-free particles. Thus, quasi-continuum electronic states are formed in the SCH and the RS. The total densities of states for the SCH (per volume) and that for the RS (per area) are, respectively, given by [71]

$$\rho_{SCH} = 2 \left(2 \frac{m_{SCH}^*}{\hbar^2} \pi k_B T \right)^{3/2} \tag{3.1}$$

$$\rho_{RS} = \frac{m_{RS}^*}{\pi \hbar^2} k_B T \tag{3.2}$$

with m^* the effective mass of either electrons or holes. The quasi-continuum carrier reservoir coupling with the localized energy states of the dots results in smaller energy separations and thus overlapping states at higher energies [72]. The discrete states lying at lower energies are separated by a few tens of millielectronvolts in the conduction band (CB), while it is smaller in the valence band (VB) due to higher effective hole mass.

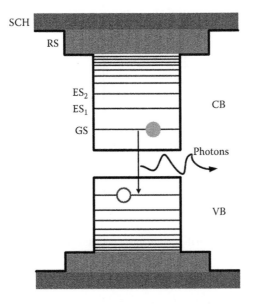

FIGURE 3.3   Schematic of a QD laser electronic band structure of electrons and holes.

Accurate simulation of the electronic structure of an InAs/InP quantum dot can be performed using multiband $\mathbf{k} \cdot \mathbf{p}$ theory including strain and piezoelectric effects [73]. Based on the eight-band $\mathbf{k} \cdot \mathbf{p}$ theory, Figure 3.4a through c shows the confinement potential of the InAs/InP Qdot in comparison with the InAs/GaAs Qdot system [33]. In the absence of strain (Figure 3.4b), the band edge of InP is different from that of GaAs relative to the active material InAs. InP confines the holes more strongly, whereas GaAs does the same for the electrons. In the presence of strain, the band-edge energies are altered as shown in Figure 3.4a for InAs/InP and in Figure 3.4c for InAs/GaAs, by mainly hydrostatic strain in the conduction band and biaxial strain in the valence band. Consequently, the heavy/light hole degeneracy is lifted at the Γ point. The change of band edge for InAs embedded in GaAs is stronger because of the larger lattice mismatch (6.6% compared to only 3.1% for InP). The most striking feature is the smaller (strained) bandgap of InAs in InP than in GaAs, which enables the former to reach the 1.55 μm emission, which is hard to achieve for InAs/GaAs Qdots. Figure 3.4a and c shows that the depth of the electron confinement potential is similar in both systems, and therefore one can expect a comparable spectrum for confined electron states provided the Qdots share the same morphological properties. However, this does not hold for hole states since the confinement potential for InAs/InP Qdots is much deeper and the heavy–light hole splitting is smaller. Figure 3.4d and e illustrates the first three electron and hole wave functions (70% isosurface) of a single particle for both (100) (Figure 3.4d) and (311)B InP (Figure 3.4e) substrates, which is obtained by solving the Schrödinger equation. The symbols e0 and h0 stand for 1Se and 1Sh states, and e1, e2, and h1, h2 stand for 1Pe and 1Ph states. The single-particle states provide a basis for the configuration interaction model, which can be applied to calculate the excitonic properties, including correlation and exchange. Finally, the excitonic optical absorption spectra can be computed; the details can be found in [33].

Because of the Coulomb interaction, electrons and holes in the semiconductor can be bound into electron–hole pairs, known as *excitons*. The distance between the electron and the hole within an exciton is called the Bohr radius of the exciton. Typical exciton Bohr radius of semiconductors is a few nanometers [74]. The exciton's nature can be modified by the confinement structure, and thus it can exhibit different optical properties. The assumption holds as long as the electron and hole populations do not show significant deviations. Such a simplified picture was capable of describing the basic optical properties.

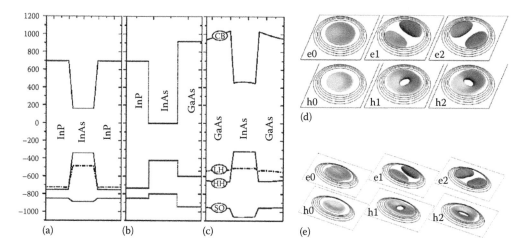

FIGURE 3.4 (a) Vertical scan through the confinement potential for an InAs Qdot embedded in InP, (b) energetic positions of the unstrained band edges of InP and GaAs relative to InAs, (c) vertical scan through the confinement potential for an InAs Qdot embedded in GaAs. Wave function representations for electrons and holes for the first InAs/InP Qdot states (d) on (100) substrate and (e) on (311)B substrate, with a Qdot height of 2.93 nm. e0, e1, and e2 stand for electronic states, and h0, h1, and h2 stand for hole states. The (311)B substrate induces an anisotropy of the wave function. (From Cornet, C. et al., *Phys. Rev. B*, 74, 035312, 2006.)

3.2.2   CARRIER SCATTERING PROCESSES IN QUANTUM DOT

In a Qdot device, once the current injection generates charge carriers in the 3D barrier, the carriers will be transported into the 2D RS, which acts as a carrier reservoir for the localized discrete Qdot states. The carrier capture process refers to the subsequent carrier capture from the RS to the ESs of the dots. In the dots, the carriers relax from high energetic ESs down to the GS level. Finally, radiative recombination of electrons and holes takes place and lasing occurs often on the GS. These processes are well reflected by the time-resolved photoluminescence (TRPL) of an InAs/InP Qdot device, as shown in Figure 3.5a. The spectra are fitted with three Gaussian peaks, which, respectively, correspond to the Qdot excitonic GS at 0.94 eV, the first Qdot ES at 0.99 eV, and the RS ~1.05 eV. The peak of the spectra shifts from RS at 10 ps via ES at 600 ps toward GS at 1500 ps with time evolution after the optical excitation [75]. In addition, carrier capture directly from the RS into the GS is also possible. This direct channel accelerates the carrier indirect process (via ES) to the lasing GS, and plays an important role on the dual (GS and ES) lasing process, as pointed out in InAs/InP Qdot lasers [76]. Moreover, those electronic states also exhibit inter-dot electronic coupling [72]. As in Qwell lasers, the carrier transport process plays an important role in determining the Qdot laser's dynamics as well, which induces a parasitic-like roll-off that is indistinguishable from an RC roll-off in the modulation response, and thus limits the modulation bandwidth [77,78]. TPRL shows that the carrier transport time across barrier to the RS is several picoseconds (1–5 ps) depending on the thickness of the SCH layer [79,80]. The carrier capture and relaxation transition processes are supported mainly by two physical mechanisms: Coulomb-interaction-induced carrier–carrier scattering (Auger process), and carrier–LO phonon scattering. The scattering behavior is different at low and high excitation carrier densities. At low excitation density, the carrier interaction with LO phonons can provide efficient scattering channels provided that energy conservation is fulfilled. While the energy separation of Qdot states typically does not match the LO phonon energy, this scattering mechanism is often possible for holes due to their dense states [81]. The carrier–LO phonon scattering process is found to be temperature-dependent: high temperature accelerates this scattering rate [82,83]. When a high-density carrier plasma is created in the carrier reservoir, carrier–carrier scattering accounts for the efficient capture from the RS into the localized Qdot states as well as the relaxation between

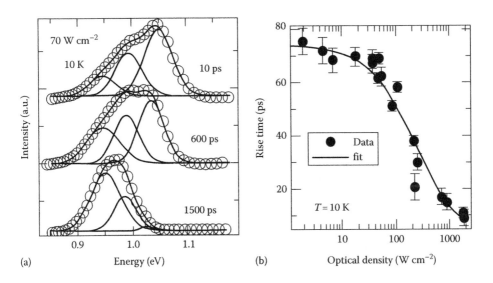

FIGURE 3.5   (a) TRPL spectra recorded at 10 K for 10, 600, and 1500 ps after the optical excitation at 790 nm with an optical excitation density of 70 W cm^{-2}. Spectra are fitted with three Gaussian curves. (b) TRPL analysis of the rise time of the Qdot as a function of the excitation intensity at 10 K. (a: From Miska, P. et al., *Appl. Phys. Lett.*, 92, 191103, 2008; b: From Miska, P. et al., *Appl. Phys. Lett.*, 94, 061916, 2009.)

the discrete Qdot states [84]. Auger scattering can be categorized into three types according to the initial electronic states of the carriers involved. One type involves two carriers in the RS states [85]; another type is with one carrier in an ES state while the other in the RS [86]; and the third type has both carriers occupied by the ES [87]. In contrast to the carrier–phonon interaction process, the Auger process is carrier-density-dependent, as shown in Figure 3.5b [88]. At 10 K, the carrier relaxation time into the GS of the Qdots reduces from ~75 ps under an optical excitation of 1 W cm^{-2} down to about 10 ps for an excitation density of 2000 W cm^{-2}.

From the analysis of TRPL rise time of Qdot devices, both the capture and relaxation times are found to vary over a wide range from 1 up to 100 ps depending on the excitation intensity [82,89]. However, for a moderate RS carrier density of 10^{11}–10^{12} cm^{-2}, the typical carrier scattering times are on the order of 1–10 ps [79,90]. For the processes related to Coulomb many-body interactions, relaxation within the Qdot is typically on a faster timescale than the carrier capture from the RS into the Qdot. Processes involving holes are typically faster than the corresponding processes involving electrons, and capture to the excited states is faster than capture to the ground states. Hence, in a dynamical scenario, first the holes are captured to the excited Qdot states and immediately scattered via relaxation to the Qdot ground states. Capture of electrons is somewhat slower; the subsequent relaxation for electrons is only slightly slower than for holes [81].

From the aspect of rate equation modeling of Qdot lasers, carrier distributions in all states are usually assumed to be under the quasi-equilibrium condition with Fermi–Dirac distribution, which is suitable for sufficiently rapid intraband relaxation processes [91]. Carrier occupations, at least in the GS and in the RS, must be modeled in order to distinguish the Qdot laser from the Qwell laser. However, for achieving moderate accuracy and correlating with experimental data in the InAs/InP Qdot system, it is necessary to consider the population in the first ES, which can have significant influence on the laser's static and dynamic characteristics. Inclusion of more states would be more accurate but at a price of losing simplicity as well as the intuitive physical image.

Rigorous calculation of carrier scattering rates is a stiff task, which requires sophisticated many-body quantum theory that treats intraband collision processes. In [91–93], a phenomenological formula was proposed to take into account the carrier-dependent capture and relaxation time in a semiempirical model:

$$\tau_i = \frac{1}{A_i + C_i N_{RS}} \qquad (3.3)$$

where
    i denotes the capture or relaxation process
    A_i is the phonon-assisted scattering rate
    C_i is the coefficient determining the Auger-assisted scattering by carriers in the RS (N_{RS})

Although this expression leads to good agreement with the TRPL experiments in [75], parameters A_i and C_i can be quite different from device to device [76], thereby limiting the applicability of this expression. Because of the fact that once the laser is operated above threshold there is a large density of carriers in the RS, which does not vary much with the bias current, it is a reasonable approximation to assume the carrier scattering time as constant, which simplifies the rate equation model for the study of Qdot laser dynamics.

3.2.3 GAIN, REFRACTIVE INDEX, AND LINE-WIDTH ENHANCEMENT FACTOR

The laser field and the semiconductor gain medium are coupled by the gain and the carrier-induced refractive index, or equivalently, by the complex optical susceptibility. To determine these quantities, it is necessary to solve the quantum mechanical gain medium equations of motion for the microscopic polarization. In principle, these dynamical equations should be derived using the full system Hamiltonian, which includes contributions from the kinetic energies, the many-body Coulomb interactions, the electric–dipole interaction between the carriers and the laser field, as well as the interactions between the carriers and phonons. The effects of injection current pumping should also be included [92].

The connection between the classical electrodynamics and quantum mechanics is effected through the macroscopic polarization P and the microscopic polarization p_α:

$$P = \frac{1}{V} \sum_\alpha \mu_\alpha p_\alpha \tag{3.4}$$

In Qdot lasers, the processes include the GS, ES, RS, and barrier transitions. V is the active region volume, and the polarization summation is performed over all interband optical transitions. The complex optical susceptibility is connected with the polarization via

$$\chi = \frac{1}{\varepsilon_0 n_b^2} \frac{P}{E} \tag{3.5}$$

where ε_0 and E are the vacuum permittivity of light and the amplitude of the electric field, respectively. The gain g and the carrier-induced refractive index δn in the model are defined by

$$\frac{d}{dt} E(t) = \Gamma_P \frac{cg}{2n_b} E(t) + j \frac{\omega \delta n}{n_b} E(t) \tag{3.6}$$

where
   ω and c are the laser frequency and the velocity of light, respectively
   n_b is the refractive index
   Γ_P is the optical confinement factor

We can obtain the following relation between the gain, refractive index, susceptibility, and the polarization [94]:

$$g = -\frac{\omega n_b}{c} \text{Im}\{\chi\} = -\frac{\omega}{\varepsilon_0 n_b c} \frac{\text{Im}\{P\}}{E} \tag{3.7}$$

$$\delta n = \frac{n_b}{2} \text{Re}\{\chi\} = \frac{1}{2\varepsilon_0 n_b} \frac{\text{Re}\{P\}}{E} \tag{3.8}$$

Introducing the differential gain a, the phenomenological gain can be expressed as

$$g = a(N - N_{tr}) \tag{3.9}$$

with N and N_{tr} being the injected carrier density and transparency carrier density of zero gain, respectively.

In semiconductor lasers, the nonlinear gain phenomenon plays an important role in both static and dynamic characteristics such as spectral properties, modulation bandwidth, and frequency chirping [95]. The main physical mechanisms behind nonlinear gain are attributed to the spectral hole-burning, spatial hole-burning, and carrier heating [96,97]. Furthermore, the gain nonlinearity was found to enhance the quantum confinement of carriers and carrier relaxation processes [98].

Spectral hole-burning is the formation of a dip in the gain spectrum due to stimulated emission. The dip occurs by the recombination of electrons and holes at a specific energy and the subsequent redistribution of carrier energies due to carrier–carrier scattering. The scattering process takes place on the timescale of the order of 50–100 fs and leads to a dip width of about 20–40 meV. It also ensures that temperature equilibrium is established among the carriers within the same timescale. Carrier heating is related to the fact that the carrier temperatures can be different from the lattice temperature due to the stimulated emission and free-carrier absorption [96]. The carrier temperatures relax toward the lattice temperature by electron–phonon scattering processes within a timescale of 0.5–1 ps. The nonlinear gain effect is usually characterized by a phenomenological gain compression factor, as

$$g_{nl} = \frac{g}{1+\xi S} \approx g(1-\xi S)$$

(3.10)

with S being the photon density. This expression shows that the linear gain g is reduced at high power density.

The refractive index change related to the optical interband transition as expressed in Equation 3.8 is known as *anomalous dispersion* [99]. Another important contribution to the index change is the free-carrier plasma originating from intraband transitions [100]. In Qwell lasers, this contribution to the differential index is well described by the Drude model [101]:

$$\delta n_{fc} = -\frac{\Gamma_p e^2 N}{2n_b \varepsilon_0 m^* \omega^2}$$

(3.11)

with e the electronic charge. Analogous transitions in Qdot lasers can be envisaged between bound Qdot states and the continuum levels of the RS and the barrier. It has been shown that the Drude formula can also be applied to the case of Qdot lasers when the Qdot carriers are not tightly confined and when working at photon energies in the 0.8–1.0 eV region [100].

In semiconductor lasers, it is well known that any change in the imaginary part of the susceptibility (gain) will be accompanied by a corresponding change in its real part (refractive index) via the Kramers–Kronig relations. The line-width enhancement factor (or α-factor) describes the coupling between the carrier-induced variation of real and imaginary parts of susceptibility, and is defined as [102]

$$\alpha_H = \frac{\partial \operatorname{Re}\{\chi\}/\partial N}{\partial \operatorname{Im}\{\chi\}/\partial N}$$

(3.12)

Employing Equations 3.7 and 3.8, the above definition is equivalent to the following often-used expression:

$$\alpha_H = -2\frac{\omega}{c}\frac{\partial n/\partial N}{\partial g/\partial N}$$

(3.13)

In a practical case, the variation of the carrier concentration is usually small, which justifies taking the derivatives at the operating point and assuming a linear dependence of $g(N)$ and $n(N)$. Equation 3.13 can therefore be written as

$$\alpha_H = -2\frac{\omega}{c}\frac{\Delta n}{\Delta g}$$

(3.14)

Through the relation between the refractive index and frequency variation, we can obtain the following equivalent formula:

$$\alpha_H = 2\frac{n_b}{c}\frac{\Delta\omega}{\Delta g} \tag{3.15}$$

The α-factor plays a crucial role in driving fundamental features of semiconductor lasers such as the spectral line-width broadening [103], frequency chirp [104], mode stability [105], and nonlinear dynamics subject to optical injection [106] or optical feedback [107–109]. Typical Qwell lasers often exhibit α-factor values in the order of 2–5 [78]. For Qdot lasers, earlier analyses have suggested a zero or near-zero α-factor due to the delta-function-like discrete density of states. A symmetrical gain curve indeed leads to a dispersive curve of the refractive index with a zero value at the gain peak. However, experimental α-factor values in Qdot lasers vary over a wide range from zero up to more than 10, particularly a giant value (as high as 60) was also reported [110–112]. The nonzero α-factor is attributed to the large inhomogeneous broadening, the off-resonant bound and continuum states, as well as the free-carrier plasma effect.

3.3 FREE-RUNNING QUANTUM DOT LASERS

3.3.1 AMPLITUDE MODULATION RESPONSE

High-speed, energy-effective, and low-cost optical communication networks primarily require semiconductor laser sources of broad modulation bandwidth. In order to theoretically discuss the amplitude modulation (AM) performance of Qdot lasers, we employ a semiclassical rate equation model [76]. The model will be analyzed in a semianalytical approach, which has the merit of giving an intuitive physical image. This numerical model of the Qdot laser holds under the assumption that the active region consists of only one Qdot ensemble; that is, the inhomogeneous broadening due to the dot size fluctuation is not considered. The electrons and holes are treated as electron–hole (e–h) pairs, meaning that the system is in excitonic energy states. Two discrete states in Qdots are taken into account: a twofold degenerate GS and a fourfold degenerate first ES. The Qdots are interconnected by the 2D RS. This simplified picture corresponds to the TRPL experimental observations in Figure 3.5 [75]. Carriers are supposed to be injected directly from the contacts into the RS, so the carrier dynamics in the 3D barrier are not taken into account in the model.

Figure 3.6 shows the schematic of the carrier dynamics in the exiton framework. First, the externally injected carrier fills directly the RS reservoir; some of the carriers are then either captured

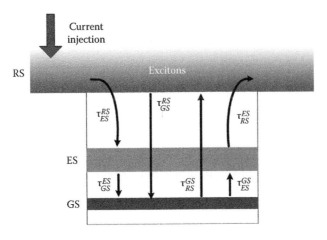

FIGURE 3.6   Sketch of carrier dynamics model including a direct relaxation channel.

into the ES within time τ_{ES}^{RS} or directly into the GS within time τ_{GS}^{RS}, and some of them recombine spontaneously with a spontaneous emission time τ_{RS}^{spon}. Once in the ES, carriers can relax into the GS within time τ_{GS}^{ES} or recombine spontaneously. On the other hand, carriers can also be thermally re-emitted from the ES to the RS with an escape time τ_{RS}^{ES}, which is governed by the Fermi distribution for the quasi-thermal equilibrium without external excitation [113]. A similar dynamic behavior is followed for the carrier population on the GS level with regard to the ES. Direct injection from the RS to the GS was introduced to reproduce the experimental results [76]. Stimulated emission occurs from the GS when the threshold is reached, and that from the ES is not taken into account in the model. Following Figure 3.6, the four coupled rate equations on carrier and photon densities are described as follows:

$$\frac{dN_{RS}}{dt} = \frac{I}{qV} + \frac{N_{ES}}{\tau_{RS}^{ES}} - \frac{N_{RS}}{\tau_{ES}^{RS}}(1-\rho_{ES}) - \frac{N_{RS}}{\tau_{GS}^{RS}}(1-\rho_{GS}) - \frac{N_{RS}}{\tau_{RS}^{spon}} + \frac{N_{GS}}{\tau_{RS}^{GS}} \tag{3.16}$$

$$\frac{dN_{ES}}{dt} = \frac{N_{RS}}{\tau_{ES}^{RS}}(1-\rho_{ES}) + \frac{N_{GS}}{\tau_{ES}^{GS}}(1-\rho_{ES}) - \frac{N_{ES}}{\tau_{RS}^{ES}} - \frac{N_{ES}}{\tau_{GS}^{ES}}(1-\rho_{GS}) - \frac{N_{ES}}{\tau_{ES}^{spon}} \tag{3.17}$$

$$\frac{dN_{GS}}{dt} = \frac{N_{RS}}{\tau_{GS}^{RS}}(1-\rho_{GS}) + \frac{N_{ES}}{\tau_{GS}^{ES}}(1-\rho_{GS}) - \frac{N_{GS}}{\tau_{ES}^{GS}}(1-\rho_{ES}) - \frac{N_{GS}}{\tau_{GS}^{spon}} - \frac{N_{GS}}{\tau_{WL}^{GS}} - g_{GS}v_gS_{GS} \tag{3.18}$$

$$\frac{dS_{GS}}{dt} = \Gamma_P g_{GS}v_gS_{GS} - \frac{S_{GS}}{\tau_p} + \beta_{SP}\frac{N_{GS}}{\tau_{GS}^{spon}} \tag{3.19}$$

where
  N_{RS}, N_{ES}, and N_{GS} are the carrier densities in the RS, ES, GS, respectively
  S_{GS} is the photon density in the cavity with GS resonance energy
  β_{SP} is the spontaneous emission factor
  Γ_P is the optical confinement factor
  τ_p is the photon lifetime
  v_g is the group velocity
  V is the volume of the laser's active region

The GS gain is given by

$$g_{GS} = a_{GS}\frac{N_B}{H_B}\left(\frac{2N_{GS}}{2N_B/H_B} - 1\right) \tag{3.20}$$

where
  a_{GS} is the differential gain
  N_B is the total Qdot surface density
  H_B is the height of the dots.

In what follows, it is important to stress that the effects of gain compression are not taken into account. In Equations 3.16 through 3.18, $\rho_{GS,ES}$ are the carrier occupation probabilities in the GS and the ES, respectively:

$$\rho_{GS} = \frac{N_{GS}}{2N_B/H_B}; \quad \rho_{ES} = \frac{N_{ES}}{4N_B/H_B} \tag{3.21}$$

Since the carrier escape from the GS to the RS has little effect on lasing properties [76], the N_{GS}/τ_{RS}^{GS} term in Equations 3.16 and 3.18 can be neglected.

The rate equations can be linearized through a small-signal analysis. Assuming a sinusoidal current modulation $dI = I_1 e^{j\omega t}$ with modulation frequency ω, the corresponding carrier and photon variations are of the form

$$dN_{RS, ES, GS} = N_{RS1, ES1, GS1} e^{j\omega t}$$

$$dS_{GS} = S_{GS1} e^{j\omega t}$$

(3.22)

Substituting the above formulas into the rate equations (Equations 3.16 through 3.18), we obtain the linearized differential rate equation in matrix form:

$$
\begin{bmatrix}
\gamma_{11} + j\omega & -\gamma_{12} & 0 & 0 \\
-\gamma_{21} & \gamma_{22} + j\omega & -\gamma_{23} & 0 \\
-\gamma_{31} & -\gamma_{32} & \gamma_{33} + j\omega & -\gamma_{34} \\
0 & 0 & -\gamma_{43} & \gamma_{44} + j\omega
\end{bmatrix}
\begin{bmatrix}
N_{RS1} \\
N_{ES1} \\
N_{GS1} \\
S_{GS1}
\end{bmatrix}
= \frac{I_1}{qV}
\begin{bmatrix}
1 \\
0 \\
0 \\
0
\end{bmatrix}
$$

(3.23)

with the following elements:

$$\gamma_{11} = \frac{1-\rho_{ES}}{\tau_{ES}^{RS}} + \frac{1-\rho_{GS}}{\tau_{GS}^{RS}} + \frac{1}{\tau_{RS}^{spon}}; \quad \gamma_{12} = \frac{1}{\tau_{RS}^{ES}} + \frac{N_{RS}}{\tau_{ES}^{RS}} \frac{1}{4N_B/H_B}$$

$$\gamma_{21} = \frac{1-\rho_{ES}}{\tau_{ES}^{RS}}; \quad \gamma_{22} = \frac{1-\rho_{GS}}{\tau_{GS}^{ES}} + \frac{1}{\tau_{RS}^{ES}} + \frac{N_{RS}}{\tau_{ES}^{RS}} \frac{1}{4N_B/H_B} + \frac{N_{GS}}{\tau_{ES}^{GS}} \frac{1}{4N_B/H_B} + \frac{1}{\tau_{ES}^{spon}}$$

$$\gamma_{23} = \frac{1-\rho_{ES}}{\tau_{ES}^{GS}} + \frac{N_{ES}}{\tau_{GS}^{ES}} \frac{1}{2N_B/H_B}; \quad \gamma_{31} = \frac{1-\rho_{GS}}{\tau_{GS}^{RS}}; \quad \gamma_{32} = \frac{1-\rho_{GS}}{\tau_{GS}^{ES}} + \frac{N_{GS}}{\tau_{ES}^{GS}} \frac{1}{4N_B/H_B}$$

(3.24)

$$\gamma_{33} = \frac{1-\rho_{ES}}{\tau_{ES}^{GS}} + \frac{N_{ES}}{\tau_{GS}^{ES}} \frac{1}{2N_B/H_B} + v_g a_{GS} S_{GS} + \frac{1}{\tau_{GS}^{spon}}; \quad \gamma_{34} = -v_g g_{GS}$$

$$\gamma_{43} = \frac{\Gamma_p \beta_{SP}}{\tau_{GS}^{spon}} + \Gamma_p v_g a_{GS} S_{GS}; \quad \gamma_{44} = \frac{1}{\tau_p} - \Gamma_p v_g g_{GS}$$

Finally, the AM or intensity modulation response of the Qdot laser is calculated by

$$H_{Qdot}(\omega) = \frac{S_{GS1}}{I_1/(qV)}$$

(3.25)

This equation is also known as the *modulation transfer function*. Through proper approximation, the AM response is given by [114]

$$H_{Qdot}(\omega) \approx H_1(\omega) H_0(\omega)$$

$$= \left(\frac{\omega_R^2}{\omega_R^2 - \omega^2 + j\omega\Gamma} \right) \left(\frac{\omega_{R0}^2}{\omega_{R0}^2 - \omega^2 + j\omega\Gamma_0} \right)$$

(3.26)

where the first part, $H_1(\omega)$, is dominated by the carrier–photon interaction processes. The resonance frequency ω_R and the damping factor Γ of the Qdot laser are, respectively, expressed as

$$\omega_R^2 = \frac{v_g a_{GS} S_{GS}}{\tau_p} + \frac{\Gamma_p \beta_{SP} N_{GS}}{\tau_{GS}^{spon} S_{GS}}\left[\left(\frac{H_B}{2N_B}\frac{N_{ES}}{\tau_{GS}^{ES}} + \frac{1-\rho_{ES}}{\tau_{ES}^{GS}}\right) + \frac{1-\beta_{SP}}{\tau_{GS}^{spon}}\right] + \frac{\beta_{SP}}{\tau_{GS}^{spon}\tau_p} \tag{3.27}$$

$$\Gamma = v_g a_{GS} S_{GS} + \left(\frac{H_B}{2N_B}\frac{N_{ES}}{\tau_{GS}^{ES}} + \frac{1-\rho_{ES}}{\tau_{ES}^{GS}}\right) + \frac{1}{\tau_{GS}^{spon}} + \frac{\Gamma_p \beta_{SP} N_{GS}}{\tau_{GS}^{spon} S_{GS}} \tag{3.28}$$

where the steady-state relationship $1/\tau_p - \Gamma_p v_g g_{GS} = \Gamma_p \beta_{SP} N_{GS}/(\tau_{GS}^{spon} S_{GS})$ has been used. In comparison with Qwell lasers [78], both expressions have an additional term $(H_B N_{ES}/(2N_B \tau_{GS}^{ES}) + (1-\rho_{ES})/\tau_{ES}^{GS})$, which describes the effective carrier scattering rate into and out of the GS. The second part, $H_0(\omega)$, is dominated by the carrier capture and relaxation processes, and the introduced two parameters ω_{R0} and Γ_0 are given by

$$\omega_{R0}^2 = \left(\frac{1-\rho_{ES}}{\tau_{ES}^{RS}} + \frac{1}{\tau_{RS}^{spon}}\right)\left(\frac{1-\rho_{GS}}{\tau_{GS}^{ES}} + \frac{1}{\tau_{ES}^{spon}} + \frac{H_B}{4N_B}\frac{N_{RS}}{\tau_{ES}^{RS}} + \frac{H_B}{4N_B}\frac{N_{GS}}{\tau_{ES}^{GS}}\right)$$
$$+ \frac{1}{\tau_{RS}^{ES}}\frac{1}{\tau_{RS}^{spon}} - \frac{H_B}{4N_B}\frac{N_{RS}}{\tau_{ES}^{RS}}\frac{1-\rho_{ES}}{\tau_{ES}^{RS}} \tag{3.29}$$

$$\Gamma_0 = \left(\frac{1-\rho_{ES}}{\tau_{ES}^{RS}} + \frac{1-\rho_{GS}}{\tau_{GS}^{RS}} + \frac{1}{\tau_{RS}^{spon}}\right)\left(\frac{1-\rho_{GS}}{\tau_{GS}^{ES}} + \frac{1}{\tau_{ES}^{RS}} + \frac{N_{RS}}{\tau_{ES}^{RS}}\frac{1}{4N_B/H_B} + \frac{N_{GS}}{\tau_{ES}^{GS}}\frac{1}{4N_B/H_B} + \frac{1}{\tau_{ES}^{spon}}\right) \tag{3.30}$$

Figure 3.7 presents an example of the calculated AM response of a Qdot laser. The solid curve $H_{Qdot}(\omega)$ shows that the response is strongly damped as usually observed in experiments [115–117], which can be attributed to the carrier relaxation and escape process of the GS as described in

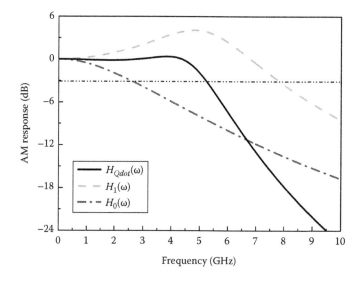

FIGURE 3.7 AM response $H_{Qdot}(\omega)$ (solid curve) of the Qdot laser. $H_1(\omega)$ (dash curve) describes mainly the contribution of carrier–photon interactions, and $H_0(\omega)$ (dash-dot curve) gives the contribution of carrier capture and relaxation processes. (From Wang, C. et al., *IEEE J. Quantum Electron.*, 48, 1144, 2012.)

Equation 3.28. On the other hand, $H_0(\omega)$ (dash-dot) exhibits a much smaller bandwidth than $H_1(\omega)$ (dash), which demonstrates that the finite carrier capture and relaxation times limit the modulation bandwidth of Qdot lasers.

3.3.2 Line-Width Enhancement Factor (α-Factor)

The line-width enhancement factor (α-factor) plays an important role in determining fundamental features of semiconductor lasers. Although the Qdot laser is predicted to have a delta-function-like discrete density of states, the measured α-factor values in experiments vary over a wide range from zero up to >10. The nonzero α-factor in Qdot lasers is partly attributed to the asymmetric gain spectrum because of the inhomogeneous broadening [118] and the carrier population in off-resonant states [119]. On the other hand, the free carrier plasma effect in the barrier and in the RS is reported to contribute almost half of the total refractive index change [100,101]. In this section, we describe an improved electric field model for Qdot lasers taking into account the contribution of off-resonant states (ES and RS) on the refractive index change, which allows a semianalytical study of the α-factor features in the Qdots. The model is capable of exploring the crucial physical mechanisms driving the Qdot laser's α-factor.

The conventional model describing the complex electric field of semiconductor lasers is given by

$$\frac{dE(t)}{dt} = \frac{1}{2}\left(\Gamma_P v_g g - \frac{1}{\tau_P}\right)E(t) + j\Delta\omega_N E(t) \qquad (3.31)$$

where the first term on the right-hand side gives the gain and the photon loss of the laser cavity. The second term, $\Delta\omega_N$, describes the carrier-induced frequency shift of the laser field with respect to the frequency at the lasing threshold. $\Delta\omega_N$ is usually expressed by the α-factor α_H as

$$\Delta\omega_N = \frac{1}{2}\left(\Gamma_P v_g g - \frac{1}{\tau_P}\right)\alpha_H \qquad (3.32)$$

This model is able to study the impacts of a nonzero α-factor on the modulation dynamics and nonlinear dynamics of Qdot lasers [120] but does not allow the study of α_H itself under different operating conditions. In order to investigate the α-factor, we need to obtain the expression of the gain and the refractive index separately. In the semiclassical theory, the semiconductor laser system can be fully described by the optical Bloch equations together with Maxell's equations [121]. The RS is treated as a discrete energy state of degeneracy $D_{RS} = k_B T m^* A_{RS}/(\pi\hbar^2)$, with m^* the reduced carrier mass and A_{RS} the RS surface area (see Section 3.1) [122]. The active region consists of only one Qdot ensemble. In addition, the electrons and holes are treated as neutral pairs (excitons). Two discrete states—the ground state (GS) and the first excited state (ES)—are considered in the dots. With these assumptions, the slowly varying electric field $E(t)$ is given by [123]

$$\frac{d}{dt}E(t) = \frac{j\omega_{LS}^0 \Gamma_P}{2\varepsilon_{bg}\varepsilon_0}\frac{1}{A_{RS}H_B}\sum_{X=GS,\,ES,\,RS}\left(\mu_X^* P_X\right) - \frac{1}{2\tau_P}E(t) \qquad (3.33)$$

where
  ω_{LS}^0 is the lasing frequency in the cold cavity
  ε_{bg} and ε_0 are the background and vacuum permittivity, respectively
  Γ_P is the optical confinement factor
  H_B is the height (equal to the dot's height)

The sum over X (X = GS, ES, RS) includes all possible optical transitions, with μ_X being the corresponding dipole transition matrix element and P_X the microscopic polarization. τ_p is the photon lifetime in the laser cavity. Assuming a sufficiently short dephasing time T_D and adiabatically eliminating the interband polarization yields the quasi-static relation

$$P_X(t) = -j\frac{\mu_X T_D}{2\hbar}(2\rho_X - 1)\frac{1 + j(\omega_X - \omega_{LS}^0)T_D}{1 + (\omega_X - \omega_{LS}^0)^2 T_D^2}E(t) \tag{3.34}$$

where

    ρ_X denotes the carrier occupation probability

    $\hbar\omega_X$ gives the transition energy of each state

Inserting Equation 3.34 into Equation 3.33, we obtain the complex gain

$$\tilde{G}(\omega_{LS}^0, t) = \left(2\mu_{GS}^*\frac{j\omega_{LS}^0\Gamma_P}{2\varepsilon_0\varepsilon_{bg}}\frac{2N_B}{H_B}\frac{P_{GS}(t)}{E(t)}\right) + \left(4\mu_{ES}^*\frac{j\omega_{LS}^0\Gamma_P}{2\varepsilon_0\varepsilon_{bg}}\frac{2N_B}{H_B}\frac{P_{ES}(t)}{E(t)}\right)$$
$$+ \left(\frac{j\omega_{LS}^0\Gamma_P}{2\varepsilon_0\varepsilon_{bg}}\frac{2}{A_{RS}H_B}D_{RS}\mu_{RS}^*\frac{P_{RS}(t)}{E(t)}\right) \tag{3.35}$$

The optical susceptibility can be derived from this complex gain expression through the relationship $\chi(\omega_{LS}^0, t) = 2\varepsilon_{bg}\tilde{G}(\omega_{LS}^0, t)/(j\omega\Gamma_P)$. The real part of Equation 3.35 is related to the laser gain, while the imaginary part gives the instantaneous frequency shift of the electric field. The three terms on the right-hand side give contributions of the GS, ES, and RS, respectively. Introducing the differential gain (a_X), we have

$$a_{GS} = \frac{2\mu_{GS}\mu_{GS}^*\omega_{GS}T_D}{\hbar v_g\varepsilon_0\varepsilon_{bg}}$$

$$a_{ES} = \frac{4\mu_{ES}\mu_{ES}^*\omega_{ES}T_D}{\hbar v_g\varepsilon_0\varepsilon_{bg}} \tag{3.36}$$

$$a_{RS} = \frac{\mu_{RS}\mu_{RS}^*\omega_{RS}T_D}{\hbar v_g\varepsilon_0\varepsilon_{bg}}$$

with v_g being the group velocity of the light. The material gain of each state is then given by

$$g_{GS} = \frac{a_{GS}}{1 + \xi S_{GS}}\frac{N_B}{H_B}\left(\frac{2N_{GS}}{2N_B/H_B} - 1\right)$$

$$g_{ES} = a_{ES}\frac{N_B}{H_B}\left(\frac{2N_{ES}}{4N_B/H_B} - 1\right) \tag{3.37}$$

$$g_{RS} = a_{RS}\frac{D_{RS}}{A_{RS}H_B}\left(\frac{2N_{RS}}{D_{RS}/(A_{RS}H_B)} - 1\right)$$

where

    N_X is the carrier density in each state

    S_{GS} is the photon density in the GS

    ξ denotes the gain compression factor

Because the real part of the complex gain approaches zero very quickly when off resonance, the field gain originates mainly from the resonant state. Considering the lasing emission in resonance with the GS transition $\omega_{LS}^0 = \omega_{GS}$, we obtain

$$\text{Re}[\tilde{G}(\omega_{GS})] \approx \Gamma_P v_g g_{GS} \tag{3.38}$$

In contrast, the imaginary part of the complex gain decays slowly for off-resonant frequencies. Thus, the off-resonant states can significantly influence the refractive index change, even though their gain contribution to the GS lasing is almost nil. Carrier populations in the off-resonant ES and RS induced frequency shifts of the laser field respectively are

$$\Delta\omega_N^{ES} = \frac{1}{2}\Gamma_P v_g g_{ES} F_{ES}^{GS} \tag{3.39}$$

$$\Delta\omega_N^{RS} = \frac{1}{2}\Gamma_P v_g g_{RS} F_{RS}^{GS} \tag{3.40}$$

with coefficients

$$F_{ES,RS}^{GS} = \frac{\omega_{GS}}{\omega_{ES,RS}} \frac{(\omega_{ES,RS} - \omega_{GS})T_D}{1 + (\omega_{ES,RS} - \omega_{GS})^2 T_D^2} \tag{3.41}$$

From Equation 3.35, it is seen that the resonant GS has no contribution to the refractive index change, which is the case when the laser is operated at the gain peak with a symmetric gain distribution. Nevertheless, as mentioned in the introduction, due to the asymmetric Qdot size dispersion, the resonant state induces a finite α-factor α_H^{GS}, and the corresponding frequency shift with respect to the cold cavity can be expressed by

$$\Delta\omega_N^{GS} = \frac{1}{2}\Gamma_P v_g g_{GS}\alpha_H^{GS} \tag{3.42}$$

Employing Equations 3.37 through 3.42, the electric field (Equation 3.33) is re-expressed as

$$\frac{dE(t)}{dt} = \frac{1}{2}\left(\Gamma_P v_g g_{GS} - \frac{1}{\tau_P}\right)E(t) + j\left(\Delta\omega_N^{GS} + \Delta\omega_N^{ES} + \Delta\omega_N^{RS}\right)E(t) \tag{3.43}$$

With carrier injection, the lasing frequency becomes $\omega_{LS} = \omega_{th}^{LS} + \Delta\omega_N^{LS}$, where $\Delta\omega_N^{LS} = \Delta\omega_N^{GS} + \Delta\omega_N^{ES} + \Delta\omega_N^{RS}$ gives the total frequency shift of the electric field from its threshold value (ω_{th}^{LS}). Through the $E(t) = \sqrt{S(t)V/\Gamma_P}\,e^{j\phi(t)}$ relationship, the photon density $S(t)$ and the phase $\phi(t)$ can be separately described. Combining with the equations describing the carrier dynamics in Qdot lasers, the laser system is finally given by

$$\frac{dN_{RS}}{dt} = \frac{I}{qV} + \frac{N_{ES}}{\tau_{RS}^{ES}} - \frac{N_{RS}}{\tau_{ES}^{RS}}(1-\rho_{ES}) - \frac{N_{RS}}{\tau_{RS}^{spon}} \tag{3.44}$$

$$\frac{dN_{ES}}{dt} = \left(\frac{N_{RS}}{\tau_{ES}^{RS}} + \frac{N_{GS}}{\tau_{ES}^{GS}}\right)(1-\rho_{ES}) - \frac{N_{ES}}{\tau_{GS}^{ES}}(1-\rho_{GS}) - \frac{N_{ES}}{\tau_{RS}^{ES}} - \frac{N_{ES}}{\tau_{ES}^{spon}} \tag{3.45}$$

$$\frac{dN_{GS}}{dt} = \frac{N_{ES}}{\tau_{GS}^{ES}}(1-\rho_{GS}) - \frac{N_{GS}}{\tau_{ES}^{GS}}(1-\rho_{ES}) - v_g g_{GS} S_{GS} - \frac{N_{GS}}{\tau_{GS}^{spon}}$$ (3.46)

$$\frac{dS_{GS}}{dt} = \left(\Gamma_p v_g g_{GS} - \frac{1}{\tau_P}\right) S_{GS} + \beta_{SP} \frac{N_{GS}}{\tau_{GS}^{spon}}$$ (3.47)

$$\frac{d\phi}{dt} = \Delta\omega_N^{GS} + \Delta\omega_N^{ES} + \Delta\omega_N^{RS}$$ (3.48)

where

τ_{GS}^{spon} is the spontaneous emission time

β_{SP} is the spontaneous emission factor

Carriers in the RS are scattered into the dots through phonon-assisted and Auger-assisted processes [82,83]. The latter makes the scattering rates nonlinearly depend on the carrier density in the RS. However, for the sake of simplicity, the carrier capture time τ_{ES}^{RS} and the relaxation time τ_{GS}^{ES} are both treated as constants in this work. On the other hand, the carrier escape times ($\tau_{RS}^{ES}, \tau_{ES}^{GS}$) are governed by the Fermi distribution for a quasi-thermal equilibrium system [124]. For semiconductor lasers operating under small-signal modulation with frequency ω, the bias current change δI induces variations of the carriers δN_X, the photon δS_{GS}, and the phase $\delta\phi$. In order to perform the analyses, the differential rate equations are derived as follows:

$$\begin{bmatrix} \gamma_{11}+j\omega & -\gamma_{12} & 0 & 0 & 0 \\ -\gamma_{21} & \gamma_{22}+j\omega & -\gamma_{23} & 0 & 0 \\ 0 & -\gamma_{32} & \gamma_{33}+j\omega & -\gamma_{34} & 0 \\ 0 & 0 & -\gamma_{43} & \gamma_{44}+j\omega & 0 \\ -\gamma_{51} & -\gamma_{52} & -\gamma_{53} & -\gamma_{54} & j\omega \end{bmatrix} \begin{bmatrix} \delta N_{RS} \\ \delta N_{ES} \\ \delta N_{GS} \\ \delta S_{GS} \\ \delta\phi \end{bmatrix} = \frac{\delta I}{qV} \begin{bmatrix} 1 \\ 0 \\ 0 \\ 0 \\ 0 \end{bmatrix}$$ (3.49)

where

$$\gamma_{11} = \frac{1-\rho_{ES}}{\tau_{ES}^{RS}} + \frac{1}{\tau_{RS}^{spon}}; \quad \gamma_{12} = \frac{1}{\tau_{RS}^{ES}} + \frac{1}{4N_B}\frac{N_{RS}}{\tau_{ES}^{RS}}; \quad \gamma_{21} = \frac{1-\rho_{ES}}{\tau_{ES}^{RS}};$$

$$\gamma_{22} = \frac{1-\rho_{GS}}{\tau_{GS}^{ES}} + \frac{1}{\tau_{RS}^{ES}} + \frac{1}{\tau_{ES}^{spon}} + \frac{1}{4N_B}\frac{N_{RS}}{\tau_{ES}^{RS}} + \frac{1}{4N_B}\frac{N_{GS}}{\tau_{ES}^{GS}}$$

$$\gamma_{23} = \frac{1-\rho_{ES}}{\tau_{ES}^{GS}} + \frac{1}{2N_B}\frac{N_{ES}}{\tau_{GS}^{ES}}; \quad \gamma_{32} = \frac{1-\rho_{GS}}{\tau_{GS}^{ES}} + \frac{1}{4N_B}\frac{N_{GS}}{\tau_{ES}^{GS}}; \quad \gamma_{33} = \frac{1-\rho_{ES}}{\tau_{ES}^{GS}} + \frac{1}{\tau_{GS}^{spon}} + \frac{1}{2N_B}\frac{N_{ES}}{\tau_{GS}^{ES}} + v_g a S_{GS}$$

$$\gamma_{34} = -v_g g_{GS} + v_g a_P S_{GS}; \quad \gamma_{43} = \Gamma_p v_g a S_{GS} + \frac{\Gamma_p \beta_{SP}}{\tau_{GS}^{spon}};$$

$$\gamma_{44} = -\Gamma_p v_g g_{GS} + \frac{1}{\tau_p} + \Gamma_p v_g a_P S_{GS}; \quad \gamma_{51} = \Gamma_p v_g a_{RS} F_{RS}$$

$$\gamma_{52} = \frac{1}{4}\Gamma_p v_g a_{ES} F_{ES}; \quad \gamma_{53} = \frac{1}{2}\Gamma_p v_g a\alpha_H^{GS}; \quad \gamma_{54} = -\frac{1}{2}\Gamma_p v_g a_P \alpha_H^{GS}$$ (3.50)

with

$$a = \frac{\partial g_{GS}}{\partial N_{GS}} = \frac{a_{GS}}{1 + \xi S_{GS}}; \quad a_P = -\frac{\partial g_{GS}}{\partial S_{GS}} = \frac{\xi}{1 + \xi S_{GS}} g_{GS} \tag{3.51}$$

Based on the above differential rate equations, the α-factor of the Qdot laser is derived as

$$\alpha_{H,QD}^{GS}(\omega) = \frac{2}{\Gamma_P v_g} \frac{\delta\left[\Delta\omega_N^{LS}(N)\right]}{\delta g_{GS}(N)} \equiv \alpha_H^{GS} + \frac{1}{2} F_{ES} \frac{a_{ES}\delta N_{ES}}{a\delta N_{GS}} + 2F_{RS} \frac{a_{RS}\delta N_{RS}}{a\delta N_{GS}} \tag{3.52}$$

Following the definition in Equation 3.13, it is noted that only the carrier contribution (δN) is included in the above equation, while the photon contribution (δS) is excluded. In the following, it will be shown that the α-factor of Qdot lasers presents peculiar characteristics under direct modulation.

Over the last decades, various techniques have been proposed for the measurement of the α-factor. In this work, we employ the well-known "FM/AM" technique for the above-threshold analysis and the widely used "Hakki-Paoli" method for the below-threshold analysis [102]. The FM/AM technique relies on the direct current modulation of the laser, which generates both the optical frequency (FM) and amplitude (AM) modulations [125]. With respect to the linearized rate equations, the ratio of the FM/AM index is derived as

$$2\frac{\beta(\omega)}{m(\omega)} = 2\frac{\delta\omega_{LS}/\omega}{\delta S_{GS}/S_{GS}}$$

$$\equiv \frac{j\omega + (1/\tau_P - \Gamma_P v_g g_{GS} + \Gamma_P v_g a_P S_{GS})}{j\omega} \times \left[\alpha_H^{GS}\left(1 - \frac{a_P \delta S_{GS}}{a\delta N_{GS}}\right) + \frac{1}{2} F_{ES} \frac{a_{ES}\delta N_{ES}}{a\delta N_{GS}} + 2F_{RS} \frac{a_{RS}\delta N_{RS}}{a\delta N_{GS}}\right] \tag{3.53}$$

where the relation $\delta\omega_{LS} = j\omega\delta\phi$ is used in the above derivation. In this approach, the laser's α-factor is usually extracted through the formula $\alpha_{H,QD}^{FM/AM} = \min\{2\beta(\omega)/m(\omega)\}$. This is indeed true for Qwell or bulk lasers; however, we show that the α-factor of Qdot lasers is dependent on the modulation frequency but we still take the minimum value to characterize the Qdot laser as reported in [119].

For semiconductor lasers operating below threshold, the Hakki-Paoli method relies on the direct measurement of the optical spectra of amplified spontaneous emission (ASE) in the laser cavity. Tuning the pump current slightly step by step (ΔI), the gain change can be extracted by the Hakki-Paoli method, and the wavelength variation can be directly recorded using an optical spectrum analyzer. Correspondingly, the below-threshold α-factor is calculated as

$$\alpha_{H,QD}^{ASE} = \alpha_H^{GS} + \frac{1}{2} F_{ES}^{GS} \frac{a_{ES}}{a_{GS}} \frac{\Delta N_{ES}}{\Delta N_{GS}} + 2F_{RS}^{GS} \frac{a_{RS}}{a_{GS}} \frac{\Delta N_{RS}}{\Delta N_{GS}} \tag{3.54}$$

The laser parameters used in the simulation are listed in Table 3.2 [57,75,126]. It is noted that the carrier occupation in the GS has a small contribution to the α-factor (<1) [127]; hence we assume the value $\alpha_H^{GS} = 0.5$ in the simulation. Figure 3.8a depicts the carrier density variations in the three states under small-signal modulation. For low frequencies (smaller than 0.1 GHz), all the carrier density variations remain almost constant, but the variations of the ES (δN_{ES}) and RS (δN_{RS}) populations are 15 dB larger than that of the GS (δN_{GS}) one. The small variation of the GS carrier population is associated with the gain-clamping above the threshold. Both δN_{GS} and δN_{ES} exhibit resonances at ~7 GHz. Beyond the resonance frequency, δN_{GS} decays faster than δN_{ES} and δN_{RS}. These features significantly impact the behavior of the α-factor, as described in Equation 3.52.

TABLE 3.2

Qdot Material and Laser Parameters for the Study of the α-Factor

Symbol	Description	Value	Symbol	Description	Value
L	Active region length	5×10^{-2} cm	τ_{ES}^{RS}	Capture time from RS to ES	6.3 ps
W	Active region width	4×10^{-4} cm	τ_{GS}^{ES}	Relaxation time from ES to GS	2.9 ps
$R_1 = R_2$	Mirror reflectivity	0.32	a_{GS}	GS differential gain	5×10^{-15} cm^2
n_r	Refractive index	3.5	a_{ES}	ES differential gain	10×10^{-15} cm^2
α_i	Internal modal loss	6 cm^{-1}	a_{RS}	RS differential gain	2.5×10^{-15} cm^2
N_B	Dot density	10×10^{10} cm^{-2}	ξ	Gain compression factor	2×10^{-16} cm^3
H_B	Dot height	5×10^{-7} cm	α_H^{GS}	GS induced α-factor	0.5
E_{RS}	RS transition energy	0.97 eV	T_D	Dephasing time	0.1 ps
E_{ES}	ES transition energy	0.87 eV	Γ_P	Optical confinement factor	0.06
E_{GS}	GS transition energy	0.82 eV	β_{SP}	Spontaneous emission factor	1×10^{-4}

Figure 3.8b compares the difference between the α-factor $\left| \alpha_{H,QD}^{GS}(\omega) \right|$ and the ratio $\left| 2\beta(\omega)/m(\omega) \right|$ as a function of the modulation frequency. At low frequencies (<0.1 GHz), there is a large discrepancy between the two parameters. As expected, $\left| 2\beta(\omega)/m(\omega) \right|$ exhibits large values due to the gain compression and the large carrier variations in the ES and in the RS. Nevertheless, $\left| \alpha_{H,QD}^{GS}(\omega) \right|$ remains constant. On increasing the modulation frequency beyond several gigahertz, the two values of both parameters decrease down to a plateau, which gives the conventional α-factor, indicated by the horizontal line. As can be seen, $\alpha_{H,QD}^{FM/AM}$ is almost the same as $\left| \alpha_{H,QD}^{GS} \right|$, which indicates that the FM/AM method is a reliable technique for the measurement of Qdot laser's α-factor. Further increase of the modulation frequency raises again both the values, as observed experimentally in a Qdot laser (inset of Figure 3.8b) [129]. It is emphasized that such a situation is not encountered in Qwell lasers [125]. This behavior is attributed to the different decay rates (versus modulation frequency) of carrier variations in each state, as shown in Figure 3.8a. In addition, Figure 3.8b shows that the ES (dash-dot curves) contributes more to the α-factor $\left| \alpha_{H,QD}^{GS} \right|$ than the RS due to its smaller energy separation with the resonant GS.

Based on the ASE and the FM/AM methods, Figure 3.9 illustrates the α-factor as a function of the normalized pump current I/I_{th}. Below threshold, carrier populations in both the resonant and off-resonant states increase with the pump current. In consequence, the α-factor increases nonlinearly. Above threshold, the carrier population in the GS is clamped, while the off-resonant state populations keep increasing. Thus, the α-factor varies almost linearly above threshold as usually measured in experiments [57,71]. At threshold, the α-factor extracted from the ASE method is similar to that using the FM/AM technique. In addition, the α-factor is larger than the sole GS-induced value of $\alpha_H^{GS} = 0.5$ both below and above threshold, which means the off-resonant ES and RS contribute to the increase of the α-factor in the Qdot laser. This is explained by the fact that the coefficients F_{ES}^{GS} and F_{RS}^{GS} are both positive since the ES and RS have higher energies than the GS (see Equation 3.41).

3.3.3 IMPACTS OF CARRIER CAPTURE AND RELAXATION PROCESSES

As discussed previously, the Qdot laser involves a carrier capture process from the 2D RS to the localized ES and a carrier relaxation process from the ES to the GS inside the dots. This section discusses the influences of these processes on the laser's modulation dynamics.

In order to study the impacts of the carrier capture process, the carrier relaxation time τ_{GS}^{ES} is fixed at 2.9 ps, while the carrier capture time τ_{ES}^{RS} is varied from 0.1 up to 50 ps. Figure 3.10a shows the variation of the AM response for different capture times. Slow capture reduces the 3 dB

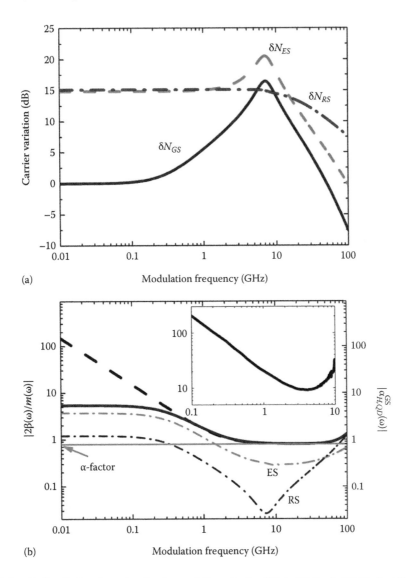

(a)

(b)

FIGURE 3.8 (a) Small-signal carrier density variations in the GS (solid line), ES (dashed line), and RS (dash-dot line) versus the modulation frequency. The bias current is $I = 1.2 \times I_{th}$, with the threshold current $I_{th} = 49$ mA. The carrier variation is normalized to the value δN_{GS} of 0.01 GHz. (b) Modulation-frequency dependence of the FM/AM ratio (dash) and of the α-factor (thick solid). The minimum level indicated by the horizontal line gives the laser's conventional α-factor. The thin dash-dotted curve represents the sole contribution of the ES or the RS to the α-factor, respectively. The inset shows an experimental curve of the FM/AM ratio for a Qdot laser. (From Wang, C. et al., *Appl. Phys. Lett.*, 105, 221114, 2014.)

modulation bandwidth from 11 GHz down to 7 GHz. Besides, the resonance peak is also slightly reduced. Interestingly, for capture times larger than 30 ps, a parasitic-like roll-off (dip) appears in the response, which is similar to the effect of the slow carrier transport process from the 3D barrier to the 2D RS [78]. In the same approach, by fixing the capture time at 6.3 ps, Figure 3.10b shows the impact of carrier relaxation time on the AM response. The modulation bandwidth is significantly reduced by the slow relaxation process from 10.8 GHz for $\tau_{GS}^{ES} = 0.1$ ps to 1.6 GHz for $\tau_{GS}^{ES} = 50$ ps. In addition, the response is strongly damped for large relaxation times. It is noted that the evolution of the AM response shape is quite different from that for the capture process in Figure 3.10a.

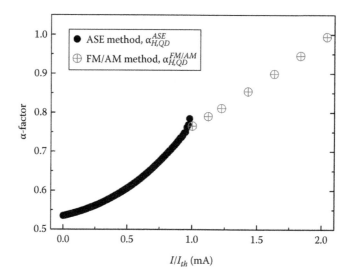

FIGURE 3.9 α-Factor as a function of the normalized bias current I/I_{th}. (From Wang, C. et al., *Appl. Phys. Lett.*, 105, 221114, 2014.)

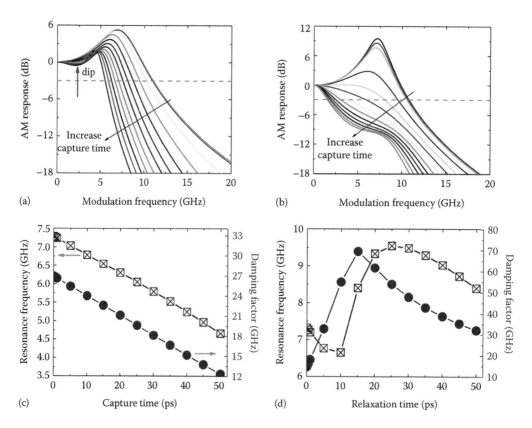

FIGURE 3.10 Influence of carrier scattering times on the AM response, resonance frequency, and damping factor. (a) and (c) are for the variation of capture time with fixed relaxation time at 2.9 ps. (b) and (d) are for the variation of relaxation time with fixed capture time at 6.3 ps.

For relaxation times $\tau_{GS}^{ES} > 25$ ps, the response shows a clear resonance at ~8 GHz. From the eigen-value analysis of the Qdot laser system [130], the resonance frequency f_R and the damping factor Γ are extracted for various carrier scattering times. It is shown that both f_R and Γ decrease linearly with increase in carrier capture time (Figure 3.10c). In contrast, for the carrier relaxation process (Figure 3.10d) the behavior is much more complex. The resonance frequency first decreases for $\tau_{GS}^{ES} < 10$ ps, while the damping factor increases with the relaxation time. At $\tau_{GS}^{ES} = 10$ ps, the AM response is rather flat. However, for $\tau_{GS}^{ES} > 10$ ps the resonance again increases until $\tau_{GS}^{ES} = 25$ ps, while the damping factor reaches the maximum at $\tau_{GS}^{ES} = 15$ ps. Beyond the peak values, both the resonance and damping decrease as a function of the relaxation time. Lastly, we note that the damping factors of the Qdot laser in both Figure 3.10c and d are much larger than those of Qwell lasers, which is attributed to the carrier occupation in the off-resonant states as well as the carrier scattering processes [131].

Figure 3.11a illustrates that the FM/AM index ratio $2\beta/m$ exhibits a significant re-increase beyond 10 GHz for a slow carrier capture process (large capture time). With respect to Equation 3.52, this can be attributed to the larger carrier variation in the RS, δN_{RS}, under high-frequency modulation

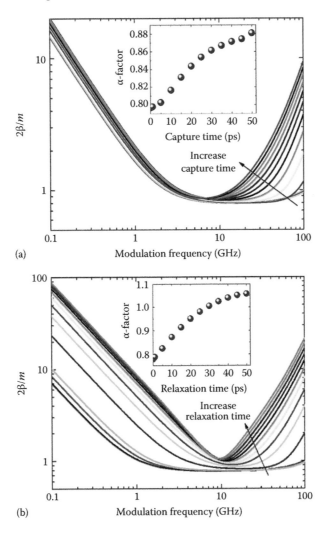

FIGURE 3.11   Influence of (a) the carrier capture time and (b) relaxation time on the ratio of FM/AM index. Insets show the α-factor variation extracted from the minimum of the FM/AM index ratio.

since the number of available carriers in the RS is larger. The inset of Figure 3.11a indicates that increasing the carrier capture time enhances the α-factor by 11% from 0.80 for $\tau_{ES}^{RS} = 0.10$ ps to 0.88 for $\tau_{ES}^{RS} = 50$ ps. Figure 3.11b depicts that a slow carrier relaxation process induces a steep re-increase of $2\beta/m$ for modulation frequencies larger than 10 GHz. This is due to the increased carrier populations and variations in the ES and RS. The inset of Figure 3.11b shows the α-factor extracted from the minimum value of $2\beta/m$, which increases by about 36% from 0.78 for $\tau_{GS}^{ES} = 0.1$ ps to 1.06 for $\tau_{GS}^{ES} = 50$ ps.

3.4   CONCLUSION

In this chapter, we discussed the electronic and optical features of InP-based nanostructure semiconductor lasers. In contrast to conventional Qwell lasers, in Qdot and Qdash lasers the existence of a carrier reservoir, discrete excited states, and the consequent carrier scattering processes bring unique characteristics to the dynamic modulation response and the line-width enhancement factor. Regarding the dynamical performance, it can be well improved by the excited-state lasing instead of ground-state lasing [132]. On the other hand, nonlinear photonic techniques such as optical injection [133,134], optical feedback [135], and optoelectronic feedback [136] can be employed for further enhancement of the dynamical performance of nanostructure lasers.

ACKNOWLEDGMENTS

This work was supported in part by the Partenariat Hubert Curien under Grant No. 30794RC (Campus France/DAAD) and by the European Office for Aerospace Research (EOARD) under grant FA9550-15-1-0104.

REFERENCES

1. E. Murphy, Enabling optical communication, *Nat. Photon.* 4, 287 (2010).
2. H. Kroemer, Theory of a wide-gap emitter for transistors, *Proc. IRE* 45, 1535 (1957).
3. Z. I. Alferov and R. F. Kazarinov, Semiconductor laser with electric pumping, Inventor's Certificate 181737 in Russian, Application 950840, priority as of March 30, 1963.
4. R. Dingle and C. H. Henry, Quantum effects in heterostructure lasers, U.S. Patent No. 3,982,207, filed on March 7, 1975, issued September 21, 1976.
5. J. P. van der Ziel, R. Dingle, R. C. Miller, W. Wiegmann, and W. A. Nordland, Laser oscillations from quantum states in very thin GaAs-Al$_{0.2}$Ga$_{0.8}$As multilayer structures, *Appl. Phys. Lett.* 26, 463 (1975).
6. Y. Arakawa and H. Sakaki, Multidimensional quantum well laser and temperature dependence of its threshold current, *Appl. Phys. Lett.* 40, 939 (1982).
7. M. Asada, Y. Miyamoto, and Y. Suematsu, Gain and the threshold of three-dimensional quantum-box lasers, *IEEE J. Quantum Electron.* 22, 1915 (1986).
8. D. Bimberg, M. Grundmann, and N. N. Ledentsov, *Quantum Dot Heterostructures*, New York: Wiley, 1998.
9. Y. Arakawa, Progress in growth and physics of nitride-based quantum dots, *Phys. Status Solidi (a)* 188, 37 (2001).
10. N. Kirstaedter, N. N. Ledentsov, M. Grundmann, D. Bimberg, V. M. Ustinov, S. S. Ruvimov, M. V. Maximov, P. S. Kop'ev, and Zh. I. Alferov, Low threshold, large T_0 injection laser emission from (InGa)As quantum dots, *Electron. Lett.* 30, 1416 (1994).
11. D. Bimberg et al., InAs-GaAs quantum pyramid lasers: In situ growth, radiative lifetimes and polarization properties, *Jpn. J. Appl. Phys.* 35, 1311 (1996).
12. N. N. Ledentsov et al., Direct formation of vertically coupled quantum dots in Stranski–Krastanow growth, *Phys. Rev. B* 54, 8743 (1996).
13. M. T. Crowley, N. A. Naderi, H. Su, F. Grillot, and L. F. Lester, GaAs based quantum dot lasers, in *Semiconductors and Semimetals: Advances in Semiconductor Lasers*, J. J. Coleman and A. C. Bryce (Eds.), 86 (2012).

14. H. Y. Liu, K. M. Groom, D. T. D. Childs, D. J. Robbins, T. J. Badcock, M. Hopkinson, D. J. Mowbray, and M. S. Skolnick, 1.3 μm InAs/GaAs multilayer quantum-dot laser with extremely low room temperature threshold current density, *Electron. Lett.* 40, 1412 (2004).

15. M. V. Maksimov et al., High-power 1.5 μm InAs-InGaAs quantum dot lasers on GaAs substrates, *Semiconductors* 38, 732 (2004).

16. A. Zhukov, M. Maksimov, and A. Kovsh, Device characteristics of long-wavelength lasers based on self-organized quantum dots, *Semiconductors* 46, 1225 (2012).

17. P. B. Joyce, T. J. Krzzewski, P. H. Steans, G. R. Bell, J. H. Neave, and T. S. Jones, Variations in critical coverage for InAs/GaAs quantum dot formation in bilayer structures, *J. Cryst. Growth* 244, 39 (2002).

18. F. Y. Chang, C. C. Wu, and H. H. Lin, Effect of InGaAs capping layer on the properties of InAs/InGaAs quantum dots and lasers, *Appl. Phys. Lett.* 82, 4477 (2003).

19. J. Oshinowo, M. Nishioka, S. Ishida, and Y. Arakawa, Highly uniform InGaAs/GaAs quantum dots (15 nm) by metal organic chemical vapor deposition, *Appl. Phys. Lett.* 65, 1421 (1994).

20. P. J. Poole, K. Kaminska, P. Barrios, Z. Lu, and J. Liu, Growth of InAs/InP-based quantum dots for 1.55 μm laser applications, *J. Cryst. Growth* 311, 1482 (2009).

21. N. Bertru et al., QD laser on InP substrate for 1.55 μm emission and beyond, *Proc. SPIE* 7608, 76081B (2010).

22. M. Gong, K. Duan, C. F. Li, R. Magri, A. Narvaez, and L. He, Electronic structure of self-assembled InAs/InP quantum dots: Comparison with self-assembled InAs/GaAs quantum dots, *Phys. Rev. B* 77, 045326 (2008).

23. R. H. Wang, A. Stintz, P. M. Varangis, T. C. Newell, H. Li, K. J. Malloy, and L. F. Lester, Room-temperature operation of InAs quantum-dash lasers on InP(001), *IEEE Photon. Technol. Lett.* 13, 767 (2001).

24. P. Miska, J. Even, C. Platz, B. Salem, and T. Benyattou, Experimental and theoretical investigation of carrier confinement in InAs quantum dashes grown on InP(001), *J. Appl. Phys.* 95, 1074 (2004).

25. M. Z. M. Khan, T. K. Ng, and B. S. Ooi, Self-assembled InAs/InP quantum dots and quantum dashes: Material structures and devices, *Prog. Quantum Electron.* 38, 6, 237 (2014).

26. J. P. Reithmaier, G. Eisenstein, and A. Forchel, InAs/InP quantum-dash lasers and amplifiers, *Proc. IEEE* 95, 1779 (2007).

27. D. Zhou, R. Piron, F. Grillot, O. Dehaese, E. Homeyer, M. Dontabactouny, T. Batte, K. Tavernier, J. Even, and S. Loualiche, Study of the characteristics of 1.55 μm quantum dash/dot semiconductor lasers on InP substrate, *Appl. Phys. Lett.* 93, 161104 (2008).

28. D. Zhou, R. Piron, M. Dontabactouny, O. Dehaese, F. Grillot, and T. Batte, Low threshold current density of InAs quantum dash laser on InP(100) through optimizing double cap technique, *Appl. Phys. Lett.* 94, 081107 (2009).

29. V. Sichkovskyi, M. Waniczek, and J. Reithmaier, High-gain wavelength-stabilized 1.55 μm InAs/InP(100) based lasers with reduced number of quantum dot active layers, *Appl. Phys. Lett.* 102, 221117 (2013).

30. C. Paranthoën et al., Height dispersion control of InAs/InP(113)B quantum dots emitting at 1.55 μm, *Appl. Phys. Lett.* 78, 1751 (2001).

31. J. Kotani, P. J. van Veldhoven, T. de Vries, B. Smalbrugge, E. A. J. M. Bente, M. K. Smit, and R. Notzel, First demonstration of single-layer InAs/InP (100) quantum-dot laser: Continuous wave, room temperature, ground state, *Electron. Lett.* 45, 1317 (2009).

32. E. Homeyer, R. Piron, F. Grillot, O. Dehaese, K. Tavernier, E. Macé, A. Le Corre, and S. Loualiche, First demonstration of a 1.52 μm RT InAs/InP (311)B laser with an active zone based on a single QD layer, *Semicond. Sci. Technol.* 22, 827 (2007).

33. C. Cornet et al., Electronic and optical properties of InAs/InP quantum dots on InP(100) and InP(311)B substrates: Theory and experiment, *Phys. Rev. B* 74, 035312 (2006).

34. V. Ustinov, A. Zhukov, A. Y. Egorov, A. Kovsh, S. Zaitsev, and N. Y. Gordeev, Low threshold quantum dot injection laser emitting at 1.9 μm, *Electron. Lett.* 34, 670 (1998).

35. V. Ustinov, A. Kovsh, A. Zhukov, A. Y. Egorov, N. N. Ledentsov, and A. V. Lunev, Low-threshold quantum-dot injection heterolaser emitting at 1.84 μm, *Tech. Phys. Lett.* 24, 22 (1998).

36. K. Nishi, M. Yamada, T. Anan, A. Gomyo, and S. Sugou, Long-wavelength lasing from InAs self-assembled quantum dots on (311)B InP, *Appl. Phys. Lett.* 73, 526 (1998).

37. N. Bertru et al., Two-dimensional ordering of self-assembled InAs quantum dots grown on (311)B InP substrate, in *Proceedings of SPIE, QD laser on InP substrate for 1.55 um emission and beyond,* San Francisco, CA, Vol. 7608, p. 76081B (2010).

38. P. Miska, J. Even, C. Paranthoën, O. Dehaese, H. Folliot, S. Loualiche, M. Senes, and X. Marie, Optical properties and carrier dynamics of InAs/InP(113)B quantum dots emitting between 1.3 and 1.55 μm for laser applications, *Physica E* 17, 56 (2003).

39. C. Paranthoën et al., Growth and optical characterizations on InAs quantum dots on InP substrate: Toward 1.55 μm quantum dot laser, *J. Cryst. Growth* 251, 230 (2003).

40. P. Caroff, C. Paranthoen, C. Platz, O. Dehaese, H. Folliot, and N. Bertru, High-gain and low-threshold InAs quantum-dot lasers on InP, *Appl. Phys. Lett.* 87, 243107 (2005).

41. E. Homeyer, R. Piron, F. Grillot, O. Dehaese, K. Tavernier, and E. Macé, Demonstration of a low threshold current in 1.54 μm InAs/InP (311)B quantum dot laser with reduced quantum dot stacks, *Jpn. J. Appl. Phys.* 46, 6903 (2007).

42. K. Klaime, C. Clo, R. Piron, C. Paranthoen, D. Thiam, and T. Batte, 23 and 39 GHz low phase noise monosection InAs/InP (113)B quantum dots mode-locked lasers, *Opt. Express* 21, 29000 (2013).

43. H. Saito, K. Nishi, and S. Sugou, Ground-state lasing at room temperature in long-wavelength InAs quantum-dot lasers on InP(311)B substrates, *Appl. Phys. Lett.* 78, 267 (2001).

44. K. Akahane, N. Yamamoto, and T. Kawanishi, Wavelength tunability of highly stacked quantum dot laser fabricated by a strain compensation technique, in *Proceedings of the 22nd IEEE International Semiconductor Laser Conference (ISLC)*, Cardiff, U.K., Vol. 37 (2010).

45. K. Akahane, N. Yamamoto, and T. Kawanishi, The dependence of the characteristic temperature of highly stacked InAs quantum dot laser diodes fabricated using a strain-compensation technique on stacking layer number, in *Proceedings of the 22nd IEEE International Semiconductor Laser Conference (ISLC)*, Cardiff, U.K., cc Vol. 82 (2012).

46. C. N. Allen, P. Poole, P. Barrios, P. Marshall, G. Pakulski, and S. Raymond, External cavity quantum dot tunable laser through 1.55 μm, *Physica E* 26, 372 (2005).

47. F. Lelarge, B. Rousseau, B. Dagens, F. Poingt, F. Pommereau, and A. Accard, Room temperature continuous-wave operation of buried ridge stripe lasers using InAs-InP(100) quantum dots as active core, *IEEE Photon. Technol. Lett.* 17, 1369 (2005).

48. S. Anantathanasarn et al., Lasing of wavelength-tunable (1.55 μm region) InAs/InGaAsP/InP (100) quantum dots grown by metal organic vapor-phase epitaxy, *Appl. Phys. Lett.* 89, 073115 (2006).

49. J. S. Kim, J. H. Lee, S. U. Hong, W. S. Han, H. S. Kwack, and C. W. Lee, Long-wavelength laser based on self-assembled InAs quantum dots in InAlGaAs on InP(001), *Appl. Phys. Lett.* 85, 1033 (2004).

50. J. S. Kim, J. H. Lee, S. U. Hong, W. S. Han, H. S. Kwack, and C. W. Lee, Room-temperature operation of InP-based quantum dot laser, *IEEE Photon. Technol. Lett.* 16, 1607 (2004).

51. C. Gilfert, V. Ivanov, N. Oehl, M. Yacob, and J. Reithmaier, High gain 1.55 μm diode lasers based on InAs quantum dot like active regions, *Appl. Phys. Lett.* 98, 201102 (2011).

52. O. Mollet, A. Martinez, K. Merghem, S. Joshi, J.-G. Provost, F. Lelarge, and A. Ramdane, Dynamic characteristics of undoped and p-doped Fabry-Perot InAs/InP quantum dash based ridge waveguide lasers for access/metro networks, *Appl. Phys. Lett.* 105, 141113 (2014).

53. C. Peucheret, *Direct and External Modulation of Light*, Technical University of Denmark, Kongens Lyngby, Denmark, 2009.

54. B. O. Seraphin and N. Bottka, Franz-Keldysh effect of the refractive index in semiconductors, *Phys. Rev.* 139, A560 (1965).

55. K. Kechaou, T. Anfray, K. Merghem, C. Aupetit-Berthelemot, G. Aubin, C. Kazmierski, C. Jany, P. Chanclou, and D. Erasme, Improved NRZ transmission distance at 20 Gbit/s using dual electro-absorption modulated laser, *Electron. Lett.* 48, 335 (2012).

56. D. Erasme et al., The dual-electroabsorption modulated laser, a flexible solution for amplified and dispersion uncompensated networks over standard fiber, *J. Lightwave Technol.* 32, 4068 (2014).

57. A. Martinez et al., Dynamic properties of InAs/InP(311B) quantum dot Fabry-Perot lasers emitting at 1.52-μm, *Appl. Phys. Lett.* 93, 021101 (2008).

58. D. Gready, G. Eisenstein, C. Gilfert, V. Ivanov, and J. P. Reithmaier, High-speed low-noise InAs/InAlGaAs/InP 1.55-μm quantum-dot lasers, *IEEE Photon. Technol. Lett.* 24, 809 (2012).

59. D. Gready, G. Eisenstein, V. Ivanov, C. Gilfert, F. Schnabel, A. Rippien, J. P. Reithmaier, and C. Bornholdt, High speed 1.55 μm InAs/InAlGaAs/InP quantum dot laser, *IEEE Photon. Technol. Lett.* 26, 11 (2014).

60. S. Bhowmick, M. Z. Baten, T. Frost, B. S. Ooi, and P. Bhattacharya, High performance InAs/In0.53Ga0.23Al0.24As/InP quantum dot 1.55 μm tunnel injection laser, *IEEE J. Quantum Electron.* 50, 7 (2014).

61. W. Kaiser, K. Mathwig, S. Deubert, J. P. Reithmaier, F. Forchel, O. Parillaud, M. Krakowski, D. Hadass, V. Mikhelashvili, and G. Eisenstein, Static and dynamic properties of laterally coupled DFB lasers based on InAs/InP Qdash structures, *Electron. Lett.* 41 (2005).

62. Z. Mi and P. Bhattacharya, DC and dynamic characteristics of p-doped and tunnel injection 1.65 μm InAs quantum-dash lasers grown on InP(001), *IEEE J. Quantum Electron.* 42, 1224 (2006).

63. S. Hein, V. von Hinten, W. Kaiser, S. Hofling, and A. Forchel, Dynamic properties of 1.5 μm quantum dash lasers on (100) InP, *Electron. Lett.* 43 (2007).

64. F. Lelarge et al., Recent advances on InAs/InP quantum dash based semiconductor lasers and optical amplifiers operating at 1.55 μm, *IEEE J. Sel. Top. Quantum Electron.* 13, 111 (2007).

65. Q. Zou, K. Merghem, S. Azouigui, A. Martinez, A. Accard, N. Chimot, F. Lelarge, and A. Ramdane, Feedback-resistant p-type doped InAs/InP quantum-dash distributed feedback lasers for isolator-free 10 Gb/s transmission at 1.55 μm, *Appl. Phys. Lett.* 97, 231115 (1010).

66. N. Chimot, S. Joshi, G. Aubin, K. Merghem, S. barbet, A. Accard, A. Ramdane, and F. Lelarge, 1550 nm InAs/InP quantum dash based directly modulated lasers for next generation passive optical network, IEEE, *Int. Conf. Indium Phosphide and Related Materials (IPRM)*, Santa Barbara, CA, p. 177 (2012).

67. S. Joshi, N. Chimot, L. A. Neto, A. Accard, and J. G. Provost, Quantum dash based directly modulated lasers for long-reach access networks, *Electron. Lett.* 50, 534 (2014).

68. R. S. Tucker, Green optical communications—Part I: Energy limitations in transport, *IEEE J. Sel. Top. Quantum Electron.* 17, 245 (2011).

69. D. Gready, G. Eisenstein, M. Gioannini, I. Montrosset, D. Arsenijevic, H. Schmeckebier, M. Stubenrauch, and D. Bimberg, On the relationship between small and large signal modulation capabilities in highly nonlinear quantum dot lasers, *Appl. Phys. Lett.* 102, 101107 (2013).

70. S. Joshi, N. Chimot, A. Ramdane, and F. Lelarge, On the nature of the linewidth enhancement factor in p-doped quantum dash based lasers, *Appl. Phys. Lett.* 105, 241117 (2014).

71. M. Gioannini and I. Montrosset, Numerical analysis of the frequency chirp in quantum-dot semiconductor lasers, *IEEE J. Quantum Electron.* 43, 941 (2007).

72. C. Cornet, C. Platz, P. Caroff, J. Even, C. Labbé, H. Folliot, and A. Le Corre, Approach to wetting-layer-assisted lateral coupling of InAs/InP quantum dots, *Phys. Rev. B* 72, 035342 (2005).

73. R. Heitz, F. Guffarth, K. Poetschke, A. Schliwa, D. Bimberg, N. D. Zakharov, and P. Werner, Shell-like formation of self-organized InAs/GaAs quantum dots, *Phys. Rev. B* 71, 045325 (2005).

74. A. J. Nozik, Multiple exciton generation in semiconductor quantum dots, *Chem. Phys. Lett.* 457, 3 (2008).

75. P. Miska, J. Even, O. Dehaese, and X. Marie, Carrier relaxation dynamics in InAs/InP quantum dots, *Appl. Phys. Lett.* 92, 191103 (2008).

76. K. Veselinov, F. Grillot, C. Cornet, J. Even, A. Bekiarski, M. Gioannini, and S. Loualiche, Analysis of the double laser emission occurring in 1.55-μm InAs-InP(113)B quantum-dot lasers, *IEEE J. Quantum Electron.* 43, 810 (2007).

77. R. Nagarajan, M. Ishikawa, T. Fukushima, R. Geels, and J. Bowers, High speed quantum well lasers and carrier transport effects, *IEEE J. Quantum Electron.* 28, 1990 (1992).

78. L. A. Coldren and S. W. Corzine, *Diode Lasers and Photonic Integrated Circuits*, New York: Wiley, 1995.

79. J. Siegert, S. Marcinkevicius, and Q. X. Zhao, Carrier dynamics in modulation-doped InAs/GaAs quantum dots, *Phys. Rev. B* 72, 085316 (2005).

80. S. Marcinkevicius and R. Leon, Carrier capture and escape in $In_xGa_{1-x}As$/GaAs quantum dots: Effects of intermixing, *Phys. Rev. B* 59, 4630 (1999).

81. T. R. Nielsen, P. Gartner, and F. Jahnke, Many-body theory of carrier capture and relaxation in semiconductor quantum-dot lasers, *Phys. Rev. B* 69, 235314 (2004).

82. B. Ohnesorge, M. Albrecht, J. Oshinowo, A. Forchel, and Y. Arakawa, Rapid carrier relaxation in self-assembled InxGa1-xAs/GaAs quantum dots, *Phys. Rev. B* 54, 11532 (1996).

83. I. V. Ignatiev, I. E. Kozin, S. V. Nair, H. W. Ren, S. Sugou, and Y. Masumoto, Carrier relaxation dynamics in InP quantum dots studied by artificial control of nonradiative losses, *Phys. Rev. B* 61, 15633 (2000).

84. M. Lorke, T. R. Nielsen, J. Seebeck, P. Gartner, and F. Jahnke, Influence of carrier-carrier and carrier-phonon correlations on optical absorption and gain in quantum-dot systems, *Phys. Rev. B* 73, 085324 (2006).

85. I. Magnusdottir, S. Bischoff, A. V. Uskov, and J. Mork, Geometry dependence of Auger carrier capture rates into cone-shaped self-assembled quantum dots, *Phys. Rev. B* 67, 205326 (2003).

86. U. Bockelmann and T. Egeler, Electron relaxation in quantum dots by means of Auger processes, *Phys. Rev. B* 46, 15574 (1992).

87. P. Ferreira and G. Bastard, Phonon assisted capture and intra-dot Auger relaxation in quantum dots, *Appl. Phys. Lett.* 74, 2818 (1999).

88. P. Miska, J. Even, X. Marie, and O. Dehaese, Electronic structure and carrier dynamics in InAs/InP double-cap quantum dots, *Appl. Phys. Lett.* 94, 061916 (2009).

89. A. V. Uskov, J. McInerney, F. Adler, H. Schweizer, and M. H. Pilkuhn, Auger carrier capture kinetics in self-assembled quantum dot structures, *Appl. Phys. Lett.* 72, 58 (1998).

90. A. V. Uskov, F. Adler, H. Schweizer, and M. H. Pilkuhn, Auger carrier relaxation in self-assembled quantum dots by collisions with two-dimensional carriers, *J. Appl. Phys.* 81, 7895 (1997).

91. W. W. Chow and S. W. Koch, Theory of semiconductor quantum-dot laser dynamics, *IEEE J. Quantum Electron.* 41, 495 (2005).

92. W. W. Chow and S. W. Koch, *Semiconductor-Laser Fundamentals*, Berlin, Germany: Springer, 1999.

93. T. W. Berg, S. Bischoff, I. Magnusdottir, and J. Mørk, Ultrafast gain recovery and modulation limitations in self-assembled quantum-dot devices, *IEEE Photon. Technol. Lett.* 13, 541 (2001).

94. W. W. Chow, M. Lorke, and F. Jahnke, Will quantum dots replace quantum wells as the active medium of choice in future semiconductor lasers, *IEEE J. Sel. Top. Quantum Electron.* 17, 1349 (2011).

95. T. Takahashi and Y. Arakawa, Nonlinear gain effects in quantum well, quantum well wire, and quantum well box lasers, *IEEE J. Quantum Electron.* 27, 1824 (1991).

96. M. Willatzen, A. Uskov, J. Mork, H. Olesen, B. Tromborg, and A. P. Jauho, Nonlinear gain suppression in semiconductor lasers due to carrier heating, *IEEE Photon. Technol. Lett.* 3, 606 (1991).

97. D. J. Klotzkin, *Introduction to Semiconductor Lasers for Optical Communications*, Springer, New York, 2014.

98. Y. Lam and J. Singh, Monte Carlo simulation of gain compression effects in GRINSCH quantum well laser structures, *IEEE J. Quantum Electron,* New York, 30, 2435 (1994).

99. H. C. Schneider, W. W. Chow, and S. W. Koch, Anomalous carrier-induced dispersion in quantum-dot active media, *Phys. Rev. B* 66, 041310(R) (2002).

100. A. V. Uskov, E. P. O'Reilly, D. McPeake, N. N. Ledentsov, D. Bimberg, and G. Huyet, Carrier-induced refractive index in quantum dot structures due to transitions from discrete quantum dot levels to continuum states, *Appl. Phys. Lett.* 84, 272 (2004).

101. S. P. Hegarty, B. Corbett, J. G. McInerney, and G. Huyet, Free-carrier effect on index change in 1.3 μm quantum-dot lasers, *Electron. Lett.* 41, 416 (2005).

102. M. Osiński and J. Buus, Linewidth broadening factor in semiconductor lasers—An overview, *IEEE J. Quantum Electron.* QE-23, 9 (1987).

103. G. H. Duan, P. Gallion, and G. Debarge, Analysis of the phase-amplitude coupling factor and spectral linewidth of distributed feedback and composite-cavity semiconductor lasers, *IEEE J. Quantum Electron.* 26, 32 (1990).

104. G. Duan, P. Gallion, and G. Gebarge, Analysis of frequency chirping of semiconductor lasers in presence of optical feedback, *Opt. Lett.* 12, 800 (1987).

105. G. P. Agrawal, Intensity dependence of the linewidth enhancement factor and its implications for semiconductor lasers, *IEEE Photon. Technol. Lett.* 1, 212 (1989).

106. S. Wieczorek, B. Krauskopf, and D. Lenstra, Multipulse excitability in a semiconductor laser with optical injection, *Phys. Rev. Lett.* 88, 063901 (2002).

107. B. Haegeman, K. Engelborghs, D. Roose, D. Pierous, and T. Erneux, Stability and rupture of bifurcation bridges in semiconductor lasers subject to optical feedback, *Phys. Rev. E* 66, 046216 (2002).

108. M. Sciamanna, P. Mégret, and M. Blondel, Hopft bifurcation cascade in small-α laser diodes subject to optical feedback, *Phys. Rev. E* 69, 046209 (2004).

109. K. Panajotov, M. Sciamanna, M. Arteaga, and H. Thienpont, Optical feedback in vertical-cavity surface-emitting lasers, *IEEE J. Sel. Top. Quantum Electron.* 19, 1700312 (2012).

110. T. C. Newell, D. J. Bossert, A. Stintz, B. Fuchs, K. J. Malloy, and L. F. Lester, Gain and linewidth enhancement factor in InAs quantum-dot laser diodes, *IEEE Photon. Technol. Lett.* 11, 1527(1999).

111. Z. Mi, P. Bhattacharya, and S. Fathpour, High-speed 1.3 μm tunnel injection quantum-dot lasers, *Appl. Phys. Lett.* 86, 153109 (2005).

112. B. Dagens, A. Markus, J. X. Chen, J. G. Provost, D. Make, O. Le Goueziou, J. Landreau, A. Foire, and B. Thedrez, Gaint linewidth enhancement factor and purely frequency modulated emission from quantum dot laser, *Electron. Lett.* 41, 323 (2005).

113. F. Grillot, K. Veselinov, M. Gioannini, I. Montrosset, J. Even, R. Piron, E. Homeyer, and S. Loualiche, Spectral analysis of 1.55 μm InAs–InP(113)B quantum-dot lasers based on a multipopulation rate equations model, *IEEE J. Quantum Electron.* 45, 872 (2009).

114. C. Wang, F. Grillot, and J. Even, Impacts of wetting layer and excited state on the modulation response of quantum-dot lasers, *IEEE J. Quantum Electron.* 48, 1144 (2012).

115. M. Kuntz, Modulated InGaAs/GaAs quantum dot lasers, PhD thesis, Berlin, Germany, 2006.

116. N. A. Naderi, External control of semiconductor nanostructure lasers, PhD thesis, University of New Mexico, Albuquerque, NM, 2011.

117. A. E. Zhukov, M. V. Maximov, A. V. Savelyev, Yu. M. Shernyakov, F. I. Zubov, V. V. Korenev, A. Martinez, A. Ramdane, J.-G. Provost, and D. A. Livshits, Gain compression and its dependence on output power in quantum dot lasers, *J. Appl. Phys.* 113, 233103 (2013).

118. J. Oksanen and J. Tulkki, Linewidth enhancement factor and chirp in quantum dot lasers, *J. Appl. Phys.* 94, 1983 (2003).

119. S. Melnik, G. Huyet, and A. V. Uskov, The linewidth enhancement factor α of quantum dot semiconductor lasers, *Opt. Express* 14, 2950 (2006).

120. S. Wieczorek, B. Krauskopf, T. B. Simpson, and D. Lenstra, The dynamical complexity of optically injected semiconductor lasers, *Phys. Rep.* 416, 1 (2005).

121. W. W. Chow and F. Jahnke, On the physics of semiconductor quantum dots for applications in lasers and quantum optics, *Prog. Quantum Electron.* 37, 109 (2013).

122. A. Markus, J. X. Chen, O. Gauthier-Lafaye, J. G. Provost, C. Paranthoen, and A. Fiore, Impact of intraband relaxation on the performance of a quantum-dot laser, *IEEE J. Sel. Top. Quantum Electron.* 9, 1308 (2003).

123. B. Lingnau, W. W. Chow, E. Schöll, and K. Lüdge, Feedback and injection locking instabilities in quantum-dot lasers: A microscopically based bifurcation analysis, *New J. Phys.* 15, 093031 (2013).

124. F. Grillot, B. Dagens, J. G. Provost, H. Su, and L. F. Lester, Gain compression and above-threshold linewidth enhancement factor in 1.3 μm InAs-GaAs quantum-dot lasers, *IEEE J. Quantum Electron.* 44, 946 (2008).

125. J.-G. Provost and F. Grillot, Measuring the chirp and the linewidth enhancement factor of optoelectronic devices with a Mach-Zehnder interferometer, *IEEE Photon. J.* 3, 476 (2011).

126. C. Cornet, C. Labbé, H. Folliot, N. Bertru, O. Dehaese, J. Even, A. Le Corre, C. Paranthoën, C. Platz, and S. Loualiche, Quantitative investigations of optical absorption in InAs/InP (311)B quantum dots emitting at 1.55 μm wavelength, *Appl. Phys. Lett.* 85, 5685 (2004).

127. Z. Mi and P. Bhattacharya, Analysis of the linewidth-enhancement factor of long-wavelength tunnel-injection quantum-dot lasers, *IEEE J. Quantum Electron.* 43, 363 (2007).

128. C. Wang, M. Osiński, J. Even, and F. Grillot, Phase-amplitude coupling characteristics in directly modulated quantum dot lasers, *Appl. Phys. Lett.* 105, 221114 (2014).

129. S. Gerhard, C. Schilling, F. Gerschutz, M. Fischer, J. Koeth, I. Krestnikov, A. Kovsh, M. Kamp, S. Hofling, and A. Forchel, Frequency-dependent linewidth enhancement factor of quantum-dot lasers, *IEEE Photon. Technol. Lett.* 20, 1736 (2008).

130. K. Lüdge and H. G. Schuster, *Nonlinear Laser Dynamics: From Quantum Dots to Cryptography*, New York: Wiley, 2011.

131. B. Lingnau, K. Lüdge, W. W. Chow, and E. Schöll, Influencing modulation properties of quantum-dot semiconductor lasers by carrier lifetime engineering, *Appl. Phys. Lett.* 101, 131107 (2012).

132. C. Wang, B. Lingnau, K. Lüdge, J. Even, and F. Grillot, Enhanced dynamic performance of quantum dot semiconductor lasers operating on the excited state, *IEEE J. Quantum Electron.* 50, 723 (2014).

133. T. B. Simpon, J. M. Liu, and A. Gavrielides, Bandwidth enhancement and broadband noise reduction in injection-locked semiconductor lasers, *IEEE Photon. Technol. Lett.* 7, 709 (1995).

134. A. Murakami, K. Kawashima, and K. Atsuki, Cavity resonance shift and bandwidth enhancement in semiconductor lasers with strong light injection, *IEEE J. Quantum Electron.* 39, 1196 (2003).

135. F. Grillot, C. Wang, N. A. Naderi, and J. Even, Modulation properties of self-injected quantum-dot semiconductor diode lasers, *IEEE J. Sel. Top. Quantum Electron.* 19, 1900812 (2013).

136. J.-P. Zhuang and S.-C. Chan, Phase noise characteristics of microwave signals generated by semiconductor laser dynamics, *Opt. Express* 23, 2777 (2015).

4 Optical, Photoluminescence, and Vibrational Spectroscopy of Metal Nanoparticles

P. Gangopadhyay

CONTENTS

4.1 INTRODUCTION

Historically, centuries ago fine precipitates of noble metals (gold, silver, and copper) were used in tinted glasses to produce beautiful colors of windowpanes used for decorations and in Lycurgus cups (developed in the fourth century AD by Roman glass workers; visit http://www.bmimages.com for the images). The first insight into the beautiful colors of stained glasses was provided by Michael Faraday in 1857, and he concluded thus: "I think that in all these cases the ruby tint is due simply to the presence of diffused finely-divided gold" (Faraday 1857). In 1908, about 50 years later, Gustav Mie theoretically explained the appearance of the striking colors as due to the resonant absorption of light in these fine metal particles (Mie 1908). This resonant absorption of light, which arises from the collective oscillations of the conduction electrons in metal nanoparticles, is being described as localized surface-plasmon resonance (SPR) in the current literature (Kreibig and Vollmer 1995, Quinten 2011). Plasmons are defined as quanta of collective oscillations of conduction electrons in metal nanoparticles vis-à-vis metal nanostructures (Ghosh and Pal 2007) and also in doped semiconductors (Luther et al. 2011). In this chapter, we discuss the interesting features of the optical responses of various metal nanoparticles and metal–semiconductor hybrid nanostructures in detail.

Scientifically interesting and enriching phenomena, for instance, strong optical absorption and photoluminescence (PL) responses in the visible range of wavelength (Bohren and Huffman 1983, Gangopadhyay et al. 2005, Xu and Suslick 2010), local field enhancements and Raman spectroscopy (Moskovits 1985, Kneipp et al. 2002, Haynes and Van Duyne 2003), sub-wavelength optics (Salerno et al. 2002), fast optical switching (Maier et al. 2001), plasmonics (Maier 2007, Murray and Barnes 2007), and so on, are fundamentally associated with various nanoscale metal particles and metal nanostructures. These exciting physical properties have been driving experimental and theoretical research for a long time (Roco et al. 1999, Allegrini et al. 2001, Hosokawa et al. 2012). Quantum-size behavior of metal nanoparticles, for example, has opened new directions in many fields of current research and modern technologies, such as electronic logic operations (Lee and Dickson 2003), catalytic and photocatalytic reactions (Bell 2003, Prieto et al. 2013), nanotweezers (Boggild et al. 2001), nanosensors (Luo et al. 2006, Saha et al. 2012), integrated biosensors (Hoa et al. 2007), and others. The field has witnessed great momentum of research studies and innovative developments during the last two decades. With the help of the transmission electron microscope, and particularly after the advent of sophisticated instruments like scanning tunneling microscope (Binnig et al. 1982a,b), the synthesis (e.g., nanolithography) and characterization (e.g., high-resolution surface microscopy and spectroscopy) of various nanomaterials have advanced. Topographic images of surfaces with atomic-scale resolution have been obtained with the tunneling microscope.

In fact, prior to the inventions of the scanning tunneling microscope, the lecture "There is plenty of room at the bottom—An Invitation to Enter a New Field of Physics" in 1959 by R. P. Feynman at Caltech had motivated and thrown many challenges at contemporary scientists and engineers (Feynman 1992, 1993). The bottom-room (i.e., the scope of science and technology with nanometer-scale materials) has really expanded with time globally and is continually being established by the multidisciplinary researchers across the boundaries of various scientific disciplines and technologists of diverse interests. These innovations and perspective developments, in effect, have touched human lives with all the benefits that could be envisaged (Xiao et al. 2003, Atwater 2007, Jie et al. 2007, Pradeep and Anshup 2009). Almost everyday, the present world is observing advanced and breakthrough discoveries that lead to miniaturization of devices. Technological advancements are being realized through the knowledge of nanoscience and nanotechnology. The science of very small things is catching up in very big ways. Nanotechnology is providing the handle to manipulate materials at the nanoscale regimes. Persistent global research activities and the large number of articles appearing in important international journals are the real testimony on this subject. "Nano" is a common buzzword in this context. According to the definition, *nanomaterials* (metals, semiconductors, or insulators) are no different from the parent phase of the bulk materials but for their physical dimensions, which range between 1 and 100 nm (1 nm = 10^{-9} m), at least in one direction. At nanoscale dimensions, the physical, chemical, and biological properties of materials differ in fundamental and favorable ways from those of individual atoms and molecules or bulk matter. That brings up spiraling applications of various nanomaterials in almost every sphere of life. Hence, understanding the sciences of various synthetic methods and novel physical properties of matter at the nanoscale is one of the most active areas of research today (Nalwa 2004, Hosokawa et al. 2012). Extensions of these scientific studies are again very relevant to significant developments in multidisciplinary fields spanning physics, chemistry, and biology. It brings out many unexpected results to ponder over. In this perspective, nanoscale metal particles, quantum dots (for semiconductors), and metal–semiconductor hybrid nanostructures, which form important classes in the world of nanomaterials, deserve special attention for comprehensive research studies (Chang et al. 2006, Akimov et al. 2007).

It is well known now that nanoscale metal structures and metal nanoparticles support localized surface-plasmon modes. Additionally, the planar surface of a metal (e.g., smooth surface of metal nanowires, Sanders et al. 2006) also supports plasmon modes, known as surface-plasmon polaritons (SPPs). These are the propagating excitations of charge-density waves. They propagate, typically, over a few tens of micrometers (Murray and Barnes 2007). The propagation length depends on the diameter of the metal nanowires or the width of nanostripes (Salerno et al. 2002). In close proximity of an optical emitter (e.g., quantum dot) with a metal nanowire, emission from the quantum dot can be guided in the nanowire because of the strong coupling of the emitted photon with the SPP modes (Chang et al. 2007, Fedutik et al. 2007). It is like seeing lights at the end of a normal dielectric-based optical fiber, making the way for highly efficient quantum optoelectronic devices. In a suitably designed nanostructured system of hybrid materials, it may be interesting to study the photon correlations among the quantum dots and the guiding SPP modes of a metal nanowire.

The prospects of a wide range of synthesis methods, basic research, and related applications of the novel nanomaterials are great. In recent times, researchers have demonstrated that semiconductor quantum dots are also useful in quantum communication and computation devices (Luther et al. 2011). These novel materials can facilitate exchanging information at the maximum communication speed (nearly the speed of light), overcoming the bottleneck of electronic transport of data in present-day conventional computers. Their scope is unlimited, but to mention a few in a limited domain, they may include, for example, high-sensitivity nanosensors for chemical and biological detectors (Alivisatos 2005), enhanced chemical and photocatalysis using plasmonic nanostructures (Moser 1996, Campbell et al. 2002, Christopher et al. 2012), surface-enhanced Raman spectroscopy (Moskovits 1985, Xu et al. 1999, Tao et al. 2003), efficiency improvements of solar cells (Nakayama et al. 2008, Atwater and Polman 2010, Bora et al. 2013, Gao et al. 2013), photothermal effects (Xiong et al. 2012), optical cloaking (invisibility) (Atwater 2007, Alù and Engheta 2008), tumor imaging and therapy (Loo et al. 2005, Atwater 2007), and so on. Plain silicate glasses containing metallic particles of few nanometers in size are also attractive materials for optical switching applications in integrated-optical devices and fast optical computers (Hughes et al. 2014). Large enhancement (a few orders of magnitudes higher than in their bulk counterpart) of the nonlinear optical susceptibility ($\chi^{(3)}$), for instance, near the SPR frequency, with a fast optical response time (on the order of a few picoseconds) makes the metal–glass nanocomposite materials technologically attractive for plasmonic optical devices (Faccio et al. 1998, Allegrini et al. 2001, Hamanaka et al. 2004, Bozhevolnyi et al. 2006).

For the most part, this chapter is designed to provide a review of the growth of metal–semiconductor nanoparticles, as well as various synthesis processes and experimental techniques to study their optical, vibrational (Raman), and PL properties at the nanoscale. Optical properties of noble metal nanoparticles, for instance, of sizes ~10 nm, show expected results that could be explained using Drude's theory for free-electron metals (Kreibig and Vollmer 1995). However, different and interesting optical responses have been observed in the case of silver particles while reducing their size to the range of ~2 nm (Srivastava et al. 2014). Quantum mechanical models have been employed to explain these results. Surface-confined acoustic vibration modes in metal or semiconductor nanoparticles are usually observed in the vibrational (phonon) spectra of these materials (Saviot et al. 2011). They reveal the nature of spheroidal (radial and quadrupolar) oscillations of these nanoparticles. A thorough understanding of the vibrations of nanoparticles is necessary to elucidate some of the novel properties that these nanomaterials exhibit, such as melting (Bottani et al. 2001), electron–phonon coupling (Bachelier and Mlayah 2004), and so on. Surface-plasmon-enhanced Raman scattering of metal nanoparticles will be discussed here in detail to explain the plasmon–phonon coupling in such cases (Gangopadhyay et al. 2007). Examples of metal–semiconductor hybrid nanostructures will be discussed to elaborate quasi-particle interactions, possible energy transfers, and interesting PL behaviors of these novel materials. Compared to the bandgap emission, PL of semiconductors may be quenched or enhanced in metal–semiconductor hybrid nanostructures (Wang et al. 2009, Walters et al. 2010, Haridas et al. 2011). Different views on the mechanisms of these processes, such as resonant coupling of surface plasmons with excitons, local field enhancements in metal nanoparticles, and electron transfers from semiconductors to metal nanoparticles, will also be discussed. Understanding and control of such interactions in weak and strong coupling regimes among different metal–semiconductor hybrid nanomaterials are impera-tive. The field has attracted much interest among the researchers for fundamental understanding of the phenomena as well as for their potential applications, though ambiguities exist among the differ-ent results and published reports (Hosoki et al. 2008, Zhou et al. 2011, Nahm et al. 2013).

4.2 NANO-BASICS

Compared to that of bulk materials, the science of nanomaterials differs considerably. Accordingly, the properties of materials vary at the nanoscale. For example, the optical, vibrational, chemical, and electrical behaviors of nanomaterials are significantly different. This is due to the following reasons: (1) for the same quantity of material, nanomaterials possess a larger surface area in com-parison to their bulk counterparts because of the large surface-to-volume ratio of the nanoscale geometry (consequently, nanomaterials are more chemically reactive); (2) due to absence of long-range translational invariance compared to the bulk crystals, nanoparticles and nanostructures may display noncrystalline structures; and (3) effects of quantum confinement (i.e., a reduction in the degrees of freedom of free electrons or charge carriers in metal nanoparticles or quantum dots, respectively) are predominant at the nanoscale dimensions of materials. These variations arise entirely from the imposed size restrictions. The size limit renders nanomaterials attractive for fundamental understanding and for technological applications.

4.2.1 PARTICLE SIZES VERSUS PHYSICAL PROPERTIES

Physical properties mentioned above depend largely on the size of the nanoscale material and the size-dependent material parameters. Standard parameters such as the particle radius (R) or diameter (d) are used to compare the sizes of the nanoscale particles of various length scales. For exam-ple, while studying the characteristic changes of optical properties with particle growth, the very first length scale to be compared is the mean free path (average distance traveled by the electrons between two successive collisions) of electrons in the metal. In bulk noble metals (silver or gold), the mean free path (l_f) of electrons is ~30 nm at room temperature (Kittel 1985). We shall classify the nanoparticles as small when $d \ll l_f$. The second characteristic length scale is the wavelength of light (λ); this scale becomes very important while exploring optical spectroscopy of relatively large metal nanoparticles in the UV–visible to near-infrared wavelength range. Accordingly, in this size regime, where $l_f \ll d \ll \lambda$, the particles are called large nanoparticles. However, for complete-ness, the finest length scale also needs to be mentioned: that is, the Fermi wavelength (de Broglie's wavelength of an electron at Fermi energy ~0.5 nm for silver or gold, for instance) of an electron. In this finest size range, particles behave as molecular types (Link et al. 2002, Zheng et al. 2004). The physical properties of such metal species are significantly different from those of the other two types of metal nanoparticles (Peyser et al. 2001, Thomas et al. 2002). Going by the above descrip-tions, two different kinds of particle-size effects are defined: intrinsic and extrinsic. Intrinsic effects are concerned with specific changes in volume and surface properties of nanomaterials. Extrinsic effects are size-dependent responses to external fields irrespective of their intrinsic variations. Experiments on intrinsic effects particularly focus on the question of how electronic and structural particle properties such as chemical reactivity, binding energy, crystallographic structure, melting temperature, and optical material property ($\varepsilon\,(\omega, R)$) vary as a function of particle size, composi-tion, and geometry. Examples of intrinsic particle size effects that will be studied in detail are the collective electronic excitations and vibrational properties. The size-dependent optical response ($\varepsilon\,(\omega, R)$) of nanoscale metal particles is an intrinsic property, for instance, which is governed only by the dimension and volume fraction (defined as the ratio of the particle volume to the sample vol-ume) of the particle materials. For the experimental studies discussed here, two length scales are of importance while elaborating the intrinsic size effects: the mean free path of electrons (l_f) and the wavelength of light (λ).

4.2.2 THEORETICAL FRAMEWORK

To understand a number of unusual experimental results of nanomaterials (e.g., optical and vibrational spectra), development of some theoretical concepts is necessary. In this section, we outline these to understand the fundamentals at the nanoscale and phenomena of nanoscale materials. Although the responses of materials are different at the nanoscale and in the bulk, in many cases experimental data measured on bulk samples provide significant clues to explain the behaviors of nanomaterials. We first discuss the optical response of metal nanoparticles.

4.2.2.1 Optical Response of Metal Nanoparticles

Before going on to detailed discussions on the optical response of nanoscale metal particles, it is imperative to revisit the experimental results on respective metals. For example, a "good" metal like silver (Ag) is chosen here for the discussions on collective excitations. Among the three metals Cu, Ag, and Au, which display SPRs in the visible wavelength of light, Ag exhibits the highest efficiency of plasmon excitation (Kreibig and Vollmer 1995). Also, light–matter interaction cross-sections for Ag can be ~10 times the geometric cross-section, implying that nanoparticles may capture much more light than is physically incident on them (Evanoff and Chumanov 2004). During optical processes, intraband (which includes Drude or free-electron absorptions) and interband transitions arising from collective excitations of an electron gas are excited in the nanoscale metal particles. In the optical wavelength ranges, the dielectric functions $\varepsilon(\omega,0) \sim \varepsilon(\omega)$ describe the collective oscillations or responses of free electrons against the positive ion core in metals, driven by the applied electric field (schematically shown in Figure 4.1).

4.2.2.1.1 Mie Theory

Spherical and well-isolated metal nanoparticles of uniform size in a dielectric matrix form the basis of the simplest model system. It facilitates understanding the major optical phenomena in practical cases. It is ideal to express the optical properties in terms of the absorption (σ_{abs}) and scattering (σ_{sca}) cross-sections. These quantities are related to the intensity loss $\Delta I(z)$ of a parallel beam of incident light due to absorption (generation of heat) or elastic scattering (changes of propagation direction). According to the Lambert–Beer law

$$\Delta I_{abs}(z) = I_o(1 - e^{-N_a \sigma_{abs} z})$$

$$\Delta I_{sca}(z) = I_o(1 - e^{-N_a \sigma_{sca} z})$$

for a purely absorbing and purely scattering system of nanoparticles (N_a being the number density of the nanoparticles), respectively. In practice, the resulting extinction cross-section is given by, $\sigma_{ext} = \sigma_{abs} + \sigma_{sca}$. The extinction, absorption, and scattering cross-sections are calculated from the

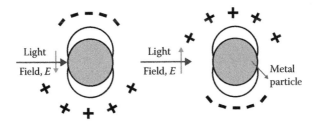

FIGURE 4.1 Schematic diagram illustrating the excitation of a dipolar surface plasmon by the electric field of the incident light at some instant of time t and after time $t + T/2$ for a spherical metal particle in the quasi-static size limit. T is the time period of the oscillating field.

Mie theory (following the common notations, see Kreibig and Vollmer 1995) by series expansion of the involved fields as

$$\sigma_{ext} = \frac{2\pi}{|k|^2} \sum_{L=1}^{\infty} (2L+1) \operatorname{Re}[a_L + b_L] \tag{4.1}$$

$$\sigma_{sca} = \frac{2\pi}{|k|^2} \sum_{L=1}^{\infty} (2L+1) \left(|a_L|^2 + |b_L|^2 \right) \tag{4.2}$$

and the resulting absorption cross-section is given as, $\sigma_{abs} = \sigma_{ext} - \sigma_{sca}$. Here, k is the wave vector ($k = (2\pi/\lambda) = (\omega/c)$) and the summation index L gives the order of multipole excitations in the nanoparticle materials. For example, $L = 1, 2, 3,...$ corresponds to the dipole, quadrupole, octupole fields, and so on, for higher-order optical excitations. The amplitudes a_L and b_L in Equations 4.1 and 4.2 are proportional to $(|k|R)^{2L+1}$. The total Mie extinction spectra consist of dipolar, quadrupolar, and higher modes of excitations. Each multipole is contributed by electric and magnetic modes, that is, plasmons and eddy currents, respectively. Each mode consists of absorption and scattering losses. For the lowest order term ($L = 1$), the simplified Mie formula for the dipolar absorption is given by

$$\sigma_{ext}(\omega) = 9\frac{\omega}{c} \varepsilon_m^{3/2} V_o \frac{\varepsilon_2(\omega)}{[\varepsilon_1(\omega) + 2\varepsilon_m]^2 + \varepsilon_2(\omega)^2} \tag{4.3}$$

where
  $\varepsilon_1(\omega)$ and $\varepsilon_2(\omega)$ are the frequency-dependent real and imaginary dielectric functions of the metal nanoparticles
  ε_m is the dielectric constant of the embedding medium
  $V_o = (4/3)\pi R^3$ is the particle volume
  c is the velocity of light in vacuum

The Mie formula (Equation 4.3) shows that the extinction is directly proportional to the volume of the nanoparticles present. However, it indirectly provides the frequency of resonance through the condition $|\varepsilon + 2\varepsilon_m|$ to be minimum. To know the exact resonance position, one needs to have knowledge about the frequency dependence of $\varepsilon_1(\omega)$ and $\varepsilon_2(\omega)$. We shall discuss this shortly with an example. However, it is important to know that in the quasi-static size regime (where $2R \ll \lambda$), the dipolar scattering cross-section which is proportional to $(|k|R)^6/|k|^2$ and higher multipolar contributions (e.g., the quadrupolar extinction $(|k|R)^5/|k|^2$ and the quadrupolar scattering $(|k|R)^{10}/|k|^2$) are strongly suppressed due to the size restrictions. Thus, in the quasi-static size range, $\sigma_{ext} \approx \sigma_{abs}$.

4.2.2.1.2 Drude–Lorentz–Sommerfeld (DLS) Model

According to free-electron theory of metals, most of the electronic and optical properties in "good" metals (particularly, alkali and noble metals) are due to the free electrons in conduction band alone. These metals have completely filled valence bands and partially filled conduction bands (e.g., the electronic configuration of Ag is $[Kr]^{36}4d^{10}5s^1$). Linear response of the metal to the interacting electromagnetic waves is described by the dielectric function $\varepsilon(\omega)$ of the free valence electron gas of metals. For alkali metals, the role of $\varepsilon(\omega)$ is primarily governed by the transitions within the conduction band alone. In noble metals, substantial contribution of interband transitions from lower lying energy levels into the conduction band or from the conduction band into higher energy unoccupied levels is quite possible. As a result, the role played by the dielectric function $\varepsilon(\omega)$ of the electron gas in noble metals in controlling the optical responses is rather complex.

This model of free-electron optical responses assumes that the response of a metal can be found out by first considering the influence of external forces on one free conduction electron alone. The macroscopic effective response to the external electromagnetic field can be found by multiplying the effect on a single electron by the number of total electrons in the given system. Thus, according to this model, the response of a free electron of mass (m) and charge (e) to an external electric field (E) is described by the equation of drift motion superimposed on the motion of an electron in the field-free case. Therefore, the kinematics of the electron in such a case can be expressed as

$$m\frac{d^2r}{dt^2} + m\Gamma\frac{dr}{dt} = eE_o e^{-\iota\omega t} \tag{4.4}$$

where Γ denotes the phenomenological damping constant. Equation 4.4 is solved to calculate the induced dipole moment p ($p = er_o$) and the polarization density P ($P = np$), (n is the number of electrons per unit volume in the system).

Now, let us define $\varepsilon(\omega) = 1 + (P/\varepsilon_o E)$ and $P = n\alpha E$, where ε_o is the permittivity of free space and α is the polarizability tensor. Assuming a solution for r as $r = r_o e^{-\iota\omega t}$, the above equation (Equation 4.4) may be solved. Using the definitions of $\varepsilon(\omega)$ and P (as above), the dielectric function $\varepsilon(\omega)$ of the free-electron gas may be obtained, which can be written as

$$\varepsilon(\omega) = 1 - \frac{\omega_P^2}{\omega^2 + \Gamma^2} + \iota\frac{\omega_P^2\Gamma}{\omega(\omega^2 + \Gamma^2)} \tag{4.5}$$

Expressing $\varepsilon(\omega) = \varepsilon_1(\omega) + \iota\varepsilon_2(\omega)$, the frequency-dependent real and imaginary parts of the dielectric function may be written as

$$\varepsilon_1(\omega) = 1 - \frac{\omega_P^2}{\omega^2 + \Gamma^2} \quad \text{and} \quad \varepsilon_2(\omega) = \frac{\omega_P^2\Gamma}{\omega(\omega^2 + \Gamma^2)} \tag{4.6}$$

where $\omega_P = \sqrt{ne^2/\varepsilon_o m}$ is the Drude frequency of plasma oscillation in metals (Kittel 1985). The damping constant Γ (also known as scattering frequency of electrons) is related to the mean free path (l_f) of the conduction electrons by $\Gamma = V_F/l_f$, where V_F is the Fermi velocity of electrons in metals. In real metals, $V_F \sim 10^6$ m s^{-1} and $\Gamma \sim 10^{13}$ s^{-1}. Now, in the UV–visible wavelength range of light, for which $\omega \gg \Gamma$, the real and imaginary parts of the dielectric function of the free-electron gas of metals can be written in simplified form as $\varepsilon_1(\omega) \approx 1 - (\omega_P^2/\omega^2)$ and $\varepsilon_2(\omega) \approx \omega_P^2\Gamma/\omega^3$ (from Equation 4.6). Here, $\varepsilon_1(\omega)$ and $\varepsilon_2(\omega)$ of the dielectric function describe the polarization and energy dissipation of light in matter, respectively.

4.2.2.1.3 Quasi-Static Responses

The laws of electrostatics facilitates the understanding of the optical response of metal nanoparticles in the size range $2R \ll \lambda$ (λ being the wavelength of light). In this size regime, the field applied is assumed to be nearly static. It is known as the quasi-static response. The positive charges in the metal nanoparticles are assumed to be immobile, and the negative charges, that is, the conduction electrons, are allowed to move under the influence of an external electric field. A schematic diagram illustrates the collective oscillations of the conduction electrons against the positive ionic core in a metal nanoparticle (see Figure 4.1). Using the electrostatic boundary conditions at the particle–dielectric interface, one can calculate the internal electric field as

$$E_i = E_o \frac{3\varepsilon_m}{\varepsilon + 2\varepsilon_m} \tag{4.7}$$

where
  ε_m is the dielectric constant of the embedding medium
  E_o is the applied field
  ε is the dielectric function of the free electrons in the metal

In the quasi-static size regime, solution of electrostatics applies as well to small metal nanoparticles in oscillating electromagnetic fields. Because of their small size, the nanoparticles experience a field that is spatially uniform. Under this special condition of size restriction, the internal electric field (E_i) shows resonance whenever $|\varepsilon + 2\varepsilon_m|$ is minimum (from Equation 4.7); to say mathematically, $(\varepsilon_1 + 2\varepsilon_m)^2 + \varepsilon_2^2 \approx$ minimum. (It may be noted that an identical condition for resonance was used while discussing the Mie theory for the dipolar absorption [see Equation 4.3] of light.) This further shows that a negative ε_1 is necessary ($\varepsilon_1 \approx -2\varepsilon_m$) for the plasmon resonance condition to be satisfied. Using Equation 4.6 for ε_1, we may further write (neglecting Γ^2 in Equation 4.6)

$$\varepsilon_1 \approx 1 - \frac{\omega_P^2}{\omega_{PR}^2} = -2\varepsilon_m \tag{4.8}$$

$$\Rightarrow \omega_{PR} = \frac{\omega_P}{\sqrt{1 + 2\varepsilon_m}} \tag{4.9}$$

where ω_{PR} is the plasmon resonance frequency. For metal nanoparticles, it is also equivalently known as the SPR frequency. The word "surface" implies that polarization of charges is mostly taking place on the surface of metal nanoparticles. Also, one can see from the resonance frequency (Equation 4.9) that the optical response of surface plasmons in metal nanoparticles is very sensitive to the local environment through the dielectric constant of the medium, ε_m. Therefore, a change in the value of ε_m alters the position of ε_1 at which the resonance occurs (see Equation 4.8).

Let us examine a practical case of knowing the localized SPR frequency of spherical metal nanoparticles in a dielectric medium, say finding the SPR frequency if silver nanoparticles are embedded in a silica glass matrix. As we know, the refractive index (μ) of silica glass in the visible wavelength of light is 1.5, and the dielectric constant ε_m is 2.25 (as $\varepsilon_m = \mu^2$). So, the resonance condition (i.e., $\varepsilon_1 = -2\varepsilon_m$) would be satisfied only when $\varepsilon_1 = -4.5$. For silver, this occurs (as shown in Figure 4.2a) at ~3 eV (415 nm). Experimental result for silver nanoparticles in a silica glass matrix is shown in Figure 4.2b. Optical response peak at 3 eV is clearly observed. Similarly, for gold nanoparticles embedded in a silica glass matrix, the SPR occurs at ~520 nm.

In addition, the size of the metal nanoparticles may also be found out from the full-width at half-maximum (FWHM) of the SPR absorption band. Assuming Drude-like free-particle behavior for the conduction electrons in the nanoscale particles within the quasi-static size range, one may write $R = V_F \tau$, where R is the average radius of the metal nanoparticles, V_F is the Fermi velocity of electrons in the metal, and τ is the time between two successive collisions (i.e., mean free time) of conduction electrons in the metal nanoparticles. Quantum confinement and frequent scattering of these electrons from the nanoscale particles in the dielectric matrix can lead to quantum

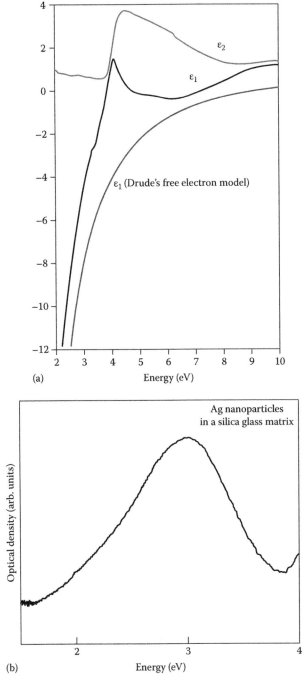

FIGURE 4.2   (a) Dielectric functions $\varepsilon_1(\hbar\omega)$, $\varepsilon_2(\hbar\omega)$ (real, imaginary parts) and the calculated free-electron contribution $\varepsilon_1^{free}(\hbar\omega)$ (Drude) for bulk Ag. Optical data were collected from the handbook. (From Palik, E.D. ed., *Handbook of Optical Constants of Solids*, Academic Press, Orlando, FL, 1985.) (b) Optical absorption spectrum of Ag nanoparticles (average size ~2.4 nm) in a silica glass matrix revealing the SPR absorption peak at ~3 eV. (Reprinted from Gangopadhyay, P. et al., *J. Appl. Phys.*, 88, 4975, 2000. With permission.)

fluctuations (ΔE) of the average energy of the free electrons around the SPR energy. Next, applying the Heisenberg's principle of uncertainty relation ($\Delta E \Delta \tau = \hbar$) and using the expression for $R = V_F \tau$, we may further write

$$R = \frac{\hbar V_F}{\Delta E} \tag{4.10}$$

where $\hbar$ is the reduced Planck's constant. This equation may be used to calculate the dimensions of metallic nanoparticles as long as the size (d) of the nanoparticles is much less than the mean free path (l_f) of electrons in the metal and the nanoparticles are of uniform size. In the dilute limit of noninteracting nanoparticles and under the quasi-static size-limiting conditions, size of nanoscale metal particles may be determined from the spectral width of the SPR absorption band. However, simple explanations like this may become increasingly inapplicable in some cases, for example, if the number density of the metal nanoparticles is beyond the dilute limit. Interactions among the nanoscale metal particles are important in such cases. In addition, simultaneous influence of other parameters (e.g., size of the nanoparticles) may become important while elaborating the optical response behaviors of an assembly of nanoscale metal particles embedded in a dielectric matrix. We shall come back to this discussion later while elaborating on the experimental results of silver nanoparticles (with a high number density) in a soda glass matrix.

4.2.2.1.4   Interband Transitions and Core Effects

Intraband transitions like the Drude or free-carrier absorption arising from collective excitations of an electron gas in nanoscale metal particles are excited by optical means. Apart from the free conduction electrons, other electrons in the core levels also contribute to the optical response of metal nanoparticles through the dielectric function $\varepsilon(\omega)$. It is common to express the dielectric function in terms of the electric susceptibility χ as $\varepsilon(\omega) = 1 + \chi^{DS}(\omega)$, where χ^{DS} denotes the free electron Drude–Sommerfeld susceptibility. Based on the optical data (Palik 1985), the calculated dielectric functions $\varepsilon_1(\hbar\omega)$, $\varepsilon_2(\hbar\omega)$ (real and imaginary parts) and the calculated free-electron contribution $\varepsilon_1^{free}(\hbar\omega)$ (Drude) for bulk silver are displayed in Figure 4.2a. The figure shows that below ~4 eV, dielectric functions are dominated by the Drude's free-electron behavior, but above 4 eV the interband transitions become dominant. Thus, from the data and the graph, we arrive at the important result that the influence of electrons that undergo interband transitions gives an additive complex contribution $\chi^{IB} = \chi_1^{IB} + \iota\chi_2^{IB}$ to the Drude–Sommerfeld (DS) susceptibility. The complex dielectric function $\varepsilon(\omega)$, incorporating all optical material properties around the visible region, is thus given as $\varepsilon(\omega) = 1 + \chi^{DS}(\omega) + \chi^{IB}(\omega)$. The imaginary part χ_2^{IB}, describing the direct energy dissipation, becomes large only for frequencies where interband transitions occur. The real part χ_1^{IB} is sometimes replaced by an average frequency independent value resembling the polarization of the ion core, χ^{core}. An alternative way to define χ^{core} is to separate the total of all interband excitations into contributions of lower and higher lying electrons; the former is then denoted as χ^{core}.

Energetically deeper lying electrons (e.g., 3d electrons in Ag) and the ion cores contribute to the dielectric function $\varepsilon(\omega)$ through this χ^{core}. Noble metal atoms such as Cu, Ag, or Au have completely filled 3d, 4d, and 5d shells, and just have one electron in the 4s, 5s, and 6s bands, respectively. It has been shown that there is an intermediate region between the pure Drude and the interband transition behavior in these noble metals (Hollstein et al. 1977). For example, whereas direct excitation of the 4s electrons (in Cu) to higher levels requires an energy of 4 eV, the threshold for direct excitation of 3d-band electrons into the conduction bands (4sp) is only about 2 eV. Similarly, the threshold for interband transition in other noble metals such as Ag and Au is found to be ~3.9 eV (Ehrenreich and Philipp 1962) and 2.38 eV (Christensen and Seraphin 1971), respectively.

4.2.2.1.5 Size-Dependent Dielectric Functions

Following the Drude's formalism of free-electron plasma oscillation in metals, the dielectric functions for nanoscale metallic particles (of radius R) may be calculated (from Equation 4.6) as

$$\varepsilon_1(\omega) = \omega_P^2 \left(\frac{1}{\omega^2 + \Gamma_\infty^2} - \frac{1}{\omega^2 + \Gamma(R)^2} \right) \tag{4.11}$$

$$\varepsilon_2(\omega) = \frac{\omega_P^2}{\omega} \left(\frac{\Gamma(R)}{\omega^2 + \Gamma(R)^2} - \frac{\Gamma_\infty}{\omega^2 + \Gamma_\infty^2} \right) \tag{4.12}$$

where
$\Gamma_\infty = V_F / l_\infty$
$\Gamma(R) = \Gamma_\infty + A(V_F / R)$

In Equations 4.11 and 4.12, Γ_∞ refers to the damping constant in bulk metals. $\Gamma(R)$ assumes surface scattering as the only size-dependent contribution to the overall damping. This simplified approach may hold approximately since the electron–phonon relaxation is known to be different in metal nanoparticles. Restriction arises because, first, the phonon spectrum of nanoparticles differs from the bulk spectrum (Dickey and Paskin 1968), and, second, the external-field-induced surface charges give rise to strong electron–phonon coupling. In metal nanoparticles, this effect would be larger than that of bulk surfaces, due to the high surface-to-volume ratio. Electromagnetic field enhancement effects in surface-enhanced resonance Raman scattering from the plasmonic properties of metal nanoparticles or nanostructures have been observed (Bachelier and Mlayah 2004, Yoshida et al. 2010). The field enhancement in nanoscale metal particles becomes an important point of discussion, and will be discussed further in this chapter with a few illustrative examples.

Thus, the above-mentioned effects change the magnitude of the dielectric functions $\varepsilon_1(\omega)$ and $\varepsilon_2(\omega)$ of the nanoparticle materials. So, while describing the optical responses of such materials, it becomes essential to define a size-dependent optical material function $\varepsilon(\omega) = \varepsilon(\omega, R)$. Thus, using Equations 4.11 and 4.12, defined above, the size-dependent dielectric function, $\varepsilon(\omega, R)$, may be expressed as

$$\varepsilon(\omega, R) = \varepsilon_b + \omega_P^2 \left(\frac{1}{\omega^2 + \Gamma_\infty^2} - \frac{1}{\omega^2 + \Gamma(R)^2} \right) + \iota \frac{\omega_P^2}{\omega} \left(\frac{\Gamma(R)}{\omega^2 + \Gamma(R)^2} - \frac{\Gamma_\infty}{\omega^2 + \Gamma_\infty^2} \right) \tag{4.13}$$

where ε_b is the dielectric function of the bulk metal. Calculations of the size-dependent optical responses (i.e., optical absorbance) for nanoscale silver particles have been performed using the $\varepsilon(\omega, R)$ (Equation 4.13) and compared with the experimental results (discussed later in this chapter).

4.2.2.1.6 Quantum Surface Plasmon Resonances

According to the quasi-static response of Mie theory for small metal nanoparticles, as discussed, the optical resonance takes place whenever the condition $|\varepsilon + 2\varepsilon_m| \approx$ minimum is satisfied. So, the optical response of the nanoparticles is strongly dependent on dielectric functions of metal and the embedding medium. Although the Drude–Sommerfeld model accurately predicts the surface-plasmon spectra for larger metal nanoparticles ($2R > 10$ nm), the error increases for nanoparticles in the quantum-size (~2 nm) regimes. Because of the quantum-confinement effects and increased surface scattering/collisions of electrons with phonons or other electrons or with defects, motions of free electrons are affected. Thus, following the Mathiessen rule, a term that is inversely proportional to the size of a metal particle has been added to the surface scattering or damping constant of bulk metals. For example, mathematically, this is expressed as $\Gamma(R) = \Gamma_\infty + A(V_F/R)$, where A is an empirical dimensional parameter usually close to 1, which used to account for factors affecting the width of the SPR in specific cases. The $1/R$ term in the expression for $\Gamma(R)$ actually reflects the ratio of the surface scattering probability (which is proportional to the surface area, $\sim\pi R^2$) and the number of electrons (being proportional to the volume of the particle, $\sim\pi R^3$).

To understand the optical responses of quantum-size metal nanoparticles, different models have been proposed. For example, based on the quasi-static optical polarizability of a sphere embedded in a homogeneous medium, Raza et al. proposed a nonlocal hydrodynamic model and a generalized local model incorporating the inhomogeneity of the electron density induced by the quantum wave nature of the electrons (Raza et al. 2013). However, here we discuss more about the well-established analytic model (Thomas–Reiche–Kuhn energy-weighted f-sum rule) that is commonly used for absorption processes. In very fine nanoparticles, due to quantum-confinement effects, the energy levels in the conduction band are discrete and only specific electronic transitions are allowed. Thus, considering quantum mechanical effects, Scholl et al. started with the modified Drude model to account for these transitions (Scholl et al. 2012). They modeled the conduction electrons as a free-electron gas constrained by infinite potential barriers at the physical boundaries of the particle. The transition frequencies ω_{if} correspond to the allowed quantum energies of transitions of conduction electrons from occupied states i within the k-space Fermi sphere to unoccupied states f immediately outside it. Here, the technique is employed in the form of oscillator strengths (S_{if}), which correspond to each transition frequency. The oscillator strength terms, as dictated by the f-sum rule, are described using the standard harmonic oscillator quantum mechanical definition

$$S_{if} = \frac{2m\omega_{if}}{\hbar N} \left| \langle f|z|i \rangle \right|^2 \tag{4.14}$$

where N is the number of conduction electrons in the nanoparticle. The matrix element term is determined by the allowed wave functions of the spherical-well model. In total, the particle's dielectric function can be expressed as

$$\varepsilon(\omega) = \varepsilon_{IB} + \omega_P^2 \sum_i \sum_f \frac{S_{if}}{\omega_{if}^2 - \omega^2 - \iota\Gamma\omega} \tag{4.15}$$

where the sum is taken over all initial and final states of the electrons. The frequency of the transition from the occupied to excited states can be described as $\hbar\omega_{if} = E_f - E_i$. The energy levels E_f and E_i depend on the geometry and potential of the system. Treating the conduction electrons as particles in an infinite spherical well, the energy eigenvalue may be calculated as $E = (\hbar^2\pi^2/8mR^2)(2n + l + 2)^2$, where m is the mass of the electron, and n and l are the principal and azimuthal quantum numbers, respectively. Figure 4.3a shows the real and imaginary parts of the dielectric function, accounting for both the quantum-confinement effects and the surface scattering of free electrons. Here, the damping term was allowed to vary in accordance with the expression $\Gamma = \Gamma_\infty + A(V_F/R)$. The calculation assumed the value of A to be 0.25. Using the dielectric function values (shown in Figure 4.3a) as input to the electromagnetic Mie theory, the absorption efficiencies (absorption cross-sections divided by physical cross-sections) of silver nanoparticles with diameters 2, 4, 6, and 8 nm were calculated. Figure 4.3b shows the results, revealing decreasing absorption efficiency and blue-shifting of the SPR peaks as the silver particle size decreases. Afterward, while discussing the optical responses of fine silver nanoparticles in a silica glass matrix, we will compare the observed experimental results with the theoretical calculations and comment on it.

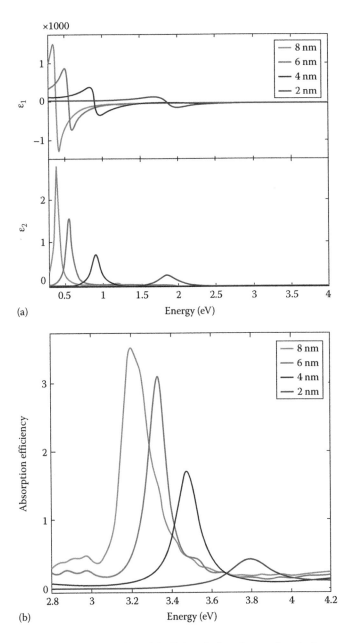

FIGURE 4.3 Analytic quantum theory of particle dielectric functions and optical spectra. (a) Real (ε_1) and imaginary (ε_2) components of the dielectric functions of Ag nanoparticles with diameters of 2, 4, 6, and 8 nm, calculated using the analytic quantum model (Equation 4.15). (b) Corresponding absorption spectra of the particles generated by Mie theory. The refractive index of the surrounding medium is set to 1.3. As the particle size decreases, absorption efficiencies reduce and surface-plasmon peak energies get blue-shifted. (Reprinted by permission from Macmillan Publishers Ltd. *Nature*, Scholl, J.A. et al., Quantum plasmon resonances of individual metallic nanoparticles, 483, 421, © 2012.)

4.2.2.2 Vibrational Spectroscopy of Nanoparticles

Solids in the form of tiny particles or consisting of nanoscale materials have particular interesting properties originating from confinements of lattice vibrations (phonons) and surface effects. A considerable amount of research (theory and experiments) has been reported during the past years regarding the surface acoustic confinements of phonons in nanomaterials (Bottani et al. 2001, Saviot et al. 2011). Although other vibrational spectroscopic studies, for example, infrared absorption (Murray et al. 2006) or neutron scattering (Saviot et al. 2008), have been carried out with a limited success, inelastic low-frequency Raman scattering (LFRS) measurements for such nanomaterials embedded in different amorphous matrices have provided good understanding about the confinement of acoustic phonons in the nanomaterials. The calculated sizes of nanoparticles using this nondestructive experimental technique have shown good agreement with direct and standard measurement procedures (e.g., transmission electron microscopy). The LFRS technique will be elaborated in the next section with illustrative examples of nanoscale metal particles embedded in different dielectric matrices. A theoretical description on the vibration of spherical bodies (arising out of Lamb's theory) will be provided here, though briefly. Nanoscale materials are generally considered as elastic bodies of spherical or ellipsoidal shape. According to Lamb's theory, two types of vibrational modes may exist for such elastic bodies: torsional and spheroidal (Lamb 1882). For the low-energy modes, torsional motion of a spherical surface can be visualized as two different hemispheres rotating in opposite directions (as shown in Figure 4.4). The torsional modes correspond to movement without any volume change. Torsional displacements induce shear. On the contrary, spheroidal oscillations (e.g., breathing modes, shown in Figure 4.5) are with dilation (i.e., with volume change). According to group-theoretical analysis, low-frequency Raman scattering from spherical nanoparticles are possible in the quasi-static size limit. Further, the theory also confirms that the observable Raman transitions are possible in this class of materials for radial ($l = 0$) as well as quadrupolar ($l = 2$) modes (Duval 1992). The characteristic vibrational frequencies associated with the spheroidal modes in nanoscale particles are in the range of few cm^{-1} to a few tens of cm^{-1}. In the vibrational spectra of nanomaterials, low-frequency Raman modes are usually observed in this range of frequencies (Duval et al. 1986).

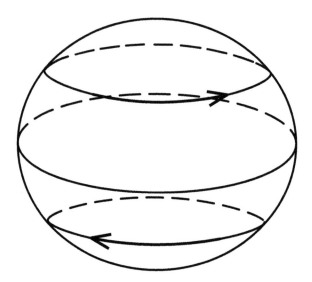

FIGURE 4.4 Relative displacements of low-energy torsional modes in a spherical body. Torsional motion of a spherical surface is visualized as two different hemispheres rotating in opposite directions without any volume change. (Reprinted from Duval, E., *Phys. Rev. B*, 46, R5795, 1992. With permission.)

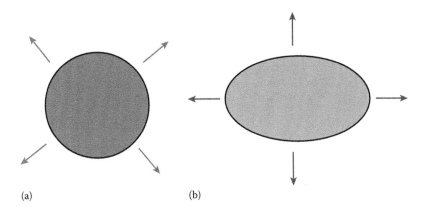

FIGURE 4.5 Different types of spheroidal oscillation modes confined to the surface of nanoscale particles. (a) Radial mode. (b) Quadrupolar mode. The spheroidal oscillations correspond to movement with volume change of materials.

4.2.2.2.1 Calculations of Vibrational Frequencies

Lamb (1882) and Tamura et al. (1982) have calculated the vibrational frequencies (Ω) of a homogeneous, elastic spherical particle under no-stress condition. The particle is free to vibrate. These calculations provide the size dependence of the peak frequency of vibrational eigen frequencies for an elastic spherical body. Following the notations used by Tamura et al., the eigenvalue equation can be derived for the spheroidal modes ($l \geq 0$) as

$$2\left[\eta^2 + (l-1)(l+2)\left\langle\frac{\eta j_{l+1}(\eta)}{j_l(\eta)} - (l+1)\right\rangle\right]\frac{\zeta j_{l+1}(\zeta)}{j_l(\zeta)} - \frac{\eta^4}{2} + (l-1)(2l+1)\eta^2$$

$$+[\eta^2 - 2l(l-1)(l+2)]\frac{\eta j_{l+1}(\eta)}{j_l(\eta)} = 0 \tag{4.16}$$

where
    $\eta = \Omega R/V_t$ and $\zeta = \Omega R/V_l$ are the nondimensional eigen frequencies
    l is an angular quantum number
    $j_l(\eta)$ is the spherical Bessel function of the first kind

The eigenvalue equation (Equation 4.16) is solved by setting the material parameter V_t/V_l, where V_l and V_t are the longitudinal and transverse sound velocities, respectively. The nondimensionalized eigen frequencies (η_{ln}^s and ζ_{ln}^s) of the spheroidal modes strongly depend on the material through the ratio V_t/V_l. For any set of l and n, it may be noted here that the eigen frequencies η_{ln}^s and ζ_{ln}^s are not independent but are related to each other by the relation $\eta/\zeta = V_t/V_l$. Only the lowest vibration modes ($n = 0$) are the surface modes, whereas the higher modes ($n \geq 1$) are the inner modes of a vibrating elastic spherical particle. Here, η_{ln}^s are the $(n + 1)$th eigenvalues corresponding to the angular momentum l. Group-theoretical analysis predicts that spheroidal modes with $l = 0$, 2 alone are Raman active. These surface acoustic vibrations are known as spheroidal quadrupolar ($l = 2$) modes and spheroidal breathing or radial ($l = 0$) modes. In agreement with the Raman selection rules, quadrupolar ($l = 2$) modes are to be observed when the polarizations of the incident and Raman scattered light are crossed. For breathing or radial ($l = 0$) modes, Raman intensity vanishes in cross-polarized conditions.

The eigen frequencies for surface modes ($n = 0$) for various metals (with suitable values of V_t/V_l) have been calculated numerically (using MATLAB®) by employing Equation 4.16. Knowing the sound velocity ratio (for example, $V_l/V_t = 2.19$ and 1.89 for silver and cobalt metals, respectively, Simmons and Wang 1971), the vibrational eigen frequencies ($\eta_{\ln}^s$ and $\zeta_{\ln}^s$) of these metal nanoparticles have been calculated. The size of nanoparticles can be calculated from these eigen frequency equations:

$$\eta = \frac{\Omega R}{V_t}, \quad \zeta = \frac{\Omega R}{V_l} \tag{4.17}$$

The experimentally measured Raman frequencies have been used in Equation 4.17 to determine average sizes of nanoparticles of interest. This method has been used to study coarsening or thermal growth of nanoparticles in different dielectric matrices. Let us discuss the experimental Raman results with a few examples in this context.

4.2.2.2.2 Illustrative LFRS Studies

4.2.2.2.2.1 Silver Nanoparticles
For example, dissolution of pristine Ag nanoparticles in a silica glass matrix due to the Si ion irradiations (see Srivastava et al. 2014 for details) has been studied by the inelastic LFRS measurements. Vibrational spectra of nanomaterials appear in the low-frequency region (from few cm^{-1} to few tens of cm^{-1}). The recorded low-frequency Raman spectra for Ag nanoparticles in a silica glass matrix are displayed in Figure 4.6. In this study, the surface acoustic vibrations corresponding to quadrupolar ($l = 2$) spheroidal modes have been observed. Te eigen frequency (η_2^s) corresponding to the quadrupolar mode is calculated from Equation 4.16, which works out to be ~2.65. Te vibrational mode frequency has been obtained after substituting the value for η in Equation 4.17. Experimentally, this vibration frequency is measured from

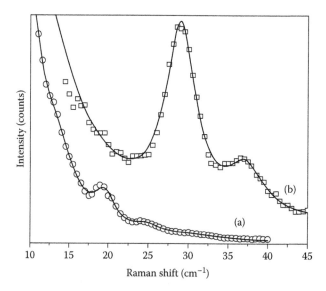

FIGURE 4.6 Polarized low-frequency Raman spectra of Ag nanoparticles of different sizes in the SiO$_2$ matrix. (a) Average size of nanoparticles 7.7 nm. (b) Average size 2.7 nm. Solid lines are the fittings through the data points (shown as symbols). Compared to spectra (a), shifts of Raman modes to higher frequencies (spectra (b)) verify the decrease in size of Ag nanoparticles. The estimated sizes agree with those from direct HRTEM measurements. (Reprinted from Srivastava, S.K. et al., *Chem. Phys. Lett.*, 607, 100, 2014. With permission.)

low-frequency Raman scattering measurements. The Raman peak frequency (v_2^s, in cm^{-1}) of the spheroidal mode corresponding to the quadrupolar vibration mode may thus be expressed as

$$v_2^s = 0.85 \frac{V_t}{dc} \tag{4.18}$$

where
   d is the diameter of silver nanoparticles
   V_t is the transverse sound velocity (1660 m s^{-1}) in bulk silver
   c is the velocity of light in vacuum

By measuring the Raman peak frequencies, the sizes of the silver nanoparticles have been estimated using Equation 4.18. In this case, the experimental data have been fitted to a Lorentzian line-shape function with an exponential Rayleigh background to estimate the Raman peak frequencies. The directly measured values (from the high-resolution transmission electron microscopy) agree well with the results obtained from the LFRS measurements.

Equation 4.18 further demonstrates that the vibration frequency is inversely proportional to the diameter of the nanoparticles. As a result, with the growth of the nanoparticles, the Raman mode frequency would further shift to lower frequencies. This condition, in particular, makes it difficult to measure sizes of larger nanoparticles through the LFRS technique. For example, typically, nanoparticles of sizes <10 nm may be measured. For larger nanoparticles, the Raman frequency shift usually merges with the huge background of Rayleigh scattering (elastic), which makes LFRS measurements, and thus obtaining the vibrational spectra of larger nanomaterials, difficult.

4.2.2.2.2.2   Cobalt Nanoparticles   LFRS studies have been carried out on cobalt nanoparticles in a silica glass matrix. For more details, see Gangopadhyay et al. (2007). Apart from finding the size of cobalt nanoparticles, the study also focused on elucidating the coupling of electrons and phonons in nanoscale metals. Raman spectra were recorded using different excitation wavelengths of the argon ion laser. As shown in Figure 4.7, intensity of the low-frequency vibrational modes depends significantly on the wavelength of laser. The strong local field due to the SPR in a cobalt nanoparticle with 351.1 nm excitation is expected to be the reason for the observed enhancement of Raman scattering intensity. As the characteristic SPR absorption for cobalt nanoparticles in a silica matrix is at ~350 nm (see Figure 4.8), the increase in the intensity of vibrational mode at 26.5 cm^{-1} with the excitation wavelength in the vicinity of SPR is thought to be due to an enhanced electric field in the nanoparticles of cobalt atoms. The results thus obtained from the low-frequency Raman scattering spectroscopy experiments emphasize the importance of surface plasmon–phonon coupling in metallic nanoparticles.

   As shown previously, strong light scattering is achieved with excitation close to the SPR wavelength. With the resonance wavelength, mainly spherical nanoparticles are excited. The dependence of Raman scattering on the incident and scattered photon polarizations can be used to characterize the observed mode according to the Raman selection rule. Since the scattering due to pure radial vibrations is totally polarized, the Raman mode observed at 26.5 cm^{-1} should disappear in depolarized spectra. To corroborate this, cross-polarized Raman measurements have been carried out, and the polarization dependence of the Raman spectra is displayed in Figure 4.9. The Raman mode observed in the polarized spectra has disappeared in the depolarized configuration. Measurements thus confirm the mode of vibration to be of radial ($l = 0$) type in accordance with the vibrational selection rule (Duval 1992). Assuming a homogeneous elastic sphere with a free surface, the nondimensionalized eigen frequency corresponding to the radial mode has been worked out to be ~2.46. The assumption of free vibration is justified because sound velocity and mass density values are

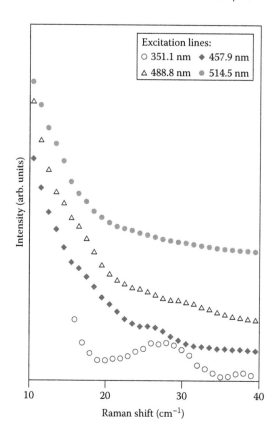

FIGURE 4.7 Enhancement of intensity of the surface-acoustic vibrational modes confined to the Co nano-particles in the SiO$_2$ matrix as the excitation wavelength from the argon-ion laser source is varied. Spectra are offset in the intensity axis for a better clarity. (Reprinted from Gangopadhyay, P. et al., *Appl. Phys. Lett.*, 90, 063108, 2007. With permission.)

very much different in cobalt compared to those in the silica glass matrix. Thus, the Raman peak frequency (ν_o^s) corresponding to the radial surface modes may be expressed as

$$\nu_o^s = 0.78 \frac{V_l}{dc} \tag{4.19}$$

where

    d is the diameter of the cobalt nanoparticles
    V_l is the longitudinal sound velocity (5856 m s^{-1}) in bulk cobalt metal
    c is the velocity of light in vacuum

From the measured Raman mode frequencies, sizes of the nanoscale cobalt particles in the silica glass matrix have been estimated using Equation 4.19.

Although confined surface-acoustic phonons in nanoscale metal particles can give rise to radial (symmetric) and quadrupolar Raman modes, only quadrupolar modes are usually observed in low-frequency Raman experiments. Quadrupolar modes are easily detected in Raman scattering experiments. This is probably because the radial modes ($l = 0$), for which size changes preserve the shape, produce weak modulations of the electric dipole (Gersten et al. 1980). However, in addition to the quadrupolar mode, radial modes have also been observed by other groups of

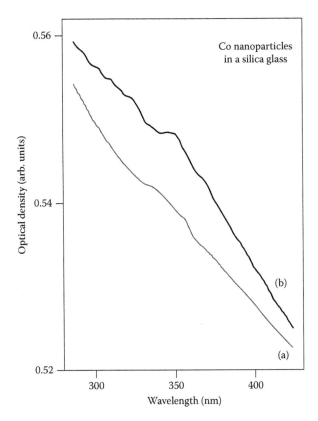

FIGURE 4.8 Optical absorption spectra of Co ion–implanted (fluence 2×10^{16} ions cm^{-2}) silica glass samples. (a) Before and (b) after annealing in high vacuum (for 573 K, 90 min). Absorbance has increased (spectra b) due to thermal growth of the Co nanoparticles in the matrix. (Reprinted with permission from Gangopadhyay, P., *AIP Conference Proceedings*, Bikaner, Rajasthan, India, Vol. 1536, p. 7. © 2013, American Institute of Physics.)

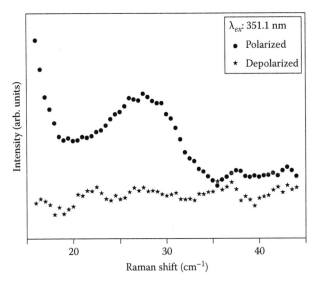

FIGURE 4.9 Polarization dependence of low-frequency Raman spectra of Co nanoparticles embedded in the silica glass matrix, confirming the vibration mode of the nanoparticles to be of radial type. (Reprinted from Gangopadhyay, P. et al., *Appl. Phys. Lett.*, 90, 063108, 2007. With permission.)

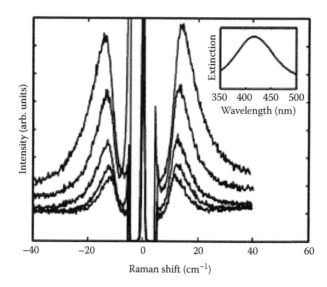

FIGURE 4.10 Raman spectra measured on Ag particles with 1.4 nm mean radius and excited at 413, 458, 476, 488, and 514 nm (from top to bottom). The corresponding extinction spectrum showing the surface-plasmon resonance is plotted in the inset. (Reprinted from Bachelier, G. et al., *Phys. Rev. B*, 76, 235419, 2007. With permission.)

researchers in their resonant Raman scattering experiments (Portales et al. 2001, Courty et al. 2002). During these measurements, the quadrupolar mode appeared very close to the direct Rayleigh (elastic scattering) line. Compared to this, radial modes were observed with lower intensities and relatively far away from the Rayleigh line. Bachelier and Mlayah have explained these experimental observations after taking into account of various coupling mechanisms between confined acoustic vibrations and surface plasmon–polariton states (Bachelier and Mlayah 2004). In another experimental study, the maximum Raman scattering intensity was observed from fine silver nanoparticles while exciting with a line close to the SPR absorption wavelength of the metal nanoparticles (see Figure 4.10).

4.3 SYNTHESIS OF METAL NANOPARTICLES

Nanomaterials may be prepared in one dimension (e.g., very thin surface coatings), in two dimensions (nanowires and nanotubes), or in three dimensions (quantum dots and nanoparticles). Some selected methods for the synthesis of metal nanoparticles are enlisted in Table 4.1. Depending on the processing conditions, preparations are generally classified into bottom-up or top-down approaches. As the name suggests, in bottom-up methods, the nanomaterials are grown from the state of atoms or ions of the materials. Growth behaviors differ depending on the processing conditions. We can observe this while dealing with different methods of preparations. On the contrary, top-down processing starts with bulk materials. The top-down techniques (mechanical attrition, lithography, or ion-beam sputtering) gradually transform the bulk into different forms of nanostructures. We will elaborate both synthesis approaches later on with a few examples. It may be restated that precise control of the synthesis of nanomaterials is very important as many of the interesting properties depend crucially on their size, shape, and uniformity. Controlling the level of impurities is again vital all through the processing using chemical methods, for example. Necessarily, this requires a good understanding of the science of the synthesis of nanomaterials.

TABLE 4.1

Selected Methods for the Synthesis of Metal Nanoparticles

Methods of Synthesis	Materials	References
Mechanical attrition (e.g., ball milling)	Fe nanoparticles	Muñoz et al. (2007)
Nanosphere lithography	Ag nanostructures on various substrates	Haynes and Van Duyne (2001)
Dip-pen nanolithography	Supported Au nanostructures	Zhang et al. (2003)
Ion exchange followed by thermal annealing, laser irradiation, light ion irradiation	Ag nanoparticles in a soda glass matrix	Gangopadhyay et al. (2005), Miotello et al. (2001), and Magudapathy et al. (2001)
Ion implantation (direct and recoil) techniques	Ag nanoparticles in a SiO_2 glass matrix; Au nanoparticles in sapphire	Srivastava et al. (2014), Gangopadhyay et al. (2000), and Zhou et al. (2012)
Ion-beam sputtering	Pd nanostructures on SiO_2; Ag and Au nanoparticles on SiO_2 glass samples	Ruffino et al. (2009b) and Gangopadhyay et al. (2010)
Melt-quenching technique	Cu nanoparticles in barium phosphate glass	Sendova et al. (2015)
RF co-sputtering	Ag nanoparticles in a SiO_2 glass matrix	Fujii et al. (1991)
Ion-beam mixing	Alloy nanoparticles (Co–Pt) in a-SiO_2	Balaji et al. (2014)
Pulsed-laser deposition	Au nanoparticles in Al_2O_3, Co nanoparticles in ZrO_2 and Al_2O_3	Gonzalo et al. (2005) and Clavero et al. (2007)
Atomic layer deposition	Alloy (Ru–Pt) nanoparticles	Kim et al. (2009)
Gas-phase preparation	Ag–PbS nanocomposites	Maisels et al. (2000)
Microemulsion technique	Alloy (Co–Pt) nanoparticles	Kumbhar et al. (2001)
Spark discharge technique	Au–Ge nanocomposites	Kala et al. (2013)

4.3.1 BOTTOM-UP METHODS

Bottom-up methods are more prevalent for the preparation of nanomaterials (see Table 4.1). Precise control on the nucleation and growth of metal nanoparticles through the bottom-up synthesis is of prime importance. Most of the novel optoelectronic properties of nanocomposite materials depend on physical dimensions and morphologies of these entities (Stamplecoskie and Scaian 2010). To achieve a good control on the preparation, a basic understanding of mechanisms governing the formation and evolution of metal nanoparticles while processing these materials is a must. Hence, attention is drawn first to elucidating the nucleation and growth characteristics of metal nanoparticles through thermal annealing of ion-implanted samples, for example. In addition, the thermochemical stability of optical materials of this kind is an important and pertinent issue to tailor the nanomaterials to suit technological demands. Ion implantation and ion exchange (chemical route) will be discussed thoroughly to elucidate the nucleation and growth of nanoparticles in different matrices. These methods are taken as illustrative examples of bottom-up approaches.

4.3.1.1 Ion Implantation

Metal particles nucleate when there is a supersaturation of the metal atoms in a given matrix. The only prerequisite is that the metal should be insoluble in the host matrix. By ion implantation, for example, metal atoms are introduced into a dielectric matrix at concentrations well above the solubility limit. Depending on the energy of metal ions, particles may be formed at different depths in the given matrix. Being a nonequilibrium process, good combinations of ion projectiles and target matrices are feasible. The bottom-up technique provides chemically clean nanoparticles with reasonable sizes and depth distributions in a matrix. Many researchers have used ion implantation/irradiation to control and tailor the size and spatial distributions of metal nanoparticles (Ren et al. 2006, Rizza et al. 2007, Bernas 2010, Ramjauny et al. 2010). Ion implantation techniques are thus highly suitable to study the nucleation and growth of metal nanoparticles in insoluble matrices (Mazzoldi and Mattei 2005, Bernas 2009). In fact, ion beams are also used as a top-down approach to prepare plasmonic nanostructures on a substrate through sputtering (Gangopadhyay et al. 2010). The basic processes behind the nucleation and growth of nanoparticles formed through the ion implantation are shown schematically in Figure 4.11. Supersaturation of metal atoms in the glass matrix leads to the nucleation of metal particles in the form of fine precipitates. According to the theory of precipitate growth (Lifshitz and Slyozov 1961, Wagner 1961), particles with radii exceeding a critical radius can grow during further implantation of the ions or during subsequent thermal annealing, and lead to the growth or coarsening of the nanoparticles. It may take place following three different stages. In the first stage, nuclei of the precipitates are formed with an average radius equal to the critical value. Homogeneous or heterogeneous nucleation may occur, but in an ion-implanted system, where copious defects are produced, it is likely that the implantation-produced defects may act as nucleating centers (i.e., heterogeneous nucleation). In the second stage, diffusional growth of the precipitates occurs by solute depletion of the surrounding matrix. Here the growth is governed by the diffusion coefficient of the precipitating element, and, more importantly, the average radius of the particle scales as $\sqrt{Dt}$, where D is the diffusion constant and t is the time of diffusion. In the third stage, the coarsening regime, the particle distribution established during the nucleation coarsens by the growth of larger nanoparticles. In this process, growth of larger nanoparticles occurs at the expense of smaller ones. It is a slow process. Here, the average radius of the growing particle scales as $(Dt)^{1/3}$. This coarsening stage is known as Ostwald ripening, where the mass of the material is conserved. The nucleation

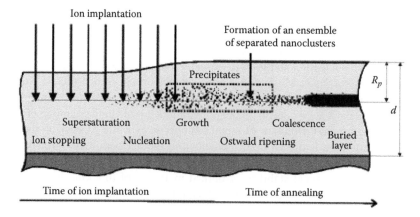

FIGURE 4.11 Ion-beam synthesis of nanostructures shown schematically. High fluence of ion implantation into a solid leads to supersaturation of the impurity atoms. Nanoscale particles nucleate and grow during ion implantation or during the subsequent thermal annealing. Average size of the nanoparticles as well as their spatial and size distributions change during the Ostwald ripening. At very high fluences, buried layers can form by coalescence of the embedded nanoparticles. (Reprinted from Reiss, S. and Heinig, K., *Nucl. Instrum. Methods Phys. Res. B*, 102, 256, 1995. With permission.)

and growth of nanoparticles of the implanted atoms take place through controlled post-annealing processes. The size of the embedded nanoparticles can be altered by adjusting the annealing parameters and the implantation fluences (concentration of the implanted atoms). In fact, one may end up with a continuous buried layer (see Figure 4.11) with very a high fluence of implanted ions or at a high annealing temperature.

Now let us discuss a few examples. First, we consider the case where Ag atoms are recoil-implanted in a silica glass matrix. During this study, the top Ag layer (~45 nm thick) of the Ag/SiO$_2$ sample was irradiated at room temperature by argon ions (energy 105 keV, total fluence 1×10^{16} ions cm^{-2}). Because of the energetic impacts of the argon ions, Ag atoms get recoil-implanted in the silica glass. After the ion irradiation, the unmixed Ag layer is removed from the surface of the silica substrates by dipping it for 2 min in a hydrochloric acid bath. The silica matrix embedded with Ag atoms was post-annealed at different temperatures between 333 and 523 K for 1 h at each annealing step. Optical absorption and Raman scattering spectroscopy carried out on these samples revealed thermal growth of Ag nanoparticles in the matrix (Gangopadhyay et al. 2000). According to Mie theory, optical absorption increases with increase in the volume of the nanoparticles (Equation 4.3). Precisely, this is what has been observed (see Figure 4.12). The Raman scattering results are shown in Figure 4.13. The systematic shift of the Raman mode positions to lower frequencies with the increase of post-annealing temperature may be noted. As discussed earlier, vibration frequency scales inversely with the diameter (Equation 4.18) of nanoparticles. Raman measurements also confirm the continual thermal growth of the embedded Ag nanoparticles in the matrix. Thus, from the low-frequency Raman scattering spectroscopy experiments, the size of Ag nanoparticles was measured, which was in the range 1–4 nm. The sizes of Ag nanoparticles are seen to increase with increase of the annealing temperature (see Figure 4.13).

Here, two stages are identified with the nucleation and growth of Ag nanoparticles in the silica glass matrix. First, because of concentration fluctuations, nucleation of Ag atoms takes place, and

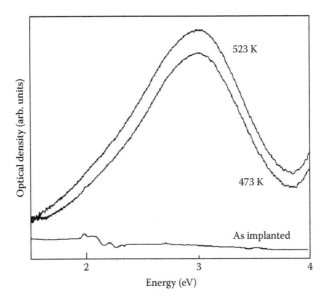

FIGURE 4.12 Optical absorption spectra for Ag nanoparticles embedded in a silica glass matrix, prior to thermal annealing and after isochronal annealing for 1 h at 473 and 523 K. Coarsening of Ag nanoparticles increases the optical absorption intensity. (Reprinted from Gangopadhyay, P. et al., *J. Appl. Phys.*, 88, 4975, 2000. With permission.)

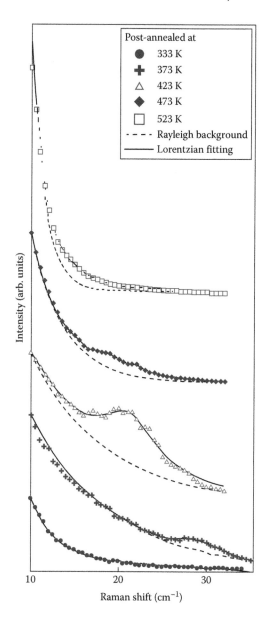

FIGURE 4.13   Polarized low-frequency Raman spectra of Ag nanoparticles in the SiO_2 glass matrix. Post-annealing temperature is indicated for each spectrum. Symbols are data, and solid lines are the Lorentzian fitting of data with the Rayleigh background (shown as dashed lines). (Reprinted from Gangopadhyay, P. et al., *J. Appl. Phys.*, 88, 4975, 2000. With permission.)

these nuclei grow directly as a result of the supersaturation. In the second stage of the process, nuclei that have reached a critical size grow according to the coarsening or Oswald ripening mechanism. In the coarsening regime, nucleation of new particles is negligible. According to the theory of ripening (Lifshitz and Slyozov 1961, Wagner 1961), if the total mass of clustering matter is conserved, the following growth behavior may be observed at asymptotic times:

$$R_t^3 - R_o^3 = \frac{4}{9} D\alpha t \qquad (4.20)$$

where

R_t (after annealing for time *t*) and R_o (prior to annealing) are the average radii of the nanoparticles

D is the diffusion constant of clustering atoms in the matrix

t is the time of annealing

α characterizes the kinetics of precipitation

By applying this growth formula (Equation 4.20), the activation energy for the diffusion of Ag atoms in the silica glass matrix was determined from these experimental results (Gangopadhyay et al. 2000).

Moving on to the study of the higher-order modes of optical responses of metal nanoparticles, next we consider an experimental case where Ag nanoparticles are formed in a plain soda glass matrix. Implantation of Ag^+ ions of 1.0 MeV energy with various fluences (maximum fluence being 5×10^{16} ions cm^{-2}) was carried out using the Tandetron (1.7 MV, High Voltage Engineering Europe (HVEE)) accelerator at Materials Science Group (MSG), Indira Gandhi Centre for Atomic Research (IGCAR), for the synthesis of Ag nanoparticles in a soda glass matrix. Although dipolar SPR is predominant in the recorded optical absorption spectra (see Figure 4.14), sample with the maximum fluence of 5×10^{16} Ag^+ ions cm^{-2} shows some interesting features (spectra "d", Figure 4.14). In this particular sample, Ag nanoparticles of sizes ~100–200 nm have been observed (SEM image, Figure 4.15). In the sample with the highest fluences of Ag^+ ions, defect-enhanced mobility of Ag atoms in the soda glass led to segregation of Ag atoms close to the glass surface. Higher mobility of Ag atoms on the glass surface further helps the growth of Ag nanoparticles. Large metal nanoparticles play a dominant role in modifying their optical properties. As observed here, compared to other samples, the intensity and width of the SPR absorption at 420 nm are greatly modified for the sample with the maximum ion fluence. Phase retardation of light and additional

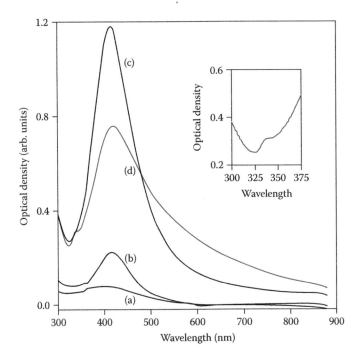

FIGURE 4.14 Optical absorption spectra for Ag^+ ion–implanted soda glass samples with various fluences: (a) 5×10^{15}, (b) 1×10^{16}, (c) 3×10^{16}, and (d) 5×10^{16} ions cm^{-2}. Quadrupolar response for larger Ag nanoparticles is shown in the inset. (Reprinted with permission from Gangopadhyay, P., *AIP Conference Proceedings*, Bikaner, Rajasthan, India, Vol. 1536, p. 7. © 2013, American Institute of Physics.)

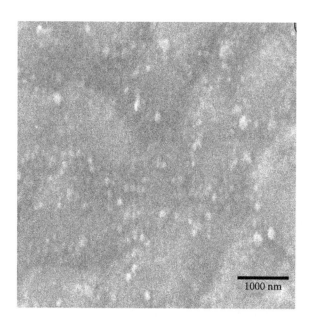

FIGURE 4.15 Scanning electron microscope (SEM) image showing large Ag nanoparticles that appear close to the surface of soda glass. Various sizes of Ag nanoparticles appear as bright dots in the image. (Reprinted with permission from Gangopadhyay, P., *AIP Conference Proceedings*, Bikaner, Rajasthan, India, Vol. 1536, p. 7. © 2013, American Institute of Physics.)

higher-order optical resonances have become particularly important in this sample. Because of the nonzero field gradient across large Ag nanoparticles in the glass matrix, phase retardation may lead to larger widths of the optical resonances. Further, the optical spectra reveal a shoulder on the low-wavelength side (~340 nm) of the dipolar peak (inset, Figure 4.14) for the sample with larger Ag nanoparticles. This is assigned to the quadrupolar SPR absorption in these Ag nanoparticles (Kreibig and Vollmer 1995).

Next we discuss the optical absorption and growth of Au nanoparticles in a silica glass matrix (De Marchi et al. 2002). Interestingly, the study has attempted to identify the difference between a pure diffusional and a coarsening growth stage of the precipitation process. From the analysis of experimental results, two distinct growth regimes have been observed, namely $t^{1/2}$ and $t^{1/3}$ (as shown in Figure 4.16), which correspond to growth due to precipitation of supersaturated solution and coarsening, respectively. Optical absorption spectra of Au ion–implanted silica samples at different annealing times in air atmosphere are shown in Figure 4.17. The characteristic dipolar SPR near 530 nm, due to Au nanoparticles in the matrix, is evident. A gradual increase of optical absorption intensity around the SPR wavelength from the as-implanted to the 12 h-annealed sample has been observed. This is expected with the thermal growth of the Au nanoparticles (according to Mie theory, Equation 4.3).

4.3.1.2 Ion-Exchange Processing

Next we describe the experimental observations on the nucleation and growth of Ag nanoparticles, for instance, in plain soda glass samples. Ion exchange of various metal ions (silver, copper, gold) in soda glasses followed by light ion irradiation is an established method to modify the linear and nonlinear optical properties of glasses. The method is advantageous because it is easy to prepare and commercially viable for applications in optoelectronic devices. Silver ion-exchange processing ($Ag^+ \leftrightarrow Na^+$) in the soda glass occurs because the difference in ionic radii of Ag^+ and Na^+ ions is ~29%.

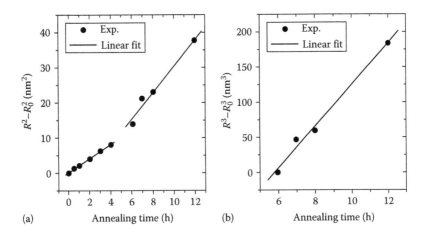

FIGURE 4.16 (a) $[R^2(t) - R_0^2]$ and (b) $[R^3(t) - R_0^3]$ evolutions in Au-implanted silica glass samples annealed in air at 1173 K for different time intervals. Solid lines are the linear fits to the experimental data (filled circles) from TEM and optical fit results. (Reprinted with permission from De Marchi, G. et al., Two stages in the kinetics of gold cluster growth in ion-implanted silica during isothermal annealing in oxidizing atmosphere, *J. Appl. Phys.*, 92, 4249. © 2002, American Institute of Physics.)

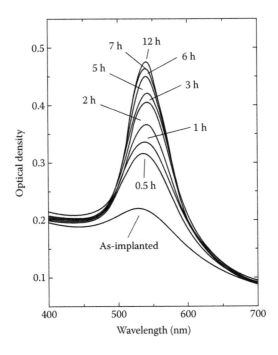

FIGURE 4.17 Optical absorption spectra of Au-implanted silica glass samples (at fluence 4×10^{16} Au^{+1} cm^{-2}, energy 190 keV), annealed in air at 1173 K for different time intervals. (Reprinted with permission from De Marchi, G. et al., Two stages in the kinetics of gold cluster growth in ion-implanted silica during isothermal annealing in oxidizing atmosphere, *J. Appl. Phys.*, 92, 4249. © 2002, American Institute of Physics.)

Composition of the glass is Si 21.49%, Na 7.1%, Ca 5.78%, Mg 0.34%, and Al 0.15% by weight. Silver-exchanged glass samples are prepared by immersing preheated soda glass slides for 2 min in a molten-salt bath of $AgNO_3$ and $NaNO_3$ (1:4 weight ratio) mixture at 593 K.

Subsequent evolution of Ag nanoparticles in the ion-exchanged soda glass samples occur under different processing conditions: (1) irradiation with light ions (Arnold et al. 1996), (2) thermal annealing (Wang 1996), (3) laser irradiation (Miotello et al. 2001), and (4) gamma irradiation (Farah et al. 2014). In this chapter, most of our discussions would be based on the first two processes. Let us first discuss the case of light-ion irradiation, where 100 keV He^+ ions bombard the silver-exchanged soda glass samples with a low beam current density (~1 μA cm^{-2}) to minimize the ion beam's heating effect. The ion-irradiated samples exhibit an optical absorption peak at ~415 nm (Figure 4.18a). This characteristic absorption is due to the SPR in silver nanoparticles in the glass matrix. In this case, defects due to the ion irradiation are expected to mediate the diffusion of silver ions in the ion-exchanged soda glass samples and control the formation and growth of silver nanoparticles. The effect due to the ion-beam heating might be less significant. Here, optical absorption results are primarily looked into, as they show interesting features. For example, optical absorbance increases and, more importantly, the absorbance band broadens considerably with the increase of helium ion fluence (Figure 4.18). The resonance peak at 415 nm apparently vanishes and the spectra appear flat over a range of optical wavelengths with higher fluences. It may imply that SPR occurs over a broad range of photon energies in the samples irradiated with higher fluences (Figure 4.18c and d). The systematic increase of optical absorbance with the increase of helium ion fluences in samples confirms that the volume fractions of the nanoparticles have increased. It is inferred from detailed studies, reported elsewhere (Magudapathy et al. 2001), that the sizes of silver nanoparticles remained unchanged but number densities have increased with the increase of helium

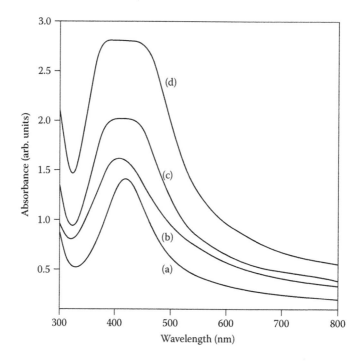

FIGURE 4.18 Optical absorption spectra for the nanoscale Ag particles embedded in the ion-exchanged soda glass samples irradiated with the 100 keV He^+ ions. Irradiation was carried out on Ag ion–exchanged soda glass samples with various fluences of He^+ ions: (a) 1×10^{16}, (b) 5×10^{16}, (c) 5×10^{17}, and (d) 1×10^{18} ions cm^{-2}. Optical absorbance has increased with the increase of ion fluence. Optical band-broadening for the samples with higher fluences is seen.

ion fluence. The reason for this additional broadening cannot be explained within the framework of Drude's free-electron model or from Mie scattering theory. These theories explain the optical absorption results for noninteracting and a low volume fraction of nonoverlapping metal nanoparticles. However, now we understand that with increase in the number density of silver nanoparticles, multiple scattering events among the silver nanoparticles play an important role with respect to their optical response functions. In this regard, a recent study based on the effective dielectric function formalism (Sancho-Parramon 2009) may be of relevance here. The author has calculated the optical absorption spectra by taking into account varying concentrations and multiple scattering events of silver nanoparticles in a silica glass matrix. Based on this computational study, the observed additional broadening of the SPR absorption is attributed to an increase of the local field fluctuations due to multiple interactions among the high-density silver nanoparticles that are embedded randomly in the soda glass matrix. As a result of these interactions, silver nanoparticles present in the glass may support SPR absorptions over a broad range of incident photon energies.

Next we discuss the nucleation and thermal growth of Ag nanoparticles in the ion-exchanged soda glass samples. Rutherford backscattering spectrometry (RBS) and various precise spectroscopy measurements have elucidated these studies. Silver ion–exchanged soda glass samples, annealed for 1 h in vacuum (1×10^{-6} mbar) up to 873 K, were studied. We would like to mention here that ion exchange and post-annealing leave the silver with different charge states in the glass matrix. This will be confirmed in subsequent sections through a range of spectroscopy studies. Relevant experimental results would now be analyzed and discussed in detail.

4.3.1.2.1 RBS Measurements

Rutherford backscattering experiments have been carried out on silver ion–exchanged soda glass samples to measure the concentration profiles of Ag before and after the thermal annealing (isochronal for 1 h in vacuum) at various temperatures. An analyzing beam of He$^+$ ions at the energy of 2.0 MeV, backscattered by an angle of 160°, was used during the backscattering measurements. RBS spectra for the silver ion–exchanged soda glass samples before and after annealing are displayed in Figure 4.19. A complete spectrum (inset, Figure 4.19) of the silver ion–exchanged soda glass sample (prior to annealing) reveals other elements (e.g., Ca, Si, Na, and O) present in the glass matrix. In the silver ion–exchanged soda glass sample, silver has the highest atomic mass among the other elements present. So, energetic He$^+$ ions transfer minimum energy to the Ag atoms during elastic collisions and backscatter almost with the same incident energy (Chu et al. 1978). Thus, the He$^+$ ions detected with the highest energy correspond to Ag-edge in the obtained backscattering spectra, and elements with lower atomic masses appear at respective lower energy channels (inset, Figure 4.19). For the annealed samples, partial spectra (high-energy channels) have been shown for better clarity and to emphasize the backscattering from silver atoms only. It is observed from the backscattering spectra that silver atoms have got accumulated near the surface of the glass during annealing. Accumulation of silver atoms is higher for the samples annealed at higher temperatures (Figure 4.19). Near-surface accumulation may be due to the thermal diffusion of silver ions in the silver-exchanged soda-glass matrix. Outward diffusion of silver ions may help to relax the stress (possibly arising from size difference of Ag$^+$ and Na$^+$ ions; ratio of ionic sizes of Ag to Na is ~1.29; Wang 1996), thus minimizing the total energy of the system (Dubiel et al. 2003). Out-diffusion of silver ions at elevated temperatures (723–873 K) explains the observed increase in the backscattering intensity from Ag atoms near the glass surface. (One may wonder how these Ag$^+$ ions are generated during annealing at elevated temperatures. To get an insight into this, better surface-sensitive experiments (e.g., X-ray photoelectron spectroscopy) have been performed on these samples. So, this would be answered or get clarified a little later when we analyze XPS as well as PL spectroscopy results.) The obtained backscattering data have been analyzed with the help of RUMP software program (Doolittle 1986) for calculating the depth distribution of silver atoms, starting near the surface up to a certain depth inside the silver-exchanged glass samples. A depth of about 100 nm over which the silver atoms got

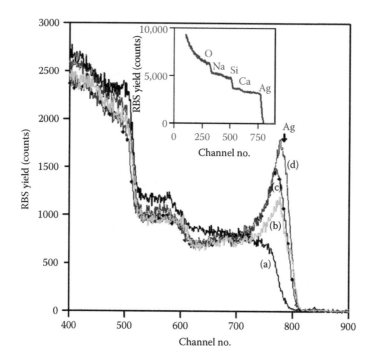

FIGURE 4.19 RBS spectra of Ag ion–exchanged soda glass samples after isochronal annealing for 1 h at various temperatures: (a) 593, (b) 723, (c) 823, and (d) 873 K. Accumulation of Ag atoms is higher for samples annealed at higher temperatures. Inset shows the complete spectrum for Ag ion–exchanged soda glass sample prior to annealing. (Reprinted with permission from Gangopadhyay, P. et al., *Chem. Phys. Lett.*, 388, 416, 2004.)

accumulated during annealing has been calculated from the backscattering data. This calculated depth is significantly smaller than the depth (~5000 nm) to which silver diffused during the silver ion–exchange process (De Marchi et al. 1996). Thus, the observed phenomenon may be due to the short-range diffusion of silver ions during the annealing the silver-exchanged soda glass samples. This result allows considering the case as a semi-infinite diffusion system with the bulk acting as a constant source of silver ions. By applying diffusion theory for a semi-infinite system, the accumulated mass of silver atoms per unit area (say, m) near the surface may be calculated from the following equation (Jacobs 1967):

$$m = c_o \sqrt{\frac{Dt}{\pi}} \tag{4.21}$$

where
  c_o is the source density
  t is the time of diffusion in the sample held at temperature T (i.e., 1 h of annealing time)
  D is the diffusion coefficient of silver ions in the soda glass matrix

For the samples annealed at different temperatures (Figure 4.19), the accumulated mass of silver atoms per unit area is estimated using the expression $m = N_A M / N_{Avo}$ (Chu et al. 1978), where N_A is the number of Ag atoms per cm², obtained from the backscattering data through the RUMP analysis, M is the atomic weight of silver (~108 atomic mass unit), and N_{Avo} is the Avogadro number (6.023 × 10²³). Assuming the diffusion coefficient (D) to vary as $D \approx e^{-\varepsilon_a/k_B T}$, where ε_a is the activation energy for thermal diffusion of silver ions in the soda glass matrix, and k_B is the Boltzmann constant, Equation 4.21 may be simplified as

$$\ln(m) = K_o - \frac{\varepsilon_a}{2k_B T} \tag{4.22}$$

where $K_o = \ln(c_o \sqrt{t}/\pi)$ is a constant under the present experimental conditions of isochronal annealing ($t = 1$ h) as c_o remains constant for the fixed time of the ion-exchange (2 min here) process. Equation 4.22 is in the Arrhenius form. From the experimentally calculated values of m (using $m = N_A M/N_{Avo}$) and corresponding temperature of annealing, the data (with symbols ♦) has been plotted in Figure 4.20. From the Arrhenius plot (the best linear fit to the data) and comparing with Equation 4.22, the slope of the fitted line is calculated. From the slope of the line, the activation energy for thermal diffusion of silver ions in the soda glass samples is estimated, which works out to be ~0.74 eV. Note that the activation energy obtained is much higher than that of diffusion of silver atoms in the soda glass matrix (Miotello et al. 2000). Diffusion of silver ions would require a higher activation energy than silver atoms. Hence, it is more than likely that thermal diffusion of silver ions has been predominant in the present study.

4.3.1.2.2 XPS and Auger Spectroscopy

4.3.1.2.2.1 Principle   X-ray photoelectron spectroscopy (XPS) is based on a single photon in–electron out process (as shown in Figure 4.21). The kinetic energy of the emitted photoelectrons is given by

$$E_k = h\nu - E_b - \varphi \tag{4.23}$$

where
  E_k is the measured kinetic energy
  $h\nu$ is the energy of the X-ray source
  E_b is the binding energy of the electrons in the solid
  φ is the work function of the spectrometer

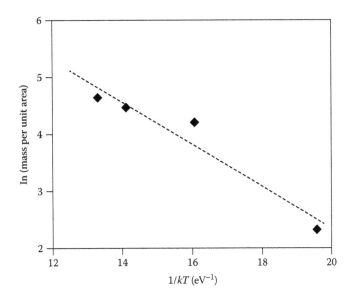

FIGURE 4.20   Temperature dependence of the mass of Ag atoms per unit area near the soda glass surface, showing an Arrhenius behavior. Symbols (♦) are the data estimated from RBS spectra (see Figure 4.19), and the solid line is the best linear fit to the analyzed backscattering data. Activation energy for thermal diffusion of Ag ions in the soda glass matrix is estimated from the slope of the line. It works out to be ~0.74 eV. (Reprinted with permission from Gangopadhyay, P. et al., *Chem. Phys. Lett.*, 388, 416, 2004.)

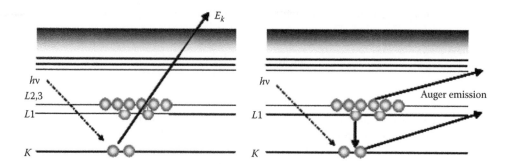

FIGURE 4.21 Schematic of photoelectron (XPS) and Auger electron emission processes. The material is being excited by the X-ray photons of energy $h\nu$.

The spectrometer work function value (~4.5 eV) and the X-ray source energy are incorporated in the data acquisition software. In commercial spectrometers, X-ray sources Al Kα (1486.6 eV) and Mg Kα (1253.6 eV) are normally used. The kinetic energy distribution (or the binding energy) of the emitted photoelectrons is measured using a hemispherical electron energy analyzer, and a photoelectron spectrum is recorded. Binding energy values of photoelectrons are characteristic of the elements. The quantification of the elements is possible by taking their peak area and the sensitivity of the peaks. Atomic concentration of an element A, for instance, in an alloy composition AB is given by

$$c(A) = \frac{I_A/S_A}{(I_A/S_A) + (I_B/S_B)} \tag{4.24}$$

where
   I is the peak intensity (area under the peak)
   S stands for sensitivity of the respective element

The binding energy and shape of the photoelectron peak can be analyzed precisely to find the chemical states of elements and their quantitative composition within the electron penetration depth of the solid (Briggs and Seah 1990).

4.3.1.2.2.2 Modified Auger Parameters   In a photoionization process, Auger electron emission also takes place (Figure 4.21). A change in the chemical state giving rise to a chemical shift in photoelectron lines will also produce a chemical shift in the Auger lines. However, often the magnitude of the Auger chemical shift is significantly greater than that of the photoelectron chemical shift. The modified Auger parameter (MAP, $\dot{\alpha}$) is defined as the sum of the binding energy of the prominent photoelectrons and the kinetic energy of the prominent Auger electrons of an element. So, by definition, $\dot{\alpha}$ is always a positive number. It may be noted here that MAP can be experimentally found out only through the XPS system. This parameter is a good characteristic of the chemical state of the element concerned. Use of this parameter while identifying chemical phases of materials is advantageous, because (1) any systematic error in the photoelectron peak shift (arising as a result of sample charging) is removed, and (2) because of a higher Auger chemical shift (compared to the photoelectron binding energy shift), the modified Auger parameter ($\dot{\alpha}$) values are distinctly different.

XPS measurements have provided chemical phase information of silver in the ion-exchanged glass samples before and after different annealing treatments. The binding energy peak of the Ag $3d_{5/2}$ photoelectrons is at 367.5 eV in the as-exchanged glass sample (Figure 4.22). However,

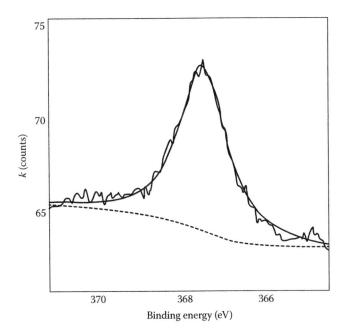

FIGURE 4.22 Photoelectron spectra of Ag $3d_{5/2}$ electrons from Ag in an ion-exchanged soda glass sample. Data are fitted with Gaussian–Lorentzian spectral function and Shirley background. Binding energy peak of Ag $3d_{5/2}$ photoelectrons is at 367.5 eV. Photoelectron spectroscopy analysis reveals the chemical phase of Ag in ion-exchanged soda glass sample as AgO. Modified Auger parameter (MAP) was calculated to confirm the chemical phase of the Ag in this sample (details are in Table 4.2). (Reprinted from Gangopadhyay, P. et al., *Phys. Rev. Lett.*, 94, 047403, 2005. With permission)

chemical analysis using XPS alone is insufficient, particularly in case of silver, where the shift in the electronic binding energies due to chemical binding is comparable to the resolution of instrument. In order to confirm further the chemical state of silver in the as-exchanged sample, the modified Auger parameter (α) has been calculated analytically taking into account the Ag $3d_{5/2}$ photoelectron-binding energy and the Ag $M_4N_{45}N_{45}$ Auger transition energy (Figure 4.23). Values of binding energies, α, are listed in Table 4.2. The modified Auger parameter is estimated to be 724.9 eV in the as-exchanged sample. This value agrees well with the reported value (724.8 eV) in the literature for the AgO phase (Hoflund et al. 2000). Thus, the chemical phase has been confirmed as AgO in the as-exchanged sample.

For comparison, the parameter α for silver has been calculated to be 726.2 and 724.1 eV in pure bulk silver and pure Ag_2O powders, respectively (Table 4.2). As seen from this table, the parameter values (α) are well separated for different chemical phases of the element (silver here). For example, with reference to that in bulk silver, shift in the Auger parameter values for silver in AgO and pure Ag_2O powder is 1.3 and 2.1 eV, respectively. These shifts are much higher compared to the binding energy shift alone (~0.4 eV, Table 4.2), providing much confidence in identifying the chemical phases of these compounds. Next, in thermally annealed samples, similar analyses confirm the existence of a mixture of two chemical phases: pure Ag and Ag_2O. Details may be found in Table 4.2 and the corresponding XPS spectra are shown in Figure 4.24. Near the surface, the ratio of Ag_2O to Ag remains comparable as the annealing temperature increases. However, this ratio reduces drastically inside the matrix (Table 4.2 and Figure 4.25). It clearly shows a significant increase in the volume fractions of pure Ag atoms in the glass samples with increasing annealing temperatures. Because of post-annealing, silver ions may be released in the matrix as a result of thermal

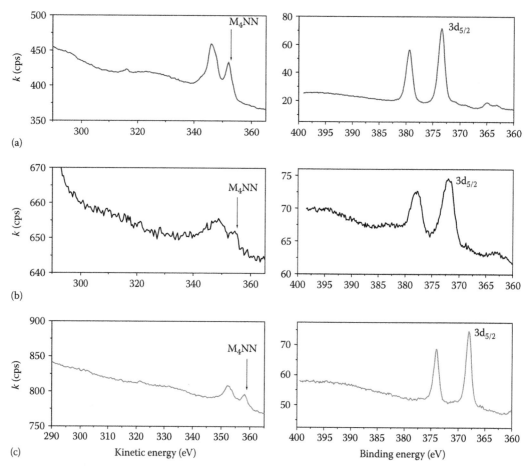

FIGURE 4.23 Ag $M_4N_{45}N_{45}$ Auger spectra and photoelectron spectra of the Ag 3d electrons. (a) Pure bulk Ag sample. (b) Pure Ag_2O standard powders. (c) Ag ion-exchanged soda glass sample.

TABLE 4.2

Binding Energies of Ag $3d_{5/2}$ Photoelectrons at Different Stages of Sample Treatment and the Corresponding Modified Auger Parameters (MAP, α)

Sample Treatments	BE (FWHM) of Ag $3d_{5/2}$ (eV)	Deconvoluted BE (FWHM) (eV)	MAP, α (eV)	Sample Composition	Ag_2O/Ag Ratio
Bulk Ag	368.4 (1.1)	—	726.2	—	—
Ag_2O powders	367.8 (1.2)	—	724.1	—	—
As-exchanged	367.5 (1.3)	—	724.9	AgO	—
Annealed at 723 K	367.9 (1.6)	367.8 (1.3)	—	Ag_2O	2.7
		368.4 (1.2)		Ag	
Annealed at 873 K	367.8 (1.6)	367.9 (1.3)	725.6	Ag_2O	2.4
		368.4 (1.3)		Ag	
Annealed at 823 K and polished	368.3 (1.9)	367.9 (1.3)	726.0	Ag_2O	0.6
		368.5 (1.5)		Ag	

Note: Experimentally obtained binding energy of Ag $3d_{5/2}$ photoelectrons and α values for bulk Ag and Ag_2O powders are also shown.

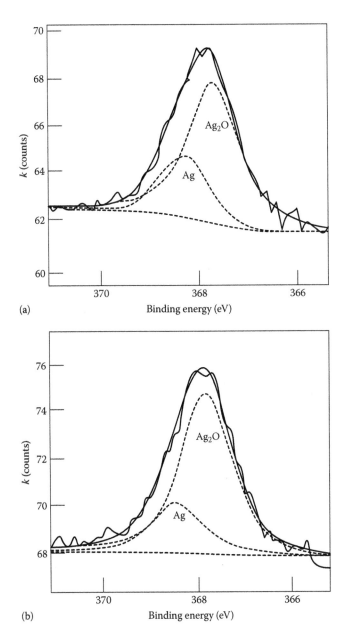

FIGURE 4.24   Deconvoluted photoelectron spectra of Ag $3d_{5/2}$ electrons from Ag in the ion-exchanged soda glass samples post-annealed at (a) 723 and (b) 873 K. Data are fitted with Gaussian–Lorentzian spectral function and Shirley background. Low (367.8 eV) and high (368.4 eV) energy peaks in the spectra correspond to Ag_2O and Ag materials, respectively. Modified Auger parameters and the intensity ratios of Ag_2O/Ag in these samples are listed in Table 4.2. (Reprinted from Gangopadhyay, P. et al., *Phys. Rev. Lett.*, 94, 047403, 2005. With permission.)

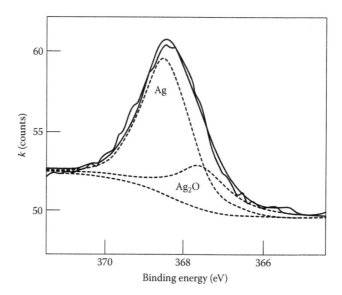

FIGURE 4.25   Deconvoluted photoelectron spectra of Ag $3d_{5/2}$ electrons from Ag in the ion-exchanged soda glass samples post-annealed at 823 K followed by mechanical polishing. XPS data are fitted with Gaussian–Lorentzian function and Shirley background. Low (367.9 eV) and high (368.5 eV) energy peaks in the spectra correspond to Ag_2O and Ag materials, respectively. Modified Auger parameter and the intensity ratio of Ag_2O/Ag in this sample are listed in Table 4.2. Polishing reduced the intensity ratio of Ag_2O/Ag drastically.

decomposition of oxides of silver (AgO and Ag_2O). Decomposition reactions of the oxides due to the thermal annealing may proceed as follows:

$$3\,AgO \rightarrow Ag + Ag_2O + O_2 \quad \text{(for annealing temperatures} \leq 723\ K) \quad \text{(Reaction 4.1)}$$

$$2\,Ag_2O \rightarrow 4\,Ag + O_2 \quad \text{(for annealing temperatures} \geq 723\ K) \quad \text{(Reaction 4.2)}$$

Because of these thermochemical reactions in vacuum, copious amounts of silver ions are released into the glass matrix. Nucleated silver particles formed during the ion-exchange process can act as further nucleation sites. Thus, during the thermal annealing, growth of these nucleated Ag particles may be expected due to diffusion of the silver species (atoms/ions) in the glass matrix. In fact, the conjecture about the growth of Ag nanoparticles is indeed true and has been confirmed with optical absorption, LFRS and PL spectroscopy experiments. As we will see in subsequent sections, thermal growth of the nanoparticles has significantly altered the optical properties of theses samples.

4.3.1.2.3   Optical Absorption Studies

Before discussing the optical absorption spectroscopy results, a visual inspection of these ion-exchanged glass samples before and after the thermal annealing may be interesting. A simple photographic image is shown in Figure 4.26 to demonstrate how the colors have changed as a result of the heat treatment. Glass samples are changing hue towards deeper shades of yellowish-brown with increase of the annealing temperature. This change has increased opacity of the samples. In other words, it implies that optical color density has increased consistently with the increase of annealing temperature. This qualitative visual observation is in conformity with the precise experimental results of optical absorption spectroscopy.

Experimental optical absorption spectrum for the silver ion-exchanged soda glass sample taken at room temperature is shown in Figure 4.27. The absorption peak structures observed in this sample are assigned to excitonic absorptions in silver monoxide (AgO) nanostructures. The chemical

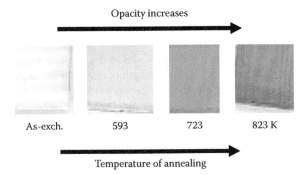

FIGURE 4.26 Photographic images showing Ag ion-exchanged soda glass samples after different stages of vacuum annealing for 1 h. Darkening is due to increase of optical absorbance in these samples. (Reprinted with permission from Gangopadhyay, P., *AIP Conference Proceedings*, Bikaner, India, Vol. 1536, p. 7. © 2013, American Institute of Physics.)

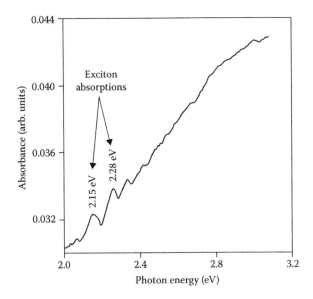

FIGURE 4.27 Room-temperature optical absorption spectrum of the Ag ion-exchanged soda glass sample showing peaks at 2.15 and 2.28 eV (marked with arrows) due to the formation of excitons in AgO. It is to be noted that PL emissions are also observed at these energies in this sample.

identification of this phase was performed by XPS studies, which were discussed in the previous section. Silver monoxide is known to be an unstable semiconductor compound. Not many related studies have been performed with nanostructured AgO materials. However, assuming a two-band (valence and conduction band) model, the observed peak structures in the optical absorption spectrum (Figure 4.27) are explained as follows: photoexcitation of electrons may induce selective transitions from the valence band (filled states) of the AgO nanostructures to higher energy states (unoccupied), leaving holes behind, and the excited electrons thus occupy different energy levels very close to the bottom of the conduction band of the material. The excited electron and the hole may be bound together by their attractive Coulomb interaction, just as an electron bound to a proton in an atom. The bound electron–hole pair, which is electrically neutral, is known as the exciton (Kittel 1985). Excitons are quasi-particle excitations. They can mediate optical absorptions and PL processes in semiconductor materials. The signature of the formation of excitons in AgO nanostructures has been experimentally observed here. Further confirmation regarding the possible existence of the exciton would be supplemented by the experimental results of PL studies in this sample. Excitons are usually unstable and decay into free charge carriers (electrons and holes). Ultimately, the electrons recombine with the holes, and in the recombination process there may be a radiative emission of photons at particular wavelength or energy. The emission is measured in PL spectroscopy experiments. To call these optical processes as excitonic in nature, the optical absorption and emission energies ought to match. In the present sample, the exciton absorption peaks are observed at 2.15 and 2.28 eV, respectively (as shown in Figure 4.27). The absorption peaks are fairly broadened. This may be because the optical absorption measurements are carried out on the samples at room temperature. Apart from normal lifetime broadening, quasi-continuous exciton energy bands just below the conduction band of the AgO nanostructures might be a possible reason for the large absorption widths at room temperature in this material.

The recorded optical absorption spectra of thermally annealed samples are displayed in Figure 4.28. Metal nanoparticles are characterized through the characteristic SPR absorptions. This is performed with optical absorption spectroscopy experiments. As discussed before,

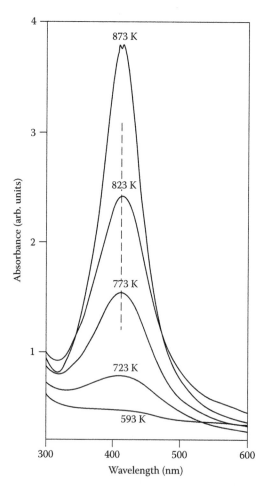

FIGURE 4.28 Recorded optical absorption spectra of Ag nanoparticles in the ion-exchanged soda glass samples before and after isochronal annealing for 1 h at different temperatures (indicated in the figure). Observed increase of optical absorbance and concomitant reduction in the width of the SPR band with the rise of annealing temperature indicates thermal growth of Ag nanoparticles in the matrix. (Reprinted from Gangopadhyay, P. et al., *Chem. Phys. Lett.*, 388, 416, 2004. With permission)

the position of the resonance wavelength in the optical absorption spectra is specific to the metal, and it depends on the dielectric constants of the metal and the host matrix. As an example, for Ag metal nanoparticles in a glass matrix the SPR wavelength position is ~415 nm. Silver is known to form a supersaturated solid solution because of its low solubility in the soda glass matrix (Pretorious et al. 1978). So, nucleation of silver atoms to form very minute silver particles can take place because of the fluctuations in the local concentrations of diffusing silver ions at the onset of the ion-exchange ($Ag^+ \leftrightarrow Na^+$) process. Nucleated silver particles formed during the ion-exchange process can act as further nucleation sites. During annealing, growth of these nucleated particles may be expected because of thermal diffusion of silver ions in the glass matrix. Growth of the silver nanoparticles may take place directly from the supersaturation. Assuming the metal nanoparticles to be of spherical type, average size may increase with the annealing time t as (Miotello et al. 2000, Christian 2002)

$$d^2(t) = d_o^2 + 8\frac{c_s - c_e}{c_p - c_e} Dt \tag{4.25}$$

where
    d_o is the value of d (diameter of metal nanoparticles) at $t = 0$ (i.e., prior to growth)
    c_p is the concentration of limiting reactants in the particle
    D is the diffusion coefficient for the limiting reactants

The role of temperature is contained in D, the diffusion coefficient. c_s and c_e are the concentration of the limiting reactant prior to growth and the equilibrium concentration in the matrix, respectively. The degree of supersaturation ($c_s - c_e$) decreases during this stage of growth. In the present experimental studies, the system is still in the early stage of growth because of the possible diffusion and reduction processes of the available silver ions. During annealing, the electron capture reaction may proceed as $Ag^+ + e^- \rightarrow Ag^o$. Thermal decomposition of silver oxides (according to Reactions 4.1 and 4.2) provides the required silver ions in the glass medium. As a result, the supply of c_s would maintain the degree of supersaturation in the system, and this could also prevent coarsening (where larger nanoparticles grow at the expense of the smaller ones) to be effective during the time of heat treatments. Growth of the silver nanoparticles observed here occurs as a result of temperature-mediated, reaction-controlled supersaturation processes, as prescribed by Equation 4.25. Thus, the observed increase in the absorption intensity (Figure 4.28) with the increase in the annealing temperature is attributed to the supersaturation growth of the silver nanoparticles in the glass matrix. The observed growth may also be explained from the Mie formula of light scattering. As can be seen from Equation 4.3, the SPR absorption intensity is proportional to the volume of the metal nanoparticles. So, with the growth of the silver nanoparticles, because of thermal annealing, the resonance optical absorption intensity is expected to increase. In particular, for samples annealed beyond 723 K, a drastic increase in the optical absorption intensity at 420 nm has been observed (Figure 4.28). This is because, according to the thermochemical reaction (Reaction 4.2), a large number of silver ions get released into the soda glass matrix through the thermal decomposition of Ag_2O. (Ag_2O is a semiconducting material with a direct bandgap of ~2.23 eV (Varkey and Fort 1993) and decomposes at ~698 K.)

The observed optical absorption features have been discussed further based on theoretical calculations. Details of the calculation have been discussed earlier. In brief, taking into account of size-dependent optical material functions ($\varepsilon(\omega, R)$) and applying the Mie's formula for dipolar absorption, the optical absorption spectra for nanoscale silver particles in a soda glass matrix have been calculated. For comparison, the calculated and experimental spectra are shown in Figure 4.29 for different average sizes of silver nanoparticles. The average sizes of silver nanoparticles are obtained from LFRS experiments. Details of the Raman measurements will be discussed shortly. The comparison between the calculated and experimental absorption spectra shows good agreement for the SPR position for the silver nanoparticles in the soda glass matrix (as shown in Figure 4.29). The result

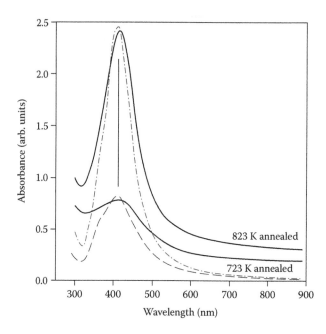

FIGURE 4.29 Calculated (broken lines) and experimental (continuous lines) optical absorption spectra of Ag nanoparticles for two different sizes (3.6 and 6.4 nm) in the soda glass matrix. Vertical line indicates the SPR position. (Reprinted with permission from Gangopadhyay, P., *AIP Conference Proceedings*, Bikaner, India, Vol. 1536, p. 7. © 2013, American Institute of Physics.)

shows that the resonance position does not vary with the size of the silver nanoparticles. Further, it is observed that the agreement is not very good for the spectral width of the resonance band. This disagreement arises because of the presence of an abrupt interface between the metal nanoparticles and the dielectric matrix. The damping (Γ) is certainly different at the interface compared to that at the pure nanoparticles. This has been discussed as due to chemical interface damping (Hovel et al. 1993). The damping depends on the chemical properties of the interface and the energy transfer between the metal nanoparticles and the surrounding matrix by temporary charge-transfer reactions. Chemical interface damping has not been properly accounted for in this simplified calculation. However, calculations properly estimate the trend of line broadening with the decrease of particle sizes.

As the increase in optical absorbance is directly related to the growth of the metal nanoparticles, it is possible to estimate the activation energy for the diffusion of Ag ions in the glass matrix. From the experimental data, the logarithm of the peak absorbance versus $(k_B T)^{-1}$ is plotted (Figure 4.30) to estimate the diffusion parameter. The graph shows an Arrhenius behavior for the growth of silver nanoparticles in this case. The slope of the line directly gives the activation energy for the diffusion of silver ions in the soda glass matrix, and it works out to be about 0.76 eV. It may be noted here that the diffusion activation energy of silver ions estimated from the optical absorption study compares well with values obtained from RBS experiments.

4.3.1.2.4 Low-Frequency Raman Scattering Studies

LFRS measurements have been carried out on silver ion–exchanged glass samples before and after thermal annealing. The recorded Raman spectra are shown in Figure 4.31. Details about the Raman measurements are provided in Gangopadhyay et al. (2004). As discussed earlier, confined surface-acoustic phonons in metal nanoparticles give rise to low-frequency Raman modes in the vibrational spectra of the nanomaterials. In this case, as the Raman modes appeared in both polarized and depolarized geometries, the observed surface acoustic vibrations is assigned to the quadrupolar ($l = 2$) Raman mode. By knowing the Raman peak frequencies from experiments, the size of the

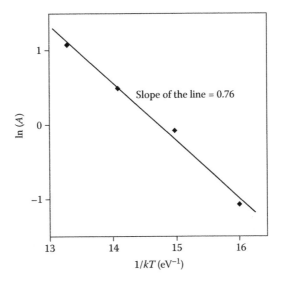

FIGURE 4.30 Arrhenius graph showing the dependence of optical peak absorbance of Ag nanoparticles on the annealing temperature. Line through the experimental data points is the best linear fit. Slope of the line gives the activation energy (~0.76 eV) for the diffusion of Ag ions in the soda glass matrix.

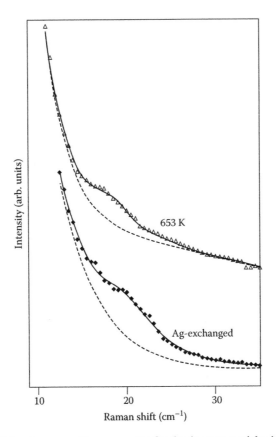

FIGURE 4.31 Polarized low-frequency Raman spectra for the Ag nanoparticles in the ion-exchanged soda glass samples and after vacuum annealing at 653 K. Symbols are the experimental data, and solid lines are the best fit to data points with the Lorentzian line-shape function and an exponential background. (Spectra are shifted along the intensity axis for better clarity.)

silver nanoparticles is estimated using the Equation 4.18. The sample details and average sizes of the silver nanoparticles are given in Table 4.3. The measurements show thermal growth of the silver nanoparticles in the soda glass matrix. Thermal growth of the silver nanoparticles observed here is related to the reaction-controlled processes. The growth dynamics here is best described by Equation 4.25. From the growth of silver nanoparticles and applying Equation 4.25 to the Raman scattering data (from Table 4.3), the activation energy for the diffusion of silver ions in the soda glass matrix is estimated. The slope of the Arrhenius plot of $\ln(d^2 - d_o^2)$ versus $(k_B T)^{-1}$ (Figure 4.32) is a measure of the activation energy, and it works out to be ~0.72 eV. The value obtained compares

TABLE 4.3
Sizes of the Silver Nanoparticles in the Silver Ion–Exchanged Soda Glass Samples Before and After Thermal Annealing at Various Temperatures

Annealing Temperature (K) of the Ion-Exchanged Glass	Measured Raman Peak Frequency (cm⁻¹)	Average Sizes of Ag Nanoparticles, d (nm)
As-exchanged	20.0	2.35
593	18.8	2.49
653	18.2	2.58
723	12.9	3.63
773	10.2	4.62
823	7.4	6.35
873	6.0	7.82

Note: Sizes have been calculated (using Equation 4.18) from the LFRS measurements. Growth of the silver nanoparticles with increase of annealing temperature is observed here.

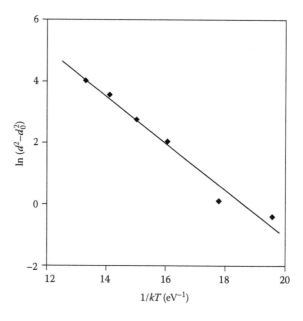

FIGURE 4.32 Arrhenius plot for the calculation of the activation energy for the diffusion of Ag ions in the soda glass matrix. d_o and d are the sizes of the Ag nanoparticles in the as-exchanged and in the post-annealed soda glass samples, respectively. Solid line through the experimental data points is the best linear fit. Slope (~0.72 eV) of the fitted line estimates the activation energy for the diffusion of Ag ions in the matrix. (Reprinted from Gangopadhyay, P. et al., *Chem. Phys. Lett.*, 388, 416, 2004. With permission.)

well with those estimated from other studies (RBS and optical absorption spectroscopy). According to a previous study, the activation energy for diffusion of silver atoms in the soda glass matrix is ~0.28 eV (Miotello et al. 2000). This value is quite low compared to the value obtained here. Since thermal diffusion of silver ions in the soda glass matrix is expected to take place with higher activation energy, it is quite likely that the diffusing species here are silver ions.

4.3.1.2.5 Photoluminescence Studies

The PL spectra of the silver-exchanged soda glass sample and after annealing the samples in high vacuum at different temperatures are displayed in Figure 4.33. Drastic changes of the PL intensity are observed on annealing (Figure 4.33). For instance, the PL intensity reaches a maximum and sharply falls to a minimum value after annealing at 723 and 873 K, respectively. Through a proper analysis of the PL data, explanations for the observed quenching of PL intensity are provided with the help of optical absorption and XPS results. A dominant PL peak at 2.15 eV (577 nm) along with the other less intense peak at 2.28 eV (545 nm) have been observed in the silver ion–exchanged soda glass sample (Figure 4.34). It may be noticed here that optical absorptions are also observed exactly at the same energy values (as shown in Figure 4.27). Thus, these emissions may be ascribed to the excitonic PL in the AgO nanostructures in this sample. Photoexcited (by the 488 nm laser line) electrons may undergo nonradiative thermal scattering processes before forming bound excitons

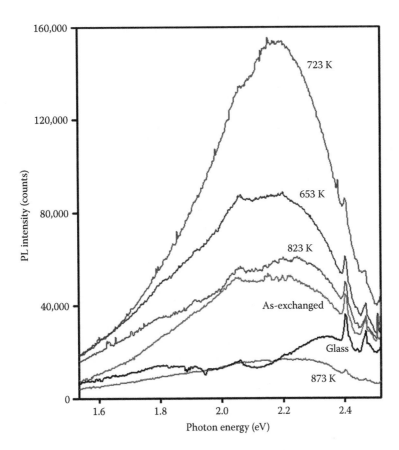

FIGURE 4.33 Room-temperature PL spectra of the Ag ion-exchanged soda glass samples before and after annealing at temperatures 653, 723, 823, and 873 K. The soda-glass substrate is also shown. Laser excitation is at 488 nm (~2.54 eV). Drastic quenching of the PL intensity is observed after annealing samples beyond 723 K. This happens because of the thermal growth of Ag nanoparticles in the matrix. (Reprinted from Gangopadhyay, P. et al., *Phys. Rev. Lett.*, 94, 047403, 2005. With permission.)

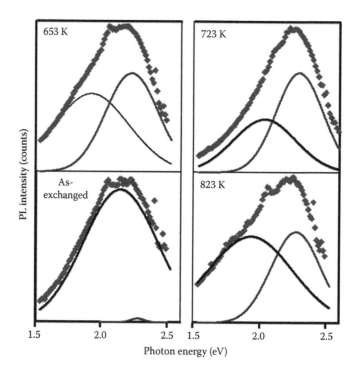

FIGURE 4.34 Experimental PL spectra as a convolution of two Gaussian functions. Symbols are the experimental data, and lines are the fitted functions to the data. Excitonic PL peaks in AgO nanostructures are shown in the as-exchanged sample. In annealed samples (653 to 823 K), the lower and higher energy PL peaks are from the Ag nanoparticles and Ag_2O materials, respectively. Observed PL bands centered around 1.95 eV are attributed to Ag nanoparticles, and the emission at 2.23 eV is assigned to the band-to-band radiative transition in Ag_2O due to annealing of the ion-exchanged soda glass samples. (Reprinted from Gangopadhyay, P. et al., *Phys. Rev. Lett.*, 94, 047403, 2005. With permission.)

(e–h pairs). Excitons are confined in space by Coulomb force and have enhanced probability to recombine radiatively. Broad PL bands centered around 1.95 eV (637 nm) and 2.23 eV (557 nm) are displayed for the samples annealed at 653 K (Figure 4.34b) and 723 K (Figure 4.34c). While the emissions lower than 2.0 eV are attributed to the presence of silver nanoparticles in the soda glass samples, the emission at 2.23 eV is ascribed to the band-to-band radiative transition in Ag_2O nanostructures (Peyser et al. 2001). The measured PL energy (at 2.23 eV) agrees well with the known optical bandgap (2.25 eV) of Ag_2O materials (Varkey and Fort 1993). It may be noted that observed PL is from the hybrid nanostructures of Ag and Ag_2O in all the annealed samples.

A similar work, in which a time-resolved PL study was carried out, was reported for Ag nanoparticles in a soda lime glass matrix (Karthikeyan 2008). Ag ion–exchanged soda glass slides were heated in air for 1 h at different temperatures (maximum 823 K) for the Ag nanoparticles to grow. With the growth of Ag nanoparticles, reduction or quenching of PL intensity and clear red shift of the emission peak were observed. The shift and quenching of PL intensity was explained as due to the Kubo effect. According to this model, mean level spacing $\delta(E_F)$ near the Fermi level (E_F) is given by $\delta(E_F) \approx E_F / z N_{At}$, where N_{At} is the number of atoms in the nanoparticle and z is the valence of the atom (Kubo 1962). The number of Ag atoms are increased in Ag nanoparticles due to thermal growth, which reduces the energy gap $\delta(E_F)$ between the sp band and d band of the metallic nanoparticles. This causes the observed shift and quenching of PL intensity emanating from Ag nanoparticles (see Figure 4 of Karthikeyan 2008). Nonradiative decay is favored when $\delta \ll k_B T$. So, electron–hole recombination is mostly nonradiative in larger Ag nanoparticles, whereas radiative emissions are more probable for smaller nanoparticles.

Time-resolved PL decay at 530 nm from the annealed silver-exchanged soda glasses with 400 nm excitation is shown in Figure 4.35. The excitation wavelength is chosen to be close to the SPR of Ag nanoparticles in the matrix. One observes a sharp rise in the time-resolved PL, followed by a fast decay and a long-lived component. In metal nanoparticles, the excited-state electrons undergo electron–electron and electron–phonon scattering and get de-excited through the radiative recombination of electrons and holes (Adelt et al. 1998). The scattering events are estimated to be fast (a few picoseconds). Radiative recombination is long lived and lasts up to 40 ns (Figure 4.35). Measurements suggest that the optical excitation excites the valence band d electrons to the sp conduction band through the interband transition, and, after initial fast electronic scattering processes, radiative recombination follows, giving rise to the visible PL.

Metal particles of very fine sizes (e.g., ~0.7 nm, the Fermi wavelength of an electron) behave like multielectron artificial atoms (Link et al. 2002). These particles do exhibit PL; however, the PL in such cases arises from the intraband transitions of the free electrons in the metal particles (Zheng et al. 2004). PL of noble metals was first observed and explained by Mooradian (1969). In the case of noble metals, in particular for silver ($[Kr]4d^{10}5s^1$) and gold ($[Xe]\ 4f^{14}5d^{10}6s^1$), the optical properties are due to the d (valence) and the sp (conduction) electrons. As for the origin of PL in silver nanoparticles in thermally annealed samples, the explanation provided by Mooradian is more pertinent. Schematic of the optically excited radiative interband recombination of sp electrons and holes in the d band is shown in Figure 4.36. The band structure of a typical noble metal has been represented by a simple model, which includes an sp conduction band and two sets of d bands (illustrated with the hatches). Excitation takes place from filled states in the upper d bands to levels at and above the Fermi energy. It is quite likely in this case that PL occurs from the direct recombination of conduction band electrons below the Fermi energy with d band holes that have been scattered to momentum states less than the Fermi momentum k_F. Band structure calculations of noble metals suggest that the d bands can be relatively flat at some regions in the momentum space less than the k_F. This might allow photoexcited holes to be scattered within a sufficient range of momentum values to account for the observed broadening of the PL spectra (as shown in Figure 4.34).

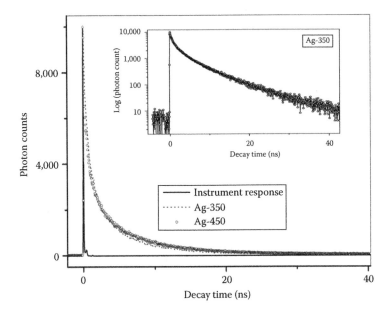

FIGURE 4.35 Time-resolved PL spectra of heat-treated Ag ion-exchanged soda glasses revealing fast electronic scattering and long-lived radiative recombination processes in Ag nanoparticles. (Reprinted from Karthikeyan, B., *J. Appl. Phys.*, 103, 114313, 2008. With permission.)

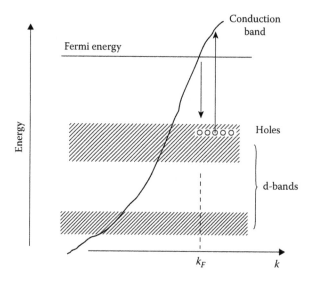

FIGURE 4.36 Schematic band structure of a noble metal showing the excitation and radiative interband recombination between the sp electrons and d-band holes. (Reprinted from Mooradian, A., *Phys. Rev. Lett.*, 22, 185, 1969. With permission.)

With increase in the annealing temperature, silver may be released in the matrix as a result of thermal decomposition of AgO. The species AgO formed in the ion-exchanged soda glass sample is unstable against heating, and decomposes into Ag_2O and Ag when annealed under high vacuum. The reaction may proceed as $3AgO \rightarrow Ag + Ag_2O + O_2$ (Reaction 4.1). As a result of this thermochemical reaction, volume fractions of both nanomaterials (Ag and Ag_2O) increase. The presence of these chemicals or moieties are confirmed by XPS and Auger spectroscopy measurements (as shown in Figure 4.23). Ag_2O phase is thermodynamically more stable than AgO (Najafi 1992); further, being a direct bandgap material, it is highly photoluminescent. Thus, increased amount of the Ag_2O causes the PL intensity to reach the highest level at ~723 K. However, around this temperature, Ag_2O thermally decomposes. Decomposition reaction of Ag_2O due to thermal annealing at higher temperatures may proceed as follows: $2Ag_2O \rightarrow 4Ag + O_2$ (Reaction 4.2). Further increase of the annealing temperature leads to the rapid growth of the silver nanoparticles (as a result of thermal decomposition of Ag_2O) within a depth scale of about 100 nm from the soda glass surface (backscattering results, Figure 4.19). Thus, thermal decomposition of Ag_2O and rapid growth of silver nanoparticles simultaneously may have resulted in the drastic quenching of the PL intensity observed for the samples post-annealed at 823 and 873 K (as shown in Figure 4.33). Photoelectron spectroscopy (XPS) results reveal that the Ag_2O phase is dominant near the surface of the post-annealed samples (see Figure 4.24). However, the estimated ratio of Ag_2O/Ag (within a sensitive depth of ~5 nm in the photoelectron spectroscopy) has not really changed with the increase of annealing temperature (shown in Table 4.2). So, to explain the observed quenching phenomena further, photoelectron spectroscopy measurements have been carried out on a specially prepared sample: the exchanged soda glass sample annealed at 823 K was mechanically polished using 0.25 μm diamond lapping for 5 min to remove a few layers from the surface of the material. This was performed to avoid the unwanted sputtering effects in the XPS system for obtaining depth information. Removal of few layers in this sample reasonably ensured that PL and XPS measurements are carried out deep inside and on identical materials in the soda glass sample. Interestingly, the recorded photoelectron spectra of this sample show significant changes in the peak profiles (Figure 4.25). Most markedly, the estimated ratio of Ag_2O/Ag in the polished sample reduced drastically to 0.6 from 2.7 (see Table 4.2) when compared to the most photoluminescent sample

(723 K annealed, as shown in Figure 4.33) in the present study. Particularly, the photoelectron spectroscopy results confirm that high-temperature-annealed samples have a higher volume fraction of Ag than Ag_2O. Following the same line of argument, successive quenching of the PL intensity for the 873 K post-annealed sample has been explained. Thus, an important correlation between temperature-induced changes of the PL intensity and thermal growth of the nanoscale silver particles in the soda glass matrix has been established through various precise spectroscopic studies.

4.3.1.2.6 Exciton and Surface-Plasmon Interactions

Studying quasi-particle interactions such as between surface plasmons and optical excitons in metal–semiconductor hybrid nanostructures have attracted great research interest in the last decade (Achermann 2010, and references therein). Nanoscale silicon structures, for instance, have been combined with metal nanostructures to explore exciton–plasmon interactions (Walters et al. 2010). Silicon nanostructures are photoluminescent (though they are indirect bandgap materials) and have attracted much attention as they can be integrated into existing semiconductor processing technologies (Heitmann et al. 2005). Hybridized quantum states of plasmons–excitons (named "plexcitons") in the core–shell geometry of hybrid nanomaterials have been obtained and studied by many researchers (Fofang et al. 2008, Antosiewicz et al. 2014). Naturally, optical emission and absorption properties of such a complex system depend on interactions of the quasi-particles, surface plasmons, and excitons. Core–shell nanostructures of Ag or Au metal nanoparticles, for example, coated with J-aggregates (a class of organic semiconductors, Jelley 1936) exhibit absolutely different optical absorption properties due to the interactions of surface plasmons with excitons (Wiederrecht et al. 2008). Variations in their interactions are attributed to several reasons: spectral overlap of semiconductor emission and surface plasmon energy of metal nanoparticles, their hybrid geometry, and the interparticle spacing between quantum dots and metal nanoparticles in hybrid nanomaterials (Hosoki et al. 2008, Haridas et al. 2011, Zhou et al. 2011). In close proximities of metal and semiconductor nanostructures, effects of weak and strong coupling may be discussed. In a weak coupling situation, exciton wave functions and electromagnetic modes of surface plasmons are considered unperturbed, and exciton–plasmon interactions are described by the coupling of the exciton's dipole with the field of the surface plasmons. Phenomena like enhanced absorption cross-sections, increased radiative rates, and exciton–plasmon energy transfer are described in the weak coupling regime. In the strong coupling regime, resonant exciton–plasmon interactions modify exciton wave functions and the surface-plasmon modes. This can also lead to change of exciton and surface-plasmon resonance energies. In this range of coupling, the excitation energy is shared, and it oscillates between the plasmonic and excitonic systems (known as Rabi oscillations, Bellessa et al. 2004). Investigating the coupling mechanisms of surface plasmons with excitons in hybrid nanomaterials is important due to a wide range of possible applications such as light-emitting diodes, quantum information and communication technology, and biosensing (Okamoto et al. 2005, Lee et al. 2007, Zhang et al. 2010, Belacel et al. 2013) among others. Theoretical studies, advancing the basic understanding of the underlying physical processes such as the resonant energy transfer between excitons and plasmons, strong electromagnetic coupling interactions with coherent exciton–plasmon, and so on, play vital roles in designing and modifying the optical properties of the semiconductor–metal coupled nanosystems of interest (Vasa et al. 2008, Marinica et al. 2013).

With this short introduction, let us now move on to discuss an experimental observation of exciton–surface plasmon interactions in a hybrid nanomaterial. Here, for example, the core–shell hybrid nanostructure of Ag–Ag_2O (Ag as core and Ag_2O as shell) is considered. The core–shell hybrid nanomaterial is formed during the thermal annealing of silver ion–exchanged soda glass samples. From the relative intensities of Ag $3d_{5/2}$ photoelectron peaks of Ag and Ag_2O (see Figure 4.25), the thickness of the Ag_2O layer has been estimated. Assuming photoelectrons are emitted from the Ag_2O layer of uniform thickness ς over the metallic Ag nanoparticle, the thickness of the oxide layer is calculated from $\varsigma = -\rho \ln(1 - \%Ag_2O)\cos\varphi$ (Laibinis and Whitesides 1992), where ρ is the mean free path and φ is the take-off angle of photoelectrons. Using this expression, the thickness of the Ag$_2$O layer works out to be ~2.2 nm for the sample annealed at 823 K. The shell thickness is small compared to the average core size of Ag nanoparticles of diameter ~6.4 nm (Table 4.3). Schematically, the core–shell nanostructure is shown in Figure 4.37. Thus, on the surface of Ag nanoparticles, a shell of Ag$_2$O has formed during thermal annealing of the samples in a high-vacuum atmosphere (Bera et al. 2006). This observation plays an important role in understanding the optical results shown in Figure 4.38. For example, signatures of excitonic absorptions and optical emissions from the silver oxide (Ag$_2$O) nanomaterials are absent in the annealed soda glass samples. This may be explained as follows: because of the incoherent and out-of-phase dipolar interactions of the surface plasmons with the excitons, in close proximity of Ag nanoparticles and Ag$_2$O shells, exciton absorptions and emissions in the Ag$_2$O nanomaterial may be suppressed. Possible excitonic optical absorptions in the semiconducting Ag$_2$O have not been observed in the annealed samples (Figure 4.38). This is probably due to the destructive interferences of excitons (having dipole moment $\vec{p}_{exi}$) with the surface-plasmon oscillations (with a dipole moment $\vec{p}_{SP}$) in the Ag$_2$O/Ag hybrid nanostructures during optical excitations (see Figure 4.37).

4.3.2   TOP-DOWN METHODS

Alternative fabrications of nanoparticles and nanostructures have become vitally important as sizes of devices reach nanoscale dimensions and the resolution of conventional optical lithography approaches its physical limit. A variety of alternative approaches, such as thermal dewetting, laser beam irradiation, ion-beam-induced dewetting, and direct writing by e-beam (Bischof et al. 1996, Lian et al. 2006, Vorobyev et al. 2007, Elbadawi et al. 2013), have been developed for the synthesis of desired supported nanostructures. Apart from these different successful sample processing techniques, synthesis (top-down approach) by ion-beam sputtering is also a promising

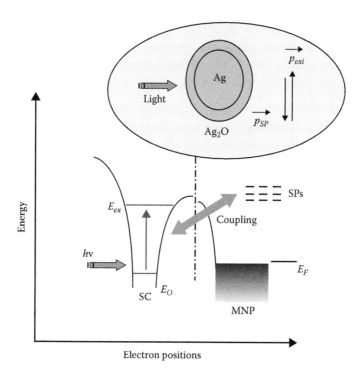

FIGURE 4.37   Schematic diagram illustrating the surface plasmon–exciton coupling in the core–shell metal–semiconductor hybrid nanostructure of Ag/Ag$_2$O. In close proximity of the nanoparticles, possible excitonic optical absorption in Ag$_2$O has not been observed in the annealed samples (also see Figure 4.38).

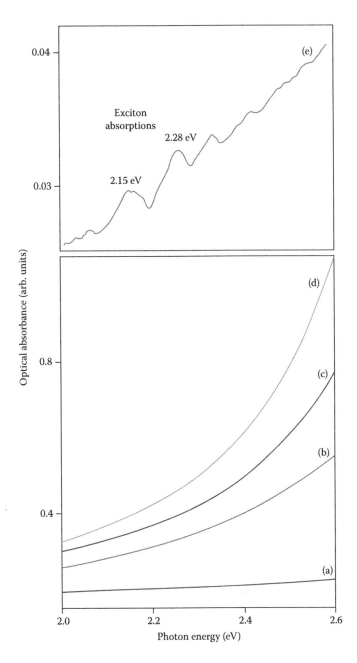

FIGURE 4.38 Optical absorption spectrum revealing excitonic signatures in AgO (e) in the silver ion–exchanged soda glass sample. Optical absorption spectra of samples are shown after isochronal annealing at temperatures (a) 653 (b) 723 (c) 773, and (d) 873 K. On annealing between 653 and 873 K, no significant feature between 2 and 2.6 eV is observed in the absorption spectra except for the known excitation energy dependence (see spectra (a)–(d)). (Reprinted from Gangopadhyay, P. et al., *Phys. Rev. Lett.*, 94, 047403, 2005. With permission.) Thermal growth of Ag nanoparticles in the soda-glass matrix has been noticed due to the annealing (also see Figure 4.28).

method leading to the fabrication of tailored nanostructures on given substrates (Facsko et al. 1999, Datta and Bhattacharyya 2003, Ruffino et al. 2009a). In fact, the production of submicrometric and nanometric features on the surfaces of solid metal targets using argon ion beam etching (sputtering) at low energy was reported first by Cunningham et al. (1960). In recent times, there have been several attempts to develop experimental techniques that would allow low-cost and large-area fabrication of nanoscale structures on supported functional substrates (Li et al. 2001, Chan and Chason 2007). During ion-beam-induced sputtering, self-organized nanostructures are often observed. Nanostructuring occurs spontaneously from the interaction of ions with the surface and near-surface target atoms. The main underlying processes are ion-induced sputtering, generation of point defects, and diffusion of surface defects (including adatoms). Nonequilibrium processing based on ion-beam sputtering is also of interest because it is relatively simple and cost effective compared to direct-write techniques, such as optical or electron beam lithography. The preparation of nanostructure materials on different substrates is possible using the ion-beam sputtering on various thin films (e.g., metals, semiconductors, insulators). The process can be suitably controlled through the selection of the mass of the projectile ions, fluences, ion-beam and target materials parameters, and the substrate temperature. Ion-beam-induced surface diffusion of the atoms of interest plays a key role in determining the ultimate topography of ion-irradiated samples. Formation of such supported novel structures by ion-beam irradiation techniques is thus a technology of choice for the benefit of seamless integration with the fabrication lines. Nanomaterials consisting of metallic nanoparticles (particularly noble metals) on insulating dielectric or glass surfaces are technologically important materials in many areas, ranging from the field of plasmonics (Smolyaninov et al. 2002, Maier and Atwater 2005, Maier 2007) to chemical reaction catalysis (Moser 1996, Moriarty 2001, Cortie and van der Lingen 2002). It has been demonstrated that noble metal nanostructures of fairly uniform size distribution could be produced, for example, on a silica glass substrate by the ion-beam sputtering (Gangopadhyay et al. 2010). Morphological differences between different target metal films do arise because of various effects of the ion-beam-induced sputtering. Factors that lead to morphological differences are (1) variations of sputtering yields of targets, (2) activation energies for atomic surface diffusion on the substrate being different for different target metals, and (3) anisotropic growth.

4.3.2.1 Ion-Beam Sputtering Phenomena

Sputtering or erosion from the surface occurs when energetic ions impinge on a solid surface. Sputtering is a physical process whereby atoms are ejected from a solid surface as a result of bombardment of the target by energetic particles. It is driven by the momentum transfer between the ions and atoms in the materials during ion–atom collisions (Sigmund 1969, Behrisch and Eckstein 2007). The incident ions initiate collision cascades in the target. Moving atoms in the cascade may eventually reach the surface and, if kinetic energy of the atoms is large enough to overcome the attractive potential of the solid, they may escape from the solid and get sputtered (Thompson 1969). Two distinct scenarios occur depending on the density of the moving atoms: When the density of the moving atoms is small, so that the probability of collision between two moving atoms is negligible, the sputtering is said to be linear. Figure 4.39 schematically describes the sputtering from a linear collision cascade. Consequence of such linearity is that the sputtering yield, that is, the average number of ejected atoms per incoming particle, is a linear function of the energy deposited at the surface. The sputtering yield depends on the ion's incident angle, the energy of the ion, the masses of the ion and target atoms, and the surface binding energy of the atoms in the target.

Since the positions of entry of the ions into the solid are statistically distributed, individual ion stopping positions and details of the collision volume are also statistically distributed. In the linear cascade regime, the collision volume is approximately of ellipsoidal shape. Contours of equal energy deposition are ellipsoids of revolution, with the major axis along the incident beam direction and centered at a depth below the surface, which increases with increasing ion energy. As a result of this distribution of energy, atoms on/and close to the surface may receive sufficient energy and

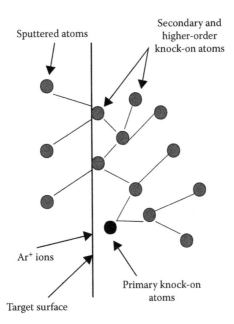

FIGURE 4.39 Schematic of the Ar$^+$ ion beam–induced sputtering of target atoms due to a linear collision cascade close to the surface of the target.

momentum, so they are likely to be ejected or sputtered from the surface of a target. A schematic diagram is shown in Figure 4.39 to show the ion-beam-induced sputtering of target atoms due to a linear displacement cascade process near the surface of the target.

Point defects (vacancies and interstitials) are created on/and below the target surface as a result of ion irradiation. Some of these point defects may recombine and annihilate or cluster to form extended defects; others may migrate to the surface. Radiation-induced release of interstitials that survive on the surface form the adatoms. Because of the statistical nature of incident ion beam and the nonoverlapping character of linear collision cascades in the interaction timescale ($\approx 10^{-13}$ s), these defects are expected to be randomly distributed on the surface, resulting in a rough surface at the early stage of ion bombardment or prior to any structure formation. Ion-beam-induced migration or diffusion of these defects plays a significant role in developing the surface morphologies of ion-irradiated samples.

4.3.2.2 Ion-Beam Sputtering Experiments

Examples of Ar$^+$ ion-beam sputtering experiments on gold and silver metal films on silica glass substrates are discussed here. A thermal evaporation setup was used to deposit gold and silver metallic films separately on cleaned silica glass substrates. Subsequently, the as-deposited metal films were irradiated at room temperature using 100 keV Ar$^+$ ions with various fluences in the range of 5×10^{15} and 1×10^{17} ions cm^{-2}. Mass-analyzed Ar$^+$ ion beam impinged at normal incidence and scanned over the metallic films on an area ≈ 1 cm^2 for uniform irradiation. The ion beam current density was kept at ~1 μA cm^{-2} on the samples during the irradiation to minimize heating due to the ion beam. The evaporated thicknesses of the silver and gold films were ~60 and ~40 nm, respectively. The thickness of the silver and gold metal films and energy of the Ar$^+$ ions were chosen (based on the TRIM simulation results, Ziegler et al. 1985) primarily to ensure the stopping of the ions within the metal/glass interface. The TRIM-calculated projected range and straggling values of 100 keV Ar$^+$ ions in gold are ~34 and ~20 nm, respectively. Thicknesses of the as-deposited metal films on silica glass substrates were more than the projected ranges of the energetic Ar$^+$ ions in the respective metals. Thus, the probability of ballistic transport or mixing of the metal atoms in the underlying glass substrate could be minimized. This further ensured that the observed experimental results are only due to the metal nanoparticles and the nanostructured materials supported on the silica glass substrates. The metallic nanostructures were observed here after a significant fraction of the metals was ejected or sputtered out as a result of the impact of the Ar+ ions (Gangopadhyay et al. 2010).

In the next sections, we discuss various experimental (e.g., optical absorption spectroscopy, RBS spectrometry, scanning electron microscopy) results to highlight the effects of the ion-beam sputtering on metallic films. Development of various types of nanostructures and nanoparticles of noble metals on the silica glass substrates has been observed with the increase of fluence of the Ar+ ions. These morphological transitions are solely due to the ion-beam-induced sputtering of the thin metallic films.

4.3.2.3 Rutherford Backscattering Results

RBS experiments were carried out to measure the areal density of gold atoms in the as-deposited film (thickness ~40 nm) as well as in the 100 keV Ar+ ion–irradiated samples with various ion fluences. An analyzing $^4He_2{}^+$ ion beam of energy 2.0 MeV was used during RBS experiments. A surface barrier detector kept at a scattering angle of 165° was utilized for detection of the backscattered particles. The experimentally obtained backscattering spectra are displayed in Figure 4.40. The figure displays the RBS spectra of the Au film and of films sputtered by 100 keV Ar+ ions with various fluences ranging from 5×10^{15} to 7×10^{16} Ar+ ions cm^{-2}. The spectra (Figure 4.40) show the systematic decrease of the backscattering yield from the Au atoms with the increase of the Ar+ ion fluence. The decrease in the area under the Au peak (which is proportional to the Au areal density) with increase of Ar+ ion fluence confirms the loss of Au atoms from the films due to the ion-beam sputtering.

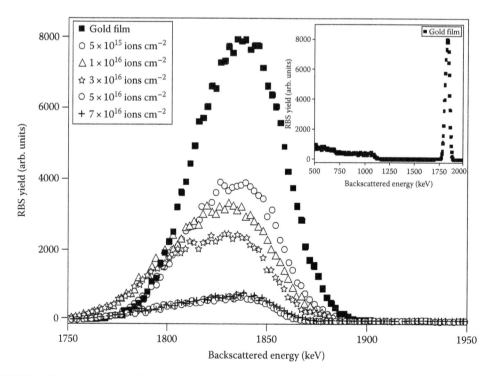

FIGURE 4.40 Rutherford backscattering spectra of the Au film and Ar+ ion–irradiated Au films as a function of the ion fluence. Partial spectra to emphasize the backscattering from Au atoms are shown. Systematic decrease of the area under the Au peak with increase of the ion fluence confirms the ion-beam-induced sputtering of the Au film. The inset shows the complete backscattering spectra of the Au film on the silica-glass substrate. (Reprinted from Gangopadhyay, P. et al., *Vacuum*, 84, 1411, 2010. With permission.)

4.3.2.4 Electron Microscopy Results

Scanning electron microscopy measurements were carried out using a Philips XL 30 scanning electron microscope (SEM). Microscopy experiments were carried out primarily to observe the morphology of the top surface of the as-deposited gold and silver metallic films in real space, as well as to study the morphological changes in the post-irradiated samples as a function of the Ar$^+$ ion fluence. Also, the approximate sizes of nanoparticles of gold and silver atoms were estimated using electron microscopy in the ion-beam-sputtered samples with higher fluences of Ar$^+$ ions. As discussed before, ion-beam sputtering is known to create many hills and valleys on the surface of the metals. According to the sputtering model due to Bradley and Harper, atoms in the valley regions are sputtered preferentially compared to those on the hills (Bradley and Harper 1988). So, compared to the valleys, sputtering is low from the tops of the hills. Because of the differences in sputtering efficiencies, hills are more stable against the ion-beam-induced sputtering. This may lead to morphological contrasts among the hills and the valleys at the microscopic level. This contrast has been actually observed in the electron micrographs (displayed in Figure 4.41). As discussed in the earlier section, argon ion-beam-induced sputtering and radiation-induced surface diffusion may be identified as the underlying process of the observed microscopic features on the surface of the glass substrate. The scanning electron micrographs (Figure 4.41) display surface morphologies of the as-deposited gold film on the silica glass substrate as well as the variation of surface morphologies of the gold film as a function of the fluence of the Ar$^+$ ions.

Interestingly, as shown in electron micrographs, an initial continuous and smooth film of gold (Figure 4.41a) turns into a discontinuous or network-like film (Figure 4.41b) structure (a kind of labyrinth nanostructures) because of the ion-beam sputtering. On continuing the sputtering experiments with higher fluences (e.g., 5×10^{16} and 7×10^{16} ions cm^{-2}) of Ar$^+$ ions, drastic changes in the surface morphology of the gold film were observed (see Figure 4.41c and d). The micrographs clearly reveal that the gold metal film has been transformed into various sizes of gold nanoparticles

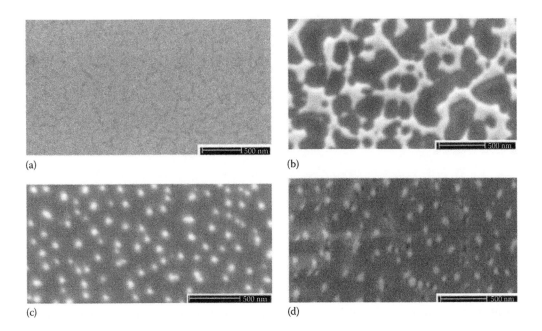

(a)

(b)

(c)

(d)

FIGURE 4.41 Scanning electron micrographs comparing the surface morphologies of the (a) as-deposited Au film (thickness ~40 nm) on the silica glass substrate, and the evolution of the Au nanostructures as a function of the Ar$^+$ ion fluence of (b) 1×10^{16}, (c) 5×10^{16}, and (d) 7×10^{16} ions cm^{-2}. (Reprinted from Gangopadhyay, P. et al., *Vacuum*, 84, 1411, 2010. With permission.)

on the surface of the glass substrate as a result of the extended ion-beam sputtering in association with the radiation-induced surface diffusion of the gold atoms.

Because of the ion-beam-induced sputtering of Au atoms, the electrical conductivity of the sample decreases at higher ion fluences. This has made the SEM images blurred or of poorer quality (as seen in Figure 4.41d) compared to Figure 4.41c. In corroboration with the RBS results (shown in Figure 4.40), it only means that the area occupied by the gold atoms on the silica glass surface has decreased with the increase of fluence of the Ar^+ ions. In these ion-beam-sputtered samples, the average sizes of the Au nanoparticles are measured to be in the range of 50–60 nm (estimated from Figure 4.41c and d). Nearly uniform sizes of Au nanoparticles are observed in the samples sputtered with higher fluences of the Ar^+ ions. Also, as shown, the Au nanoparticles are quite isolated. Although data on radiation-induced adatom diffusion of Au on the surface of the substrate is not available, maybe the low surface diffusivity of Au atoms on the substrate forces the nanoparticles to be isolated. This also possibly explains why the Au nanoparticles do not coarsen during the ion-beam-induced sputtering. Generally speaking, nanoparticles are expected to be isolated and of uniform size provided diffusion-controlled coarsening is inhibited. This is particularly observed during the irradiation of Au films with higher fluences of Ar^+ ions.

Next, in the case of ion-beam sputtering of Ag metallic films with Ar^+ ions of energy 100 keV, different surface morphological results are observed. For example, a qualitative comparison of the surface morphologies between the ion-beam-sputtered samples of Au and Ag films can be made. As shown in the micrographs in Figures 4.41 and 4.42, Ar^+ ions of same energy and with similar ion fluences have sputtered the metal films quite differently. This difference may be expected. Ion simulation (TRIM) calculations show higher sputtering yield for the Au film

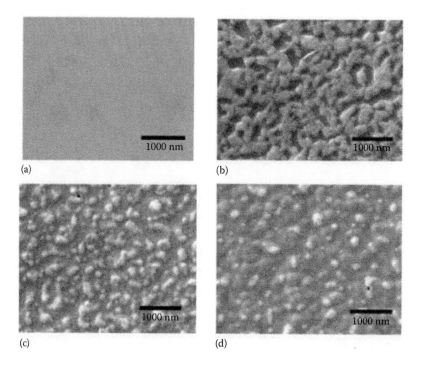

(a)   (b)

(c)   (d)

FIGURE 4.42 Scanning electron micrographs displaying (a) as-deposited Ag film (thickness ~60 nm) on silica glass, and Ar^+ ion–irradiated Ag films with ion fluence (b) 1×10^{16}, (c) 3×10^{16}, and (d) 5×10^{16} ions cm^{-2}. Network structures of Ag at lower ion fluences (see image (b)) give rise to various sizes of Ag nanoparticles at higher ion fluences (see images (c) and (d)) due to the ion-beam sputtering and radiation-induced surface diffusion of the Ag atoms on the substrate. (Reprinted from Gangopadhyay, P. et al., *Vacuum*, 84, 1411, 2010. With permission.)

(about nine atoms per ion) compared to the Ag film (about seven atoms per ion) upon Ar^+ ion irradiation (Ziegler et al. 1985). According to Sigmund's theory of linear cascade sputtering, the yield is more for heavier atoms (Nastasi et al. 1996). The surface morphologies of the Au and Ag films sputtered with higher ion fluences are shown in these micrographs (Figures 4.41c and 4.42d) for comparison. For equal fluences of the Ar^+ ions, these micrographs reveal completely different surface morphologies. As a consequence of higher sputtering yield of Au compared to Ag, and also due to a low surface diffusivity of Au atoms on the substrate (Levine et al. 1991), nearly uniform sizes (average size ~50 nm) of Au nanoparticles are produced on the surface (see Figure 4.41c and d). On the contrary, Ag nanoparticles are found to be of various sizes (the bright speckles in Figure 4.42d), the maximum size being ~200 nm. The following points need to be considered while explaining the observed differences in the surface morphologies (shown in Figures 4.41 and 4.42) of the ion-beam-sputtered samples. First, according to the TRIM calculations and Sigmund's theory, sputtering yields are different. Second, according to thermal diffusion studies, the experimentally found activation energy for the diffusion of Ag atoms was lower than that of Au atoms on silica glass (Levine et al. 1991, Kim and Alford 2002). These differences are likely to enhance the ripening of Ag nanoparticles on the silica glass substrate compared to Au nanoparticles.

4.3.2.5 Optical Absorption Results

The presence of the nanoscale metal particles and nanostructures of Au and Ag atoms in the ion-beam-sputtered metallic films is elaborated further through the optical absorption results. The optical absorption spectra of the supported Au nanoparticles on the silica glass substrates are displayed in Figure 4.43. As seen in these spectra, the characteristic optical absorption peak of Au nanoparticles on the substrate has developed with the increase of Ar^+ ion fluence. The observed optical absorption peak is in the range 550–570 nm following the increase of Ar^+ ion fluences from 5×10^{16} to 1×10^{17} ions cm^{-2}, respectively. The response peak is attributed to the SPR absorption in Au nanoparticles (Kreibig and Vollmer 1995). Moreover, as shown in the optical absorption spectra for the supported Au nanoparticles on the substrate, the peak position of the SPR absorption band is shifted by ~20 nm toward longer wavelengths (shown by the arrow in Figure 4.43) in the sample

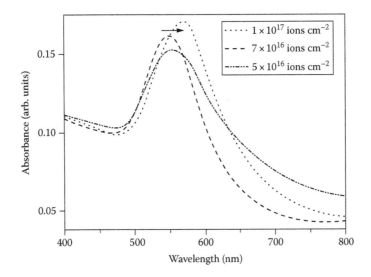

FIGURE 4.43 Optical absorption spectra of Au nanoparticles on silica glass substrates. Au nanoparticles are formed during the ion beam sputtering of Au films with various Ar^+ ion fluences (indicated in the figure). The optical absorption peak around 550 nm is due to the surface-plasmon resonance in Au nanoparticles. The arrow indicates the shift of the absorption band. (Reprinted from Gangopadhyay, P. et al., *Vacuum*, 84, 1411, 2010. With permission.)

sputtered with the maximum Ar$^+$ ion fluence of 1×10^{17} ions cm^{-2}. The observed results must be related to the size, shape, and morphology of the ion-beam-sputtered Au samples on the silica substrate. The observed red shift (~20 nm) may be due to the reduction in the average sizes of the Au nanoparticles. For fully embedded Au nanoparticles in a dielectric matrix, size reduction shifts the absorption band toward lower wavelengths (blue shift) (Kreibig and Genzel 1985, Kreibig and Vollmer 1995). However, using the discrete dipole approximation calculations, the red shift of SPR has been theoretically calculated for 10 nm Ag nanoparticles on a mica substrate to highlight the substrate effects (Kelly et al. 2003). In the present case of supported Au nanoparticles on the substrate, it adds a new level of complexity to the electromagnetic effects on metal nanoparticles due to dissimilar environments. In the present case, a plausible explanation of the observed red shift with the decrease in size of the Au nanoparticles may be understood in the following way: Because of size effects, the number of polarized electrons on the surface of the Au nanoparticles is reduced (as polarizability is proportional to volume of nanoparticles). Additionally, there could be a finite polarization imbalance at the metal–dielectric interface for supported metal nanoparticles because of interaction of surface charges with the glass substrate (Persson 1993). This may lead to further reduction of the overall charge density. The decreased surface charge corresponds to a smaller restoring force, which in turn may shift the resonance toward higher wavelengths.

Experimental optical absorption spectrum of supported Ag nanoparticles on the silica glass substrate is displayed (solid line) in Figure 4.44. The sample is obtained after Ag thin film samples were sputtered with the maximum fluence of 5×10^{16} Ar$^+$ ions cm^{-2} (energy 100 keV). The recorded spectrum was deconvoluted to show the optical absorption peaks of Ag nanoparticles on the substrate. The observed peak at 410 nm is attributed to the characteristic dipolar SPR absorption in nanoscale Ag particles: dipolar absorption is the most dominant. In addition to the dipolar absorption peak, a low-intensity peak on lower wavelength side (~375 nm) is also observed. Earlier studies have reported on the appearance of such a shoulder for metal nanoparticles on LiF crystal and silica glass substrates (Hoheisel et al. 1993, Kreibig and Vollmer 1995, Kelly et al. 2003). This peak is assigned to the quadrupole resonance absorption in large-sized Ag nanoparticles on the substrate.

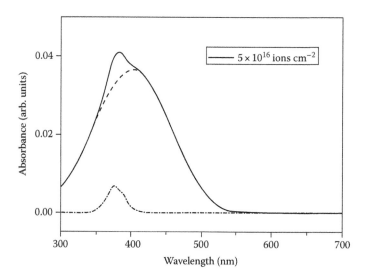

FIGURE 4.44 Optical absorption spectra of the supported Ag nanoparticles on the silica glass substrate. Various sizes of Ag nanoparticles are formed during the ion beam sputtering of the Ag metallic film with high fluence (5×10^{16} ions cm^{-2}) of Ar$^+$ ions. In addition to the dipolar absorption peak at 410 nm, a low-intensity peak on the lower wavelength side (~375 nm) is also observed. This peak is assigned to the quadrupole resonance absorption in large Ag nanoparticles on the silica glass substrate. (Reprinted from Gangopadhyay, P. et al., *Vacuum*, 84, 1411, 2010. With permission.)

In the ion-beam-sputtered sample, large-scale Ag nanoparticles are observed (see the SEM micrograph, Figure 4.42d). Related effects, such as phase retardation, field becoming inhomogeneous across the nanoparticles, and presence of higher multipole optical resonances, are usually observed in relatively larger metal nanoparticles (Kreibig and Vollmer 1995, Ovchinnikov 2008).

4.4 SUMMARY

Tremendous advances have taken place during last few decades in synthesis of materials of nanoscale dimensions. In this chapter, we discussed various metal nanoparticles and metal–semiconductor hybrid nanomaterials with a few examples of some interesting optical (absorption and emission) and vibrational properties. For instance, collective oscillations of electrons in nanoscale metals give rise to SPR absorptions. Further, electromagnetic interactions of surface plasmons with excitons in hybrid nanomaterials show interesting optical features. Because of the phonon confinement on nanomaterials, their vibrational properties are drastically different from those of bulk materials. Vibrational modes measured through LFRS can quantify the size of nanomaterials very fast and in a nondestructive way. Estimation of the thermal growth of metal nanoparticles and activation energy for the diffusion of metal species in different glass matrices are feasible. Vibrational spectroscopy has established the existence of electron–phonon coupling in nanoscale metal particles. These studies help us to explain plasmon-field-induced enhancements of Raman intensity in metal nanoparticles.

Ion-beam-induced sputtering (a top-down approach) has ample potential for the synthesis of nanoparticles and nanostructures of different materials on functional substrates. Depending on the ion-beam parameters, very interesting surface morphologies may be obtained. For example, uniform size of Au nanoparticles has been observed at relatively high fluences of argon ions. The study also suggests that the higher mobility of Ag atoms compared to Au atoms on the glass surface can lead to the synthesis of large Ag nanoparticles. Quadrupolar optical absorption is possible in such large-sized metal nanoparticles.

ACKNOWLEDGMENTS

The author thanks the collaborators and colleagues for many scientific discussions and their help in performing various experiments that were discussed here. The author is particularly grateful to the management of IGCAR for the support in pursuing this research study.

REFERENCES

Achermann, M. 2010. Exciton–plasmon interactions in metal-semiconductor nanostructures. *J. Phys. Chem. Lett.* 1: 2837.

Adelt, M., S. Nepijko, W. Drachsel, and H.-J. Freund. 1998. Size-dependent luminescence of small palladium particles. *Chem. Phys. Lett.* 291: 425.

Akimov, A. V., A. Mukherjee, C. L. Yu et al. 2007. Generation of single optical plasmons in metallic nanowires coupled to quantum dots. *Nature* 450: 402.

Allegrini, M., N. Garci, and O. Marti (eds.). 2001. *Nanometer Scale Science and Technology*. Amsterdam, the Netherlands: IOS Press.

Alivisatos, A. P. 2005. The use of nanocrystals in biological detection. *Nat. Biotechnol.* 22: 47.

Alù, A. and N. Engheta. 2008. Multifrequency optical invisibility cloak with layered plasmonic shells. *Phys. Rev. Lett.* 100: 113901.

Antosiewicz, T. J., S. P. Apell, and T. Shegai. 2014. Plasmon-exciton interactions in a core–shell geometry: From enhanced absorption to strong coupling. *ACS Photon.* 1: 454.

Arnold, G. W., G. De Marchi, F. Gonella et al. 1996. Formation of nonlinear optical waveguides by using ion-exchange and implantation techniques. *Nucl. Instrum. Methods Phys. Res. B* 116: 507.

Atwater, H. A. 2007. The promise of plasmonics. *Sci. Am.* 296: 56.

Atwater, H. A. and A. Polman. 2010. Plasmonics for improved photovoltaic devices. *Nat. Mater.* 9: 205.

Bachelier, G., J. Margueritat, A. Mlayah et al. 2007. Size dispersion effects on the low-frequency Raman scattering of quasispherical silver nanoparticles: Experiment and Theory. *Phys. Rev. B* 76: 235419.

Bachelier, G. and A. Mlayah. 2004. Surface plasmon mediated Raman scattering in metal nanoparticles. *Phys. Rev. B* 69: 205408.

Balaji, S., B. K. Panigrahi, K. Saravanan et al. 2014. Ion beam shaping of embedded metal nanoparticles by Si^+ ion irradiation. *Appl. Phys. A* 116: 1595.

Behrisch, R. and W. Eckstein. 2007. *Sputtering by Particle Bombardment: Experiments and Computer Calculations from Threshold to MeV Energies*. Berlin, Germany: Springer-Verlag.

Belacel, C., B. Habert, F. Bigourdan et al. 2013. Controlling spontaneous emission with plasmonic optical patch antennas. *Nano Lett.* 13: 1516.

Bell, A. T. 2003. The impact of nanoscience on heterogeneous catalysis. *Science* 299: 1688.

Bellessa, J., C. Bonnand, J. C. Plenet, and J. Mugnier. 2004. Strong coupling between surface plasmons and excitons in an organic semiconductor. *Phys. Rev. Lett.* 93: 036404.

Bera, S., P. Gangopadhyay, K. G. M. Nair, B. K. Panigrahi, and S. V. Narasimhan. 2006. Electron spectroscopic analysis of silver nanoparticles in a soda-glass matrix. *J. Electron Spectrosc. Relat. Phenom.* 152: 91.

Bernas, H. (ed.). 2009. *Materials Science with Ion Beams*. Berlin, Germany: Springer.

Bernas, H. 2010. Can ion beams control nanostructures in insulators? *Nucl. Instrum. Methods Phys. Res. B* 268: 3171.

Binnig, G., H. Rohrer, Ch. Gerber, and E. Weibel. 1982a. Tunneling through a controllable vacuum gap. *Appl. Phys. Lett.* 40: 178.

Binnig, G., H. Rohrer, Ch. Gerber, and E. Weibel. 1982b. Surface studies by scanning tunneling microscopy. *Phys. Rev. Lett.* 49: 57.

Bischof, J., D. Scherer, S. Herminghaus, and P. Leiderer. 1996. Dewetting modes of thin metallic films: Nucleation of holes and spinodal dewetting. *Phys. Rev. Lett.* 77: 1536.

Boggild, P., T. M. Hansen, C. Tanasa, and F. Grey. 2001. Fabrication and actuation of customized nanotweezers with a 25 nm gap. *Nanotechnology* 12: 331.

Bohren, C. F. and D. R. Huffman. 1983. *Absorption and Scattering of Light by Small Particles*. New York: Wiley.

Bora, M., E. M. Behymer, D. A. Dehlinger et al. 2013. Plasmonic black metals in resonant nanocavities. *Appl. Phys. Lett.* 102: 251105.

Bottani, C. E., A. Li Bassi, A. Stella, P. Cheyssac, and R. Kofman. 2001. Investigation of confined acoustic phonons of tin nanoparticles during melting. *Europhys. Lett.* 56: 386.

Bozhevolnyi, S. I., V. S. Volkov, E. Devaux, J.-Y. Laluet, and T. W. Ebbesen. 2006. Channel plasmon subwavelength waveguide components including interferometers and ring resonators. *Nature* 440: 508.

Bradley, R. M. and J. M. E. Harper. 1988. Theory of ripple topography induced by ion bombardment. *J. Vac. Sci. Technol. A* 6: 2390.

Briggs, D. and M. P. Seah (eds.). 1990. *Practical Surface Analysis—Auger and X-Ray Photoelectron Spectroscopy*, 2nd edn. Chichester, U.K.: Wiley Interscience.

Campbell, C. T., S. C. Parker, and D. E. Starr. 2002. The effect of size-dependent nanoparticle energetics on catalyst sintering. *Science* 298: 811.

Chan, W. L. and E. Chason. 2007. Making waves: Kinetic processes controlling surface evolution during low energy ion sputtering. *J. Appl. Phys.* 101: 121301.

Chang, D. E., A. S. Sørensen, P. R. Hemmer, and M. D. Lukin. 2006. Quantum optics with surface plasmons. *Phys. Rev. Lett.* 97: 053002.

Chang, D. E., A. S. Sørensen, P. R. Hemmer, and M. D. Lukin. 2007. Strong coupling of single emitters to surface plasmons. *Phys. Rev. B* 76: 035420.

Christensen, N. E. and B. O. Seraphin. 1971. Relativistic band calculation and the optical properties of gold. *Phys. Rev. B* 4: 3321.

Christian, J. W. 2002. *The Theory of Transformations in Metals and Alloys*. Amsterdam, the Netherlands: Pergamon.

Christopher, P., H. Xin, A. Marimuthu, and S. Linic. 2012. Singular characteristics and unique chemical bond activation mechanisms of photocatalytic reactions on plasmonic nanostructures. *Nat. Mater.* 12: 1044.

Chu, W. K., J. W. Mayer, and M.-A. Nicolet. 1978. *Backscattering Spectrometry*. New York: Academic Press.

Clavero, C., G. Armelles, J. Margueritat et al. 2007. Interface effects in the magneto-optical properties of Co nanoparticles in dielectric matrix. *Appl. Phys. Lett.* 90: 182506.

Cortie, M. B. and E. van der Lingen. 2002. Catalytic gold nano-particles. *Mater. Forum* 26: 1.

Courty, A., I. Lisiecki, and M. P. Pileni. 2002. Vibration of self-organized silver nanocrystals. *J. Chem. Phys.* 116: 8074.

Cunningham, R. L., P. Haymann, C. Lecomte, W. J. Moore, and J. J. Trillat. 1960. Etching of surfaces with 8-keV Argon ions. *J. Appl. Phys.* 31: 839.

Datta, D. and S. R. Bhattacharyya. 2003. Role of interface modifications in ion-sputtering mechanism of gold thin films. *Nucl. Instrum. Methods Phys. Res. B* 212: 201.

De Marchi, G., F. Caccavale, F. Gonella et al. 1996. Silver nanoclusters formation in ion-exchanged waveguides by annealing in hydrogen atmosphere. *Appl. Phys. A* 63: 403.

De Marchi, G., G. Mattei, P. Mazzoldi, C. Sada, and A. Miotello. 2002. Two stages in the kinetics of gold cluster growth in ion-implanted silica during isothermal annealing in oxidizing atmosphere. *J. Appl. Phys.* 92: 4249.

Dickey, J. M. and A. Paskin. 1968. Phonon spectrum changes in small particles and their implications for superconductivity. *Phys. Rev. Lett.* 21: 1441.

Doolittle, L. R. 1986. A semiautomatic algorithm for Rutherford backscattering analysis. *Nucl. Instrum. Methods Phys. Res. B* 15: 227.

Dubiel, M., H. Hofmeister, G. L. Tan, K.-D. Schicke, and E. Wendler. 2003. Silver diffusion and precipitation of nanoparticles in glass by ion implantation. *Eur. Phys. J. D* 24: 361.

Duval, E. 1992. Far-infrared and Raman vibrational transitions of a solid sphere: Selection rules. *Phys. Rev. B* 46: 5795.

Duval, E., A. Boukenter, and B. Champagnon. 1986. Vibration eigenmodes and size of microcrystallites in glass: Observation by very-low-frequency Raman scattering. *Phys. Rev. Lett.* 56: 2052.

Ehrenreich, H. and H. R. Philipp. 1962. Optical properties of Ag and Cu. *Phys. Rev.* 128: 1622.

Elbadawi, C., M. Toth, and C. J. Lobo. 2013. Pure platinum nanostructures grown by electron beam induced deposition. *ACS Appl. Mater. Interfaces* 5: 9372.

Evanoff, D. D. and Jr. G. Chumanov. 2004. Size-controlled synthesis of nanoparticles. Measurement of extinction, scattering, and absorption cross sections. *J. Phys. Chem. B* 108: 13957.

Faccio, D., P. Di Trapani, E. Borsella, F. Gonella, P. Mazzoldi, and A. M. Malvezzi. 1998. Measurement of the third-order nonlinear susceptibility of Ag nanoparticles in glass in a wide spectral range. *Europhys. Lett.* 43: 213.

Facsko, S., T. Dekorsy, C. Koerdt et al. 1999. Formation of ordered nanoscale semiconductor dots by ion sputtering. *Science* 285: 1551.

Faraday, M. 1857. Experimental relations of gold (and other metals) to light. *Trans. R. Soc. Lond.* 147: 145.

Farah, K., F. Hosni, A. Mejri, B. Boizot, A. H. Hamzaoui, and H. B. Ouada. 2014. Effect of gamma rays absorbed doses and heat treatment on the optical absorption spectra of silver ion-exchanged silicate glass. *Nucl. Instrum. Methods Phys. Res. B* 323: 36.

Fedutik, Y., V. V. Temnov, O. Schops, and U. Woggon. 2007. Exciton-plasmon-photon conversion in plasmonic nanostructures. *Phys. Rev. Lett.* 99: 136802.

Feynman, R. P. 1992. There's plenty of room at the bottom. *J. Microelectromech. Syst.* 1: 60.

Feynman, R. P. 1993. Infinitesimal machinery. *J. Microelectromech. Syst.* 2: 4.

Fofang, N. T., T.-H. Park, O. Neumann, N. A. Mirin, P. Nordlander, and N. J. Halas. 2008. Plexcitonic nanoparticles: Plasmon-exciton coupling in nanoshell-J aggregate complexes. *Nano Lett.* 8: 3481.

Fujii, M., T. Nagareda, S. Hayashi, and K. Yamamoto. 1991. Low-frequency Raman scattering from small silver particles embedded in SiO_2 thin films. *Phys. Rev. B* 44: 6243.

Gangopadhyay, P. 2013. *AIP Conference Proceedings*, Bikaner, Rajasthan, India, Vol. 1536, p. 7.

Gangopadhyay, P., R. Kesavamoorthy, S. Bera et al. 2005. Optical absorption and photoluminescence spectroscopy of the growth of silver nanoparticles. *Phys. Rev. Lett.* 94: 047403.

Gangopadhyay, P., R. Kesavamoorthy, K. G. M. Nair, and R. Dhandapani. 2000. Raman scattering studies on silver nanoclusters in a silica matrix formed by ion-beam mixing. *J. Appl. Phys.* 88: 4975.

Gangopadhyay, P., P. Magudapathy, R. Kesavamoorthy, B. K. Panigrahi, K. G. M. Nair, and P. V. Satyam. 2004. Growth of silver nanoclusters embedded in soda glass matrix. *Chem. Phys. Lett.* 388: 416.

Gangopadhyay, P., T. R. Ravindran, K. G. M. Nair, S. Kalavathi, B. Sundaravel, and B. K. Panigrahi. 2007. Raman scattering studies of cobalt nanoclusters formed during high energy implantation of cobalt ions in a silica matrix. *Appl. Phys. Lett.* 90: 063108.

Gangopadhyay, P., S. K. Srivastava, P. Magudapathy, T. N. Sairam, K. G. M. Nair, and B. K. Panigrahi. 2010. Ion-beam sputtering and nanostructures of noble metals. *Vacuum* 84: 1411.

Gao, H. L., X. W. Zhang, Z. G. Yin, S. G. Zhang, J. H. Meng, and X. Liu. 2013. Efficiency enhancement of polymer solar cells by localized surface plasmon of Au nanoparticles. *J. Appl. Phys.* 114: 163102.

Gersten, J. I., D. A. Weitz, T. J. Gramila, and A. Z. Genack. 1980. Inelastic Mie scattering from rough metal surfaces: Theory and experiment. *Phys. Rev. B* 22: 4562.

Ghosh, S. K. and T. Pal. 2007. Interparticle coupling effect on the surface plasmon resonance of gold nanoparticles: From theory to applications. *Chem. Rev.* 107: 4797.

Gonzalo, J., A. Perea, D. Babonneau et al. 2005. Competing processes during the production of metal nanoparticles by pulsed laser deposition. *Phys. Rev. B* 71: 125420.

Hamanaka, Y., K. Fukuta, A. Nakamura, L. M. Liz-Marzán, and P. Mulvaney. 2004. Enhancement of third-order nonlinear optical susceptibilities in silica-capped Au nanoparticle films with very high concentrations. *Appl. Phys. Lett.* 84: 4938.

Haridas, M., L. N. Tripathi, and J. K. Basu. 2011. Photoluminescence enhancement and quenching in metal-semiconductor quantum dot hybrid arrays. *Appl. Phys. Lett.* 98: 063305.

Haynes, C. L. and R. P. Van Duyne. 2001. Nanosphere lithography: A versatile nanofabrication tool for studies of size-dependent nanoparticle optics. *J. Phys. Chem. B* 105: 5599.

Haynes, C. L. and R. P. Van Duyne. 2003. Plasmon-sampled surface-enhanced Raman excitation spectroscopy. *J. Phys. Chem. B* 107: 7426.

Heitmann, J., F. Muller, M. Zacharias, and U. Gosele. 2005. Silicon nanocrystals: Size matters. *Adv. Mater.* 17: 795.

Hoa, X. D., A. G. Kirk, and M. Tabrizian. 2007. Towards integrated and sensitive surface plasmon resonance biosensors: A review of recent progress. *Biosens. Bioelectron.* 23: 151.

Hoflund, G. B., Z. F. Hazos, and G. N. Salaita. 2000. Surface characterization study of Ag, AgO, and Ag_2O using x-ray photoelectron spectroscopy and electron energy-loss spectroscopy. *Phys. Rev. B* 62: 11126.

Hoheisel, W., M. Vollmer, and F. Trager. 1993. Desorption of metal atoms with laser light: Mechanistic studies. *Phys. Rev. B* 48: 17463.

Hollstein, T., U. Kreibig, and F. Leis. 1977. Optical properties of Cu and Ag in the intermediate region between pure Drude and interband absorption. *Phys. Status Solidi (b)* 82: 545.

Hosokawa, M., K. Nogi, M. Naito, and T. Yokoyama (eds.). 2012. *Nanoparticle Technology Handbook*, 2nd edn. Oxford, U.K.: Elsevier.

Hosoki, K., T. Tayagaki, S. Yamamoto, K. Matsuda, and Y. Kanemitsu. 2008. Direct and stepwise energy transfer from excitons to plasmons in close-packed metal and semiconductor nanoparticle monolayer films. *Phys. Rev. Lett.* 100: 207404.

Hovel, H., S. Fritz, A. Hilger, U. Kreibig, and M. Vollmer. 1993. Width of cluster plasmon resonances: Bulk dielectric functions and chemical interface damping. *Phys. Rev. B* 48: 18178.

Hughes, M. A., Y. Fedorenko, B. Gholipour et al. 2014. n-Type chalcogenides by ion implantation. *Nat. Commun.* 5: 5346.

Jacobs, M. H. 1967. *Diffusion Processes*. New York: Springer-Verlag.

Jelley, E. E. 1936. Spectral absorption and fluorescence of dyes in the molecular state. *Nature* 138: 1009.

Jie, G., B. Liu, H. Pan, J.-J. Zhu, and H.-Y. Chen. 2007. CdS nanocrystal-based electrochemiluminescence biosensor for the detection of low-density lipoprotein by increasing sensitivity with gold nanoparticle amplification. *Anal. Chem.* 79: 5574.

Kala, S., R. Theissmann, and F. E. Kruis. 2013. Generation of AuGe nanocomposites by co-sparking technique and their photoluminescence properties. *J. Nanopart. Res.* 15: 1963.

Karthikeyan, B. 2008. Fluorescent glass embedded silver nanoclusters: An optical study. *J. Appl. Phys.* 103: 114313.

Kelly, K. L., E. Coronado, L. L. Zhao, and G. C. Schatz. 2003. The optical properties of metal nanoparticles: The influence of size, shape, and dielectric environment. *J. Phys. Chem. B* 107: 668.

Kim, H., H.-B.-R. Lee, and W.-J. Maeng. 2009. Applications of atomic layer deposition to nanofabrication and emerging nanodevices. *Thin Solid Films* 517: 2563 (a review article).

Kim, H. C. and T. L. Alford. 2002. Thickness dependence on the thermal stability of silver thin films. *Appl. Phys. Lett.* 81: 4287.

Kittel, C. 1985. *Introduction to Solid State Physics*. New Delhi, India: Wiley Eastern Limited.

Kneipp, K., H. Kneipp, I. Itzkan, R. R. Dasari, and M. S. Feld. 2002. Surface-enhanced Raman scattering and biophysics. *J. Phys.: Condens. Matter* 14: R597.

Kreibig, U. and L. Genzel. 1985. Optical absorption of small metallic particles. *Surf. Sci.* 156: 678.

Kreibig, U. and M. Vollmer. 1995. *Optical Properties of Metal Clusters*. Berlin, Germany: Springer.

Kubo, R. 1962. Electronic properties of metallic fine particles. *J. Phys. Soc. Jpn.* 17: 975.

Kumbhar, A., L. Spinu, F. Agnoli, K.-Y. Wang, W. Zhou, and C. J. O'Connor. 2001. Magnetic properties of cobalt and cobalt-platinum alloy nanoparticles synthesized via microemulsion technique. *IEEE Trans. Magnet.* 37: 2216.

Lamb, H. 1882. On the vibrations of an elastic sphere. *Proc. Lond. Math. Soc.*13: 187.

Laibinis, P. E. and G. M. Whitesides. 1992. Self-assembled monolayers of n-alkanethiolates on copper are barrier films that protect the metal against oxidation by air. *J. Am. Chem. Soc.* 114: 9022.

Lee, J., P. Hernandez, J. Lee, A. O. Govorov, and N. A. Kotov. 2007. Exciton-plasmon interactions in molecular spring assemblies of nanowires and wavelength-based protein detection. *Nat. Mater.* 6: 291.

Lee, T.-H. and R. M. Dickson. 2003. Discrete two-terminal single nanocluster quantum optoelectronic logic operations at room temperature. *Proc. Natl. Acad. Sci. U.S.A.* 100: 3043.

Levine, J. R., J. B. Cohen, and Y. W. Chung. 1991. Thin film island growth kinetics: A grazing incidence small angle x-ray scattering study of gold on glass. *Surf. Sci.* 248: 215.

Li, J., D. Stein, C. McMullan, D. Branton, M. J. Aziz, and J. A. Golovchenko. 2001. Ion-beam sculpting at nanometre length scales. *Nature* 412: 166.

Lian, J., L. Wang, X. Sun, Q. Yu, and R. C. Ewing. 2006. Patterning metallic nanostructures by ion-beam-induced dewetting and Rayleigh instability. *Nano Lett.* 6: 1047.

Lifshitz, I. M. and V. V. Slyozov. 1961. The kinetics of precipitation from supersaturated solid solutions. *J. Phys. Chem. Solids* 19: 35.

Link, S., A. Beeby, S. FitzGerald, M. A. El-Sayed, T. G. Schaaff, and R. L. Whetten. 2002. Visible to infrared luminescence from a 28-atom gold cluster. *J. Phys. Chem. B* 106: 3410.

Loo, C., A. Lowery, N. Halas, J. West, and R. Drezek. 2005. Immunotargeted nanoshells for integrated cancer imaging and therapy. *Nano Lett.* 5: 709.

Luo, X., A. Morrin, A. J. Killard, and M. R. Smyth. 2006. Application of nanoparticles in electrochemical sensors and biosensors. *Electroanalysis* 18: 319.

Luther, J. L., P. K. Jain, T. Ewers, and A. P. Alivisatos. 2011. Localized surface plasmon resonances arising from free carriers in doped quantum dots. *Nat. Mater.* 10: 361.

Magudapathy, P., P. Gangopadhyay, B. K. Panigrahi, K. G. M. Nair, and S. Dhara. 2001. Electrical transport studies of Ag nanoclusters embedded in a glass matrix. *Physica B* 299: 142.

Maier, S. A. 2007. *Plasmonics: Fundamentals and Applications.* Berlin, Germany: Springer-Verlag.

Maier, S. A. and H. A. Atwater. 2005. Plasmonics: Localization and guiding of electromagnetic energy in metal/dielectric structures. *J. Appl. Phys.* 98: 011101.

Maier, S. A., M. L. Brongersma, P. G. Kik, S. Meltzer, A. A. G. Requicha, and H. A. Atwater. 2001. Plasmonics—A route to nanoscale optical devices. *Adv. Mater.* 13: 1501.

Maisels, A., F. E. Kruis, H. Fissan, B. Rellinghaus, and H. Zähres. 2000. Synthesis of tailored composite nanoparticles in the gas phase. *Appl. Phys. Lett.* 77: 4431.

Marinica, D. C., H. Lourenco-Martins, J. Aizpurua, and A. G. Borisov. 2013. Plexciton quenching by resonant electron transfer from quantum emitter to metallic nanoantenna. *Nano Lett.* 13: 5972.

Mazzoldi, P. and G. Mattei. 2005. Potentialities of ion implantation for the synthesis and modification of metal nanoclusters. *Riv Nuovo Cimento* 28: 1.

Mie, G. 1908. Beitrage zur optik truber medien speziell kolloidaler metallosungen. *Ann. Phys. Lpz.* 25: 377.

Miotello, A., M. Bonelli, G. De Marchi, G. Mattei, P. Mazzoldi, and C. Sada. 2001. Formation of silver nanoclusters by excimer laser interaction in silver-exchanged soda-lime glass. *Appl. Phys. Lett.* 79: 2456.

Miotello, A., G. De Marchi, G. Mattei, P. Mazzoldi, and A. Quaranta. 2000. Clustering of silver atoms in hydrogenated silver-sodium exchanged glasses. *Appl. Phys. A* 70: 415.

Mooradian, A. 1969. Photoluminescence of metals. *Phys. Rev. Lett.* 22: 185.

Moriarty, P. 2001. Nanostructured materials. *Rep. Prog. Phys.* 64: 297.

Moser, W. R. 1996. *Advanced Catalysts and Nanostructured Materials.* San Diego, CA: Academic Press.

Moskovits, M. 1985. Surface-enhanced spectroscopy. *Rev. Mod. Phys.* 57: 783.

Muñoz, J. E., J. Cervantes, R. Esparza, and G. Rosas. 2007. Iron nanoparticles produced by high-energy ball milling. *J. Nanopart. Res.* 9: 945.

Murray, D. B., C. H. Netting, L. Saviot et al. 2006. Far-infrared absorption by acoustic phonons in titanium dioxide nanopowders. *J. Nanoelectron. Optoelectron.* 1: 92.

Murray, W. A. and W. L. Barnes. 2007. Plasmonic materials. *Adv. Mater.* 19: 3771.

Nahm, C., D.-R. Jung, J. Kim et al. 2013. Photoluminescence enhancement by surface-plasmon resonance: Recombination-rate theory and experiments. *Appl. Phys. Exp.* 6: 052001.

Najafi, S. I. (ed.). 1992. *Introduction to Glass Integrated Optics.* Norwood, MA: Artech House.

Nakayama, K., K. Tanabe, and H. A. Atwater. 2008. Plasmonic nanoparticle enhanced light absorption in GaAs solar cells. *Appl. Phys. Lett.* 93: 121904.

Nalwa, H. S. (ed.). 2004. *Encyclopedia of Nanoscience and Nanotechnology.* Los Angeles, CA: American Scientific Publishers.

Nastasi, M., J. W. Mayer, and J. K. Hirvonen. 1996. *Ion-Solid Interactions: Fundamentals and Applications.* Cambridge, U.K.: Cambridge University Press.

Okamoto, K., I. Niki, A. Scherer, Y. Narukawa, T. Mukai, and Y. Kawakami. 2005. Surface plasmon enhanced spontaneous emission rate of InGaN/GaN quantum wells probed by time-resolved photoluminescence spectroscopy. *Appl. Phys. Lett.* 87: 071102.

Ovchinnikov, V. 2008. Formation and characterization of surface metal nanostructures with tunable optical properties. *Microelectron. J.* 39: 664.

Palik, E. D. (ed.). 1985. *Handbook of Optical Constants of Solids.* Orlando, FL: Academic Press Inc.

Persson, B. N. J. 1993. Polarizability of small spherical metal particles: Influence of the matrix environment. *Surf. Sci.* 281: 153.

Peyser, L. A., A. E. Vinson, A. P. Bartko, and R. M. Dickson. 2001. Photoactivated fluorescence from individual silver nanoclusters. *Science* 291: 103.

Portales, H., L. Saviot, E. Duval et al. 2001. Resonant Raman scattering by breathing modes of metal nanoparticles. *J. Chem. Phys.* 115: 3444.

Pradeep, T. and Anshup. 2009. Noble metal nanoparticles for water purification: A critical review. *Thin Solid Films* 517: 6441.

Pretorius, R., J. M. Harris, and M.-A. Nicolet. 1978. Reaction of thin metal films with SiO_2 substrates. *Solid-State Electron.* 21: 667.

Prieto, G., J. Zecevic, H. Friedrich, K. P. de Jong, and P. E. de Jongh. 2013. Towards stable catalysts by controlling collective properties of supported metal nanoparticles. *Nat. Mater.* 12: 34.

Quinten, M. 2011. *Optical Properties of Nanoparticle Systems: Mie and Beyond.* Weinheim, Germany: Wiley-VCH.

Ramjauny, Y., G. Rizza, S. Perruchas, T. Gacoin, and R. Botha. 2010. Controlling the size distribution of embedded Au nanoparticles using ion irradiation. *J. Appl. Phys.* 107: 104303.

Raza, S., N. Stenger, S. Kadkhodazadeh et al. 2013. Blueshift of the surface plasmon resonance in silver nanoparticles studied with EELS. *Nanophotonics* 2: 131.

Reiss, S. and K. Heinig. 1995. Computer simulation of mechanisms of the SIMOX process. *Nucl. Instrum. Methods Phys. Res. B* 102: 256.

Ren F., C. Jiang, C. Liu, J. Wang, and T. Oku. 2006. Controlling the morphology of Ag nanoclusters by ion implantation to different doses and subsequent annealing. *Phys. Rev. Lett.* 97: 165501.

Rizza, G., H. Cheverry, T. Gacoin, A. Lamasson, and S. Henry. 2007. Ion beam irradiation of embedded nanoparticles: Toward an in situ control of size and spatial distribution. *J. Appl. Phys.* 101: 014321.

Roco, M. C., R. S. Williams, and A. P. Alivisatos (eds.). 1999. *Nanotechnology Research Directions: IWGN Workshop Report.* Dordrecht, the Netherlands: Kluwer Academic Publishers.

Ruffino, F., M. G. Grimaldi, C. Bongiorno et al. 2009a. Normal and abnormal grain growth in nanostructured gold film. *J. Appl. Phys.* 105: 054311.

Ruffino, F., A. Irrera, R. De Bastiani, and M. G. Grimaldi. 2009b. Room-temperature grain growth in sputtered nanoscale Pd thin films: Dynamic scaling behaviour on SiO_2. *J. Appl. Phys.* 106: 084309.

Saha, K., S. S. Agasti, C. Kim, X. Li, and V. M. Rotello. 2012. Gold nanoparticles in chemical and biological sensing. *Chem. Rev.* 112: 2739.

Salerno, M., J. R. Krenn, B. Lamprecht et al. 2002. Plasmon polaritons in metal nanostructures: The optoelectronic route to nanotechnology. *Optoelectron. Rev.* 10: 217.

Sancho-Parramon, J. 2009. Surface plasmon resonance broadening of metallic particles in the quasi-static approximation: A numerical study of size confinement and interparticle interaction effects. *Nanotechnology* 20: 235706.

Sanders, A. W., D. A. Routenberg, B. J. Wiley, Y. Xia, E. R. Dufresne, and M. A. Reed. 2006. Observation of plasmon propagation, redirection, and fan-out in silver nanowires. *Nano Lett.* 6: 1822.

Saviot, L., A. Mermet, and E. Duval. 2011. Acoustic vibrations in nanoparticles. In *Handbook of Nanophysics*, Vol. 3, K. D. Sattler (ed.). Boca Raton, FL: Taylor & Francis.

Saviot, L., C. H. Netting, and D. B. Murray. 2008. Inelastic neutron scattering due to acoustic vibrations confined in nanoparticles: Theory and experiment. *Phys. Rev. B* 78: 245426.

Scholl, J. A., A. L. Koh, and J. A. Dionne. 2012. Quantum plasmon resonances of individual metallic nanoparticles. *Nature* 483: 421.

Sendova, M., J. A. Jiménez, R. Smitha, and N. Rudawskic. 2015. Kinetics of copper nanoparticle precipitation in phosphate glass: an isothermal plasmonic approach. *Phys. Chem. Chem. Phys.* 17: 1241.

Sigmund, P. 1969. Theory of sputtering. I. Sputtering yield of amorphous and polycrystalline targets. *Phys. Rev.* 184: 383.

Simmons, G. and H. Wang (eds.). 1971. *Single Crystal Elastic Constants and Calculated Aggregate Properties: A Handbook*, 2nd edn. Cambridge, U.K.: MIT.

Smolyaninov, I. I., A. V. Zayats, A. Gungor, and C. C. Davis. 2002. Single-photon tunneling via localized surface plasmons. *Phys. Rev. Lett.* 88: 187402.

Srivastava, S. K., P. Gangopadhyay, S. Amirthapandian et al. 2014. Effects of high-energy Si ion-irradiations on optical responses of Ag metal nanoparticles in a SiO_2 matrix. *Chem. Phys. Lett.* 607: 100.

Stamplecoskie, K. G. and J. C. Scaian. 2010. Light emitting diode irradiation can control the morphology and optical properties of silver nanoparticles. *J. Am. Chem. Soc.* 132: 1825.

Tao, A., F. Kim, C. Hess et al. 2003. Langmuir-Blodgett silver nanowire monolayers for molecular sensing using surface-enhanced Raman spectroscopy. *Nano Lett.* 3: 1229.

Tamura, A., K. Higeta, and T. Ichinokawa. 1982. Lattice vibrations and specific heat of a small particle. *J. Phys. C* 15: 4975.

Thomas, O. C., W. Zheng, S. Xu, and Jr. K. H. Bowen. 2002. Onset of metallic behavior in magnesium clusters. *Phys. Rev. Lett.* 89: 213403.

Thompson, M. W. 1969. *Defects and Radiation Damage in Metals*. Cambridge, U.K.: Cambridge University Press.

Varkey, A. J. and A. F. Fort. 1993. Some optical properties of silver peroxide (AgO) and silver oxide (Ag_2O) films produced by chemical-bath deposition. *Sol. Energy Mater. Sol. Cells* 29: 253.

Vasa, P., R. Pomraenke, S. Schwieger et al. 2008. Coherent exciton-surface-plasmon-polariton interaction in hybrid metal-semiconductor nanostructures. *Phys. Rev. Lett.* 101: 116801.

Vorobyev, A. Y., V. S. Makin, and C. Guo. 2007. Periodic ordering of random surface nanostructures induced by femtosecond laser pulses on metals. *J. Appl. Phys.* 101: 034903.

Wagner, C. 1961. Theory of transformation in sludge throw the resolution. *Z. Elektrochemie.* 65: 581.

Walters, R. J., R. V. A. van Loon, I. Brunets, J. Schmitz, and A. Polman. 2010. A silicon-based electrical source of surface plasmon polaritons. *Nat. Mater.* 9: 21.

Wang, P. W. 1996. Thermal stability of silver in ion-exchanged soda lime glasses. *J. Vac. Sci. Technol. A* 14: 465.

Wang, Y., T. Yang, M. T. Tuominen, and M. Achermann. 2009. Radiative rate enhancements in ensembles of hybrid metal-semiconductor nanostructures. *Phys. Rev. Lett.* 102: 163001.

Wiederrecht, G. P., G. A. Wurtz, and A. Bouhelier. 2008. Ultrafast hybrid plasmonics. *Chem. Phys. Lett.* 461: 171.

Xiao, Y., F. Patolsky, E. Katz, J. F. Hainfeld, and I. Willner. 2003. Plugging into enzymes: Nanowiring of redox enzymes by a gold nanoparticle. *Science* 299: 1877.

Xiong, Y., R. Long, D. Liu et al. 2012. Solar energy conversion with tunable plasmonic nanostructures for thermoelectric devices. *Nanoscale* 4: 4416.

Xu, H., E. J. Bjerneld, M. Kall, and L. Borjesson. 1999. Spectroscopy of single hemoglobin molecules by surface enhanced Raman scattering. *Phys. Rev. Lett.* 83: 4357.

Xu, H. and K. S. Suslick. 2010. Water-soluble fluorescent silver nanoclusters. *Adv. Mater.* 22: 1078.

Yoshida, K., T. Itoh, H. Tamaru, V. Biju, M. Ishikawa, and Y. Ozaki. 2010. Quantitative evaluation of electromagnetic enhancement in surface-enhanced resonance Raman scattering from plasmonic properties and morphologies of individual Ag nanostructures. *Phys. Rev. B* 81: 115406.

Zhang, H., K.-B. Lee, Z. Li, and C. A Mirkin. 2003. Biofunctionalized nanoarrays of inorganic structures prepared by dip-pen nanolithography. *Nanotechnology* 14: 1113.

Zhang, J. T., Y. Tang, K. Lee, and M. Ouyang. 2010. Tailoring light-matter-spin interactions in colloidal hetero-nanostructures. *Nature* 466: 91.

Zheng, J., C. Zhang, and R. M. Dickson. 2004. Highly fluorescent, water-soluble, size-tunable gold quantum dots. *Phys. Rev. Lett.* 93: 077402.

Zhou, L., C. Zhang, Y. Yang, B. Li, and L. Zhang. 2012. Narrow size distribution of Au nanocrystals formed in sapphire by utilizing Ar ion irradiation and thermal annealing. *Nucl. Instrum. Methods Phys. Res. B* 278: 42.

Zhou, X. D., X. H. Xiao, J. X. Xu, G. X. Cai, F. Ren, and C. Z. Jiang. 2011. Mechanism of the enhancement and quenching of ZnO photoluminescence by ZnO–Ag coupling. *Europhys. Lett.* 93: 57009.

Ziegler, J. F., J. P. Biersack, and U. Littmark. 1985. *The Stopping and Range of Ions in Solids*. New York: Pergamon.

5 Silicon Nanocrystals
Properties and Potential Applications

Spiros Gardelis

CONTENTS

5.1 OVERVIEW

Silicon nanocrystals exhibit unique optical and electrical properties due to their quantum size and charging effects, respectively. In the 1990s, for the first time, it was observed that nanostructured silicon in the form of porous silicon emitted efficiently visible light even at room temperature. Later, it was also observed that silicon nanocrystals with sizes smaller than the Bohr radius of the exciton in bulk silicon could also emit visible light. These observations triggered enormous efforts by the scientific community to use silicon nanostructures for optoelectronic applications, as this material could be integrated in an all-silicon chip containing optoelectronic components. Beyond their applications in optoelectronics, silicon nanocrystals showed potential use in sensors, photonics, and solar cells. The most recent report on 100% internal quantum efficiency of light-emitting silicon nanocrystals has revived interest in silicon nanocrystals as an efficient light-emitting material that could replace the toxic Cd- and Pb-based quantum dots in many biological and medical applications. On the other end, the charging effects observed in silicon nanocrystals have already been exploited in embedded nonvolatile memories in microcontrollers. In this chapter, we give a brief historical account on this field, present methods of growing silicon nanocrystals, and review their optical and electrical properties as well as the prospects for their applications in electronics, sensors, photonics, photovoltaic devices, biology, and medicine.

5.2 INTRODUCTION

After the discovery of the transistor in 1948, there was a revolution in the electronics industry, and semiconductors have been the key materials used extensively in microelectronics, optoelectronics, and solar cells. An important property of a semiconductor that makes it useful for the aforementioned applications is its energy bandgap. The existence of a bandgap poses a barrier in the carrier transport. At energies above this barrier, carriers can flow as in the case of a metal resistor, whereas below it the carriers cannot flow and the material acts as an insulator. The presence of the bandgap results in the existence of two types of carriers in the semiconductor: electrons and holes. Suitable doping of the semiconductor with acceptors or donors can make the material p-type or n-type, respectively. Depending on the doping concentration, the resistivity can be tailored over a broad range. This makes possible the fabrication of diodes and transistors for various applications. Thus, the rectifying behavior of a diode and the property of switching in a transistor—the two properties microelectronics is based on—are due to the existence of the energy bandgap.

Another important consequence of the existence of energy bandgap in semiconductors is light emission and photoconductivity, which are essential in optoelectronics and solar cell applications, respectively. Miniaturization and integration are two essential properties of semiconductors that have helped the semiconductor industry contribute significantly to the world economic growth steadily in the last 40 years with significant consequences in technology and society. Both these properties have driven the race for increasing densities of semiconductor devices that can be incorporated in a single chip in order to make processors faster and carry out more complicated tasks. Of course, the leading semiconducting material for this race is silicon. The reason for this is that silicon is abundant in nature and thus cheaper to fabricate. Although the first transistor was fabricated using germanium, this material was quickly abandoned in the microelectronics industry as it has a lower cutoff current with smaller temperature variation compared to silicon and a poorer ability to withstand heat than silicon. Besides, silicon is more abundant and thus cheaper than germanium.

Although silicon is a semiconductor material with very good electrical properties, it is an indirect bandgap semiconductor, and as such it is not very useful for optoelectronic applications. The idea of all-silicon optoelectronics, that is, the fabrication of a silicon chip where lasers, photodetectors, and transistors are fully integrated, has been the driving force for finding materials with good optoelectronic properties that could be integrated with silicon. Many significant efforts have been made in this direction. SiGe, erbium-doped silicon, and silicon nanocrystals were the best candidates to fulfill this idea of an all-silicon superchip. Prompted by this important potential application of silicon nanocrystals, we will review the methods of their fabrication, their optoelectronic properties, and, finally, their applications that span from optoelectronics to biology and medicine.

This chapter is structured as follows: First, we will give a historic review of this subject. Then we will present the fabrication methods of silicon nanocrystals. Then we will review the optical, electrical, and optoelectronic properties of silicon nanocrystals, which are different from those of bulk silicon as a result of quantum size effects. Finally, we will review the possible applications of silicon nanocrystals.

5.3   A HISTORIC OVERVIEW

In 1990, Canham observed efficient room-temperature photoluminescence in the visible region from porous silicon [1]. This was strange, as silicon is an indirect energy bandgap material and as such it cannot emit efficiently. This was literally the starting point of rigorous research on the origin of light emission from porous silicon. After many contradicting theories put forward to explain the origin of the effect, there is now consensus that light emission is a result of quantum and spatial confinement of electrons and holes, which recombine radiatively within the nanostructured silicon that exists between the pores of porous silicon. For light emission to occur, the silicon nanostructures must have dimensions smaller than the Bohr radius of exciton in bulk silicon (~4.3 nm) [2]. The idea that quantum confinement within the silicon nanostructures in porous silicon is the origin of light emission led the researchers to fabricate silicon nanocrystals with controllable sizes able to emit in different energies within the visible and near-infrared region [3–5].

In addition to the light emission properties of the silicon nanocrystals, unique electrical properties were also observed because of their charging effects [6–8]. This initiated rigorous research to use this property in nonvolatile memory structures [6,9–11].

5.4   METHODS OF FABRICATION OF SILICON NANOCRYSTALS

Various routes of fabrication have been used to produce silicon nanocrystals such as electrochemical, sputtering, ablation, and colloidal chemistry. We review briefly each of these methods.

5.4.1   ELECTROCHEMISTRY: POROUS SILICON AND STAIN ETCHING

Electrochemical anodization of crystalline silicon in aqueous or ethanolic solutions of hydrofluoric acid (HF) can cause the selective etching of silicon and modify it into a porous material. Between the pores, the remaining silicon retains its crystallinity. Depending on the conditions of the anodic reaction, the material can be nanoporous to microporous (pore size < 2 nm), mesoporous (pore size 2–50 nm), or macroporous (pore size > 50 nm). Figure 5.1 shows an example of

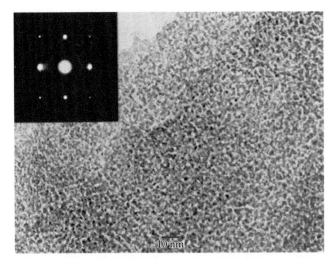

FIGURE 5.1   Transmission electron microscopy (TEM) image of a microporous silicon layer. (Reproduced with permission from Cullis, A.G., Canham, L.T., and Calcott, P.D.J., The structural and luminescence properties of porous silicon, *J. Appl. Phys.*, 82, 909. © 1997, American Institute of Physics.)

microporous silicon [12]. Specifically, the morphology of porous silicon depends on the doping and type of the starting bulk silicon, the concentration of hydrofluoric acid in the electrolyte, and the anodic current density [12,13]. The etching takes place with the help of holes. Thus, if the starting material is p-type, the anodic reaction is carried out in the dark. In the case where the starting material is of n-type and the majority carriers are electrons, etching can occur only under illumination so that the necessary holes can be generated for the reaction. The thickness of the porous silicon layer depends on the anodization reaction time.

Stain etching is another method to fabricate nanocrystalline silicon. In this method, crystalline Si is dipped in aqueous solutions containing HF and HNO_3 [14]. Selective etching of Si occurs via its oxidation by hole injection from HNO_3 and subsequent etching of the oxide by the HF. The thickness of the film is limited by the diffusion of the ionic species and not by the reaction time [15]. Increasing the impurity doping of the substrate results in thicker films [16].

Porous Si produced by both electrochemical and chemical etching of silicon is hydrogen-terminated and very vulnerable to environmental oxidation, which affects its stability with regard to its light emission properties. In order to stabilize the porous structure against degradation in light emission, controlled oxidation was carried out by chemical, electrochemical, or thermal oxidation processes. The next logic step was to consider fabricating silicon nanocrystals embedded in a dielectric matrix.

5.4.2 Silicon Nanocrystals Embedded in Dielectric Matrices

Two dielectric matrices compatible with Si technology were investigated in order to produce Si nanocrystals of well-controlled sizes and to stabilize mainly their light emission properties: they were stoichiometric silicon dioxide and silicon nitride. In the following, we review the fabrication methods of these structures.

5.4.2.1 Oxidation of Porous Silicon

Oxidation of porous silicon can produce silicon nanocrystals that are interconnected and covered by a silicon dioxide layer. We now review the different methods of oxidation of porous silicon. Four different methods were used to oxidize the pore walls intentionally and stabilize them against further oxidation due to exposure to the ambient air. This can be done chemically, electrochemically, by thermal oxidation, or by wet oxidation. Chemical oxidation can be done in oxidants such as hydrogen peroxide, nitric acid, and dimethylsulfoxide [17,18]. Electrochemical oxidation takes place in electrolytes such as KNO_3 with the use of current in an electrochemical cell [19]. Thermal oxidation is performed in dry oxygen ambient in a furnace at temperatures between 300°C and 900°C. At moderate temperatures between 300°C and 600°C, oxidation causes loss of hydrogen from the pore walls, resulting in poor electronic and light emission properties. To resolve this problem, a subsequent high-temperature oxidation (between 800°C and 900°C)—usually a rapid one in order to avoid total oxidation of porous silicon—must be performed. This process leads to a good passivation of the pore walls with stoichiometric silicon dioxide and consequently improves and stabilizes the light emission and electronic properties of the material [20–22]. Finally, wet oxidation at high pressures [23] or at ambient pressure [24] has also been used to passivate porous silicon with a good quality layer of silicon dioxide.

5.4.2.2 Silicon Nanocrystals Embedded in Silicon Dioxide

The luminescent properties of porous silicon led various research groups to develop new methods of producing silicon nanostructures. In these methods, the silicon nanocrystals are literally embedded in silicon dioxide. We will review these methods, which produce silicon nanocrystals with controllable sizes. One approach is ion implantation of silicon in a silicon dioxide matrix. Ion doses above the equilibrium solid solubility limit results in phase separation during ion implantation and aggregation of silicon ions, which eventually form clusters. Post-annealing, depending on the temperature, and the ion dose determine the final size and the crystallinity of the resulting silicon clusters. Also, acceleration energies determine the depth to which these ions can penetrate and eventually form the silicon clusters [25–27]. Another approach of silicon nanocrystal growth in silicon dioxide is by sputtering. One example is sputtering of silicon suboxides and subsequent annealing. Annealing at temperatures as high as 1100°C in a nitrogen ambient separates the silicon suboxide into two different phases of silicon and silicon dioxide, forming silicon nanocrystals embedded in a silicon dioxide matrix [3]. Cosputtering of crystalline silicon and silicon dioxide targets and subsequent annealing at similar temperatures in nitrogen also leads to aggregation of silicon in silicon nanocrystals within a silicon dioxide matrix [7].

Growth of silicon nanocrystals embedded in silicon dioxide out of the gas phase and subsequent thermal oxidation is another well-known method. The method has two variations. The first one is called low-pressure chemical vapor deposition (LPCVD) and the other one is plasma-enhanced chemical vapor deposition (PECVD). Silane is used in both techniques as a precursor. In LPCVD, the temperature of the reaction can be as high as 600°C, whereas in PECVD the deposition can be done at lower temperatures. For example, the growth of silicon on a layer of silicon dioxide grown by LPCVD at a growth temperature 580°C is amorphous with nanocrystalline nucleation sites [4,5], whereas at growth temperatures above 600°C the silicon layers grown are nanocrystalline [28,29]. In PECVD, where the deposition is performed at a much lower growth temperature (e.g., 250°C), the deposited silicon layer is amorphous. In this case, the crystallization temperature needs to be as high as 1000°C [30]. In either LPCVD or PECVD, to grow well-separated silicon nanocrystals by silicon dioxide, a subsequent step of thermal oxidation at elevated temperatures, usually between 800°C and 900°C, is necessary. The oxidation time and the initial thickness of the nanocrystalline silicon layer determine the final size of the silicon nanocrystals. In the PECVD method, an annealing step is necessary before thermal oxidation to crystallize the initially amorphous silicon layer. The size of the silicon nanocrystals grown by these CVD techniques after thermal oxidation is well controlled and the size dispersion narrow. However, in order to fabricate silicon nanocrystals well separated by silicon dioxide barriers, the initial nanocrystalline layer thickness must be small enough, that is, a few nanometers. Silicon nanocrystalline layers have also been grown on quartz substrates by these techniques. In the case of LPCVD growth, the nanocrystalline layer is columnar, consisting of silicon nanocrystals with a certain lateral size dispersion, and their dimension in the growth direction is equal to the thickness of the nanocrystalline layer [28,29]. In this case, silicon nanocrystals embedded in silicon dioxide can be fabricated after thermal oxidation but the initial nanocrystalline layer must be thin enough so that the nanocrystals become separated by a silicon dioxide barrier [28,29]. Figure 5.2 shows two examples of multiple layers of silicon nanocrystals embedded in silicon dioxide [3,5].

5.4.2.3 Silicon Nanocrystals Embedded in Other Dielectric Matrices (Silicon Nitride–Silicon Oxynitride)

One drawback of silicon dioxide is its high barrier, which makes carrier injection and tunneling through it more difficult and thus electroluminescence poor, and the prospect of using this system as a light-emitting diode (LED) is challenging [31,32]. Another important drawback is that the color tunability of such a structure is limited by defects at the silicon dioxide–silicon nanocrystal interface, which introduce energy states within the energy bandgap of the silicon nanocrystals [33,34]. To resolve these problems and make silicon nanocrystals more useful for optoelectronic application, silicon dioxide has been replaced by silicon nitride, which has a lower energy bandgap barrier. This has been realized by growing silicon-rich silicon nitride layers either by PECVD or LPCVD and subsequently annealing them at elevated temperatures so that silicon within these layers can agglomerate forming silicon nanocrystals within a stoichiometric silicon nitride matrix [35–39].

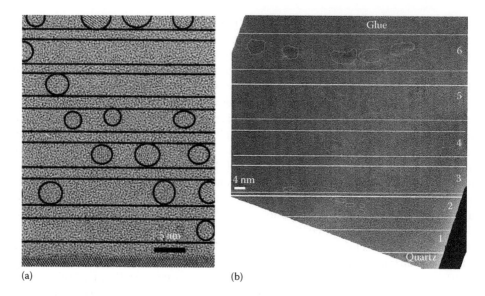

FIGURE 5.2 Silicon nanocrystals/silicon dioxide superlattices fabricated (a) by silicon suboxide sputtering and subsequent annealing, and (b) by LPCVD deposition of silicon and subsequent thermal oxidation. (a: Reproduced with permission from Zacharias, M., Heitmann, J., Scholz, R., Kahler, U., Schmidt, M., and Bläsing, J., Size-controlled highly luminescent silicon nanocrystals: A SiO/SiO$_2$ superlattice approach, *Appl. Phys. Lett.*, 80, 661. © 2002, American Institute of Physics; b: Reproduced with permission from Gardelis, S., Nassiopoulou, A.G., Vouroutzis, N., and Frangis, N., Effect of exciton migration on the light emission properties in silicon nanocrystal ensembles, *J. Appl. Phys.*, 105, 113509. © 2009, American Institute of Physics.)

5.4.2.4 Growth of Silicon Nanocrystals by Laser Ablation

Another technique of silicon nanocrystal growth is laser ablation. In this technique, the target can be crystalline silicon onto which a pulsed laser is focused. The laser heats locally the silicon target, generating a plasma, which nucleates on the substrate and forms silicon nanocrystals that are unpassivated. Controlled passivation with either silicon dioxide or hydrogen makes the silicon nanocrystals to emit visible light [40,41].

5.4.2.5 Colloidal Chemistry

Chemistry offers versatile routes to produce silicon nanocrystals with suitable surface functionalization for various applications, mainly biological ones [42]. These chemical routes of silicon nanocrystal fabrication are based on the reduction of halide salts [43,44], reaction of alkali silicides [45–47], or thermal treatment of sol–gel precursors [48,49]. Finally, it has been demonstrated that silicon nanocrystals can be doped. Such an example is phosphorous doping using PCl$_3$ [50].

5.5 OPTICAL PROPERTIES OF SILICON NANOCRYSTALS

5.5.1 Light Emission from Silicon Nanocrystals—General

The first observation of light emission from silicon nanostructures was reported by Canham in 1990 and was from porous silicon. This was attributed from the very beginning to the quantum confinement of the excitons within the silicon nanocrystals or nanowires of porous silicon [1,51]. Other groups also confirmed efficient room-temperature luminescence from other silicon nanostructures such as porous SiGe [51], silicon nanocrystals embedded in silicon dioxide [3,4], or silicon nitride [36–39]. There has been a very vivid discussion on the origin of the light emission, which was sparked by the fact that luminescence from porous silicon was vulnerable to the chemical termination of the surface. For example, the fact that luminescence was quenched after desorption of hydrogen in the freshly anodized porous silicon and recovered after its immersion in hydrofluoric acid solutions led some researchers to claim that silicon hydride [52] or siloxane [53] was the origin of the light emission. However, observations complying with the quantum confinement of excitons within the silicon nanocrystals in the porous silicon such as the blue shift of the luminescence [1,51,54] and of the absorption edge with decreasing size of the silicon nanocrystals, which were also confirmed by theoretical calculations [55], led to the firm conclusion that quantum confinement of excitons is the origin of luminescence. Surface termination and passivation of silicon nanostructures determine the emission properties, but are not the origin of luminescence [56]. For example, hydrogen passivation does not limit blue shift of the luminescence with decreasing size, and the luminescence follows the bandgap opening. On the contrary, silicon oxide termination limits the blue shift, as it introduces energy states within the energy bandgap of the silicon nanocrystals that are involved in the emission [33]. Different chemicals on the surface can quench luminescence as they introduce nonradiative centers. Specifically, a number of organic and inorganic molecules adsorbed on the surface of the silicon nanocrystals exhibited a quenching behavior for the light emission. This effect has been mainly explained by the energy or electron transfer between the intrinsic energy states of the nanocrystals and those of the quencher [57–59]. For example, in the presence of aromatic nitro compounds, luminescence is quenched because of the transfer of an electron from the conduction band of the nanocrystal to a vacant orbital of the quencher [60,61]. Moreover, it was reported that suitable surface passivation of the silicon nanocrystals can result in quantum yields as high as 40%–70% for silicon nanocrystals [62–64]. More recently, colloidal silicon nanocrystals with suitable ligand passivation showed quantum yields near 100% [65].

Why then can silicon nanocrystals be so bright when silicon itself is an indirect bandgap semiconductor? Two preconditions must be fulfilled in order to observe light emission from a silicon nanocrystal. The most important one is that the size of the silicon nanocrystal must be smaller than the Bohr radius of exciton in bulk silicon, which has been calculated to be 4.3 nm [2,66]. The second one is the passivation of any dangling bonds, which could act as nonradiative centers and quench the luminescence, with suitable atoms such as hydrogen. We will examine each of these conditions separately. The spatial confinement of the generated excitons within a silicon nanocrystal smaller than 4.3 nm results in a significant overlap, which increases the oscillator strength of the radiative emission even though silicon is an indirect bandgap semiconductor and as such any transition needs phonons to be fulfilled. Because of the Heisenberg uncertainty principle, spatial confinement enhances the uncertainty in the momentum space k. Thus, the selection rule for phonon-assisted transitions breaks down, resulting in a much larger number of phonon modes fulfilling the radiative recombination of the excitons within the silicon nanocrystal [67]. The second precondition for efficient luminescence is the passivation of the dangling bonds. This can be done by hydrogen passivation, which protects silicon nanocrystals from oxidation that results in the degradation of luminescence efficiency and also in the change in the energetics of the luminescence [33,68]. Other than hydrogen passivation, additional passivation is needed in the case where silicon nanocrystals are to be used as luminescent labels in biological applications where cadmium-based quantum dots can be toxic. Thus, the need for a suitable functionalization of their surface, so that silicon nanocrystals could be linked to biomolecules, has led to the alkylation of their hydrogen-terminated surface [69–71]. This makes them chemically stable, able to retain the efficient luminescence, and also suitably functionalized for such applications.

Another important consequence of the confinement of the carriers in the silicon nanocrystals is quantization of the energy states where the carrier can be found. This occurs when the size of the nanocrystal is smaller than the de Broglie length, or in other words, the wavelength that corresponds to the carrier. In the case of silicon, the wavelength of the electron is 8 nm. When the size of the nanocrystal decreases below the electron wavelength, then quantum size effects and a widening of the energy bandgap occur. Thus, for nanocrystal sizes smaller than the exciton Bohr radius, where efficient light emission is observed, the energy states of the nanocrystal become quantized and the energy bandgap increases with decreasing nanocrystal size.

5.5.2 Light Emission from Silicon Nanocrystals Embedded in Silicon Dioxide and Silicon Nitride

From the very beginning of the investigation of light emission from silicon nanostructures, the role of oxygen in the degradation and the modification of the energetics of the luminescence from these structures were acknowledged. This was first observed in luminescent porous silicon. Freshly prepared porous silicon is very reactive to the oxygen of the atmosphere, and as it ages in atmosphere, hydrogen is replaced gradually by silicon oxides. This causes quenching and spectral shift of the initial efficient luminescence of the freshly prepared porous silicon. It has been observed that for sizes of silicon nanocrystals smaller than 3 nm, although oxidation would cause decrease of the silicon size and thus a blue shift to the luminescence spectrum, the luminescence peak could not shift to energies higher than 2.1 eV. Instead, after air exposure, a red shift of the luminescence as large as 1 eV was observed for sizes less than 2 nm. Wolkin et al. [33] were the first to resolve this controversy by introducing a model in which silicon oxide was assumed to introduce energy states in the energy bandgap of the smaller silicon nanocrystals (smaller than 3 nm), which are involved in the radiative recombination process, thus determining the energetics of the luminescence. Figure 5.3 shows the evolution of the electronic states in oxidized silicon nanocrystals as a function of their size, as proposed by Wolkin et al. [33].

Theoretical work supported this idea and explained further the role of oxidation of the silicon nanostructures in light emission. It was confirmed that silicon oxides introduce interface states within the energy bandgap of the silicon nanocrystals or even strain, which modify luminescence spectrum from the case of the hydrogen termination of the nanocrystals [72–75].

However, a deliberate passivation of the surface of porous silicon with a good-quality thermal silicon dioxide formed at elevated temperatures (800°C–900°C) has been proved to decrease the number of nonradiative centers and stabilize the photoluminescence properties of porous silicon

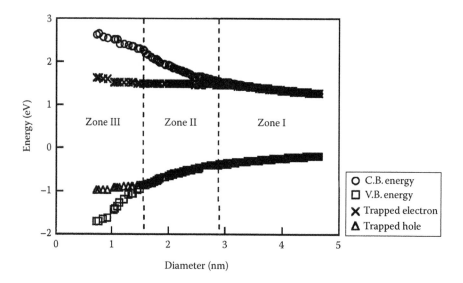

FIGURE 5.3 Electronic states in oxidized silicon nanocrystals as a function of their size. The trapped electron state is a p-state localized on the Si atom of the Si=O bond, and the trapped hole state is a p-state localized on the oxygen atom. (Reprinted with permission from Wolkin, M.V., Jorne, J., Fauchet, P.M., Allan, G., and Delerue, C., *Phys. Rev. Lett.*, 82, 197, 1999. © 1999 by the American Physical Society.)

against degradation in atmosphere. However, oxidation results in a large Stokes shift (~1 eV). The same occurs in the light emission of silicon nanocrystals embedded in silicon dioxide. An example of this Stokes shift of the photoluminescence peak relative to the absorption edge is shown in Figure 5.4 [29]. Regardless of the large Stokes shift, tunability of the energy bandgap with the size of the silicon nanocrystals has been shown. Figure 5.5 shows the blue shift of the photoluminescence spectra of silicon nanocrystals embedded in a silicon dioxide matrix with decreasing nanocrystal size [3].

Efficient luminescence has also been observed in silicon nanocrystals embedded in silicon nitride and attributed to quantum size effects within the silicon nanocrystals [32,35–39]. Colloidal silicon nanocrystals have also shown efficient luminescence, which can be tailored from the near-infrared to the whole visible spectrum depending on the size [42].

5.5.3 PHOTOLUMINESCENCE (PL) DYNAMICS

PL decay measurements have shown a stretched exponential time dependence of the PL intensity, $I_{PL}(t)$, expressed by

$$I_{PL}(t) = I_0 e^{-(t/\tau_{PL})^\beta} \tag{5.1}$$

where
I_0 is the photoluminescence intensity at $t = 0$
τ_{PL} is a characteristic decay time, or carrier recombination time
β is a dispersion factor that takes values between 0 and 1 and expresses the deviation of the photoluminescence time decay from the single exponential [5,76,77]

Figure 5.6 shows PL decay times extracted from stretched exponential fittings (insert) as a function of temperature for different films of silicon nanocrystals [5].

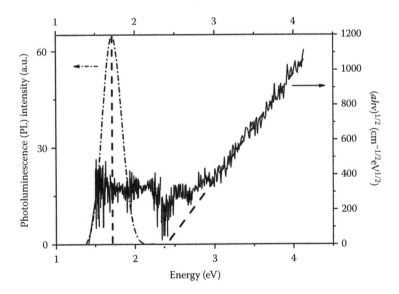

FIGURE 5.4 Example of Stokes shift of photoluminescence peak relative to the absorption edge in a layer of silicon nanocrystals embedded in silicon dioxide. The size of the nanocrystals is less than 2 nm. (Reproduced with permission from Gardelis, S., Nassiopoulou, A.G., Manousiadis, P., Milita, S., Gkanatsiou, A., Frangis, N., and Lioutas, Ch.B., Structural and optical characterization of two-dimensional arrays of Si nanocrystals embedded in SiO$_2$ for photovoltaic applications, *J. Appl. Phys.*, 111, 083536. © 2012, American Institute of Physics.)

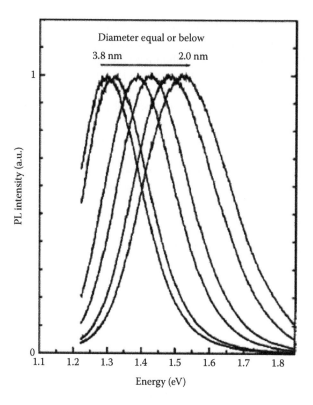

FIGURE 5.5 Normalized PL spectra from silicon nanocrystals embedded in silicon dioxide as a function of their sizes. (Reproduced with permission from Zacharias, M., Heitmann, J., Scholz, R., Kahler, U., Schmidt, M., and Bläsing, J., Size-controlled highly luminescent silicon nanocrystals: A SiO/SiO₂ superlattice approach, *Appl. Phys. Lett.*, 80, 661. © 2002, American Institute of Physics.)

The measured PL recombination rate r_{PL} is the sum of the radiative (r_r) and nonradiative (r_{nr}) recombination rates according to the following expression:

$$r_{PL} = r_r + r_{nr} = \frac{1}{\tau_{PL}} = \frac{1}{\tau_r} + \frac{1}{\tau_{nr}} \qquad (5.2)$$

where τ_r and τ_{nr} are the radiative and nonradiative recombination times, respectively. This stretched exponential behavior of PL decay is generally expected in disordered systems in which more than one pathway of carrier recombination exists. Different models have been proposed to explain this behavior. Some of these models suggest that this distribution of different times is a result of the interaction between the silicon nanocrystals in an ensemble of such crystals, which allows exciton migration from smaller to larger ones [78–82]. Other models suggest that migration of carriers is not necessary for the observed behavior of the PL decay. Instead, the excitons contributing to the PL are strongly localized, and recombination dynamics, which is a competition between radiative and nonradiative recombination of the localized excitons, is characterized by different decay times due to the dispersion in the sizes of the silicon nanocrystals in an ensemble. That is to say, the stretched exponential behavior of the PL decay is not due to the interaction between the silicon nanocrystals in an ensemble [83]. Specifically, this behavior is a result of the indirect nature of the energy bandgap of the silicon nanocrystals [84,85].

PL recombination times of up to a few microseconds have been reported for silicon nanocrystals. These long recombination times are due to the indirect nature of the energy bandgap of the

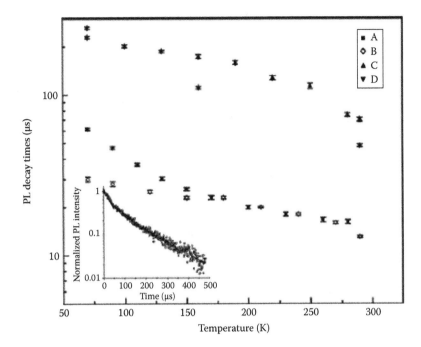

FIGURE 5.6 PL decay times extracted from stretched exponential fitting (insert) as a function of temperature. Samples A and B are porous silicon layers with porosities 85% and 62%, respectively. Sample C is a heavily oxidized porous silicon layer containing silicon nanocrystals within a silicon dioxide matrix. Sample D contains silicon nanocrystals embedded in a silicon dioxide matrix, grown by LPCVD deposition of silicon and subsequent thermal oxidation (its structure is shown in Figure 5.2b). (Reproduced with permission from Gardelis, S., Nassiopoulou, A.G., Vouroutzis, N., and Frangis, N., Effect of exciton migration on the light emission properties in silicon nanocrystal ensembles, *J. Appl. Phys.*, 105, 113509. © 2009, American Institute of Physics.)

silicon nanocrystals. It has been observed that the recombination time increases with decreasing temperature (see, e.g., Figure 5.6). Also, the PL intensity increases with decreasing temperature. This is due to the competition between radiative and nonradiative recombination. At higher temperatures, nonradiative transitions come into play, resulting in the quenching of PL and in a decrease of the measured recombination time according to Equation 5.2. Another interesting observation is that the PL intensity increases at lower temperatures up to a certain temperature and then decreases, whereas the recombination time increases. This effect was attributed to the fact that, due to quantum confinement, the orbital degeneracy of the bulk Si valence band is lifted by quantum confinement, resulting in splitting of a localized exciton state by the exchange interaction into a spin singlet and a spin triplet, with the triplet lower in energy. At low enough temperatures (e.g., 2 K), the generated excitons should be in the lower energy triplet state. Optical transitions from the triplet state are forbidden due to spin selection rules. On the contrary, transitions from the singlet state are allowed [12,86–88]. Thus, the radiative lifetime corresponding to transitions form triplet state should be infinite. However, due to spin–orbit scattering, there is some mixing between triplet and singlet states, and thus the transition from the triplet state becomes weakly allowed. At higher temperatures, singlet states become thermally populated and transitions take place from these states. Thus, the radiative lifetime becomes smaller, and the PL intensity increases with temperature. At some temperature, the PL intensity reaches its maximum value and at higher temperatures it decreases again due to the nonradiative transitions, as we have already mentioned.

Recombination lifetime has also a spectral dependence. It increases with increasing wavelength or decreasing detection energy. PL at lower wavelengths originates from the smaller nanocrystals within the ensemble, and thus the oscillator strength of the recombination transitions is larger. This is a consequence of the breakdown of the k-conservation rule and explains the faster recombination times at the lower wavelengths in the PL spectrum [89].

In an ensemble of silicon nanocrystals, recombination rates can be affected by the degree of electrical interconnection between the silicon nanocrystals. For example, in mesoporous and nanoporous silicon films, one measures recombination rates faster than in ensembles of silicon nanocrystals embedded in a silicon dioxide matrix. For example, in the first case, excitons can migrate and recombine away from their generation locations, whereas in the latter case the excitons are more localized. This can be understood if we introduce in the measured recombination rates a rate r_m, which is due to this exciton migration. Thus, the recombination rate r_{PL} of equation becomes $r_{PL} = r_r + r_{nr} + r_m$ [5,78–82]. Exciton migration in porous silicon results in a red shift of the luminescence with increasing temperature, which is larger than that expected from the temperature reduction of the energy bandgap of the nanocrystals. This occurs because at higher temperatures excitons are less localized and can recombine via lower energy states, that is, within larger silicon nanocrystals, than those in which they initially were generated. On the contrary, in the case of silicon nanocrystals embedded in the silicon dioxide, the excitons are more localized and there is almost no shift of the PL with the temperature [5].

5.5.4 Other Issues Regarding Light Emission

5.5.4.1 Optical Gain

Optical gain is necessary for laser applications. Pavesi et al. claimed that they observed optical gain in silicon nanocrystals embedded in a silicon dioxide matrix [90]. They explained this effect by introducing a three-level model. Specifically, they considered that the generated carriers, before their recombination, populate intermediate energy states, within the energy bandgap of the silicon nanocrystals, associated with interface states of the silicon nanocrystals with the surrounding silicon nanocrystals. This created a population inversion, which resulted in stimulated emission and thus optical gain. Other researchers have also reported optical gain in similar systems of silicon nanocrystals [91,92].

However, when population inversion is needed to achieve lasing, high excitation is needed. Under these conditions, fast nonradiative processes, such as Auger recombination (participation of three particles in nonradiative processes or free carrier absorption), come into play. These processes provide loss mechanisms, as they deplete the excited population. Thus, gain competes with these loss mechanisms. Due to these competing loss mechanisms, optical gain from silicon nanocrystals still remains an open question. Suppression of these two loss mechanisms would result in the exploitation of gain for lasing applications [93].

5.5.4.2 Multiple Exciton Generation (MEG) in Silicon Nanocrystals

MEG was first demonstrated in colloidal silicon nanocrystals. It was found that the threshold for MEG in silicon nanocrystals with a size of 9.5 nm, which corresponds in an energy bandgap of 1.2 eV, is 2.4 times the energy bandgap of the silicon nanocrystal. Also, a quantum yield of 2.6 per absorbed photon was found in silicon at 3.4 times the energy bandgap [94]. Later, the effect was demonstrated in silicon nanocrystals embedded in silicon dioxide. The multiple excitons are generated by the absorption of one photon of energy larger than twice the energy bandgap of the nanocrystals, not in the same nanocrystal but in adjacent ones, so that multiple excitons do not undergo nonradiative Auger recombination. This effect is very important for solar cell applications, as the generation of multiple excitons could improve the conversion efficiencies of silicon-based solar cells [95].

5.5.4.3   Silicon Nanocrystals as Sensitizers

Silicon nanostructures have been used as optical amplifiers in order to enhance erbium luminescence (1.54 μm), which is used in optical communications due to the low attenuation of this emission in long-haul optical fibers. Also, this optical amplification provided by the silicon nanocrystals is promising for the development of erbium lasers. The first efforts were made in porous silicon [96,97]. For example, an order of magnitude more intense luminescence from erbium was observed in porous silicon compared to that of erbium in crystalline bulk silicon. This was one of the first observations that silicon nanostructures could provide amplification of the erbium luminescence [96]. More recently, erbium-doped silicon nanocrystals showed remarkable enhancement of the erbium luminescence [98–102]. This enhancement originates from the higher absorption cross-section of silicon nanocrystals in combination with the efficient energy transfer to the adjacent erbium ions.

5.5.5   ELECTROLUMINESCENCE

The significant result of light emission from silicon nanostructures initiated significant effort to demonstrate the effect electrically, that is, with electrical injection of electrons and holes within the silicon nanostructures. Electroluminescence was first demonstrated by a redox reaction in a solution during the anodic oxidation of porous silicon and then in various types of solid-state devices, such as Schottky [103–109], p–n, or p–i–n diodes [110–116]. The interest and the effort to fabricate efficient electroluminescent devices of silicon nanostructures passed from porous silicon to other types of silicon nanostructures. A demonstration of electroluminescence from silicon nanostructures other than porous silicon was reported in 1996 for silicon nanopillars (nanowires) fabricated in a vertical direction on a crystalline silicon wafer. The device was fabricated as a Schottky diode [117]. The main interest now for silicon nanocrystal–based electroluminescent devices has shifted to silicon nanocrystals embedded in dielectric matrices. Although these structures show very stable and efficient photoluminescence, the presence of the electrically insulating dielectric matrix is an obstacle for efficient carrier injection into the silicon nanocrystals, making electroluminescence very inefficient. Silicon nitride or silicon carbide has been suggested as a dielectric matrix alternative to silicon dioxide because of their smaller energy bandgap compared to that of silicon dioxide. This would ease tunneling or injection of the carriers into the silicon nanocrystals and consequently improve the efficiency of the electroluminescent device. Also, thinner dielectric matrices would make carrier injection into the silicon nanocrystals easy [118]. Earlier works showed electroluminescence in MOS devices in which the silicon nanocrystals were embedded in the silicon dioxide of the gate. These devices work under high voltages, higher than 5 V. Under these voltages, unipolar field-enhanced Fowler–Nordheim tunneling is the main mechanism of carrier injection. Under these conditions, charge transport occurs in the oxide via hot electrons, whereas the electron–hole pairs are generated within the silicon nanocrystals by impact ionization [119–127]. Quantum efficiencies of these devices were generally small, 10^{-3}% to 10^{-4}%. A record high of 0.16% under pulsed operation has been reported [112]. A previous record high was 0.01% [114]. Figure 5.7 shows an example of electroluminescence from silicon nanocrystals embedded in the silicon dioxide of the gate of an MOS device [122].

Hybrid devices with silicon nanocrystals and organic polymers have shown efficient white-light electroluminescence or tunable electroluminescence depending on the size of the silicon nanocrystals. External quantum efficiencies as high as 1.1% or even 3.6% have been reported [128,129]. A significant drawback of the silicon nanocrystal–based electroluminescent devices, apart from their low efficiencies, is that they suffer from long emission decays, which limit the modulation frequency of the electroluminescence only to applications for light-emitting display diodes, thus making them unsuitable for laser applications.

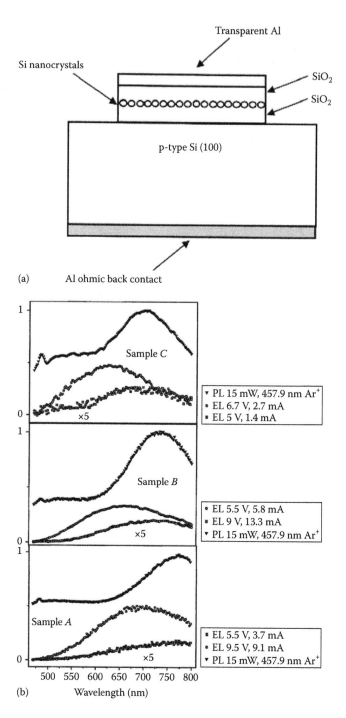

FIGURE 5.7 (a) Light-emitting diode based on silicon nanocrystals embedded in a silicon dioxide matrix. Hot electrons are injected into the silicon nanocrystals generating electron–hole pairs by impact ionization, which eventually recombine giving rise to the observed electroluminescence (EL). (b) Comparison of PL and EL from samples containing different sizes of silicon nanocrystals (A = 2 nm, B = 1.5 nm, and C = 1 nm). (Reproduced with permission from Photopoulos, P. and Nassiopoulou, A.G., Room- and low-temperature voltage tunable electroluminescence from a single layer of silicon quantum dots in between two thin SiO_2 layers, *Appl. Phys. Lett.*, 77, 1816. © 2000, American Institute of Physics.)

5.6 CHARGING EFFECTS IN SILICON NANOCRYSTALS—NONVOLATILE MEMORIES

Silicon nanocrystals show charging effects when their sizes are of the order of a few nanometers [6–9]. Specifically, if the silicon nanocrystal is charged, extra energy is necessary to inject an extra electron into it because of Coulomb interactions. This extra energy is called the "charging energy" of the nanocrystal. This effect can be exploited in nonvolatile memories. The use of embedded flash nonvolatile memories in microcontrollers, which have found their use in consumer, automotive, and other industrial markets, is the driving force for more scalable and thus less energy-consuming flash memories. However, the existing technology of flash memories has its limitations in scalability and energy consumption. For example, floating memories, which dominate the industry, suffer from scalability as the tunnel oxide cannot be made thinner than 10 nm, as stress-induced leakage currents come into play and reduce the retention times of the memory. Other types of memories, such as resistive, ferroelectric RAM, or phase change memories, are not as mature as the silicon-based ones. Silicon nanocrystal–based nonvolatile memories can solve these problems of flash memories. The silicon nanocrystal nonvolatile memory is fabricated as a field-effect transistor with its gate formed by silicon nanocrystals embedded in silicon dioxide. This memory is better than the standard floating-gate flash memory as the gate is not a continuous nanocrystalline silicon layer but consists of discrete nanocrystals that can be charged separately. In the standard flash memory, the continuity of the gate results in leakages, which destroy the memory effect. In the case of the silicon nanocrystal, if any defect at the silicon nanocrystal–silicon dioxide interface causes any leakage of the charge out of the nanocrystal, this causes only a minimal shift in the threshold voltage without destroying the nonvolatile behavior of the memory. Also, this allows scaling the bottom tunnel oxide by 40%–50% compared to a floating gate and consequently for lower energy consumption [11]. The concept of using discrete silicon nanocrystals for storing charge was introduced in 1996 [130]. Since then, different approaches of a single layer of silicon nanocrystals embedded in the gate oxide of a field-effect transistor have been implemented. For example, LPCVD [4,28,29] and PECVD [131] growth of nanocrystalline or amorphous silicon and subsequent thermal oxidation results in the formation of a single layer of silicon nanocrystals embedded in a silicon dioxide matrix. Another method is by implantation of silicon ions in the gate oxide of a field-effect transistor at low silicon ion energies and the subsequent annealing, which results in the formation of a single layer of silicon nanocrystals in the gate oxide [10,132]. Today, microcontrollers with embedded silicon nanocrystal nonvolatile memories are commercially available [11,133].

5.7 APPLICATIONS OF SILICON NANOCRYSTALS

Silicon has been the dominant material in the microelectronics industry for over half a century now. Advancements in silicon CMOS (complementary metal–oxide–semiconductor) technology, integration, and miniaturization have enabled the fabrication of complex, fast, and low-power-consumption devices. However, silicon as an indirect bandgap semiconductor cannot be used for optical amplifiers and light sources. It would be ideal if such devices could be fabricated and integrated with the existing CMOS technology in integrated circuits where all components are made out of silicon, that is, transistors, optical amplifiers, light sources, and photodetectors. Efficient light emission from silicon nanostructures was the driving force to investigate this material as a potential light source that could be integrated with the existing CMOS technology. Unfortunately, the slow switching and the low electroluminescence efficiency of the silicon nanocrystals make this material unsuitable for laser applications. On the other hand, the observed optical gain that is achieved at high pumping powers must compete with the nonradiative Auger recombination, and thus optical gain is not sufficient to support lasing. Nevertheless, efficient light-emitting diodes for display purposes could be produced using silicon nanocrystals and could replace other toxic materials such as Cd or Pb [128,129]. However, silicon nanocrystals could be used as sensitizers for erbium light emission, resulting in a considerable enhancement of erbium emission, which could be used for lasing applications or optical amplifiers [96,101]. Thus, silicon nanocrystals indirectly can be very useful for realizing a silicon-based laser source or an optical amplifier that could be integrated.

The recently observed 100% internal quantum efficiency of suitably passivated silicon nanocrystals shows that silicon nanocrystals, which are biocompatible, can replace the toxic Cd- and Pb-based quantum dots in biomedical imaging and diagnostic applications [65]. Another interesting application of such highly efficient light-emitting silicon nanocrystals could be their use in silicon-based solar cell application for downshifting [134,135]. As silicon is an indirect bandgap material, the photons of the solar spectrum with energies much higher than the energy bandgap of silicon can be lost as heat before being collected by the electrodes of the solar cell. By depositing a suitable layer of light-emitting silicon nanocrystals on the surface of the solar cell, the high-energy photons of the solar spectrum can be converted efficiently into photons of lower energies, which could be absorbed from the silicon absorber underneath and converted into electricity more efficiently, thus improving the conversion efficiency of the solar cell. This effect of downshifting could be enhanced even further by the use of metal nanoparticles deposited in the vicinity of the silicon nanocrystals, which can couple with the plasmon modes of the metal nanoparticles.

Silicon nanocrystals, as a result of their energy bandgap tailoring, could be used in tandem solar cell configurations [136]. These cells can be fabricated as multiple p–i–n junctions in a tandem configuration, where i is a layer of silicon nanocrystals of a certain size that are embedded in a dielectric matrix. Such solar cells are expected to show higher conversion efficiencies compared to the conventional silicon-based solar cells. Apart from silicon dioxide, other dielectric matrices such as silicon nitride and silicon carbide have been suggested, as these dielectrics form smaller barriers with the embedded silicon nanocrystals, thus making it easier for the carriers generated within the silicon nanocrystals to overcome the barrier and be collected by the electrodes of the cell [137]. Much effort has been made to produce suitable solar cell absorber layers of silicon nanocrystals embedded in a dielectric matrix [138–140]. In another approach, a hybrid silicon nanocrystal–organic solar cell has been realized [141]. Also a Schottky-type solar cell containing a layer of silicon nanocrystals has been demonstrated. The conversion efficiency, though, was very low, only 0.02% [142]. All these approaches have resulted in very small conversion efficiencies compared to conventional silicon-based cells, although silicon nanocrystals show higher absorbance than silicon [29,143]. One drawback of using layers of nanocrystals is the difficulty to extract the generated carriers out of the dots and eventually to collect them by the electrodes of the cell. The cell is resistive due to the separation of the silicon nanocrystals. One way to reduce this effect is by embedding higher densities of silicon nanocrystals that are very close to each other, thus increasing conductivity and absorption. Finally, multiple exciton generation in silicon nanocrystals, as already described previously, could be exploited in silicon nanocrystal–based solar cells to improve conversion efficiencies in comparison with bulk silicon solar cells.

5.8 CONCLUSIONS

In conclusion, silicon nanocrystals show unique optical and electronic properties due to quantum (for sizes less than the electron wavelength of ~8 nm in silicon) and spatial confinement of carriers (for sizes less than the Bohr radius of excitons in bulk silicon of ~5 nm). The result of quantum confinement is the increase of the energy bandgap of the silicon nanocrystals as their size decreases. On the other hand, the result of the spatial confinement is the increase of the absorption and radiative emission oscillator strength due to the increasing overlap of the electron and hole wave functions of the generated excitons within the silicon nanocrystals after the absorption of photons with energies larger than the energy bandgap. Both these effects result in the efficient light emission from silicon nanocrystals for sizes less than 5 nm, which shifts to higher energies with decreasing size. With suitable surface passivation, the internal quantum efficiency can reach almost 100%, although silicon nanocrystals retain the indirect nature of the energy bandgap.

However, the indirectness of the bandgap results in slow recombination mechanisms that govern the light emission, making silicon nanocrystals unsuitable for fast-switching optical applications. Also, the optical gain, which is necessary for laser applications, and multiple exciton generation, which have been observed in silicon nanocrystals, compete with nonradiative Auger recombination mechanisms, thus making these effects very inefficient. However, light emission can be exploited in sensing and particularly in biological and medical applications, as silicon nanocrystals are compatible with living organisms and can replace the toxic Cd- and Pb-based quantum dots in many bioapplications. Also, they can be used for downshifting in solar cell applications. Finally, spatial confinement results in charging effects within the silicon nanocrystals, thus making them useful in nonvolatile memory applications.

REFERENCES

1. L. T. Canham, Silicon quantum wire array fabrication by electrochemical and chemical dissolution of wafers, *Appl. Phys. Lett.* 57, 1046 (1990).
2. A. D. Yoffe, Low-dimensional systems: Quantum size effects and electronic properties of semiconductor microcrystallites (zero-dimensional systems) and some quasi-two-dimensional systems, *Adv. Phys.* 42, 173 (1993).
3. M. Zacharias, J. Heitmann, R. Scholz, U. Kahler, M. Schmidt, and J. Bläsing, Size-controlled highly luminescent silicon nanocrystals: A SiO/SiO_2 superlattice approach, *Appl. Phys. Lett.* 80, 661 (2002).
4. P. Photopoulos, A. G. Nassiopoulou, D. N. Kouvatsos, and A. Travlos, Photoluminescence from nanocrystalline silicon in Si/SiO_2 superlattices, *Appl. Phys. Lett.* 76, 3588 (2000).
5. S. Gardelis, A. G. Nassiopoulou, N. Vouroutzis, and N. Frangis, Effect of exciton migration on the light emission properties in silicon nanocrystal ensembles, *J. Appl. Phys.* 105, 113509 (2009).
6. A. G. Nassiopoulou, Silicon nanocrystals in SiO_2 thin layers, in *Encyclopedia of Nanosciences and Nanotechnology*, H. S. Nalwa (ed.), Vol. 9, pp. 793–813 (2004).
7. I. Balberg, E. Savir, J. Jedrzejewski, A. G. Nassiopoulou, and S. Gardelis, Fundamental transport processes in ensembles of silicon quantum dots, *Phys. Rev. B* 75, 235329 (2007).
8. P. Manousiadis, S. Gardelis, and A. G. Nassiopoulou, Lateral electronic transport in 2D arrays of oxidized Si nanocrystals on quartz: Coulomb blockade effect and role of hydrogen passivation, *J. Appl. Phys.* 109, 083718 (2011).
9. D. N. Kouvatsos, V. Ioannou-Sougleridis, and A. G. Nassiopoulou, Charging effects in silicon nanocrystals within SiO_2 layers, fabricated by chemical vapor deposition, oxidation and annealing, *Appl. Phys. Lett.* 82(3), 397 (2003).
10. P. Dimitrakis, E. Kapetanakis, D. Tsoukalas, D. Skarlatos, C. Bonafos, G. Ben Asssayag, A. Claverie et al., Silicon nanocrystal memory devices obtained by ultra-low-energy ion-beam-synthesis, *Solid State Electron.* 48, 1511 (2004).
11. J. A. Yater, Implementation of Si nanocrystals in non-volatile memory devices, *Phys. Status Solidi A* 210(8), 1505–1511 (2013).
12. A. G. Cullis, L. T. Canham, and P. D. J. Calcott, The structural and luminescence properties of porous silicon, *J. Appl. Phys.* 82, 909 (1997).
13. S. Gardelis, U. Bangert, A. J. Harvey, and B. Hamilton, Double crystal x-ray diffraction, electron diffraction and high resolution electron microscopy of luminescent porous silicon, *J. Electrochem. Soc.* 142, 2094 (1995).
14. É. Váazsonyi, E. Szilágyi, P. Petrik, Z. E. Horvátha, T. Lohner, M. Fried, and G. Jalsovszky, Porous silicon formation by stain etching, *Thin Solid Films* 388, 295 (2001).
15. C. J. M. Eukel, J. Branebjerg, M. Elwenspoek, and F. C. M. Van De Pol, A new technology for micromachining of silicon: Dopant selective HF anodic etching for the realization of low-doped monocrystalline silicon structures, *IEEE Electron Dev. Lett.* 11, 588 (1990).
16. M. T. Kelly, J. K. M. Chun, and A. B. Bocarsly, High efficiency chemical etchant for the formation of luminescent porous silicon, *Appl. Phys. Lett.* 64, 1693 (1994).
17. C. A. Caras, J. M. Reynard, and F. V. Bright, An in-depth study linking the infrared spectroscopy and photoluminescence of porous silicon during ambient hydrogen peroxide oxidation, *Appl. Spectrosc.*, 67, 570 (2013).
18. M. Steinert, J. Acker, A. Henssge, and K. Wetzig, Experimental studies on the mechanism of wet chemical etching of silicon in HF/HNO_3 mixtures, *J. Electrochem. Soc.* 152, C843 (2005).

19. M. A. Hory, R. Hkrino, M. Ligeon, F. Muller, F. Gaspard, I. Mihalcescu, and J. C. Vial, Fourier transform IR monitoring of porous silicon passivation during post-treatments such as anodic oxidation and contact with organic solvents, *Thin Solid Films* 255, 200 (1995).

20. V. Petrova-Koch, T. Muschik, A. Kux, B. K. Meyer, F. Koch, and V. Lehmann, Rapid-thermal-oxidized porous Si—The superior photoluminescent Si, *Appl. Phys. Lett.* 61, 943 (1992).

21. S. Gardelis and B. Hamilton, The effect of surface modification on the luminescence of porous silicon, *J. Appl. Phys.* 76, 5327 (1994).

22. L. Debarge, J. P. Stoquert, A Slaoui, L. Stalmans, and J. Poortmans, Rapid thermal oxidation of porous silicon for surface passivation, *Mater. Sci. Semicond. Process.* 1, 281 (1998).

23. B. Gelloz and N. Koshida, Highly enhanced photoluminescence of as-anodized and electrochemically oxidized nanocrystalline p-type porous silicon treated by high-pressure water vapor annealing, *Thin Solid Films* 508, 406 (2006).

24. Y. Ishikawa, A. V. Vasin, J. Salonen, S. Muto, V. S. Lysenko, A. N. Nazarov, N. Shibata, and V.-P. Lehto, Color control of white photoluminescence from carbon-incorporated silicon oxide, *J. Appl. Phys.* 104, 083522 (2008).

25. T. Shimizu-Iwayama, N. Kurumado, D. E. Hole, and P. D. Townsend, Optical properties of silicon nanoclusters fabricated by ion implantation, *J. Appl. Phys.* 83, 6018 (1998).

26. H. Z. Song and X. M. Bao, Visible photoluminescence from silicon-ion-implanted SiO_2 film and its multiple mechanisms, *Phys. Rev. B* 55, 6988 (1997).

27. T. Shimizu-Iwayama, S. Nakao, and K. Saitoh, Visible photoluminescence in Si^+-implanted thermal oxide films on crystalline Si, *Appl. Phys. Lett.* 65, 1814 (1994).

28. Ch. B. Lioutas, N. Vouroutzis, I. Tsiaoussis, N. Frangis, S. Gardelis, and A. G. Nassiopoulou, Columnar growth of ultra-thin nanocrystalline Si films on quartz by low pressure chemical vapor deposition: Accurate control of vertical size, *Phys. Status Solidi A* 205, 2615 (2008).

29. S. Gardelis, A. G. Nassiopoulou, P. Manousiadis, A. Gkanatsiou, N. Frangis, and Ch. B. Lioutas, Structural and optical characterization of two-dimensional arrays of Si nanocrystals embedded in SiO_2 for photovoltaic applications, *J. Appl. Phys.* 111, 083536 (2012).

30. L. Tsybeskov, K. D. Hirschman, S. P. Duttagupta, M. Zacharias, P. M. Fauchet, J. P. McCaffrey, and D. J. Lockwood, Nanocrystalline-silicon superlattice produced by controlled recrystallization, *Appl. Phys. Lett.* 72, 43 (1998).

31. L.-Y. Chen, W.-H. Chen, and F. C.-N. Hong, Visible electroluminescence from silicon nanocrystals embedded in amorphous silicon nitride matrix, *Appl. Phys. Lett.* 86, 193506 (2005).

32. C. Liu, C. Li, A. Ji, L. Ma, Y. Wang, and Z. Cao, Intense blue photoluminescence from Si-in-SiNx thin film with high-density nanoparticles, *Nanotechnology* 16, 940 (2005).

33. M. V. Wolkin, J. Jorne, P. M. Fauchet, G. Allan, and C. Delerue, Electronic states and luminescence in porous silicon quantum dots: The role of oxygen, *Phys. Rev. Lett.* 82, 197 (1999).

34. F. Iacona, G. Franzò, and C. Spinella, Correlation between luminescence and structural properties of Si nanocrystals, *J. Appl. Phys.* 87, 1295 (2000).

35. D. S. Chao and J. H. Liang, Annealing temperature dependence of photoluminescent characteristics of silicon nanocrystals embedded in silicon-rich silicon nitride films grown by PECVD, *Nucl. Instrum. Methods Phys. Res. B: Beam Interact. Mater. Atoms* 307, 344 (2013).

36. L. Dal Negro, J. H. Yi, L. C. Kimerling, S. Hamel, A. Williamson, and G. Galli, Light emission from silicon-rich nitride nanostructures, *Appl. Phys. Lett.* 88, 183103 (2006).

37. T.-Y. Kim, N.-M. Park, K.-H. Kim, G. Y. Sung, Y.-W. Ok, T.-Y. Seong, and C.-J. Choi, Quantum confinement effect of silicon nanocrystals in situ grown in silicon nitride films, *Appl. Phys. Lett.* 85, 5355 (2004).

38. V. Em. Vamvakas, N. Vourdas, and S. Gardelis, Optical characterization of Si-rich silicon nitride films prepared by low pressure chemical vapor deposition, *Microelectron. Reliab.* 47, 794 (2007).

39. V. Em. Vamvakas and S. Gardelis, FTIR characterization of light emitting Si-rich nitride films prepared by low pressure chemical vapor deposition, *Surf. Coat. Technol.* 201, 9359 (2007).

40. E. Werwa, A. A. Seraphin, L. A. Chiu, C. Zhou, and K. D. Kolenbrander, Synthesis and processing of silicon nanocrystallites using a pulsed laser ablation supersonic expansion method, *Appl. Phys. Lett.* 64, 1821 (1994).

41. T. A. Burr, A. A. Seraphin, E. Werwa, and K. D. Kolenbrander, Carrier transport in thin films of silicon nanoparticles, *Phys. Rev. B* 56, 4818 (1997).

42. B. F. P. McVey and R. D. Tilley, Solution synthesis, optical properties, and bioimaging applications of silicon nanocrystals, *Acc. Chem. Res.* 47, 3045 (2014).

43. J. H. Warner, A. Hoshino, K. Yamamoto, and R. D. Tilley, Water-soluble photoluminescent silicon quantum dots, *Angew. Chem., Int. Ed.* 44, 4550 (2005).

44. J. P. Wilcoxon, G. A. Samara, and P. N. Provencio, Optical and electronic properties of Si nanoclusters synthesized in inverse micelles, *Phys. Rev. B* 60, 2704 (1999).

45. C.-S. Yang, R. A. Bley, S. M. Kauzlarich, H. W. H. Lee, and G. R. Delgado, Synthesis of alkyl-terminated silicon nanoclusters by a solution route, *J. Am. Chem. Soc.* 121, 5191 (1999).

46. M. P. Singh, T. M. Atkins, E. Muthuswamy, S. Kamali, C. Tu, A. Y. Louie, and S. M. Kauzlarich, Development of iron-doped silicon nanoparticles as bimodal imaging agents, *ACS Nano* 6, 5596 (2012).

47. C. Tu, X. Ma, P. Pantazis, S. M. Kauzlarich, and A. Y. Louie, Paramagnetic, silicon quantum dots for magnetic resonance and two-photon imaging of macrophages, *J. Am. Chem. Soc.* 132, 2016 (2010).

48. C. M. Hessel, E. J. Henderson, and J. G. C. Veinot, Hydrogen silsesquioxane: A molecular precursor for nanocrystalline Si–SiO$_2$ composites and freestanding hydride-surface-terminated silicon nanoparticles, *Chem. Mater.* 18, 6139 (2006).

49. E. J. Henderson, J. A. Kelly, and J. G. C. Veinot, Influence of HSiO$_{1.5}$ sol–gel polymer structure and composition on the size and luminescent properties of silicon nanocrystals, *Chem. Mater.* 21, 5426 (2009).

50. R. K. Baldwin, J. Zou, K. A. Pettigrew, G. J. Yeagle, R. D. Britt, and S. M. Kauzlarich, The preparation of a phosphorus doped silicon film from phosphorus containing silicon nanoparticles, *Chem. Commun.* 6, 658–660 (2006).

51. S. Gardelis, J. S. Rimmer, P. Dawson, B. Hamilton, R. A. Kubiak, T. E. Whall, and E. H. C. Parker, Evidence for quantum confinement in the photoluminescence of porous Si and SiGe, *Appl. Phys. Lett.* 59, 2118 (1991).

52. S. M. Prokes, O. J. Glembocki, V. M. Bermudez, R. Kaplan, L. E. Friedersdorf, and P. C. Searson, SiH$_x$ excitation: An alternate mechanism for porous Si photoluminescence, *Phys. Rev. B* 45, 13788(R) (1992).

53. H. D. Fuchs, M. Stutzmann, M. S. Brandt, M. Rosenbauer, J. Weber, A. Breitschwerdt, P. Deák, and M. Cardona, Porous silicon and siloxene: Vibrational and structural properties, *Phys. Rev. B* 48, 8172 (1993).

54. D. J. Lockwood and A. G. Wang, Quantum confinement induced photoluminescence in porous silicon, *Solid State Commun.* 94, 905 (1995).

55. C. Delerue, G. Allan, and M. Lannoo, Theoretical aspects of the luminescence of porous silicon, *Phys. Rev. B* 48, 11024 (1993).

56. H. Mizuno, H. Koyama, and N. Koshida, Oxide-free blue photoluminescence from photochemically etched porous silicon, *Appl. Phys. Lett.* 69, 3779 (1996).

57. J. M. Lauerhaas and M. J. Sailor, Chemical modification of the photoluminescence quenching of porous silicon, *Science* 261, 1567 (1993).

58. M. T. Kelly, J. K. Chun, and A. B. Bocarsly, General Brönsted acid behavior of porous silicon: A mechanistic evaluation of proton-gated quenching of photoemission from oxide-coated porous silicon, *J. Phys. Chem. B* 101, 2702 (1997).

59. J. L. Coffer, S. C. Lilley, R. A. Martin, and L. Files-Sesler, Surface reactivity of luminescent porous silicon, *J. Appl. Phys.* 74, 2094 (1993).

60. J. M. Rehm, G. L. McLendon, and P. M. Fauchet, Conduction and valence band edges of porous silicon determined by electron transfer, *J. Am. Chem. Soc.* 118, 4490 (1996).

61. I. N. Germanenko, S. Li, and M. S. El-Shall, Decay dynamics and quenching of photoluminescence from silicon nanocrystals by aromatic nitro compounds, *J. Phys. Chem. B* 105, 59 (2001).

62. D. Jurbergs, E. Rogojina, L. Mangolini, and U. Kortshagen, Silicon nanocrystals with ensemble quantum yields exceeding 60%, *Appl. Phys. Lett.* 88, 233116 (2006).

63. R. J. Anthony, D. J. Rowe, M. Stein, J. Yang, and U. Kortshagen, Routes to achieving high quantum yield luminescence from gas-phase-produced silicon nanocrystals, *Adv. Funct. Mater.* 21, 4042 (2011).

64. M. L. Mastronardi, F. Maier-Flaig, D. Faulkner, E. J. Henderson, C. Kübel, U. Lemmer, and G. A. Ozin, Size-dependent absolute quantum yields for size-separated colloidally-stable silicon nanocrystals, *Nano Lett.* 12, 337 (2012).

65. F. Sangghaleh, I. Sychugov, Z. Yang, J. G. C. Veinot, and J. Linnros, Near-unity internal quantum efficiency of luminescent silicon nanocrystals with ligand passivation, *ACS Nano* 9(7), 7097–7104 (2015).

66. D. J. Lockwood and L. Pavesi, Silicon fundamentals for photonic applications, in *Silicon Photonics*, Topics in Applied Physics, Vol. 94, p. 13, L. Pavesi and D. J. Lockwood (eds.), Springer-Verlag, Berlin, Germany (2004).

67. D. Kovalev, H. Heckler, M. Ben-Chorin, G. Polisski, M. Schwartzkopff, and F. Koch, Breakdown of the k-conservation rule in Si nanocrystals, *Phys. Rev. Lett.* 81, 2803 (1998).

68. S. Godefroo, M. Hayne, M. Jivanescu, A. Stesmans, M. Zacharias, O. I. Lebedev, G. Van Tendeloo, and V. V. Moshchalkov, Classification and control of the origin of photoluminescence from Si nanocrystals, *Nat. Nanotechnol.* 2, 486–489 (2007).

69. L. H. Lie, M. Duerdin, E. M. Tuite, A. Houlton, and B. R. Horrocks, Preparation and characterization of luminescent alkylated silicon quantum dots, *J. Electroanal. Chem.* 538, 183 (2002).

70. L. H. Lie, S. N. Patole, A. R. Pike, L. C. Ryder, B. A. Connolly, A. D. Ward, E. M. Tuite, A. Houlton, and B. R. Horrocks, Immobilisation and synthesis of DNA on Si(111), nanocrystalline porous silicon and silicon nanoparticles, *Faraday Discuss.* 125, 235 (2004).

71. Y. Chao, A. Houlton, B. R. Horrocks, M. R. C. Hunt, N. R. J. Poolton, J. Yang, and L. Šillera, Optical luminescence from alkyl-passivated Si nanocrystals under vacuum ultraviolet excitation: Origin and temperature dependence of the blue and orange emissions, *Appl. Phys. Lett.* 88, 263119 (2006).

72. M. Luppi and S. Ossicini, Ab initio study on oxidized silicon clusters and silicon nanocrystals embedded in SiO_2: Beyond the quantum confinement effect, *Phys. Rev. B* 71, 035340 (2005).

73. L. E. Ramos, J. Furthmüller, and F. Bechstedt, Effect of backbond oxidation on silicon nanocrystallites, *Phys. Rev. B* 70, 033311 (2004).

74. I. Vasiliev, J. R. Chelikowsky, and R. M. Martin, Surface oxidation effects on the optical properties of silicon nanocrystals, *Phys. Rev. B* 65, 121302(R) (2002).

75. G. Hadjisavvas and P. C. Kelires, Structure and energetics of Si nanocrystals embedded in a-SiO_2, *Phys. Rev. Lett.* 93, 226104 (2004).

76. L. Pavesi and M. Ceschini, Stretched-exponential decay of the luminescence in porous silicon, *Phys. Rev. B* 48, 17625(R) (1993).

77. P. J. Ventura, M. C. do Carmo, and K. P. O'Donnell, Excitation dynamics of luminescence from porous silicon, *J. Appl. Phys.* 77, 323 (1995).

78. L. Pavesi, Influence of dispersive exciton motion on the recombination dynamics in porous silicon, *J. Appl. Phys.* 80, 216 (1996).

79. F. Priolo, G. Franzò, D. Pacifici, V. Vinciguerra, F. Iacona, and A. Irrera, Role of the energy transfer in the optical properties of undoped and Er-doped interacting Si nanocrystals, *J. Appl. Phys.* 89, 264 (2001).

80. J. Linnros, N. Lalic, A. Galeckas, and V. Grivickas, Analysis of the stretched exponential photoluminescence decay from nanometer-sized silicon crystals in SiO_2, *J. Appl. Phys.* 86, 6128 (1999).

81. J. Heitmann, F. Müller, L. Yi, M. Zacharias, D. Kovalev, and F. Eichhorn, Excitons in Si nanocrystals: Confinement and migration effects, *Phys. Rev. B* 69, 195309 (2004).

82. I. V. Antonova, M. Gulyaev, E. Savir, J. Jedrzejewski, and I. Balberg, Charge storage, photoluminescence, and cluster statistics in ensembles of Si quantum dots, *Phys. Rev. B* 77, 125318 (2008).

83. I. Mihalcescu, J. C. Vial, and R. Romestain, Absence of carrier hopping in porous silicon, *Phys. Rev. Lett.* 80, 3392 (1998).

84. C. Delerue, G. Allan, C. Reynaud, O. Guillois, G. Ledoux, and F. Huisken, Multiexponential photoluminescence decay in indirect-gap semiconductor nanocrystals, *Phys. Rev. B* 73, 235318 (2006).

85. X. Zianni and A. G. Nassiopoulou, Photoluminescence lifetimes of Si quantum dots, *J. Appl. Phys.* 100, 074312 (2006).

86. P. D. J. Calcott, K. J. Nash, L. T. Canham, M. J. Kane, and D. Brumhead, Identification of radiative transitions in highly porous silicon, *J. Phys.: Condens. Matter* 5, L91 (1993).

87. O. Bisi, S. Ossicini, and L. Pavesi, Porous silicon: A quantum sponge structure for silicon based optoelectronics, *Surf. Sci. Rep.* 9, 126 (2000).

88. N. Arad-Vosk and A. Sa'ar, Radiative and nonradiative relaxation phenomena in hydrogen- and oxygen-terminated porous silicon, *Nanoscale Res. Lett.* 9(1), 47 (2014).

89. D. Kovalev, J. Diener, H. Heckler, G. Polisski, N. Künzner, and F. Koch, Optical absorption cross sections of Si nanocrystals, *Phys. Rev. B* 61, 4485 (2000).

90. L. Pavesi, L. Dal Negro, C. Mazzoleni, G. Franzó, and F. Priolo, Optical gain in silicon nanocrystals, *Nature* 408, 440 (2000).

91. L. Khriachtchev, M. Räsänen, S. Novikov, and J. Sinkkonen, Optical gain in SiO/SiO_2 lattice: Experimental evidence with nanosecond pulses, *Appl. Phys. Lett.* 79, 1249 (2001).

92. K. Luterová, I. Pelant, I. Mikulskas, R. Tomasiunas, D. Muller, J.-J. Grob, J.-L. Rehspringer, and B. Hönerlage, Stimulated emission in blue-emitting Si+-implanted SiO_2 films, *J. Appl. Phys.* 91, 2896 (2002).

93. I. Pelant, Optical gain in silicon nanocrystals: Current status and perspectives, *Phys. Status Solidi A* 208, 625 (2011).
94. M. C. Beard, K. P. Knutsen, P. Yu, J. M. Luther, Q. Song, W. K. Metzger, R. J. Ellingson, and A. J. Nozik, Multiple exciton generation in colloidal silicon nanocrystals, *Nano Lett.* 7(8), 2506 (2007).
95. M. Tuan Trinh, R. Limpens, W. D. A. M. de Boer, J. M. Schins, L. D. A. Siebbeles, and T. Gregorkiewicz, Direct generation of multiple excitons in adjacent silicon nanocrystals revealed by induced absorption, *Nat. Photon.* 6, 316 (2012).
96. T. Taskin, S. Gardelis, J. H. Evans, B. Hamilton, and A. R. Peaker, Sharp 1.54 µm luminescence of porous, erbium doped silicon, *Electron. Lett.* 31, 2132 (1995).
97. L. Gu, Z. Xiong, G. Chen, Z. Xiao, D. Gong, X. Hou, and X. Wang, Luminescent erbium-doped porous silicon bilayer structures, *Adv. Mater.* 13, 1402 (2001).
98. M. Fujii, M. Yoshida, S. Hayashi, and K. Yamamoto, Photoluminescence from SiO_2 films containing Si nanocrystals and Er: Effects of nanocrystalline size on the photoluminescence efficiency of Er^{3+}, *J. Appl. Phys.* 84, 4525 (1998).
99. C. E. Chryssou, A. J. Kenyon, T. S. Iwayama, C. W. Pitt, and D. E. Hole, Evidence of energy coupling between Si nanocrystals and Er^{3+} in ion-implanted silica thin films, *Appl. Phys. Lett.* 75, 2011 (1999).
100. G. Franzò, D. Pacifici, V. Vinciguerra, F. Priolo, and F. Iacona, Er^{3+} ions–Si nanocrystals interactions and their effects on the luminescence properties, *Appl. Phys. Lett.* 76, 2167 (2000).
101. M. Makarova, V. Sih, J. Warga, R. Li, L. Dal Negro, and J. Vuckovic, Enhanced light emission in photonic crystal nanocavities with Erbium-doped silicon nanocrystals, *Appl. Phys. Lett.* 92, 161107 (2008).
102. R. Hoffmann, J. Beyer, V. Klemm, D. Rafaja, B. C. Johnson, J. C. McCallum, and J. Heitmann, Erbium-doped slot waveguides containing size-controlled silicon nanocrystals, *J. Appl. Phys.* 17, 163106 (2015).
103. N. Koshida and H. Koyama, Visible electroluminescence from porous silicon, *Appl. Phys. Lett.* 60, 347 (1992).
104. H. Shi, Y. Zheng, Y. Wang, and R. Yuan, Electrically induced light emission and novel photocurrent response of a porous silicon device, *Appl. Phys. Lett.* 63, 770 (1993).
105. A. Richter, P. Steiner, F. Kozlowski, and W. Lang, Current induced light-emission from a porous silicon device, *IEEE Electron. Device Lett.* 12, 691 (1991).
106. V. A. Kuznetsov, I. Andrienko, and D. Haneman, High efficiency blue–green electroluminescence and scanning tunneling microscopy studies of porous silicon, *Appl. Phys. Lett.* 72, 3323 (1998).
107. T. Oguro, H. Koyama, T. Ozaki, and N. Koshida, Mechanism of the visible electroluminescence from metal/porous silicon/n-Si devices, *J. Appl. Phys.* 81, 1407 (1997).
108. N. Koshida, H. Koyama, Y. Yamamoto, and G. J. Collins, Visible electroluminescence from porous silicon diodes with an electropolymerized contact, *Appl. Phys. Lett.* 63, 2655 (1993).
109. Y. Yang, Q. Li, and X. Liu, Study on electroluminescence from porous silicon light-emitting diode, *Chin. Opt. Lett.* 4, 297 (2006).
110. F. Namavar, H. P. Maruska, and N. M. Kalkhoran, Visible electroluminescence from porous silicon np heterojunction diodes, *Appl. Phys. Lett.* 60, 2514 (1992).
111. H. Li, B. Huang, D. Yi, H. Cui, Y. He, and J. Peng, Efficient visible electroluminescence from porous silicon diodes with low driven voltage, *Chin. Opt. Lett.* 2, 171 (2004).
112. J. Linnros and N. Lalic, High quantum efficiency for a porous silicon light emitting diode under pulsed operation, *Appl. Phys. Lett.* 66, 3048 (1995).
113. K. Nishimura, Y. Nagao, and N. Ikeda, High external quantum efficiency of electroluminescence from photoanodized porous silicon, *Jpn. J. Appl. Phys.* 37, L303 (1998).
114. P. Steiner, F. Kozlowski, and W. Lang, Light-emitting porous silicon diode with an increased electroluminescence quantum efficiency, *Appl. Phys. Lett.* 62, 2700 (1993).
115. G. Barillaro, A. Diligent, F. Pieri, F. Fuso, and M. Allegrini, Integrated porous-silicon light-emitting diodes: A fabrication process using graded doping profiles, *Appl. Phys. Lett.* 78, 4154 (2001).
116. F. Kozlowski, P. Steiner, W. Lang, and H. Sandmaier, Light-emitting diodes in porous silicon, *Sens. Actuators A* 43, 153 (1994).
117. A. G. Nassiopoulos, S. Grigoropoulos, and D. Papadimitriou, Electroluminescent device based on silicon nanopillars, *Appl. Phys. Lett.* 69, 2267 (1996).
118. G. Conibeer, M. Green, E.-C. Cho, D. König, Y.-H. Cho, T. Fangsuwannarak, G. Scardera et al., Silicon quantum dot nanostructures for tandem photovoltaic cells, *Thin Solid Films* 516, 6748 (2008).

119. H.-Z. Song, X.-M. Bao, N.-S. Li, and J.-Y. Zhang, Relation between electroluminescence and photoluminescence of Si⁺-implanted SiO₂, *J. Appl. Phys.* 82, 4028 (1997).

120. A. G. Nassiopoulou, V. Ioannou-Sougleridis, P. Photopoulos, A. Travlos, V. Tsakiri, and D. Papadimitriou, Stable visible photo- and electroluminescence from nanocrystalline silicon thin films fabricated on thin SiO₂ layers by low pressure chemical vapour deposition, *Phys. Status Solidi A* 165, 79 (1998).

121. P. Photopoulos, A. G. Nassiopoulou, D. N. Kouvatsos, and A. Travlos, Photo- and electroluminescence from nanocrystalline silicon single and multilayer structures, *Mater. Sci. Eng.* B69–B70, 345 (2000).

122. P. Photopoulos and A. G. Nassiopoulou, Room- and low-temperature voltage tunable electroluminescence from a single layer of silicon quantum dots in between two thin SiO₂ layers, *Appl. Phys. Lett.* 77, 1816 (2000).

123. J. Valenta, N. Lalic, and J. Linnros, Electroluminescence microscopy and spectroscopy of silicon nanocrystals in thin SiO₂ layers, *Opt. Mater.* 17, 45 (2001).

124. G. Franzò, A. Irrera, E. C. Moreira, M. Miritello, F. Iacona, D. Sanfilippo, G. Di Stefano, P. G. Fallica, and F. Priolo, Electroluminescence of silicon nanocrystals in MOS structures, *Appl. Phys. A* 74, 1–5 (2002).

125. R. J. Walters, G. I. Bourianoff, and H. A. Atwater, Field-effect electroluminescence in silicon nanocrystals, *Nat. Mater.* 4, 143 (2005).

126. T. Creazzo, B. Redding, E. Marchena, J. Murakowski, and D. W. Prather, Tunable photoluminescence and electroluminescence of size-controlled silicon nanocrystals in nanocrystalline-Si/SiO₂ superlattices, *J. Lumin.* 130, 631 (2010).

127. J. López-Vidrier, Y. Berencén, S. Hernández, B. Mundet, S. Gutsch, J. Laube, D. Hiller et al., Structural parameters effect on the electrical and electroluminescence properties of silicon nanocrystals/SiO₂ superlattices, *Nanotechnology* 26, 185704 (2015).

128. B. Ghosh, Y. Masuda, Y. Wakayama, Y. Imanaka, J.-I. Inoue, K. Hashi, K. Deguchi et al., Hybrid white light emitting diode based on silicon nanocrystals, *Adv. Funct. Mater.* 24, 7151 (2014).

129. F. Maier-Flaig, J. Rinck, M. Stephan, T. Bocksrocker, M. Bruns, C. Kübel, A. K. Powell, G. A. Ozin, and U. Lemmer, Multicolor silicon light-emitting diodes (SiLEDs), *Nano Lett.* 13, 475 (2013).

130. S. Tiwari, F. Rana, H. Hanafi, A. Hartstein, E. F. Crabbé, and K. Chan, A silicon nanocrystals based memory, *Appl. Phys. Lett.* 68, 1377 (1996).

131. H. Qin, X. Gu, H. Lu, J. Liu, X. Huang, and K. Chen, Observation of Coulomb-blockade in a field-effect transistor with silicon nanocrystal floating gate at room temperature, *Solid State Commun.* 111, 171 (1999).

132. E. Kapetanakis, P. Normand, D. Tsoukalas, and K. Beltsios, Room-temperature single-electron charging phenomena in large-area nanocrystal memory obtained by low-energy ion beam synthesis, *Appl. Phys. Lett.* 80, 2794 (2002).

133. S.-T. Kang, B. Winstead, J. Yater, M. Suhail, G. Zhang, C.-M. Hong, H. Gasquet et al., High performance nanocrystal based embedded flash microcontrollers with exceptional endurance and nanocrystal scaling capability, in *2012 Fourth IEEE International Memory Workshop (IMW)*, Milan, Italy, pp. 1–4, doi:10.1109/IMW.2012.6213668.

134. W. R. Taube, A. Kumar, R. Saravanan, P. B. Agarwal, P. Kothari, B. C. Joshi, and D. Kumar, Efficiency enhancement of silicon solar cells with silicon nanocrystals embedded in PECVD silicon nitride matrix, *Solar Energy Mater. Solar Cells* 101, 32 (2012).

135. F. Sgrignuoli, P. Ingenhoven, G. Pucker, V. D. Mihailetchi, E. Froner, Y. Jestin, E. Moser, G. Sànchez, and L. Pavesi, Purcell effect and luminescent down shifting in silicon nanocrystals coated back-contact solar cells, *Solar Energy Mater. Solar Cells* 132, 267 (2015).

136. E.-C. Cho, M. A. Green, G. Conibeer, D. Song, Y.-H. Cho, G. Scardera, S. Huang et al., Silicon quantum dots in a dielectric matrix for all-silicon tandem solar cells, *Adv. OptoElectron.* 2007, Article ID 69578 (2007).

137. G. Conibeer, M. Green, R. Corkish, Y. Cho, E.-C. Cho, C.-W. Jiang, T. Fangsuwannarak et al., Silicon nanostructures for third generation photovoltaic solar cells, *Thin Solid Films* 511–512, 654 (2006).

138. W. R. Taube, A. Kumar, R. Saravanan, P. B. Agarwal, P. Kothari, B. C. Joshi, and D. Kumar, Efficiency enhancement of silicon solar cells with silicon nanocrystals embedded in PECVD silicon nitride matrix, *Solar Energy Mater. Solar Cells* 101, 32–35 (2012).

139. P.-J. Wu, Y.-C. Wang, and I-C. Chen, Fabrication of Si heterojunction solar cells using P-doped Si nanocrystals embedded in SiNₓ films as emitters, *Nanoscale Res. Lett.* 8, 457 (2013).

140. P. Löper, A. Witzky, A. Hartel, S. Gutsch, D. Hiller, J. C. Goldschmidt, S. Janz, S. W. Glunz, and M. Zacharias, Photovoltaic properties of silicon nanocrystals in silicon carbide, in *Physics, Simulation, and Photonic Engineering of Photovoltaic Devices*, Proceedings of SPIE, Vol. 8256, p. 82560G, A. Freundlich and J.-F. F. Guillemoles (eds.) (February 9, 2012), San Francisco, CA, doi: 10.1117/12.906669.

141. Y. Ding, R. Gresback, R. Yamada, K. Okazaki, and T. Nozaki, Hybrid silicon nanocrystal/poly(3-hexylthiophene-2,5-diyl) solar cells from a chlorinated silicon precursor, *Jpn. J. Appl. Phys.* 52, 11NM04 (2013).

142. C.-Y. Liu and U. R. Kortshagen, A silicon nanocrystal Schottky junction solar cell produced from colloidal silicon nanocrystals, *Nanoscale Res. Lett.* 5(8), 1253 (2010).

143. S.-K. Kim, C.-H. Cho, B.-H. Kim, S.-J. Park, and J. W. Lee, Electrical and optical characteristics of silicon nanocrystal solar cells, *Appl. Phys. Lett.* 95, 143120 (2009).

6 Electronic and Optical Properties of Si and Ge Nanocrystals

Tupei Chen

CONTENTS

6.1 INTRODUCTION

Bulk crystalline silicon (Si) and germanium (Ge) are the two most important indirect bandgap semiconductors. Bulk Si crystal is the main material of today's microelectronic/nanoelectronic, photovoltaic, and microelectromechanical system (MEMS) technologies. In particular, the ubiquitous "Si chip" is taken for granted in today's society. Germanium is a semiconductor material that formed the basis for the development of transistor technology (the first solid-state transistor was made with Ge). But by the late 1950s, silicon had emerged as the favored semiconductor, and it has remained ever since because of the breakthrough of planar technology and integrated circuit (IC) technology. Nevertheless, Ge is an important material for optoelectronic devices such as photodetectors and solar cells, and in recent years there has been renewed interest in Ge, which has been triggered by its strong potential for deep submicrometer (sub-45 nm) IC technologies.

Optical properties of a semiconductor can be any of the properties that involve the interaction between light and the semiconductor, including absorption, diffraction, polarization, reflection, refraction, and so on. Most optical properties of semiconductors are related to the particular nature of the electronic band structures of the semiconductors, which are in turn determined by the crystallographic structure, the particular atoms, and their bonding. The full symmetry of the space groups is also essential in determining the structure of the energy bands. From the macroscopic viewpoint, the interaction of matter with light is described by Maxwell's equations, in which the optical properties of matter are introduced as the constants characterizing the medium, such as the dielectric constant. The dispersion of the dielectric constant over a particular range of photon energy is called the "dielectric function" $(\varepsilon) \cdot \varepsilon$ is a complex number, which is often described by its real (ε_1) and imaginary (ε_2) parts, that is, $\varepsilon = \varepsilon_1 - i\varepsilon_2$. The dielectric function can be expressed also in terms of the complex refractive index (N) with $\varepsilon = N^2 = (n-ik)^2$ where n and k are the refractive index and extinction coefficient of the semiconductor, respectively. Both n and k are important optical constants; they are real and positive numbers.

The band structure of a semiconductor can be significantly modified when its dimensions are reduced to the scale of the exciton Bohr radius (e.g., ~4.9 nm for bulk crystalline silicon) of the bulk crystalline semiconductor, which is due to the quantum confinement effect. Therefore, the electronic and optical properties of a semiconductor in the nanoscale (e.g., Si and Ge nanocrystals) would be significantly different from those of its bulk counterpart. In this chapter, we review the electronic and optical properties of Si and Ge nanocrystals. The organization of this chapter is based on the following two main sections: in Section 6.2, some theoretical calculations of the electronic and optical properties of small Si and Ge nanocrystals with up to several hundreds of atoms are introduced; in Section 6.3, some experimental studies of the optical properties (with a focus on the dielectric functions and optical constants) of Si nanocrystals embedded in a dielectric matrix (SiO_2 and Si_3N_4) and of self-assembled Ge nanocrystals, which are based on spectroscopic ellipsometry (SE), are presented.

6.2 THEORETICAL STUDIES

The calculation of the electronic and optical properties of Si and Ge nanocrystals is a difficult task. The nanocrystals that can be observed in experiments are usually too small to be described by those theoretical techniques (e.g., the k-space techniques) that are used for large crystals. On the other hand, they are too big to be described by real-space, high-level *ab initio* quantum chemistry techniques that are used for small molecules. For example, a small Si nanocrystal with a diameter of 2.5 nm consists of 281 Si atoms and 172 H atoms on the surface including 4106 electrons [1]. This is a huge many-body system. The solution to the time-independent Schrödinger equation for the system is a formidable task. It is therefore necessary to resort to approximations, or to different approaches.

The theoretical *ab initio* investigation of the optical properties demands an accurate account of electron correlation for both the ground and the excited states and involves high computational cost, which scales at least as the fifth or sixth power of the diameter d of the nanocrystal [2]. The *ab initio* approaches are limited by their applicability to systems of less than 1000 atoms with the current computer power. Calculations of the optical properties can be performed using an *ab initio* technique based on density functional theory (DFT) and local density approximation (LDA) [3–5] or using time-dependent LDA [6,7]. Many-body effects such as self-energy corrections and excitonic effects have been taken into account [6,8–10]. However, the correct treatment of the many-body effects in the calculation of the electron–hole pair excitation energies is a concern [6,11,12]. Many theoretical calculations on the electronic and optical properties of Si and Ge nanocrystals are semiempirical. For example, semiempirical tight-binding approaches [13,14] and the empirical pseudopotential approach [15,16] allow the treatment of much larger nanocrystalline systems. Such approaches are based on the knowledge of the electronic structure of bulk crystals, but the transferability of bulk electronic interaction parameters to a nanocrystalline environment is questionable. This holds true for both the tight-binding approximation and the empirical-pseudopotential approach [5]. Various approaches have been used to calculate the electronic and optical properties of small nanocrystals, such as hydrogenated Si and Ge nanocrystals [2–4,6–8,10,14,17–20], oxidized Si nanocrystals [2,18,21–23], and Si nanocrystals embedded in a SiO_2 matrix [22,24–27].

In this section, we present theoretical analysis of the structural and electronic properties and optical gaps of various nanocrystal systems including hydrogenated Si and Ge nanocrystals, oxidized Si nanocrystals, and Si nanocrystals embedded in SiO_2 matrix.

6.2.1 HYDROGENATED Si NANOCRYSTALS (nc-Si)

Simple cubic supercells can be used to model the nc-Si surrounded by vacuum and with different passivation techniques (e.g., hydrogen passivation) of the surface dangling bonds. For hydrogen passivation, with the relaxed Si_mH_n structures, Degoli et al. performed an *ab initio* calculation of the electronic and structural properties of small hydrogenated nc-Si as a function of dimension [17]. The calculation was done within DFT using a pseudopotential plane-wave approach. They first examined how the ground-state structural properties of the nc-Si change as a function of the cluster dimension and how the creation of an electron–hole pair modifies the overall structures. Then they showed that the structural modifications are immediately reflected into the electronic properties of nc-Si.

In the study reported by Degoli et al. [17], the calculations for each cluster were performed both in the ground and the excited states, and the structural properties were determined by allowing full relaxation of each hydrogenated nc-Si. The starting configuration for the clusters was fixed with all Si atoms occupying the same position as in the bulk crystal, and passivation of the surface was done with H atoms placed along the bulk crystal directions. The formation of an electron–hole pair under excitation was taken into account by forcing one electron to occupy the the lowest unoccupied single-particle state (LUMO), thus leaving a hole in the highest occupied single-particle state (HOMO) [17].

Figure 6.1 shows the average Si–Si bond lengths of the relaxed Si_5H_{12}, $Si_{10}H_{16}$, $Si_{29}H_{36}$, and $Si_{35}H_{36}$ clusters in their ground- and excited state configurations [17]. It can be clearly seen in the figure that there is a contraction of the hydrogenated nc-Si in its ground-state configuration with respect to bulk silicon. The contraction becomes smaller and the average interatomic distances tend to the calculated bulk value as the size of the cluster increases (the average bond lengths approach the bulk value for hydrogenated nc-Si with diameter of about 20 Å). However, no clear trend can be identified for the excited-state configuration because of heavy clusters distortions.

Figure 6.2 demonstrates that the cluster relaxation in the excited configuration causes structural distortions [17]. As shown in Figure 6.2a for the $Si_{10}H_{16}$ cluster, the first-, second-, third-, and fourth-neighbor Si–Si distances are almost unchanged, with respect to bulk values in the ground-state configuration, whereas significant deviations are induced by the excitation. The same situation is observed for the $Si_{29}H_{36}$ cluster (see Figure 6.2b). It has been concluded that the presence of an electron–hole pair in the clusters causes a strong deformation of the structures with respect to their ground-state configuration, and this is more evident for smaller clusters [17].

The structural modifications due to the cluster dimension effect or to the creation of an electron–hole pair lead to changes in the electronic structures. Figure 6.3 shows the Kohn–Sham levels

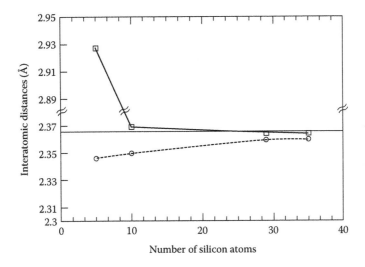

FIGURE 6.1 Average Si–Si interatomic distances for the relaxed Si_5H_{12}, $Si_{10}H_{16}$, $Si_{29}H_{36}$, and $Si_{35}H_{36}$ clusters in their ground-state (circle) and excited-state (square) configurations. The horizontal line indicates the calculated interatomic distances in bulk Si. (From Degoli, E. et al., *Phys. Rev. B*, 69, 155411, 2004. With permission.)

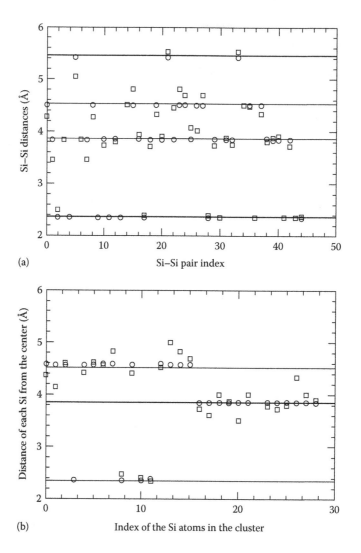

FIGURE 6.2 Calculated Si–Si distances (a) for each pair of Si atoms for the $Si_{10}H_{16}$ cluster, and (b) of each Si atom with respect to the central Si of the $Si_{29}H_{36}$ cluster. Circles and squares represent ground- and excited-state configurations, respectively. Straight lines represent the calculated bulk silicon values for the first-, second-, third-, and fourth-neighbor distances. (From Degoli, E. et al., *Phys. Rev. B*, 69, 155411, 2004. With permission.)

for the Si_5H_{12}, $Si_{10}H_{16}$, $Si_{29}H_{36}$, and $Si_{35}H_{36}$ clusters in both the ground- (left panel) and excited-state (right panel) configurations [17]. It can be concluded from the figure that the ground-state energy gap increases as the cluster dimension decreases. In addition, the excitation of the electron–hole pair causes a reduction of the energy gap, which is more significant for smaller clusters. For the excited clusters, the HOMO and LUMO become strongly localized in correspondence to the structural distortion, giving rise to defect-like states that reduce the gap; and the effect is stronger for a smaller cluster [17].

The nanocrystal size influences both the energetic positions of the optical transitions and their oscillator strengths. This has been demonstrated by many calculations [5,18]. Figure 6.4 shows one such calculation of the transition energies and the oscillator strengths of the HOMO–LUMO

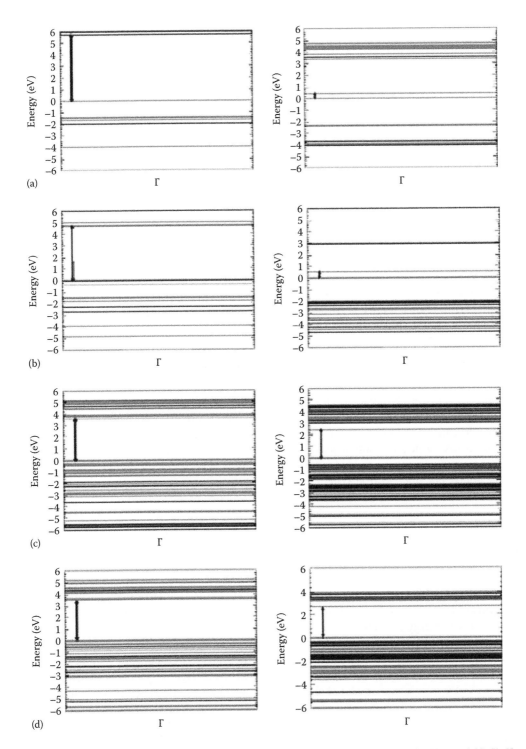

FIGURE 6.3 Calculated energy levels at the Γ point for (a) Si$_5$H$_{12}$, (b) Si$_{10}$H$_{16}$, (c) Si$_{29}$H$_{36}$, and (d) Si$_{35}$H$_{36}$ clusters in ground-state (left panel) and excited-state (right panel) configuration. The energies are referred to the highest valence level. (From Degoli, E. et al., *Phys. Rev. B*, 69, 155411, 2004. With permission.)

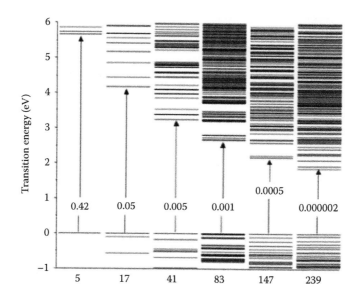

FIGURE 6.4 Energy-level schemes for the Kohn–Sham eigenvalues of Si nanocrystals passivated with hydrogen. The number of Si atoms is indicated by integers. The arrows indicate the HOMO–LUMO transition with the oscillator strength given by the real numbers. (From Ramos, L.E. et al., *Phys. Stat. Sol.* (*b*), 242, 3053, 2005. With permission.)

transitions for nc-Si passivated with hydrogen [18]. In this figure, the quantum confinement effects on both the energy levels and the oscillator strengths are clearly visible.

Öğüt et al. performed the calculation on the quantum size effect on the quasiparticle gaps, self-energy corrections, exciton Coulomb energies, and optical gaps in Si nanocrystals (i.e., quantum dots) from first principles using a real-space pseudopotential method [8]. The calculations were performed on hydrogen-passivated spherical Si clusters with diameters up to 27.2 Å (~800 Si and H atoms). For an n-electron system, the quasiparticle gap E_g^{qp} can be expressed in terms of the ground-state total energies E of the $(n + 1)$-, $(n - 1)$-, and n-electron systems as [8]

$$E_g^{qp} = E(n+1) + E(n-1) - 2E(n) = E_g^{band} + \Sigma \qquad (6.1)$$

where
   E_g^{band} is the usual single-particle LDA (local density approximation) band gap (defined as the eigenvalue difference between the lowest unoccupied and the highest occupied orbitals)
   Σ is the self-energy correction

Figure 6.5 shows the size dependence of the quasiparticle and LDA band gaps, and the self-energy corrections [8]. The quasiparticle bandgap ($E_g^{qp}(d)$), the LDA bandgap ($E_g^{band}(d)$), and the self-energy correction ($\Sigma(d)$) for a nanocrystal of size d are enhanced substantially with respect to their bulk values ($E_g^{qp}(bulk), E_g^{band}(bulk)$), and ($\Sigma(bulk)$), respectively) as a result of quantum confinement. Power-law fitting to the calculated data in Figure 6.5 shows that $E_g^{qp}(d) - E_g^{qp}(bulk), E_g^{qp}(d) - E_g^{band}(bulk)$, and $\Sigma(d) - \Sigma(bulk)$ scale as $d^{-1.2}$, $d^{-1.1}$, and $d^{-1.5}$, respectively [8].

For direct comparison with experimental absorption data, the Coulomb and exchange-correlation energies of the exciton need to be included. Compared to the Coulomb energy, exciton

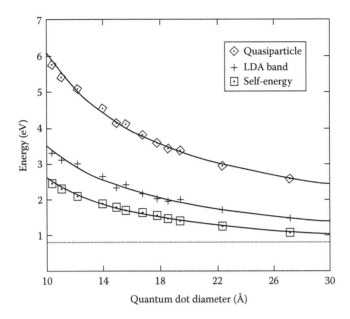

FIGURE 6.5 Calculated quasiparticle and LDA bandgaps and self-energy corrections as a function of the nanocrystal (quantum dot) diameter d (in Å). The solid lines are power-law fits to the calculated data. The horizontal dotted line is the bulk limit of the self-energy correction (0.68 eV). (From Öğüt, S. et al., *Phys. Rev. Lett.*, 79, 1770, 1997. With permission.)

exchange-correlation energies are much smaller for the nanocrystals, and can therefore be neglected [8]. Quantum confinement (QC) in nanostructures enhances the bare exciton Coulomb interaction and also reduces electronic screening, so that the exciton Coulomb energy E_{Coul} can be comparable to the quasiparticle gap. E_{Coul} can be approximately calculated from the effective mass approximation (EMA), which yields (in atomic units) [8]

$$E_{Coul} = \frac{3.572}{\varepsilon d} \tag{6.2}$$

where ε and d are the dielectric constant and diameter of the nanocrystal, respectively [8]. Though EMA is not able to yield accurate exciton Coulomb energies, it can provide a simple picture of the quantum size dependence of the exciton Coulomb energy. Figure 6.6 shows the unscreened exciton Coulomb energies obtained from various calculations, including the direct *ab initio* pseudopotential calculation, the EMA calculation, and the semiempirical pseudopotential calculation [8]. A strong size dependence of the unscreened Coulomb energy can be observed from Figure 6.6. The unscreened exciton Coulomb energies are inversely proportional to the nanocrystal diameter d, that is, there is a power-law dependence $d^{-\alpha}$ with $\alpha = 0.7$ for the *ab initio* pseudopotential calculation, 0.8 for the semiempirical calculation, and 1 for the EMA calculation.

An accurate calculation of the exciton Coulomb energy requires the knowledge of the dielectric constant of the nanocrystal. If E_{Coul} can be calculated accurately, the optical gap can be extracted as $E_g^{opt} = E_g^{qp} - E_{Coul}$ [8]. Figure 6.7 shows the calculated optical gaps along with the quasiparticle gaps and experimental absorption data as a function of the nanocrystal diameter [8]. A strong quantum size dependence of the optical gap can be observed from the figure.

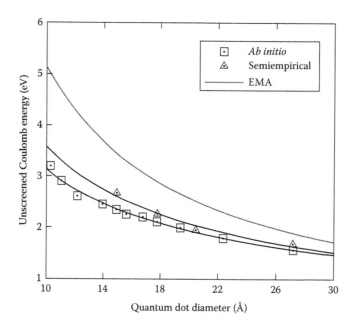

FIGURE 6.6   Unscreened exciton Coulomb energies as a function of the nanocrystal (quantum dot) diameter d (in Å) calculated by (1) effective mass approximation (EMA) (dot line), (2) direct semiempirical pseudopotential calculations [28], and (3) direct *ab initio* pseudopotential calculations. The solid lines are power-law fits to the calculated data. (From Öğüt, S. et al., *Phys. Rev. Lett.,* 79, 1770, 1997. With permission.)

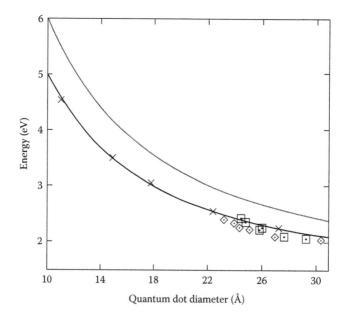

FIGURE 6.7   Calculated quasiparticle gaps (dotted line), optical gaps (shown by × fitted to the solid line), and experimental absorption data from Si nanocrystals (□ and ◇) as a function of the nanocrystal (quantum dot) diameter d. (From Öğüt, S. et al., *Phys. Rev. Lett.,* 79, 1770, 1997. With permission.)

A simple expression for the size dependence of the optical band gap can be obtained from the power-law fitting to the calculated optical gaps. The expression can be written as

$$E_g^{opt}(d) = E_g^{opt}(\infty) + \frac{c}{d^m} \tag{6.3}$$

where

$E_g^{opt}(d)$ is the optical gap (in eV)
$E_g^{opt}(\infty)$ is the bandgap of the bulk material (here $E_g^{opt}(\infty)$ is 1.12 eV for bulk crystalline silicon)
d is the nanocrystal diameter (in nm)

The constants c and m are usually different for different calculations. Weissker et al. reported the size dependence with $c = 2.96$ and $m = 1.0$ [6]. The calculation of Delerue et al. yields $c = 3.73$ and $m = 1.39$ [9]. Lehtonen and Sundholm proposed $c = 3.00$ and $m = 1.575$ [29].

The optical absorption spectra of Si_nH_m nanoclusters up to ~250 atoms were computed by Vasiliev et al. using a linear response theory within the time-dependent local density approximation (TDLDA) [7]. The TDLDA formalism incorporates the electronic screening and correlation effects, which determine exciton binding energies, and thus it represents a full *ab initio* formalism for the excited states. Figure 6.8 shows the calculated absorption spectra of Si_nH_m clusters with various sizes (up to 147 Si atoms and 100 H atoms) [7]. It can be observed from the figure that, as the size of the cluster increases, the absorption gaps gradually decrease, and the discrete spectra for small clusters evolve into quasi-continuous spectra for silicon nanocrystals. At the same time, the oscillator strength of dipole-allowed transitions near the absorption edge decreases with increasing cluster size. This fact is consistent with the formation of an *indirect* bandgap in the limit of bulk silicon [7].

Delerue et al. calculated the absorption coefficients of bigger silicon nanocrystals [9]. Figure 6.9 shows the absorption coefficient of a silicon nanocrystal with the size of 3.86 nm, which has a calculated band gap of 1.67 eV (without exciton binding energy) [9]. As shown in Figure 6.9a, the absorption edge is shifted to near 3.5 eV, which corresponds to the direct-gap absorption of bulk silicon. Nothing is visible in the figure between 1.67 and ~3.0 eV at that scale. The reason is that the optical matrix element for transitions with energy between 1.67 and 3.0 eV is several orders of magnitude lower than for transitions above 3.0 eV. The optical absorption becomes very close to that of bulk silicon—but with a blue shift of the absorption edge, that is, it is very close to the absorption of an indirect semiconductor. In order to have a close view of the absorption near the band edge (1.67 eV), the absorption coefficient in the energy range 1.4–2.4 eV is plotted in Figure 6.9b at a different scale and in the form of a bar chart (the amplitudes of the bars represent the integrated absorption coefficient over the width of the bar). Figure 6.9b shows that the absorption threshold is at the bandgap energy (1.67 eV) because the transition is dipole-allowed even if this is only with fairly weak oscillator strength. The optical threshold is also subject to a blue shift depending on the size of the nanocrystals. It is interesting to note that the absorption coefficient follows approximately a square-law variation with photon energy [9].

Weissker et al. performed parameter-free calculations of the frequency-dependent dielectric functions of Si nanocrystals [6]. The calculations are based upon the independent-particle approximation and a pseudopotential plane wave method. The nanocrystals are described by clusters of up to 363 atoms. Their surfaces are passivated by hydrogen atoms. The electronic structure calculations are based on the DFT-LDA. Each of the Si nanocrystals is situated at the center of a supercell. The supercells form an artificial simple cubic (sc) crystal. They considered large sc supercells with nominally 1000 atoms in the bulk limit. The supercells allow the treatment of nearly spherical nanocrystals with 5, 17, 41, 83, 147, 239, and 363 Si atoms and a corresponding number of passivating hydrogen atoms. The atoms are assumed to be tetrahedrally coordinated with a distance 2.34 Å, which was taken from the bulk crystal. Some of the calculation results are briefly described in the following.

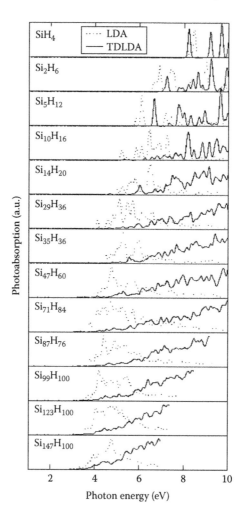

FIGURE 6.8 Calculated TDLDA absorption spectra of Si_nH_m clusters (solid lines). Spectra of time-independent Kohn–Sham LDA eigenvalues (dotted lines) are shown for comparison. (From Vasiliev, I. et al., *Phys. Rev. Lett.*, 86, 1813, 2001. With permission.)

Figure 6.10 shows the calculated dielectric functions of spherical nanocrystals with 5, 17, 41, 83, 147, and 239 Si atoms [6]. Although the model atomic structure used is somewhat unrealistic, and important many-body effects are not included, the absorption spectrum for the smallest Si cluster of five atoms shows the same basic features as obtained in more sophisticated calculations. The spectra in Figure 6.10, in particular, those representing the imaginary part of the dielectric function, are strongly influenced by quantum-confinement effects. The magnitude of the imaginary part of the dielectric function decreases with decreasing nanocrystal size. On the other hand, the absorption threshold moves to lower energies with increasing nanocrystal size. In the imaginary part of the dielectric function in Figure 6.10b, only one broad structure appears with a maximum between 4 and 6 eV. It exhibits a small shift of the main peak toward smaller photon energies with increasing nanocrystal size. The structure seems to develop into the bulk E_2 peak. In the case of the largest nanocrystals considered here, the almost complete absence of the E_1 structure may be related to neglecting the excitonic effects [6].

It has been mentioned previously that the magnitude of the imaginary part of the dielectric function decreases with decreasing nanocrystal size. Actually, an important finding from many

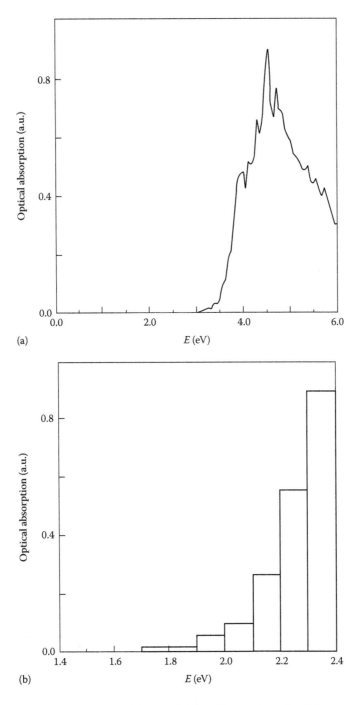

FIGURE 6.9   (a) Calculated optical absorption coefficient with respect to the photon energy E for a silicon nanocrystal with a diameter of 3.86 nm (the bandgap is calculated at 1.67 eV). (b) Same as (a) but only the energy region near the bandgap is plotted (amplitudes of the bars represent the integrated absorption coefficient over the width of the bar). (Adapted from Delerue, C. et al., *Phys. Rev. B*, 48, 11024, 1993. With permission.)

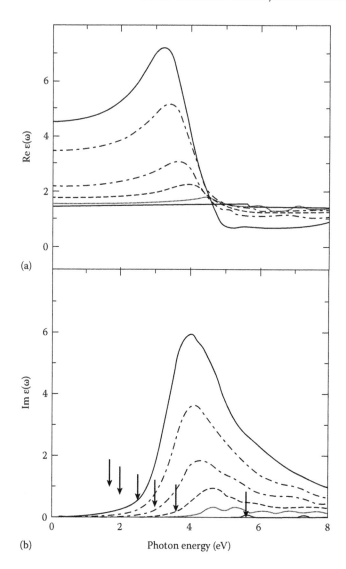

FIGURE 6.10 (a) Real and (b) imaginary part of dielectric functions of Si nanocrystallites with a varying number N of atoms. $N = 5$, solid line; $N = 17$, dotted line; $N = 41$, dashed line; $N = 83$, long-dashed line; $N = 147$, dot-dashed line; and $N = 239$, solid line. The vertical arrows in the absorption spectra indicate the single-particle HOMO–LUMO gaps. (Adapted from Weissker, H.-Ch. et al., *Phys. Rev. B*, 65, 155328-7, 2002. With permission.)

investigations on semiconductor nanocrystallites is the drastic reduction in the static dielectric constant when the size of nanocrystals approaches less than 10 nm. Reduction in the static dielectric constant causes an increase in the Coulomb interaction energy between electrons, holes, and ionized shallow impurities, and therefore can significantly modify the optical absorption and the transport phenomenon of a nanometer-sized device [30]. The existing theoretical investigations on the size dependence of static dielectric constant can be divided into two categories: (1) calculations performed with the use of the semiempirical version of the Penn model, referred to as the modified or generalized Penn model [31,32], and (2) numerically computed results employing empirical pseudo-potentials and the semiempirical linear combination of atomic orbitals [15,33,34].

Tsu et al. presented a simple single-oscillator model for the size-dependent reduction of the static dielectric constant of silicon nanocrystals [32]. The modified Penn model, taking into account the quantum confinement–induced discrete-energy states, leads to the size-dependent dielectric constant

$$\varepsilon_s(R) = 1 + \frac{\varepsilon_b - 1}{1 + (\Delta E / \Delta E_g)^2} \tag{6.4}$$

where
    R is the radius of a spherical silicon nanocrystal
    ε_b is the bulk dielectric constant
    ΔE is an energy separation
    E_g is an energy gap, as shown in Figure 6.11 [32]

Quantum confinement increases the separations of the discrete states, resulting in an increase in the energy denominator and a subsequent reduction in the static dielectric constant. Referring to Figure 6.11, the energy separation is given by [32]

$$\Delta E = \frac{\pi E_F}{k_F R} \tag{6.5}$$

Taking the parameters for Si, $\varepsilon_b = 12$ (or 11.3), $E_g = 4$ eV (note that the fundamental Γ–Δ gap at 1.1 eV plays almost no role in the dielectric function), and filling the energy bands up to E_F (=12.6 eV), the computed $\varepsilon_s(R)$ according to the modified Penn model is shown in Figure 6.12 [32].

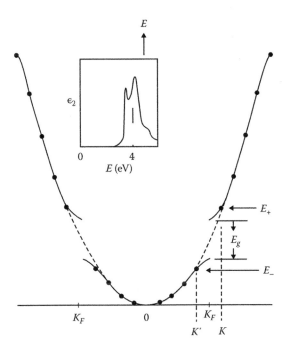

FIGURE 6.11 Electron energy versus k for an isotropic, three-dimensional, nearly free electron model. The inset shows the absorption spectra (ε_2) versus photon energy, giving justification for setting $E_g = 4$ eV for Si. Round dots indicate the discrete energies and momenta. E_+ and E_- define the positions of the new gap. (From Tsu, R. et al., *J. Appl. Phys.*, 82, 1327, 1997. With permission.)

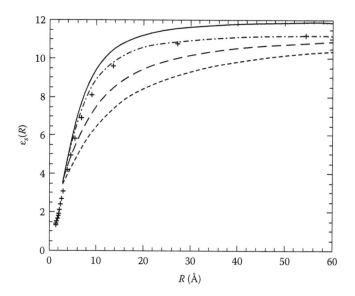

FIGURE 6.12 Size dependence of the static dielectric constants $\varepsilon_s(R)$ of silicon nanocrystals. (a) The solid line delineates $\varepsilon_s(R)$ from the modified Penn model using $\varepsilon_b = 12$. (b) The dash-dot line delineates $\varepsilon(R)$ from the modified Penn model using $\varepsilon_b = 11.3$. (c) The crosses depict $\varepsilon_s(q)$ from [35] converted into $\varepsilon_s(R)$ using $q = \pi/R$. (d) The dashed line with shorter dashes delineates the screening dielectric constant from [15]. (e) The dashed line with longer dashes delineates the dielectric constant from [36]. (From Tsu, R. et al., *J. Appl. Phys.*, 82, 1327, 1997. With permission.)

It has been suggested that the static dielectric constant $\varepsilon_s(R)$ for a spherical silicon nanocrystal with radius R can approximately be represented by

$$\varepsilon_s(R) = 1 + \frac{\varepsilon_b - 1}{1 + (\alpha/R)^m} \tag{6.6}$$

where $m = 2$, $\alpha = 10.93$ Å in Si, and $\varepsilon_b = 11.4$ is the bulk dielectric constant [15]. This generalized Penn model (GPM) is simple, but may be of questionable validity. It has been often used to explain experimental results on photoluminescence, photoabsorbance, and other optical properties. However, the size dependence of the static dielectric constant is still not fully understood. The existing calculations based on GPM are semiempirical in nature, in which the size dependence of the static dielectric constant has mainly been introduced through the size-dependent energy separation between two discrete energy levels. The oscillator strength and the energy bandgap have been treated as independent of the size in most of the existing calculations [30].

On the other hand, Wang et al. performed a "microscopic" calculation of the static dielectric constant of Si nanocrystals based on the quantum mechanical pseudopotential calculation of the absorption spectra $\varepsilon_2(E)$ [15]. The absorption spectra $\varepsilon_2(E)$ of Si nanocrystals containing up to ~1300 Si atoms were calculated fully quantum mechanically using an empirical pseudopotential plane wave representation and a novel moments method. The static dielectric constant ε_s is given by the integral of the absorption spectra $\varepsilon_2(E)$:

$$\varepsilon_s = 1 + \frac{2}{\pi} \int_0^\infty \varepsilon_2(E)/E \, dE \tag{6.7}$$

The calculation according to Equation 6.7 gives the total polarization dielectric constant depicted in Figure 6.13 [15]. Fitting the calculation result to the analytic form of Equation 6.6 (see the solid line shown in Figure 6.13) yields $\alpha = 4.25$ Å, $m = 1.25$ for the total polarization dielectric constant ε_s. Both α and m are significantly smaller than the results of the GPM, which gives $\alpha = 10.93$ Å and $m = 2$.

6.2.2 OXIDIZED nc-Si

It should be clear from the above discussions that the electronic and optical properties of nc-Si are determined by the nc-Si size. In addition to the size, the surface and interface properties of nc-Si should also play an important role in the electronic and optical properties. For instance, it is generally accepted that the quantum confinement, caused by the restricted space of the nanometer sizes, is essential for visible light emission in Si nanostructures, but some controversial interpretations of the photoluminescence (PL) properties of low-dimensional Si structures still exist [22]. In particular, the effect of oxidation of nc-Si surface is a question that needs to be addressed. Both theoretical calculations and experiments have been carried out to examine the role of the surface/interface on the electronic and optical properties. Wolkin et al. observed that oxidation introduces defects in the nc-Si bandgap, which pin the transition energy [37]. They claimed the formation of a Si=O double bond as the pinning state. However, Vasiliev et al. pointed out that similar results could also be obtained for O connecting two Si atoms (single bond) at the nc-Si surface [23]. The assistance of Si–O vibrations at the interface was also proposed as the dominant path for recombination [38]. On the other hand, interface radiative states have been suggested to play a key role in the mechanism of population inversion at the origin of the optical gain observed in nc-Si [39,40].

Garoufalis and Zdetsis reported accurate high-level calculations of the optical gap and absorption spectrum of small Si nanocrystals, with hydrogen and oxygen at the surface [2]. The calculations were performed in the framework of time-dependent density functional theory (TDDFT) using the hybrid nonlocal exchange and correlation functional of Becke and Lee, Yang and Parr (B3LYP). The oxygen contamination is introduced with several different bonding configurations (e.g., Si=O double bonds, Si–O–Si bridging bonds, and hydroxyl passivation Si–OH) [2].

Table 6.1 shows the calculated optical gaps of some oxygen-free Si nanocrystals and oxygen-contaminated Si nanocrystals containing Si=O bonds, Si–O–Si bridging bonds, and both bridging and double bonds [2]. As can be seen in the table, oxygen contamination leads to a significant reduction in the optical gaps. There is a red shift of 0.94 eV in the optical gap when the nanocrystal structure is changed from the oxygen-free $Si_{99}H_{100}$ to oxygen-rich $Si_{99}H_{76}O_{12}$ with Si=O bonds;

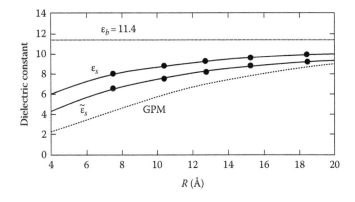

FIGURE 6.13 Dielectric constants as a function of the radius R of Si nanocrystals. Here, ε_s is for the total polarization and $\tilde{\varepsilon}_s$ is for exciton screening. The symbols denote the results calculated with Equation 6.7, while the solid lines are the fitted curves. The dashed curve corresponds to the GPM of Equation 6.6. (From Wang, L.-W. and Zunger, A., *Phys. Rev. Lett.*, 73, 1039, 1994. With permission.)

TABLE 6.1
Optical Gap of Oxygen-Free and Oxygen-Rich Si Nanocrystals

Oxygen-Free Nanocrystal	Optical Gap (eV)	Oxygenated Nanocrystal	Optical Gap (eV)
$Si_{99}H_{100}$	3.39	$Si_{99}H_{76}O_{12}$	2.45
$Si_{147}H_{100}$	3.19	$Si_{147}H_{52}O_{24}$	1.85
		$Si_{47}H_{48}O_6$	2.70
		$Si_{47}H_{36}O_{12}$	3.7
		$Si_{47}H_{24}O_{18}$	2.5

Source: Adapted from Garoufalis, C.S. and Zdetsis, A.D., *Phys. Chem. Chem. Phys.*, 8, 808, 2006. With permission.

while the red shift is 1.34 eV when the structure is changed from the larger oxygen-free $Si_{147}H_{100}$ to oxygen-rich $Si_{99}H_{52}O_{24}$ with Si=O bonds. Figure 6.14 shows a comparison of the nanocrystal's size dependence of the optical gap between oxygen-free Si nanocrystal and oxygen-containing Si nanocrystal [2]. The calculated red shift ranges from ~2 eV for the smaller nanocrystals down to ~1 eV for the larger nanocrystals considered here, which is in agreement with Wolkin et al., who have shown that even a 3 min exposure of the samples in air produces a red shift as large as 1 eV (for relatively large nanocrystals) due to the formation of oxygen bonds [37]. It seems that the oxygen effect is at least an important factor responsible for the conflicting experimental results reported in the literature about the exact dependence of the optical gap on the size of the nanocrystals and in particular the critical size for visible PL. Wolkin et al. obtained the optical gaps as small as 2.2 eV, for nanoclusters with a diameter of 18 Å [37]. For nanoclusters of about the same size, Wilcoxon et al. showed a much larger gap (larger than 3.2 eV) for highly purified samples of the same diameter [41].

The critical parameter for the change of the optical gap is not just the amount of oxygen but the bonding environment and the way the oxygen atoms are accumulated on the surface [2]. The role of the bonding type of the surface can be highlighted by the comparison of optical gap among the

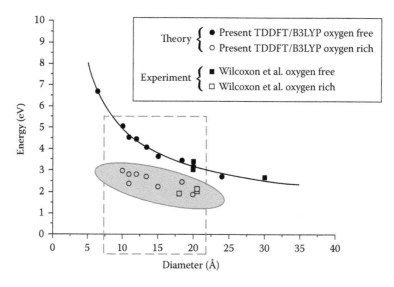

FIGURE 6.14 Variation of the optical gap as a function of the Si nanocrystal diameter (in Å). The shaded area corresponds to oxygen-contaminated nanocrystals containing Si=O bonds. The experimental result is from Wilcoxon et al. [41]. For diameters $10 < d < 20$ Å, the average calculated optical gap for oxygenated nanocrystals ranges from 2.5 to 1.85 eV. (From Garoufalis, C.S. and Zdetsis, A.D., *Phys. Chem. Chem. Phys.*, 8, 808, 2006. With permission.)

following similar nanocrystals with different bonding structures: $Si_{47}H_{48}O_6$ (with Si=O double bonds), $Si_{47}H_{36}O_{12}$ (with Si–O–Si bridging oxygen bonds), and $Si_{47}H_{24}O_{18}$ nanocrystal (with both Si=O and Si–O–Si bonds). As can be seen in Table 6.1, the optical gaps of these nanoparticles are different. The $Si_{47}H_{48}O_6$ nanocrystal containing Si=O double bonds has an optical gap of 2.70 eV; but the optical gap of the $Si_{47}H_{36}O_{12}$ nanocrystal containing Si–O–Si bridging oxygen bonds is blue-shifted up to 3.7 eV. However, in the $Si_{47}H_{24}O_{18}$ nanocrystal, which contains both double (Si=O) and bridging (Si–O–Si) oxygen bonds, the fundamental optical gap is red-shifted back to the lower energy side (2.5 eV, which is close to the optical gap of the $Si_{47}H_{48}O_6$ nanocrystal containing only Si=O double bonds).

To examine the possible role of single-bonded surface oxygen atoms, Garoufalis and Zdetsis also performed calculations on the Si_{35} nanocrystals with the surface dangling bonds passivated by both hydrogen atoms and hydroxyl groups [2]. For the fully hydroxyl-passivated $Si_{35}(OH)_{36}$ nanocrystal, the HOMO–LUMO gap obtained is 2.6 eV. However, for the similar partially "hydroxylated" nanocrystals such as $Si_{35}H_{24}(OH)_{12}$ and $Si_{35}H_{34}(OH)_2$, the gap is increased by as much as 1.8 eV for $Si_{35}H_{24}(OH)_{12}$ and 2.2 eV for $Si_{35}H_{34}(OH)_2$ (the HOMO–LUMO gaps of $Si_{35}H_{24}(OH)_{12}$ and $Si_{35}H_{34}(OH)_2$ are 4.4 and 4.8 eV, respectively). This indicates that the effect of hydroxyl passivation is largely dependent on the number of hydroxyl groups deposited on the surface and probably on their spatial distribution. This observation highlights the role of the relative positions of the oxygen atoms. The large difference in the behavior between the hydroxyl and double-bonded oxygen is that, although a few (even two or three) double-bonded oxygen atoms can significantly reduce the HOMO–LUMO gap, to accomplish the same reduction of the gap by hydroxyl groups, a much larger number of hydroxyls is needed. Therefore, the presence of double-bonded oxygen "contaminants" is most effective to reduce the gap [2].

Garoufalis and Zdetsis also calculated the absorption spectra of oxygen-free and oxygen-rich Si nanocrystals with emphasis on the lower part of the absorption spectra [2]. Figure 6.15 shows the comparison of the calculated absorption spectrum between the oxygen-free nanocrystals ($Si_{35}H_{36}$ and $Si_{71}H_{84}$) and oxygen-rich nanocrystals ($Si_{35}H_{24}O_6$ and $Si_{71}H_{72}O_6$) containing Si=O double bonds [2]. The optical gaps of the oxygen-free $Si_{35}H_{36}$ and $Si_{71}H_{84}$ nanocrystals are 4.42 and 3.64 eV, respectively; the optical gaps of the oxygen-rich $Si_{35}H_{24}O_6$ and $Si_{71}H_{72}O_6$ nanocrystals are red-shifted to 2.82 and 2.21 eV, respectively. As we can see in Figure 6.15, the presence of surface oxygen not only affects the value of the fundamental optical gap but also produces a significant enhancement of the absorbance (large oscillator strengths). The result shown in Figure 6.15 suggests that, in addition to reducing the optical gap, the Si=O bonds could also enhance the light absorption in the tail of the spectrum.

6.2.3 Ge Nanocrystals

The extension of research interest to Ge nanocrystals (nc-Ge) is rather straightforward, as both Si and Ge are group IV semiconductors. PL has been experimentally observed in the wavelength range 350–700 nm from Ge nanocrystals with the sizes of 2–5 nm [42], while there have also been studies concerning much larger nanoparticles [43]. Various researchers have observed evidence for quantum confinement effects [42,44–46] and size-dependent PL in the near-infrared region, which could possibly be due to a radiative recombination of excitons confined in the nanoclusters [44].

Theoretical studies on nc-Ge are relatively limited as compared to those on nc-Si. There have been calculations reported using various techniques including tight-binding [47], empirical pseudopotentials [48,49], DFT in the local density approximation (LDA) [50,51], a combination of DFT in the LDA or local spin density approximation (LSDA) with self-consistent field theory (SCF) [5,6,52,53], and TDDFT [54,55]. The qualitative trend of all of these reports is a significant increase in the optical gap as the cluster diameter decreases; however, there are disagreements between the theoretical results. Some results of the theoretical calculations are briefly described in the following.

Garoufalis et al. conducted *ab initio* calculations of the electronic structures and optical gaps of Ge nanocrystals based on TDDFT employing the hybrid nonlocal exchange-correlation functional of Becke, Lee, Yang, and Parr (B3LYP) [56]. The size of the nanocrystals considered in the

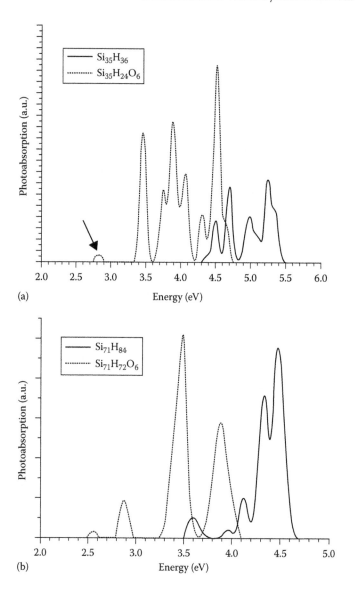

FIGURE 6.15 Comparison of the absorption spectrum of oxygen-free and oxygen-rich Si nanocrystals. The small arrows indicate a weak peak at ~2.8 eV. (From Garoufalis, C.S. and Zdetsis, A.D., *Phys. Chem. Chem. Phys.*, 8, 808, 2006. With permission.)

calculations is rather small, ranging from 5 to 99 Ge atoms, with 12–100 H atoms (the diameter of the largest nanocrystal with $Ge_{99}H_{100}$ is 19 Å). Figure 6.16 shows the calculated DOS (density of states) for two Ge nanocrystals with different sizes ($Ge_{47}H_{60}$ and $Ge_{99}H_{100}$) [56]. As can be observed from the figure, the gap of the nanocrystals increases with decreasing size of the crystals, which is practically "symmetrical" with respect to the conduction and valence band edges. On the other hand, the similarity of the electronic structure of Si and Ge nanocrystals can be demonstrated by Figure 6.17, in which the DOS curves for Ge_{47}:H_{60} and Si_{47}:H_{60} are shown [56]. The electronic structures of the two nanocrystals are fully homologous. This could imply that their optical properties are homologous also. However, Garoufalis et al. showed that, due to the smaller bandgaps of Ge nanocrystals, the diameter of the smallest Ge nanocrystal that can emit in the visible region is ~19–20 Å, which is smaller than the corresponding "critical" diameter of Si nanocrystals (22 Å) [56].

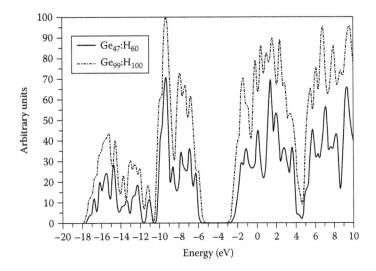

FIGURE 6.16 Comparison of the density of states of $Ge_{47}H_{60}$ and $Ge_{99}H_{100}$ nanocrystals. (From Garoufalis, C.S. et al., *J. Phys.: Conf. Ser.*, 10, 97, 2005. With permission.)

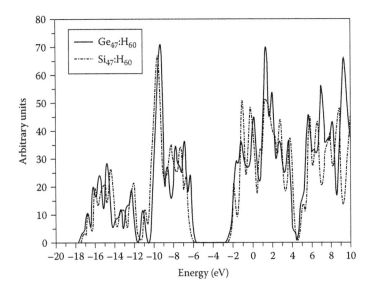

FIGURE 6.17 Comparison of density of states of $Ge_{47}H_{60}$ and $Si_{47}H_{60}$. (From Garoufalis, C.S. et al., *J. Phys.: Conf. Ser.*, 10, 97, 2005. With permission.)

Weissker et al. calculated the energy levels and optical transitions of Ge and Si nanocrystals using the DFT-LDA technique [6]. In their calculation, the nanocrystals are described by clusters of up to 363 atoms, and their surfaces are passivated by hydrogen atoms. Figure 6.18 shows the energy level schemes and optical transitions for Ge nanocrystals with 41, 147, and 239 atoms obtained from the calculation [6]. In this figure, the quantum confinement effects on the energy levels are clearly visible, and the effect of confinement is also obvious for the oscillator strengths of the optical transitions. Note that the HOMO–LUMO transition is forbidden by symmetry; in contrast, the transition from the threefold degenerate second-highest state into the nondegenerate LUMO state possesses extremely large oscillator strength. Very strong transitions occur close to the absorption edge, even for clusters

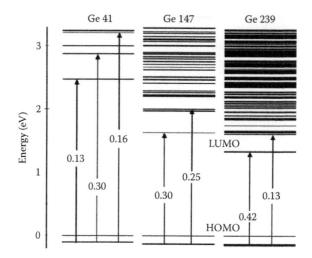

FIGURE 6.18 Level scheme for Ge nanocrystals with 41, 147, and 239 Ge atoms. The HOMO level defines zero energy. The allowed optical transitions are indicated by vertical arrows. The oscillator strength of a given transition is indicated by the number at the corresponding arrow. (From Weissker, H.-Ch. et al., *Phys. Rev. B*, 65, 155328-7, 2002. With permission.)

with a diameter of ~2.2 nm (239 Ge atoms). These transitions could be related to the formation of an E_0-like absorption feature in the more extended Ge nanocrystallites. The behavior of the oscillator strengths near the absorption edges of Ge nanocrystals is very different from that of Si nanocrystals. In the case of Si nanocrystals with diameters above 1.5 nm (more than 83 Si atoms), a tail of weak transitions appears just above the HOMO–LUMO gap. The oscillator strengths of these transitions are much smaller than the maximum oscillator strengths of ~0.4. The occurrence of the tail can be interpreted as an indication of the development of bulk properties with increasing nanocrystal size [6].

On the other hand, Nesher et al. also conducted a TDLDA (time-dependent local density approximation) calculation of average oscillator strengths of near-gap transitions for Ge nanocrystals [55]. They pointed out that, because of the large number of near-gap transitions, theoretical predictions for near-gap optical activity are meaningful only if one averages the oscillator strengths over a relatively narrow energy window around the optical gap. The size dependence of the calculated average oscillator strengths for Ge nanocrystals is shown in Figure 6.19 [55]. It can be clearly seen from the figure that the average oscillator strength of the optical transitions near the absorption edge decreases strongly with increasing nanocrystal size. The diminishing optical activity with increasing size is consistent with a shift from direct absorption for the smallest molecules and clusters to the formation of an indirect bandgap in the limit of bulk Ge.

As can be seen in Figure 6.18, the nanocrystal size influences not only the oscillator strengths but also the energetic positions of the optical transitions. Garoufalis et al. showed that both the HOMO–LUMO gap and the fundamental optical gap increase as the diameter of the Ge nanocrystals decreases, exhibiting the quantum-size dependence similar to that of Si nanocrystals [56]. They also showed that, as compared to the Si nanocrystals with the same sizes, the Ge nanocrystals have a smaller HOMO–LUMO gap and optical gap. The optical gap for the nanocrystal of $Ge_{99}H_{100}$ (about 19 Å in diameter) obtained from their calculation is 2.95 eV (420 nm in wavelength). Weissker et al. also calculated the lowest electron–hole pair excitation energies of Ge and Si nanocrystals with various sizes, and the result is shown in Figure 6.20 [6]. The pair excitation energies are not much larger than the single-particle HOMO–LUMO gaps estimated by means of the Kohn–Sham eigenvalues. For the considered crystallite sizes, the pair excitation energy varies between 5 (6) and 1 (2) eV for Ge (Si) nanocrystals. The decrease of the transition energies with the nanocrystal size is rather rapid. The pair excitation energies are smaller in Ge nanocrystals than in Si nanocrystals.

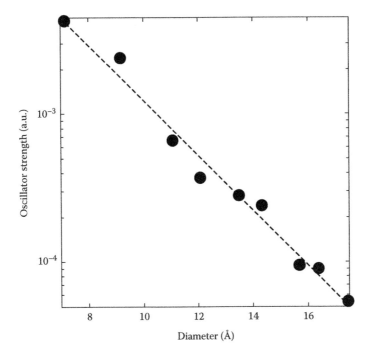

FIGURE 6.19 Average oscillator strengths for near-edge optical transitions in Ge nanocrystals as a function of nanocrystal size. The dashed line is a linear fit. (From Nesher, G. et al., *Phys. Rev. B*, 71, 035344-5, 2005. With permission.)

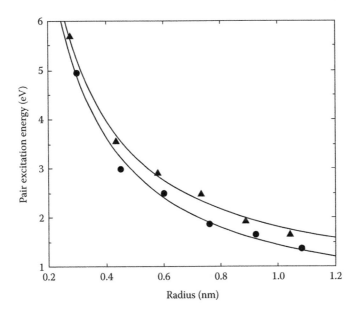

FIGURE 6.20 Lowest electron–hole pair excitation energies of Ge (dots) and Si (triangles) nanocrystals. The solid lines are the fitted curves according to $E_g(R) = E_g(\text{bulk}) + \alpha(\text{Å}/R)^l$ with $\alpha = 14.7$ eV and $l = 1.0$ for Ge and $\alpha = 14.8$ eV and $l = 1.0$ for Si. (From Weissker, H.-Ch. et al., *Phys. Rev. B*, 65, 155328-7, 2002. With permission.)

In the interested size range, the Si energies are larger by about 0.5 eV. This value corresponds roughly to the difference of the fundamental energy gaps in the bulk limit. It has been reported that the size dependence of the bandgap of Ge nanocrystals also follows Equation 6.3. The fitting to the calculated pair excitation energies of Ge nanocrystals shown in Figure 6.20 with Equation 6.3 yields $c = 2.94$ and $m = 1.0$ [6]. This means that with $m = 1.0$ the gap energy varies approximately like the inverse nanocrystalline diameter. This is much weaker than expected from the quantum mechanics of the three-dimensional spherical potential well.

Garoufalis et al. also estimated the binding energy (E_B) of the exciton formed as a result of the electronic excitation, which is approximately the difference between the HOMO–LUMO gap and the fundamental optical gap, and the result is shown in Figure 6.21 [56]. As expected by the quantum confinement hypothesis, the binding energy increases as the diameter of the Ge nanocrystals decreases. For diameters of ~20 Å, the value of E_B is about 0.45 eV. It is interesting to note that, although the Si and Ge nanocrystals with the diameter of ~20 Å contain different numbers of atoms, their E_B values are about the same.

Nesher et al. calculated *ab initio* absorption spectra for hydrogen-passivated Ge nanocrystals using TDDFT within the adiabatic local density approximation [55]. The computed time-dependent LDA (TDLDA) spectra are shown in Figure 6.22. Each spectrum is compared with a time-independent LDA spectrum obtained by considering filled and empty Kohn–Sham orbitals as true one-electron wave functions. As can be observed in the figure, with increasing nanocrystal size, both LDA and TDLDA spectra feature a gradual decrease in the absorption and a gradual evolution from discrete spectra for the smaller nanocrystals to quasicontinuous spectra for the larger ones.

The confinement effect is found to have a strong influence on the frequency-dependent dielectric function for Ge nanocrystals. Figure 6.23 shows the dielectric function as a function of the number of Ge atoms calculated by Ramos et al. in the framework of the independent-particle approach [18]. As can be observed in the figure, the effect of quantum confinement on the HOMO–LUMO (corresponding to E_0 in bulk Ge) transition is clearly visible in the imaginary part, though it becomes weaker for high-energy optical transitions. The peak corresponding to the E_1 peak in the bulk spectrum shows a red shift with increasing nanocrystal size. However, the size effect is practically not observed in the case of the E_2-like transitions near 4.2 eV.

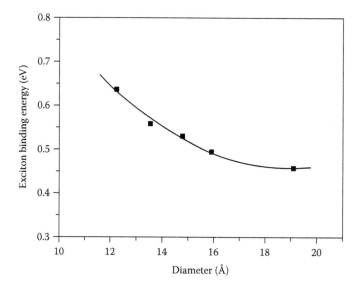

FIGURE 6.21 Variation of the exciton binding energy of Ge nanocrystals with the crystal size. (From Garoufalis, C.S. et al., *J. Phys.: Conf. Ser.*, 10, 97, 2005. With permission.)

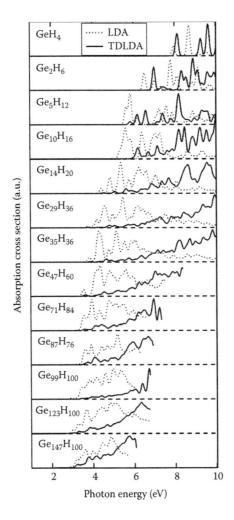

FIGURE 6.22 Absorption spectra of Ge nanocrystals calculated using TDLDA (solid lines) and LDA (dashed lines). (From Nesher, G. et al., *Phys. Rev. B*, 71, 035344-5, 2005. With permission.)

The optical gap can be extracted from the low-energy optical spectrum. Figure 6.24 shows the size dependence of the optical gap of spherical Ge nanocrystals calculated using tight binding [47] and TDLDA [54]. In the same figure, the PL peak as function of the cluster diameter, measured by Takeoka et al. [44] and Kanemitsu et al. [57], is included also. As can be seen in Figure 6.24, the optical gap increases with decreasing nanocrystal size. The dependence of the optical gap on the size of the nanocrystals can be explained with the help of the quantum confinement model [54]. According to this model, PL comes from the recombination of electron–hole pairs confined in the nanocrystals. By considering the simplified picture of an exciton confined into an infinite spherical potential, the lowest energy of the electron–hole pair E_g is given by [54]

$$E_g = E_{g0} + \frac{\pi^2 \hbar^2}{2\mu R^2} \tag{6.8}$$

where
E_{g0} is the optical gap for the bulk crystalline material
μ is the reduced mass of the exciton
R is the radius of the spherical potential

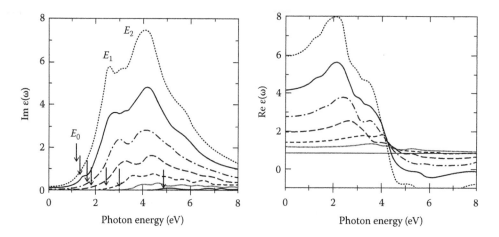

FIGURE 6.23 (a) Imaginary and (b) real part of the frequency-dependent dielectric functions of Ge nanocrystals passivated with H, containing 363 (dotted line), 239 (solid line), 147 (dash-dotted line), 83 (long-dashed line), 41 (short-dashed line), 17 (dotted line), and 5 (solid line) Ge atoms. The arrows indicate the single-particle HOMO–LUMO gaps. (From Ramos, L.E. et al., *Phys. Stat. Sol. (b)*, 242, 3053, 2005. With permission.)

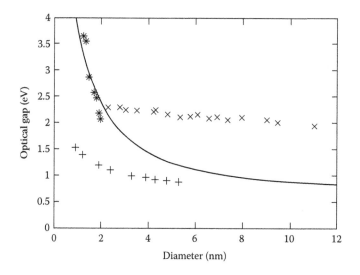

FIGURE 6.24 Optical gap of spherical Ge nanocrystals calculated using tight binding by Niquet et al. (dotted line) [47] and TDLDA by Tsolakidis et al. (asterisks) [54]. Experimental photoluminescence energies are from Kanemitsu et al. [57] (×) and Takeoka et al. [44] (+). (From Tsolakidis, A. and Martin, R.M., *Phys. Rev. B*, 71, 125319, 2005. With permission.)

This simplified model can provide a qualitative explanation for the size dependence of the optical gap. However, as can be seen in Figure 6.24, the optical gap does not have the simple size dependence of $1/R^2$ predicted by Equation 6.8.

6.3 EXPERIMENTAL STUDIES OF OPTICAL PROPERTIES

As discussed previously, there have been many theoretical calculations on the optical properties of Si and Ge nanocrystals. Experimental works on the optical properties were mainly focused on the luminescence properties of Si and Ge nanocrystals. Although some experimental studies of the optical properties of a continuous Si nanocrystal thin film [58] and SiO_2/nanocrystalline Si multilayers [59] have been reported recently, there have been relatively few optical studies so far to experimentally determine the dielectric functions or optical constants of Si and Ge nanocrystals. In particular, a comprehensive experimental study of the optical properties of Si and Ge nanocrystals embedded in various dielectric matrixes over a wider photon energy range is still lacking. In the material system of dielectric films embedded with silicon or Ge nanocrystals, the dielectric functions or optical constants of the embedded nanocrystals should be different from those of the bulk crystalline semiconductor due to the size effect, and should be also different from those of a continuous nanocrystal thin film. In this section, we present some experimental studies of the dielectric functions/optical constants of both Si nanocrystals embedded in SiO_2 and Si_3N_4 matrixes and self-assembled Ge nanocrystals using spectroscopic ellipsometry (SE) [60–69]; in addition, we also describe briefly an experimental study of size-dependent near-infrared photoluminescence from Ge nanocrystals embedded in SiO_2 matrices [44].

6.3.1 nc-Si Embedded in SiO_2 Thin Films

SE has been used to study the optical properties of silicon nanocrystals (nc-Si) embedded in SiO_2 matrix [63–67]. For example, Ding et al. reported such a study on the determination of the optical properties of nc-Si in the photon energy range 1.1–5.0 eV [63]. nc-Si with a mean size of ~4 nm embedded in a SiO_2 matrix was synthesized by Si^+ implantation with a dose of 1×10^{17} atoms/cm^2 at the energy of 100 keV into a 550-nm-thick SiO_2 film thermally grown on a p-type Si substrate. Stable nc-Si was formed after thermal annealing at 1000°C for 30 min in nitrogen gas. The dielectric function of the nc-Si is found to be well described by both the Lorentz oscillator model [70] and the Forouhi–Bloomer (FB) model [71].

The dielectric functions of the nc-Si obtained from the SE analysis is shown in Figure 6.25 [63]. As can be observed in the figure, the overall spectral features of optical properties of the nc-Si are similar to those of bulk crystalline Si. However, the nc-Si shows a significant reduction in the dielectric function compared to bulk crystalline Si.

Using the same SE methodology, Ding et al. obtained the dielectric functions and optical constants of the nc-Si with various sizes embedded in SiO_2, which are shown in Figures 6.26 and 6.27, respectively [66]. As can be observed in the two figures, nc-Si exhibits a significant reduction in the dielectric functions and optical constants with respect to bulk crystalline silicon, and the size of nc-Si has a large influence on both the magnitude and shape of the spectra of the dielectric functions and optical constants. It is well known that dielectric functions of a crystalline material are closely associated with its electronic band structure, which is often described by the joint DOS [66]. The critical points observed in the dielectric spectra of crystalline material are believed to originate from singularities in the joint DOS. As can be observed in Figure 6.26, the imaginary part of the dielectric function of bulk crystalline silicon has main peaks at the transition energies E_1 (~3.4 eV) and E_2 (~4.3 eV) as its critical points. The main peaks are responsible for the high absorption of the light wave by the material. In the case of nc-Si embedded in SiO_2, the magnitude of the imaginary part of the dielectric functions decreases with decreasing nc-Si size. There is not much change in the transition energy of E_1 for different nc-Si sizes, but a large red shift (~0.3 eV) in the transition energy of E_2 is observed for the nc-Si sizes of 4.6 and 5.3 nm [66].

The static dielectric constant of the widely distributed nc-Si embedded in SiO_2 matrix is found to be 9.7, which is obtained from the SE analysis based on the four-term FB model by setting the photon energy to zero [60]. Compared to the static dielectric constant (11.4) of bulk crystalline silicon, there is a significant reduction in that of nc-Si. It has been well established that reduction of the static dielectric constant becomes significant as the size of the quantum-confined physical systems, such as quantum dots and wires, approaches the nanometer range [15,32,33,66,72]. However, the origin of the reduction in static dielectric constant with the size is still not fully understood. It is often attributed to the opening of the gap, which should lower the polarizability, but it has also been shown that the

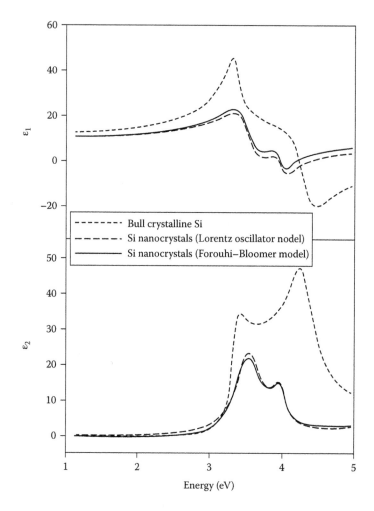

FIGURE 6.25   Real (ε_1) and imaginary (ε_2) part of the complex dielectric function of the nc-Si obtained from the spectral fittings based on the Lorentz oscillator model and the FB model. The dielectric function of bulk crystalline silicon is also included for comparison. (From Ding, L. et al., *Phys. Rev. B*, 72, 125419, 2005. With permission.)

reduction is due to the breaking of polarizable bonds at the surface and not to the opening of the bandgap induced by the confinement [34]. Taking the screening effect by the medium into account, Equation 6.6 can be expressed for the screening static dielectric constant of the nc-Si [15,60,66], as

$$\varepsilon_r(D) = 1 + \frac{\varepsilon_r(\infty) - 1}{1 + (6.9/R)^{1.37}} \qquad (6.9)$$

where $\varepsilon_r(\infty)$ (=11.4) is the static dielectric constant of bulk crystalline silicon, and R is the radius of nc-Si (in Å). The static dielectric constant of nc-Si with the diameter of 4.5 nm calculated with Equation 6.9 is 9.7, which is equal to the value obtained from the SE analysis based on the FB model [60]. Figure 6.28 shows the size dependence of the static dielectric constant of silicon nanocrystal calculated with Equation 6.9.

     The FB model can also yield the bandgap of the isolated nc-Si embedded in the SiO_2 matrix. The bandgap of the nc-Si obtained from the SE spectral fitting based on the FB model is 1.74 eV, which is ~0.6 eV larger than the bandgap of bulk crystalline Si [63]. On the other hand, based on the plot of $(\alpha E)\gamma$ versus E, where $\alpha = 4\pi k/\lambda$ (where k is the extinction coefficient of the nc-Si obtained

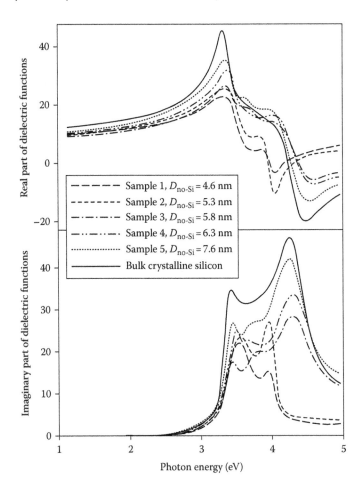

FIGURE 6.26   Real (ε_r) and imaginary (ε_i) part of the complex dielectric function of the nc-Si with various sizes obtained from the spectral fittings. The dielectric function of bulk crystalline silicon is also included for comparison. (From Ding, L. et al., *J. Appl. Phys.*, 101, 103525, 2007. With permission.)

from the SE analysis and λ is the wavelength) is the absorption coefficient of the nc-Si and E is the photon energy, one can examine whether the nc-Si is a direct ($\gamma = 2$) or indirect ($\gamma = 1/2$) bandgap semiconductor [60,73]. The Tauc plots shown in [73] indicate that nc-Si has an indirect bandgap structure with the value of 1.75 eV, which is almost the same as the value (1.74 eV) obtained from the SE analysis based on the FB model. The bandgap obtained here is in good agreement with that obtained from the first-principles calculation of the optical gap of Si nanocrystals based on quantum confinement [8]. A fit using Equation 6.3 to the calculation result (Figure 6.7) obtained by Öğüt et al. [8] yields the band gap expansion [66]

$$E_g(D) - E_{g0} = \frac{C}{D^n} \tag{6.10}$$

where

D is the nanocrystal size (in nm)
$E_g(D)$ is the bandgap (in eV) of the nanocrystal
$E_{g0} = 1.12$ eV is the bandgap of bulk crystalline Si
$C = 3.9$
$n = 1.22$

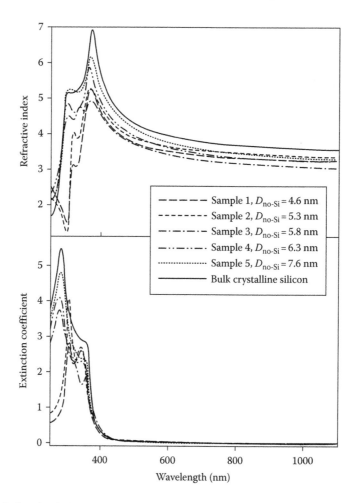

FIGURE 6.27 Refractive index (n) and extinction coefficient (k) of nc-Si of various sizes. The optical constants of bulk crystalline silicon are also included for comparison. (From Ding, L. et al., *J. Appl. Phys.*, 101, 103525, 2007. With permission.)

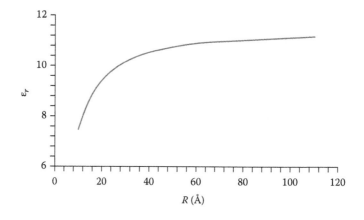

FIGURE 6.28 Size dependence of the static dielectric constant of silicon nanocrystal calculated with Equation 6.9.

For the nc-Si size of ~4.2 nm, Equation 6.10 gives a bandgap expansion of ~0.67 eV, which agrees with the values mentioned previously [73].

From the SE spectral fitting based on the FB model, Ding et al. obtained the bandgaps of nc-Si with various sizes embedded in SiO_2 [66]. nc-Si exhibits a large expansion in the band gap as compared to that of the bulk crystalline silicon, and the bandgap of the nc-Si increases when the nc-Si size is reduced. The bandgap expansion as a function of the nc-Si size is shown in Figure 6.29 [66]. For comparison, a calculation of the bandgap expansion with Equation 6.10 is also shown in the figure. A good agreement can be seen in the figure. The bandgap expansion is the most direct evidence of quantum confinement effect of nc-Si. The bandgap expansion of nc-Si and its dependence on the nanocrystal size have been demonstrated by many theoretical calculations, as discussed in Section 6.2.

6.3.2 nc-Si EMBEDDED IN Si_3N_4 THIN FILMS

Cen et al. reported an SE study on the optical properties of Si nanocrystals dispersed in a silicon nitride (Si_3N_4) thin film synthesized with the Si ion implantation [68]. The study is summarized here.

A Si_3N_4 thin film with thickness of ~120 nm was deposited by low-pressure chemical vapor deposition (LPCVD) onto a 30 nm SiO_2 thin film (i.e., a stress-relief oxide layer) thermally grown on a Si (100) substrate. The Si_3N_4 film was subsequently implanted with Si ions with a dose of 3.5×10^{16} atoms/cm^2 at the energy of 30 keV. Afterward, thermal annealing was carried out in nitrogen ambient at different temperatures (i.e., 800°C, 900°C, 1000°C, and 1100°C) for 1 h. The multilayer optical model shown in Figure 6.30a was used in the SE analysis [68]. Transmission electron microscopy (TEM) measurement showed that the nc-Si size was in the order of 2 nm. Figure 6.30b shows a typical TEM image.

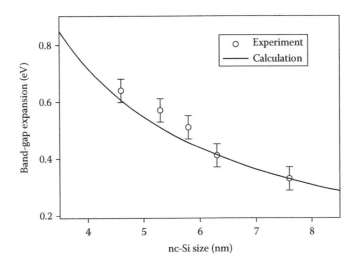

FIGURE 6.29 Bandgap expansion of nc-Si as a function of the crystal size. The symbols are the data obtained from the SE spectral fitting based on the FB model, and the line represents the calculation with Equation 6.10. (From Ding, L. et al., *J. Appl. Phys.*, 101, 103525, 2007. With permission.)

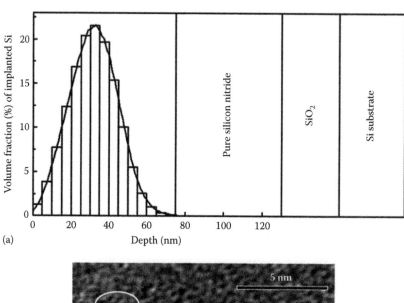

(a)

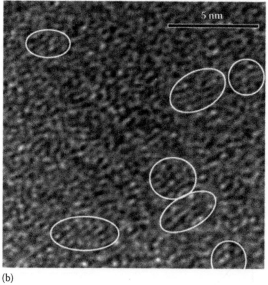

(b)

FIGURE 6.30 (a) Multilayer model used in the SE analysis and (b) TEM image of the Si nanoparticles or nanoclusters embedded in the silicon nitride thin film for the sample annealed at 1100°C. (From Cen, Z.H. et al., *Appl. Phys. Lett.*, 93, 023122, 2008. With permission.)

In the SE analysis, the Kramers–Kronig-consistent Tauc–Lorentz (TL) model was used [68]. In the model, the imaginary part (ε_2) of dielectric function was written as

$$
\varepsilon_2(E) = \begin{cases} 0, & \text{for } 0 < E \le E_g \\ \left[\dfrac{A E_0 C (E - E_g)^2}{(E^2 - E_0^2)^2 + C^2 E^2} \cdot \dfrac{1}{E} \right], & \text{for } E > E_g \end{cases} \tag{6.11}
$$

where
  A is the amplitude of a Lorentz oscillator
  E_0 is the peak transition energy
  C is the broadening term
  E_g is the bandgap energy

The real part (ε_1) of dielectric function is obtained by Kramers–Kronig integration of ε_2 with a fitting constant $\varepsilon_1(\infty)$. In the SE spectral fitting, A, E_0, C, and E_g as well as the thicknesses of both the silicon nitride layer and the SiO_2 layer are the fitting parameters. Figure 6.31 shows the spectral fittings for the sample annealed at 1100°C [68]. The dielectric functions of the nc-Si distributed in Si_3N_4 obtained from the SE fittings are shown in Figure 6.32 for different annealing temperatures [68].

It is observed from Figure 6.32 that the dielectric functions for all the samples are generally similar. The spectral features of the dielectric functions are also similar to those of amorphous Si. The broadened peak structures, particularly the broad peaks of the as-implanted sample, suggest that the implanted Si is in an amorphous or disordered state to a certain extent. The nc-Si distributed in Si_3N_4 shows a significant reduction in the dielectric functions as compared to bulk crystalline silicon, and the situation is similar to that of nc-Si distributed in SiO_2 [68].

6.3.3 SELF-ASSEMBLED Ge NANOCRYSTALS

Goh et al. reported a study on the bandgap and optical properties (dielectric functions and optical constants) of self-assembled Ge nanocrystals (nc-Ge) using SE based on the FB optical dispersion model [69]. They showed that, as compared to bulk crystalline Ge, the nc-Ge exhibited a large bandgap expansion and a significant reduction in the dielectric function. The information presented in this section is adapted from [69].

The synthesis procedure of the self-assembled nc-Ge is as follows. A 3-nm-thick Ge layer was deposited onto a 3 nm SiO_2 layer, which was thermally grown on p-type $\langle 100 \rangle$ Si substrate, using electron beam evaporation under the base pressure of 5×10^{-6} mbar with the deposition rate of 0.07 nm/s. The sample underwent a rapid thermal annealing (RTA) in nitrogen ambient at 450°C for 50 s to form nc-Ge. With the existence of the SiO_2 layer, possible reaction between the deposited Ge layer and the Si substrate during the annealing process could be avoided. The formation of the nc-Ge is clearly demonstrated by the comparison of the atomic force microscopy (AFM) image (planar view) of the Ge layer before the RTA with that after the RTA, as shown in Figure 6.33a and b [69]. In addition, as revealed by the TEM image shown in Figure 6.33c, the

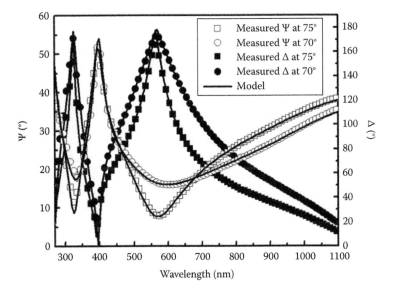

FIGURE 6.31 SE spectral fittings for the sample annealed at 1100°C. (From Cen, Z.H. et al., *Appl. Phys. Lett.*, 93, 023122, 2008. With permission.)

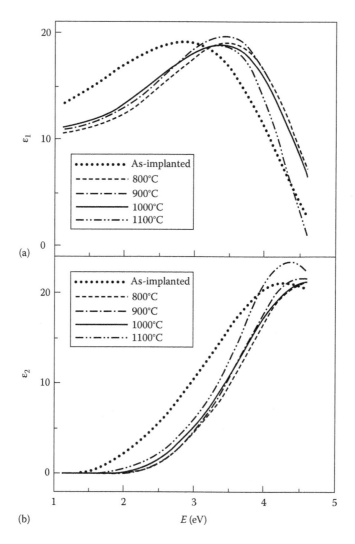

FIGURE 6.32 Annealing effect on the dielectric functions of the implanted Si. (a) Real part (ε_1) and (b) imaginary part (ε_2) of the dielectric functions. (From Cen, Z.H. et al., *Appl. Phys. Lett.*, 93, 023122, 2008. With permission.)

self-assembled nc-Ge has a dome shape with average size of ~6 nm in height and ~13 nm in diameter [69]. The formation of nc-Ge is explained by stress relaxation, dispersion force, and surface energy minimization during the annealing process [69].

In the SE analysis for the nc-Ge on the SiO$_2$ layer thermally grown on the Si substrate, the four-phase model, that is, air/nc-Ge layer/SiO$_2$ layer/Si substrate, which is shown in the inset of Figure 6.34, was used. The nc-Ge layer comprised of voids and nc-Ge. Its effective dielectric function ε_i ($= N_i^2$, where N_i is the complex refractive index of the layer) was modeled with the Bruggeman effective medium approximation (EMA) [69]:

$$\frac{\varepsilon_{\text{nc-Ge}} - \varepsilon_i}{\varepsilon_{\text{nc-Ge}} + 2\varepsilon_i} f + \frac{\varepsilon_{void} - \varepsilon_i}{\varepsilon_{void} + 2\varepsilon_i}(1 - f) = 0 \tag{6.12}$$

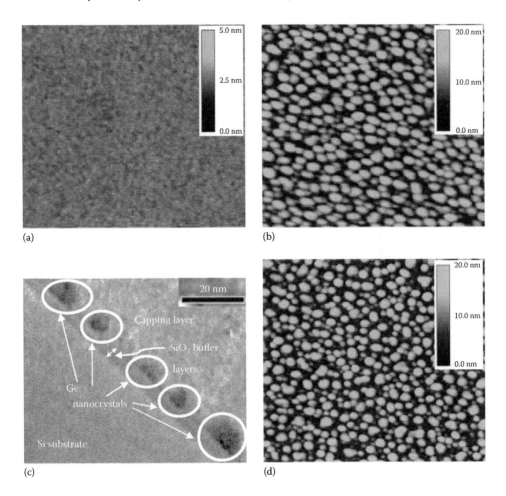

FIGURE 6.33 AFM images (plane view) of the 3 nm Ge thin film sample on Si substrate (a) before anneal-ing and (b) after the annealing to form nc-Ge. (c) TEM image of the nc-Ge formed on Si substrate (note that a SiO$_2$ capping layer was used for the TEM experiment). (d) Planar view AFM image of the nc-Ge formed on fused silica substrate. (From Goh, E.S.M. et al., *J. Appl. Phys.*, 109, 064307, 2011. With permission.)

where $\varepsilon_{\text{nc-Ge}}$ ($= (n_{\text{nc-Ge}} - k_{\text{nc-Ge}})^2, n_{\text{nc-Ge}}$ and $k_{\text{nc-Ge}}$ are the refractive index and extinction coefficient of the nc-Ge, respectively) is the effective dielectric function of the nc-Ge, f is the volume fraction of the voids in the layer, and the dielectric function ($\varepsilon_{\text{void}}$) of the voids is 1. It was found that the FB model satisfactorily describes the optical constants of the nc-Ge [69].

Figure 6.34 shows the fittings based on the Bruggeman EMA and the FB model to the SE spectra measured at three angles of incidence [69]. As can be seen in this figure, the spectra of both Ψ and Δ in the whole wavelength range for all the three angles of incidence fit excel-lently. The volume fraction of nc-Ge in the nc-Ge layer and the effective thickness of the layer obtained from the fittings are 55.79% and ~6 nm, respectively. The result agrees with the TEM measurement, from which the effective thickness and the volume fraction were esti-mated based on the average diameter and height of nc-Ge. This indicates that the FB model describes the dispersion of the optical constants of the nc-Ge well and the fitting procedure is effective [69].

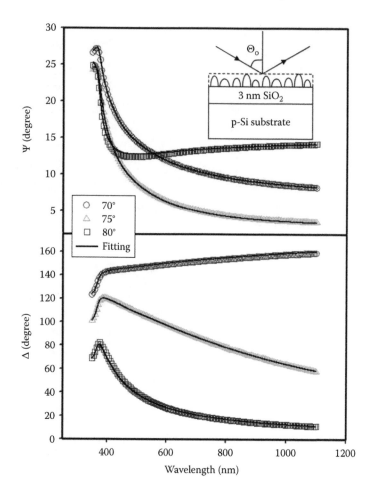

FIGURE 6.34 Spectral fittings of (a) Ψ and (b) Δ at the angles of incidence of 70°, 75°, and 80°. The inset shows the four-phase model used in the SE analysis. (From Goh, E.S.M. et al., *J. Appl. Phys.*, 109, 064307, 2011. With permission.)

Figure 6.35 shows the real and imaginary parts of the complex dielectric function of the nc-Ge [69]. For comparison, the dielectric function of bulk crystalline Ge is also included in the figure. A strong reduction in the dielectric function is observed for the nc-Ge with respect to bulk crystalline Ge. For example, at 400 nm, the real and the imaginary parts of dielectric function of nc-Ge are lower than those of bulk crystalline Ge by 30% and 76%, respectively.

Perhaps the most important parameter obtained from the spectral fitting based on the FB model is the bandgap of nc-Ge. This value of the nc-Ge formed on Si substrate obtained is 0.92 eV, which is consistent with the absorption measurement of the nc-Ge formed on a fused silica substrate [69]. Figure 6.36 shows the Tauc plot of $(\alpha E)^{1/2}$ versus E obtained from the absorption measurement [69]. The bandgap of the nc-Ge formed on fused silica substrate obtained from the linear extrapolation of the Tauc plot is 0.86 eV. Therefore, it could be concluded that the nc-Ge has a bandgap expansion of ~0.2 eV compared to bulk crystalline Ge (note that the bandgap of bulk crystalline Ge is 0.66 eV). On the other hand, the good linearity of the Tauc plot of $(\alpha E)^{1/2}$ versus E could suggests that nc-Ge has an indirect bandgap structure.

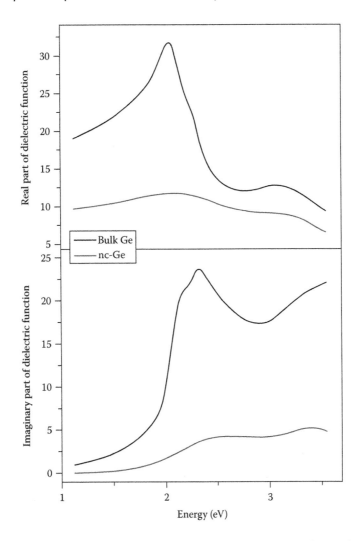

FIGURE 6.35 Real (ε_1) and imaginary (ε_2) part of the dielectric function of nc-Ge, obtained from the spectral fittings based on the FB model. The dielectric function of bulk crystalline Ge is included for comparison. (From Goh, E.S.M. et al., *J. Appl. Phys.*, 109, 064307, 2011. With permission.)

6.3.4 Ge Nanocrystals Embedded in SiO$_2$ Matrices

Takeoka et al. reported an experimental study of the PL of nc-Ge with 0.9–5.3 nm average diameter (D) in the near-infrared region [44]. Ge nanocrystals were fabricated by rf cosputtering of Ge and SiO$_2$ and post-annealing at 800°C for 30 min. The size nc-Ge was controlled by changing the volume fraction (f_{Ge}) of Ge in the films. Figure 6.37 shows a typical cross-sectional high-resolution TEM (HRTEM) image of the sample with f_{Ge} = 3.6%. Figure 6.38 shows the PL spectra for various nc-Ge sizes. As can be observed in the figure, the sample with D = 5.3 nm shows a PL peak at ~0.88 eV, which is slightly larger than the bandgap of bulk Ge crystal. As D decreases, the PL peak shifts monotonously to higher energies and reaches ~1.54 eV as D decreases to 0.9 nm [44]. The strong size dependence of the PL peak energy indicates that the bandgap of the nc-Ge widens when the nc-Ge size decreases as a result of quantum confinement and that the light emission originates from the recombination of electron–hole pairs confined in nc-Ge.

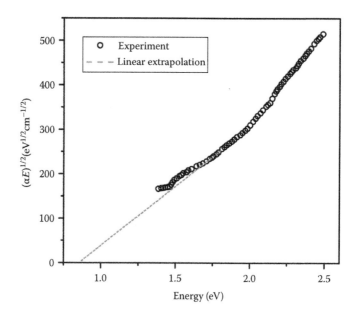

FIGURE 6.36 Tauc plot of $(\alpha E)^{1/2}$ versus E yielded from the absorption measurement for the nc-Ge formed on a fused silica substrate. The bandgap of the nc-Ge obtained from the linear extrapolation of the Tauc plot is 0.86 eV. (From Goh, E.S.M. et al., *J. Appl. Phys.*, 109, 064307, 2011. With permission.)

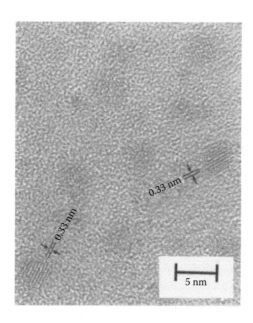

FIGURE 6.37 Cross-sectional HRTEM image of the sample with $f_{Ge} = 3.6\%$. Lattice fringes correspond to {111} planes of Ge. The diamond structure can clearly be seen. (From Takeoka, S. et al., *Phys. Rev. B*, 58, 7921, 1998. With permission.)

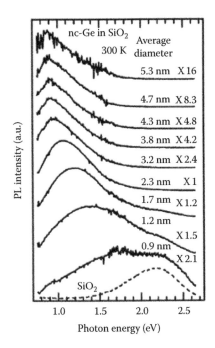

FIGURE 6.38 Dependence of PL spectra on the average diameter of nc-Ge. A PL spectrum of a SiO_2 film is also shown. (From Takeoka, S. et al., *Phys. Rev. B*, 58, 7921, 1998. With permission.)

ACKNOWLEDGMENT

The author wishes to thank Y. Liu, L. Ding, Z. H. Cen, E. S. M. Goh, and others for their contributions to the results from the group at Nanyang Technological University presented here.

REFERENCES

1. Zdetsis, A. D. 2006. Optical and electronic properties of small size semiconductor nanocrystals and nanoclusters. *Rev. Adv. Mater. Sci.* 11: 56–78.
2. Garoufalis, C. S. and Zdetsis, A. D. 2006. High accuracy calculations of the optical gap and absorption spectrum of oxygen contaminated Si nanocrystals. *Phys. Chem. Chem. Phys.* 8: 808–813.
3. Delley, B. and Steigmeier, E. F. 1995. Size dependence of band gaps in silicon nanostructures. *Appl. Phys. Lett.* 67: 2370–2372.
4. Delley, B. and Steigmeier, E. F. 1993. Quantum confinement in Si nanocrystals. *Phys. Rev. B* 47: 1397–1400.
5. Weissker, H.-Ch., Furthmüller, J., and Bechstedt, F. 2002. Optical properties of Ge and Si nanocrystallites from *ab initio* calculations. I. Embedded nanocrystallites. *Phys. Rev. B* 65: 155327(1-9).
6. Weissker, H.-Ch., Furthmüller, J., and Bechstedt, F. 2002. Optical properties of Ge and Si nanocrystallites from *ab initio* calculations. II. Hydrogenated nanocrystallites. *Phys. Rev. B* 65: 155328(1-7).
7. Vasiliev, I., Öğüt, S., and Chelikowsky, J. R. 2001. *Ab initio* absorption spectra and optical gaps in nanocrystalline silicon. *Phys. Rev. Lett.* 86: 1813–1816.
8. Öğüt, S., Chelikowsky, J. R., and Louie, S. G. 1997. Quantum confinement and optical gaps in Si nanocrystals. *Phys. Rev. Lett.* 79: 1770–1773.
9. Delerue, C., Allan, G., and Lannoo M. 1993. Theoretical aspects of the luminescence of porous silicon. *Phys. Rev. B* 48: 11024–11036.
10. Delerue, C., Lannoo, M., and Allan, G. 2000. Excitonic and quasiparticle gaps in Si nanocrystals. *Phys. Rev. Lett.* 84: 2457–2460.

11. Godby, R. W. and White, I. D. 1998. Density-relaxation part of the self-energy. *Phys. Rev. Lett.* 80: 3161.

12. Franceschetti, A., Wang, L. W., and Zunger, A. 1999. Comment on "quantum confinement and optical gaps in Si nanocrystals. *Phys. Rev. Lett.* 83: 1269.

13. Baierle, R. J., Caldas, M. J., Molinari, E., and Ossicini, S. 1997. Optical emission from small Si particles. *Solid State Commun.* 102: 545–549.

14. Palummo, M., Onida, G., and Sole, R. D. 1999. Optical properties of germanium nanocrystals. *Phys. Stat. Sol. (a)* 175: 23–31.

15. Wang, L.-W. and Zunger, A. 1994. Dielectric constants of silicon quantum dots. *Phys. Rev. Lett.* 73: 1039–1042.

16. Zunger, A. 2001. Pseudopotential theory of semiconductor quantum dots. *Phys. Stat. Sol. (b)* 224: 727–734.

17. Degoli, E., Cantele, G., Luppi, E. et al. 2004. *Ab initio* structural and electronic properties of hydrogenated silicon nanoclusters in the ground and excited state. *Phys. Rev. B* 69: 155411(1-10).

18. Ramos, L. E., Weissker, H.-Ch., Furthmüller, J., and Bechstedt, F. 2005. Optical properties of Si and Ge nanocrystals: Parameter-free calculations. *Phys. Stat. Sol. (b)* 242: 3053–3063.

19. Vincent, J. E., Kim, J., and Martin, R. M. 2007. Quantum Monte Carlo calculations of the optical gaps of Ge nanoclusters using core-polarization potentials. *Phys. Rev. B* 75: 045302(1-10).

20. Weissker, H.-Ch., Furthmüller, J., and Bechstedt, F. 2003. Oscillator strengths and excitation energies of Ge and Si nanocrystals from ab initio supercell calculations. *Mater. Sci. Eng. B* 101: 39–42.

21. Luppi, E., Iori, F., Magri, R. et al. 2007. Excitons in silicon nanocrystallites: The nature of luminescence. *Phys. Rev. B* 75: 033303(1-4).

22. Luppi, M. and Ossicini, S. 2005. *Ab initio* study on oxidized silicon clusters and silicon nanocrystals embedded in SiO_2: Beyond the quantum confinement effect. *Phys. Rev. B* 71: 035340(1-15).

23. Vasiliev, I., Chelikowsky, J. R., and Martin, R. M. 2002. Surface oxidation effects on the optical properties of silicon nanocrystals. *Phys. Rev. B* 65: 121302(1-4).

24. Guerra, R. and Ossicini, S. 2010. High luminescence in small Si/SiO_2 nanocrystals: A theoretical study. *Phys. Rev. B* 81: 245307(1-6).

25. Guerra, R., Degoli, E., and Ossicini, S. 2009. Size, oxidation, and strain in small Si/SiO_2 nanocrystals. *Phys. Rev. B* 80: 155332(1-5).

26. Guerra, R., Marri, I., Magri, R. et al. 2009. Silicon nanocrystallites in a SiO_2 matrix: Role of disorder and size. *Phys. Rev. B* 79: 155320(1-9).

27. Guerra, R., Marri, I., Magri, R. et al., 2009. Optical properties of silicon nanocrystallites in SiO_2 matrix: Crystalline vs. amorphous case. *Superlatt. Microstruct.* 46: 246–252.

28. Francheschetti, A. and Zunger, A. 1997. Direct pseudopotential calculation of exciton coulomb and exchange energies in semiconductor quantum dots. *Phys. Rev. Lett.* 78: 915–918.

29. Lehtonen, O. and Sundholm, D. 2009. Computational studies of free-standing silicon nanoclusters. In *Silicon Nanophotonics*. Khriachtchev, L. (ed.), pp. 61–88. World Scientific Publishing, Singapore.

30. Sharma, A. C. 2006. Size-dependent energy band gap and dielectric constant within the generalized Penn model applied to a semiconductor nanocrystallite. *J. Appl. Phys.* 100: 084301(1-8).

31. Tsu, R. and Babić, D. 1994. Doping of a quantum dot. *Appl. Phys. Lett.* 64: 1806–1808.

32. Tsu, R., Babić, D., and Ioriatti, Jr. L. 1997. Simple model for the dielectric constant of nanoscale silicon particle. *J. Appl. Phys.* 82: 1327–1329.

33. Wang, L. W. and Zunger, A. 1996. Pseudopotential calculations of nanoscale CdSe quantum dots. *Phys. Rev. B* 53: 9579–9582.

34. Delerue, C., Lannoo, M., and Allen, G. 2003. Concept of dielectric constant for nanosized systems. *Phys. Rev. B* 68: 15411(1-4).

35. Walter, J. P. and Cohen, M. L. 1970. Wave-vector-dependent dielectric function for Si, Ge, GaAs, and ZnSe. *Phys. Rev. B* 2: 1821–1826.

36. Lannoo, M., Delerue, C., and Allan, G. 1995. Screening in semiconductor nanocrystallites and its consequences for porous silicon. *Phys. Rev. Lett.* 74: 3415–3418.

37. Wolkin, M. V., Jorne, J., Fauchet, P. M., Allan, G., and Delerue, C. 1999. Electronic states and luminescence in porous silicon quantum dots: The role of oxygen. *Phys. Rev. Lett.* 82: 197–200.

38. Garrido, B., Lopez, M., Gonzalez, O., Perez-Rodriguez, A., Morante, J. R., and Bonafos, C. 2000. Correlation between structural and optical properties of Si nanocrystals embedded in SiO_2: The mechanism of visible light emission. *Appl. Phys. Lett.* 77: 3143–3145.

39. Pavesi, L., Negro, L. D., Mazzoleni, C., Franzò, G., and Priolo, F. 2000. Optical gain in silicon nanocrystals. *Nature (London)* 408: 440–444.

40. Negro, L. D., Cazzanelli, M., Pavesi, L., Ossicini, S., Pacifici, D., Franzò, G., Priolo, F., and Iacona, F. 2003. Dynamics of stimulated emission in silicon nanocrystals. *Appl. Phys. Lett.* 82: 4636–4638.

41. Wilcoxon, J. P., Samara, G. A., and Provencio, P. N. 1999. Optical and electronic properties of Si nanoclusters synthesized in inverse micelles. *Phys. Rev. B* 60: 2704–2714.

42. Wilcoxon, J. P., Provencio, P. P., and Samara, G. A. 2001. Synthesis and optical properties of colloidal germanium nanocrystals. *Phys. Rev. B* 64: 035417(1-9).

43. Heath, J. R., Shiang, J. J., and Alivisatos, A. P. 1994. Germanium quantum dots: Optical properties and synthesis. *J. Chem. Phys.* 101: 1607–1615.

44. Takeoka, S., Fujii, M., Hayashi, S., and Yamamoto, K. 1998. Size-dependent near-infrared photoluminescence from Ge nanocrystals embedded in SiO_2 matrices. *Phys. Rev. B* 58: 7921–7925.

45. Taraschi, G., Saini, S., Fan, W. W., Kimerling, L. C., and Fitzgerald, E. A. 2003. Nanostructure and infrared photoluminescence of nanocrystalline Ge formed by reduction of $Si_{0.75}Ge_{0.25}O_2/Si_{0.75}Ge_{0.25}$ using various H_2 pressures. *J. Appl. Phys.* 93: 9988–9996.

46. Konchenko, A., Nakayama, Y., Matsuda, I., Hasegawa, S., Nakamura, Y., and Ichikawa, M. 2006. Quantum confinement observed in Ge nanodots on an oxidized Si surface. *Phys. Rev. B* 73: 113311(1-4).

47. Niquet, Y., Allan, G., Delerue, C., and Lannoo, M. 2000. Quantum confinement in germanium nanocrystals. *Appl. Phys. Lett.* 77: 1182–1183.

48. Reboredo, F. A. and Zunger, A. 2000. L-to-X crossover in the conduction-band minimum of Ge quantum dots. *Phys. Rev. B* 62: R2275–R2278.

49. Reboredo, F. A. and Zunger, A. 2001. Surface-passivation-induced optical changes in Ge quantum dots. *Phys. Rev. B* 63: 235314(1-7).

50. Melnikov, D. V. and Chelikowsky, J. R. 2003. Absorption spectra of germanium nanocrystals. *Solid State Commun.* 127: 361–365.

51. Melnikov, D. V. and Chelikowsky, J. R. 2004. Electron affinities and ionization energies in Si and Ge nanocrystals. *Phys. Rev. B* 69: 113305(1-4).

52. Weissker, H.-Ch., Furthmüller, J., and Bechstedt, F. 2003. Structural relaxation in Si and Ge nanocrystallites: Influence on the electronic and optical properties. *Phys. Rev. B* 67: 245304(1-7).

53. Weissker, H.-Ch., Furthmüller, J., and Bechstedt, F. 2004. Structure- and spin-dependent excitation energies and lifetimes of Si and Ge nanocrystals from ab initio calculations. *Phys. Rev. B* 69: 15310(1-8).

54. Tsolakidis, A. and Martin, R. M. 2005. Comparison of the optical response of hydrogen-passivated germanium and silicon clusters. *Phys. Rev. B* 71: 125319(1-8).

55. Nesher, G., Kronik, L., and Chelikowsky, J. R. 2005. Ab initio absorption spectra of Ge nanocrystals. *Phys. Rev. B* 71: 035344(1-5).

56. Garoufalis, C. S., Skaperda, M. S., and Zdetsis, A. D. 2005. The optical gap of small Ge nanocrystals. *J. Phys.: Conf. Ser.* 10: 97–100.

57. Kanemitsu, Y., Uto, H., Masumoto, Y., and Maeda, Y. 1992. On the origin of visible photoluminescence in nanometer size Ge crystallites. *Appl. Phys. Lett.* 61: 2187–2189.

58. Amans, D., Callard, S., Gagnaire, A. et al. 2003. Ellipsometric study of silicon nanocrystal optical constants. *J. Appl. Phys.* 93: 4173–4179.

59. Lee, K.-J., Kang, T.-D., Lee, H. et al. 2005. Optical properties of SiO_2/nanocrystalline Si multilayers studied using spectroscopic ellipsometry. *Thin Solid Films* 476: 196–200.

60. Chen T. P. and Ding, L. 2010. Optical and optoelectronic properties of silicon nanocrystals embedded in SiO_2 matrix. In *Nanostructured Thin Films and Coatings: Functional Properties*, Zhang, S. (ed.), pp. 113–165. CRC Press, Boca Raton, FL.

61. Cen Z. H. 2011. Studies of optical properties and light emission of silicon-rich silicon nitride thin films. PhD thesis, Nanyang Technological University, Singapore.

62. Ding L. 2009. Optical and optoelectronic properties of Si nanocrystals embedded in dielectric matrix. PhD thesis, Nanyang Technological University, Singapore.

63. Ding, L., Chen, T. P., Liu, Y., Ng, C. Y., and Fung, S. 2005. Optical properties of silicon nanocrystals embedded in a SiO_2 matrix. *Phys. Rev. B* 72: 125419(1-7).

64. Ding, L., Chen, T. P., Liu, Y. et al. 2005. Thermal annealing effect on the band gap and dielectric functions of silicon nanocrystals embedded in SiO_2 matrix. *Appl. Phys. Lett.* 87: 121903(1-3).

65. Chen, T. P., Liu, Y., Tse, M. S. et al. 2003. Dielectric functions of Si nanocrystals embedded in a SiO_2 matrix. *Phys. Rev. B* 68: 153301(1-4).

66. Ding, L., Chen, T. P., Liu, Y. et al. 2007. Influence of nanocrystal size on optical properties of Si nanocrystals embedded in SiO_2 synthesized by Si ion implantation. *J. Appl. Phys.* 101: 103525(1-6).

67. Ding, L., Chen, T. P., Wong, J. I. et al. 2006. Dielectric functions of densely stacked Si nanocrystal layer embedded in SiO_2 thin films. *Appl. Phys. Lett.* 89: 251910(1-3).

68. Cen, Z. H., Chen, T. P., Ding, L. et al. 2008. Annealing effect on the optical properties of implanted silicon in a silicon nitride matrix. *Appl. Phys. Lett.* 93: 023122(1-3).
69. Goh, E. S. M., Chen, T. P., Huang, S. F., Liu, Y. C., and Sun, C. Q. 2011. Band gap expansion and dielectric suppression of self-assembled Ge nanocrystals. *J. Appl. Phys.* 109: 064307(1-4).
70. Azzam, R. M. A. and Basharra, N. M. 1977. *Ellipsometry and Polarized Light.* Amsterdam, the Netherlands: North-Holland.
71. Forouhi, A. R. and Bloomer, I. 1988. Optical properties of crystalline semiconductors and dielectrics. *Phys. Rev. B* 38: 1865–1874.
72. Penn, D. R. 1962. Wave-number-dependent dielectric function of semiconductors. *Phys. Rev.* 128: 2093–2097.
73. Ding, L., Chen, T. P., Liu, Y. et al. 2008. Evolution of photoluminescence mechanisms of Si+-implanted SiO$_2$ films with thermal annealing. *J. Nanosci. Nanotechnol.* 8: 3555–3560.

7 Light Emission Properties of Si Nanocrystals Embedded in a Dielectric Matrix

Tupei Chen

CONTENTS

7.1 INTRODUCTION

A semiconductor could emit photons under either optical or electrical pumping. Photoluminescence (PL) and electroluminescence (EL) refer to light emission due to the excitation by optical pumping (i.e., optical radiation) and electrical pumping (i.e., application of electric field or current injection), respectively. PL and EL are usually concerned with the following optical transitions: (1) interband transition with intrinsic emission corresponding very close in energy to the bandgap of the semiconductor, where phonons and excitons may be involved, and (2) the transition involving states in the bandgap due to impurities or defects. Not all transitions can occur in the same material or under the same conditions, and not all transitions are radiative. The light emission process consists of two optical transitions: the excitation transition, and the recombination transition. In the excitation transition, the electron is excited from the ground state (e.g., a state in the valence band of the semiconductor) to the excited state (e.g., a state in the conduction band of the semiconductor) by absorption of a photon under optical pumping, or is electrically injected into the excited state under electrical pumping. In the radiative recombination transition, the electron in the excited state recombines with the hole at the ground state to emit a photon. For a given input excitation energy, the radiative recombination process competes with the nonradiative process (such as the Auger recombination). To have efficient light emission, the radiative transitions must predominate over the nonradiative ones [1–3].

The excitation and recombination optical transitions are different in the direct and indirect bandgap semiconductors. If the conduction band minimum and the valence band maximum have the same k-vectors (wave vectors) in the Brillouin zone, the semiconductor is called a "direct bandgap" material; and if the k-vectors are different, it is called an "indirect bandgap" material [1].

In the direct bandgap semiconductor, an electron in the valence band can be excited into the conduction band after the absorption of a photon with an energy equal to or larger than the energy bandgap (E_g). However, such a vertical absorption cannot happen in the indirect bandgap materials. To conserve momentum, the photon absorption process must be assisted by either the absorption or emission of a phonon, because the change in the electron momentum cannot be provided by the absorption of a photon [2,3]. The minimum photon energy required for the indirect transition involving absorption of a phonon with energy of E_p is ($E_g - E_p$), and it is ($E_g + E_p$) for the indirect transition involving phonon emission. In a direct bandgap material, any photon with energy equal to or larger than the bandgap energy can participate in a direct transition and excite an electron from the top of the filled valence band to one of the states at the bottom of the conduction band. However, nonvertical absorptions in an indirect bandgap material occur via an intermediate virtual state, whereby photon absorption is accompanied by either the creation or annihilation of a phonon to achieve momentum conservation [1]. In a perfect semiconductor, electron–hole pairs thermalize and accumulate at the conduction and valence band extremes. Fundamental radiative transitions in a semiconductor are those occurring at or near the band edges, namely, band-to-band or excitonic transitions. In a direct bandgap material, momentum-conserving transitions connect states with the same k-value. Accordingly, the emitted photon has a low-energy threshold at $h\nu = E_g$ [1]. Indirect band-to-band radiative transitions could be realized by the assistance of phonon emission or absorption. With the involvement of a phonon, both the requirements of energy and momentum conservations are met for the band-to-band transitions in the indirect bandgap semiconductor [1]. However, the light emission efficiency of the indirect bandgap semiconductor would be much lower due to the need for the involvement of phonons as compared to that of the direct bandgap semiconductor.

Silicon is the most important semiconductor material in microelectronics industry, because nowadays, silicon devices constitute over 95% of the market of semiconductor devices [1,3]. On the other hand, as electronic device dimensions become smaller and smaller, the traditional electrical interconnects that are used for intra- and inter-chip communication become increasingly impractical due to the RC delay and heat dissipation associated with the metal interconnects. When the device dimensions decrease to the nanoscale, the device will suffer a lot from the traditional electrical interconnects due to propagation delays, high power consumption, and low bandwidth [4]. Fortunately, optical interconnects would probably provide us with a promising alternative strategy for overcoming these challenges. Since communication via optical interconnects requires an on-chip emitter and detector, an important challenge on the materials and the integration of photonic devices into the main-stream Si process has triggered a new research subject, namely, Si photonics, recently [1,3]. Most of the photonic components fabricated with Si technology have been demonstrated, such as optical modulators [5,6], switches [7,8], detectors [9,10], and low-loss waveguides [11,12]. However, bulk crystalline silicon cannot be used for light-emitting diodes (LEDs) because it is an indirect bandgap semiconductor and thus the probability for a radiative transition is very low. In this regard, the most challenging task is to look for Si-based materials with high emission efficiency for realizing Si-based LEDs.

Many strategies have been proposed to realize Si-based LEDs. The most successful ones are based on the exploitation of low-dimensional silicon, where the silicon material is nanostrucured such that its optical and electronic properties are modified by quantum confinement effects [1,3]. Of all the types of low-dimensional silicon, porous Si has received the most intensive research attention in the early years of exploitation of nanoscaled Si. The discovery of the visible PL from porous Si in 1990 [13] has triggered large research efforts in realizing luminescence from porous silicon. However, it is not reliable to utilize porous silicon in optoelectronic devices due to various issues such as its instability in light emission, structural fragility, and incompatibility with conventional complementary metal–oxide–semiconductor (CMOS) technology [14–16]. Another nanostructured silicon material, namely, the silicon nanocrystal (nc-Si), is considered to be a promising material system for application in Si LEDs [15–17]. In particular, nc-Si embedded in a dielectric matrix such as SiO_2 or Si_3N_4 thin films is a promising candidate for realizing Si-based LEDs with the advantages of chemical stability and full compatibility with the mainstream silicon technology that is used in the microelectronics industry.

7.2 FUNDAMENTAL LUMINESCENCE PROPERTIES OF SILICON NANOCRYSTALS

The band structure of crystalline silicon can be significantly modified when its dimensions are reduced to the scale of the exciton Bohr radius (~4.9 nm) of bulk crystalline silicon, which is due to the quantum confinement effect. In bulk crystalline silicon, since the radiative time of indirect transitions is extremely long and the transport of excitons is efficient, the main decay channel for free excitons is their capture in bound exciton states or nonradiative recombination. This results in a very low quantum yield of free exciton emission. The situation changes in a silicon nanocrystal with a dimension comparable to the exciton Bohr radius [1]. The following changes significantly affect the luminescence properties of silicon: the spatial confinement in the nanocrystal results in shifts of both the absorbing and luminescing states to higher energies; the geometrical change in the nanocrystal causes delocalization of carriers in k-space, thus allowing zero phonon optical transitions and significantly enhancing the oscillator strength of the zero phonon optical transitions in extremely small silicon nanocrystals; and due to the better overlap of electron and hole envelope wavefunctions, a strong enhancement of the electron–hole (e–h) exchange interaction would be expected [2].

Spatial localization of the electron and hole wavefunctions in a Si nanocrystal of nanometer scale increases their spread in the momentum space, thus increasing their overlap in the Brillouin zone. Due to the spatial confinement of the electrons and holes in the nanocrystal, their motion and energy states will be quantized in three dimensions. As a result of the broadening of the electron–hole pair state in momentum space, the momentum conservation rule is relaxed for the radiative transitions in Si nanocrystals, and more phonons with satisfied momentum can be involved in the radiative transitions. Thus, the probability of radiative recombination is increased with the lifetime reduced to microseconds or even nanoseconds. Moreover, the bandgap of Si nanocrystals expands (i.e., the nc-Si bandgap is larger than that of bulk crystalline Si) as a result of the quantum confinement effect. Therefore, the luminescence wavelength can be manipulated from near-infrared (NIR) (the energy close to 1.12 eV, the bandgap of bulk crystalline Si) to the visible range by controlling the size of the Si nanocrystal [2,3].

Light emission from Si nanocrystal is frequently concerned with the system of Si nanocrystal dispersed in a dielectric matrix such as SiO_2 or Si_3N_4. For such system, besides the quantum confinement effect of the Si nanocrystal, the interfaces between the Si nanocrystals and the host dielectric matrix play a significant role in the light emission because of the large surface-to-volume ratio of Si nanocrystal and the existence of interface states. Therefore, effects related to both the interfaces and the matrix, such as surface passivation and the luminescence centers at the interfaces or even in the matrix, should be taken into account for the light emission from Si nanocrystal embedded in a dielectric matrix.

In spite of the intensive research progress made toward understanding the nature of efficient light emission from a silicon nanocrystal, the mechanism is still controversial. This is because of the extreme complexity of the material system, as silicon nanocrystal ensembles are complicated heterogeneous systems that include various parameters such as crystalline network properties, matrix properties, surface passivation conditions, nanocrystal size and shape distributions, and so on, affecting the optical properties of the nanocrystal [2,3]. Figure 7.1 shows some possible radiative electron–hole recombination processes in the system of nc-Si embedded in a dielectric matrix [3]. The quantum confinement effect of nc-Si, luminescent surface/interface states in the interface regions between nc-Si and the dielectric matrix, and luminescent defect states in the dielectric matrix could be involved in the light emission [3].

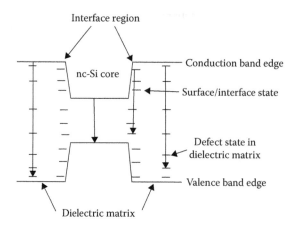

FIGURE 7.1 Schematic of some possible radiative electron–hole recombination processes in the system of nc-Si embedded in a dielectric matrix.

As pointed out previously, quantum confinement effect is likely to be a source of photon emission in structures with one or more dimensions smaller than the Bohr radius of an exciton (excited electron–hole pair). When injected electrons and holes by electrical or optical stimulation are confined into an extremely small Si nanocrystal, the potential barrier formed around each silicon nanocrystal inhibits the diffusion of electrons and holes. This effect significantly increases the radiative recombination possibility. If quantum confinement effect is responsible for the light emission, a blue shift in the luminescence wavelength with reduction of nc-Si size should be observed. This is because the quantum confinement effect causes an expansion of the nc-Si bandgap when the nc-Si size is reduced. Figure 7.2 shows the bandgap of nc-Si as a function of the nc-Si diameter [16]. As shown in the figure, the bandgap increases with decreasing nc-Si size. The expanded bandgap results in a blue shift of the emitted wavelength. The blue shift of the luminescence peak wavelength with reduction of nc-Si size has been frequently observed. Figure 7.3 shows PL spectra of nc-Si embedded in SiO_2 prepared by cosputtering Si and SiO_2 and annealing at temperatures higher than 1100°C [18]. The sample with

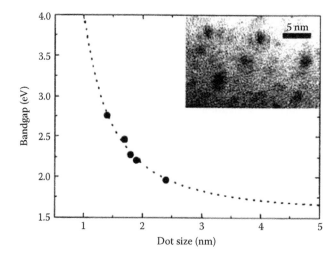

FIGURE 7.2 Bandgap (i.e., PL peak energy) of amorphous silicon quantum dot as a function of the dot size. The dashed line was obtained from the effective mass theory for three-dimensionally confined Si dots (i.e., nc-Si). (From Park, N.-M. et al., *Appl. Phys. Lett.*, 78, 2575, 2001. With permission.)

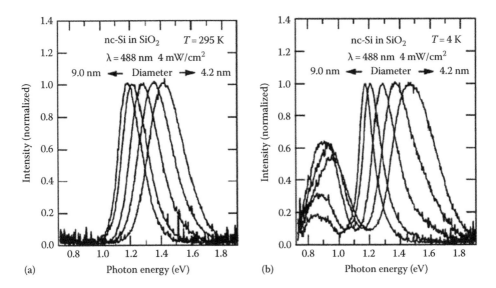

FIGURE 7.3 PL spectra of nc-Si embedded in SiO_2 thin films (a) at room temperature and (b) at 4 K. (From Takeoka, S. et al., *Phys. Rev. B*, 62, 16820, 2000. With permission.)

the average nc-Si diameter of 59.0 nm exhibits PL at ~1.19 eV. The peak energy is slightly larger than the bandgap energy of bulk Si crystal (1.12 eV). With decreasing average nc-Si diameter from 9.0 to 4.2 nm, the PL peak shifts monotonously to higher energy and reaches 1.42 eV [18]. Figure 7.4 shows the PL peak energy at room temperature as a function of the nc-Si size [18]. The PL energy varies from the vicinity of the bulk bandgap to 1.9 eV when decreasing the size from 9 to 1 nm.

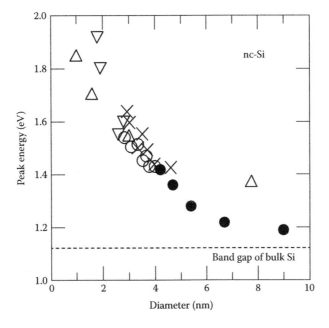

FIGURE 7.4 PL peak energy versus average diameter of nc-Si embedded in SiO_2 thin films. (From Takeoka, S. et al., *Phys. Rev. B*, 62, 16820, 2000. With permission.)

Figure 7.5 shows PL lifetime for two samples of nc-Si embedded in SiO_2 thin films with different nc-Si sizes (average diameters of 4.2 and 9.0 nm) [18]. We can see that the temperature dependence of the PL lifetime is much different between the two samples. For the sample with the average diameter of 4.2 nm, the PL lifetime changes abruptly at 10–20 K. On the other hand, for the average diameter of 9.0 nm, the PL lifetime gradually decreases until 150 K. The temperature dependence can be well explained by the model proposed by Calcott et al. [19]; the exciton state is split into a singlet state and a triplet state due to the electron–hole exchange interaction. In this model (see the inset of Figure 7.5), the excitonic levels are split by the energy (ΔE) due to the exchange interaction of an electron and a hole. The lower (upper) level is a triplet (singlet) state with a radiative lifetime τ_t (τ_s). The overall temperature dependence of the PL lifetime can be calculated on the basis of Boltzmann statistics with a weight factor of 3 for the triplet state [18]. The broken curves in Figure 7.5 represent the results of the fitting based on the model. We can see that this model can well reproduce the observed temperature dependence of the PL lifetime below 150 K. Both the lifetimes (τ_t and τ_s,) of the triplet and singlet states and the splitting energy (ΔE) can be obtained from the fitting. As the PL energy increases, τ_t and τ_s become shorter and ΔE becomes larger. The values of τ_t, τ_s, and ΔE represent the degree of carrier confinement; τ_t and τ_s depend on the optical transition oscillator strength, and ΔE is determined by the spatial separation between an electron and a hole in nc-Si. The continuous change of these parameters suggests that the PL mechanism is the same in the wide PL energy range. Furthermore, the PL peak energy and the splitting energy of excitons change continuously from the bulk values. This continuity from the bulk values indicates that the PL is due to the recombination of electron–hole pairs confined in nc-Si, that is, a quantum confinement model [18].

The presence of luminescent surface states in the interface between nc-Si and the dielectric matrix can result in photon emission also. Figure 7.3b shows the PL spectra of nc-Si embedded in SiO_2 at 4 K [18]. At this temperature, in addition to the PL peak at ~1.2–1.4 eV, which can be observed at room temperature, another peak appears at ~0.8–0.9 eV. The low-energy peak position shows a size dependence similar to that of the high-energy one, although the dependence is weaker.

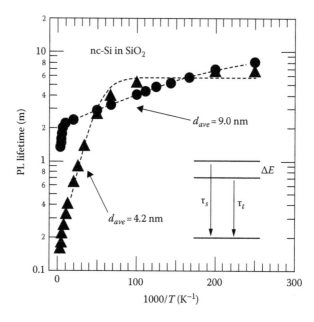

FIGURE 7.5 Temperature dependence of PL lifetime for two samples of nc-Si embedded in SiO_2 thin films with different nc-Si sizes [18]. The inset shows a schematic illustration of singlet and triplet splitting of an exciton state. τ_s and τ_t represent singlet and triplet lifetimes, respectively, and ΔE the splitting energy. (From Takeoka, S. et al., *Phys. Rev. B*, 62, 16820, 2000. With permission.)

The low-energy peak is commonly observed for oxidized nc-Si, for example, nc-Si embedded in SiO_2 matrices and surface-oxidized porous Si. This peak is assigned to the recombination of carriers trapped at P_b centers at nc-Si/SiO_2 interfaces [18], or to the recombination of a conduction band electron with a hole in a deep surface trap at nc-Si/SiO_2 interfaces. The size-dependent shift of the low-energy peak is considered to reflect that of the conduction band edge.

The component of defect luminescence can be diminished or removed with passivating techniques. It can also be distinguished by its rapid decay: in this indirect bandgap semiconductor system, a long lifetime (longer than ~1 μs) is usually expected for excitonic recombination, while defect-related luminescence could have a lifetime of less than 100 ns. The defects can locate at the interface between nc-Si and the dielectric matrix. However, the recombination of carriers trapped at radiative recombination centers that form in the defects in the dielectric matrix itself can result in luminescence also [20–23]. Defect-related PL from SiO_2 has been observed; for example, Figure 7.6 shows the two PL bands at 4.3 and 2.6 eV due to oxygen-deficiency-type defects in a boron-implanted thermal SiO_2 film and oxygen deficiency-type bulk SiO_2 [22].

Figures 7.7 through 7.9 present a study of PL from Si-implanted SiO_2 layers [24]. Figure 7.7 shows the PL spectra for a sample implanted at a dose of 2×10^{17} cm^{-2} and after vacuum annealing at 500°C and 700°C for 30 min. On the as-implanted sample, a broad band around 550–600 nm (a green band) is observed. The intensity of this band at ~560 nm increases as a function of the annealing temperature, up to 700°C. No NIR band around 750–850 nm is detected. An NIR band is observed after 1000°C annealing, as shown in Figure 7.8. The PL intensity for the NIR band grows with increasing annealing time (Figure 7.9). As can be observed in the three figures, the green band exists in the as-implanted sample and also after 1000°C annealing and its intensity increases as a function of the annealing temperature. The intensity reaches its maximum after 1000°C annealing. For increasing annealing time at 1000°C, the green band intensity decreases (Figure 7.9). The emission times for the green band are in the nanosecond range, while the NIR emission is characterized by times from 1 to 0.3 ms. The short emission time and the annealing behavior suggest that the green emission is related to extended defects like silicon chains in the oxide matrix while the NIR band is due to the emission from the nc-Si (the quantum confinement effect) [24].

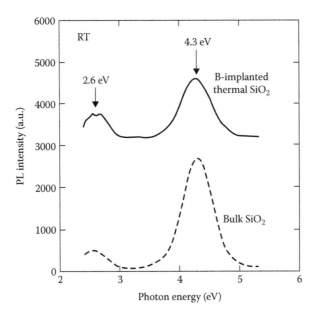

FIGURE 7.6 PL spectrum at room temperature at a B-implanted thermal SiO_2 film (acceleration energy: 30 keV, dose: 10^{16} cm^{-2}) and oxygen-deficient bulk SiO_2 when excited at 5.0 eV photons from a KrF excimer laser. (From Nishikawa, H. et al., *J. Appl. Phys.*, 78, 842, 1995. With permission.)

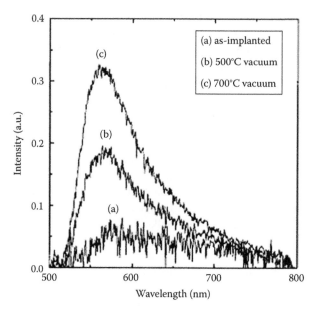

FIGURE 7.7 (a) PL spectra for a sample implanted at the Si^+ dose of 2×10^{17} cm^{-2}. (b) Spectra after 500°C and (c) 700°C 30 min vacuum annealing. (From Ghislotti, G. et al., *J. Appl. Phys.*, 79, 8660, 1996. With permission.)

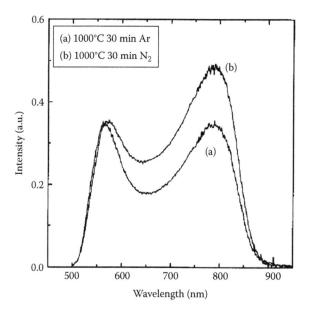

FIGURE 7.8 PL spectra for the sample already annealed in vacuum (reported in Figure 7.7) after 1000°C annealing in (a) argon and (b) nitrogen annealing. (From Ghislotti, G. et al., *J. Appl. Phys.*, 79, 8660, 1996. With permission.)

For silicon-rich silicon nitride (SRN) materials (the excess Si can be in the form of nanoparticle or nanocrystal distributed in the nitride matrix), various localized states in the band tail of the SRN materials can assist the radiative recombination [2]. In an amorphous SRN film, the static disorder of the structural randomness can cause potential fluctuations within the exponential band tails [25–28]. According to the Anderson localization principle, if the average additional disorder-induced potential is strong enough to suppress the interaction between atoms, frequent electron

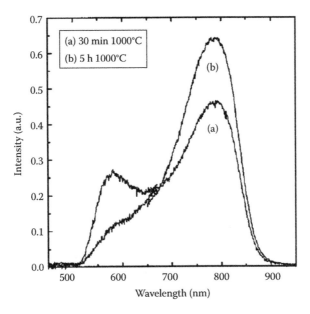

FIGURE 7.9 PL emission from a sample implanted at 3×10^{17} cm^{-2} fluence and annealed in nitrogen at 1000°C for different times: (a) 30 min and (b) 5 h. (From Ghislotti, G. et al., *J. Appl. Phys.*, 79, 8660, 1996. With permission.)

scattering results in the loss of the phase coherence in the electron wavefunction and the formation of strong localization of the scattered electrons [2]. In this case, a large uncertainty in the electron momentum is caused by the strong scattering, which leads to the loss of the momentum conservation in the electronic transitions. As a result, an optical transition is allowed between two states for which energy conservation applies. Thus, the radiative probability is increased in the amorphous SRN system [2]. Therefore, the light emission from a SRN material may be related to the quantum confinement effect of the Si nanocrystal or the localized bandtail states.

7.3 PHOTOLUMINESCENCE AND ELECTROLUMINESCENCE FROM Si$^+$-IMPLANTED SiO$_2$ THIN FILMS

7.3.1 PHOTOLUMINESCENCE

Efficient room-temperature PL from SiO$_2$ thin films embedded with nc–Si synthesized by various techniques has been frequently reported, and various mechanisms for the PL have been proposed. For the nc–Si/SiO$_2$ system synthesized with Si ion implantation into the SiO$_2$ thin films, the visible-light (in particular the red light) luminescence could be due to the quantum confinement effect of the nc–Si [15,29–33]. However, it was also suggested that the short-wavelength luminescence was related to the presence of the oxygen vacancies in the Si$^+$-implanted dielectric matrix [24,34]. It is now generally agreed that there are multiple mechanisms responsible for the luminescence. As an example, the study of PL properties of silicon nanocrystals embedded in SiO$_2$ films synthesized by the Si ion implantation technique reported by Ding et al. [35] is summarized in the following.

In their study, Si ion implantation into a 550 nm SiO$_2$ thin film thermally grown on p-type Si wafer was carried out with the dose of 1×10^{17} cm^{-2} at the energy of 100 keV. Thermal annealing was carried out in nitrogen ambient at the temperatures of 500°C, 600°C, 700°C, 850°C, 900°C, 1000°C, and 1100°C for 20 min. The PL spectra of the samples consist of six PL bands peaked at the wavelengths of ~415, ~460, ~520, ~630, ~760, and ~845 nm. Such PL bands have been frequently reported and widely accepted in the literature. The PL bands do not show a significant change in the full-width at half-maximum (FWHM) with the annealing temperature. The PL bands at ~415, ~460, and ~520 nm have a similar dependence on the annealing temperature. The integrated intensity of the three bands first increases slightly and then decreases dramatically with the increase of the annealing temperature. The three bands almost disappear when the annealing temperature is higher than 900°C. The ~630 nm band shows no change with annealing temperatures up to 1000°C but disappears at the annealing temperature of 1100°C. On the other hand, the integrated PL intensity of both the ~760 and ~845 nm bands increases slightly with the annealing temperature up to 1000°C and then shows a dramatic increase by ~30 times when the annealing temperature reaches 1100°C [35].

The above results suggest that not all the PL bands originate from the quantum confinement effect of the nc-Si and some PL bands can be ascribed to the oxide matrix. It is well known that various defects in the oxide matrix can serve as visible luminescent centers. The weak oxygen bond (WOB) defects in silicon oxide could be responsible for the 415 nm (~3 eV) PL band [1,36–38]. Ion implantation produces oxygen vacancies in the SiO_2 network, and oxygen interstitials (the precursors of the WOB defects) are created at the same time. The oxygen interstitials change into the WOB defects immediately after annealing, and the reaction is reversible subject to an excessive thermal annealing energy. The reverse reaction can explain the decrease in the intensity of the 415 nm PL band with the increase of the annealing temperature when the temperature is higher than 700°C. Pure silica glass [39] and Si-rich silicon oxides [34] have been reported to have a PL band at ~460 nm (~2.7 eV), which is associated with the neutral oxygen vacancy (NOV) defect represented by $O_3 \equiv Si–Si \equiv O_3$ [34,36]. The NOV defects are actually generated in the same process as the WOB defects in the Si^+-implanted SiO_2 film. This explains why the ~415 (WOB) and ~460 nm (NOV) PL bands have a similar dependence on the annealing temperature. The ~520 nm band is attributed to the E_δ' defects [23,35,36], which can be generated by extrinsic ion-implantation-induced dissociation process [36,37] and/or intrinsic UV photon absorption during the PL measurement [41]. An NOV defect can be transformed to an E_δ' center [23,36]. The ~630 nm (~2.0 eV) PL band could be due to another radiative defect, the non-bridging oxygen hole center (NBOHC) that has been observed in both pure silica glass and ion-irradiated SiO_2 [37,42–45].

The drastic increase in the peak intensity of the ~760 and ~845 nm bands when the annealing temperature approaches 1100°C indicates that the luminescence mechanisms of the two bands are different from that of the defects in the oxide matrix discussed above. The ~760 nm band is attributed to the transitions in the nc-Si due to the quantum confinement (QC) effect [1,35]. However, there is an energy difference of ~0.13 eV (i.e., the Stokes shift) between the energy of the ~760 nm PL peak (~1.63 eV) and the nc-Si bandgap (~1.76 eV). As the nc-Si has an indirect band structure [1,35], the probability of direct transitions is very low. Here the nc-Si/SiO_2 interface is thought to play an important role in the emission of the ~760 nm PL band. The Stokes shift is about the same as the energy (~0.13 eV) of the Si–O vibration with a stretching frequency of ~1083 cm^{-1} in the system of nc-Si embedded in SiO_2 [46]. It was shown in a previous study that the coupling of the confined excitons and the Si–O stretching vibrations dominate the PL process in nc-Si embedded in SiO_2 [47]. The phonons (i.e., the Si–O vibration at the nc-Si/SiO_2 interface) can provide the means required for both the energy dissipation due to the energy conservation requirement and the momentum conservation in the PL process [1,35]. Therefore, the ~760 nm band can be attributed to the band-to-band transition of the nc-Si assisted by the emission of a phonon (i.e., the involvement of the Si–O vibration at the nc-Si/SiO_2 interface). On the other hand, the 845 nm band could originate from the localized luminescent centers at the nc-Si/SiO_2 interface [1,35,48,49].

7.3.2 ELECTROLUMINESCENCE

Visible EL from Si^+-implanted SiO_2 films has been frequently reported [20,50–59]. As an example, the EL from nc-Si light-emitting structures with a 30 nm SiO_2 thin film embedded with nc-Si fabricated with low-energy Si ion implantation is briefly discussed in the following [1,57].

In the study reported in [57], the nc-Si-distributed SiO_2 light-emitting layer was fabricated by the implantation of Si ions with various doses into thermally grown 30 nm SiO_2 thin films on p-type Si wafers at low implantation energies (2–8 keV). A post-implantation thermal annealing was conducted at 1000°C or 1100°C in N_2 for 1 h to induce nanocrystallization of excess Si in SiO_2. Visible and NIR EL has been observed from the EL structures under a negative gate voltage (i.e., a negative bias is applied to the indium–tin–oxide (ITO) electrode).

It was shown that the EL spectrum consisted of the four EL bands at ~460, ~610, ~740, and ~1260 nm [57]. The ~610 nm band dominates the major EL peak, and the ~460 and ~740 nm bands also have a large contribution to the major EL peak; and the ~1260 nm band is obviously responsible for the weak luminescence in the NIR region. The peak wavelengths of the four EL bands are independent of the gate voltage. However, the contribution of each EL band (i.e., the ratio of the integrated intensity of each EL band to the integrated intensity of the EL spectrum) changes with the gate voltage. As the gate voltage increases, the contribution of the ~610 nm band decreases, but the contributions of the ~460 and ~740 nm bands increase and the contribution of the ~1260 nm band remains small [57].

Both the implantation dose and energy have a large impact on the EL. The EL intensity increases with the implantation dose; the EL intensity also increases with the implantation energy [57]. The increase in the EL intensity with either the implanted Si ion dose or the implantation energy can be explained by the enhancement in the current conduction of the Si^+-implanted oxide. Although the integrated EL intensity increases with both implanted Si ion dose and implantation energy, the contribution of each EL band shows a different picture of its dependence on the implanted Si ion dose and implantation energy. The contribution of each EL band is almost independent of the Si ion dose. However, the contribution of each EL band changes with the implantation energy. As the implantation energy increases from 2 to 8 keV, the contribution of the ~600 nm band significantly increases from ~55% to ~61%, the contributions of both the ~740 nm band and the ~1260 nm band remain unchanged, and the contribution of the ~460 nm EL band significantly decreases from ~23% to ~13% [1,57].

Similar EL bands have been reported by different groups (e.g., see Song et al. [56]). Various luminescent defects, such as WOB (O–O) defects [20,36], neutral oxygen vacancy (NOV) (O_3≡Si–Si≡O_3) [20,34,36,40,45,53,55,56,61], non-bridging oxygen hole center (NBOHC) (O_3≡Si–Si–O·) [20,36,45,56,60,61], D center [61], and E' center [15,20,31,36,56,61] have been proposed as the mechanisms of luminescence from Si^+-implanted SiO_2 films. It has been well accepted that the defect luminescence centers play a major role in EL from Si^+-implanted SiO_2 films. Although the origins for the above-mentioned EL bands are still under debate, the ~460 nm (~2.7 eV) band and the ~600 nm (~2 eV) band have been ascribed to the NOV defect and the NBOHC defect, respectively [1,57].

As discussed earlier, both PL and EL bands at ~740 nm are observed, suggesting that they have the same emission mechanism. It has been suggested that the ~740 nm PL band (~1.7 eV) is due to the band-to-band transition in the nc-Si assisted by the emission of a phonon, which is actually the Si–O vibration at the nc-Si/SiO_2 interface. Therefore, the ~740 nm EL band can be attributed to the recombination of carries within the nc-Si assisted by the Si–O vibration at the nc-Si/SiO_2 interface [1,57]. Here the phonons (i.e., the Si–O vibration at the nc-Si/SiO_2 interface) provide the means required for both the energy dissipation due to the energy conservation requirement and the momentum conservation in the EL process [1]. As for the infrared EL band centered at ~1260 nm, it most possibly originates from the Si substrate because it is also observed from the pure, thermally grown SiO_2 thin films and the SiO_2 films implanted with different elements such as Ge and Al. It is a result of electron–hole recombination in the accumulation region formed beneath the oxide layer near the interface between the Si substrate and SiO_2 [57]. Such recombination is assisted by the Si–O vibration at the interface between SiO_2 and the Si substrate [1,57].

The excitation of the NOV defect acting as a luminescence center under electrical pumping can be enhanced by the electric field in the oxide [20,56,62]. This explains the result that the contribution of the ~460 nm band, which is attributed to the NOV defect, increases with the applied voltage [1,57]. On the other hand, for higher implantation energy, the implanted-Si-distributed region is extended, leading to a reduction in the tunnel oxide (i.e., the pure SiO_2 region) and thus an increase in the injection current. As a result, for a given applied voltage, the voltage drops in the ITO/oxide junction and the oxide/substrate junction increase, and thus the voltage drop in the implanted-Si-distributed oxide region decreases. Thus, the electric field in the implanted-Si-distributed region decreases, weakening the excitation of the NOV defect. Therefore, the contribution of the ~460 nm band decreases with the implantation energy [1,57].

7.4 PHOTOLUMINESCENCE AND ELECTROLUMINESCENCE FROM Si+-IMPLANTED Si₃N₄ THIN FILMS

7.4.1 Photoluminescence

In this section, we present a study on the PL from Si^+-implanted Si_3N_4 thin films reported by Cen et al. [63]. Silicon nitride films with the thickness of 120 nm were deposited by low-pressure chemical vapor deposition (LPCVD) on a 30 nm thick SiO_2 thin film thermally grown on Si(100) wafers. Si ions were implanted into the silicon nitride films at the energy of 30 keV with a dose of 3.5×10^{16} atoms/cm². The implanted films were thermally annealed in N_2 ambient for 1 h at temperatures ranging from 800°C to 1100°C. The PL spectrum of the Si-implanted sample annealed at 1100°C is shown in Figure 7.10 [63]. The PL from the control sample (i.e., without the Si ion implantation) is also included in the figure for comparison. Two emission bands located at ~435 nm (2.8 eV, blue PL band) and ~680 nm (1.8 eV, red PL band) can be observed for both samples, but the red light emission from pure silicon nitride is much weaker compared to that of the Si-implanted sample. The blue PL band is attributed to the radiative transition between the defect states related to the nitrogen dangling bonds [63]. Although this blue emission is believed to originate from the silicon nitride host matrix, it is affected by the Si ion implantation also. As shown in Figure 7.10, compared to the blue PL band from the pure silicon nitride, the blue PL band from the Si-implanted sample shows a large reduction in the peak intensity and is broadened and red-shifted. This is somewhat similar to the situation of Ar^+ implantation into silicon nitride [64]. This reduced intensity of the blue PL band from the Si-implanted silicon nitride film could be due to the nonradiative defect centers

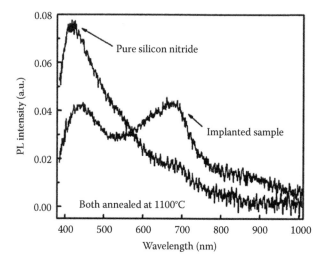

FIGURE 7.10 PL spectra of the Si-implanted silicon nitride and the pure silicon nitride thin film (i.e., the control sample), both annealed at 1100°C. (From Cen, Z.H. et al., *Electrochem. Solid-State Lett.*, 12, H38, 2009. With permission.)

introduced during the Si ion implantation [2,64,65]. In contrast to the blue PL band, the red PL band is enhanced by the Si ion implantation. As shown in Figure 7.10, the red PL band at ~680 nm is much stronger due to the Si implantation. The origin of the red PL band is still controversial. It is often attributed to the recombination via the defect states, such as the Si–Si defects in the silicon nitride matrix (note that Si implantation can generate defects); however, the radiative transition due to the quantum confinement in the nc-Si formed in the Si-implanted silicon nitride could be responsible for the red PL band also [63].

More experiments are carried out to understand the red PL band. It can be observed from the PL measurements shown in Figure 7.11 [63] that the red PL band from the as-implanted sample is insignificant, but it emerges after a thermal annealing. The nonradiative centers introduced by the ion implantation in the as-implanted sample, which suppress the light emission, can explain the absence of the red PL band. With the elimination of these nonradiative centers by thermal annealing, the intensity of the red and blue PL bands increases. When the annealing temperature is higher than 800°C, the light emission of the two bands is quenched gradually, in general, as shown in the inset of Figure 7.11. The evolution of the PL intensity with annealing temperature is similar for the two bands, and such evolution can also be observed for the control sample. It can be interpreted in terms of the nonradiative defects due to the hydrogen desorption [63]. On the other hand, with the 1100°C annealing, the red PL band is as strong as or even stronger than the blue PL band, as shown in Figure 7.10. Indeed, for a sample with a thinner Si-implanted silicon nitride film and annealed at a high temperature, the red PL band is the major contributor to the light emission, because the contribution of the blue PL band originating from the silicon nitride host matrix is relatively small due to the thin silicon nitride film. There is no significant shift in the red PL band for different annealing temperatures, which is a characteristic of defect-related PL [63].

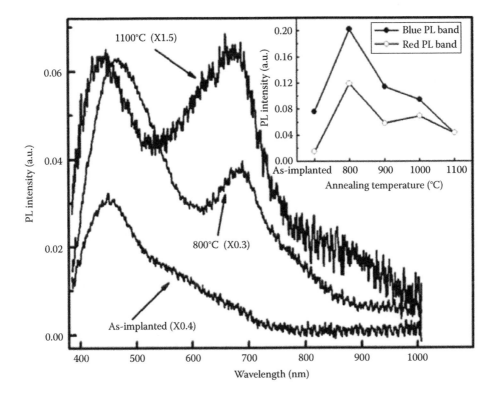

FIGURE 7.11 PL spectra of the Si-implanted silicon nitride thin films annealed at different temperatures. The inset shows the evolution of the PL intensity with the annealing temperature for the blue (435 nm) and red (680 nm) PL bands. (From Cen, Z.H. et al., *Electrochem. Solid-State Lett.*, 12, H38, 2009. With permission.)

The emission mechanism of the red PL band can be further investigated with the PL excitation (PLE) measurement. The PLE spectra monitored at 680 nm (the red PL band) for the as-implanted and the annealed Si-implanted samples as well as the pure silicon nitride are shown in Figure 7.12 [63]. As the red PL band is weak for the as-implanted sample and the control sample, there is no obvious feature or structure in the PLE spectra of these samples. As shown in Figure 7.12a, for the short excitation wavelengths, a broad PLE peak at around 290 nm exists for the annealed samples. The excitation transition process begins at the excitation wavelength of ~310 nm (4 eV), which is close to the Tauc bandgaps (4.1 eV) of the annealed Si-implanted samples derived from the optical transmissions [63]. Therefore, it is suggested that the major contribution for the red band emission is the excitation transition between the Si-bond-related states of σ and σ*, which are near the valence and conduction band edges of silicon nitride, respectively [63].

On the other hand, another PLE peak at around 425 nm (2.9 eV) emerges after the annealing at 1100°C, and it is not observed in other Si-implanted samples annealed at lower temperatures, as shown in Figure 7.12b [63]. The emergence of this PLE peak after the annealing at 1100°C indicates that annealing introduces another excitation transition for the red PL band. It has been well established that annealing at 1100°C leads to the formation of stable nc-Si or Si nanoclusters in the silicon nitride layer. Therefore, the PLE peak at 425 nm could originate from the formation of stable nc-Si or Si nanoclusters. To further understand the PLE peak at 425 nm, PLE was conducted at different monitored wavelengths of the red PL band for the sample annealed at 1100°C, and the result is shown in Figure 7.13 [63]. It can be seen from the figure that the PLE peak at ~425 nm

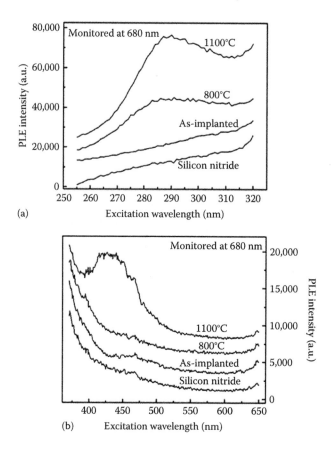

FIGURE 7.12 (a) PLE spectra monitored at 680 nm of the Si-implanted samples annealed at different temperatures. (b) Individual PLE spectra are shifted along the vertical axis for a clear comparison. (From Cen, Z.H. et al., *Electrochem. Solid-State Lett.*, 12, H38, 2009. With permission.)

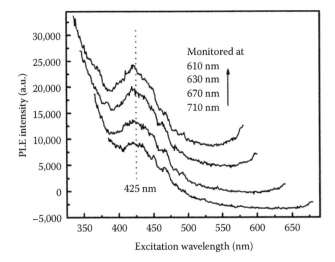

FIGURE 7.13 PLE spectra monitored at different wavelengths for the Si-implanted sample annealed at 1100°C. Individual PLE spectra are shifted along the vertical axis for a clear comparison. (From Cen, Z.H. et al., *Electrochem. Solid-State Lett.*, 12, H38, 2009. With permission.)

is independent of the monitored wavelength. It is expected that the PLE peak shifts toward lower energy as the monitored wavelength is increased for the transitions associated with the quantum confinement effect [63]. Therefore, it is believed that the quantum confinement effect does not play a major role for the red PL band. As the red PL band is strongly influenced by the Si implantation and cannot be observed from the Ge-implanted silicon nitride films [65], it is reasonable to argue that the emission of the red PL band originates from the radiative recombination associated with the defect states related to the nc-Si or Si nanoclusters embedded in the silicon nitride matrix [63]. The large Stokes shifts observed for the two PLE peaks (i.e., at 290 and 425 nm, respectively) suggest that the PL primarily occurs via the relaxed radiative recombination states [63].

7.4.2 ELECTROLUMINESCENCE

In this section, we present a systematic study of the EL from Si^+-implanted Si_3N_4 thin films reported by Cen et al. [66–70]. The EL mechanisms, thermal annealing effect, and EL color tunability are discussed.

7.4.2.1 Fabrication of EL Structures

A silicon nitride thin film with the thickness of 30 nm was deposited on a p-type (100) Si wafer by the LPCVD technique at 800°C. The reactant gases were dichlorosilane (DCS) and ammonia (NH_3) with flow rates of 50 and 200 sccm, respectively, and the total deposition pressure was 300 mTorr. In order to enhance the injection current and improve the light-emission performance, a nearly uniform distribution of the implanted Si ions was achieved by multiple implantations with the following implantation recipe [2,66]: the first implantation at an energy of 25 keV with a dose of 4×10^{16} atoms/cm²; the second implantation at 8 keV with a dose of 8×10^{15} atoms/cm²; and the third implantation at 2 keV with a dose of 3×10^{15} atoms/cm². As revealed by the stopping and range of ions in matter (SRIM) simulation, a nearly uniform distribution of the implanted Si in the silicon nitride thin film is achieved, and the average implanted Si concentration throughout the Si-implanted film is ~1.2×10^{22} cm⁻³ (the corresponding volume fraction of the implanted Si is ~25%), except for the area close to the film surface. The multiple implantations were used for the studies on the EL mechanism and the annealing effect. For other studies discussed later, a single implantation was used, and the details will be given in the relevant discussions. After the implantations, thermal annealing was conducted in nitrogen ambient at various temperatures for 1 h. To fabricate the EL structure, a 200 nm thick Al layer was deposited onto the backside of the Si substrate to form the ohmic contact, and circular transparent ITO electrodes with a thickness of 130 nm were deposited on the Si-implanted silicon nitride thin film through a hard shadow mask with a pad radius of 1.2 mm by the sputtering technique [66].

7.4.2.2 EL Properties and Mechanisms

Figure 7.14 shows the EL spectra from the 1100°C-annealed Si-implanted silicon nitride thin film measured at various gate biases [66]. A primary EL peak at ~415 nm can be observed in the EL spectra. In addition to the main EL peak, there is an apparent shoulder at ~560 nm in the EL spectra. For low applied voltages, the EL intensity increases with the applied voltage, but large voltages cause EL degradation. Despite the variation of the EL intensity, no significant changes in the spectral shape can be observed under different voltages.

The current conduction of the silicon nitride thin film is greatly enhanced by the nearly uniform distribution of the implanted Si in the silicon nitride thin film. This means that a large injection current can be easily achieved with a relatively low voltage. Figure 7.15 shows the current–voltage (I–V) characteristic of the EL structure [66]. As can be observed in the figure, the current conduction in the Si$^+$-implanted nitride film follows a power law that can be described by the equation shown in Figure 7.15. The power-law fitting to the experimental I–V data yielded a ζ value of 2.06, suggesting that the current conduction is close to 2D transport. The power-law behavior indicates that the

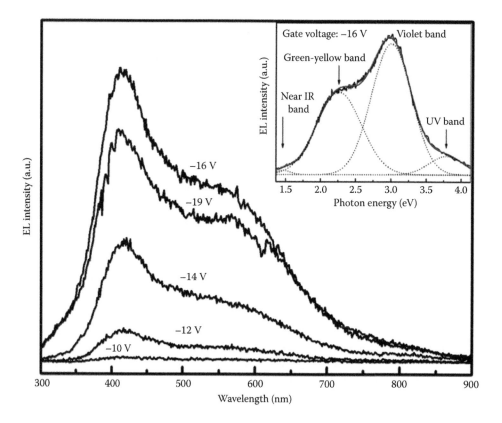

FIGURE 7.14 EL spectra measured at different gate voltages. The inset shows the deconvolution of the EL spectrum measured at the gate voltage of −16 V. (From Cen, Z.H. et al., *Appl. Phys. Lett.*, 94, 041102, 2009. With permission.)

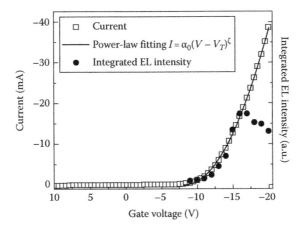

FIGURE 7.15 Gate current and integrated EL intensity as funtions of the gate voltage. The power-law fitting yields $\zeta = 2.06$, $\alpha_0 = 0.23$ mAV$^{-\zeta}$, and $V_T = -8.1$ V. (From Cen, Z.H. et al., *Appl. Phys. Lett.*, 94, 041102, 2009. With permission.)

enhancement in the current conduction can be explained in terms of the formation of conductive percolation paths by the defects, that is, nc-Si and Si nanoparticles distributed in the nitride matrix due to the Si implantation [66].

The dependence of the integrated EL intensity on the voltage is also shown in Figure 7.15 [66]. There is no light emission under a positive voltage due to insufficient hole injection from the ITO electrode. EL is observed under a negative voltage with the voltage magnitude larger than ~10 V. As can be seen in Figure 7.15, in the voltage range of −10 to −16 V, the integrated EL intensity is proportional to the injection current, implying that the EL originates from the recombination of the electrons and holes injected from the ITO gate and the p-type Si substrate, respectively. However, when the magnitude of the voltage is larger than ~16 V, the integrated EL intensity is quenched, and it decreases gradually as the magnitude of the voltage further increases. It has been found that such an EL quenching is related to the charge trapping in the Si-implanted nitride thin film, which reduces both the electron injection from the ITO gate and the hole injection from the p-Si substrate [2]. Moreover, the high electric field in the silicon nitride film, which is estimated to be ~6 MV/cm for the voltages of −17 to −20 V, could have an impact on the radiative lifetime of the luminescent centers, leading to a quenching of the radiative recombination [71]. In addition, nonradiative Auger recombination may also affect the light emission when multiple carriers are present in the same nanostructures [72].

The EL spectra can be deconvolved into two primary Gaussian-shaped EL bands, that is, the violet EL band (~3.0 eV or 415 nm) and the green-yellow EL band (~2.2 eV or 560 nm). When the magnitude of the voltage is larger than 15 V, an ultraviolet (UV) band (~3.8 eV or 327 nm) and an NIR band (~1.45 eV or 850 nm) emerge, but their contributions to the total EL intensity are much smaller compared to the two main EL bands. As an example, the deconvolution of the EL spectrum measured at an applied voltage of −16 V is shown in the inset of Figure 7.14 [66]. The two primary EL bands, that is, the violet and green-yellow bands, together contribute up to 90% of the total EL intensity. The intensity of these two EL bands (in particular, the violet band) decreases with the voltage when its magnitude is larger than ~16 V, which is responsible for the quenching of the total EL intensity at high voltage. Some of the possible radiative recombination processes responsible for the four EL bands (i.e., the violet band, green-yellow band, UV band, and NIR band) are schematically shown in Figure 7.16 and discussed below [2,67].

The strong violet EL band at 3.0 eV, which can also be observed from the Ge-implanted SiO$_2$ layer [55], has seldom been reported for silicon nitride–based materials, although violet PL has been

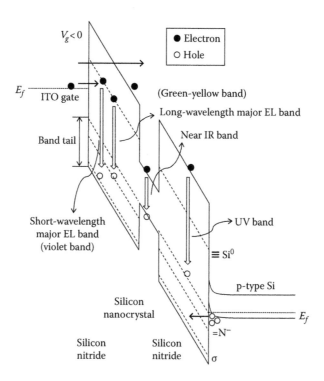

FIGURE 7.16 Schematic band diagram of the EL structure under forward bias. Carrier transport, defect states, and Si nanocrystal in the nitride film, as well as various recombination processes are also shown. (From Cen, Z.H. et al., *J. Appl. Phys.*, 105, 123101, 2009. With permission.)

obtained from Si nanostructures in silicon nitride films [16,73]. PL with energies close to this violet EL band from thick Si-implanted silicon nitride films, which are fabricated with the implantation recipe similar to that of the EL structures, has been observed. One of the mechanisms usually used to explain the short wavelength light emission from silicon nitride films is radiative recombination through the defect states within the silicon nitride bandgap [74,75]. The defect-related mechanism should particularly play an important role in the case of Si-implanted silicon nitride thin films because various point defects can be introduced into the dielectric matrix during the Si implantation [61]. This can explain the difference in light emissions between the Si-rich silicon nitride materials fabricated by ion implantation and those by CVD. The violet emission at ~3.0 eV could be attributed to the radiative transition from the defect state related to the silicon dangling bonds ($\equiv Si^0$) located at 3.1 eV above the valence band maximum (VBM) of silicon nitride to the bonding state (σ) of the $\equiv Si–Si\equiv$ unit that is close to the valence band [75–77] or the radiative transition from the defect state ($\equiv Si^0$) to the VBM [66]. On the other hand, similar defect-related luminescence mechanisms could be also used to explain another primary band, the green-yellow EL band. Band tail radiative recombination has been used to account for the emission of the green-yellow band [78]; nevertheless the radiative recombination assisted by defect states could also be the origin [74,79]. The green-yellow band could be due to the transition between the $\equiv Si^0$ state and the state related to the nitrogen dangling bonds ($=N^-$) in the valence band tail, which is located at 0.8 eV above the VBM [2,75,77].

As mentioned earlier, the two minor EL bands, that is, the UV band and the NIR band, appear at a sufficiently high voltage (high-level injection). The mechanism responsible for the UV band peak at ~3.8 eV is still not clear. Two possible mechanisms are explained in the following [2,66]. Compared to the stoichiometric silicon nitride, Si-implanted silicon nitride thin film has been found to have a large trap density of shallow energy levels below the silicon nitride conduction band minimum (CBM). A significant number of electrons can be captured by these shallow traps during the electron injection from the ITO gate under a high voltage. The radiative recombination between the electrons captured by the shallow traps and the holes in the valence band of silicon nitride injected from the p-type Si substrate produces a UV emission [66]. It is also possible that under a high bias voltage, electrons can be injected into the conduction band of silicon nitride and then recombine with the holes in the band tail states above the silicon nitride VBM, leading to the UV band emission [2,67], as shown in Figure 7.16. It is worth mentioning that PL near the UV region has been observed from silicon-rich silicon nitride thin films prepared by LPCVD also [80]. On the other hand, at a high voltage, significant electron or hole injection into the conduction or valence bands of the nc-Si embedded in the silicon nitride matrix, respectively, could also occur. The radiative recombination of the injected electrons and holes via nc-Si can produce EL in the NIR range as a result of the quantum confinement effect. This argument is supported by the fact that the peak energy of the NIR band is close to the bandgap of the nc-Si (~1.5 eV) [66].

7.4.2.3 Annealing Effect

Figure 7.17 shows the effect of annealing on the I–V characteristics of the EL structure [67]. Compared to the as-implanted film, the film thermally annealed at 800°C shows a reduction in the current conduction at gate voltages larger than ~12 V; and the current reduction becomes more significant and over a wider voltage range for annealing at 1100°C. The annealing-induced current reduction can be attributed to the recovery of the silicon nitride matrix during the thermal annealing process [67]. As shown in Figure 7.17, the I–V characteristics also follow a power law, which can be explained in terms of the formation of conductive percolation tunneling paths by the Si nanocrystals as well as of the defect states created by the ion implantation [67]. With a large number of nc-Si or Si nanoclusters and defect states distributed in the thin film, the injected carriers can be transported through tunneling via the Si nanostructures and the defect states. As the defect states can be removed by annealing, the tunneling via the defect states will decrease after annealing. This explains the increase of the global threshold voltage or the reduction of the conduction current with increase in annealing temperature [67]. The annealing effect on the current conduction will, of course, affect the EL properties.

The integrated EL intensity for the as-implanted film and the films annealed at 800°C and 1100°C as a function of the gate voltage is shown in Figure 7.18 [67]. For both the as-implanted film

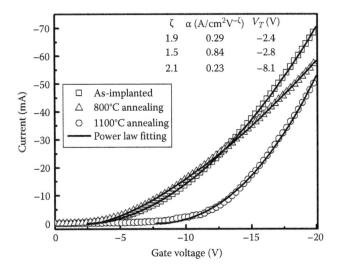

FIGURE 7.17 Annealing effect on the I–V characteristics. (From Cen, Z.H. et al., *J. Appl. Phys.*, 105, 123101, 2009. With permission.)

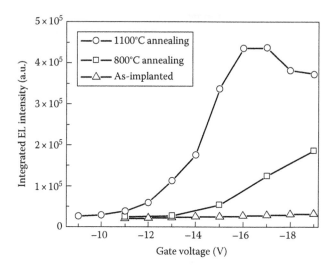

FIGURE 7.18 Integrated EL intensity for the as-implanted and annealed films as a function of the gate voltages. (From Cen, Z.H. et al., *J. Appl. Phys.*, 105, 123101, 2009. With permission.)

and the film annealed at 800°C, the integrated EL intensity increases with the voltage, and there is a linear relationship between the integrated EL intensity and the injection current, implying that the EL originates from the recombination of electrons and holes injected from the ITO gate and the Si substrate, respectively. However, for the sample annealed at 1100°C, as can be seen in Figure 7.18, the integrated EL intensity decreases when the voltage is larger than ~16 V. The quenching phenomenon can be explained by different mechanisms, as pointed out earlier. On the other hand, the EL quantum efficiency can be enhanced largely by annealing because annealing can lead to a large increase in the EL intensity (Figure 7.18) and a large reduction in the injection current (Figure 7.17). In addition, as can be seen in Figure 7.18, the EL turn-on voltage decreases significantly as a result of annealing. The nonradiative defects can be removed by annealing, and nc-Si and other luminescent centers can also be formed during annealing, particularly at a high temperature. Thus, annealing can increase the light emission quantum efficiency [67].

In addition to the EL intensity, the EL spectrum is also significantly affected by annealing. The EL spectrum of the as-implanted film shows a broad peak located at an orange wavelength of ~600 nm (~2.0 eV). However, in the films annealed at 800°C or 1100°C, a dominant peak at ~400 nm (i.e., the violet band) and a shoulder band at ~550 nm (i.e., the green-yellow band), which make up a nearly white EL emission, dominate the EL spectra. Figure 7.19 shows the comparison of the EL deconvolution between annealing at 800°C and 1100°C [67]. All the EL spectra of the as-implanted and annealed films consist of two major bands including the violet band (named the "short-wavelength major band" in [67]) and the green-yellow band (i.e., the "long-wavelength major band" in [67]). The minor UV band emerges after annealing at both 800°C and 1100°C, and the minor NIR band appears after annealing at 1100°C, as shown in Figure 7.19b.

Figure 7.20 shows the evolution of the peak position, intensity, and FWHM of the four EL bands with annealing temperature [67]. Annealing leads to blue shifts of the two major bands (i.e., violet and green-yellow). As the annealing temperature is increased from 800°C to 1100°C, the UV band becomes remarkable, and the NIR band is detectable after annealing at 1100°C [67]. The blue shift of the two major bands with increase in annealing temperature may be due to the evolution of the band-tail states with annealing. Due to annealing, the implantation-induced damage is recovered and the Si-implanted film becomes more ordered, thus reducing the band tail. Therefore, for a higher annealing temperature, the radiative recombination of the electrons from the luminescence

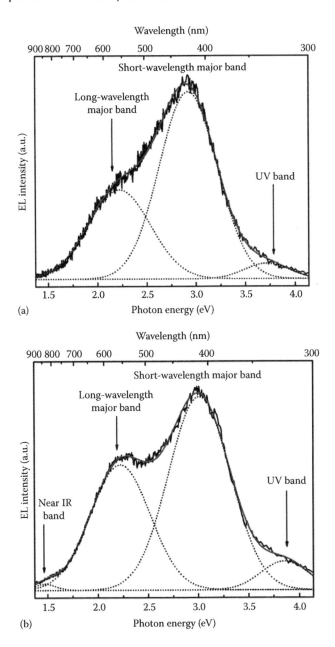

FIGURE 7.19 Deconvolution of the EL spectrum at the gate voltage of −17 V for annealing at (a) 800°C and (b) 1100°C. (From Cen, Z.H. et al., *J. Appl. Phys.*, 105, 123101, 2009. With permission.)

centers with the holes in the states in the reduced band tail can produce light emission with higher energies [2,67]. If the UV band is due to the recombination of the injected electrons in the conduction band of silicon nitride with the holes in the band tail above the silicon nitride valence band, its slight blue shift with annealing temperature shown in Figure 7.20a can be explained by the reduced valence band tail at a higher annealing temperature [2,67]. On the other hand, as the NIR band is attributed to the band-to-band transition in the nc-Si, its emergence should be related to the formation of stable nc-Si as a result of annealing at the high temperature [2,67].

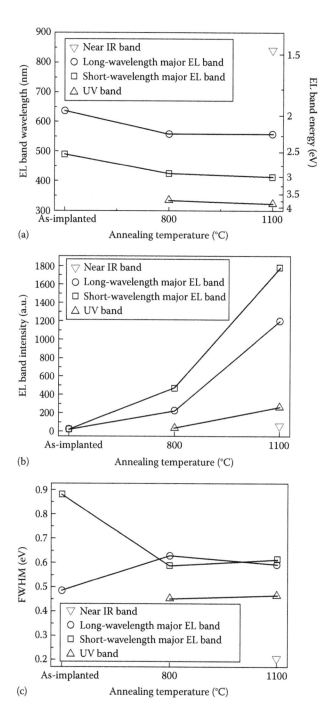

FIGURE 7.20 Evolution of each EL band with annealing temperature at the gate voltage of −17 V. (a) Peak energy. (b) Intensity. (c) FWHM. (From Cen, Z.H. et al., *J. Appl. Phys.*, 105, 123101, 2009. With permission.)

7.4.2.4 Electrically Tunable EL

It has been shown that the EL wavelength can be tuned by varying the injection current [69]. In the study reported in [69], a single Si ion implantation into a 30 nm Si_3N_4 thin film at the energy of 12 keV with the implantation dose of 4×10^{16} atoms/cm^2 was carried out, and the implanted film was annealed at 1100°C in N_2 atmosphere for 1 h. Figure 7.21 shows EL spectra and the photographs of the corresponding light emissions under various injection currents [69]. As can be seen in the figure, the spectral shape or the emission color changes with the injection current, and even bright white light emission can be achieved at a relatively high injection current. The change in the emission color is actually due to the changes in both the intensities and FWHM of the two major bands (i.e., the violet band and the green-yellow band).

The peak positions (i.e., the peak wavelength λ_{peak} or the peak energy E_{peak}), the FWHM of the two major bands, the ratio (I_{max-GY}/I_{max-V}) of the peak intensity of the green-yellow band to that of the violet band, and the integrated EL intensity ratio (S_{GY}/S_V) of the green-yellow band to the violet band as a function of the injection current are shown in Figure 7.22a through d, respectively [69]. As can be seen in Figure 7.22a, the peak positions of the two EL bands do not shift with the injection current. As shown in Figure 7.22b, the FWHM of the green-yellow band is larger than that of the violet band, and as the injection current is increased from −26 mA/cm^2, the FWHM of the green-yellow band decreases significantly but that of the violet band increases gently. The peak intensities of both bands increase with injection current simultaneously. However, the peak intensity of the green-yellow band increases faster than that of the violet band, leading to an increase in the peak intensity ratio (I_{max-GY}/I_{max-V}) with injection current, as shown in Figure 7.22c. For low injection currents (e.g., less than −61 mA cm^{-2}), $I_{max-GY}/I_{max-V} < 0.55$, the violet band is dominant. However, at high injection current of −265 mA/cm^2, $I_{max-GY}/I_{max-V} = 1.1$, showing that the peak intensity of the green-yellow band

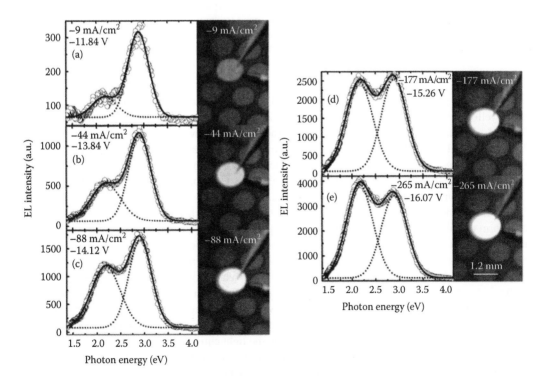

FIGURE 7.21 (a–c) EL spectra and photographs of the corresponding light emissions under different injection currents. (d,e) The EL spectra consist of the two major bands, namely, the violet band and the green-yellow band. (From Cen, Z.H. et al., *Opt. Express*, 18, 20439, 2010. With permission.)

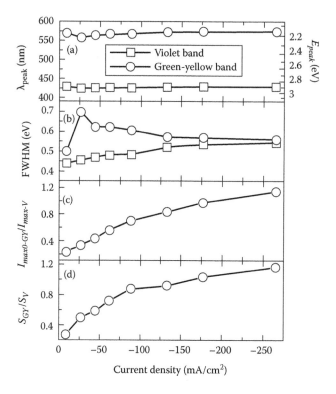

FIGURE 7.22 (a) Peak positions (λ_{peak} or E_{peak}), (b) peak FWHM of the violet and green-yellow bands, (c) peak-intensity ratio $I_{max\text{-}GY}/I_{max\text{-}V}$ of the green-yellow band to the violet band, and (d) the integrated-EL-intensity ratio S_{GY}/S_V of the green-yellow band to the violet band as a function of the injection current. (From Cen, Z.H. et al., *Opt. Express*, 18, 20439, 2010. With permission.)

is even slightly greater than that of the violet band. The integrated-EL-intensity ratio (S_{GY}/S_V) of the green-yellow band to the violet band, which includes the effects of both the peak intensity and the FWHM, also increases as the injection current is increased, as shown in Figure 7.22d. The changes in $I_{max\text{-}GY}/I_{max\text{-}V}$ and FWHM with injection current are responsible for the observed changes in the spectral shape or the color shown in Figure 7.21. In fact, the color transition from violet at low injection currents to white at high injection currents is mainly due to the changes in $I_{max\text{-}GY}/I_{max\text{-}V}$ [69].

The evolution of the dominant EL bands (the violet and green-yellow) with injection current of the Si-implanted film with a single implantation cannot be observed in the multiply Si-implanted films with a uniform distribution of the implanted Si ions [69]. The difference in EL evolution between a single implantation (most of the implanted Si atoms are located in the middle of the silicon nitride film) and multiple implantations (the implanted Si is distributed almost uniformly throughout the entire film) suggests that the evolution of the two major EL bands could be related to the distribution of the implanted Si in the silicon nitride thin film. In the case of the single implantation, the major radiative recombination can occur in the region near the Si substrate for low negative voltages (small injection current) and in middle of the Si-implanted film for high negative voltages (large injection current) [69]. Therefore, the luminescence centers from the Si_3N_4 host determine the light emission under small injection currents, which is due to the low implanted Si concentration near the Si substrate, while the large concentration of implanted Si in the middle of the film should play a more important role in EL for large injection currents. As discussed previously, the violet band originates from the silicon nitride matrix and the green-yellow band is related to the implanted Si. Therefore, if the major radiative recombination region shifts from the lightly implanted Si areas to the highly implanted Si areas due the increase in the voltage, a transition in the dominant EL band from violet to green-yellow could occur. On the other hand, the EL quenching of Si-implanted Si_3N_4 films can also affect the evolution of the intensity ratio of the violet band to the green-yellow band with increasing injection current [69].

ACKNOWLEDGMENTS

The author wishes to thank Y. Liu, L. Ding, Z. H. Cen, and others at the Nanyang Technological University for their contribution to the results presented here.

REFERENCES

1. Chen, T. P. and Ding, L. 2010. Optical and optoelectronic properties of silicon nanocrystals embedded in SiO_2 matrix. In *Nanostructured Thin Films and Coatings: Functional Properties*, Zhang, S. (Ed.), pp. 113–165. CRC Press, Boca Raton, FL.
2. Cen, Z. H. 2011. Studies of optical properties and light emission of silicon-rich silicon nitride thin films. PhD thesis, Nanyang Technological University, Singapore.
3. Ding, L. 2009. Optical and optoelectronic properties of Si nanocrystals embedded in dielectric matrix. PhD thesis, Nanyang Technological University, Singapore.
4. Pavesi, L. and Lockwood, D. J. (Eds.) 2004. *Silicon Photonics*. Springer, Berlin, Germany.
5. Cutolo, A., Iodice, M., Spirito, P., and Zeni, L. 1997. Silicon electro-optic modulator based on a three terminal device integrated in a low-loss single-mode SOI waveguide. *Journal of Lightwave Technology* 15:505–518.
6. Liu, A., Jones, R., Liao, L. et al. 2004. A high-speed silicon optical modulator based on a metal-oxide-semiconductor capacitor. *Nature* 427:615–618.
7. Liu, Y., Liu, E., Li, G. et al. 1994. Novel silicon waveguide switch based on total internal reflection. *Applied Physics Letters* 64:2079–2080.
8. Zhao, C. Z., Chen, A. H., Liu, E. K., and Li, G. Z. 1997. Silicon-on-insulator asymmetric optical switch based on total internal reflection. *IEEE Photonics Technology Letters* 9:1113–1115.
9. Ghioni, M., Zappa, F., Kesan, V. P., and Warnock, J. 1996. A VLSI-compatible high-speed silicon photodetector for optical data link applications. *IEEE Transactions on Electron Devices* 43:1054–1060.
10. Hawkins, A. R., Wu, W., Abraham, P., Streubel, K., and Bowers, J. E. 1997. High gain-bandwidth-product silicon heterointerface photodetector. *Applied Physics Letters* 70:303–305.
11. Pellegrino, P., Garrido, B., Garcia, C. et al. 2005. Low-loss rib waveguides containing Si nanocrystals embedded in SiO_2. *Journal of Applied Physics* 97:074312(1-8).
12. Valenta, J., Pelant, I., and Linnros, J. 2002. Waveguiding effects in the measurement of optical gain in a layer of Si nanocrystals. *Applied Physics Letters* 81:1396–1398.
13. Canham, L. T. 1990. Silicon quantum wire array fabrication by electrochemical and chemical dissolution of wafers. *Applied Physics Letters* 57:1046–1048.
14. Tischler, M. A., Collins, R. T., Stathis, J. H., and Tsang, J. C. 1992. Luminescence degradation in porous silicon. *Applied Physics Letters* 60:639–641.
15. Mutti, P., Ghislotti, G., Bertoni, S. et al. 1995. Room-temperature visible luminescence from silicon nanocrystals in silicon implanted SiO_2 layers. *Applied Physics Letters* 66:851–853.
16. Park, N.-M., Kim, T.-S., and Park, S.-J. 2001. Band gap engineering of amorphous silicon quantum dots for light-emitting diodes. *Applied Physics Letters* 78:2575–2577.
17. Belyakov, V. A., Burdov, V. A., Lockwood, R., and Meldrum, A. 2008. Silicon nanocrystals: Fundamental theory and implications for stimulated emission. *Advances in Optical Technologies* 2008: Article ID 279502, 32pp.
18. Takeoka, S., Fujii, M., and Hayashi, S. 2000. Size-dependent photoluminescence from surface-oxidized Si nanocrystals in a weak confinement regime. *Physical Review B* 62:16820–16825.
19. Calcott, P. D. J., Nash, K. J., Canham, L. T., Kane, M. J., and Brumhead, D. 1993 Identification of radiative transitions in highly porous silicon. *Journal of Physics: Condensed Matter* 5:L91–L98.
20. Lin, C. J. and Lin, G. R. 2005. Defect-enhanced visible electroluminescence of multi-energy silicon-implanted silicon dioxide film. *IEEE Journal of Quantum Electronics* 41:441–447.
21. Guha, S. 1998. Characterization of Si^+ ion-implanted SiO_2 films and silica glasses. *Journal of Applied Physics* 84:5210–5217.
22. Nishikawa, H., Watanabe, E., Ito, D. et al. 1995. Photoluminescence study of defects in ion-implanted thermal SiO_2 films. *Journal of Applied Physics* 78:842–846.

23. Nishikawa, H., Stahlbush, R. E., and Stathis, J. H. 1999. Oxygen-deficient centers and excess Si in buried oxide using photoluminescence spectroscopy. *Physical Review B* 60:15910–15918.
24. Ghislotti, G., Nielsen, B., Asoka-Kumar, P. et al. 1996. Effect of different preparation conditions on light emission from silicon implanted SiO₂ layers. *Journal of Applied Physics* 79:8660–8663.
25. Giorgis, F., Vinegoni, C., and Pavesi, L. 2000. Optical absorption and photoluminescence properties of a-Si$_{1-x}$N$_x$:H films deposited by plasma-enhanced CVD. *Physical Review B* 61:4693–4698.
26. Yerci, S., Li, R., Kucheyev, S. O. et al. 2010. Visible and 1.54 µm emission from amorphous silicon nitride films by reactive cosputtering. *IEEE Journal of Selected Topics in Quantum Electronics* 16:114–123.
27. Kato, H., Kashio, N., Ohki, Y., Seol, K. S., and Noma, T. 2003. Band-tail photoluminescence in hydrogenated amorphous silicon oxynitride and silicon nitride films. *Journal of Applied Physics* 93:239–244.
28. Shen, J. M., Palsule, C., Gangopadhyay, S., Naseem, H. A., Kizzar, S., and Goh, F. H. C. 1994. Characterization of fluorinated hydrogenated amorphous-silicon nitride (a-SiN$_x$:H) alloys. *Journal of Applied Physics* 76:1055–1061.
29. Guha, S., Pace, M. D., Dunn, D. N., and Singer, I. L. 1997. Visible light emission from Si nanocrystals grown by ion implantation and subsequent annealing. *Applied Physics Letters* 70:1207–1209.
30. Wilkinson, A. R. and Elliman, R. G. 2004. The effect of annealing environment on the luminescence of silicon nanocrystals in silica. *Journal of Applied Physics* 96:4018–4020.
31. Shimizu-Iwayama, T., Fujita, K., Nakao, S. et al. 1994. Visible photoluminescence in Si⁺-implanted silica glass. *Journal of Applied Physics* 75:7779–7783.
32. Guha, S., Qadri, S. B., Musket, R. G., Wall, M. A., and Shimizu-Iwayama, T. 2000. Characterization of Si nanocrystals grown by annealing SiO₂ films with uniform concentrations of implanted Si. *Journal of Applied Physics* 88:3954–3961.
33. Cheylan, S. and Elliman, R. G. 2001. Effect of particle size on the photoluminescence from hydrogen passivated Si nanocrystals in SiO₂. *Applied Physics Letters* 78:1912–1914.
34. Liao, L.-S., Bao, X.-M., Zheng, X.-Q., Li, N.-S., and Min, N.-B. 1996. Blue luminescence from Si⁺-implanted SiO₂ films thermally grown on crystalline silicon. *Applied Physics Letters* 68:850–852.
35. Ding, L., Chen, T. P., Liu, Y. et al. 2008. Evolution of photoluminescence mechanisms of Si⁺-implanted SiO₂ films with thermal annealing. *Journal of Nanoscience and Nanotechnology* 8:3555–3560.
36. Lin, C.-J., Lee, C.-K., Diau, E. W.-G., and Lin, G.-R. 2006. Time-resolved photoluminescence analysis of multidose Si-ion-implanted SiO₂. *Journal of the Electrochemical Society* 153:E25–E32.
37. Fischer, T., Petrova-Koch, V., Shcheglov, K., Brandt, M. S., and Koch, F. 1996. Continuously tunable photoluminescence from Si⁺-implanted and thermally annealed SiO₂ films. *Thin Solid Films* 276:100–103.
38. Cheang-Wong, J. C., Oliver, A., Roiz, J. et al. 2001. Optical properties of Ir²⁺-implanted silica glass. *Nuclear Instruments and Methods in Physics Research B* 175–177:490–494.
39. Tohmon, R., Shimogaichi, Y., Mizuno, H. et al. 1989. 2.7-eV luminescence in as-manufactured high-purity silica glass. *Physical Review Letters* 62:1388–1391.
40. Bae, H. S., Kim, T. G., Whang, C. N. et al. 2002. Electroluminescence mechanism in SiOx layers containing radiative centers. *Journal of Applied Physics* 91:4078–4081.
41. Stathis, J. H. and Kastner, M. A. 1984. Photoinduced paramagnetic defects in amorphous silicon dioxide. *Physical Review B* 29:7079–7081.
42. Skuja, L. 1992. Time-resolved low temperature luminescence of non-bridging oxygen hole centers in silica glass. *Solid State Communications* 84:613–616.
43. Munekuni, S., Yamanaka, T., Shimogaichi, Y. et al. 1990. Various types of nonbridging oxygen hole center in high-purity silica glass. *Journal of Applied Physics* 68:1212–1217.
44. Bakos, T., Rashkeev, S. N., and Pantelides, S. T. 2002. The origin of photoluminescence lines in irradiated amorphous SiO₂. *IEEE Transactions on Nuclear Science* 49:2713–2717.
45. Valakh, M. Y., Yukhimchuk, V. A., Bratus, V. Y. et al. 1999. Optical and electron paramagnetic resonance study of light-emitting Si⁺ ion implanted silicon dioxide layers. *Journal of Applied Physics* 85:168–173.
46. Liu, Y., Chen, T. P., Fu, Y. Q. et al. 2003. A study on Si nanocrystal formation in Si-implanted SiO₂ films by x-ray photoelectron spectroscopy. *Journal of Physics D: Applied Physics* 36:L97–L100.
47. Kanemitsu, Y., Shimizu, N., Komoda, T., Hemment, P. L. F., and Sealy, B. J. 1996. Photoluminescent spectrum and dynamics of Si⁺-ion-implanted and thermally annealed SiO₂ glasses. *Physical Review B* 54:14329–14332.
48. Allan, G., Delerue, C., and Lannoo, M. 1996. Nature of luminescent surface states of semiconductor nanocrystallites. *Physical Review Letters* 76:2961–2964.
49. Zhuravlev, K. S., Gilinsky, A. M., and Kobitsky, A. Y. 1998. Mechanism of photoluminescence of Si nanocrystals fabricated in a SiO₂ matrix. *Applied Physics Letters* 73:2962–2964.

50. Kulakci, M., Serincan, U., and Turan, R. 2006. Electroluminescence generated by a metal oxide semiconductor light emitting diode (MOS-LED) with Si nanocrystals embedded in SiO_2 layers by ion implantation. *Semiconductor Science and Technology* 21:1527–1532.

51. Lalic, N. and Linnros, J. 1999. Light emitting diode structure based on Si nanocrystals formed by implantation into thermal oxide. *Journal of Luminescence* 80:263–267.

52. Luterova, K., Pelant, I., Valenta, J. et al. 2000. Red electroluminescence in Si^+-implanted sol–gel-derived SiO_2 films. *Applied Physics Letters* 77:2952–2954.

53. Matsuda, T., Nishihara, K., Kawabe, M. et al. 2004. Blue electroluminescence from MOS capacitors with Si-implanted SiO_2. *Solid-State Electronics* 48:1933–1941.

54. Muller, D., Knapek, P., Faure, J. et al. 1999. Blue electroluminescence from high dose Si^+ implantation in SiO_2. *Nuclear Instruments and Methods in Physics Research B* 148:997–1001.

55. Rebohle, L., von Borany, J., Yankov, R. A. et al. 1997. Strong blue and violet photoluminescence and electroluminescence from germanium-implanted and silicon-implanted silicon-dioxide layers. *Applied Physics Letters* 71:2809–2811.

56. Song, H.-Z., Bao, X.-M., Li, N.-S., and Zhang, J.-Y. 1997. Relation between electroluminescence and photoluminescence of Si^+-implanted SiO_2. *Journal of Applied Physics* 82:4028–4032.

57. Ding, L., Chen, T. P., Liu, Y. et al. 2007. The influence of the implantation dose and energy on the electroluminescence of Si^+-implanted amorphous SiO_2 thin films. *Nanotechnology* 18:455306(1-6).

58. Ding, L., Chen, T. P., Yang, M. et al. 2009. Relationship between current transport and electroluminescence in Si^+-implanted SiO_2 thin films. *IEEE Transactions on Electron Devices* 56:2785–2791.

59. Garrido, B., Lopez, M., Perez-Rodriguez, A. et al. 2004. Optical and electrical properties of Si-nanocrystals ion beam synthesized in SiO_2. *Nuclear Instruments and Methods in Physics Research B* 216:213–221.

60. Jeong, J. Y., Im, S., Oh, M. S., Kim, H. B., Chae, K. H., Whang, C. N., and Song, J. H. 1999. Defect versus nanocrystal luminescence emitted from room temperature and hot-implanted SiO_2 layers. *Journal of Luminescence* 80:285–289.

61. Song, H. Z. and Bao, X. M. 1997. Visible photoluminescence from silicon-ion-implanted SiO_2 film and its multiple mechanisms. *Physical Review B* 55:6988–6993.

62. Sahoo, P. K., Gasiorek, S., Dhar, S., Lieb, K. P., and Schaaf, P. 2006. Cathodoluminescence and epitaxy after laser annealing of Cs^+-irradiated α-quartz. *Applied Surface Science* 252:4477–4480.

63. Cen, Z. H., Chen, T. P., Ding, L. et al. 2009. Optical transmission and photoluminescence of silicon nitride thin films implanted with Si ions. *Electrochemical and Solid-State Letters* 12:H38–H40.

64. Seol, K. S., Futami, T., Watanabe, T., Ohki, Y., and Takiyama, M. 1999. Effects of ion implantation and thermal annealing on the photoluminescence in amorphous silicon nitride. *Journal of Applied Physics* 85:6746–6750.

65. Tyschenko, I. E., Volodin, V. A., Rebohle, L., Voelskov, M., and Skorupa, V. 1999. Photoluminescence of Si_3N_4 films implanted with Ge^+ and Ar^+ ions. *Semiconductors* 33:523–528.

66. Cen, Z. H., Chen, T. P., Ding, L. et al. 2009. Strong violet and green-yellow electroluminescence from silicon nitride thin films multiply implanted with Si ions. *Applied Physics Letters* 94:041102(1-3).

67. Cen, Z. H., Chen, T. P., Ding, L. et al. 2009. Evolution of electroluminescence from multiple Si-implanted silicon nitride films with thermal annealing. *Journal of Applied Physics* 105:123101(1-5).

68. Cen, Z. H., Chen, T. P., Ding, L. et al. 2011. Influence of implantation dose on electroluminescence from Si-implanted silicon nitride thin films. *Applied Physics A: Materials Science and Processing* 104:239–245.

69. Cen, Z. H., Chen, T. P., Liu, Z. et al. 2010. Electrically tunable white-color electroluminescence from Si-implanted Silicon nitride thin film. *Optics Express* 18:20439–20444.

70. Cen, Z. H., Chen, T. P., Ding, L. et al. 2009. Quenching and reactivation of electroluminescence by charge trapping and detrapping in Si-implanted silicon nitride thin film. *IEEE Transactions on Electron Devices* 56:3212–3217.

71. Photopoulos, P. and Nassiopoulou, A. G. 2000. Room- and low-temperature voltage tunable electroluminescence from a single layer of silicon quantum dots in between two thin SiO_2 layers. *Applied Physics Letters* 77:1816–1818.

72. Delerue, C., Lannoo, M., Allan, G. et al. 1995 Auger and Coulomb charging effects in semiconductor nanocrystallites. *Physical Review Letters* 75:2228–2231.

73. Ma, L. B., Song, R., Miao, Y. M. et al. 2006. Blue-violet photoluminescence from amorphous Si-in-SiN_x thin films with external quantum efficiency in percentages. *Applied Physics Letters* 88:093102(1-3).

74. Mo, C. M., Zhang, L. D., Xie, C. Y., and Wang, T. 1993. Luminescence of nanometer-sized amorphous silicon nitride solids. *Journal of Applied Physics* 73:5185–5188.

75. Robertson, J. and Powell, M. J. 1984. Gap states in silicon nitride. *Applied Physics Letters* 44:415–417.
76. Chen, L. Y., Chen, W. H., and Hong, F. C. N. 2005. Visible electroluminescence from silicon nanocrystals embedded in amorphous silicon nitride matrix. *Applied Physics Letters* 86:193506(1-3).
77. Deshpande, S. V., Gulari, E., Brown, S. W., and Rand, S. C. 1995. Optical properties of silicon nitride films deposited by hot filament chemical vapor deposition. *Journal of Applied Physics* 77:6534–6541.
78. Hao, H. L., Wu, L. K., Shen, W. Z., and Dekkers, H. F. W. 2007. Origin of visible luminescence in hydrogenated amorphous silicon nitride. *Applied Physics Letters* 91:201922(1-3).
79. Pei, Z. W., Chang, Y. R., and Hwang, H. L. 2002. White electroluminescence from hydrogenated amorphous-SiN_x thin films. *Applied Physics Letters* 80:2839–2841.
80. Gritsenko, V. A., Zhuravlev, K. S., Milov, A. D. et al. 1999. Silicon dots/clusters in silicon nitride: Photoluminescence and electron spin resonance. *Thin Solid Films* 353:20–24.

8 Optical Properties of Semiconductor Nanoparticles in Photoelectrochemical Cells

M.H. Buraidah, S.N.F. Yusuf, I.M. Noor, and A.K. Arof

CONTENTS

8.1 INTRODUCTION

Photoelectrochemical cells are devices that can harness energy from the sun and do not need moving parts. These devices are environmentally friendly, noiseless, and nonpolluting. One of the components of a photoelectrochemical cell is a semiconductor layer. The dye-sensitized solar cell (DSSC) and the quantum dot–sensitized solar cell (QDSSC) are photoelectrochemical cells that convert energy from sunlight into electricity. In DSSCs and QDSSCs, the semiconducting layer is usually a metal oxide (MO). Apart from the MO semiconductor, which has its own role to play, metal chalcogenide semiconductors are used as sensitizers in QDSSCs. Examples of MO semiconductors used in DSSCs and QDSSCs include TiO_2 (Gratzel 2001), SnO_2 (Chappel and Zaban 2002), and ZnO (Zhang et al. 2009), whereas examples of sensitizers or quantum dots used in QDSSCs are CdS (Schaller and Klimov 2004), CdSe (Fuke et al. 2010), CdTe (Xiaoyan et al. 2014), and PbS (Lee et al. 2009, Ju et al. 2010) metal chalcolgenide semiconductors. Knowledge of the semiconductors is important especially for the design of photoelectrochemical cells with good performance. In this chapter, we focus on these materials and their application in DSSCs and QDSSCs.

8.2 STRUCTURE AND MECHANISM OF DSSC AND QDSSC

The mesoporous MO film is an important component of DSSC and QDSSC devices. This oxide film, which consists of an MO network, is coated over a dense blocking layer deposited on a transparent conducting oxide (TCO) substrate. The two layers are sintered to establish electronic conduction (Hagfeldt et al. 2010). Typically, the mesoporous layer thickness is ~20 μm (Arof et al. 2013, 2014, Aziz et al. 2014, Bandara et al. 2015a, Yusuf et al. 2014) and the nanoparticle is ~20 nm (Arof et al. 2013, 2014, Aziz et al. 2014, Bandara et al. 2015b, Yusuf et al. 2014). The TCO substrate can be made of glass or plastic. The most common TCO substrate is fluorine-doped tin oxide (FTO), although indium-doped tin oxide (ITO) substrate is also used. The charge-transfer dye is attached to the surface of the mesoporous film.

8.2.1 STRUCTURE AND MECHANISM OF A DSSC

Figure 8.1 depicts the DSSC structure. It consists of a transparent substrate that has been coated with FTO or ITO, an MO semiconductor layer, a dye, an electrolyte or ionic conductor, and a counter electrode (CE) coated on another substrate.

The mesoporous MO film has pores that will provide additional surface area for dye loading so that more of the incident light can be absorbed. The MO can be, but not limited to, TiO_2 (Kim et al. 2013), ZnO (Wang et al. 2013), SnO_2 (Lee et al. 2011), Nb_2O_5 (Ghosh et al. 2011), and $SrTiO_3$ (Yang et al. 2010) only. The mesoporous layer or film ensures direct contact of the dye molecule with the electrolyte, which also fills the pores in the film. The CE is usually a platinum-coated transparent substrate. Apart from platinum, binary and ternary metal chalcogenides such as $NiSe_2$ (Gong et al. 2013), $CoSe_2$ (Dong et al. 2015), MoS_2 (Kim et al. 2015, Wu et al. 2011a), CoS (Huo et al. 2015, Wang et al. 2009), CuS (Kim et al. 2015, Savariraj et al. 2014), Cu_2S (Feng et al. 2015), WS_2 (Wu et al. 2011b), and $NiCo_2S_4$ (Xiao et al. 2013); MOs, for example, V_2O_3 (Wu et al. 2012), ZnO (Wang et al. 2013), SnO_2 (Bu and Zheng 2015), and MoO_2 (Wu et al. 2012); metal carbides such as TiC (Wu et al. 2013), VC (Wu et al. 2012), WC (Vijayakumar et al. 2015), and MoC (Wu et al. 2011a); and metal nitrides such as MoN (Song et al. 2012), Mo_2N (Wu et al. 2011c), Nb_4N_5 (Cui et al. 2015), and TiN (Li et al. 2010) can also be used as CE. In addition, Au has also been reported as a CE (Lee and Lo 2009, Seol et al. 2010). An electrolyte is sandwiched between the two electrodes. The electrolyte can be solid, quasi-solid, or liquid containing a redox mediator. For liquid electrolytes, a thermoplastic film is placed in between the photoanode and the CE to act as a spacer so that the space created can be filled by the electrolyte. The film also serves as a sealant.

The working of a DSSC is also shown in Figure 8.1 and has been discussed by many researchers (Aziz et al. 2014, Bandara et al. 2014, Buraidah et al. 2011, Gratzel 2001, Hagfeldt et al. 2010, Le Bahers et al. 2013, Maçaira et al. 2014, Nazeeruddi et al. 2004, Noor et al. 2011, Peter 2011, Yusuf et al. 2014). A photocurrent is generated when the dye sensitizer absorbs the incident light. The MO semiconductor nanoparticles play an important role in allowing the light to pass through to the dye. The dye should strongly adhere to the mesoporous semiconductor. Thus, MO semiconductor nanoparticles with a large energy gap are needed for DSSC. This shows that the optical properties of the semiconducting MO nanoparticles are important for photoelectrochemical cells. Solar energy

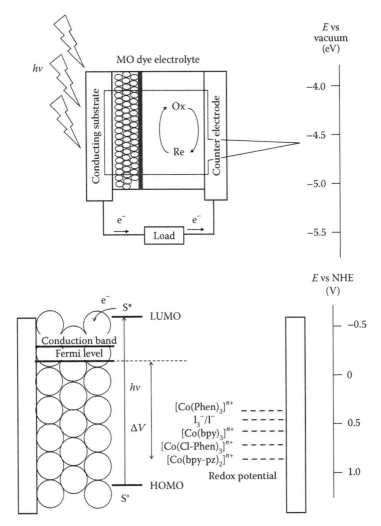

FIGURE 8.1 Operating principles of dye-sensitized solar cells.

absorbed by the dye should be more than or equal to the energy gap of the dye. This results in the excitation of the dye molecules. This excitation process is represented by Equation 8.1:

$$S + h\nu \rightarrow S^* \quad \text{light absorption by the dye} \tag{8.1}$$

According to Nazeeruddin et al. (2004), the dye to be used in DSSCs should have the following properties: it should harvest light in the visible spectrum (i.e., from 400 to 700 nm); it must have good adhesion to the semiconductor MO surface; and the O^- anions of the dye should be electrostatically bound to the H^+ on the MO surface (Figure 8.2).

The energy level of the dye's excited state must be above the semiconductor's conduction band (CB) edge. For rapid dye regeneration, the redox potential of the mediators must be sufficiently less positive (with reference to the E vs vacuum scale) than the potential of the dye in the ground state. The dye should be stable over a long period (i.e., between 20 and 25 years). The excited dye molecule (S^*) releases an electron into the semiconducting MO. The dye molecule is now in an oxidized state (S^+) (Equation 8.2):

$$S^* \rightarrow e^-_{(MO)} + S^+ \quad \text{electron injection} \tag{8.2}$$

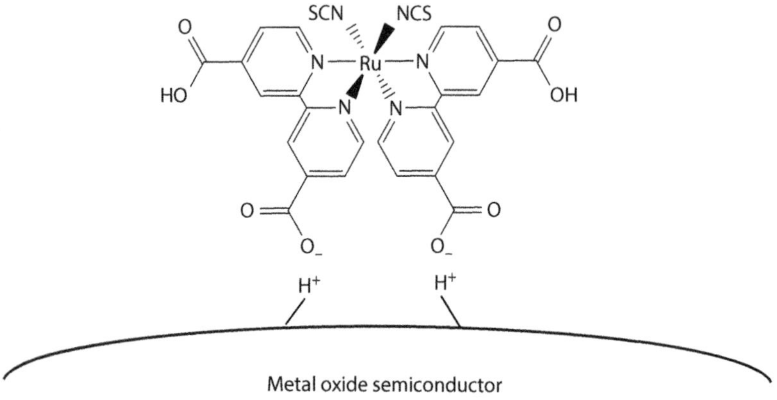

FIGURE 8.2 Attachment of the dye to the TiO$_2$ surface.

The transport of electrons to the TCO occurs through the dye/MO interface. This can be attributed to the coupling between the dye and the MO (Thomas et al. 2014), as shown in Figure 8.2. The electron pathway begins from the highest occupied molecular orbital (HOMO) of the dye, for example, the Ru-SCN moieties, to the lowest unoccupied molecular orbital (LUMO) at the carboxylated bipyridyl ligand into the CB of the MO semiconductor within a few femtoseconds (Gratzel 1991). The transferred electrons percolate through the interconnected MOs till they reach the TCO substrate. The electrons then pass through a load, and finally reach the CE. The circuit or pathway of the electrons is completed when they are returned to the holes in the dye molecules via the redox couple in the electrolyte. If the redox mediator is an I$^-$/I$_3^-$ couple, then it is the I$_3^-$ ion that accepts the electron from the CE and is reduced to the I$^-$ ion. The I$^-$ ions release the electrons to the holes in the dye. The reduction of I$_3^-$ is shown in Equation 8.3:

$$\frac{1}{2}I_3^- + e^-_{(CE)} \rightarrow \frac{3}{2}I^- \quad \text{overall charge transfer reaction} \tag{8.3}$$

Upon regeneration of the dye molecules, the I$^-$ ion is oxidized to I$_3^-$ ion. The process is shown in Equation 8.4. This process occurs continuously.

$$S^+ + \frac{3}{2}I^- \rightarrow S + \frac{1}{2}I_3^- \quad \text{dye regeneration} \tag{8.4}$$

8.2.2 STRUCTURE AND MECHANISM OF QDSSC

The structure and working principle of a QDSSC (Figure 8.3) are similar to those of the DSSC. The difference between QDSSC and DSSC is in their sensitizer (quantum dots and dyes, respectively) and the electrolyte component. For QDSSCs, instead of the dye, the mesoporous layer is covered by quantum dot (QD) semiconductors such as CdS (Schaller and Klimov 2004), CdSe (Fuke et al. 2010), CdTe (Xiaoyan et al. 2014), PbS (Lee et al. 2009), and PbSe (Schaller and Klimov 2004). A typical redox mediator is the I$^-$/I$_3^-$ couple and is used in DSSC. However, I$^-$/I$_3^-$ mediators can corrode the QDs. Hence, a polysulfide electrolyte consisting of S^{2-}/S$_x^{2-}$ redox couple (x = 2–5) is used in QDSSCs (Jun et al. 2013). The semiconductor nanocrystals mentioned above are able to absorb visible light from the solar spectrum. This is why these QDs can serve as sensitizers. The structure of a QDSSC is illustrated in Figure 8.3. The working principles of QDSSCs have been discussed by many researchers (Brown and Kamat 2008, Choi et al. 2014, Fuke et al. 2010, Jabbour and Doderer 2010, Jun et al. 2013, Kamat 2008, 2013, Kim et al. 2015, Lee and Lo 2009, Lee et al. 2008, 2009,

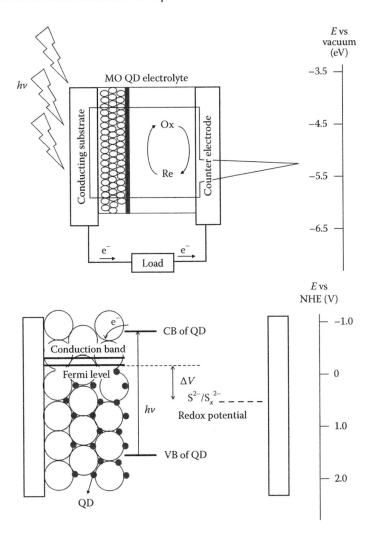

FIGURE 8.3   Structure and working of a QDSSC.

Parsi Benehkohal et al. 2012, Rhee et al. 2013, Salant et al. 2010, Santra and Kamat 2012, Santra et al. 2013, Savariraj et al. 2014, Tian and Cao 2013, Zhao et al. 2014).

Under illumination, photons will be absorbed by the QDs, and electron–hole (e–h) pairs (electrons in the CB and holes in the valence band (VB)) will be produced.

$$QD + photons \rightarrow QD(e+h) \tag{8.5}$$

The electrons in the CB of QD will then be injected into the CB of the MO semiconductor (Jun et al. 2013) such as TiO_2, ZnO, and SnO_2 (Figure 8.4).

$$QD(e+h) + MO \rightarrow QD(h) + MO(e) \tag{8.6}$$

The injected electrons then reach the transparent conducting glass through the semiconductor MO network and travel to the CE through the external circuit (Hassan et al. 2014). At the electrolyte/CE interface, S_x^{2-} is reduced by the electrons to S^{2-}, which then releases the electrons to the holes in the VB of the QDs (Jun et al. 2013) to complete the circuit, as shown in Equations 8.7 through 8.9 and Figure 8.4.

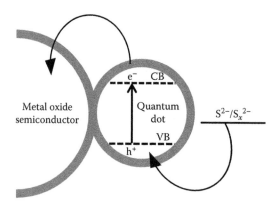

FIGURE 8.4   Electron injection in TiO_2 and CdS QD regeneration by the redox mediator S^{2-}/S_x^{2-}.

$$S_x^{2-} + 2e \rightarrow S_{x-1}^{2-} + S^{2-} \qquad (8.7)$$

$$CdX(h) + S^{2-} \rightarrow CdX + S \qquad (8.8)$$

$$S + S_{x-1}^{2-} \rightarrow S_x^{2-} \qquad (8.9)$$

The voltage produced depends on the difference between the Fermi level of the MO semiconductor and the redox potential of the polysulfide electrolyte.

The issues of concern in this chapter with respect to DSSCs and QDSSCs are (1) light harvesting and (2) electron transfer from the MO layer to the TCO and finally to the external circuit. Due to the size of the TiO_2 nanoparticles, which is ~20 nm in diameter, scattering of the incident light is negligible. Electron transfer through the network of nanoparticles in the mesoporous and blocking layers is quite slow and may lead to electron recombination to the electrolyte and/or dye. We will show in the following section what researchers have done to address these issues.

8.3   SEMICONDUCTING METAL OXIDES FOR DSSCs AND QDSSCs AND THEIR OPTICAL PROPERTIES

Many types of MOs have been used as the blocking layer between the TCO and the electrolyte. MOs with wide bandgaps have been widely used because these materials are able to control the transfer of electrons from the dye's excited state to the substrate (Krüger et al. 2011). These MOs include TiO_2, ZnO, and Nb_2O_5 (Sangiorgi et al. 2014). A blocking layer acts as a barrier at the interface between the conducting substrate and electrolyte in order to prevent electron recombination during the transport and collection process of electrons in DSSCs. Electron recombination that occurs at the conducting substrate/electrolyte interface is a factor responsible for the limitation of DSSC's performance.

8.3.1   TiO₂

There are three forms of TiO_2, namely, anatase, rutile, and brookite (Tang et al. 1994). Although these are isostructural, the open-circuit voltage of the DSSC using anatase TiO_2 is higher than that using rutile TiO_2. This is due to the smaller difference in energy between the CB and the redox potential in rutile TiO_2. The bandgap for rutile is 3.0 eV and for anatase it is 3.2 eV (Tang et al. 1994). As it is thermodynamically less stable, the brookite form of TiO_2 can be transformed into other phases at high temperatures (Koyama et al. 2006). According to Zallen and Moret (2006), the absorption spectrum of brookite extends throughout the visible region with a broad and gradual edge. These authors have reported that the lowest bandgap for brookite should be at least 3.54 eV. Among the MOs shown in Figure 8.5, TiO_2 has the highest refractive index in the wavelength region from 430 to 1530 nm (Bond 1965, Devore 1951, Dodge 1986, Gao et al. 2012, Pan et al. 2008). The anatase form of TiO_2 is usually used in DSSCs and QDSSCs.

Table 8.1 lists the refractive index of TiO_2 films versus wavelength at several oxidation temperatures (Ting et al. 2000). The values of the temperature and refractive index are estimated from the original results. It can be seen that the refractive index increases with the oxidation temperature for all wavelengths in the visible region. However, as the wavelength increases, the refractive index decreases. This is true at all temperatures.

The absorption coefficient of TiO_2 films does not show much change in the visible range. However, the absorption coefficient, α (cm^{-1}), increases abruptly as the wavelength moves toward the UV region (Devore 1951). In the visible region, the absorption coefficient shows noticeable changes, and at 400 nm at 900°C, the absorption is 2.53×10^4 cm^{-1}. As the temperature increases from 700°C to 900°C, the energy gap decreases by 0.80 eV (Ting et al. 2000).

TiO_2 is said to be the most effective electrolyte blocker among the many MOs tried. It blocks electron recombination to the electrolyte (Barea and Bisquert 2013). The blocking layer can be prepared by grinding TiO_2 powder (particle size ~15 nm) with 0.1 M nitric acid, spin-coating the

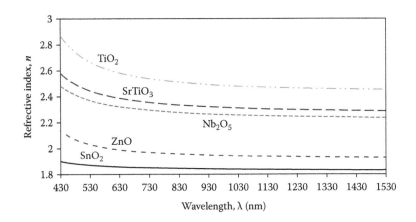

FIGURE 8.5 Refractive index for TiO_2, ZnO, $SrTiO_3$, Nb_2O_5, and SnO_2 metal oxide semiconductor. (From Gao, L. et al., *Opt. Express,* 20(14), 15734, 2012; Pan, S.S. et al., *J. Appl. Phys.*, 103(9), 093103, 2008; Devore, J.R., *J. Opt. Soc. Am.,* 41(6), 416, 1951; Bond, W.L., *J. Appl. Phys.*, 36(5), 1674, 1965; Dodge, M.J., Refractive index, in M.J. Weber, ed., *Handbook of Laser Science and Technology*, CRC Press, Boca Raton, FL, 1986.)

TABLE 8.1

Effect of Oxidation Temperature on Refractive Index of the TiO_2 Films in Terms of Wavelength

Wavelength (nm) → Temperature (°C) ↓	400	500	600	700	800
700	2.70	2.61	2.55	2.51	2.47
800	2.76	2.64	2.57	2.53	2.49
900	2.77	2.68	2.62	2.58	2.57

Source: Devore, J.R., *J. Opt. Soc. Am.*, 41(6), 416, 1951.

mixture on the conducting glass substrate, and sintering at ~450°C for 30 min (Arof et al. 2014, Aziz et al. 2013, Bandara et al. 2013, Hassan et al. 2014, Yusuf et al. 2014). The blocking layer is dense. The dense blocking layer generates effective electron pathways to the FTO and finally to the external circuit. The blocking layer should not exceed a certain thickness because this will lead to the presence of more trap states in the thicker layer, which can hinder the passage of electrons to the external circuit (Berger et al. 2007, Choi et al. 2012). According to Seo et al. (2011), the performance of the DSSC with the blocking layer is 30% higher than that of the cell without the blocking layer. A mesoporous TiO_2 film is then applied over the blocking layer, which can be prepared by grinding TiO_2 powder (particle size ~20 nm) with 0.1 M nitric acid, a low molecular weight polymer, and a few drops of a surfactant (Arof et al. 2014, Aziz et al. 2014, Bandara et al. 2013, Hassan et al. 2014, Yusuf et al. 2014). Using the doctor-blade technique, the TiO_2 layer is then deposited on the blocking layer and heated at ~450°C for 30 min. The evaporation of the low molecular weight polymer in the composition produces mesopores in the MO layer, which is necessary for sufficient dye loading. The MO layer in Figures 8.1 and 8.3 comprises a blocking layer and a mesoporous layer. This layout was reported in various works (Arof et al. 2013, 2014, Aziz et al. 2013, Bandara et al. 2012, 2013, 2014, 2015a,b, Yusuf et al. 2014).

According to Lee et al. (2012), since the size of the TiO_2 crystallites in the mesoporous layer is ~20 nm, light cannot be scattered, which results in photons not being totally absorbed by the dye-sensitized photoanode. The photons will be either lost through the counter electrode or partially absorbed by the electrolyte solution. Hence part of the light is wasted. In order to optimize harnessing of the incoming light, the escaping photons must be collected. To achieve this, a highly diffusive reflecting layer consisting of submicrometer-sized TiO_2 spheres is applied over the mesoporous film. The scattering intensity depends on the size and refractive index of the particles. High-refractive-index, submicrometer TiO_2 particles of size between 0.3 and 1 μm can be used as efficient scatters.

In an effort to enhance DSSC performance, the mesoporous TiO_2 electrodes have been coated with MOs and used in DSSCs (Diamant et al. 2004). These MOs have wide a bandgap, and include Nb_2O_5, ZnO, $SrTiO_3$, ZrO_2, Al_2O_3, SnO_2, and V_2O_5, which act as the shell materials. The Nb_2O_5 coating delays or prevents electron recombination by forming a surface energy barrier between the transparent conducting oxide and the electrolyte. The other shell materials, namely, ZnO, $SrTiO_3$, ZrO_2, Al_2O_3, and SnO_2, shift the conduction band of the TiO_2 by forming a dipole layer at the core–shell interface. For example (Diamant et al. 2004), surface dipole generation at the interface of the $TiO_2/SrTiO_3$ core/shell shifts the CB to a more negative level. This is due to quantum confinement. The surface dipole is created by the difference in electron affinity or isolectic point of the $SrTiO_3$ and TiO_2 semiconductors. Utilizing the core/shell structure in DSSCs, the open-circuit voltage (V_{OC}) increased but the short-circuit current (J_{SC}) decreased. However, the overall conversion efficiency increased by ~15%. Elbohy et al. (2015) proved that DSSCs with mesoporous TiO_2 coated with V_2O_5 exhibited enhanced efficiency by ~10%. This indicates that V_2O_5 functions effectively as a blocking layer at the TiO_2/electrolyte interface, thus preventing reactions.

8.3.2 SnO_2

SnO_2 is another MO semiconductor that has been used as a blocking or compact layer. However, among the many MOs presented in Figure 8.5, SnO_2 has the lowest refractive index between 1.8 and 2 in the wavelength range 430–530 nm. The bandgap of SnO_2 is 3.6 eV (Dou et al. 2011, Okuya et al. 2001, Ramasamy and Lee 2011). SnO_2 is transparent, chemically stable, and has high electron mobility between 100 and 200 cm²/V s (Dou et al. 2011).

Yong et al. (2014) introduced an ultrathin SnO_2 blocking layer in DSSCs in the effort to improve the cell's performance. The SnO_2 blocking layer was deposited on FTO-coated glass before TiO_2 was pasted over it using the doctor-blade method. The SnO_2 blocking layer reduced the recombination of charges at the FTO/electrolyte interface. The cascading band structure of TiO_2 and SnO_2, shown in Figure 8.6, increased the electron lifetime, resulting in an effective charge collection.

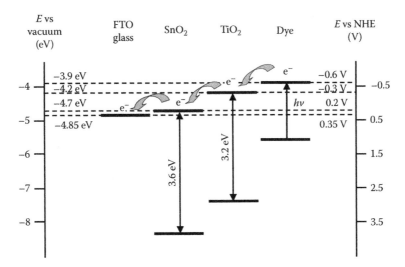

FIGURE 8.6 Schematic of the cascading band structure formed by introducing an SnO_2 blocking layer between FTO glass and TiO_2 photoelectrode. (From Yong, S.-M. et al., *Thin Solid Films*, 556(0), 503, 2014.)

The bottom of the CB and the top of the VB of TiO_2 were both higher than those of SnO_2, enabling the photogenerated electrons to easily migrate from TiO_2 into SnO_2, with fewer photogenerated electrons undergoing recombination.

Duong et al. (2013) have also employed SnO_2 thin films to reduce the number of electrons entering the electrolyte and recombining with the mediator. The authors observed that the electron lifetime is longer in samples with SnO_2 blocking layers. The performance of the DSSC with the SnO_2 blocking layer improved as compared to the DSSC without the SnO_2 blocking layer.

8.3.3 ZnO

ZnO is another wide bandgap semiconductor used in DSSCs (Qiu et al. 2011, Wong et al. 2012). Among the MOs shown in Figure 8.5, only ZnO has a refractive index larger than that of SnO_2. Since the refractive index $n > 1$, the refractive angle is always smaller than the incident angle and the refracted light is closer to the normal, implying that the incident light will mostly be directed into the solar cell.

Al-Kahlout (2015) has used ZnO nanoparticles in the photoanode of the DSSC. The ZnO nanopowder was wetted to make it into a paste. The ZnO paste was deposited on a conducting substrate using the doctor-blade technique to form several films. The films were dried at 100°C for 1 h, followed by sintering in air at temperatures ranging from 200°C to 500°C for a specified length of time. DSSCs using the ZnO films exhibited different performances, as shown in Table 8.2.

It can be seen that the DSSC using ZnO layers sintered at 400°C exhibited the best cell performance with an efficiency of 3.01% and a high photocurrent density of 15.6 mA/cm². The improved performance may be attributed to increased light harvesting and better charge transport channel with less charge recombination at FTO/ZnO/electrolyte interface. These results indicate that the overall performance of the DSSC can be optimized by sintering the ZnO film at a suitable temperature.

8.3.4 Nb$_2$O$_5$

Xia et al. (2007) have used Nb_2O_5 as the blocking layer or a compact film between the FTO glass and mesoporous MO film for application in DSSC. The use of Nb_2O_5 as blocking layer improved the fill factor (FF) and open-circuit voltage (V_{OC}). The probable reason for the improvement is the

TABLE 8.2

ZnO-DSSC Performance with Photoelectrodes Annealed at Different Temperatures

Temperature (°C)	V_{OC} (V)	J_{SC} (mA/cm²)	FF (%)	Efficiency, η (%)
200	0.545	4.42	44.42	1.07
300	0.544	12.51	29.68	2.02
350	0.550	12.87	33.62	2.38
400	0.552	15.66	34.82	3.01
450	0.554	13.38	30.49	2.26
500	0.573	8.64	29.69	1.47

formation of a barrier at the interface between the FTO conducting oxide and the mesoporous TiO$_2$ layer, which effectively suppresses electron leakage (Figure 8.7). This results in the increase of V_{OC}. From the figure, it is seen that the Nb$_2$O$_5$ blocking layer potential is higher than the CB edge of TiO$_2$. This will impose difficulty to the electrons and prevent the charge carriers from crossing over to the electrolyte. However, electron injection can take place via electron tunneling because of the nanometer thickness of the Nb$_2$O$_5$ compact layer.

Sacco et al. (2015) have also shown that a Nb$_2$O$_5$ blocking layer can increase the photovoltaic efficiency by reducing electron leakage or recombination at substrate/MO semiconductor interface. Nb$_2$O$_5$ has a permittivity of ~53. Its bandgap is 3.3 eV (Hashemzadeh et al. 2014, Hu and Liu 2015). Since it is a good insulator, it is expected to act as a barrier to block back electron reaction from the TCO to the redox couple without lowering the mobility of the injected electrons. This makes Nb$_2$O$_5$ very suitable as a blocking layer. DSSC with a Nb$_2$O$_5$ blocking layer has shown a short-circuit current density of 124 A/m², which is more than 40% increase compared to that of the cell without Nb$_2$O$_5$ blocking layer (Sacco et al. 2015).

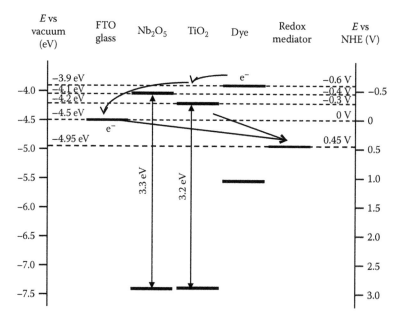

FIGURE 8.7 Schematic view of the electron transfer of the new structured electrode. (From Xia, J. et al., *J. Photochem. Photobiol. A: Chem.*, 188(1), 120, 2007.)

8.3.5 SrTiO₃

SrTiO$_3$ has a bandgap of ~3.2 eV (Burnside et al. 1999). Its CB is ~0.2 eV higher than the CB of the anatase form of TiO$_2$. SrTiO$_3$, which has structural similarities with anatase TiO$_2$, is expected to produce good photovoltage, thereby making it suitable for the development of DSSCs. According to Jayabal et al. (2014), porous SrTiO$_3$ can be produced via hydrothermal interaction of strontium acetate and titanium isopropoxide. SrTiO$_3$ is a perovskite material with a cubic structure. Although SrTiO$_3$ and TiO$_2$ are electronically isostructural, the flat-band potential of SrTiO$_3$ is larger than that of anatase TiO$_2$. SrTiO$_3$ absorbs more at wavelengths less than 400 nm and shows a maximum at 350 nm. SrTiO$_3$ has high refractive index, below that of TiO$_2$ as shown in Figure 8.5. Using the synthesized SrTiO$_3$ and organic Eosin yellow dye as a sensitizer, a DSSC has been fabricated. The cell showed a V_{OC} of 730 mV. The value of J_{SC}, FF, and η were 44 A/m^2, 55%, and 0.51%, respectively.

8.4 OPTICAL PROPERTIES OF METAL CHALCOLGENIDES FOR QUANTUM DOTS (QDs)

QDs of different sizes can emit light of different wavelengths or colors although made up of the same material. This is attributed to quantum confinement. Large QDs exhibit redder fluorescence as their size increases and bluer as their size decreases. Thus, larger QDs emit light of lower energy compared to smaller QDs. The coloration produced is directly related to the bandgap energy. This is because, as the particle size decreases (and may become too small to be comparable to the electron or exciton wavelength), the decreased confining dimension makes the energy levels discrete and this widens the bandgap, resulting in an increase in the bandgap energy. Larger QDs have more closely spaced energy levels, allowing the QDs to absorb low-energy photons. The closely spaced energy levels of large QDs can also trap the excitons or electron–hole quasiparticles, and therefore electron–hole pairs in larger QDs have longer life. QDs can increase the efficiency of QDSSCs through multiple exciton generation (Schaller and Klimov 2004), as shown in Figure 8.8.

8.4.1 LEAD CHALCOGENIDE QUANTUM DOTS

As we already mentioned, the optical properties of QDs are dependent on their size. For a QD of size d, its molar extinction coefficient ε increases according to $d^{1.3}$ (Moreels et al. 2009). For PbS QDs, the exciton lifetime τ is between 1 and 1.18 μs. Bulk lead chalcogenide semiconductors such as PbS,

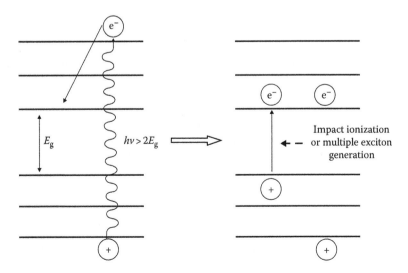

FIGURE 8.8 Schematic diagram of multiple exciton generation (MEG) or impact ionization.

PbSe, and PbTe have energy bandgaps of 0.41, 0.28, and 0.31 eV, respectively, and are suitable as sensitizers for QDSSCs (Onicha et al. 2012). The exciton Bohr radius of PbS, PbSe, and PbTe is 18, 46, and 150 nm, respectively (Fu and Tsang 2012, Zhao et al. 2014). The energy bandgaps of PbS, PbSe, and PbTe QDs can be tailored to be in the range 0.9–1.1, 0.7–1.7, and 0.6–1.1 eV, respectively (Murphy et al. 2006, Rhee et al. 2013). Therefore, for PbS, PbSe, and PbTe, the optical absorption edge can be extended to ~1300, ~1500, and ~2000 nm, respectively, which is in the infrared region. As mentioned earlier, QDs can produce multiple excitons from one high-energy photon through impact ionization. The threshold photon energy hv_{th} for impact ionization or multiple exciton generation (MEG) to occur is given by

$$hv_{th} = E_g \left(2 + \frac{m_e^*}{m_h^*} \right) \tag{8.10}$$

where

m_e^* is the effective electron mass

m_h^* is the effective mass of the hole

The impact ionization threshold for PbSe QD was observed to be close to $3E_g$, whereas the threshold of MEG was found to be ~$2.5E_g$ and ~$2E_g$ for PbS and PbTe, respectively (Hardman et al. 2011, Murphy et al. 2006, Schaller and Klimov 2004). As an example, when PbS QDs with the energy bandgap of 1.1 eV is irradiated with light, energy from wavelength ~1300 nm to the UV region will be absorbed. Each photon absorbed by the PbS QD produces one exciton until the threshold wavelength ~451 nm is reached, when more than one exciton will be produced. The threshold wavelength of 451 nm is based on the PbS bandgap of 1.1 eV. Since the energy bandgap of PbS can be tailored from 0.9 to 1.1 eV, the smaller tailored bandgap MEG will occur at the longer wavelength and more excitons or electron–hole quasiparticles will be produced. The increase in the number of such exciton in the visible region will lead to an increase in the electron injection rate, current density, and, finally, the solar conversion efficiency.

These also apply to PbSe and PbTe QDs, and therefore lead chalcogenide can help enhance the performance of solar cells.

Figure 8.9 shows the dependence of the energy gap and CB edge on the size of the PbS QDs. The figure suggests that the electrons in the CB of bulk PbS cannot easily cross over into the CB of TiO_2 due to the CB edge of bulk PbS being at a lower level than that of TiO_2. The CB edge of bulk PbS is at −4.7 eV. The CB edge of bulk TiO_2 is 0.5 eV above that of PbS (Tian and Cao 2013). To enable electrons to jump into the CB of TiO_2, the energy gap has to be increased, and this can be done by decreasing the size of PbS. From Figure 8.9, it can be seen that electron injection will increase with decreasing PbS size in the PbS-TiO_2 photoelectrode.

Doping in QDs with transition metals such as Cu^{2+} and Mn^{2+}, which are optically active, creates electronic states in the bandgap of the QDs, enabling the tuning of their properties. This can alter the dynamics of charge separation and recombination. The CB of the PbS QDs was found to be shifted upward with Hg^{2+} doping. This can lead to higher J_{SC}. The efficiency of 2.01% and J_{SC} of 210 A/m² have been achieved when using a Cu-doped PbS/CdS photoelectrode (Rhee et al. 2013).

There are several factors limiting the use of lead chalcogenides in DSSCs. The structure of lead chalcogenide QDs is unstable at temperatures above 100°C. The optical properties of PbSe QDs are also sensitive to how they are dispersed in a medium. There are also other factors that can reduce the quantum yield and lead to band shift in the photoluminescence (PL) spectrum (Zhao et al. 2014). By coating the MO semiconductor with suitable QDs to form a core/shell structure, stability can be

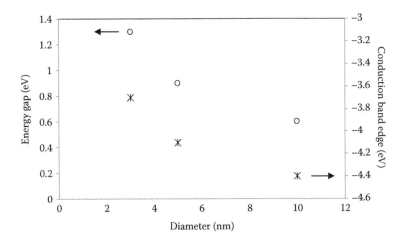

FIGURE 8.9 Energy gap and conduction band edge of PbS QDs with different diameters. (From Tian, J. and Cao, G., *Nano Rev.*, 4, 8, 2013.)

improved and maximization of fluorescence of the QD core can be achieved. The core can be protected by the shell against oxidation by acting as a barrier and minimize surface defects by providing better passivation. At the same time, thermal and photostability of QDs can also be improved.

8.4.2 CADMIUM CHALCOGENIDE QUANTUM DOTS

Over the years, cadmium chalcogenide QDs have served as sensitizers in photoelectrochemical cells to enhance device performance and stability. The sensitizers include CdS, CdSe, and CdTe. The optical bandgap of CdS of 5 nm size is 2.3 eV, enabling light absorption until ~540 nm (Rhee et al. 2013). For CdSe with 3 nm size, the bandgap is 1.7 eV; and thus light can be absorbed up to ~731 nm wavelength. CdTe with 3.8 nm size has a bandgap of 1.45 eV, leading to light absorption at the wavelengths up to ~887 nm. Fluorescent intensity and Stokes shift are known to increase and decrease, respectively, with temperature when CdS QDs are embedded in a silica matrix (Reda 2008). These results indicate that low temperature is favored for the preparation high-efficiency QDs. The size of CdS-doped silica crystallites increases with the annealing temperature from 373 to 673 K. Absorption and emission maxima are also red-shifted. Photostability studies have shown that QDs prepared at low temperatures have high fluorescent intensity and Stokes shift, indicating that the QDs are efficient for solar cells. In another study, Manna et al. (2002) showed that the size of CdS and CdS/ZnS semiconducting QDs is independent of the growth time. CdS/ZnS core–shell QDs exhibit more intense PL than that of CdS QDs without the ZnS shell. According to Lee and Lo (2009), the efficiency of QDSSCs using CdS and CdSe is 1.15% and 1.24%, respectively. QDSSCs with CdSe exhibited a higher efficiency due to the smaller energy gap of CdSe compared to CdS. However, since the CB edge of CdSe is almost at the same level as that of TiO$_2$ (−4.2 eV), electron injection occurs less forcefully or at a slower rate. The CB edge position of CdS is higher than that of both TiO$_2$ and CdSe (−3.9 eV). Thus, by co-sensitizing CdS/CdSe QDs on a TiO$_2$ mesoporous layer, adjustment of the Fermi level resulted in the increase in injection rate from CdSe to TiO$_2$, as proven by the efficiency increase to 2.9%. Figure 8.10 shows the cascadal potential of TiO$_2$, CdS, and CdSe after the Fermi level adjustment.

The efficiency of QDSSCs using Mn-doped CdS/CdSe electrode was observed to increase up to 5.4% (Kamat 2013). The midgap states created by Mn doping prevent electron leakage so that electron–hole and electron–oxidized polysulfide electrolyte recombination could not occur (Santra and Kamat 2012).

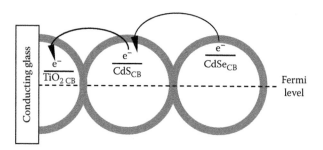

FIGURE 8.10 Possible band edge structure after the adjustment of Fermi level for TiO$_2$/CdS/CdSe electrode. (From Lee, Y.-L. and Lo, Y.-S., *Advanced Functional Materials*, 19, 604, 2009.)

8.4.3 ZINC CHALCOGENIDE QUANTUM DOTS

Among the II–VI semiconductor QDs, ZnS, ZnSe, and ZnTe are the safe choices compared to toxic cadmium and lead. However, the larger energy bandgap may limit their performance in QDSSCs. The optical bandgap for ZnS, ZnSe, and ZnTe is 3.6, 2.7, and 2.4 eV, respectively. Zinc chalcogenide, particularly ZnS, has been extensively studied as a surface passivation layer for photoelectrodes in QDSSCs. ZnS can protect QD materials from photocorrosion. In addition, ZnS can prevent electrons recombining at the electrode/electrolyte interface. An increase in QDSSC's efficiency was observed from 1.5% using TiO$_2$/CdSe electrode to 1.9% using TiO$_2$/CdSe/ZnS electrode. A similar observation was made by Lee and Lo (2009) when ZnS was coated on TiO$_2$/CdS/CdSe photoelectrode. The efficiency increased from 2.9% to 3.7%. A schematic diagram of surface passivation layer is shown in Figure 8.11.

8.4.4 CADMIUM SELENIUM TELLURIDE

Alloyed semiconducting QDs (cadmium selenium telluride) have been prepared to tailor the optical properties of individual binary QDs and maintain the particle size (Bailey and Nie 2003). The absorption/emission energies and composition of these alloyed semiconducting QDs have a nonlinear relationship. This leads to new properties different from those of the parent binary systems. This new QDs opens more avenues for bandgap tailoring.

Figure 8.12 shows the bandgaps and energy levels of some semiconductors (metal oxides, CdS, and CdSe) used in DSSCs and QDSSCs.

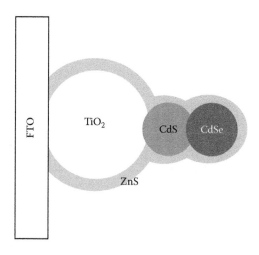

FIGURE 8.11 Surface passivation of ZnS on TiO$_2$/CdS/CdSe photoelectrode.

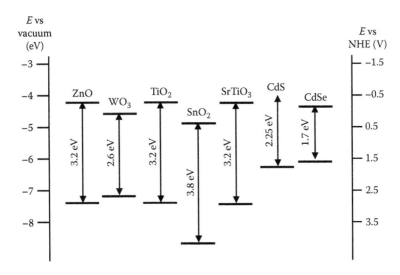

FIGURE 8.12   Energy diagram for some metal oxides, CdS, and CdSe.

8.5   NANOSTRUCTURES

In the attempt to address the issues concerning DSSCs and QDSSCs, for example, preventing electron recombination at the interface by improving the electron transport from the MO nanoparticle
network to the TCO, nanostructures have been synthesized. Structures with at least one dimension
less than 100 nm are called "nanostructures." Nanostructures can be classified into zero-dimensional
(0D nanoparticle), one-dimensional (1D, nanotube or nanowire), two-dimensional (2D, nanofilm),
and three-dimensional (3D) spheres and helix structures.

8.5.1   1D NANOSTRUCTURE

One of the strategies to address the transport-limiting process is to use 1D nanostructures (Besra
and Liu 2007). Nanowires (Bendall et al. 2011, Fan et al. 2013a), nanorodes (Jeng et al. 2013, Lai
et al. 2010, 2011, Shaikh et al. 2013, Wang et al. 2015), and nanotubes (Dembele et al. 2013,
Jingbin et al. 2010, Pugliese et al. 2014, Song et al. 2014) have been prepared. According to Lin
et al. (2003), 1D nanostructures can be composed from various MOs such as TiO_2, ZnO, or SnO.
Diffusion coefficients of these materials are higher than those of unordered nanostructures. The
electron diffusion lengths of these films are also longer than their thickness. These nanostructures
can serve as photoelectrodes in DSSCs and QDSSCs. The bandgap of TiO_2 nanowires embedded in
anodic alumina membranes annealed at 500°C is ~3.35 eV. Because of the quantum size effect, the
optical absorption band edge of these nanowire arrays shows a blue shift compared to that of bulk
anatase TiO_2, with a bandgap of 3.2 eV.

Using ZnO in the form of nanowires and the cobalt complex $[Co(bpy)_3]^{2+/3+}$ as redox mediators,
the fabricated DSSCs exhibited a V_{OC} of ~0.2 V higher than that of the DSSC with I^-/I_3^- redox couple
(Fan et al. 2013b), see Figure 8.13. Barpuzary et al. (2014) investigated the performance of a DSSC
consisting of 1D ZnO nanowire, donor−π−acceptor type carbazole dye, or more specifically 2-cyano-
3-(4-(2-(9-p-tolyl-9H-fluoren-6-yl)vinyl)phenyl)acrylic acid, and cobalt tris(2,2′-bipyridyl) redox
mediator. A solar conversion efficiency of 5.7% with J_{SC} of 122 A/m², V_{OC} of 0.72 V, and FF of 0.65
were achieved. The higher V_{OC} with the cobalt complex mediator is due to the more positive redox
potential (E vs NHE scale) of the cobalt complex compared to that of I^-/I_3^-, as shown in Figure 8.13.

Although DSSCs with these 1D nanostructures showed improved electron transport properties
and assist electrolyte diffusion throughout the photoelectrode, the performance of DSSCs using

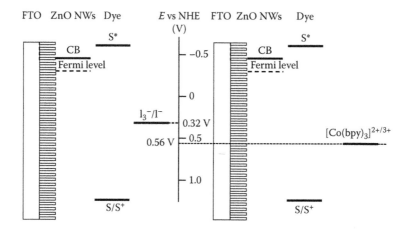

FIGURE 8.13 Schematic energy diagram of DSSC using ZnO nanowire with I^-/I_3^- or cobalt complex redox couples. (From Fan et al., *ACS Applied Materials & Interfaces*, 5, 1902, 2013b.)

TiO$_2$ nanoparticles is still better. This is because of the low surface area of the 1D nanostructures, which causes less absorption of the dye and the QDs. In other words, the light harvesting issue is still not addressed. Hence, although vertically aligned nanostructures do help to address the transport-limiting problems and assist the electrolyte diffusion throughout the photoelectrode, they still could not exhibit high device performance (Lee et al. 2014a). Therefore, further efforts have to be made to improve the performance of solar cells by using 2D and 3D nanostructures.

8.5.2 2D NANOSTRUCTURE

A 2D semiconductor nanostructure is a type of natural semiconductor with thicknesses on the atomic scale. An example is the 2D semiconducting graphene. This material is composed of one layer of C atoms arranged in a honeycomb lattice. Because of their efficient stacking, the 2D nanostructures can provide good light reflecting ability and good electron diffusion paths (Chen et al. 2012). The organized 2D nanostructure should be a suitable candidate for scattering layers in DSSCs.

As we now know, although nanoparticles can provide high surface area for dye and QD adsorption, such nanoparticle-based cells still exhibit poor particle–particle interconnectivity, leading to high electron leakage and low fill factor. Hence the nanoparticle networks should be replaced by nanostructures with improved interconnectivity, better electron pathways, and good light harvesting capability. Xu et al. (2014) have synthesized porous SnO$_2$ nanosheets with surface area ~5 times that of a single-crystalline SnO$_2$ nanosheet, which led to improved conversion efficiency. The efficiency enhancement could be attributed to the porous architecture of the 2D nanostructure, which provides an efficient electron pathway and good light scattering ability.

8.5.3 3D NANOSTRUCTURE

According to Maçaira et al. (2013), 3D photoelectrode nanostructure should have a large surface area and large interconnected pores for efficient electrolyte diffusion. Apart from that, hindrance to electron transport by defect levels, particle–particle boundaries, and recombination losses should be minimized. Lee et al. (2014b) fabricated a CdSe QDSSC with a 3D TiO$_2$ nanohelix to address the light scattering issue. The absorbance of nanohelix array was greater than that of TiO$_2$ nanoparticles in the reference DSSC. This work has shown that the 3D TiO$_2$ nanohelix array has improved light harnessing, leading to increased absorbance. The electron transport time also decreased.

8.5.4 1D–2D–3D Nanostructure

Kim and Yong (2013) have developed 1D ZnO nanowires and 3D ZnO nanostructures. They compared the diffuse reflectance spectra of a ZnO sputtered film, a 1D ZnO nanorod, and a 3D nanosheet-branched ZnO nanorods. The 3D nanosheet-branched ZnO nanorods demonstrated good light trapping or capturing properties and efficient light scattering capacity in the wavelength region 400–750 nm. Table 8.3 shows the comparison of the characteristics between the QDSSCs with nanosheet-branched ZnO nanorods and with ZnO nanorods.

Feng et al. (2015) have developed TiO_2 nanowires (TNWs), TiO_2 nanowire-TiO_2 nanosheet (TNW-TNS) arrays, and hyperbranched TiO_2 nanowire-TiO_2 nanorod-ZnO nanorod heterostructured array (TNW-TNS-ZNR). Light scattering increased in the order TNW < TNW-TNS < TNW-TNS-ZNR in the wavelength range 380–800 nm. The high light scattering intensity for the 3D heterostructure will be an advantage for photocurrent enhancement, which will increase the solar conversion efficiency. Impedance characteristic studies revealed that the Nyquist plot of the nanostructures consists of two depressed semicircles, which can be represented by an equivalent circuit as shown in Figure 8.14.

TABLE 8.3

Characteristics of QDSSC with 1D and 3D Nanostructures

ZnO Nanostructure	J_{SC} (mA/cm²)	V_{OC} (mV)	Fill Factor, FF (%)	Efficiency, η (%)
QDSSCs with nanosheet-branched ZnO nanorods	14.2	631	48	4.4
QDSSCs with ZnO nanorods	12.3	643	45	3.5

Source: Feng, H.-L. et al., *J. Mater. Chem. A*, 3, 7, 2015.

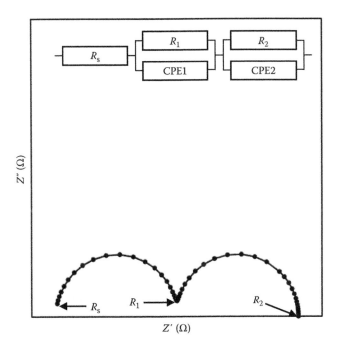

FIGURE 8.14 Nyquist plot of TiO_2 nanowire.

TABLE 8.4
R_2 **and** τ **Values**

Nanostructure	R_2 (Ω)	τ (ms)
1D	52.11	66
1D–2D	94.13	88
3D	103.9	99

Source: Feng, H.-L. et al., *J. Mater. Chem. A*, 3, 7, 2015.

In Figure 8.14, R_2 represents electron-recombination resistance at the semiconductor array/QD/electrolyte interface. The electrolyte lifetime (τ) within the cells is obtained as $2\pi/\omega$, where ω is the minimum frequency at the low-frequency side of each semicircle. Table 8.4 shows the R_2 and τ values for the DSSCs using the 1D, 1D–2D, and 3D nanostructures.

The increase in the lifetime indicates the effect of charge extraction and collection due to the branch in the nanostructure.

8.6 SEMICONDUCTOR ASSEMBLY PROCESS

8.6.1 SELF-ASSEMBLY TECHNIQUE

Self-assembly is a technique to form organized structures from a disordered system of pre-existing components without external direction. In DSSCs, self-assembly forms a compact TiO_2 under-layer film to enhance optical transmission and electron extraction properties, which could lead to improved DSSC performance. Self-assembly of TiO_2 layer results in good contact with TCO glass, increased optical transmission, and reduced charge recombination. In the self-assembly technique, the MO solution is poured on the desired substrate, and the solution is allowed to spread over the substrate (Xie et al. 2014). In the self-assembly process, the MO solution extends outward for the MO nanocrystals to gradually pack as a film. By varying the concentration and coating volume of nanocrystals forming the film, the thickness can be controlled. According to Sun et al. (2012), self-assembled TiO_2 layers when used in DSSCs exhibit better efficiency compared to DSSCs without self-assembled TiO_2.

8.6.2 LAYER-BY-LAYER TECHNIQUE

The layer-by-layer (LbL) technique is used for preparing thin films. This can be accomplished by solution immersion, spinning, or spraying. LbL is simple and can be inexpensive. Film thickness can be controlled to as fine as 1 nm resolution.

The LbL self-assembly technique has been used to deposit SnO_2 layers on an FTO substrate (Kim et al. 2012). The DSSC exhibited an increase in J_{SC} from 89.6 to 109.7 A/m². The efficiency increased from 5.43% to 6.57%. The enhancement can be attributed to the improved TiO_2-FTO adhesion by the ultrathin SnO_2 layer. The ultrathin SnO_2 layer also reduced the electron percolation time in the TiO_2 layer. The observed photovoltaic properties were attributed to the unique CB location of the LbL-assembled SnO_2, which is higher than the CB edge of the FTO and lower than that of TiO_2. The bandgap of the FTO and nano SnO_2-based film formed via the LbL-assembly technique was calculated to be 3.92 eV and that of SnO_2 was 4.23 eV (Kim et al. 2012). Hence, there is good band-match so that electrons can be transported from the dye to TiO_2, SnO_2, and the FTO collecting substrate, as shown in Figure 8.6, for the electrons to exit to the external circuit and do work. This band alignment formed upon introducing a SnO_2 ultrathin interfacial layer increases the electron diffusion from TiO_2 to the TCO, thus shortening the transit time of the electrons in TiO_2.

8.6.3 ELECTROPHORETIC DEPOSITION TECHNIQUE

Electrophoretic deposition (EPD) is a method of coating an electrode with a layer of solid particles. In the initial steps of EPD, the solid particles to be deposited are dispersed in an ion-containing liquid. Ions in the solution are selectively adsorbed on the surface of the solid particles. Hence, the surface of the solid particles will be charged (Santhanagopalan et al. 2010). The charged solid particles will move toward the corresponding electrodes and will be deposited when a voltage is applied across the electrodes. Santhanagopalan et al. (2010) have deposited 1D carbon nanostructures on TCOs at room temperature. All colloidal particles that can carry charge and form stable suspensions can be used in EDP to coat ceramics, organics, and even metals to any electrically conductive surface.

8.7 SUMMARY

We attempted to address two DSSC- and QDSSC-related issues, namely, optimization of light harvesting, and electron transport from the mesoporous MO layer to the glass or plastic collecting substrate that has been coated with fluorine-doped or indium-doped tin oxide. The blocking or compact MO layer can be TiO_2, SnO_2, ZnO, Nb_2O_5, or $SrTiO_3$. The best blocker is TiO_2. All these MOs have a wide bandgap. These blocking layers are dense and have shown the ability to reduce electron leakage, ensuring smooth electron flow to the external circuit. The optical properties of QDs were also discussed in this chapter. Several approaches have been adopted in the efforts to enhance the photoelectrochemical cell performance. TiO_2 nanoparticles were coated with blocking oxides, the size of QDs was reduced, bandgaps were tuned, and QD/MO structures were passivated. Since the dimensional geometry of MO conductors are also important, various 1D, 2D, and 3D nanostructures and heterostructures have been fabricated. In the 1D nanostructure, the smaller surface area can limit the photoelectrochemical cell performance. Even the 3D helix TiO_2 nanostructure exhibits a smaller surface area than conventional nanoparticles, but it is a good light harvester. Finally, we described briefly how nanostructures can be prepared.

REFERENCES

Al-Kahlout, A. 2015. Thermal treatment optimization of ZnO nanoparticles-photoelectrodes for high photovoltaic performance of dye-sensitized solar cells. *Journal of the Association of Arab Universities for Basic and Applied Sciences* 7:66.

Arof, A.K., Aziz, M.F., Noor, M.M. et al. 2014. Efficiency enhancement by mixed cation effect in dye-sensitized solar cells with a PVdF based gel polymer electrolyte. *International Journal of Hydrogen Energy* 39(6):2929–2935.

Arof, A.K., Naeem, M., Hameed, F. et al. 2013. Quasi solid state dye-sensitized solar cells based on polyvinyl alcohol (PVA) electrolytes containing I⁻/I₃⁻ redox couple. *Optical and Quantum Electronics* 46(1):143–154.

Aziz, M.F., Noor, I.M., Sahraoui, B. et al. 2013. Dye-sensitized solar cells with PVA-KI-EC-PC gel electrolytes. *Optical and Quantum Electronics* 46(1):133–141.

Aziz, M.F., Noor, I.M., Sahraoui, B. et al. 2014. Dye-sensitized solar cells with PVA-KI-EC-PC gel electrolytes. *Optical and Quantum Electronics* 46(1):133–141.

Bailey, R.E. and Nie, S. 2003. Alloyed semiconductor quantum dots: Tuning the optical properties without changing the particle size. *Journal of the American Chemical Society* 125:7.

Bandara, T.M.W.J., Aziz, M.F., Fernando, H.D.N.S. et al. 2015a. Efficiency enhancement in dye-sensitized solar cells with a novel PAN-based gel polymer electrolyte with ternary iodides. *Journal of Solid State Electrochemistry* 19:7.

Bandara, T.M.W.J., Jayasundara, W.J.M.J.S.R., Dissanayake, M.A.K.L. et al. 2013. Effect of cation size on the performance of dye sensitized nanocrystalline TiO_2 solar cells based on quasi-solid state PAN electrolytes containing quaternary ammonium iodides. *Electrochimica Acta* 109(0):609–616.

Bandara, T.M.W.J., Jayasundara, W.J.M.J.S.R., Fernado, H.D.N.S. et al. 2014. Efficiency enhancement of dye-sensitized solar cells with PAN:CsI:LiI quasi-solid state (gel) electrolytes. *Journal of Applied Electrochemistry* 44(8):917–926.

Bandara, T.M.W.J., Jayasundara, W.J.M.J.S.R., Fernado, H.D.N.S. et al. 2015b. Efficiency of 10% for quasi-solid state dye-sensitized solar cells under low light irradiance. *Journal of Applied Electrochemistry* 45(4):289–298.

Bandara, T.M.W.J., Svensson, T., Dissanayake, M.A.K.L. et al. 2012. Tetrahexylammonium iodide containing solid and gel polymer electrolytes for dye sensitized solar cells. *Energy Procedia* 14(0):1607–1612.

Barea, E.M. and Bisquert, J. 2013. Properties of chromophores determining recombination at the TiO_2–dye–electrolyte interface. *Langmuir* 29(28):8773–8781.

Barpuzary, D., Patra, A.S., Vaghasiya, J.V. et al. 2014. Highly efficient one-dimensional ZnO nanowire-based dye-sensitized solar cell using a metal-free, D–π–A-type, carbazole derivative with more than 5% power conversion. *ACS Applied Materials & Interfaces* 6(15):12629–12639.

Bendall, J.S., Etgar, L., Tan, S.C. et al. 2011. An efficient DSSC based on ZnO nanowire photo-anodes and a new D-[small pi]-A organic dye. *Energy & Environmental Science* 4(8):2903–2908.

Berger, T., Lana-Villarreal, T., Monllor-Satoca, D. et al. 2007. An electrochemical study on the nature of trap states in nanocrystalline rutile thin films. *Journal of Physical Chemistry C* 111(27):9936–9942.

Besra, L. and Liu, M. 2007. A review on fundamentals and applications of electrophoretic deposition (EPD). *Progress in Materials Science* 52(1):1–61.

Bond, W.L. 1965. Measurement of the refractive indices of several crystals. *Journal of Applied Physics* 36(5):1674–1677.

Brown, P. and Kamat, P.V. 2008. Quantum dot solar cells. Electrophoretic deposition of CdSe–C60 composite films and capture of photogenerated electrons with nC60 cluster shell. *Journal of the American Chemical Society* 130(28):8890–8891.

Bu, I.Y.Y. and Zheng, J. 2015. A new type of counter electrode for dye sensitized solar cells based on solution processed SnO_2 and activated carbon. *Materials Science in Semiconductor Processing* 39:223–228.

Buraidah, M.H., Teo, L.P., Yusuf, S.N.F. et al. 2011. TiO_2/Chitosan-NH4I(+I2)-BMII-based dye-sensitized solar cells with anthocyanin dyes extracted from black rice and red cabbage. *International Journal of Photoenergy* 10(1155):11.

Burnside, S., Moser, J.-E., Brooks, K. et al. 1999. Nanocrystalline mesoporous strontium titanate as photoelectrode material for photosensitized solar devices: Increasing photovoltage through flatband potential engineering. *Journal of Physical Chemistry B* 103(43):9328–9332.

Chappel, S. and Zaban, A. 2002. Nanoporous SnO_2 electrodes for dye-sensitized solar cells: Improved cell performance by the synthesis of 18 nm SnO_2 colloids. *Solar Energy Materials and Solar Cells* 71(2):141–152.

Chen, H.-Y., Kuang, D.-B., and Su, C.-Y. 2012. Hierarchically micro/nanostructured photoanode materials for dye-sensitized solar cells. *Journal of Materials Chemistry* 22:15.

Choi, H., Nahm, C., Kim, J. et al. 2012. The effect of $TiCl_4$-treated TiO_2 compact layer on the performance of dye-sensitized solar cell. *Current Applied Physics* 12(3):737–741.

Choi, H.M., Ji, I.A., and Bang, J.H. 2014. Metal selenides as a new class of electrocatalysts for quantum dot-sensitized solar cells: A tale of Cu1.8Se and PbSe. *ACS Applied Materials and Interfaces* 6(4):2335–2343.

Cui, H., Zhu, G., Liu, X. et al. 2015. Niobium nitride Nb_4N_5 as a new high-performance electrode material for supercapacitors. *Advanced Science* 2: 1500126. doi:10.1002/advs.201500126.

Dembele, K.T., Selopal, G.S., Soldano, C. et al. 2013. Hybrid carbon nanotubes–TiO_2 photoanodes for high efficiency dye-sensitized solar cells. *Journal of Physical Chemistry C* 117(28):14510–14517.

Devore, J.R. 1951. Refractive indices of rutile and sphalerite. *Journal of the Optical Society of America* 41(6):416–417.

Diamant, Y., Chappel, S., Chen, S.G. et al. 2004. Core–shell nanoporous electrode for dye sensitized solar cells: The effect of shell characteristics on the electronic properties of the electrode. *Coordination Chemistry Reviews* 248(13–14):1271–1276.

Dodge, M.J. 1986. Refractive index. In M.J. Weber (ed.), *Handbook of Laser Science and Technology*, Vol. 4. Boca Raton, FL: CRC Press.

Dong, J., Wu, J., Jia, J. et al. 2015. Cobalt selenide nanorods used as a high efficient counter electrode for dye-sensitized solar cells. *Electrochimica Acta* 168:69–75.

Dou, X., Sabba, D., Mathews, N. et al. 2011. Hydrothermal synthesis of high electron mobility Zn-doped SnO_2 nanoflowers as photoanode material for efficient dye-sensitized solar cells. *Chemistry of Materials* 23(17):3938–3945.

Duong, T.-T., Choi, H.-J., He, Q.-J. et al. 2013. Enhancing the efficiency of dye sensitized solar cells with an SnO_2 blocking layer grown by nanocluster deposition. *Journal of Alloys and Compounds* 561(0):206–210.

Elbohy, H., Thapa, A., Poudel, P. et al. 2015. Vanadium oxide as new charge recombination blocking layer for high efficiency dye-sensitized solar cells. *Nano Energy* 13:368–375.

Fan, J., Fàbrega, C., Zamani, R.R. et al. 2013a. Enhanced photovoltaic performance of nanowire dye-sensitized solar cells based on coaxial TiO_2@TiO heterostructures with a cobalt(II/III) redox electrolyte. *ACS Applied Materials & Interfaces* 5(20):9872–9877.

Fan, J., Hao, Y., Cabot, A. et al. 2013b. Cobalt(II/III) redox electrolyte in ZnO nanowire-based dye-sensitized solar cells. *ACS Applied Materials & Interfaces* 5(6):1902–1906.

Feng, H.-L., Wu, W.-Q., Rao, H.-S. et al. 2015. Three-dimensional hyperbranched TiO_2/ZnO heterostructured arrays for efficient quantum dot-sensitized solar cells. *Journal of Materials Chemistry A* 3:7.

Fu, H. and Tsang, S.-W. 2012. Infrared colloidal lead chalcogenide nanocrystals: Synthesis, properties, and photovoltaic applications. *Nanoscale* 4(7):2187–2201.

Fuke, N., Hoch, L.B., Koposov, A.Y. et al. 2010. CdSe quantum-dot-sensitized solar cell with ~100% internal quantum efficiency. *ACS Nano* 4(11):6377–6386.

Gao, L., Lemarchand, F., and Lequime, M. 2012. Exploitation of multiple incidences spectrometric measurements for thin film reverse engineering. *Optics Express* 20(14):15734–15751.

Ghosh, R., Brennaman, M.K., Uher, T. et al. 2011. Nanoforest Nb_2O_5 photoanodes for dye-sensitized solar cells by pulsed laser deposition. *ACS Applied Materials & Interfaces* 3(10):3929–3935.

Gong, F., Xu, X., Li, Z. et al. 2013. $NiSe_2$ as an efficient electrocatalyst for a Pt-free counter electrode of dye-sensitized solar cells. *Chemical Communications* 49(14):1437–1439.

Gratzel, M. 1991. Mesoporous oxide junctions and nanostructured solar cells. *Current Opinion in Colloid & Interface Science* 4:8.

Gratzel, M. 2001. Photoelectrochemical cells. *Nature* 414(6861):338–344.

Hagfeldt, A., Boschloo, G., Sun, L. et al. 2010. Dye-sensitized solar cells. *Chemical Reviews* 110(11):6595–6663.

Hardman, S.J.O., Graham, D.M., Stubbs, S.K. et al. 2011. Electronic and surface properties of PbS nanoparticles exhibiting efficient multiple exciton generation. *Physical Chemistry Chemical Physics* 13:9.

Hashemzadeh, F., Rahimi, R., and Ghaffarinejad, A. 2014. Mesoporous nanostructures of Nb_2O_5 obtained by an EISA route for the treatment of malachite green dye-contaminated aqueous solution under UV and visible light irradiation. *Ceramics International* 40(7, Part A):9817–9829.

Hassan, H.C., Abidin, Z.H.Z., Careem, M.A. et al. 2014. Chlorophyll as sensitizer in I^-/I_3^--based solar cells with quasi-solid-state electrolytes. *High Performance Polymers* 26(6):647–652.

Hu, B. and Liu, Y. 2015. Nitrogen-doped Nb_2O_5 nanobelt quasi-arrays for visible light photocatalysis. *Journal of Alloys and Compounds* 635:1–4.

Huo, J., Zheng, M., Tu, Y. et al. 2015. A high performance cobalt sulfide counter electrode for dye-sensitized solar cells. *Electrochimica Acta* 159:166–173.

Jabbour, G.E. and Doderer, D. 2010. Quantum dot solar cells: The best of both worlds. *Nature Photonics* 4(9):604–605.

Jayabal, P., Sasirekha, V., Mayandi, J. et al. 2014. A facile hydrothermal synthesis of $SrTiO_3$ for dye sensitized solar cell application. *Journal of Alloys and Compounds* 586:456–461.

Jeng, M.-J., Wung, Y.-L., Chang, L.-B. et al. 2013. Dye-sensitized solar cells with anatase TiO_2 nanorods prepared by hydrothermal method. *International Journal of Photoenergy* Article ID 280253, 2013(8 pp).

Jingbin, H., Fengru, F., Chen, X. et al. 2010. ZnO nanotube-based dye-sensitized solar cell and its application in self-powered devices. *Nanotechnology* 21(40):405203.

Ju, T., Graham, R.L., Zhai, G. et al. 2010. High efficiency mesoporous titanium oxide PbS quantum dot solar cells at low temperature. *Applied Physics Letters* 97(4):043106.

Jun, H.K., Careem, M.A., and Arof, A.K. 2013. Quantum dot-sensitized solar cells—Perspective and recent developments: A review of Cd chalcogenide quantum dots as sensitizers. *Renewable and Sustainable Energy Reviews* 22:148–167.

Kamat, P.V. 2008. Quantum dot solar cells. Semiconductor nanocrystals as light harvesters. *Journal of Physical Chemistry C* 112(48):18737–18753.

Kamat, P.V. 2013. Quantum dot solar cells. The next big thing in photovoltaics. *The Journal of Physical Chemistry Letters* 4(6):908–918.

Kim, B., Park, S.W., Kim, J.Y. et al. 2013. Rapid dye adsorption via surface modification of TiO_2 photoanodes for dye-sensitized solar cells. *ACS Applied Materials and Interfaces* 5(11):5201–5207.

Kim, H. and Yong, K. 2013. A highly efficient light capturing 2D (nanosheet)-1D (nanorod) combined hierarchical ZnO nanostructure for efficient quantum dot sensitized solar cells. *Physical Chemistry Chemical Physics* 15(6):2109–2116.

Kim, H.-J., Kim, J.-H., Pavan Kumar, C.S.S. et al. 2015. Facile chemical bath deposition of CuS nano peas like structure as a high efficient counter electrode for quantum-dot sensitized solar cells. *Journal of Electroanalytical Chemistry* 739:20–27.

Kim, Y.J., Kim, K.H., Kang, P. et al. 2012. Effect of layer-by-layer assembled SnO_2 interfacial layers in photovoltaic properties of dye-sensitized solar cells. *Langmuir* 28(28):10620–10626.

Koyama, H., Fujimoto, M., Ohno, T. et al. 2006. Effects of thermal annealing on formation of micro porous titanium oxide by the sol–gel method. *Journal of the American Ceramic Society* 89(11):3536–3540.

Krüger, S., Hickey, S.G., Tscharntke, S. et al. 2011. Study of the attachment of linker molecules and their effects on the charge carrier transfer at lead sulfide nanoparticle sensitized ZnO substrates. *Journal of Physical Chemistry C* 115(26):13047–13055.

Lai, M.H., Lee, M.W., Wang, G.-J. et al. 2011. Photovoltaic performance of new-structure ZnO-nanorod dye-sensitized solar cells. *International Journal of Electrochemical Science* 6:9.

Lai, M.-H., Tubtimtae, A., Lee, M.-W. et al. 2010. ZnO-nanorod dye-sensitized solar cells: New structure without a transparent conducting oxide layer. *International Journal of Photoenergy* 2010, Article ID 497095, 5 pp.

Le Bahers, T., Pauporté, T., Lainé, P.P. et al. 2013. Modeling dye-sensitized solar cells: From theory to experiment. *The Journal of Physical Chemistry Letters* 4(6):1044–1050.

Lee, G., Lee, H., Um, M.-H. et al. 2012. Light scattering amplification on dye sensitized solar cells assembled by Hollyhock-shaped CdS-TiO_2 composites. *Bulletin of the Korean Chemical Society* 33(9):5.

Lee, H.J., Chen, P., Moon, S.-J. et al. 2009. Regenerative PbS and CdS quantum dot sensitized solar cells with a cobalt complex as hole mediator. *Langmuir* 25(13):7602–7608.

Lee, J.-H., Park, N.-G., and Shin, Y.-J. 2011. Nano-grain SnO_2 electrodes for high conversion efficiency SnO_2–DSSC. *Solar Energy Materials and Solar Cells* 95(1):179–183.

Lee, K., Mazare, A., and Schmuki, P. 2014a. One-dimensional titanium dioxide nanomaterials: Nanotubes. *Chemical Reviews* 114(19):9385–9454.

Lee, S.H., Jin, H., Kim, D.-Y. et al. 2014b. Enhanced power conversion efficiency of quantum dot sensitized solar cells with near single-crystalline TiO_2 nanohelixes used as photoanodes. *Optics Express* 22(S3):A867–A879.

Lee, Y.-L., Huang, B.-M., and Chien, H.-T. 2008. Highly efficient CdSe-sensitized TiO_2 photoelectrode for quantum-dot-sensitized solar cell applications. *Chemistry of Materials* 20(22):6903–6905.

Lee, Y.-L. and Lo, Y.-S. 2009. Highly efficient quantum-dot-sensitized solar cell based on co-sensitization of CdS/CdSe. *Advanced Functional Materials* 19(4):604–609.

Li, G.-r., Wang, F., Jiang, Q.-W. et al. 2010. Carbon nanotubes with titanium nitride as a low-cost counter-electrode material for dye-sensitized solar cells. *Angewandte Chemie International Edition* 49(21):3653–3656.

Lin, Y., Wu, G.S., Yuan, X.Y. et al. 2003. Fabrication and optical properties of TiO_2 nanowire arrays made by sol–gel electrophoresis deposition into anodic alumina membranes. *Journal of Physics: Condensed Matter* 15:2917–2922.

Maçaira, J., Andrade, L., and Mendes, A. 2013. Review on nanostructured photoelectrodes for next generation dye-sensitized solar cells. *Renewable and Sustainable Energy Reviews* 27:334–349.

MaçAira, J., Andrade, L., and Mendes, A. 2014. Modeling, simulation and design of dye sensitized solar cells. *RSC Advances* 4(6):2830–2844.

Manna, L., Scher, E.C., Li, L.-S. et al. 2002. Epitaxial growth and photochemical annealing of graded CdS/ZnS shells on colloidal CdSe nanorods. *Journal of the American Chemical Society* 124:10.

Moreels, I., Lambert, K., Smeets, D. et al. 2009. Size-dependent optical properties of colloidal PbS quantum dots. *ACS Nano* 3(10):3023–3030.

Murphy, J.E., Beard, M.C., Norman, A.G. et al. 2006. PbTe colloidal nanocrystals: Synthesis, characterization, and multiple exciton generation. *Journal of the American Chemical Society* 128:7.

Nazeeruddin Md, K., Zakeeruddin, S.M., Lagref, J.J. et al. 2004. Stepwise assembly of amphiphilic ruthenim sensitizers and their applications in dye-sensitized solar cell. *Coordination Chemistry Reviews* 248:12.

Noor, M.M., Buraidah, M.H., Yusuf, S.N.F. et al. 2011. Performance of dye-sensitized solar cells with (PVDF-HFP)-KI-EC-PC electrolyte and different dyematerials. *International Journal of Photoenergy* 10(1155):5.

Okuya, M., Kaneko, S., Hiroshima, K. et al. 2001. Low temperature deposition of SnO_2 thin films as transparent electrodes by spray pyrolysis of tetra-n-butyltin(IV). *Journal of the European Ceramic Society* 21(10–11):2099–2102.

Onicha, A.C., Petchsang, N., Kosel, T.H. et al. 2012. Controlled synthesis of compositionally tunable ternary PbSexS1–x as well as binary PbSe and PbS nanowires. *ACS Nano* 6(3):2833–2843.

Pan, S.S., Zhang, Y.X., Teng, X.M. et al. 2008. Optical properties of nitrogen-doped SnO_2 films: Effect of the electronegativity on refractive index and bandgap. *Journal of Applied Physics* 103(9):093103.

Parsi Benehkohal, N., González-Pedro, V., Boix, P.P. et al. 2012. Colloidal PbS and PbSeS quantum dot sensitized solar cells prepared by electrophoretic deposition. *Journal of Physical Chemistry C* 116(31):16391–16397.

Peter, L.M. 2011. The Grätzel cell: Where next? *The Journal of Physical Chemistry Letters* 2(15):1861–1867.

Pugliese, D., Lamberti, A., Bella, F. et al. 2014. TiO$_2$ nanotubes as flexible photoanode for back-illuminated dye-sensitized solar cells with hemi-squaraine organic dye and iodine-free transparent electrolyte. *Organic Electronics* 15(12):3715–3722.

Qiu, J., Guo, M., and Wang, X. 2011. Electrodeposition of hierarchical ZnO nanorod-nanosheet structures and their applications in dye-sensitized solar cells. *ACS Applied Materials & Interfaces* 3(7):2358–2367.

Ramasamy, E. and Lee, J. 2011. Ordered mesoporous Zn-doped SnO$_2$ synthesized by exotemplating for efficient dye-sensitized solar cells. *Energy & Environmental Science* 4(7):2529–2536.

Reda, S.M. 2008. Synthesis and optical properties of CdS quantum dots embedded in silica matrix thin films and their applications as luminescent solar concentrators. *Acta Materialia* 56(2):259–264.

Rhee, J.H., Chung, C.-C., and Diau, E.W.-G. 2013. A perspective of mesoscopic solar cells based on metal chalcogenide quantum dots and organometal-halide perovskites. *NPG Asia Materials* 5:17.

Sacco, A., Di Bella, M.S., Gerosa, M. et al. 2015. Enhancement of photoconversion efficiency in dye-sensitized solar cells exploiting pulsed laser deposited niobium pentoxide blocking layers. *Thin Solid Films* 574(0):38–42.

Salant, A., Shalom, M., Hod, I. et al. 2010. Quantum dot sensitized solar cells with improved efficiency prepared using electrophoretic deposition. *ACS Nano* 4(10):5962–5968.

Sangiorgi, A., Bendoni, R., Sangiorgi, N. et al. 2014. Optimized TiO$_2$ blocking layer for dye-sensitized solar cells. *Ceramics International* 40(7, Part B):10727–10735.

Santhanagopalan, S., Teng, F., and Meng, D.D. 2010. High-voltage electrophoretic deposition for vertically aligned forests of one-dimensional nanoparticles. *Langmuir* 27(2):561–569.

Santra, P.K. and Kamat, P.V. 2012. Mn-doped quantum dot sensitized solar cells: A strategy to boost efficiency over 5%. *Journal of the American Chemical Society* 134(5):2508–2511.

Santra, P.K., Nair, P.V., George Thomas, K. et al. 2013. CuInS2-sensitized quantum dot solar cell. Electrophoretic deposition, excited-state dynamics, and photovoltaic performance. *The Journal of Physical Chemistry Letters* 4(5):722–729.

Savariraj, A.D., Viswanathan, K.K., and Prabakar, K. 2014. CuS nano flakes and nano platelets as counter electrode for quantum dots sensitized solar cells. *Electrochimica Acta* 149:6.

Schaller, R.D. and Klimov, V.I. 2004. High efficiency carrier multiplication in PbSe nanocrystals: Implications for solar energy conversion. *Physical Review Letters* 92(18):4.

Seo, H., Son, M.-K., Kim, J.-K. et al. 2011. Method for fabricating the compact layer in dye-sensitized solar cells by titanium sputter deposition and acid-treatments. *Solar Energy Materials and Solar Cells* 95(1):340–343.

Seol, M., Kim, H., Tak, Y. et al. 2010. Novel nanowire array based highly efficient quantum dot sensitized solar cell. *Chemical Communications* 46(30):5521–5523.

Shaikh, S.F., Kalanur, S.S., Mane, R.S. et al. 2013. Monoclinic WO$_3$ nanorods-rutile TiO$_2$ nanoparticles core–shell interface for efficient DSSCs. *Dalton Transactions* 42(28):10085–10088.

Song, C.B., Zhao, Y.L., Song, D.M. et al. 2014. Dye-sensitized solar cells based on TiO$_2$ nanotube/nanoparticle composite as photoanode and Cu$_2$SnSe$_3$ as counter electrode. *International Journal of Electrochemical Science* 9:8.

Song, J., Li, G.R., Xiong, F.Y. et al. 2012. Synergistic effect of molybdenum nitride and carbon nanotubes on electrocatalysis for dye-sensitized solar cells. *Journal of Materials Chemistry* 22(38):20580–20585.

Sun, Z., Kim, J.H., Zhou, Y. et al. 2012. Improved photovoltaic performance of dye-sensitized solar cells with modifiedself-assembling highly ordered mesoporous TiO$_2$ photoanodes. *Journal of Materials Chemistry* 22:9.

Tang, H., Prasad, K., Sanjinbs, R. et al. 1994. Electrical and optical properties of TiO$_2$ anatase thin films. *Journal of Applied Physics* 75(4):6.

Thomas, S., Deepak, T.G., Anjusree, G.S. et al. 2014. A review on counter electrode materials in dye-sensitized solar cells. *Journal of Materials Chemistry A* 2:17.

Tian, J. and Cao, G. 2013. Semiconductor quantum dot-sensitized solar cells. *Nano Reviews* 4:8.

Ting, C.-C., Chen, S.-Y., and Liu, D.-M. 2000. Structural evolution and optical properties of TiO$_2$ thin films prepared by thermal oxidation of sputtered Ti films. *Journal of Applied Physics* 88:7.

Vijayakumar, P., Senthil Pandian, M., Lim, S.P. et al. 2015. Investigations of tungsten carbide nanostructures treated with different temperatures as counter electrodes for dye sensitized solar cells (DSSC) applications. *Journal of Materials Science: Materials in Electronics* 26(10):7977–7986.

Wang, H., Wang, B., Yu, J. et al. 2015. Significant enhancement of power conversion efficiency for dye sensitized solar cell using 1D/3D network nanostructures as photoanodes. *Scientific Reports* 5:9305, 9 pp.

Wang, H., Wei, W., and Hu, Y.H. 2013. Efficient ZnO-based counter electrodes for dye-sensitized solar cells. *Journal of Materials Chemistry A* 1(22):6622–6628.

Wang, M., Anghel, A.M., Marsan, B. et al. 2009. CoS supersedes pt as efficient electrocatalyst for triiodide reduction in dye-sensitized solar cells. *Journal of the American Chemical Society* 131(44):15976–15977.

Wong, K.K., Ng, A., Chen, X.Y. et al. 2012. Effect of ZnO nanoparticle properties on dye-sensitized solar cell performance. *ACS Applied Materials & Interfaces* 4(3):1254–1261.

Wu, M., Lin, X., Hagfeldt, A. et al. 2011a. Low-cost molybdenum carbide and tungsten carbide counter electrodes for dye-sensitized solar cells. *Angewandte Chemie International Edition* 50(15):3520–3524.

Wu, M., Lin, X., Wang, Y. et al. 2012. Economical Pt-free catalysts for counter electrodes of dye-sensitized solar cells. *Journal of the American Chemical Society* 134(7):3419–3428.

Wu, M., Wang, Y., Lin, X. et al. 2011b. Economical and effective sulfide catalysts for dye-sensitized solar cells as counter electrodes. *Physical Chemistry Chemical Physics* 13:4.

Wu, M., Wang, Y., Lin, X. et al. 2013. TiC/Pt composite catalyst as counter electrode for dye-sensitized solar cells with long-term stability and high efficiency. *Journal of Materials Chemistry A* 1(34):9672–9679.

Wu, M., Zhang, Q., Xiao, J. et al. 2011c. Two flexible counter electrodes based on molybdenum and tungsten nitrides for dye-sensitized solar cells. *Journal of Materials Chemistry* 21(29):10761–10766.

Xia, J., Masaki, N., Jiang, K. et al. 2007. Fabrication and characterization of thin Nb_2O_5 blocking layers for ionic liquid-based dye-sensitized solar cells. *Journal of Photochemistry and Photobiology A: Chemistry* 188(1):120–127.

Xiao, J., Zeng, X., Chen, W. et al. 2013. High electrocatalytic activity of self-standing hollow $NiCo_2S_4$ single crystalline nanorod arrays towards sulfide redox shuttles in quantum dot-sensitized solar cells. *Chemical Communications* 49(100):11734–11736.

Xiaoyan, L., Chunlei, W., Shuhong, X. et al. 2014. Manipulation of inter-particle interactions between TiO_2 and CdTe: An effective method to enhance the performance of quantum dot sensitized solar cells. *Journal of Physics D: Applied Physics* 47(1):015103.

Xie, F., Cherng, S.-J., Lu, S. et al. 2014. Functions of self-assembled ultrafine TiO_2 nanocrystals for high efficient dye-sensitized solar cells. *ACS Applied Materials & Interfaces* 6(8):5367–5373.

Xu, X., Qiao, F., Dang, L., and Gao, F. 2014. Porous tin oxide nanosheets with enhanced conversion efficiency as dye-sensitized solar cell electrode. *Journal of Physical Chemistry C* 118:16856.

Yang, S., Kou, H., Wang, J. et al. 2010. Tunability of the band energetics of nanostructured $SrTiO_3$ electrodes for dye-sensitized solar cells. *Journal of Physical Chemistry C* 114(9):4245–4249.

Yong, S.-M., Tsvetkov, N., Larina, L. et al. 2014. Ultrathin SnO_2 layer for efficient carrier collection in dye-sensitized solar cells. *Thin Solid Films* 556(0):503–508.

Yusuf, S.N.F., Aziz, M.F., Hassan, H.C. et al. 2014. Phthaloylchitosan-based gel polymer electrolytes for efficient dye-sensitized solar cells. *Journal of Chemistry* Article ID 783023, 2014 (8 pp).

Zallen, R. and Moret, M.P. 2006. The optical absorption edge of brookite TiO_2. *Solid State Communications* 137:4.

Zhang, Q., Dandeneau, C.S., Zhou, X. et al. 2009. ZnO nanostructures for dye-sensitized solar cells. *Advanced Materials* 21:22.

Zhao, H., Liang, H., Vidal, F. et al. 2014. Size dependence of temperature-related optical properties of PbS and PbS/CdS core/shell quantum dots. *Journal of Physical Chemistry C* 118:9.

9 Second-Order Nonlinear Susceptibility in Quantum Dot Structures

M. Abdullah, Farah T. Mohammed Noori,
and Amin H. Al-Khursan

CONTENTS

9.1 INTRODUCTION

With the development of the fabrication technique of nanostructures, multiform semiconductor structures are fabricated [1]. The twenty-first century will see a dramatic change in lighting technologies. By 2025, fluorescent and incandescent illumination sources should be replaced by more efficient, long-lasting, and versatile light sources, offering more lumens per cm^2 and decreasing the consumption of energy for lighting by 29% [2]. The core of such lighting devices, in its simplest form, is a junction, a relatively simple multilayered structure formed by a semiconductor crystal between two higher bandgap semiconductors, which emits light when an electric current passes through it. The localization of carriers in all three dimensions breaks down the classical band structure of the continuous dispersion of energy as a function of momentum. Unlike quantum wells (QWs) and quantum wires (QWi's), the energy-level structure of quantum dots (QDs) is quite discrete. This unique structure of QDs opens a new chapter both in fundamental physics in which they can be regarded as artificial atoms and in potential applications as devices [3,4]. The density of states for a bulk material is a function of energy ($\sim E^{1/2}$), while in a zero-dimensional (QD) crystal, the density of states is described by a discrete δ-function, ($\delta(E)$) [5], due to the quantum confinement effect.

9.2   SEMICONDUCTOR NANOSTRUCTURES

One of the important features of semiconductor nanostructures is the flexibility of controlling and designing the properties of such materials [5]. In nanostructures, ideally 1–50 nm in scale, the dimensions commensurate with the de Broglie wavelength of the charge carriers so that quantum confinement effects become important and the properties of the semiconductors are significantly modified [5].

Advanced semiconductor growth techniques, such as molecular beam epitaxy (MBE) and metal organic chemical vapor deposition (MOCVD), allow the fabrication of various semiconductor nanostructures or low-dimensional structures. Such low-dimensional structures include (1) QWs, where the charge carriers are confined along the growth direction but are free in other two directions; (2) QWi's, where the charge carriers are confined in two directions and allow free carrier motion in only one dimension; and (3) QDs, where the charge carriers are confined in all the three directions (see Figure 9.1). QW or QWi confinement gains the electron at some degrees of freedom, but it still gives the electrons at least one direction to propagate. On the other hand, today's technology allows us to create QD nanostructures where electron subbands were quantized in all the degrees of freedom [6]. Strong interband (IB) transitions are possible in these low-dimensional structures because there is a probability of a strong overlap between the wave functions of electrons and holes.

9.3   QUANTUM DOTS

The term "quantum dot" is usually used to describe a semiconductor nanocrystal. QDs are zero-dimensional semiconductor systems created at nanoscale. This confinement results in properties that are not found in bulk materials; see Figure 9.2. One of the main differences between QDs and traditional semiconductors is the tunability by both the dot size and composition [7].

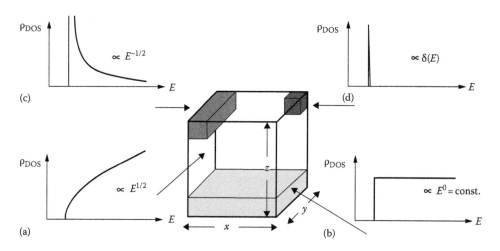

FIGURE 9.1 Electronic density of states of semiconductor crystals for (a) bulk, (b) quantum well, (c) quantum wire, and (d) quantum dot.

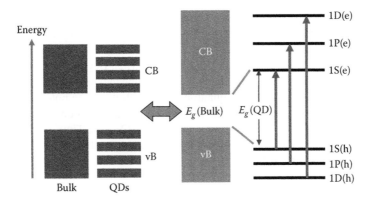

FIGURE 9.2 Comparison of energy levels between a bulk and a quantum dot semiconductor.

The importance of QDs is originated from the fact that their electrical conductivity can be altered by an external stimulus such as voltage or photon flux. QDs are very sensitive to the surface properties due to a high surface to volume ratio. Surface states play a dominant role in these systems, acting as efficient traps for electrons and holes, and thus surface passivation is essential for fabricating practical semiconductor devices based on such low-dimensional structures [5].

9.4 SEMICONDUCTOR NONLINEARITIES

Semiconductor materials play an important role in nonlinear optics because they both produce a large nonlinear optical response and lend themselves to the construction of integrated devices in which electronic devices, semiconductor lasers, and nonlinear optical components are all fabricated on a single semiconductor substrate [8].

Modern technology has allowed scientists to fabricate high-precision semiconductor QD nanostructures. In these structures, the precise engineering would enable to confine the motion of charge carriers in three dimensions. So, a great deal of work has been performed in this area [9,10]. Electronic and optical properties of QWs with an applied external electric field are of increasing interest [11]. An analytical relation for energy subband calculations is stated earlier and gives a good result compared with experimental measurements [10–12] and numerical calculations [13,14].

By adding an additional distortion to the energy subbands through the application of an external electric field, one can tune QD nonlinear properties and change the emission spectrum of photoluminescence, both in intensity and wavelength [15]. This results in a precise control of oscillator strength, which opens the way for the development of practical devices such as optical filters and color-tunable sources [16]. Thus, studying of nonlinearity in QD nanostructures under an applied electric field is important for a large number of device applications.

Dane et al. [17] studied the effect of an electric field on the binding energy of a shallow donor impurity in a spherical QD with an infinite barrier. Xie [18] studied the second-order nonlinear susceptibility (SONS) in QDs with a quantum disk shape under an applied electric field. Baskoutas et al. [19] studied nonlinear optical rectification (OR) in semiparabolic QDs, where they found that SONS depends on the type of quantum confinement. Vaseghi et al. [20] studied OR in cubic QDs with an infinite potential barrier, where they found that both the dot size and electric field strength increase with OR. Second-harmonic generation (SHG) in cubic QDs with infinite potential under an applied electric field is considered in the work of Shao et al. [21], where they found a nonmonotonic behavior of SONS with cubic length and applied field. The aforementioned works either consider parabolic confinement or infinite potential. The parabolic confinement gives an equidistant energy subband that is far from the experiment. Also, the use of an infinite barrier limits the accuracy compared with experimental data. Although this type of calculations to get an overview of the physical problem is important, it is impractical to consider the finite barrier for real calculations. The accurate description of QD states requires a multiband $k \cdot p$ calculation, which is limited by the knowledge of the exact shape and composition of QDs [22]. Beyond this tedious calculation, it is required to build a model to calculate the QD energy subbands under the applied electric field. No work deals with QDs in the shape of quantum disks with the finite potential barrier under the applied electric field. This shape of dots was considered in a large number of literatures and self-assembled QDs can be approximated to it, for example, see [22–29]. Thus, starting from the quantum disk model [12], we introduced the electric field effect on QD subbands, where the subbands are shifted to higher energies with the field.

9.5 NONLINEAR OPTICAL SUSCEPTIBILITY

Nonlinear optical properties of semiconductors have received much attention in recent years. The very large optical nonlinearities in semiconductors have offered promise for practical applications in low-power, high-speed, room-temperature optical switching, and signal processing devices [30]. Nonlinear optics has been rapidly growing as a scientific field in recent decades. It is based on the phenomena related to the interaction of intense coherent light radiation with matter. Nonlinear optics is the study of interactions of light with matter under conditions in which the nonlinear response plays an important role. During the past three decades, optics has secured a good place in application areas previously dominated by electronics. Developments in the field of nonlinear optics promise for important applications in optical information processing, telecommunications, and integrated optics. Because of the emergence of this field from solid-state physics in which inorganic semiconductors, insulators, and crystals have constituted a major part of the scientific base, the early experimental and theoretical investigations were primarily concerned with materials from these classes [31].

Nonlinear optical phenomena are "nonlinear" in the sense that they occur when the response of a material system to an applied optical field depends in a nonlinear manner on the strength of the optical field. For example, SHG occurs as a result of the part of the atomic response that scales quadratically with the strength of the applied optical field. Consequently, the intensity of light generated at the second-harmonic frequency tends to increase as the square of the intensity of the applied field [8].

Second-order nonlinear optical interactions can occur only in noncentrosymmetric crystals, that is, in crystals that do not display inversion symmetry. Since liquids, gases, amorphous solids (such as glass), and even many crystals display inversion symmetry, SONS ($\chi^{(2)}$) vanishes identically for such media, and consequently such materials cannot produce second-order nonlinear optical interactions. On the other hand, third-order nonlinear optical interactions (i.e., those described by a third-order nonlinear susceptibility $\chi^{(3)}$) can occur for both centrosymmetric and noncentrosymmetric media [8].

A linear dielectric medium is characterized by the linear relation $P = \varepsilon_0 \cdot \chi E$, between the polarization density, P, and the electric field, E, where ε_0 is the permittivity of free space and χ is the optical susceptibility of the medium. A nonlinear dielectric medium, on the other hand, is characterized by a nonlinear relation between P and E; see Figure 9.3. The nonlinearity may be of microscopic or macroscopic origin. The polarization density $P = N\rho$ is a product of the individual dipole moment ρ induced by the applied electric field E and the number density of dipole moments N. The nonlinear behavior may reside either in ρ or in N. The relation between P and E is linear when E is small but becomes nonlinear when E acquires values comparable to interatomic electric fields. Since externally applied optical electric fields are small in comparison with characteristic interatomic or crystalline fields, even when focused laser light is used, the nonlinearity is usually weak. The relation between P and E is then approximately linear for small E [32].

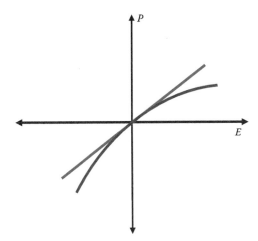

FIGURE 9.3 The *P–E* relation for a linear dielectric medium (blue line) and a nonlinear medium (red line).

9.6 SECOND-ORDER NONLINEARITY

In the regime of conventional optics, the electric polarization vector P is simply assumed to be linearly proportional to the electric field strength E of an applied optical wave, that is [33]

$$P = \varepsilon_0 \chi E \tag{9.1}$$

where
  ε_0 is the free-space permittivity
  χ is the susceptibility of a given medium

A plot of P versus E is a straight line. Equation 9.1 is valid for field strengths of conventional sources. The quantity χ is a constant only in the sense of being independent of E; its magnitude is a function of the frequency. With sufficiently intense laser radiation, this relation does not hold good and has to be generalized to Equation 9.2, which can be written in the following vector form, as by a power series [33]:

$$\tilde{P}(t) = \varepsilon_0 \left[\chi^{(1)} \tilde{E}(t) + \chi^{(2)} \tilde{E}^2(t) + \chi^{(3)} \tilde{E}^3(t) + \cdots \right] \tag{9.2}$$

The second-order nonlinear effects can occur in the noncentrosymmetrical crystals only. In the dielectric dipole approximation, isotropic media and centrosymmetrical crystals cannot be used to generate second-order nonlinear effects. Therefore, the media for SHG should be the crystals having no inversion symmetry. This requirement is the same as that for the piezoelectric effect; thus, all SHG crystals are piezoelectric crystals although the physical mechanisms for these two effects are different [33].

In SHG, the combination (addition) of two photons of the same frequency was considered to produce a single photon of twice the original frequency. Generalization of this process allows the case in which the two photons have different frequencies. SHG is a nonlinear process where a photon at frequency 2ω is generated from the interaction between intense light at frequency ω and a nonlinear medium. The interaction of weak light field with matter is dominated by a linear process.

With a very intense light field, nonlinear process such as SHG becomes observable [34]. OR in anisotropic media is well known and was first observed in a potassium dihydrogen phosphate crystal [8]. The magnitude of the induced polarization was found to be proportional to the square of the optical electric field amplitude. For the effect to arise, the medium, isotropic or anisotropic, needs to be of sufficiently low symmetry such that the optically induced electric polarization does not reverse exactly with the optical field [35]. The second-order nonlinear polarization created in the crystal is given by [36]

$$\tilde{P}^{(2)}(t) = \varepsilon_0 \chi^{(2)} \tilde{E}^2(t) = 2\chi^{(2)} EE^* + (\chi^{(2)} E^2 e^{-2iwt}) + \text{c.c.} \tag{9.3}$$

The second-order polarization combines two parts: The first part is with zero frequency that leads to the generation of a static electric field within the nonlinear crystal and is known as OR. The second part contains e^{-2iwt} frequency that also leads to the generation of the radiation at the SHG [36].

Let us consider the process of SHG, which is illustrated schematically in Figure 9.4. Here, a laser beam is incident upon a crystal for which the second-order susceptibility $\chi^{(2)}$ is nonzero. The nonlinear polarization that is created in such a crystal is given according to Equation 9.3. Under proper experimental conditions, the process of SHG can be so efficient that nearly all of the power in the incident beam at frequency ω is converted into radiation at the second-harmonic frequency 2ω. One common use of SHG is to convert the output of a fixed-frequency laser to a different spectral region. SHG can be visualized by considering the interaction in terms of the exchange of photons between the various frequency components of the field. According to this picture, which is illustrated in part (b) of Figure 9.4, two photons of frequency ω are destroyed and a photon of frequency 2ω is simultaneously created in a single quantum-mechanical process [8].

Let the optical field incident upon a second-order nonlinear optical medium consists of two distinct frequency components; it can be represented by

$$\tilde{E}(t) = E_1 e^{-iw_1 t} + E_2 e^{-iw_2 t} + \text{c.c.} \tag{9.4}$$

The nonlinear polarization, using Equation 9.2, is given by

$$P^2(t) = \varepsilon_0 \chi^{(2)} \left[E_1^2 e^{-2iw_1 t} + E_2^2 e^{-2iw_2 t} + 2E_1 E_2 e^{-i(w_1 + w_2)t} + 2E_1 E_2^* e^{-i(w_1 - w_2)t} + \text{c.c.} \right]$$

$$+ 2\varepsilon_0 \chi^{(2)} [E_1 E_1^* + E_2 E_2^*] \tag{9.5}$$

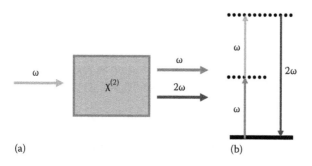

(a)   (b)

FIGURE 9.4   (a) Geometry of second-harmonic generation. (b) Energy-level diagram describing second-harmonic generation.

Each expression in this equation refers to a nonlinear physical process. The first two terms with e^{-2iwt} dependence are related to SHG, the third term powers to $(w_1 + w_2)$ refers to sum-frequency generation (SFG), the fourth term powers to $(w_1 - w_2)$ is the difference-frequency generation (DFG), while the last two terms refer to the OR process. There is also a response to the negative of each of nonzero frequencies. It is not necessary to take explicit account of both the positive and negative frequency components [8]. SHG, SFG, DFG, and OR are four different nonzero frequency components present in the nonlinear polarization. However, typically no more than one of these frequency components will be present with any appreciable intensity in the radiation generated by the nonlinear optical interaction. The reason for this behavior is that the nonlinear polarization can efficiently produce an output signal only if a certain phase-matching condition is satisfied, and usually this condition cannot be satisfied for more than one frequency component of the nonlinear polarization. Operationally, one often chooses which frequency component will be radiated by properly selecting the polarization of the input radiation and the orientation of the nonlinear crystal [8].

The process of SFG is illustrated in Figure 9.5. In many ways, it is analogous to that of SHG, except that in SFG the two input waves are at different frequencies. The process of DFG is illustrated in Figure 9.6. Superficially, DFG and SFG appear to be a very similar process [8].

Nonlinear optics has been growing rapidly as an important scientific field in recent decades. It is based on a phenomenon related to the interaction of intense coherent light radiation with matter. Development in this field is promised for important applications in the optical information processing, telecommunications, and integrated optics. Because of the emergence of this field from solid-state physics, in which inorganic semiconductors, insulators, and crystals have constituted a major part of the scientific base, the early experimental and theoretical investigations were primarily concerned with the materials from these classes [31]. Modern technology has allowed scientists to fabricate high-precision semiconductor nanostructures, QDs. Therefore, they tremendously attract the attention due to their unique physical properties and their potential applications in micro- and nano-optoelectronic devices [9,10].

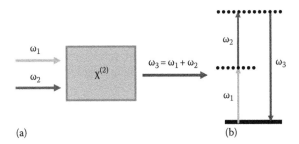

FIGURE 9.5 Sum-frequency generation. (a) Geometry of the interaction. (b) Energy-level description.

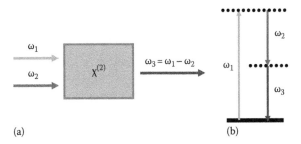

FIGURE 9.6 Difference-frequency generation. (a) Geometry of the interaction. (b) Energy-level description.

One of the most basic nonlinear processes is the SONS. It requires a symmetry breaking. Thus, it can be explored in QWs having asymmetry resulted from the modulation doping or from the application of an external electric field. QDs are considered as asymmetric structures due to their inhomogeneity, and thus SONS in QDs is calculated in some articles without consideration of an applied electric field [37]. Third-order nonlinearity in QDs was studied in detail; as an example, see [23–26,38], considering zero-diagonal matrix elements (symmetry is found).

QDs are promising candidates to achieve large nonlinear susceptibilities. In nanostructures, nonlinear susceptibility increases due to two factors: (1) large dipole matrix elements associated with intersubband (ISB) transitions and (2) these ISB transitions that can be adjusted by changing the QW size [37]. Additionally, nonlinear phenomena at nanoscale are different from that at bulk (conventional structures). The conventional and second-harmonic waves must propagate in phase to fulfill the phase-matching condition to create constructive interference and, then, increase conversion efficiency. For small photonic cavities, the classical phase-matching conditions are replaced by a spatial overlap of localized modes [39].

Earlier studies of SONS in QDs depend on the asymmetrical shape of the self-assembled QDs. In [37], a lens-shaped QD with infinite barrier potential was considered by Brunhes et al. to study SONS. They predicted a giant SONS compared with the bulk and QW response due to the achievement of resonance conditions with intraband transitions. A deviation from experimental results was observed, which can be reasoned to the infinite potential considered. The research work in [37] was developed in [40] using the finite potential barrier, and a good agreement was obtained between theoretical predictions and experimental measurements of SONS.

In addition to the asymmetrical shape of spontaneously formed QDs during the growth process, most of the works dealing with SONS in QDs under applied electric field consider an infinite barrier or a simple harmonic oscillator type. This is resultant from the complexity associated with the numerical calculations and then only small number of studies dealt with QDs under the applied electric field [15].

The purpose of this chapter is to show that a record value of SONS can be achieved using QD IB and ISB transitions. To demonstrate this effect, the chosen model system was InAs/InGaAs self-assembled QDs, which corresponds to the standard Stranski–Krastanov growth mode [37]. Although the presented results are directly connected to the shape and composition of the QDs, similar features are expected to occur for other types or shapes of QDs. The QD energy subbands are first calculated in the effective mass approximation by solving the three-dimensional Schrodinger equation. Then, SONS is computed in both the conduction and valence bands from the calculated energy dependence of the confined subbands.

Recent development in the field of nonlinear optics has been pushing nonlinear optical materials into practical applications. Nonlinear optical materials are those in which light waves can interact with each other [41]. To measure the nonlinear response of matter to electromagnetic waves in the optical region, in general, high fields are necessary, starting at about 1 kV/cm. The corresponding light intensities of some kW/cm^2 necessitate laser beams. As laser physics started with the ruby laser with its high pulse intensities, it took only few years after the invention of the laser [42] that many classical experiments in nonlinear optics were successfully performed.

9.7 QUANTUM DOT STRUCTURE UNDER STUDY

The QD structure simulated in this theoretical study is a 10-fold InAs QD layer grown by MBE at NanoSemiconductor GmbH in Germany [26,28]. Each QD array is covered with an InGaAs wetting layer (WL) and a 33 nm thick GaAs barrier layer. Each QD active layer is sandwiched between 1.5 µm thick AlGaAs cladding layers. An InAs QD is treated as a quantum disk with a radius of a and a height of h. The corresponding material parameters used are listed in Tables 9.1 and 9.2.

TABLE 9.1

Experimental Parameters

Parameter	InAs (Dot)	In$_x$Ga$_{1-x}$As (Barrier)
E_g (eV)	0.354	0.75 at $x = 0.47$
m_e^*	$0.023m_o$	$0.041m_o$
m_h^*	$0.4m_o$	$0.46m_o$
ζ	41	—
N	5×10^{10} cm^{-2}	—

TABLE 9.2

Experimental Parameters of Relaxations and Frequencies [26]

Parameter	InAs (Dot)	InGaAs (Barrier)
γ_{21} (1/ps)	6.25	—
γ_{20} (1/ps)	1	—
γ_{32} (1/ps)	0.333	—
γ_{31} (1/ps)	0.2	—
γ_{10} (1/ps)	0.3	—
γ_{12} (1/ps)	0.833	—
γ_{02} (1/ps)	0.33	—
γ_{23} (1/ps)	1	—
γ_{13} (1/ps)	0.1	—
γ_{01} (1/ps)	0.25	—
Ω_s (eV)	0.1	—
Ω_p (eV)	1	—

9.8 QUANTUM DISK MODEL UNDER APPLIED ELECTRIC FIELD

Dots here are considered as a quantum disk with a radius of a and a height of h grown on WL in the form of a QW, with a finite constant potential assumed for both quantum disk and WL. The Hamiltonian in the cylindrical coordinates (ρ, ϕ, z) is given by

$$H = -\frac{\hbar^2}{2m^*}\left[\frac{1}{\rho}\frac{\partial}{\partial\rho}\left(\frac{1}{\rho}\frac{\partial}{\partial\rho}\right) + \frac{1}{\rho^2}\frac{\partial^2}{\partial\phi^2} + \frac{\partial^2}{\partial z^2}\right] + V \tag{9.6}$$

where the effective mass is $m^* = m_d^*$ inside the disk and $m^* = m_b^*$ in the barrier. Similarly, the electric potential is $V = V_d$ inside the disk and $V = V_b$ in the barrier. For strongly confined nanostructures, the electron–hole Coulomb interaction is surpassed by the quantization energy [43]. This is used in a large number of works that deal with QDs under an applied electric field. For example, in [21], the Hamiltonian is the same, but the researchers considered infinite potential that is far from practice although it has theoretical importance in viewing the problem. In [44], the similar Hamiltonian is used, but the researchers considered a harmonic oscillator type of confinement potential that makes energy subbands as that of the simpler problem of an electron in a box where energy subbands become multipliers of $\hbar w$, with the addition of the part due to the applied field. Thus, our Hamiltonian is acceptable and more practical. Solving the Schrodinger equation under the parabolic band model gives the wave function of the quantum disk. Each state can be characterized by three integral quantum numbers (nml), where nm and l correspond to ρ–ϕ (transverse) and z-dependence, respectively. The wave function of the state (nml) at the position $r = (\rho, \phi, z)$ is expressed as [12]

$$\phi_d(r) = C_{nm} \frac{e^{im\phi}}{\sqrt{2\pi}} \begin{cases} J_m(p\rho)\cos(k_z z) \\ J_m(p\rho)\cos(k_z h/2)e^{-\alpha(|z|-h/2)} & \rho \leq a \quad \text{and} \quad |z| > h/2 \\ \dfrac{J_m(p\rho)}{K_m(q\rho)} K_m(q\rho)\cos(k_z z) & \rho > a \quad \text{and} \quad |z| \leq h/2 \\ \dfrac{J_m(p\rho)}{K_m(q\rho)} K_m(q\rho)\cos(k_z h/2)e^{-\alpha(|z|-h/2)} & \text{Otherwise} \end{cases} \quad (9.7)$$

where

$J_m(p\rho)$ and $K_m(q\rho)$ are the Bessel function of the first kind and the modified Bessel function of the second kind, respectively

C_{nm} is the normalization constant

p, q, k_z, and α are constants that are determined from the boundary conditions at the interface between the quantum disk and the surrounding material

If the separation of variables is assumed in the solution of the Hamiltonian, an approximate wave function of the quantum disk can be obtained [42] by solving the well-known problems of the two-dimensional circular potential well in the ρ–ϕ direction and the one-dimensional square potential well in the z-direction. In the ρ–ϕ direction, we have a solution of the form

$$\Psi(\rho, \phi) = \frac{e^{im\phi}}{\sqrt{2\pi}} \begin{cases} C_1 J_m(p\rho) & \rho \leq a \\ C_2 K_m(q\rho) & \rho > a \end{cases} \quad (9.8)$$

where

$$p = \frac{\sqrt{2m_d^*(E_\rho - V_d)}}{\hbar} \quad \text{and} \quad q = \frac{\sqrt{2m_b^*(V_b - E_\rho)}}{\hbar}$$

Using the boundary condition in which the wave function Ψ and its first derivative $(d\Psi/dt)$ are divided by the effective mass (i.e., $(1/m^*)(d\Psi/d\rho)$) and are continuous at the interface between the barrier and the dot, one can obtain the eigenequation. The procedure of derivation is described well in [12]. In [27], the results of the model are compared with that obtained from tight-binding calculations and are found convenient with it. For convenience, if the potential in the disk is taken as $V_d = 0$, the transverse eigenenergy E_ρ is given by [12]

$$E_\rho = \frac{\hbar^2}{2m_d^*} \frac{(p\rho)^2}{a^2} \quad (9.9)$$

The solution for the wave function for the z-dependence in a finite QW of width L and depth V_o in the presence of a constant electric field F lies along the positive direction of the well z [13]. Generally, when an electric field is applied to a QW structure as schematically illustrated in Figure 9.7, the profile of the potential will be changed. The total potential is given by [45]

$$V(z,F) = V(z,0) - ezF \tag{9.10}$$

where
   $V(z, 0)$ is the potential profile of the QW
   F is the applied electric field in (kV/cm)
   e is the electronic charge
   z is the associated spatial coordinate

We choose the origin to be at the center of the well.
   Substituting Equation 9.10 into the Schrodinger equation, we arrive at the following formula:

$$-\frac{\hbar^2}{2m^*}\frac{d^2}{dz^2}\psi(z) + |e|Fz\psi(z) = E\psi(z) \quad |z| \leq L/2$$

$$-\frac{\hbar^2}{2m^*}\frac{d^2}{dz^2}\psi(z) + (V_o + |e|Fz)\psi(z) = E\psi(z) \quad |z| \leq L/2 \tag{9.11}$$

where the potential profile of the QD in the z-direction is given by

$$V(z) = \begin{cases} 0 & \text{for } |z| \leq L/2 \\ V_o & \text{for } |z| \geq L/2 \end{cases} \tag{9.12}$$

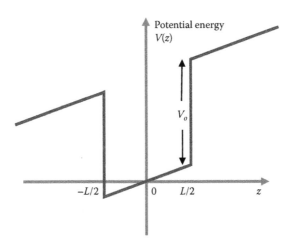

FIGURE 9.7 Potential energy profile $V(z)$ for a quantum dot in the z-direction, with depth V_o and width L subjected to an external electric field. Note that the origin is chosen at the center of the well.

where $V_o = B_{eff}[E_{gw} - E_{gd}]$, in which B_{eff} is the band offset and E_{gw} and E_{gd} are the bandgaps of WL and QD, respectively. The wave function in the well and barrier regions is described by [13]

$$\Psi(z) = \begin{cases} C_3 Ai(\eta_2) & z > -L/2 \\ C_4 Ai(\eta_1) + D_2 Bi(\eta_1) & |z| \le L/2 \\ C_5 \left[Bi(\eta_2) + iAi(\eta_2) \right] & z < -L/2 \end{cases} \tag{9.13}$$

where

C_3, C_4, C_5, and D_2 are constants
Ai and Bi are the homogeneous Airy function

From the properties of Airy function, it is clear that $Bi(\eta_2)$ increases with increasing η_2 and becomes infinity when η_2 is at infinity. In order to make the wave function well behaved in the entire region, this part is not added in the wave function in the region $z > -L/2$. Note that

$$\eta_1 = -\left[\frac{2m^*}{(e\hbar F)^2} \right]^{1/3} (E - |e| Fz)$$

$$\eta_2 = -\left[\frac{2m^*}{(e\hbar F)^2} \right]^{1/3} (E - V_o - |e| Fz) \tag{9.14}$$

The required boundary conditions for the coefficients are obtained from the current continuity conditions at the heterojunction as

$$\Psi(\eta_1)\big|_{z=z_o} = \Psi(\eta_2)\big|_{z=z_o}$$

$$\frac{1}{m_b^*} \frac{d\Psi(\eta_1)}{dz}\bigg|_{z=z_o} = \frac{1}{m_d^*} \frac{d\Psi(\eta_2)}{dz}\bigg|_{z=z_o} \tag{9.15}$$

This results in the determinant

$$\det \begin{vmatrix} Ai(\eta_1^+) & Bi(\eta_1^+) & -Ai(\eta_2^+) & 0 \\ A'i(\eta_1^+) & B'i(\eta_1^+) & -A'i(\eta_2^+) & 0 \\ Ai(\eta_1^-) & Bi(\eta_1^-) & 0 & -\left[Bi(\eta_2^-) + iAi(\eta_2^-) \right] \\ A'i(\eta_1^-) & B'i(\eta_1^-) & 0 & -\left[B'i(\eta_2^-) + iA'i(\eta_2^-) \right] \end{vmatrix} = 0 \tag{9.16}$$

where $\eta_1^\pm$ and $\eta_2^\pm$ are the values of η_1 and η_2 evaluated at $z = L/2$ and $z = -L/2$, respectively. Then, we obtain

$$A_{11} = \left\{ \left[A_i(\eta_1)B_i'(\eta_2) - \left(\frac{m_b^*}{m_d^*} \right) A_i'(\eta_1)B_i(\eta_2) \right] + i \left[A_i(\eta_1)A_i'(\eta_2) - \left(\frac{m_b^*}{m_d^*} \right) A_i'(\eta_1)A_i(\eta_2) \right] \right\} \tag{9.17}$$

$$A_{12} = \left\{ \left[B_i(\eta_1)B_i'(\eta_2) - \left(\frac{m_b^*}{m_d^*} \right) B_i'(\eta_1)B_i(\eta_2) \right] + i \left[B_i(\eta_1)A_i'(\eta_2) - \left(\frac{m_b^*}{m_d^*} \right) B_i'(\eta_1)A_i(\eta_2) \right] \right\} \tag{9.18}$$

$$A_{21} = \left\{ \left[A'i(\eta_1)Ai(\eta_2) - \left(\frac{m_b^*}{m_d^*} \right) Ai(\eta_1)A'i(\eta_2) \right] \right\} \tag{9.19}$$

$$A_{22} = -\left\{ \left[A'i(\eta_2)Bi(\eta_1) - \left(\frac{m_b^*}{m_d^*} \right) Ai(\eta_2)B'i(\eta_1) \right] \right\} \tag{9.20}$$

$$a_o A_{11} + b_o A_{12} = 0$$
$$a_o A_{21} + b_o A_{22} = 0 \tag{9.21}$$

$$\begin{vmatrix} A_{11} & A_{12} \\ A_{21} & A_{22} \end{vmatrix} = 0 \tag{9.22}$$

The eigenenergy E_z is obtained by solving Equation 9.22. The total eigenenergy of the quantum disk E_d is, approximately, the summation of the transverse and longitudinal eigenenergies and is expressed as

$$E_d = E_\rho + E_z \tag{9.23}$$

This gives the eigenenergy of the QD structure under the applied electric field.

9.9 CALCULATED QD SUBBANDS UNDER APPLIED ELECTRIC FIELD

Figure 9.8 shows the calculated QD ground (GS) and excited state (ES) conduction subbands under an applied electric field. The solid lines represent the GS conduction subband (Ec_1), while the dashed lines represent the first ES conduction subband (Ec_2). For the height $h = 2$ nm

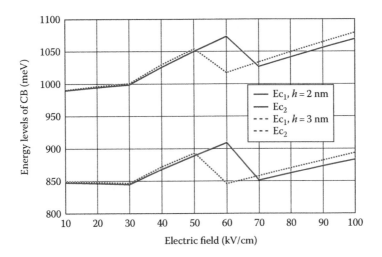

FIGURE 9.8 Quantum dot energy conduction subbands versus the applied electric field for heights $h = 2$ nm (the red line) and $h = 3$ nm (the blue line). The disk radius $a = 13$ nm. The solid lines are for ground state (Ec_1), while the dashed lines are for first excited state (Ec_2).

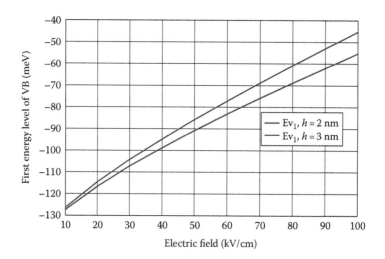

FIGURE 9.9 Quantum dot ground-state valence subband versus the applied electric field for heights $h = 2$ nm (the red line) and $h = 3$ nm (the blue line). The disk radius is $a = 13$ nm.

(the red lines) and when the field changes from 10 to 100 V/m, Ec_1 changes by 35.4 meV, while Ec_2 changes by 78.9 meV. This is a Stark shift for QD subbands. Thus, a nonmonotonic change was obtained with the applied field on the QD ES conduction subband. For $h = 3$ nm (the blue lines), approximately similar changes were obtained. A nonmonotonic change of energy subbands with an electric field is also shown in [45,46]. This behavior leads us to use Airy functions to describe wave functions under an applied electric field. From Figure 9.9, it is shown that subband energy becomes higher at some value of the applied electric field. This depends on the type of confinement, QD shape, and size. This result coincides with the conclusion drawn in [19,45,46]. This can be justified by a resonance that occurs between the applied field and the subband energy.

Figure 9.9 shows the change of QD GS valence subband for disk radius $a = 13$ nm and for two heights ($h = 2$ nm the red line and $h = 3$ nm the black line) when the field changes by the same range (10–100 V/m). For $h = 2$ nm, GS shifts by 89 meV. This shows that the valence subbands are more sensitive to the applied electric field.

Additionally, as it is shown in Figure 9.10, GS conduction subband was plotted for different quantum disk radii and at three field values. The subband energy decreased with increasing the disk radius and increased with increasing the electric field. Electrons and holes are separated under the applied electric field, and they are pushed apart by the field. Reducing the QD size increases the confinement of carrier wave functions. The shift of QD energy subbands by the Stark effect competes with the shift due to the quantum-size effect. Increasing the disk radius by 1 nm reduces the subband energy by ~3 meV, while increasing electric field by 1 kV/cm increases the subband energy by ~0.7 meV.

9.10 DENSITY MATRIX FORMULATION OF OPTICAL SUSCEPTIBILITY

To derive SONS relation, we use the density matrix approach. It has been proven very useful to study quantum electronic processes in different material systems [47]. The density operator ρ satisfies the equation of motion

$$\frac{\partial}{\partial t}\rho = \frac{-i}{\hbar}[H,\rho] \tag{9.24}$$

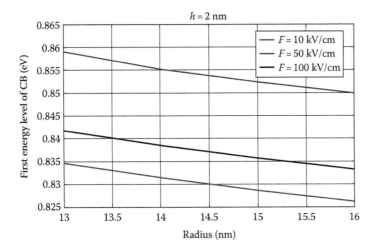

FIGURE 9.10 Quantum dot ground-state energy conduction subbands versus radius for three values of applied electric field. The disk height is $h = 2$ nm.

The Hamiltonian operator consists of three parts

$$H = H_0 + H' + H_{random} \tag{9.25}$$

where H_0 is the unperturbed Hamiltonian and H' accounts for the interaction such as the electron–photon interaction, which is given by

$$H' = -\mu \cdot E(t) = -\sum_{j=1}^{3} \mu^j E_j(t) \tag{9.26}$$

where μ is the dipole operator, $\mu = er$, and $e = -|e|$ for electrons and $+|e|$ for holes. E is the electric field. H_{random} includes the relaxation effects due to incoherent scattering processes. The equation of motion, Equation 9.24, becomes

$$\frac{\partial}{\partial t}\rho = \frac{-i}{\hbar}[H_0 + H', \rho] + \left(\frac{\partial \rho}{\partial t}\right)_{relax} \tag{9.27}$$

where

$$\left(\frac{\partial \rho}{\partial t}\right)_{relax} = \frac{-i}{\hbar}\left[H_{random}, \rho - \rho^{(0)}\right] \tag{9.28}$$

which can be written in terms of T_1 and T_2 time constants as follows:

$$\frac{\partial}{\partial t}\left(\rho_{uu} - \rho_{uu}^{(0)}\right)_{relax} = -\frac{\rho_{uu} - \rho_{uu}^{(0)}}{(T_1)_{uu}} \tag{9.29}$$

$$\left(\frac{\partial}{\partial t}\rho_{uu}\right)_{relax} = -\frac{\rho_{uu}}{(T_2)_{uv}} \quad \text{for } u = v \tag{9.30}$$

For the initial distributions, $\rho_{uv}^{(0)} = \rho_{uu}^{(0)}\delta_{uv}$, which are diagonal for each state. It is convenient to use

$$\gamma_{uu} = \frac{1}{(T_1)_{uu}}$$

$$\gamma_{uv} = \frac{1}{(T_2)_{uv}} \quad \text{for } u \neq v$$

(9.31)

Taking the uv component of equation of motion (Equation 9.27) and using

$$\left\langle u \middle| H_0\rho - \rho H_0 \middle| v \right\rangle = (E_u - E_v)\rho_{uv}$$

(9.32)

we obtain the density matrix equation in the presence of an optical excitation

$$\frac{\partial}{\partial t}\rho_{uv} = \frac{-i}{\hbar}(E_u - E_v)\rho_{uv} + \frac{i}{\hbar}\sum_{u'}\sum_j \left(\mu_{uu'}^j\rho_{u'v} - \rho_{uu'}\mu_{u'v}^j\right)E_j(t) - \gamma_{uv}\left(\rho_{uv} - \rho_{uv}^{(0)}\right)$$

(9.33)

In general, we may define

$$w_{uv} = \frac{E_u - E_v}{\hbar}$$

(9.34)

and consider the interaction term $H' = -\mu \cdot E(t)$ as a small perturbation. The perturbation series gives

$$\rho = \rho^{(0)} + \rho^{(1)}(E) + \rho^{(2)}(E^2) + \cdots$$

(9.35)

One obtains

$$\frac{\partial}{\partial t}\rho_{uv}^{(n+1)} = (-iw_{uv} - \gamma_{uv})\rho_{uv}^{(n+1)} + \frac{i}{\hbar}\sum_{u'}\sum_j \left(\mu_{uu'}^j\rho_{u'u}^{(n)} - \rho_{uu'}^{(n)}\mu_{u'v}^j\right)\tilde{E}_j(t)$$

(9.36)

For $n \geq 0$, consider an optical field given by

$$\tilde{E}(t) = \sum_{\alpha=1}^{N} \tilde{E}(w_\alpha)e^{-iw_\alpha t}$$

(9.37)

The polarization per unit volume is calculated from the trace of the matrix product of the dipole moment matrix μ and the density matrix

$$\tilde{P}(t) = \frac{1}{V}Tr[\rho(t)\mu]$$

$$= \frac{1}{V}\sum_{uv}\rho_{uv}(t)\mu_{uv}$$

$$= \frac{1}{V}(\rho_{aa}\mu_{aa} + \rho_{ab}\mu_{ba} + \rho_{ba}\mu_{ab} + \rho_{bb}\mu_{bb})$$

(9.38)

The ith component of the polarization density, to the first order in the optical electric field, is given by

$$\tilde{P}_i^{(n)}(t) = \frac{1}{V}\left(\mu_{ba}^i \rho_{ab}^{(n)}(w) + \mu_{ab}^i \rho_{ba}^{(n)}(w)\right)e^{-iwt}$$

$$+ \frac{1}{V}\left(\mu_{ba}^i \rho_{ab}^{(n)}(-w) + \mu_{ab}^i \rho_{ba}^{(n)}(-w)\right)e^{iwt} \tag{9.39}$$

The definition of the electric susceptibility χ_{ij} is obtained using

$$\tilde{P}^{(n)}(t) = \varepsilon_0 \chi^{(n)}(w)\tilde{E}(w)e^{-iwt} + \varepsilon_0 \chi^{(n)}(-w)\tilde{E}(-w)e^{+iwt} \tag{9.40}$$

By calculating the density operator $\rho^{(n)}$ for the nth order for the two- or three-level system and substituting into Equation 9.39 and comparing with Equation 9.40, we obtain the optical susceptibility $\chi^{(n)}(w)$.

In the case of conventional (i.e., linear) optics, the induced polarization depends linearly on the electric field strength in a manner that can often be described by the relationship

$$\tilde{P}^{(1)}(t) = \varepsilon_0 \chi^{(1)}\tilde{E}(t) \tag{9.41}$$

where the constant of proportionality $\chi^{(1)}$ is known as the linear susceptibility and ε_0 is the permittivity of free space. In nonlinear optics, the optical response can often be described by generalizing Equation 9.41 by expressing the polarization $P(t)$ as a power series in the field strength $E(t)$ as

$$\tilde{P}(t) = \varepsilon_0\left[\chi^{(1)}\tilde{E}(t) + \chi^{(2)}\tilde{E}^2(t) + \chi^{(3)}\tilde{E}^3(t) + \cdots\right]$$

$$\equiv \tilde{P}^{(1)}(t) + \tilde{P}^{(2)}(t) + \tilde{P}^{(3)}(t) + \cdots \tag{9.42}$$

The quantities $\chi^{(2)}$ and $\chi^{(3)}$ are known as the second- and third-order nonlinear optical susceptibilities, respectively.

In the three-level system considered (Figure 9.11) here, only $2 \rightarrow 1$ and $3 \rightarrow 2$ transitions are dipole allowed [29]. From the earlier relations, the IB transition, $\rho_{21}^{(2)}$, can be derived. Then, SONS for IB transition is written as

$$\chi^{(2)}(w_s) = \frac{1}{\hbar\varepsilon_0}\left\{\frac{\Omega_s\mu_{12}(\rho_{11}^{(0)} - \rho_{22}^{(0)})(\mu_{22} - \mu_{11})}{[(w_{21} - w_s) - i\gamma_{21}]^2}\right\} \tag{9.43}$$

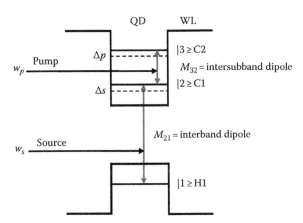

FIGURE 9.11 Schematic diagram of the potential profiles and relevant parameters in a semiconductor quantum dot system. (H1 = the heavy-hole ground state, C1 = the conduction band ground state, and C2 = the first excited state of conduction band.)

Similarly, for ISB transitions, ISB SONS was derived. It can be written as

$$\chi^{(2)}(w_p) = \frac{1}{\hbar\varepsilon_0}\left\{\frac{\Omega_p\mu_{23}(\rho_{22}^{(0)} - \rho_{33}^{(0)})(\mu_{33} - \mu_{22})}{\left[(w_{32} - w_p) - i\gamma_{32}\right]^2}\right\} \qquad (9.44)$$

where ω_s and ω_p are their frequency separations of the signal and the pump, respectively, $\Omega_s = \mu_{21}E_s/\hbar$, $w_{21} = (E_2 - E_1)/\hbar$ and $\Omega_p = \mu_{32}E_p/\hbar$, $w_{32} = (E_3 - E_2)/\hbar$. The ISB dipole moment (μ_{32}) between the GS (C1 or level 2) and the first ES (C2 or level 3) of the conduction band is calculated according to [25]

$$\mu_{32} = \left\langle \phi_{d3}(r) \left| er \right| \phi_{d2}(r) \right\rangle \qquad (9.45)$$

where

  e is the unit charge
  $|r|$ is the distance

On the other hand, the IB dipole moment (μ_{21}) is expressed as follows [25]:

$$\mu_{21} = \frac{e}{m_0 w_{21}} \zeta \left\langle u_c \left| \hat{p} \right| u_v \right\rangle \left\langle \phi_{d2}(r) \left| \phi_{d1}(r) \right\rangle \qquad (9.46)$$

where

  ζ is an enhancement factor due to excitonic effects
  m_0 is the free electron mass

9.11 RESULTS ON SONS

The real and imaginary parts of the SONS for IB transition at 10–100 kV/cm electric field strengths are shown in Figure 9.12. From Figure 9.12a through c, it is shown that both real and imaginary parts of susceptibility were reduced and the peak wavelength was shifted toward the longer wavelength by increasing the electric field strength. The susceptibility was reduced by two times and the peak wavelength was increased by 36 nm when the electric field increases to 50 kV/cm and by 1.5 times and 89 nm when the electric field increases to 100 kV/cm. This is consistent with the experimental evidence that the applied electric field quenches the photoluminescence in QDs [16]. It is a result of the reduced overlap between electron and hole wave functions, where they are separated by the electric field due to their opposite charges. Our results of SONS in Figure 9.12 are on the order of 10^{-2} m/V, which is in the range of He and Xie results [48].

For the case of conduction ISB transitions, SONS is shown in Figure 9.13, both real and imaginary parts of SONS were increased by more than two orders of magnitude, and their peak wavelengths were increased more than 7000 nm when the field changes from 10 to 100 kV/cm. Note that the peak wavelength was reduced when the field is 50 kV/cm than that at 10 and 100 kV/cm. This relates to the position of subband energy when the field was increased to 50 kV/cm as shown in Figures 9.8 and 9.10.

SONS, for valence ISB transitions, is shown in Figure 9.14. The peak values are of the same order of that of conduction ISB transitions, while their wavelengths are increased by more than one order of magnitude. Their obtained wavelengths lie between 33 and 157 µm, which are important in terahertz applications. The subband energy difference at 10 kV/cm is 0.8, 0.14, and 0.037 eV for the first IB transition, conduction ISB transition, and first valence ISB, respectively. These differences can explain the ranges of wavelengths of these structures shown in Figures 9.12 through 9.14.

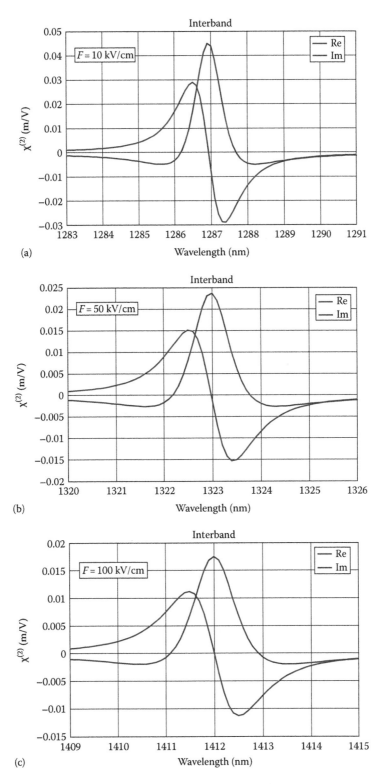

FIGURE 9.12 The imaginary (the blue line) and the real (the red line) second-order nonlinear susceptibility for interband transition under applied electric field with strengths of (a) 10 kV/cm, (b) 50 kV/cm, and (c) 100 kV/cm.

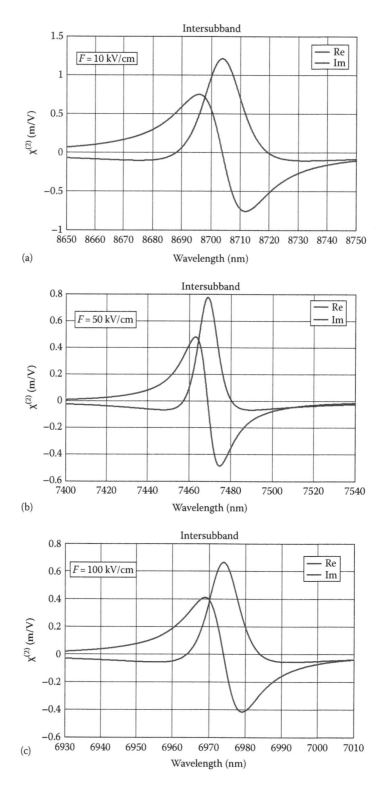

FIGURE 9.13 The imaginary (the blue line) and the real (the red line) second-order nonlinear susceptibility for conduction intersubband transition under applied electric field with strengths of (a) 10 kV/cm, (b) 50 kV/cm, and (c) 100 kV/cm.

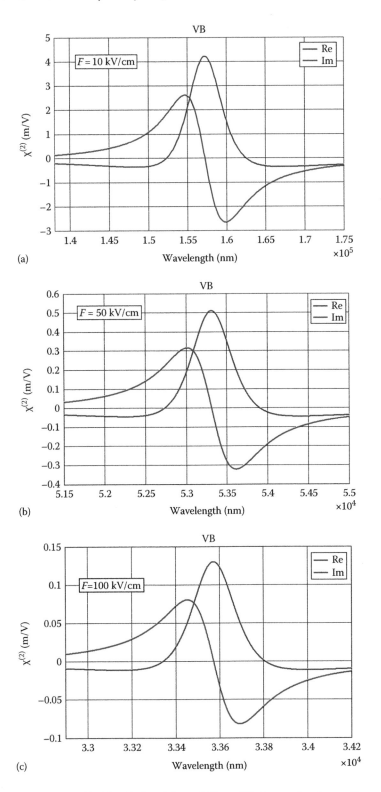

FIGURE 9.14 The imaginary (the blue line) and the real (the red line) second-order nonlinear susceptibility for valence intersubband transition under applied electric field with strengths of (a) 10 kV/cm, (b) 50 kV/cm, and (c) 100 kV/cm.

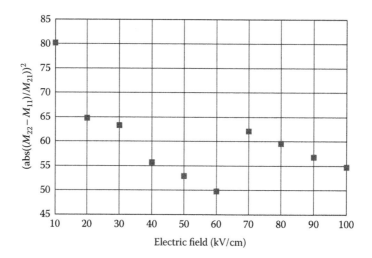

FIGURE 9.15 Ratio of momentum matrix elements for quantum dot structure shown in Figure 9.11 versus the applied electric field.

Our result of SONS for both conduction and valence ISB transitions in Figures 9.13 and 9.14 is higher by one order of magnitude than that of IB transitions shown in Figure 9.12. The main factor that controls SONS in QDs, here, is the momentum matrix element. For conduction ISB transitions, the factor $\mu_{23}(\mu_{33} - \mu_{22})$ in Equation 9.44 is higher by one order of magnitude than the factor $\mu_{12}(\mu_{22} - \mu_{11})$ in Equation 9.43 for IB transitions. This can explain why SONS of ISB transitions is higher than that of IB structure. Figure 9.15 shows the ratio $|(M_{22} - M_{11})/M_{12}|$ for the IB structure where a nonmonotonic change is shown.

9.12   SHG, SFG, DFG, AND OR

The second-order nonlinear processes SHG, SFG, DFG, and OR are the simplest examples of nonlinear molecular spectroscopes and will serve as our example of diagrams that describe nonlinear spectroscopy. Second-order spectroscopes vanish for isotropic samples and are therefore surface selective. There are various nonlinear optical processes. Some of them involve the generation of a new frequency that is different from that of the pump source, while some others do not. The former includes SHG, SFG, and DFG. They are also the topics that will be involved in this chapter. The processes of SHG, SFG, and DFG are all due to the second-order optical nonlinearity [49].

An electromagnetic field incident on a medium induces bound electrons to oscillate about their equilibrium position. In the linear regime, the resulting dielectric polarization is proportional to the applied electric field [8]. The polarization is described by Equation 9.42, where

$$\tilde{E}(t) = E_1 e^{-iw_1 t} + E_2 e^{-iw_2 t} + \text{c.c.} \tag{9.47}$$

It can be shown that the nonlinear polarization is given by

$$P^2(t) = \varepsilon_0 \chi^{(2)} \left[E_1^2 e^{-2iw_1 t} + E_2^2 e^{-2iw_2 t} + 2E_1 E_2 e^{-i(w_1 + w_2)t} + 2E_1 E_2^* e^{-i(w_1 - w_2)t} + \text{c.c.} \right]$$

$$+ 2\varepsilon_0 \chi^{(2)} \left[E_1 E_1^* + E_2 E_2^* \right] \tag{9.48}$$

It is convenient to express this result using the notation

$$P^{(2)}(t) = \sum_n P(w_n)e^{-iw_n t} \tag{9.49}$$

where the summation extends over positive and negative ω_n frequencies. The complex amplitudes of various frequency components of the nonlinear polarization are hence given by [8]

$$P(2w_1) = \varepsilon_0 \chi^{(2)} E_1^2 \quad \text{(SHG)}$$

$$P(2w_2) = \varepsilon_0 \chi^{(2)} E_2^2 \quad \text{(SHG)}$$

$$P(w_1 + w_2) = 2\varepsilon_0 \chi^{(2)} E_1 E_2 \quad \text{(SFG)} \tag{9.50}$$

$$P(w_1 - w_2) = 2\varepsilon_0 \chi^{(2)} E_1 E_2^* \quad \text{(DFG)}$$

$$P(0) = 2\varepsilon_0 \chi^{(2)} (E_1 E_1^* + E_2 E_2^*) \quad \text{(OR)}$$

Note that, in accordance with our complex notation, there is also a response at the negative of each of the nonzero frequencies just given by

$$P(-2w_1) = \varepsilon_0 \chi^{(2)} E_1^{*2} \quad \text{(SHG)}$$

$$P(-2w_2) = \varepsilon_0 \chi^{(2)} E_2^{*2} \quad \text{(SHG)}$$

$$P(-w_1 - w_2) = 2\varepsilon_0 \chi^{(2)} E_1^* E_2^* \quad \text{(SFG)} \tag{9.51}$$

$$P(w_2 - w_1) = 2\varepsilon_o \chi^{(2)} E_2 E_1^* \quad \text{(DFG)}$$

However, since each of these quantities is simply the complex conjugate of one of the quantities given in Equation 9.50, it is not necessary to take explicit account of both the positive and negative frequency components.

For SFG, according to Equation 9.51, the complex amplitude of the nonlinear polarization describing this process is given by the expression [8]

$$P(w_1 + w_2) = 2\varepsilon_0 \chi^{(2)} E_1 E_2 \tag{9.52}$$

In many ways, the process of SFG is analogous to that of SHG, except that in SFG where the two input waves are at different frequencies.

For DFG, the process of DFG is described by a nonlinear polarization of the form [8]

$$P(w_1 - w_2) = 2\varepsilon_0 \chi^{(2)} E_1 E_2^* \tag{9.53}$$

Here, the frequency of the generated wave is the difference of those of the applied fields, and DFG and SFG appear to be of very similar processes. We see that the conservation of energy requires that for every photon that is created at the difference frequency $\omega_3 = \omega_1 - \omega_2$, a photon at the higher input frequency (ω_1) must be destroyed, and a photon at the lower input frequency (ω_2) must be created.

Thus, the lower frequency input field is amplified by the process of DFG. For this reason, the process of DFG is also known as optical parametric amplification. According to the photon energy–level description of DFG, the atom, first, absorbs a photon of frequency ω_1 and jumps to the highest virtual level. This level decays by a two-photon emission process that is stimulated by the presence of the ω_2 field, which is already present. Two-photon emission can occur even if the ω_2 field is not applied. The generated fields in such a case are very much weaker, since they are created by spontaneous two-photon emission from a virtual level [8].

The definition of the electric susceptibility χ_{ij} is obtained using

$$
\tilde{P}^{(n)}(2w_s - 2w_s) = \varepsilon_0 \chi^{(n)}(2w_s)\tilde{E}(2w_s)e^{-i(2w_s)t}
$$
$$
+ \varepsilon_0 \chi^{(n)}(-2w_s)\tilde{E}(-2w_s)e^{+i(2w_s)t} \tag{9.54}
$$

It is not necessary to take explicit account of both the positive and negative frequency components [8].

Using the three-level system shown in Figure 9.11, in this configuration, only $2 \rightarrow 1$ and $3 \rightarrow 2$ dipole transitions are allowed, where ω_s and ω_p are their frequency separations of the signal and the pump, respectively, and the optical susceptibility $\chi^{(n)}(2w_s)$ can be written as

$$
\chi^{(2)}(2w_s) = \frac{N\mu_{21}\mu_{12}}{\hbar^2\varepsilon_0}\left\{\frac{(\rho_{11}^{(0)} - \rho_{22}^{(0)})(\mu_{22} - \mu_{11})}{\left[(w_{21} - 2w_s) - i\gamma_{21}\right]^2}\right\} \tag{9.55}
$$

Similarly, the SONS for ISB transitions can be written as

$$
\chi^{(2)}(2w_p) = \frac{N\mu_{32}\mu_{23}}{\hbar\varepsilon_0}\left\{\frac{(\rho_{22}^{(0)} - \rho_{33}^{(0)})(\mu_{33} - \mu_{22})}{\left[(w_{32} - 2w_p) - i\gamma_{32}\right]^2}\right\} \tag{9.56}
$$

Similarly, calculated SFG, DFG, and OR for IB and ISB transitions are derived by the same method.

9.12.1 SUM-FREQUENCY GENERATION

The SFG relation for our system can be written as

$$
\chi^{(2)}(w_s + w_p) = \frac{N\mu_{13}\mu_{31}}{2\hbar^2\varepsilon_0}\left\{\frac{\Omega_s\mu_{32}(\rho_{11}^{(0)} - \rho_{22}^{(0)})}{[(w_{21} - w_s) - i\gamma_{21}][(w_{31} - w_s - w_p) - i\gamma_{31}]}\right.
$$
$$
\left. - \frac{\Omega_p\mu_{21}(\rho_{22}^{(0)} - \rho_{33}^{(0)})}{[(w_{32} - w_p) - i\gamma_{32}][(w_{31} - w_s - w_p) - i\gamma_{31}]}\right\} \tag{9.57}
$$

where $\Omega_s = \mu_{21}E_s/\hbar$, $w_{21} = (E_2 - E_1)/\hbar$ and $\Omega_p = \mu_{32}E_p/\hbar$, $w_{32} = (E_3 - E_2)/\hbar$.

9.12.2 DIFFERENCE-FREQUENCY GENERATION

The DFG relation corresponding to our system can be written as

$$
\chi^{(2)}(w_s - w_p) = \frac{N\mu_{12}\mu_{32}}{2\hbar^2\varepsilon_0}\left\{\frac{\Omega_s\Omega_p^*(\rho_{11}^{(0)} - \rho_{22}^{(0)})(\rho_{33}^{(0)} - \rho_{22}^{(0)})(\mu_{22} - \mu_{11})(\mu_{22} - \mu_{33})}{[(w_{21} - w_s) - i\gamma_{21}]^2[(w_{23} + w_p) + i\gamma_{23}]^2}\right\} \tag{9.58}
$$

where $\Omega_p^* = \mu_{32}E_p^*/\hbar$.

9.12.3 OPTICAL RECTIFICATION

The OR for IB transitions (see Figure 9.11) is

$$\chi^{(2)}(0) = \frac{N\mu_{12}\mu_{21}}{2\hbar^2\varepsilon_0}\left\{\frac{\Omega_s\Omega_s^*(\rho_{11}^{(0)} - \rho_{22}^{(0)})(\rho_{22}^{(0)} - \rho_{11}^{(0)})(\mu_{22} - \mu_{11})(\mu_{22} - \mu_{11})}{[w_{21} - i\gamma_{21}]^2[w_{21} + i\gamma_{12}]^2}\right\} \tag{9.59}$$

The OR for ISB transitions is

$$\chi^{(2)}(0) = \frac{N\mu_{32}\mu_{23}}{2\hbar^2\varepsilon_0}\left\{\frac{\Omega_p\Omega_p^*(\rho_{22}^{(0)} - \rho_{33}^{(0)})(\rho_{33}^{(0)} - \rho_{22}^{(0)})(\mu_{33} - \mu_{22})(\mu_{22} - \mu_{33})}{[w_{32} - i\gamma_{32}]^2[w_{23} + i\gamma_{23}]^2}\right\} \tag{9.60}$$

9.13 INHOMOGENEITY IN QDs

The aforementioned works consider either parabolic confinement or infinite potential. The parabolic confinement gives an equidistant energy subband that is far from experiment. Also, the use of an infinite barrier limits the accuracy comparison with experimental data. Although this type of calculations to get an overview of the physical problem is important, it is impractical to consider a finite barrier for real calculations.

Consider that QD inhomogeneity is important from the practical point of view. The inhomogeneity was included in the study of SONS in QDs in the work of Brunhes et al. [37,40], but they took a QD distribution normalized to the dot height and not as energy distribution. The QD height is not the only factor that is causing inhomogeneous distribution. Many other factors, such as size, dot distribution, and imperfections in the QD shape, contribute to the QD inhomogeneity. Thus, one must cover the spectrum of the dots through the emission spectrum of each one. The energy is the adequate factor to represent. This is what the authors have done here. We need to deal with SONS in QDs and derive that their relations depend on the structure consideration. This is preferred than the present forms applied for any structure atom, QW, QD, or any other types of electronic transition system studied. This is done also here.

The inhomogeneous SONS in QDs is then given by

$$\chi^{(2)} = \int \chi^{(2)}(w_s + w_p, w_p, w_s)D(w)dw \tag{9.61}$$

Including QD inhomogeneity makes our formula different from all other researches that deal with SONS calculations in QDs. It is done here via the convolution over the inhomogeneous density of states, which is given by [22]

$$D(w) = \frac{s^i}{V_{dot}^{eff}}\frac{1}{\sqrt{2\pi\sigma^2}}\exp\left(\frac{-\left(\hbar w - E_{max}^i\right)^2}{2\sigma^2}\right) \tag{9.62}$$

where

s^i is the degeneracy number at the QD state ($s^i = 2$) in the quantum disk model used here

σ is the spectral variance of QDs

$V_{dot}^{eff}(= h/N_D)$ is the effective volume of QDs, h is the dot height, and N_D is the areal density of QDs

The transition energy at the QD maximum distribution of the ith optical transition is E_{max}^i. Note that although the inhomogeneity is included in the study of SONS in QDs in the work of Brunhes et al. [37], they take a QD distribution normalized to the dot height not as energy distribution as in Equation 9.62.

9.14 RESULTS AND DISCUSSION OF SHG, SFG, DFG, AND OR IN QDs

Figure 9.16 shows SONS for the first structure versus SHG wavelength at a 10 kV/cm applied field (Figure 9.16a), where 0.4 m/V SONS peak is obtained at 5095.5 nm peak wavelength. Increasing the field to 100 kV/cm reduces SONS peak to 0.12 m/V, and its peak wavelength is red shifted by approximately 200 nm as in Figure 9.16b.

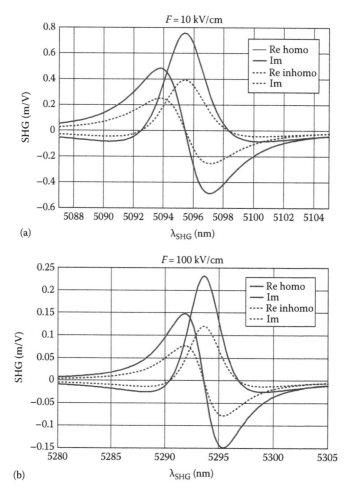

(a)

(b)

FIGURE 9.16 Second-order nonlinear susceptibility real (red lines) and imaginary (blue lines) parts by second-harmonic generation (SHG) versus SHG wavelength for homogeneous (solid lines) and inhomogeneous (dotted lines) of the first quantum dot structure at (a) 10 kV/cm and (b) 100 kV/cm electric field strengths.

Including the QD inhomogeneity in the calculations is important to cover synthesis imperfections. The SONS calculated with inhomogeneous density of states is shown as dashed curves in Figure 9.16a and b. It reduces the peak value of SONS to approximately its half value, this result is shown for both real and imaginary parts of SONS, and this is with our recent result in [24] for the third-order nonlinear susceptibility. Thus, it is important to include inhomogeneity in SONS calculations in QDs.

The ISB case (second structure) is shown in Figure 9.17. When the applied field is 10 kV/m, the spectrum is peaked at 1.18 mm as shown in Figure 9.17a, that is, it is extended by 1.175 mm as compared with Figure 9.16a, while at a 100 kV/m applied field, it is peaked at 0.28 mm as shown in Figure 9.17b, and at this latter case, SONS is reduced by more than three times (in both real and imaginary parts).

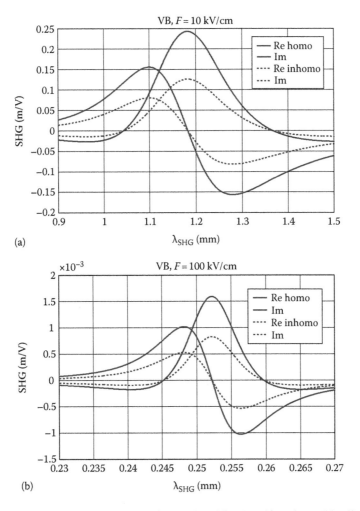

(a)

(b)

FIGURE 9.17 Second-order nonlinear susceptibility real (red lines) and imaginary (blue lines) parts by second-harmonic generation (SHG) versus SHG wavelength for homogeneous (solid lines) and inhomogeneous (dotted lines) of the second quantum dot structure at (a) 10 kV/cm and (b) 100 kV/cm electric field strengths.

SFG in the first structure is shown in Figure 9.18a, where SONS is peaked at 10 μm wavelength under the 10 kV/cm applied field and its value (imaginary SONS) is increased to 1.1 m/V. Figure 9.18b shows SONS when the applied field is increased to 100 kV/cm, where SONS peak becomes 0.35 m/V at 8 μm wavelength peak.

SFG for the second structure (valence ISB case) under 10 kV/cm applied electric field is shown in Figure 9.19a, where SONS peaks at 1.363 mm and its value is reduced by three times compared with IB case. Increasing the field to 100 kV/cm peaks SONS at 0.293 mm, as shown in Figure 9.19b, and its value was reduced by two orders compared with Figure 9.19a.

DFG in the first structure is shown in Figure 9.20a, where SONS is peaked at 7.4 μm under 10 kV/cm applied field and its value is reduced by seven orders compared with SONS of SHG and SFG of the first structure (Figures 9.16a and 9.18a). Figure 9.20b shows the case when the field is

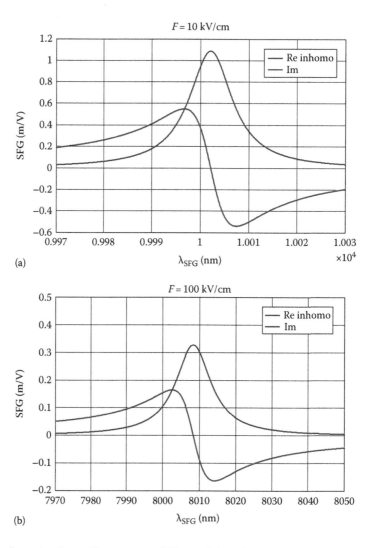

FIGURE 9.18 Second-order nonlinear susceptibility real (red lines) and imaginary (blue lines) parts by sum-frequency generation (SFG) versus SFG wavelength for inhomogeneous broadening of the first quantum dot structure at (a) 10 kV/cm and (b) 100 kV/cm electric field strengths.

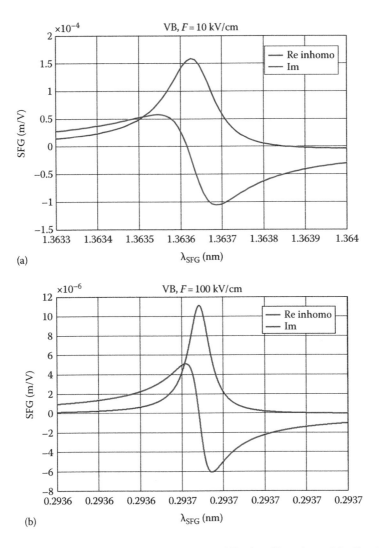

FIGURE 9.19 Second-order nonlinear susceptibility real (red lines) and imaginary (blue lines) parts by sum-frequency generation (SFG) versus SFG wavelength for inhomogeneous broadening of the second quantum dot structure at (a) 10 kV/cm and (b) 100 kV/cm electric field strengths.

increased to 100 kV/cm, where SONS is peaked at 5.3 μm and its value is reduced by one order of magnitude compared with Figure 9.20a.

Figure 9.21a shows DFG for the second structure at 10 kV/m applied electric field, where a high SONS is obtained, and it is peaked at 0.468 mm wavelength. Increasing the field to 100 kV/cm, Figure 9.21b, SONS is peaked at 0.1 mm and its value is reduced by two orders compared with Figure 9.21a.

Figure 9.22a shows OR for the first structure under the applied electric field 10 kV/m, where it is peaked at 1275 nm and its value is 7.8×10^{-5} m/V. Increasing the field to 100 kV/cm reduces SONS peak by more than two times and the peak wavelength is red shifted to 1323 nm.

For OR in the second structure, Figure 9.23a, SONS is reduced by two times compared with Figure 9.22a, while the peak wavelength is 0.285 mm. Reducing the field to 100 kV/cm as in

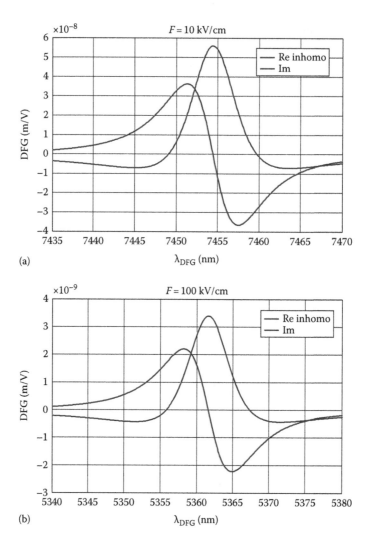

FIGURE 9.20 Second-order nonlinear susceptibility real (red lines) and imaginary (blue lines) parts by difference-frequency generation (DFG) versus DFG wavelength for inhomogeneous broadening of the first quantum dot structure at (a) 10 kV/cm and (b) 100 kV/cm electric field strengths.

Figure 9.23b reduces SONS by four time, while the peak wavelength is 64 μm. For both structures, SONS was reduced with increasing field in all SHG, SFG, DFG, and OR, which refers to a higher asymmetry at a low applied field.

This reduction in SONS is due to the Stark shift of both conduction and valence subband energies under the application of the electric field. For the first (IB) structure, the difference between the first two conduction subbands is 186.3 meV for 100 kV/cm, while this difference is reduced to 142 meV for 10 kV/cm applied field. Additionally, the momentum matrix element μ_{12} is reduced when the field was increased from 10 to 100 kV/cm. These factors cause the reduction of SONS when the field increases to 100 kV/cm. A nonmonotonic behavior of SHG in QDs with field is also noticed by Shao et al. [21]. Chen et al. [44] show the reduction of OR in QDs with an increasing electric field and their results are in the range of 10^{-6} m/V, which is on the order of our results of IB

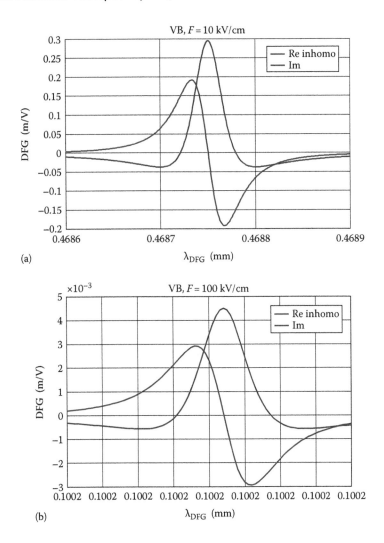

(a)

(b)

FIGURE 9.21 Second-order nonlinear susceptibility real (red lines) and imaginary (blue lines) parts by difference-frequency generation (DFG) versus DFG wavelength for inhomogeneous broadening of the second quantum dot structure at (a) 10 kV/cm and (b) 100 kV/cm electric field strengths.

case (first structure) under 100 kV/cm. Results of Baskoutas et al. range between 10^{-8} and 10^{-3} m/V depending on the type of confinement of QDs [18]. Our results here for peak OR range between 6.5×10^{-11} m/V, for the second structure under 100 kV/cm (Figure 9.23b), and 7.5×10^{-5} m/V, for the first structure under 10 kV/cm (Figure 9.22a). This shows the difference between the confinements of the two structures. The momentum matrix elements of the IB structure are higher than that of the ISB structure.

The difference between the first two valence subbands at 10 kV/cm applied field was 4.2 meV, while this difference increases to 36.7 meV when the field increases to 100 kV/cm, that is, the subbands of the second structure shift by approximately one order of magnitude, while the first (subband) structure increases by 1.3 times with an increasing field. This is with the wavelength

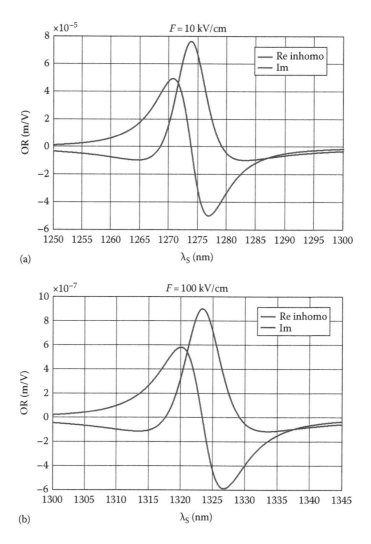

(a)

(b)

FIGURE 9.22 Second-order nonlinear susceptibility real (red lines) and imaginary (blue lines) parts by optical rectification (OR) versus OR wavelength for inhomogeneous broadening of the first quantum dot structure at (a) 10 kV/cm and (b) 100 kV/cm electric field strengths.

shift of the two structures. When the field increases from 10 to 100 kV/cm, the peak wavelength is shifted by 1–2 μm (or less) for the IB structure, while it is shifted by one order of magnitude for the ISB structure.

One of the main results that must be discussed in the study is the very long wavelengths obtained in the second structures, millimeter wavelengths. Energy subbands of the structures can also explain it. While the peak wavelength ranges between 1.2 and 10 μm for the first structure, it ranges between 0.064 and 1.363 mm for the second structure; the reason for this also lies in the values of energy difference between energy states of the two structures. This difference is only few milli–electron volts for the ISB (second) structure, while it is few electron volts for IB (first) structure. Thus, ISB structures can be used in the application of millimeter wavelength.

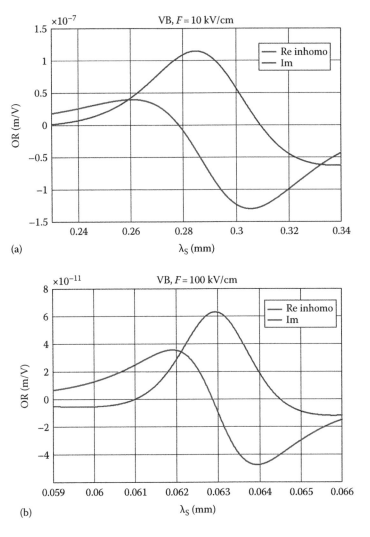

FIGURE 9.23 Second-order nonlinear susceptibility real (red lines) and imaginary (blue lines) parts by optical rectification (OR) versus OR wavelength for inhomogeneous broadening of the second quantum dot structure at (a) 10 kV/cm and (b) 100 kV/cm electric field strengths.

REFERENCES

1. X. Li, C. Zhang, Y. Tang, and B. Wang, Nonlinear optical rectification in asymmetric quantum dots with an external static magnetic field, *Physica E*, 56, 2014, 130–133.
2. A. L. Allenic, Structural, electrical and optical properties of p-type ZnO epitaxial films, PhD thesis, University of Michigan, Ann Arbor, MI, 2008.
3. T. Inoshita and H. Sakaki, Electron relaxation in a quantum dot-significance of Multiphonon Processes, *Physical Review B*, 46, 1992, 7260–7263.
4. J. Shah, *Ultrafast Spectroscopy of Semiconductors and Semiconductor Nanostructures*, Springer-Verlag, Berlin, Germany, 1999.
5. X. Wen, Ultrafast spectroscopy of semiconductor nanostructures, PhD thesis, University of Technology, Melbourne, Victoria, Australia, 2007.

6. M. Abdullah, A study on electronic and optical properties of ZnO-MgZnO quantum-dot semiconductor optical amplifiers, MSc thesis, University of Baghdad, Baghdad, Iraq, 2009.

7. Amin H. Al-Khursan, Intensity noise characteristics in quantum-dot lasers: Four-level rate equations analysis. *Journal of Luminescence*, 113, 2005, 129–136.

8. R. W. Boyd, *Nonlinear Optics*, 3rd edn., Academic Press, London, 2003.

9. M. Sabaeian and A. K. Nasab, Size-dependent intersubband optical properties of dome-shaped InAs/GaAs quantum dots with wetting layer, *Applied Optics*, 51, 2012, 4176–4185.

10. A. Enshaeian, G. Rezaei, and S. F. Taghizadeh, Investigation of an external electric field effects on diamagnetic susceptibility of an off-center hydrogenic donor in a spherical Gaussian quantum dot, in *Proceedings of the Fourth International Conference on Nanostructures (ICNS4)*, March 12–14, 2012, Kish Island, Iran.

11. D. Ahn and S. L. Chuang, Exact calculations of quasibound states of an isolated quantum well with uniform electric field: Quantum-well Stark resonance, *Physical Review B*, 34, 1986, 9034–9037.

12. J. Kim and S. L. Chuang, Theoretical and experimental study of optical gain, refractive index change and linewidth enhancement factor of p-doped quantum dot lasers, *IEEE Journal of Quantum Electronics*, 42, 2006, 942–952.

13. D. Ahn and S. L. Chuang, Calculation of linear and nonlinear intersubband optical absorptions in a quantum well model with an applied electric field, *IEEE Journal Quantum Electronics*, 23, 1987, 2196–2204.

14. Z. Li and X. Hong-Jing, Studies on the second-order nonlinear optical properties of parabolic and semi-parabolic quantum wells with applied electric fields, *Communications in Theoretical Physics*, 41, 2004, 761–766.

15. E. C. Niclescu and M. Cristea, Dielectric mismatch effect on the electronic states in ZnS/CdSe core–shell quantum dots under applied electric field, *U.P.B. The Scientific Bulletin A*, 75, 2013, 195–204.

16. R. Korlacki, R. F. Saraf, and S. Ducharme, Electrical control of photoluminescence wavelength from semiconductor quantum dots in a ferroelectric polymer matrix, *Applied Physics Letter*, 99, 2011, 153112.

17. C. Dane, H. Akbas, S. Minez, and A. Guleroglu, Electric field effect in GaAs/Al/As spherical quantum dot, *Physica E*, 41, 2008, 278.

18. W. Xie, The nonlinear optical rectification of a confined exciton in a quantum dot, *Journal of Luminescence*, 131, 2011, 943–946.

19. S. Baskoutas, E. Paspalakis, and A. F. Terzis, Effects of excitons in nonlinear optical rectification in semiparabolic quantum dots, *Physical Review B*, 74, 2006, 153306.

20. B. Vaseghi, G. Rezaei, V. Azizi, and S. M. Azami, Spin–orbit interaction effects on the optical rectification of a cubic quantum dot, *Physica E*, 44, 2012, 1241–1243.

21. S. Shao, K. Guo, Z. Zhang, N. Li, and C. Peng, Studies on the second-harmonic generations in cubical quantum dots with applied electric field, *Physica B*, 406, 2011, 393–396.

22. D. Ahn and S.-H. Park, *Engineering Quantum Mechanics*, John Wiley & Sons, Hoboken, NJ, 2011.

23. B. Al-Nashy, S. M. M. Amin, and A. H. Al-Khursan, Kerr dispersion in Y-configuration quantum dot system, *Journal of Optics*, 16, 2014, 105205.

24. B. Al-Nashy, S. M. M. Amin, and A. H. Al-Khursan, Kerr effect in Y-configuration double quantum dot System, *Journal of the Optical Society of America B*, 31, 2014, 1991–1996.

25. A. H. Al-Khursan, M. K. Al-Khakani, and K. H. Al-Mossawi, Third-order non-linear susceptibility in a three-level QD system, *Photonics and Nanostructures—Fundamentals and Applications* 7, 2009, 153–160.

26. J. Kim, M. Laemmlin, C. Meuer, D. Bimberg, and G. Eisenstein, Static gain saturation model of quantum-dot semiconductor optical amplifiers, *IEEE Journal of Quantum Electronics*, 44, 2008, 658–666.

27. H. Al-Husaini, A. H. Al-Khursan, and S. Y. Al-Dabagh, III-N QD lasers, *Open Nanoscience Journal*, 3, 2009, 1–11.

28. S. S. Mikhrin, A. R. Kovsh, I. L. Krestnikov, A. V. Kozhukhov, D. A. Livshits, N. N. Ledentsov, Y. M. Shernyakov et al., High power temperature-insensitive 1.3 μm InAs/InGaAs/GaAs quantum dot lasers, *Semiconductor Science and Technology*, 20, 2005, 340–342.

29. C. J. Chang-Hasnain, P. C. Ku, J. Kim, and S. L. Chuang, Variable optical buffer using slow light in semiconductor nanostructures, *Proceedings of IEEE*, 91, 2003, 1884–1896.

30. Y. C. Chang, Nonlinear optical properties of semiconductor superlattices, *Journal of Applied Physics*, 58, 1985, 499–509.

31. S. Suresh, A. Ramanand, D. Jayaraman, and P. Mani, Review on theoretical aspect of nonlinear optics, *Reviews on Advanced Materials Science*, 30, 2012, 175–183.
32. N. Bloembergen, Nonlinear optics, in *Fundamentals of Photonics*, Bahaa E. A. Saleh and Malvin C. Teich (Eds.), John Wiley & Sons, New York, 1991.
33. K. Semwal and S. C. Bhatt, Tuning of wavelengths for producing eye safe laser using second order nonlinear processes, *International Journal of Optics and Applications*, 2, 2012, 20–28.
34. Y. R. Shen, Optical second harmonic generation at interfaces, *Annual Review of Physical Chemistry* 40, 1989, 327–350.
35. P. Fischer and A. C. Albrecht, On optical rectification in isotropic media, *Laser Physics*, 2002, 1177–1181.
36. M. V. Farsi, Investigation of heavily doped congruent lithium tantalite for single-pass second harmonic generation, Master thesis work, Universitat Politècnica de Catalunya (UPC), Universitat Autònoma de Barcelona (UAB), Barcelona, Spain, 2013.
37. T. Brunhes, P. Boucaud, S. Sauvage, A. Lemaıtre, J. M. Gerard, F. Glotin, R. Prazeres, and J.-M. Ortega Infrared second-order optical susceptibility in InAs/GaAs self-assembled quantum dots, *Physical Review B*, 61, 2000, 5562.
38. Y. She, X. Zheng, D. Wang, and W. Zhang, Controllable double tunneling induced transparency and solitons formation in a quantum dot molecule, *Optics Express*, 21, 2013, 17392–17403.
39. P. Ginzburg, A. Krasavin, Y. Sonnefraud, A. Murphy, R. J. Pollard, S. A. Maier, and A. V. Zayats, Nonlinearly coupled localized plasmon resonances: Resonant second-harmonic generation, *Physical Review B*, 86, 2012, 085422.
40. T. Brunhes, P. Boucaud, S. Sauvage, F. Glotin, R. Prazeres, J.-M. Ortega, A. Lemaitre, and J.-M. Gerard, Midinfrared second-harmonic generation in p-type InAs/GaAs self-assembled quantum dots, *Applied Physics Letters*, 75, 1999, 835–837.
41. P. Gter (Ed.), *Nonlinear Optical Effects and Materials*, Springer, Berlin, Germany, 2000.
42. M. Imlau, *Nonlinear Optics*, Fachbereich Physic, University at Osnabruck, Osnabruck, Germany, 2005.
43. D. Bimberg, M. Grumman, and N. Ledentsov, *Quantum Dot Heterostructures*, John Wiley & Sons, Chichester, U.K., 1999.
44. T. Chen, W. Xie, and S. Liang, The nonlinear optical rectification of an ellipsoidal quantum dot with impurity in the presence of an electric field, *Physica E*, 44, 2012,786–790.
45. E. Psarakis, Simulation of performance of quantum well infrared photodetection, PhD thesis, University of California, Berkeley, CA, 2005.
46. W. Xie, Nonlinear optical rectification of a hydrogenic impurity in a disc-like quantum dot, *Physica B*, 404, 2009, 4142–4145.
47. Y. R. Shen, *The Principles of Nonlinear Optics*, John Wiley & Sons, New York, 1984.
48. L. He and W. Xie, Effects of an electric field on the confined hydrogen impurity states in a spherical parabolic quantum dot, *Superlattices and Microstructures*, 47, 2010, 266–273.
49. F. Xie, Resonant optical nonlinearities in cascade and coupled Quantum well structures, PhD thesis, Texas A&M University, College Station, TX, 2008.

10 Metallic Nanopastes for Power Electronic Packaging

Denzel Bridges, Ruozhou Li, Zhiming Gao,
Zhiqiang Wang, Zhiyong Wang, Anming Hu,
Zhili Feng, Leon M. Tolbert, and Ning-Cheng Lee

CONTENTS

10.1 INTRODUCTION

Power electronics are devices designed to handle the control and conversion of electrical power. Such devices include four types of power converters, that is, DC–DC, DC–AC, AC–DC, and AC–AC converters. Power electronics are particularly crucial in the development of renewable energy (RE) technologies, smart grid technologies, and electric/hybrid electric vehicles (HEVs) because these applications (Bose 2012; Hegazy et al. 2013) operate under high power (from tens to hundreds of kW) and/or high temperature (>200°C) (Drobnik and Jain 2013). Power semiconductor modules, the basic elements of a power electronic converter, are specifically designed to handle high power; however, high-temperature operation is still a challenge. As such, the constituent materials of power semiconductor modules (shown in Figure 10.1) must be able to sustain such harsh conditions especially for the packaging materials that bond the power semiconductors to the substrate and bond power electronic modules to the thermal management modules (Sheng and Colino 2004). At high power and high temperatures, these packaging materials face accelerated fatigue conditions. The development of wide-bandgap semiconductors such as silicon carbide (SiC) and gallium nitride (GaN) has increased the potential and feasibility of high-temperature and high-power electronics (Khazaka et al. 2015). Power electronic packaging components including substrates, bonding materials (e.g., soldering alloys), and functional components have to satisfy the requirements of the enhanced bonding strength, high-temperature operation, and resistance to harsh environments.

The bonding materials currently used for die attachment and interconnecting electronic components are generally Pb-containing solders. However, due to the well-known long-term toxic nature of Pb, many countries have placed limitations on the use of Pb-containing materials. A Pb-free substitute is greatly desired. Some Sn-based alternatives have been developed, such as Sn–Bi and Sn–Ag–Cu solder alloys. However, Sn–Bi and Sn–Ag–Cu solder alloys suffer from a melting temperature below 225°C and significant creep deformation at service temperatures (Ohnuma et al. 2000; Shen et al. 2013). Sn–Bi solder alloys sometimes contain nanometallic fillers to improve its creep resistance, hardness, and other mechanical properties. However, these soldering alloys cannot survive at higher temperatures of power electronic applications such as the conditions in exhaust sensing in an electric vehicle (EV), where the temperatures can be as high as 850°C (Greenwell et al. 2011).

Potential alternatives to conventional bulk soldering materials (e.g., Pb–Sn, Sn–Bi, and Sn–Ag–Cu) are metallic nanopastes. Several studies have been done to investigate the development and suitability of metallic nanopastes as a replacement for conventional solders (Chen et al. 2014; Siow 2014). In this chapter, we will discuss the recent progress of synthesis and sintering methods of metallic nanopastes and their potential applications in the power electronic industry. We will also discuss

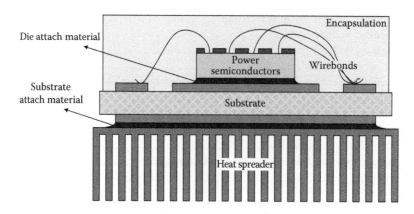

FIGURE 10.1 Basic packaging structure of a power semiconductor module. (From Sheng, W.W. and Colino, R.P., *Power Electronic Modules Design and Manufacture*, CRC Press, Boca Raton, FL, 2004.)

future trends in the development of metallic nanopastes for power electronic packaging. This chapter is organized as follows: first, we will discuss the synthesis methods for Ag-based and Cu-based metallic nanopastes, then sintering fundamentals and techniques, applications in power electronics, and finally future work in the field. We believe that this chapter is useful for advancing the relevant techniques and facilitate innovative power electronic applications.

10.2 SYNTHESIS METHODS

There are two primary approaches to metal nanomaterial synthesis: top-down and bottom-up. In a top-down approach, particles are formed by breaking a bulk metal into smaller particles and dispersed in a proper medium. These methods include mechanical grinding, laser ablation, rapid condensation of metal vapor, thermal heating/decomposition, or plasma excitation of bulk metal or metal powders. In a bottom-up approach, nanomaterials are formed from an ionic precursor by reaction with a reducing agent (Kamyshny and Magdassi 2014). As seen in Figure 10.2, the two main mechanisms of particle growth in a bottom-up approach are (1) the reduction of the precursor and (2) particle coarsening through consuming nearby smaller particles. Bottom-up approaches are among the most popular methods because this approach is promising for large-scale, low-cost metal nanomaterial synthesis. However, laser ablation of bulk metal in a liquid medium is also used to fabricate nanoparticles (NPs) (Nguyen et al. 2012).

10.2.1 CHEMICAL REDUCTION

Chemical reduction methods are one of the most common methods for synthesizing metallic nanomaterials. These processes typically involve dispersing an ionic precursor and reducing agent in a medium and heating to an elevated temperature until the reaction is complete.

In addition to an ionic precursor and reducing agent, polymers and other organic compounds can also be used as capping agents to stabilize materials such as Cu and Fe that are prone to oxidation (Jeong et al. 2013; Kim et al. 2014a; Mott et al. 2007; Peng et al. 2006; Zhang et al. 2012). Some capping agents, such as polyvinylpyrrolidone (PVP), can also be used to regulate the nanomaterial growth in certain directions to form nanomaterials with controlled shapes, such as metallic nanowires (MNWs) or nanoplates (NPLs) (Li et al. 2015). Some methods use an oxygen scavenger such as triethanolamine to prevent highly reactive NPs from oxidation (Chang et al. 2005; Yang et al. 2013). Another way to protect reactive nanomaterials from oxidation during synthesis is to fabricate them in an oxygen-free atmosphere filled with an inert gas such as nitrogen or argon (Ha et al. 2011).

A popular subset of chemical reduction methods is called polyol methods. Polyol methods use polyols (e.g., ethylene glycol, propylene glycol, and diethylene glycol) as both solvents and reducing

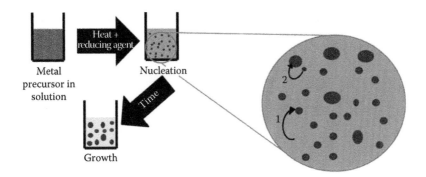

FIGURE 10.2 General process of nanoparticle synthesis using a bottom-up approach. During a typical bottom-up approach, nanoparticles typically grow after nucleation via two different pathways: (1) consuming the precursor material and (2) larger particles consume smaller particles.

agents. These methods allow reactions to be carried out at a higher temperature than aqueous solution (Li et al. 2013; Siow 2012; Yan et al. 2011). Polyols are mild reducing agents and are most suitable for obtaining nanomaterials of electropositive metals such as silver, gold, platinum, and palladium (Dzido et al. 2015; Long et al. 2010, 2011, 2013; Mottaghi et al. 2014). However, polyol methods can be used for more electronegative metals such as copper (Lee et al. 2008; Mott et al. 2007).

There are further two primary types of chemical reduction methods: standard chemical reduction (performed at ambient pressure) and solvothermal (ST)/hydrothermal (HT) reduction (performed at high pressure).

10.2.1.1 Chemical Reduction at Ambient Pressure

The main advantage of chemical reduction methods at ambient pressure is their relative simplicity. It is also easy to determine and control the experimental conditions in situ (e.g., temperature and pressure). Other methods such as ST/HT methods are more difficult to determine the experimental conditions. The chemical reduction process at ambient pressure is also scalable and easy to implement in an industrial setting (Carroll et al. 2011; Lee et al. 2008).

The main disadvantage of wet chemical methods is that the yield is generally low. Wet chemical methods require a significant amount of solvent, most of which is decanted when the synthesis is completed. This is of course quite inefficient. In addition, a conventional wet chemical method is limited to a synthesis temperature lower than or approximately equal to the boiling temperature of the solvent.

10.2.1.2 Hydrothermal/Solvothermal Synthesis

HT/ST synthesis is a chemical reduction method performed under high pressure (several MPa). HT synthesis is performed using an aqueous solution. ST synthesis is the counterpart of HT synthesis that uses a nonaqueous solvent. Usually, the solvent is organic. This method enables users to synthesize metallic nanomaterials at temperatures as high as the supercritical temperatures of the solvent (Kawai-Nakamura et al. 2008; Li et al. 2015). The high pressure nature of the reaction allows the precursor solution to be heated to supercritical temperatures.

HT and ST syntheses can be performed using either a batch reactor or a flow-cell reactor (Miyakawa et al. 2014). Using the batch reactor procedure, it is difficult to control the heat-up time of the precursor solution, which affects the rapid and homogeneous nucleation of the resulting nanomaterials (Aksomaityte et al. 2013). Flow-cell synthesis is a faster HT/ST synthesis technique because the precursor solution can be constantly fed into the apparatus such as the one in Figure 10.3 for large-scale synthesis. The other reactants that can be added via a different line in the flow cell can either be stabilizing agents or additional ionic precursors that enable core–shell NP synthesis (Miyakawa et al. 2014).

Most ST/HT methods use conventional heating methods; however, they can also be incorporated with microwave (MW)-assisted heating. MW heating is typically faster than traditional heating

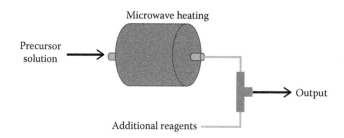

FIGURE 10.3 Basic diagram of the construction of a flow cell incorporated with microwave heating. (From Kawai-Nakamura, A. et al., *Mater. Lett.*, 62, 3471, 2008.)

methods. Traditional heating methods take minutes to hours to complete, while MW heating can take mere seconds to complete (Ramulifho et al. 2012). To efficiently convert MW energy into thermal energy, the solvent must have a high dielectric constant, so polyols and water are popular choices (Kim et al. 2014b; Yi et al. 2013).

10.2.1.3 Sonochemical and Sonoelectrochemical Methods

Sonochemical synthesis of metal nanomaterials uses ultrasonic vibrations generated in a liquid medium to generate the necessary energy and localized hot spots for chemical reduction of the precursor solution to occur (Godínez-García et al. 2012; Xu and Suslick 2010). Sonochemical methods can be quite slow compared to some of the previously mentioned synthesis techniques like HT and ST techniques, sometimes taking hours to complete.

Sonoelectrochemical methods combine sonochemical and electrodeposition methods by using ultrasonic vibrations to enhance a conventional electrodeposition process (Zin et al. 2014). Sonoelectrochemical methods can be faster than some electrodeposition techniques but have many of the same weaknesses as electrodeposition techniques (Chang et al. 2012). Both types of methods allow for synthesis temperatures as low as room temperature (Darroudi et al. 2012; Haas et al. 2006), while the localized heating can be as high as 5000 K (Sakkas et al. 2012). The center of the material is exposed to extreme conditions because the ultrasonic waves from all directions focus at the center, as seen in Figure 10.4.

10.2.2 Electrodeposition

Electrodeposition is a process that uses an electric field between two electrodes to reduce a metal precursor solution and deposit them on a substrate. A sample setup is shown in Figure 10.5. While this technique is frequently used to form thin films (Zhou et al. 2012), it can also be used to fabricate and/or align metal nanowires (NWs) using nanoporous templates (Cui et al. 2009) or microelectrodes (Wang et al. 2013a). Using a nanoporous membrane regulates the growth of NWs in one direction without the use of a capping agent (Ohgai and Hashiguchi 2013). An example of such is shown in Figure 10.5. Anodized alumina films (AAFs) are usually used for templating for NW growth. This method is cheap and simple to synthesize aligned NWs. After growth, the AAF can be easily removed by a strong base solution. However, this templating method is difficult to grow NWs directly from an arbitrary substrate.

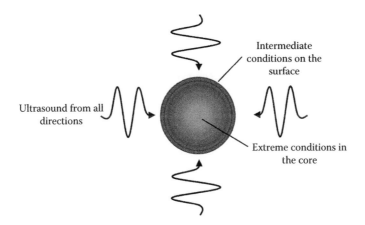

FIGURE 10.4 A metal precursor material that is exposed to ultrasonic vibrations transported through a liquid medium. Since the ultrasound is focused from all direction, all ultrasonic waves focus at the core of the precursor material.

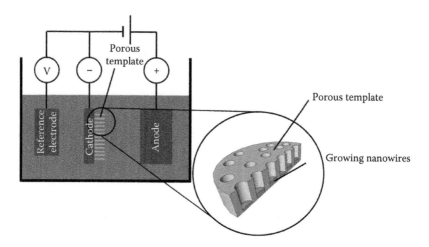

FIGURE 10.5 Schematic of electrodeposition that employs a porous template for growing nanowires. Inset: a schematic of a porous template cross section, inside which nanowires are being grown.

10.2.3 LASER ABLATION

Laser ablation is the process of removing a material from a block or plate using laser irradiation (Kamyshny and Magdassi 2014; Nguyen et al. 2012). Interaction with the laser causes the material to either sublimate or become a plasma, and then the material is rapidly quenched and condenses to form NPs in a liquid or gas medium (Mottaghi et al. 2014). As a plasma, the target material can reach temperatures of several thousand Kelvin (Saito et al. 2014).

Most laser ablation methods in the last few years use laser pulses as opposed to continuous laser irradiation because laser ablation processes with millisecond and longer laser pulses as an input energy source are often described by a thermal evaporation process. For shorter laser pulses (nanosecond or even shorter), the ablation processes is described by additional nonequilibrium processes such as thermal and mechanical fragmentation of superheated melted material, spinodal decomposition, explosive boiling, and spallation (Hu et al. 2007; Itina and Voloshko 2013). The laser wavelength, fluence, and pulse duration are chosen based on the surface plasmon resonance, desired NP size, and particle shape (Amendola and Meneghetti 2009; Becker et al. 1998; Muniz-Miranda et al. 2011). For example, Messina et al. reported creating cubic Ag NPs when a 532 nm laser radiation is used and octahedral Ag NPs when using 1064 nm laser radiation.

Laser ablation is a very versatile top-down method that can be done in an ambient atmosphere, in a vacuum or even in a liquid (Hu et al. 2006, 2007, 2008; Ossi et al. 2011; Tarasenko et al. 2005). Laser ablation can be used to produce NPs of almost any material. One of the most important advantages of laser ablation is that it is a green synthesis (GS) method and generally no additional reagents are required. The disadvantages of laser ablation are the small yield of NPs and the fact that it is difficult to use laser beams serially. These issues are particularly important for industrial applications.

10.2.4 COMPARISON

The synthesis methods discussed in this section differ in maturity, speed, and complexity. These factors play a role in whether or not they are ready for commercial or industrial implementation. In terms of complexity, chemical reduction at ambient pressure is the most widely used and well-understood synthesis technique. Chemical reduction at ambient pressure is also easily scalable for

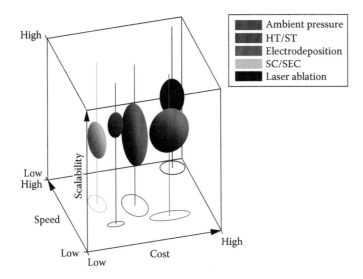

FIGURE 10.6 Comparison of the current synthesis methods.

industrial applications, but depending on the reactants used, this can introduce health hazards to workers and environmental contamination. HT/ST methods have similar health hazards depending on the technique, but when the method is MW-assisted and incorporated with a flow cell, it has the potential to be one of the fastest synthesis techniques. Electrodeposition and sonochemical methods can potentially be used in industrial applications; however, more research is needed to make these methods suitable for large-scale production. Laser ablation is inherently one of the greenest and fastest synthesis methods, but the small yield of NPs and difficulty in coupling laser energy to the target due to attenuation of plumes limits laser power efficiency. Special skills required for operating the laser also limit its application. Figure 10.6 compares five different synthesis methods based on the current cost, speed, and scalability. The cost is judged based on the current cost of equipment and operation for the technique. Scalability is based on the current feasibility of large-scale synthesis of nanomaterials using this technique. The speed is in terms of the relative NP yields per unit time. To satisfy a low-cost, large-scale synthesis for industrial application, a synthesis technique must be highly scalable and fast.

10.3 SINTERING FUNDAMENTALS

Sintering differs from other joining processes (i.e., solid-state bonding, brazing, soldering) in several ways. For instance, brazing involves the soldering/bonding material being distributed between the two bonding surfaces using capillary action after heating to a high temperature, usually above 450°C. Soldering involves melting the soldering/bonding material directly onto the bonding surface at a temperature lower than 450°C. Solid-state bonding does not involve any melting whatsoever but is usually dominated by pressure-assisted solid-state diffusion. Sintering is the process of joining of two particles through a solid-state diffusion (solid-state sintering) or fusing smaller particles together usually by surface melting of the particles (liquid-phase sintering [LPS]). Figure 10.7 shows sintering that occurs in solid state and in liquid state, separately. Sintering generally occurs at a lower temperature than soldering and brazing because the particles are not necessarily completely melted (Zhou and Hu 2011).

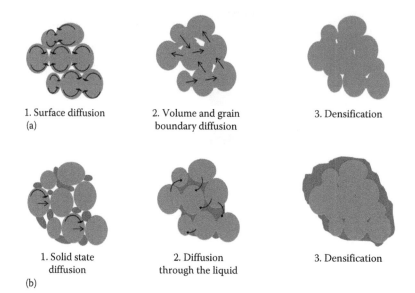

1. Surface diffusion
(a)

2. Volume and grain
boundary diffusion

3. Densification

1. Solid state
diffusion

2. Diffusion
through the liquid

3. Densification

(b)

FIGURE 10.7 Simplified comparison of (a) solid-state sintering and (b) liquid-phase sintering diffusion pathways.

The thermodynamic driving force of sintering is the reduction of surface energy. The driving force is given by

$$\sigma = \gamma \left(\frac{1}{R_1} + \frac{1}{R_2} \right) \tag{10.1}$$

where
  γ is the surface energy of the material
  R_1 and R_2 are the principal radii of curvature of the neighboring particles (Fang and Wang 2008)

10.3.1 Mass Transport Mechanisms

Surface diffusion is the initial mass transport mechanism at sintering temperatures because the activation energy of surface diffusion is smaller than that of other mass transport mechanisms such as lattice diffusion. Surfaces with high curvature typically have a high defect population that includes ledges, kinks, vacancies, and adatoms. The typical atomic movement that occurs during surface diffusion comes in three steps: (1) breaking an atom away from existing bonds, usually at a kink, (2) random motion across the surface, and (3) reattachment at an available surface site. When it comes to sintering two particles in contact with each other, this available surface site is typically on the surface of a neighboring particle.

As sintering progresses, surface diffusion becomes less important and lattice diffusion and grain boundary diffusion increase in importance. Lattice diffusion and grain boundary diffusion are both dominated by the motion of vacancies. As atoms diffuse through the volume of the particle to the grain boundary between fused particles, mass is deposited in the neck region. Grain boundary diffusion is particularly important for sintering densification of metals because grain boundaries form in the sinter bond between misaligned crystals and it is through grain boundary diffusion that mass is deposited in the neck region like the one in Figure 10.7a (German 1996). Vacancies are particularly important for these types of diffusion because generally all types of diffusion occur at a faster rate when the concentration of available lattice sites increases.

The mass transport mechanisms and driving force of LPS differ slightly from solid-state sintering. LPS involves dispersing solid grains in a liquid. The solubility causes the liquid to wet the solid, providing capillary forces, and pulls the grains together (German et al. 2009). Atoms in the liquid phase also display higher mobility, which facilitates the sintering process. Compared to solid-state diffusion, LPS provides higher diffusion rates and is potentially more time-effective because it typically occurs at a higher temperature than a solid-state diffusion procedure (Alarifi et al. 2013; Hu et al. 2013; Marzbanrad et al. 2013).

10.3.2 Thermal Effects

Sintering is usually thermally activated, so naturally increasing the temperature increases the sintering rate. The equilibrium vacancy concentration in a material is proportional to kT, where k is the Boltzmann constant and T is temperature. Therefore, when there is a temperature gradient in a material, generally a vacancy concentration gradient also exists. According to Fick's first law (Equation 10.2), the diffusive flux (J) of a substance (e.g., atoms or vacancies) is proportional to the concentration gradient ($d\Phi/dx$). When vacancies diffuse in one direction, atoms diffuse in the other direction until a steady state is reached. Therefore, when the vacancy concentration increases, the diffusive flux of vacancies increases. Since the diffusive flux is the rate at which a substance is transported per unit area per unit time, a higher diffusive flux in sintering means that atoms are diffusing to the neck region and neighboring particles faster (i.e., the sintering rate increases):

$$J = \text{Diffusivity} \times \left(\frac{d\Phi}{dx} \right) \tag{10.2}$$

In addition, the sintering temperature of silver nanopastes has been shown to be positively correlated with the shear strength of the sintered joint (Alarifi et al. 2011; Khazaka et al. 2014). Akada et al. report that the bonding strength of a sintered joint using silver NPs increases as the sintering temperature increases (Akada et al. 2008).

Chemically active materials such as Cu and Fe may oxidize at high sintering temperatures. Oxides of these materials are typically weak, brittle, and nonconductive (Hai et al. 2013). Such oxides are detrimental to power electronic packaging. Organic shells can be used to stabilize some metal NPs and protect them from oxidation. However, organic shells are not an optimized solution for high-temperature power electronic packing because organic shells typically decompose at high temperatures. In addition, removal of the organic shell is critical to sintering because a thick organic shell prevents direct contact between particle surfaces (Manikam et al. 2012). An example of this is shown in Figure 10.8 (Wang et al. 2013b).

10.3.3 Effect of Pressure

Applying pressure during the sintering process can increase the strength of the sintered joint by increasing the green density. The green density is the ratio between the actual volume of the sintered nanomaterial and the external volume of the sintered structure. A higher green density means that the particles are more close-packed. When the particles are more close-packed, there are more neighboring particles, thereby allowing a lower sintering temperature (German 1996) and stronger bonds between particles.

One drawback to increase the pressure (especially when joining areas ≥100 mm) is that increasing the green density of a nanopaste can prevent solvents and other volatile components in the nanopaste from escaping the center of the sintered region (Fu et al. 2014; Xiao et al. 2013). If the volatile components cannot be removed, these also hinder the sintering process because many of

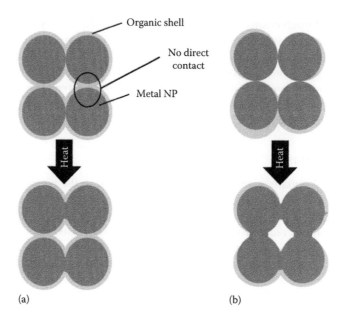

FIGURE 10.8 Example of sintering with an organic shell around the particles. (a) Schematic of sintering metal nanoparticles with a thick organic shell. (b) Schematic of sintering metal nanoparticles with a thin organic shell. (From Wang, S. et al., *Scripta Mater.*, 69, 789, 2013b.)

these components are organic. In addition, it is also a challenge to sinter some power electronic semiconductor materials under high pressure since they are quite fragile.

10.3.4 PARTICLE SIZE

It is well known that the melting point of a material decreases as the particle size decreases as seen in Figure 10.9 (Yang et al. 2013). The same is true for sintering for which the temperature can be calculated from the following equation from Fang et al.:

$$T_{is}(r) = 0.3T_m(\infty)\exp\left[-\frac{2S_m(\infty)}{3k}\frac{1}{(r/r_0)-1}\right]$$

(10.3)

where
    T_{is} is the sintering temperature
    r is the particle radius
    $T_m(\infty)$ is the melting temperature of the bulk material
    $S_m(\infty)$ is the bulk melting entropy
    k is the Boltzmann constant
    $r_0 = 3h$, where h is the atomic diameter (Fang and Wang 2008)

The decreasing sintering temperature of NPs is also due to the increasing surface energy. This higher surface energy increases the driving force, allowing large particles to grow by "consuming" smaller particles (Asoro et al. 2014). Once sintered, the nanomaterial thermal, electrical, and physical properties more closely resemble the bulk material properties (Hu et al. 2010; Lee et al. 2008; Yang et al. 2013).

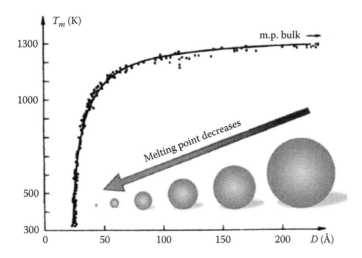

FIGURE 10.9 Relationship between melting temperature and particle diameter of gold nanoparticles. (From Yang, C. et al., *J. Mater. Chem. C*, 1(207890), 4052, 2013.)

10.3.5 SHAPE EFFECT

Sintering of convex surfaces is more energetically favorable than concave surfaces because, according to Fang et al., the curvature of a convex surface is taken to be positive, which yields a positive driving force according to Equation 10.1. When surfaces of different shapes come into contact, anisotropic sintering occurs. For example, Li et al. discuss how the morphology and anisotropic sintering in an Ag NPL/NW composite paste can change as a function of temperature and how it is different in a pure Ag NW paste. In this study, the sides of NWs mostly sinter to the edge of NPLs instead of the face of the NPLs. The surface energy of the edge of an NPL is greater than the face due to the smaller radius of curvature. This increases the driving force of the sintering process according to Equation 10.1 and is thus more energetically favorable than side-to-face sintering (Li et al. 2015).

10.4 SINTERING METHODS

10.4.1 THERMAL SINTERING

Thermal sintering is the most popular and well-understood sintering method. Thermal sintering involves heating a nanopaste to its sintering temperature using a hot plate or furnace. For nanomaterials, sintering can occur at as low as room temperature, but most of the time, this method is conducted at a temperature greater than 100°C (Alarifi et al. 2011; Hu et al. 2010) and can be done with and without additional pressure (Fu et al. 2014; Kähler et al. 2012). This method is also feasible for removing solvents and organic material from the surface of nanomaterials (Kamyshny and Magdassi 2014; Morisada et al. 2010). If the organic shell is removed, it is possible to sinter NPs at room temperature (Marzbanrad et al. 2013; Peng et al. 2013, 2015).

This sintering process is conceptually simple and relies on thermally activated diffusion. Thermal sintering is relatively cheap, but the process is relatively slow, and heating certain NPs to high temperatures can cause unwanted oxides to form.

10.4.2 JOULE HEATING

Joule heating (or current-assisted sintering) is a unique sintering method that employs a sufficiently high current to melt the surface of the nanomaterials and then sinter them together. The dissipated heat due to the internal resistance of the materials in combination with electromigration in the materials causes the materials to locally join.

This joule heating process may take only a few seconds to complete if a pulsed current is applied, and it is practical in an industrial setting. Joule heating is an attractive option for forming conductive interconnections due to its simplicity, cleanliness, and reliability. However, the data on the application of joule heating in nanojoining applications are still limited because high current flow through metal nanomaterials can cause severe structural and functional damage.

Peng et al. reported that this can be offset by including sacrificial NWs to prevent the functional NWs from being damaged (Peng et al. 2009). However, this will further increase the material costs of utilizing this technique. When using DC joule heating, Vafaei et al. report employing a voltage ramp to find a "surge voltage" (the voltage at which there is a sharp increase in current) and impose a current limit to avoid damage to Ag NWs (Vafaei et al. 2014). A recent study has also investigated replacing DC with AC to enhance the sintering process for Ag nanopaste and found that the joints can survive much longer than conventional pressure-assisted thermal sintering (Mei et al. 2014).

10.4.3 Ultrasonic Sealing

Ultrasonic sealing uses high-frequency vibrations beyond the audible range to melt or diffuse materials together. This is a versatile method because the vibrational energy can be transmitted through substrates to the joint area where the mechanical energy is converted to thermal energy through internal friction. The thermal energy is enough to induce either melting or thermally activated diffusion to join nanomaterials.

Unfortunately, the quality of the sintered joint is strongly dependent on the design of the equipment, design of the components being joined, the material properties, and the energy process (Guo 2009). Right now, ultrasonic sealing for nanomaterials has been developed to join carbon nanotubes to titanium electrodes, but there are not a lot of data on ultrasonic sealing of metal nanomaterials (Chen et al. 2006).

10.4.4 Photonic Sintering

By utilizing the plasmonic effect, millisecond light pulses can be concentrated between two adjacent nano-objects to produce "hot spots" without heating their surroundings using flash lamps and camera flash sources (Li et al. 2014). The most effective wavelength of light for photonic sintering is based on the plasmon absorption band. For instance, Ag NPs have a plasmon absorption band of ~430 nm, so blue light is the most effective to initiate the sintering of Ag NPs (Hösel and Krebs 2012; Li et al. 2014). Once the particles are sintered, their plasmonic resonant frequency will dramatically shift to the red side (a long wavelength) (Li et al. 2014; Yang et al. 2011). In this sense, a white light source with wide bands is preferred for sintering Ag NPs.

By not heating the surroundings, the risk of damaging some of the temperature sensitive components in the power semiconductor module is eliminated. Photonic sintering begins with surface atoms that are the ones that absorb the majority of the photons, and then heat transfer from the surface leads to sintering of underlying nano-objects. Unfortunately, this leads to the formation of a higher-density surface layer compared to the internal bulk. Regardless, photonic sintering is very fast and cheap and has already been commercialized by NovaCentrix (2015). However, photonic sintering is not possible for die-attach applications because the nanopaste will not effectively absorb the photons (Kamyshny and Magdassi 2014).

10.4.5 Laser Sintering

Laser sintering is a type of photonic sintering that involves using laser irradiation to selectively excite nanomaterials based on their surface plasmon absorption band, and the absorbed photons melt the nanomaterial, allowing atoms to diffuse between neighboring particles and fuse together (Cui et al. 2009). Most laser sintering methods involve multiple pulsed laser interactions and depend on the length of exposure. Figure 10.10 shows the time scale for various secondary processes in laser–matter interactions (Hu et al. 2011). A long-pulse-width laser or a continuous wave laser may induce thermal sintering, while an ultrashort laser, that is, the laser pulse width shorter than the thermal coupling time, can induce a nonthermal surface melting and thereby result in an LPS (Hu et al. 2013).

By being able to selectively excite nano-objects, laser sintering holds an advantage over normal photonic sintering by being able to form specific structures such as nanoscale porous surfaces (Cui et al. 2009; Huang et al. 2012; Xie et al. 2013). Studies that investigate femtosecond laser welding focus on nonthermal laser–matter interactions. This technique is ideal for fabricating microelectromechanical systems and nanoelectromechanical systems because it allows for precise welding of electrical interconnections.

Laser sintering can also be incorporated in an industrial setting as well. However, it is difficult to use for die-attach applications because the nanomaterials cannot effectively absorb the photons (Kähler et al. 2012; Khazaka et al. 2015; Tan et al. 2015). In addition, femtosecond laser sintering is not fully understood. Three phase changes may occur during femtosecond laser interaction: melting, evaporation, and resolidification. If the laser intensity is too high, the laser can cause fragmentation of the nanomaterial, or the particle is melted completely (Hu et al. 2011).

10.4.6 Comparison

Of the methods discussed in this section, thermal sintering is one of the most well understood, easily implementable, and conceptually simple and feasible for certain applications. For example, thermal sintering is not ideal when depositing metallic nanopastes on temperature sensitive components. Likewise, photonic and laser sintering is challenging for die-attach and other applications where the nanopaste cannot effectively absorb incoming photons. As previously mentioned, photonic sintering and ultrasonic welding have made their way into industrial applications, but laser sintering has yet to be widely used in an industrial setting for metallic nanopastes. Joule heating has potential, but there are not much data on joule heating of metallic nanopastes. In short, thermal sintering is the most mature sintering technique, but there is room in industrial applications for other methods to

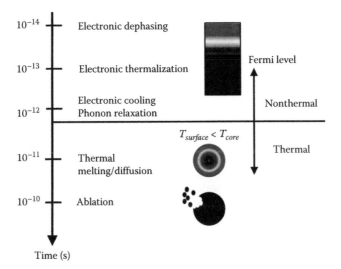

FIGURE 10.10 Time scale for various secondary processes in laser–matter interactions. (From Hu, A. et al., *Open Surf. Sci. J.*, 3, 42, 2011.)

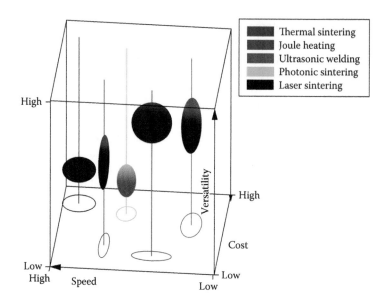

FIGURE 10.11 Comparison of current sintering methods.

make an impact. Figure 10.11 compares sintering techniques based on cost, speed, and versatility. Versatility refers to a particular technique's capability of being used on any part of the power electronic. Currently, thermal sintering, joule heating, and ultrasonic welding are the most balanced techniques in terms of cost, speed, and versatility. Photonic and laser sintering need improvements in terms of cost and versatility before they can be widely used in industrial applications.

10.5 POTENTIAL APPLICATION IN POWER ELECTRONICS

Power electronics have already been heavily integrated into many fields from telecommunications and computing power supplies to transportation to smart grid technologies (Figure 10.12). Metallic nanopastes would mostly be used to make conductive interconnections and die-attach materials (Ahmed et al. 2013; Hegazy et al. 2013).

10.5.1 Electric Vehicles

All EVs have yet to fully catch on as an attractive option for personal transportation, but more countries have started to adopt stricter carbon emission standards and have begun to turn to EVs for public transportation (Ahmed et al. 2013; Nakamura and Mufson 2014), and HEVs are becoming increasingly popular for personal transportation. However, the increasing popularity of EVs and HEVs comes with technical challenges because of new failure modes including inverter failure, sensor failure, and motor overheating (Zhuoping et al. 2014).

Current EV and HEV designs incorporate a number of power electronics including DC–DC converters, DC–AC converters, battery charging systems, and the electric motor (Drobnik and Jain 2013; Dusmez and Khaligh 2013; Madawala and Thrimawithana 2011; Onar et al. 2013). Metallic nanopastes would most likely serve as die-attach materials for high-temperature and high-power-density SiC power modules that currently have unproven high-temperature reliability in EV/HEV environments (Rajashekara 2013). Using metallic nanopastes should improve the efficiency and reliability of high-density power electronics that suffer a loss in efficiency when exposed to high electrical power (Whitaker et al. 2014).

Many of these systems operate at a temperature at or above 200°C (Xu et al. 2013); some may even operate at temperatures as high as 850°C (Greenwell et al. 2011). Electrically conductive metal

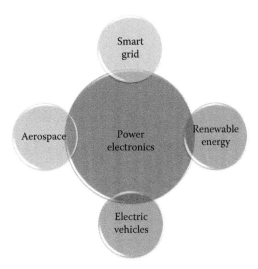

FIGURE 10.12 Applications of power electronics. (From Zhu, H. and Liu, X., *J. Chem. Pharmaceut. Res.*, 6(7), 859, 2014; Khazaka, R. et al., *IEEE Trans. Power Electron.*, 30(5), 2456, 2015; Gould, K. et al., Thermal management of silicon carbide power module for military hybrid vehicles, in *2014 Semiconductor Thermal Measurement and Management Symposium (SEMI-THERM)*, IEEE, 2014, pp. 105–108; Manojkumar, M. et al., Power electronics interface for hybrid renewable energy system—A survey, in *2014 International Conference on Green Computing Communication and Electrical Engineering (ICGCCEE)*, Las Vegas, Nevada, 2014, pp. 1–9.)

nanopastes are also typically thermally conductive, which can transfer heat away from the functional parts of the semiconductor-based components and boost efficiency (Pinkos and Guo 2013).

10.5.2 Renewable Energy

Improving the efficiency of solar cells, wind turbines, and other RE technologies is an important research topic for meeting future energy needs. Aside from losses due to the energy conversion and harvesting (e.g., light energy to electrical energy for photovoltaic cells), RE technologies suffer from DC–AC conversion losses in the power electronic converters when connected to the electrical grid. Photovoltaic cells also suffer from ohmic losses from electrical contacts.

Hybrid energy systems (HESs) and power electronic–based energy storage systems (PEBESSs) are shown in Figures 10.13 and 10.14, respectively, and are being increasingly used and investigated for the electric grid. Both systems incorporate DC–DC converters, DC–AC converters, and other power electronics into their design. HESs and PEBESSs will require more advanced power electronics to be successful (Manojkumar et al. 2014; Sirisukprasert 2014).

Ag pastes and inks are currently used in commercial photovoltaic cells and have excellent electrical properties; however, the Ag pastes and inks alone account for one quarter of the total material cost for crystalline silicon solar cells (Hai et al. 2013; Silva et al. 2011; Yang et al. 2011). Various materials have been considered as a replacement for this expensive material such as Cu–Ag core–shell particles because having a copper core greatly reduces the overall material cost (Bridges et al. 2015).

10.5.3 Smart Grid and Microgrids

A popular topic in electrical engineering is the development of the smart grid, an electric power grid that more efficiently and reliably distributes electrical power to a region based on supply, demand, location, and time. Also of interest to companies are microgrids, which can provide generation locally and continue to provide power to key loads even when there is an interruption from the main

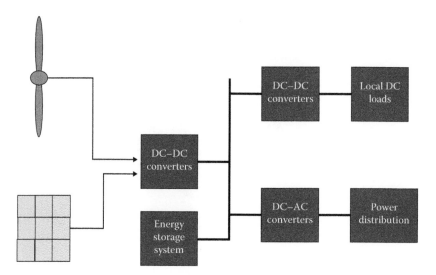

FIGURE 10.13 Schematic of a hybrid energy system and its components. (From Manojkumar, M. et al., Power electronics interface for hybrid renewable energy system—A survey, in *2014 International Conference on Green Computing Communication and Electrical Engineering (ICGCCEE)*, Las Vegas, Nevada, 2014, pp. 1–9.)

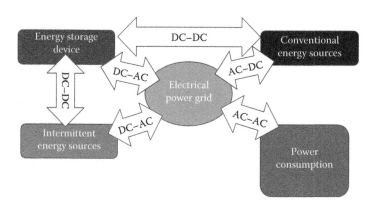

FIGURE 10.14 Block diagram of a the flow of DC–AC power between different components of a smart grid system. (From Kobayashi, T., Power electronics technology in smart grid projects—Applications and experiences, in *2014 International Power Electronics Conference*, Hiroshima, Japan, 2014, pp. 1868–1873.)

utility grid. Key components of smart grids and microgrids are intermittent RE sources such as photovoltaic and wind power, robust large-scale energy storage, regional/microenergy management system, demand-side energy management, and advanced metering infrastructure (Kobayashi 2014). Due to the interactive nature of the smart grid and microgrids, power electronics are necessary for applying varying electrical loads based on demand and location because the spatial and temporal distribution of power in a region is unbalanced (Yue et al. 2014; Zhu and Liu 2014).

The development of the smart grid is also important for high penetration levels of intermittent energy sources such as photovoltaic and wind power. The peak energy production times of intermittent energy sources do not necessarily correlate with peak energy consumption times. In tandem with a large-scale energy storage system, the smart grid would use data from the demand-side energy management system to determine when the energy generated from intermittent energy sources needs to be stored or immediately supplied to the grid. The power electronics again is the key components to implement smart electrical management.

10.5.4 MILITARY, AEROSPACE, AND HIGH-POWER SEMICONDUCTOR DEVICE PACKAGING APPLICATIONS

More advanced power electronics are also of great interest to the military, aerospace, high-power semiconductor laser (HPSL), and high-power light-emitting diodes (HPLEDs). For the military, power electronics have very different requirements. The power electronic packaging must have high resistance to pyrotechnical shocks and vibrations, thermal cycling, and corrosive atmospheres such as salt spray and fog. HEVs are also of great interest to the military (Gould et al. 2014).

In the aerospace industry, the world is moving toward "more electric" and "all electric" aircraft and hence replacing hydraulic systems with power electronic systems. Doing so will significantly improve aircraft reliability and reduce complexity, weight, and maintenance (Pathak 2012).

HPSLs are important for various scientific and industrial instruments such as laser material processing, laser projection display, and fluorescence analysis. Novel packaging materials are sought particularly for thermal management of HPSLs. The obstacles are that current die-attach materials suffer from a relatively low melting temperature and thermal conductivity and high thermal stresses due to a coefficient of thermal expansion mismatch (Higurashi et al. 2015; Okumura et al. 2014).

HPLEDs are desired for various lighting applications such as automobile headlights and other high-brightness lighting applications. Like HPSLs, novel packaging materials are desired mainly for thermal management of HPLEDs. It is well known that the luminosity of light-emitting diodes decreases as temperature increases. Without heat dissipation in the semiconductor, the temperature HPLED will continue to increase and the luminosity will deteriorate (Qu et al. 2014; Thenmozhi et al. 2014). Therefore, a thermally conductive interconnection between the high-power semiconductor module and the heat spreader in Figure 10.1 is highly desirable. Li et al. report a high luminous flux when Ag nanopaste is used as the die-attach material compared to Sn–Ag–Cu solder and silver epoxy (Li et al. 2010).

10.6 FUTURE TRENDS

Metallic nanopastes have advanced greatly in the last two decades, but there is still much room for improvement before commercialization and various applications. We will now discuss the advancements being made in metallic nanopaste technologies.

10.6.1 GREEN SYNTHESIS

As previously mentioned, several methods for synthesizing metallic nanomaterials use toxic reducing agents. As a result, GS procedures are a growing interest among researchers and the industry. GS refers to synthesis procedures that use nontoxic and environmentally friendly reactants. Most GS procedures use aqueous solutions and call for nontoxic organic reducing agents; ascorbic acid (a form of vitamin C) is the most popular one (Usman et al. 2013; Valodkar et al. 2011; Zhao et al. 2011). Other reducing agents such as starch, gelatin (Darroudi et al. 2012), and tartaric acid (Chen et al. 2013) have been used as well.

10.6.2 CORE–SHELL NANOPARTICLES

Core–shell NPs are unique nanostructured materials that are formed by first forming core NP from one metal, then coating with another metal by a transmetalation reaction such as the one in Figure 10.15 for Cu–Ag core–shell NPs (Grouchko et al. 2009) or electrodeposition (Lin-You et al. 2002). The core material is often a cheaper or chemically less stable metal and the shell is usually more chemically stable (Ger et al. 2014; Xia et al. 2010). In the case of Cu–Ag core–shell NPs, Cu and Ag have comparable thermal and electrical properties; however, Cu is prone to oxidation and pure Ag pastes are quite expensive. Therefore, a Cu core can be synthesized and then encased

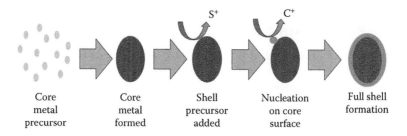

Core metal precursor | Core metal formed | Shell precursor added | Nucleation on core surface | Full shell formation

FIGURE 10.15 Simplified diagram of formation of core–shell nanoparticles by galvanic metal displacement reaction. S^+ refers to a shell material ion and C^+ refers to a core material ion.

in a Ag shell to protect the Cu core from oxidation (Bridges et al. 2015; Kirubha and Palanisamy 2014; Miyakawa et al. 2014; Peng et al. 2012).

10.6.3 NANOALLOYS AND NANOSCALE METALLIC COMPOUNDS

Some nanomaterials are mixed to lower the overall material cost by mixing a more expensive material such as Ag with a less expensive, but compatible material. As previously stated, the result of the sintered nanopaste is comparable to that of a bulk alloy. Nanomaterial mixture displays properties that are a combination of the constituent nanomaterials (Morisada et al. 2010).

Nanomaterials can also be mixed for interesting mechanical and oxidation properties. Li et al. report mixing Ag NWs and NPLs (as seen in Figure 10.16) to bond Cu wire. The result was a joint

(a)

(b)   (c)

FIGURE 10.16 The morphology of (a) Ag nanowires, (b) Ag nanoplates, and (c) Ag nanowire and nanoplate composite paste.

that had electrical conductivity resembling a pure NW paste and increased bonding strength from the added NPLs (Li et al. 2015). Mixing Ag and Cu NPs results in an oxidation-resistant Cu–Ag alloy according to several studies (Tan et al. 2015; Valodkar et al. 2011; Yan et al. 2012).

10.7 SUMMARY

Packaging with nanomaterials (i.e., nanojoining) and power electronic technology are progressing very rapidly. Both technologies will be critical in the way that we generate, transmit, store, and use energy. A number of synthesis and sintering techniques have emerged over the years, providing different metal nanopastes with different shapes and morphologies. Each nanopaste and sintering technique holds different advantages and disadvantages that are best suited depending on the intended production scale and application. The unique properties of metallic nanopastes show great potential to be a viable alternative to Pb-based solders. It is crucial that the cost of synthesizing and sintering metallic nanopastes is greatly reduced before they become widely used in power electronic applications. When improvements are made to reduce the cost and further advance the mechanical properties of metallic nanopastes, it will surely propel power electronics into the future.

ACKNOWLEDGMENTS

The authors appreciate the research initiative funding provided by the University of Tennessee as a new hire package to AH.

REFERENCES

Ahmed, S., H. Kim, T. Prohaska, T. Ronkainen, and R. Burgos. 2013. Stability study of electric vehicle power electronics based power system. In *IEEE International Electric Vehicle Conference,* Santa Clara, CA, October 23–25, 2013.

Akada, Y., H. Tatsumi, T. Yamaguchi, A. Hirose, T. Morita, and E. Ide. 2008. Interfacial bonding mechanism using silver metallo-organic nanoparticles to bulk metals and observation of sintering behavior. *Materials Transactions* 49(7): 1537–1545.

Aksomaityte, G., M. Poliakoff, and E. Lester. 2013. The production and formulation of silver nanoparticles using continuous hydrothermal synthesis. *Chemical Engineering Science* 85(January): 2–10.

Alarifi, H., A. Hu, M. Yavuz, and Y. Norman Zhou. 2011. Silver nanoparticle paste for low-temperature bonding of copper. *Journal of Electronic Materials* 40(6): 1394–1402.

Alarifi, H.A., M. Atiş, C. Özdoğan, A. Hu, M. Yavuz, and Y. Zhou. 2013. Determination of complete melting and surface premelting points of silver nanoparticles by molecular dynamics simulation. *The Journal of Physical Chemistry C* 117(23): 12289–12298.

Amendola, V. and M. Meneghetti. 2009. Laser ablation synthesis in solution and size manipulation of noble metal nanoparticles. *Physical Chemistry Chemical Physics* 11: 3805–3821.

Asoro, M.A., D. Kovar, and P.J. Ferreira. 2014. Effect of surface carbon coating on sintering of silver nanoparticles: In situ TEM observations. *Chemical Communications* 50: 4835–4838.

Becker, M.F., J.R. Brock, H. Cai, D.E. Henneke, J.W. Keto, J. Lee, W.T. Nichols, and H.D. Glicksman. 1998. Metal nanoparticles generated by laser ablation. *Nanostructured Materials* 10(5): 853–863.

Bose, B. 2012. Global energy scenario and impact of power electronics in 21st century. *IEEE Transactions on Industrial Electronics* 60(7): 1.

Bridges, D., R.-Z. Li, M. Floarea, and A. Hu. 2015. Cu–Ag core–shell nanopastes obtained by microwave-assisted solvothermal method and galvanic metal displacement for Cu to Cu bonding. Unpublished work.

Carroll, K.J., J.U. Reveles, M.D. Shultz, S.N. Khanna, and E.E. Carpenter. 2011. Preparation of elemental Cu and Ni nanoparticles by the polyol method: An experimental and theoretical approach. *The Journal of Physical Chemistry C* 115(6): 2656–2664.

Chang, C.C., K.H. Yang, Y.C. Liu, and C.C. Yu. 2012. Surface-enhanced Raman scattering-active silver nanostructures with two domains. *Analytica Chimica Acta* 709: 91–97.

Chang, Y., M.L. Lye, and H.C. Zeng. 2005. Large-scale synthesis of high-quality ultralong copper nanowires. *Langmuir* 21(20): 3746–3748.

Chen, C., L. Yan, E.S.-W. Kong, and Y. Zhang. 2006. Ultrasonic nanowelding of carbon nanotubes to metal electrodes. *Nanotechnology* 17: 2192–2197.

Chen, G., L. Yu, Y.-H. Mei, X. Li, X. Chen, and G.-Q. Lu. 2014. Reliability comparison between SAC305 joint and sintered nanosilver joint at high temperatures for power electronic packaging. *Journal of Materials Processing Technology* 214(9): 1900–1908.

Chen, K.T., D. Ray, Y.H. Peng, and Y.C. Hsu. 2013. Preparation of Cu–Ag core–shell particles with their anti-oxidation and antibacterial properties. *Current Applied Physics* 13(7): 1496–1501.

Cui, Q., F. Gao, S. Mukherjee, and Z. Gu. 2009. Joining and interconnect formation of nanowires and carbon nanotubes for nanoelectronics and nanosystems. *Small* 5(11): 1246–1257.

Darroudi, M., A.K. Zak, M.R. Muhamad, N.M. Huang, and M. Hakimi. 2012. Green synthesis of colloidal silver nanoparticles by sonochemical method. *Materials Letters* 66: 117–120.

Drobnik, J. and P. Jain. 2013. Electric and hybrid vehicle power electronics efficiency, testing and reliability. In *International Electric Vehicle Symposium and Exhibition,* Barcelona, Spain, November 17–20, 2013, pp. 1–12.

Dusmez, S. and A. Khaligh. 2013. A compact and integrated multifunctional power electronic interface for plug-in electric vehicles. *IEEE Transactions on Power Electronics* 28(12): 5690–5701.

Dzido, G., P. Markowski, A. Małachowska-Jutsz, K. Prusik, and A.B. Jarzębski. 2015. Rapid continuous microwave-assisted synthesis of silver nanoparticles to achieve very high productivity and full yield: From mechanistic study to optimal fabrication strategy. *Journal of Nanoparticle Research: An Interdisciplinary Forum for Nanoscale Science and Technology* 17(1): 27.

Fang, Z. and H. Wang. 2008. Densification and grain growth during sintering of nanosized particles. *International Materials Reviews* 53(6): 389–400.

Fu, S., Y. Mei, G.-Q. Lu, X. Li, G. Chen, and X. Chen. 2014. Pressureless sintering of nanosilver paste at low temperature to join large area (≥100 mm²) power chips for electronic packaging. *Materials Letters* 128: 42–45.

Ger, T.-R., H.-T. Huang, C.-Y. Huang, W.-C. Liu, J.-Y. Lai, B.-T. Liu, J.-Y. Chen, C.-W. Hong, P.-J. Chen, and M.-F. Lai. 2014. Comparing the magnetic property of shell thickness controlled of Ag–Ni core–shell nanoparticles. *Journal of Applied Physics* 115: 17B528.

German, R.M. 1996. *Sintering Theory and Practice.* New York: John Wiley & Sons, Inc.

German, R.M., P. Suri, and S.J. Park. 2009. Review: Liquid phase sintering. *Journal of Materials Science* 44: 1–39.

Godínez-García, A., J.F. Pérez-Robles, H.V. Martínez-Tejada, and O. Solorza-Feria. 2012. Characterization and electrocatalytic properties of sonochemical synthesized PdAg nanoparticles. *Materials Chemistry and Physics* 134(2–3): 1013–1019.

Gould, K., S.Q. Cai, C. Neft, and A. Bhunia. 2014. Thermal management of silicon carbide power module for military hybrid vehicles. In *2014, 30th annual Semiconductor Thermal Measurement and Management Symposium (SEMI-THERM),* IEEE, San Jose, CA, pp. 105–108, March 9–13, 2014.

Greenwell, R.L., B.M. McCue, L. Zuo, M.A. Huque, L.M. Tolbert, B.J. Blalock, and S.K. Islam. 2011. SOI-based integrated circuits for high-temperature power electronics applications. In *Conference Proceedings—26th annual IEEE Applied Power Electronics Conference and Exposition (APEC),* Fort Worth, TX, pp. 836–843, March 6–11, 2011.

Grouchko, M., A. Kamyshny, and S. Magdassi. 2009. Formation of air-stable copper–silver core–shell nanoparticles for inkjet printing. *Journal of Materials Chemistry* 19: 3057–3062.

Guo, K.W. 2009. A review of micro/nano welding and its future developments. *Recent Patents on Nanotechnology* 3: 53–60.

Ha, D.-H., L.M. Moreau, C.R. Bealing, H. Zhang, R.G. Hennig, and R.D. Robinson. 2011. The structural evolution and diffusion during the chemical transformation from cobalt to cobalt phosphide nanoparticles. *Journal of Materials Chemistry* 21(31): 11498–11510.

Haas, I., S. Shanmugam, and A. Gedanken. 2006. Pulsed sonoelectrochemical synthesis of size-controlled copper nanoparticles stabilized by poly(N-vinylpyrrolidone). *The Journal of Physical Chemistry B* 110(34): 16947–16952.

Hai, H.T., H. Takamura, and J. Koike. 2013. Oxidation behavior of Cu–Ag core–shell particles for solar cell applications. *Journal of Alloys and Compounds* 564: 71–77.

Hegazy, O., R. Barrero, J. Van Mierlo, P. Lataire, N. Omar, and T. Coosemans. 2013. An advanced power electronics interface for electric vehicles applications. *IEEE Transactions on Power Electronics* 28(12): 5508–5521.

Higurashi, E., K. Okumura, K. Nakasuji, and T. Suga. 2015. Surface activated bonding of GaAs and SiC wafers at room temperature for improved heat dissipation in high-power semiconductor lasers. *Japanese Journal of Applied Physics* 54(March): 030207.

Hösel, M. and F.C. Krebs. 2012. Large-scale roll-to-roll photonic sintering of flexo printed silver nanoparticle electrodes. *Journal of Materials Chemistry* 22(31): 15683.

Hu, A., I. Alkhesho, W.W. Duley, and H. Zhou. 2006. Cryogenic graphitization of submicrometer grains embedded in nanostructured tetrahedral amorphous carbon films. *Journal of Applied Physics* 100: 084319.

Hu, A., G.L. Deng, S. Courvoisier, O. Reshef, C.C. Evans, E. Mazur, and Y. Zhou. 2013. Femtosecond laser induced surface melting and nanojoining for plasmonic circuits. In *Proceedings of the SPIE Plasmonics: Metallic Nanostructures and Their Optical Properties XI,* San Diego, CA, Vol. 8809, August 25, 2013.

Hu, A., J.Y. Guo, H. Alarifi, G. Patane, Y. Zhou, G. Compagnini, and C.X. Xu. 2010. Low temperature sintering of Ag nanoparticles for flexible electronics packaging. *Applied Physics Letters* 97: 153117.

Hu, A., M. Rybachuk, I. Alkhesho, Q.-B. Lu, and W. Duley. 2008. Nanostructure and sp^1/sp^2 clustering in tetrahedral amorphous carbon thin films grown by femtosecond laser deposition. *Journal of Laser Applications* 20(1): 37.

Hu, A., M. Rybachuk, Q.B. Lu, and W.W. Duley. 2007. Direct synthesis of sp-bonded carbon chains on graphite surface by femtosecond laser irradiation. *Applied Physics Letters* 91: 131906.

Hu, A., Y. Zhou, and W.W. Duley. 2011. Femtosecond laser-induced nanowelding: Fundamentals and applications. *The Open Surface Science Journal* 3: 42–49.

Huang, C.-J., H.-T. Cheng, C.-N. Kuo, C.-W. Cheng, W.-L. Tsai, and Y.-H. Yang. 2012. Fabrication of porous ti surface by femtosecond laser sintering of Ti powder. In *2012 Conference on Lasers and Elecro-Optics,* San Jose, CA, Vol. 1, pp. 1–2, May 6–11, 2012.

Itina, T.E. and A. Voloshko. 2013. Nanoparticle formation by laser ablation in air and by spark discharges at atmospheric pressure. *Applied Physics B: Lasers and Optics* 113: 473–478.

Jeong, S., S.H. Lee, Y. Jo, S.S. Lee, Y.-H. Seo, B.W. Ahn, G. Kim et al. 2013. Air-stable, surface-oxide free Cu nanoparticles for highly conductive Cu ink and their application to printed graphene transistors. *Journal of Materials Chemistry C* 1: 2704–2710.

Kähler, J., N. Heuck, A. Wagner, A. Stranz, E. Peiner, and A. Waag. 2012. Sintering of copper particles for die attach. *IEEE Transactions on Components, Packaging and Manufacturing Technology* 2(10): 1587–1591.

Kamyshny, A. and S. Magdassi. 2014. Conductive nanomaterials for printed electronics. *Small* 10(17): 3515–3535.

Kawai-Nakamura, A., T. Sato, K. Sue, S. Tanaka, K. Saitoh, K. Aida, and T. Hiaki. 2008. Rapid and continuous hydrothermal synthesis of metal and metal oxide nanoparticles with a microtube-reactor at 523 K and 30 MPa. *Materials Letters* 62: 3471–3473.

Khazaka, R., L. Mendizabal, and D. Henry. 2014. Review on joint shear strength of nano-silver paste and its long-term high temperature reliability. *Journal of Electronic Materials* 43(7): 2459–2466.

Khazaka, R., L. Mendizabal, D. Henry, and R. Hanna. 2015. Survey of high-temperature reliability of power electronics packaging components. *IEEE Transactions on Power Electronics* 30(5): 2456–2464.

Kim, J.S., H.J. Jeon, H.W. Yoo, Y.K. Baek, K.H. Kim, D.W. Kim, and H.T. Jung. 2014a. Generation of monodisperse, shape-controlled single and hybrid core–shell nanoparticles via a simple one-step process. *Advanced Functional Materials* 24: 841–847.

Kim, M., W.-S. Son, K.H. Ahn, D.S. Kim, H.-S. Lee, and Y.-W. Lee. 2014b. Hydrothermal synthesis of metal nanoparticles using glycerol as a reducing agent. *The Journal of Supercritical Fluids* 90(June): 53–59.

Kirubha, E. and P.K. Palanisamy. 2014. Green synthesis, characterization of Au–Ag core–shell nanoparticles using gripe water and their applications in nonlinear optics and surface enhanced Raman studies. *Advances in Natural Sciences: Nanoscience and Nanotechnology* 5: 045006.

Kobayashi, T. 2014. Power electronics technology in smart grid projects—Applications and experiences. In *2014 International Power Electronics Conference*, Hiroshima, Japan, pp. 1868–1873, May 18–21, 2014.

Lee, Y., J.-R. Choi, K.J. Lee, N.E. Stott, and D. Kim. 2008. Large-scale synthesis of copper nanoparticles by chemically controlled reduction for applications of inkjet-printed electronics. *Nanotechnology* 19: 415604.

Li, R.-Z., A. Hu, T. Zhang, and K.D. Oakes. 2014. Direct writing on paper of foldable capacitive touch pads with silver nanowire inks. *Applied Materials and Interfaces* 6: 21721–21729.

Li, R.-Z., T. Zhang, A. Hu, and D. Bridges. 2015. Ag nanowire and nanoplate composite paste for low temperature bonding. *Materials Transactions* 56: 984–987.

Li, W., M. Chen, J. Wei, W. Li, and C. You. 2013. Synthesis and characterization of air-stable Cu nanoparticles for conductive pattern drawing directly on paper substrates. *Journal of Nanoparticle Research* 15: 1949.

Li, X., X. Chen, and G.Q. Lu. 2010. Effect of die-attach material on performance and reliability of high-power light-emitting diode modules. In *Proceedings of the 60th Electronic Components and Technology Conference*, Las Vegas, Nevada, pp. 1344–1346, June 1–4, 2010.

Lin-You, C., D. Peng, and L. Zhong-Fan. 2002. Preparation of Au(core)–Cu (shell) nanoparticles assembly by electrodeposition. *Acta Physico Chimica Sinica* 18(12): 1062–1067.

Long, N.V., T. Asaka, T. Matsubara, and M. Nogami. 2011. Shape-controlled synthesis of Pt–Pd core–shell nanoparticles exhibiting polyhedral morphologies by modified polyol method. *Acta Materialia* 59(7): 2901–2907.

Long, N.V., N.D. Chien, T. Hayakawa, H. Hirata, G. Lakshminarayana, and M. Nogami. 2010. The synthesis and characterization of platinum nanoparticles: A method of controlling the size and morphology. *Nanotechnology* 21(3): 035605.

Long, N.V., M. Ohtaki, M. Yuasa, S. Yoshida, T. Kuragaki, C.M. Thi, and M. Nogami. 2013. Synthesis and self-assembly of gold nanoparticles by chemically modified polyol methods under experimental control. *Journal of Nanomaterials* 2013: 793125.

Madawala, U.K. and D.J. Thrimawithana. 2011. A bidirectional inductive power interface for electric vehicles in V2G systems. *IEEE Transactions on Industrial Electronics* 58(10): 4789–4796.

Manikam, V.R., K.A. Razak, and K.Y. Cheong. 2012. Sintering of silver–aluminum nanopaste with varying aluminum weight percent for use as a high-temperature die-attach material. *IEEE Transactions on Components, Packaging and Manufacturing Technology* 2(12): 1940–1948.

Manojkumar, M., K. Porkumaran, and C. Kathirvel. 2014. Power electronics interface for hybrid renewable energy system—A survey. In *2014 International Conference on Green Computing Communication and Electrical Engineering (ICGCCEE)*, Coimbatore, India, pp. 1–9, March 6–8, 2014.

Marzbanrad, E., A. Hu, B. Zhao, and Y. Zhou. 2013. Room temperature nanojoining of triangular and hexagonal silver nanodisks. *The Journal of Physical Chemistry C* 117: 16665–16676.

Mei, Y.H., Y. Cao, G. Chen, X. Li, G.Q. Lu, and X. Chen. 2014. Characterization and reliability of sintered nanosilver joints by a rapid current-assisted method for power electronics packaging. *IEEE Transactions on Device and Materials Reliability* 14(1): 262–267.

Miyakawa, M., N. Hiyoshi, M. Nishioka, H. Koda, K. Sato, A. Miyazawa, and T.M. Suzuki. 2014. Continuous syntheses of Pd@Pt and Cu@Ag core–shell nanoparticles using microwave-assisted core particle formation coupled with galvanic metal displacement. *Nanoscale* 6: 8720–8725.

Morisada, Y., T. Nagaoka, M. Fukusumi, Y. Kashiwagi, M. Yamamoto, and M. Nakamoto. 2010. A low-temperature bonding process using mixed Cu–Ag nanoparticles. *Journal of Electronic Materials* 39(8): 1283–1288.

Mott, D., J. Galkowski, L. Wang, J. Luo, and C.J. Zhong. 2007. Synthesis of size-controlled and shaped copper nanoparticles. *Langmuir* 23(10): 5740–5745.

Mottaghi, N., M. Ranjbar, H. Farrokhpour, M. Khoshouei, A. Khoshouei, P. Kameli, H. Salamati, M. Tabrizchi, and M. Jalilian-Nosrati. 2014. Ag/Pd core–shell nanoparticles by a successive method: Pulsed laser ablation of Ag in water and reduction reaction of $PdCl_2$. *Applied Surface Science* 292: 892–897.

Muniz-Miranda, M., C. Gellini, and E. Giorgetti. 2011. Surface-enhanced Raman scattering from copper nanoparticles obtained by laser ablation. *The Journal of Physical Chemistry C* 115: 5021–5027.

Nakamura, D. and S. Mufson. 2014. China, U.S. agree to limit greenhouse gases. *The Washington Post*, November 12, 2014, Business section. http://www.washingtonpost.com/business/economy/china-us-agree-to-limit-greenhouse-gases/2014/11/11/9c768504-69e6-11e4-9fb4-a622dae742a2_story.html.

Nguyen, T.B., T.K.T. Vu, C.D. Nguyen, T.D. Nguyen, T.A. Nguyen, and T.H. Trinh. 2012. Preparation of metal nanoparticles for surface enhanced Raman scattering by laser ablation method. *Advances in Natural Sciences: Nanoscience and Nanotechnology* 3(2): 025016.

NovaCentrix. 2015. PulseForge tools for printed electronics. http://www.novacentrix.com/products/pulseforge. Accessed on May 3, 2015.

Ohgai, T. and K. Hashiguchi. 2013. Functional nanowires array electrodeposited into nano-porous membrane thin films. *Journal of Physics: Conference Series* 417: 012047.

Ohnuma, I., M. Miyashita, K. Anzai, X. J. Liu, H. Ohtani, R. Kainuma, and K. Ishida. 2000. Phase equilibria and the related properties of Sn–Ag–Cu based Pb-free solder alloys. *Journal of Electronic Materials* 29: 1137–1144.

Okumura, K., E. Higurashi, T. Suga, and K. Hagiwara. 2014. Low-temperature GaAs/SiC wafer bonding with Au thin film for high-power semiconductor lasers. In *2014 International Conference on Electronics Packaging (ICEP)*, IEEE, Toyama, Japan, pp. 716–719, April 23–25, 2014.

Onar, O.C., J. Kobayashi, and A. Khaligh. 2013. A fully directional universal power electronic interface for EV, HEV, and PHEV applications. *IEEE Transactions on Power Electronics* 28(12): 5489–5498.

Ossi, P.M., F. Neri, N. Santo, and S. Trusso. 2011. Noble metal nanoparticles produced by nanosecond laser ablation. *Applied Physics A: Materials Science and Processing* 104: 829–837.

Pathak, A.D. 2012. High reliability power electronics. In *Proceedings of the PCIM ASIA 2012 International Conference and Exhibition for Power Electronics, Intelligent Motion, Renewable Energy and Energy Management*, Shanghai, China, pp. 29–39.

Peng, P., A. Hu, A.P. Gerlich, Y. Liu, and Y.N. Zhou. 2015. Self-generated local heating induced nanojoining for room temperature pressureless flexible electronic packaging. *Scientific Reports* 5: 1–8.

Peng, P., L. Liu, A.P. Gerlich, A. Hu, and Y.N. Zhou. 2013. Self-oriented nanojoining of silver nanowires via surface selective activation. *Particle & Particle Systems Characterization* 30(5): 420–426.

Peng, S., C. Wang, J. Xie, and S. Sun. 2006. Synthesis and stabilization of monodisperse Fe nanoparticles. *Journal of the American Chemical Society* 128: 10676–10677.

Peng, Y., T. Cullis, and B. Inkson. 2009. Bottom-up nanoconstruction by the welding of individual metallic nanoobjects using nanoscale solder. *Nano Letters* 9(1): 91–96.

Peng, Y.-H., C.-H. Yang, K.-T. Chen, S.R. Popuri, C.-H. Lee, and B.-S. Tang. 2012. Study on synthesis of ultra-fine Cu–Ag core–shell powders with high electrical conductivity. *Applied Surface Science* 263: 38–44.

Pinkos, A.F. and Y. Guo. 2013. Automotive design challenges for wide-band-gap devices used in high temperature capable, scalable power vehicle electronics. In *2013 IEEE Energytech*, Cleveland, OH, May 21–23, 2013.

Qu, H.-M., X.-H. Yang, Q. Zheng, X.-T. Wang, and Q. Chen. 2014. Thermal management technology of high-power light-emitting diodes for automotive headlights. *IEICE Electronics Express* 11(23): 1–11.

Rajashekara, K. 2013. Present status and future trends in electric vehicle propulsion technologies. *IEEE Journal of Emerging and Selected Topics in Power Electronics* 1(1): 3–10.

Ramulifho, T., K.I. Ozoemena, R.M. Modibedi, C.J. Jafta, and M.K. Mathe. 2012. Fast microwave-assisted solvothermal synthesis of metal nanoparticles (Pd, Ni, Sn) supported on sulfonated MWCNTs: Pd-based bimetallic catalysts for ethanol oxidation in alkaline medium. *Electrochimica Acta* 59: 310–320.

Saito, G., Y. Nakasugi, and T. Akiyama. 2014. Excitation temperature of a solution plasma during nanoparticles formation. *Journal of Applied Physics* 116: 083301.

Sakkas, P., O. Schneider, S. Martens, P. Thanou, G. Sourkouni, and C. Argirusis. 2012. Fundamental studies of sonoelectrochemical nanomaterials preparation. *Journal of Applied Electrochemistry* 42(9): 763–777.

Shen, L., Z.Y. Tan, and Z. Chen. 2013. Nanoindentation study on the creep resistance of SnBi solder alloy with reactive nano-metallic fillers. *Materials Science and Engineering A* 561: 232–238.

Sheng, W.W. and R.P. Colino. 2004. *Power Electronic Modules Design and Manufacture*. Boca Raton, FL: CRC Press.

Silva, J.A., M. Gauthier, C. Boulord, C. Oliver, A. Kaminski, B. Semmache, and M. Lemiti. 2011. Improving front contacts of N-type solar cells. *Energy Procedia* 8: 625–634.

Siow, K.S. 2012. Mechanical properties of nano-silver joints as die attach materials. *Journal of Alloys and Compounds* 514: 6–19.

Siow, K.S. 2014. Are sintered silver joints ready for use as interconnect material in microelectronic packaging? *Journal of Electronic Materials* 43(4): 947–961.

Sirisukprasert, S. 2014. Power electronics-based energy storages: A key component for smart grid technology. In *Proceedings of the International Electrical Engineering Congress 2014*, Chonburi, Thailand, March 19–24, 2014.

Tan, K.S., Y.H. Wong, and K.Y. Cheong. 2015. Thermal characteristic of sintered Ag–Cu nanopaste for high-temperature die-attach application. *International Journal of Thermal Sciences* 87: 169–177.

Tarasenko, N.V., A.V. Butsen, and E.A. Nevar. 2005. Laser-induced modification of metal nanoparticles formed by laser ablation technique in liquids. *Applied Surface Science* 247(1–4): 418–422.

Thenmozhi, R., C. Sharmeela, P. Natarajan, and R. Velraj. 2014. Experimental investigation on performance improvement in high power LED lamps. In *2014 International Conference on Green Computing Communication and Electrical Engineering (ICGCCEE)*, Coimbatore, India, pp. 3–7.

Usman, M.S., M.E. El Zowalaty, K. Shameli, N. Zainuddin, M. Salama, and N.A. Ibrahim. 2013. Synthesis, characterization, and antimicrobial properties of copper nanoparticles. *International Journal of Nanomedicine* 8: 4467–4479.

Vafaei, A., A. Hu, and I.A. Goldthorpe. 2014. Joining of individual silver nanowires via electrical current. *Nano-Micro Letters* 6(4): 293–300.

Valodkar, M., S. Modi, A. Pal, and S. Thakore. 2011. Synthesis and anti-bacterial activity of Cu, Ag and Cu–Ag alloy nanoparticles: A green approach. *Materials Research Bulletin* 46(3): 384–389.

Wang, B., Y. Xu, K.-L. Yung, W. Chen, and C.-L. Kang. 2013a. Aligning and soldering pure-copper nanowires for nanodevice fabrication. *Journal of Microelectromechanical Systems* 22(3): 519–526.

Wang, S., M. Li, H. Ji, and C. Wang. 2013b. Rapid pressureless low-temperature sintering of Ag nanoparticles for high-power density electronic packaging. *Scripta Materialia* 69: 789–792.

Whitaker, B., Z. Cole, B. Passmore, T. Mcnutt, M.N. Ericson, S. Shane, C.L. Britton et al. 2014. High-temperature SiC power module with integrated SiC gate drivers for future high-density power electronics applications. In *IEEE Workshop on Wide Bandgap Power Devices and Applications (WiPDA)*, Knoxville, TN. IEEE, New York, pp. 36–40.

Xia, L., X. Hu, X. Kang, H. Zhao, M. Sun, and X. Cihen. 2010. A one-step facile synthesis of Ag–Ni core–shell nanoparticles in water-in-oil microemulsions. *Colloids and Surfaces A: Physiochemical Engineering Aspects* 367: 96–101.

Xiao, K., S. Luo, K. Ngo, and G.-Q. Lu. 2013. Low-temperature sintering of a nanosilver paste for attaching large-area power chips. In *15th International Symposium and Exhibition on Advanced Packaging Materials (APM 2013)*, Irvine, CA, Vol. 5, pp. 192–202.

Xie, F., X. He, X. Lu, S. Cao, and X. Qu. 2013. Preparation and properties of porous Ti–10Mo alloy by selective laser sintering. *Materials Science and Engineering C: Materials for Biological Applications* 33(3): 1085–1090.

Xu, H. and K.S. Suslick. 2010. Sonochemical synthesis of highly fluorescent Ag nanoclusters. *ACS Nano* 4(6): 3209–3214.

Xu, Z., M. Li, F. Wang, and Z. Liang. 2013. Investigation of Si IGBT operation at 200°C for traction applications. *IEEE Transactions on Power Electronics* 28: 2604–2615.

Yan, J., G. Zou, A. Hu, and Y.N. Zhou. 2011. Preparation of PVP coated Cu NPs and the application for low-temperature bonding. *Journal of Materials Chemistry* 21(1): 15981–15986.

Yan, J., G. Zou, A. Wu, J. Ren, A. Hu, and Y.N. Zhou. 2012. Polymer-protected Cu–Ag mixed NPs for low-temperature bonding application. *Journal of Electronic Materials* 41(7): 1886–1892.

Yang, C., C.P. Wong, and M.M.F. Yuen. 2013. Printed electrically conductive composites: Conductive filler designs and surface engineering. *Journal of Materials Chemistry C* 1(207890): 4052–4069.

Yang, Y., S. Seyedmohammadi, U. Kumar, D. Gnizak, E. Graddy, and A. Shaikh. 2011. Screen printable silver paste for silicon solar cells with high sheet resistance emitters. *Energy Procedia* 8: 607–613.

Yi, Z., X. Xu, K. Zhang, X. Tan, X. Li, J. Luo, X. Ye et al. 2013. Green, one-step and template-free synthesis of silver spongelike networks via a solvothermal method. *Materials Chemistry and Physics* 139(2–3): 794–801.

Yue, L., S. He, and W. Chen. 2014. On the application of advanced power electronics technology in smart grid. *Journal of Chemical and Pharmaceutical Research* 6(7): 793–797.

Zhang, J., N. Ma, F. Tang, Q. Cui, F. He, and L. Li. 2012. pH- and glucose-responsive core–shell hybrid nanoparticles with controllable metal-enhanced fluorescence effects. *ACS Applied Materials and Interfaces* 4: 1747–1751.

Zhao, J., D. Zhang, and J. Zhao. 2011. Fabrication of CuAg coreshell bimetallic superfine powders by eco-friendly reagents and structures characterization. *Journal of Solid State Chemistry* 184(9): 2339–2344.

Zhou, D., M. Zhou, M. Zhu, X. Yang, and M. Yue. 2012. Electrodeposition and magnetic properties of FeCo alloy films. *Journal of Applied Physics* 111: 07A319.

Zhou, Y. and A. Hu. 2011. From microjoining to nanojoining. *The Open Surface Science Journal* 3: 32–41.

Zhu, H. and X. Liu. 2014. On the application of advanced power electronics technology in smart grid. *Journal of Chemical and Pharmaceutical Research* 6(7): 859–863.

Zhuoping, Y., J. Wu, and L. Xiong. 2014. Research of stability control of distributed drive electric vehicles under motor failure modes. In *2014 IEEE Conference and Expo Transportation Electrification Asia-Pacific (ITEC Asia-Pacific)*, IEEE, Beijing, China, pp. 1–5, August 31–September 3, 2014.

Zin, V., K. Brunelli, and M. Dabalà. 2014. Characterization of Cu–Ni alloy electrodeposition and synthesis of nanoparticles by pulsed sonoelectrochemistry. *Materials Chemistry and Physics* 144(3): 272–279.

11 Applications of Metal Nanoparticles and Nanostructures Fabricated Using Ultrafast Laser Ablation in Liquids

S. Venugopal Rao, S. Hamad, and G. Krishna Podagatlapalli

CONTENTS

11.1 INTRODUCTION

The last few decades were dedicated for the investigation of materials that enabled scientists to understand their physical, chemical, and structural properties and utilize them in applications such as plasmonics, photonics, and biomedicine to name a few. The interrogation of materials in the nanoscale became an imperative and inevitable task since the dimensions modify the basic behavior of the material through alteration of boundary conditions (Kelly et al. 2003, Link et al. 2003). The materials whose dimensions are in the 1–100 nm range are termed as nanomaterials (Kelly et al. 2003, Link et al. 2003). The utilization of nanoscale plasmonic materials (e.g., Au, Ag, and Cu) is well known from the ancient times of history. Nanomaterials are essential building blocks of a wide range of scientific applications such as photo-induced thermal therapy, biochemical sensors, surface-enhanced Raman spectroscopy, carriers of drug delivery, nanophotonic devices, biosensing, in vivo and in vitro diagnostics, solar cells, optoelectronic devices, antibacterial agents, cancer treatment, catalysis, imaging, sensing, biology and medicine, and sensors (Arruebo et al. 2007, Guo and Wang 2011, Jain et al. 2008, Murphy et al. 2005, Pradeep and Anshup 2009). Among the various top-down and bottom-up lithographic fabrication techniques (Makaraov 2013, Narayanan and Sakthivel 2010, Pelton et al. 2008, Sakamoto et al. 2009, Sau and Rogach 2010, Xia et al. 1999), laser ablation in liquid (LAL) media is an efficient and "green" technique to produce both nanoparticles (NPs) and nanostructures (NSs) in a single experiment. This technique is deemed to be "green" since there are no chemicals (surfactants) involved during or after ablation. Ablation is the removal of fragments when a material is hit by an intense laser pulse. Fragments account for atoms, ions, molecular clusters, etc. The interaction of ultrashort laser pulses with materials conjures strong interest since their pulse duration is much less than the time required for several relaxation processes (transfer of energy between electron and lattice systems, heat diffusion, etc.) (Perez and Lewis 2003, Perez et al. 2008). These short pulse durations lead to the deposition of higher energy densities (higher peak intensities) when they focus on a bulk target surface. Consequently, higher pressures and temperatures will be attained at the point of the impact leading to surface modification of the material and effortless fragmentation. The transient state of the material after the irradiation of a focused pulsed laser can be approximated by a phase transition (Radziemski and Creamers 1989) since the absorption of laser energy by electron gas promotes the target material to a state of higher temperature by leaving the remaining target (lattice) at the initial temperatures (Radziemski and Creamers 1989). Restoration of the system to a state of equilibrium depends on the parameters of laser pulses used for ablation and the intrinsic nature of the target. Ultrafast laser ablation (ULA) minimizes the random corrugation of the target since the pulse duration is extremely short and the heat-affected zone (HAZ) provided is negligible compared to the HAZ provided by the longer laser pulses. When the fs pulses interact with materials, the energy is absorbed by inverse Bremsstrahlung followed by ionization processes typically in the ~10–100 fs time scales (Perez and Lewis 2003, Perez et al. 2008). The excited electrons collide with each other or with bound electrons, resulting in the production of high conductivity. This phenomenon is described as electron–electron scattering (or coulomb explosion), which occurs in ~1 ps time scale (Perez and Lewis 2003, Perez et al. 2008). After that, electrons transfer their energy to lattice through electron–lattice coupling and the corresponding characteristic time is ~10 ps (Perez and Lewis 2003, Perez et al. 2008). Thermal diffusion occurs inside the lattice via phonon–phonon scattering, which is typically after 10 ps and up to ~1 nanosecond (ns) time. The electron–electron, electron–phonon, and phonon–phonon relaxations can be explained by the two-temperature model (Harilal et al. 2014, Von Der Linde and Sokolowski-Tinten 2000). Further stages involved are plasma formation and plume quenching (~100 ns to 1 μs), formation and collapse of cavitation bubble (1–200 μs), and nanomaterials (NMs) formation in liquid media (~1 ms) (De Bonis et al. 2013, Harilal et al. 2014). Figure 11.1 illustrates the various mechanisms and time scales recorded for a typical ablation process.

LAL with ns pulses has been studied for more than two decades now including the pioneering work by Yang and coworkers (Liang et al. 2014, Liu et al. 2010, Wang et al. 2005, Yang 2007, 2011). There are several review articles outlining the basic physics, fabrication aspects, characterization,

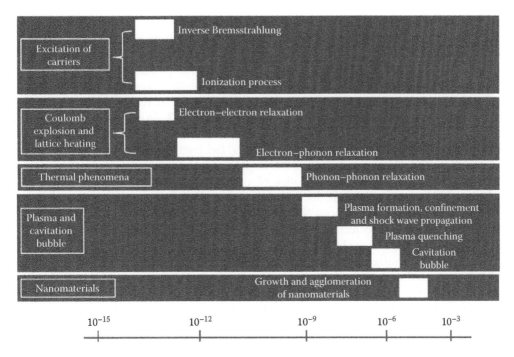

FIGURE 11.1 A typical timing diagram of fs laser ablation for various stages.

and applications of the generated NPs (Besner and Meunier 2010, Ishikawa et al. 2006, Sasaki et al. 2006, Shafeev 2008, Stratakis 2012, Yan and Chrisey 2012, Zeng et al. 2012) and NSs (Barmina et al. 2010, Tsuji et al. 2010). Several pioneering works have been published by a number of groups all over the world, and some significant contributions emanated from Meunier's group (Besner et al. 2005, 2007, 2008, 2009, 2010a,b, Boulais et al. 2013, Dallaire et al. 2012, Hatef and Meunier 2015, Kabashin and Meunier 2003, 2006, Kenth et al. 2011, Lachaine et al. 2014, Lorazo et al. 2003, Papgiannouli et al. 2015, Rioux et al. 2009, Robitaille et al. 2013, Sylvestre et al. 2004a,b, 2005, Zakharko et al. 2011), Barcikowski's group (Asahi et al. 2015, Barcikowski and Mafune 2011, Barcikowski et al. 2007, 2010, De Giacomo et al. 2013, Gökce et al. 2015, Ibrahimkutty et al. 2012, Menéndez-Manjón et al. 2011, 2013, Petersen et al. 2009, Rehbock et al. 2014, Sajti et al. 2010, Stratakis et al. 2009b, Taylor et al. 2011), Shafeev's group (Dolgaev et al. 2002, Kazakevich et al. 2004, 2006, Kuzmin et al. 2010, Menéndez-Manjón et al. 2010, Serkov et al. 2015, Simakin et al. 2001, 2004, Stratakis et al. 2009a,c, Tiedemann et al. 2014, Truong et al. 2007a,b, Zavedeev et al. 2006), Amendola and Meneghetti's group (Amendola et al. 2006, 2007, 2009a,b, 2011a,b, 2012, 2013, 2014, Bertorelle et al. 2014), Itina's group that includes the theoretical modeling of ablation phenomena (Delfour and Itina 2015, Dell'Aglio et al. 2015, Itina 2011, Noël et al. 2007, Povarnitsyn et al. 2007), Brandi's group (Bagga et al. 2013, Intartaglia et al. 2011, 2012a–c, 2013, 2014), others (Compagnini et al. 2003, Hu et al. 2011, Lalayan 2005, Medivil et al. 2013, O'Malley et al. 2014, Salminen et al. 2012, Tilaki et al. 2007, Usui et al. 2005, Jiang et al. 2011), and our group (Hamad et al. 2012, 2013, 2014a,b,c, 2015, Podagatlapalli et al. 2012, 2013, 2014a,b,c, 2015a,b, Saikiran et al. 2014, Syed et al. 2012, Vendamani et al. 2015, Venugopal Rao et al. 2014).

Our group efforts, over the last few years, have mainly been focused on ultrafast (ps or fs) ablation in a variety of liquids and investigating different plasmonic metals such as Al, Ag, Au, and alloys of Ag–Au. There are very few reports in literature (in comparison with ns ablation) on the femtosecond ablation of any targets (metallic and nonmetallic) in liquid media (Alnassar et al. 2013, Ancona et al. 2014, Arboleda et al. 2015, Barchanski et al. 2015, Chelnokov et al. 2012, De Bonis et al. 2014, Hatanaka et al. 1999, Hu et al. 2015, Liu et al. 2007, Povarnitsyn et al. 2013, Santagata et al. 2015, Semaltianos et al. 2010, Shaheen et al. 2013, Tagami et al. 2014, Tan et al. 2011, Tsuji et al. 2003). Some of the large number of applications envisaged from NPs created with LAL and other lithographic techniques include photovoltaic devices (Spyropoulus et al. 2012), photocatalytic activity (Liu et al. 2009, Zimbone et al. 2015), antibacterial properties (Guisbiers et al. 2015, Ismail et al. 2015, Kamat 2002, Nath et al. 2012, Zimbone et al. 2015), magnetic resonance imaging (MRI) (Tian et al. 2014), plasmonics (Intartaglia et al. 2013), upconversion (Onodera et al. 2013), surface-enhanced Raman scattering (SERS) studies (Fan et al. 2013, Jin et al. 2014, Nedderson et al. 1993, Nie and Emory 1997, Ou et al. 2014, Qua et al. 2013, Wang et al. 2015, Zheng et al. 2014), and photonics/nonlinear optics (Amendola et al. 2009, Del Fatti and Vallée 2001, Fan et al. 2014, Fazio and Neri 2013, Ganeev et al. 2004, 2005, Hajiesmaeilbaigi and Motamedi 2007, Karavanskii et al. 2004). LAL has enabled the fabrication of a variety of materials such as alloy NPs (Chen and Beraun 2001, Compagnini et al. 2008, Kuladeep et al. 2012, Lee et al. 2001, Olea-Mejía et al. 2015, Shah et al. 2012), semiconductor NPs (Eroshova et al. 2012, Vadavalli et al. 2014), TiO_2 NPs (Siuzdak et al. 2014), Au-poly methyl methylyacrylate (PMMA) composites (Schwenke et al. 2013), and MoS_2 NSs (Compagnini et al. 2012).

This chapter is organized as follows. We first provide a brief introduction to ultrafast ablation in air and water and explain their advantages and disadvantages. Later we discuss, in detail, the fabrication of Ag, Cu, and Al NPs and NSs. We also summarize the results obtained from different parameters such as input energy, scanning speeds, and input angle during ablation of these materials. Further, we elaborate on three different applications of NPs and NSs generated by ultrafast laser ablation in liquids (ULAL) technique: (1) SERS studies for explosives (e.g., RDX, trinitrotoluene [TNT], ANTA, FOX-7, CL-20, etc.) detection using NSs of Ag, Cu, etc., created by ULAL; (2) photonic applications of NPs created by ULAL, wherein we utilized the Z-scan and four-wave mixing techniques to study the nonlinear optical (NLO) coefficients in the ps and fs time domains; and (3) antibacterial applications of the Cu NPs generated using ULAL.

11.2 FUNDAMENTALS OF ULTRAFAST LASER ABLATION

11.2.1 ABLATION OF METALS IN AMBIENT AIR

Laser ablation of solid targets is a top-down method and attracted the attention of the scientific community for the last half century. The material cannot be fragmented from the surface immediate to the dumping of laser energy. The electric field provided by the impact of the laser pulse is sufficient to eject only the surface electrons rather than the fragments. The ejection of electrons, ions, molecular clusters, and other intermediate products is governed by two important nonlinear mechanisms, namely, multiphoton absorption and cascade ionization (Amoruso et al. 2005a,b, Batani 2010, Batani et al. 2014, Chichkov et al. 1996, Nolte et al. 1999, Upadhyay et al. 2008, Zhakhovskii et al. 2009, Zhigilei et al. 2009). These mechanisms are purely dependent on the pulse duration (Pereira et al. 2004). Initially, surface electrons of the material absorb laser energy by means of inverse Bremsstrahlung and acquire higher kinetic energies. Consequently, they maintain a temperature very much higher than that of the surrounding lattice. Gradient in the temperature of the hot electrons and lattice can be explained by the two-temperature model (Chen and Yeh 2001) governed by the following set of equations:

$$C_e \frac{\partial T_e}{\partial t} = \frac{\partial Q(z)}{\partial t} - \gamma(T_e - T_i) + S$$

$$C_i \frac{\partial T_i}{\partial t} = \gamma(T_e - T_i)$$

$$Q(z) = -k_e \frac{\partial T_e}{\partial z}$$

$$S = I(t)\alpha \exp(-\alpha z)$$

(11.1)

where

$Q(z)$ is the heat flux
$I(t)$ is the laser intensity
α is the material absorption coefficient
C_e, C_i are the electron and lattice heat capacities per unit volume
γ is the electron–phonon coupling constant
k_e is the electron thermal conductivity

In general, in the case of ULA attainable, temperature and pressure of the plasma plume are $\sim 10^3$ K and $\sim 10^{10}$ Pa, respectively (Margetic 2002). The hot electrons in the conduction band transfer the excess heat energy through the electron lattice collisions, which lasts for a few ps. After thermalization, the entire system reaches an equilibrium state. At this stage, if the attained temperature is greater than the melting threshold of the material, then the portion of the material at the impact of laser pulses starts melting immediately. The entire process lasts for several ps. In a time scale of a few microseconds, NPs are ejected from the target surface. A few articles (Amoruso et al. 2005, Gamaly et al. 2002, Ivanov and Zhigilei 2003, König et al. 2005, Noël et al. 2007, Povarnitsyn et al. 2007, Zhigilei et al. 2009) are reported in recent literature that discusses in detail the complete mechanisms of ultrafast ablation of metal and metal films in ambient air. The dynamics of ablation even in ambient air are affected by the input laser parameters to a great extent. Lorazo et al. (2006) reported that pulse duration is indeed an important parameter producing different outcomes in the case of longer pulses compared to the ultrashort pulses. Lorazo et al. (2006) made an attempt to explain the thermodynamic pathways of various stages of ablation in an extensive manner. They have observed that isochoric heating and rapid adiabatic expansion of the material (silicon) provided a natural pathway for phase explosion. This was not observed with slower, nonadiabatic cooling with ps pulses where fragmentation of the hot metallic fluid was the only pertinent ablation mechanism. Amoruso et al. (2005) reported the ultrafast ablation in vacuum, and comparison was carried out between ps and fs regime dynamics. Similarly, Balling and Schou (2006) reported the dynamics of ns laser ablation of dielectrics in air ambient with plausible explanations. Semaltianos (2010) reported the effect of medium (both air and liquid) on the fabricated NPs and underlying physics elaborately. Although there are several reports describing that ULAL fabricates a variety of metallic and nonmetallic NPs and NSs, there are very few reviews on this subject.

11.2.2 Demerits of Laser Ablation in Ambient Air

- The generated NPs during ablation process are expelled into the environment contaminating the surroundings.
- In this process, NSs can only be obtained by placing suitable wafers in the close vicinity of ablation volume, resulting in inhomogeneous (thickness) film of NPs.
- Plasma produced in ambient air expands into the surrounding rapidly, which slows down the process of NP generation. Ablation in air requires higher input energies to produce NPs.

11.3 ABLATION OF METALS IN LIQUIDS

If the target is surrounded by an aqueous medium, then the dynamics of ablation happen to be extremely complicated. When the laser beam is focused on the surface of a target material under the liquid layer, local melting of the metal target takes place. As a result, the metal portion at which the laser beam is impinged goes to the melt phase. The adjacent liquid layer of the metallic melt absorbs a part of the heat energy and attains higher temperature. At this temperature, liquid cannot sustain in its own phase and, therefore, evaporates. During the process of evaporation, the liquid layer exerts a recoil pressure on the underlying melt. This pressure is usually much higher than the atmospheric pressure. However, the surface tension forces those exist in the melt try to sustain it as a spherical entity. As a result of the higher recoil pressures exerted by the evaporating liquid layer, metallic melt splashes into fragments and each fragment of nanodimension goes into the surrounding liquid medium. At higher values of the recoil pressure, in addition to the fabrication of NPs, redistribution of the metallic melt takes place before its condensation, forming different NSs.

11.3.1 CAVITATION BUBBLE DYNAMICS

The physical processes including the generation, transformation, and condensation of plasma plume induced at the time of LAL have been investigated by many groups (Amendola et al. 2009, Besner and Meunier 2010, Semaltianos 2010, Shafeev 2008, Simakin et al. 2004, Stratakis 2012, Yan and Chrisey 2012, Zeng et al. 2012). The exact dynamics of LAL are still being debated, but according to some of the recent works, it is explained as a complicated laser–matter interaction under the liquid layer leading to the generation of the plasma plume. The created plasma expands into the surrounding liquid medium resulting in the generation of a shock wave. During the process of expansion, the plasma plume cools down and plasma transfers energy to the surrounding liquid medium. The annihilation of the plasma plume occurs in typical time scales of 10^{-8} to 10^{-7} s resulting in the formation of a cavitation bubble that tries to expand in the medium in a time scale of 10^{-7} to 10^{-6} s and continues expanding up to $\sim 10^{-4}$ s. After a certain period of time (usually of the order of few hundred microseconds) at which the inside pressure decreases compared to the surrounding liquid medium, the cavitation bubble collapses along with the generation of a second shock wave. The exact stages at which nanomaterials are generated are still under investigation. Many reports suggest that the nanomaterials are formed during the expansion of the cavitation bubble inside it on a time scale of $\sim 10^{-6}$ to 10^{-4} s. The expansion of the cavitation bubble into the liquid medium exerts a recoil pressure on the metallic melt formed under the plasma plume. As mentioned earlier, the expansion of the plasma plume and its transient dynamics can prompt the mixing of the vaporized material with the liquid medium surrounding it. During later stages, the vaporized material condenses into the liquid medium resulting in solidification (generation of NPs) in a time scale of a few hundred ns ($\ll$time taken for the condensation of vapor plume in liquid [~ 100 µs]). Nanoentities inside the cavitation bubble are at a higher temperature than the surrounding liquid environment. The gradient in temperature on both sides of the liquid–bubble interface leads to nucleation and condensation of the fabricated nanomaterials after collapse of the cavitation bubble. Fabrication of nanomaterials not only depends on the laser parameters and materials but also on the nature of the surrounding liquid. The polarity, viscosity, and refractive index of the surrounding media likewise play a key role in post ablation processes such as nucleation and growth. In general, ablation of metals in polar liquids results in the fabrication of NPs with smaller size through the formation of electrical double layers (EDLs) on the produced charged NP surface that prevents further growth. Viscosity of the liquid medium sustains the plume for a longer time and enhances the probability of the second ablation at the point of plume formation.

11.4 ADVANTAGES OF PULSED LASER ABLATION IN LIQUIDS

- Confinement of the produced plasma in aqueous media supports the ablation to occur at lower energies compared to ablation in ambient air.
- Pulsed LAL avoids the contamination of the surrounding air media since the ablated fragments (NPs) from the target surface directly enter the surrounding liquid medium to form a colloidal solution.
- Utility of surfactants is not necessary to produce dispersed NPs unlike chemical methods of NPs synthesis. Thus, ULAL can be considered as a green method.
- Simultaneous preparation of NPs and NSs can be achieved in a single experiment within a few minutes while generating NSs over an area of a few inches. Moreover, the fabricated NPs and NSs can be utilized for other experimental purposes devoid of subjecting them to a rigorous cleaning with chemicals. ULAL is a simple platform to produce not only diverse NPs of metals but also NPs of semiconductors, alloys, oxides, magnetic materials, biaxial heterostructures, core–shell type, etc.

11.5 PARAMETERS INFLUENCING THE ABLATION MECHANISM

The products of ablation depend on many parameters of the input laser pulses as well as the surrounding liquid media. Wavelength, pulse duration, fluence, beam waist, repetition rate, and number of pulses incident per spot size are the pulse parameters, whereas linear refractive index, viscosity, and the polarity of the liquid along with the thickness of the liquid layer on the target surface are the liquid parameters that influence the process of ablation to a great extent. Variation of one of the parameters to a slightest degree affects the nature of the products.

11.5.1 INPUT WAVELENGTH

Wavelength of the incident laser beam determines the skin depth and, as a consequence, ablation depth can be determined. Nichols and coworkers (2006a–c) reported that the absorption of UV photons by the surface electrons via interband transitions was more uniform and leads to nice corrugation on the surface, whereas near-infrared radiation is preferentially absorbed by the defects and impurities of the material, consequently resulting in the formation of random surface structures. For UV radiation, the tendency of scattering is higher compared to NIR. Furthermore, the wavelength of the laser beam should be chosen in such a way that it should not be absorbed by the fabricated intermediate NPs within the liquid. It has been proven that the incident laser wavelength should not be equal to the wavelength corresponding to the surface plasmon resonance (SPR) peak of the NP colloidal solution to avoid the ambiguity of the intermediate absorption of incident light. If absorption occurs, then nanomaterials are subjected to unusual modifications. It has also been documented that the shorter the wavelength, the higher the photon energies through which bond breaking and ionization processes become simpler when compared to longer wavelengths. Mortazavi and coworkers (2011) reported the observed variations in the size distribution of the fabricated Pd NPs in deionized water fabricated with the wavelengths 1064 nm (Nd:YAG) and 193 nm (Excimer laser) at an equal pulse duration of 6 ns. According to their data and analysis, laser ablation with longer wavelengths followed bottom-up behavior, which supported the aggregation of synthesized NPs due to thermal effects from plasma-induced ablation ending up with NPs of larger sizes. Laser ablation with shorter wavelengths follows top-down behavior directly fabricating smaller NPs since the photon energy is higher. Schwenke and coworkers (2011) also reported the effect of laser ablation of metal in the solvent tetrahydrofuran with fundamental (1064 nm) and second harmonic (532 nm) pulses in the ps regime and observed good rate and yield of NP fabrication with 1064 nm pulses. A linear dependence of decrease in the hydrodynamic particle size in the case of 1064 nm ablation was observed since the absorption and scattering loss is weak at this wavelength. In the case of 532 nm ablation, exponential decrement of the hydrodynamic particle size could be assigned to particle fragmentation during the postirradiation of dispersed NPs. Moreover, depending on the wavelength of the incident laser, the cross section of inverse Bremsstrahlung varies and cascade ionization will be altered since the aforementioned cross section varies directly with the second power of the incident laser (Batani 2010, Batani et al. 2014).

11.5.2 PULSE DURATION

Developments in achieving superior pulsed laser systems from ns time scales to fs time scales have further established practical applications resulting from the advantages of pulsed laser ablation. Since the short pulsed lasers support the higher peak intensities, explosive boiling of the material under consideration has become an easier task. When laser pulses of longer duration (μs to ns) are incident on the metallic targets, following the electron–lattice collisions, thermal processes such as explosive boiling and evaporation take place. Further to the absorption of laser pulse, heat diffuses into the lattice from the electrons in a very short time compared to its pulse duration (μs or ns). The process of ablation of targets with μs or ns pulses is dominated by heat conduction, melting, evaporation, and plasma formation. For the longer pulse ablation, HAZs prevail over short pulse ablation. We performed the ablation of aluminum (Al) target using ps and fs pulses, and the differences between ablation by ~7 ns and ~40 fs laser pulses in water exhibited crater formation with different properties (Podagatlapalli et al. 2014). Ablation with nanosecond pulses consists of large HAZs, while in the case of femtosecond ablation, very minimal or negligible HAZ was observed. The melting continues for a long time resulting in the boundaries of the ablated zone with no sharp boundaries. In addition, a plasma-shielding effect prevails in which the trailing part of the laser pulse interacts with the plasma generated by the leading part of the pulse, resulting in unwanted modification of the nature of previously produced nanomaterials. Recently, Niu and coworkers (2010) reported the advantage of even longer-duration pulses (ms) to promote the surface reactions of metal droplets and control over the formation of diverse NSs, hollow spheres, core–shell nanospheres, heterostructures, nanocubes, and nanowires. A vast amount of experimental work has been carried out and reported on ablation using ns laser pulses.

In the case of ultrafast pulses, ablation dynamics are explained by multiphoton-induced absorption and cascaded photoionization (avalanche). Particularly, for fs ablation, instantaneous multiphoton absorption prevails, while in the case of ps ablation, the produced hot electrons behave as seed in the buildup of avalanche. In the short pulse case, since the pulse duration τ_L is much shorter than the electron–phonon equilibrium time τ_E (10^{-12} to 10^{-11} s) and liquid–vapor equilibrium time τ_{LV} (10^{-12} to 10^{-11} s), nonthermal processes prevail, while in the ablation with longer pulses, thermal processes usually dominate. During the ablation, unwanted interaction of the pulse with the plasma plume is avoided. ULA can be considered as an isochoric process (considerable change of the volume of sample does not occur) since the irradiation of laser pulse causes local heating in very short time and lasts before the expansion of metal takes place. In the ultrafast ablation, acceleration of the ionized entities to enormous velocities leads to the development of higher pressures and temperatures. Since the interaction time is very short, the material cannot evaporate continuously but transforms into a state of overheated liquid. As a result, NSs formed on the solid targets after irradiation of ultrafast laser pulse exhibit sharp boundaries with nice corrugation since the material is evaporated with a minimal HAZ. Barmina and coworkers (2009) performed the ablation of tantalum (Ta_2O_5) with laser pulses of 350 ps, 5 ps, and 180 fs duration, and their corresponding wavelengths were 1064, 248, and 800, respectively, for fabricating metallic nanostructured substrates. According to them, surface nanostructuring was chiefly dependent on the laser pulse duration (defines the energy density) than the wavelength of incident laser. In the case of 180 fs pulses (it was not observed with the other two pulse durations), they observed small-scale periodic structures (along with hillocks) of periodicity less than the laser wavelength. The fundamental differences between the longer pulse ablation of targets with ultrafast pulse ablation were also reported, and the effect of the number of pulses on ablation along with the spot size at the focus was discussed (Leitz et al. 2011). Barcikowski and coworkers (2007) demonstrated the efficiency of ps and fs pulse ablation of a Ag strip in water flow. It was observed that the ablation with liquid flow improved the reproducibility and increased the productivity of NPs by 380% compared to ablation performed with stationary liquid. Fs ablation in water was 20% more efficient than ps laser ablation, but due to higher ps laser power (higher repetition rate), the productivity of NPs at the same fluence was three times higher for ps ablation. Further, they observed that fs (120 fs, 5 kHz) ablation was efficient to generate 2 µg of NPs/J, whereas it was 1.5 µg for ps (10 ps, 50 kHz) pulses. At the same time, the rate of generation observed was excellent for ps ablation (34 mg/h) compared to fs ablation (6 mg/h). Recent investigations (Riabinina et al. 2012) of the ablation phenomenon in water demonstrated better yields of Au NPs. They found that ~2 ps was the optimal pulse duration at 5 mJ energy per pulse at the expense of photoionization to obtain better yields of Au NPs.

11.5.3 Energy per Pulse, Spot Size, and Fluence

In general, the decomposition or material removal from a solid is the consequence of an energy input into the target, resulting in the domination of the target's binding energy. In ULAL, the total density of hot electrons depends upon the energy of the primarily ejected ballistic electrons since their kinetic energy only determines the number of secondary electrons. The mentioned electrons are the supreme initiators for the formation of metal melt through electron–phonon collisions to attain equilibrium. Thus, the entire phenomenon depends on the energy of the incident laser beam. Several groups (Upadhyay et al. 2008, Zhigilei et al. 2009) performed molecular dynamic simulations to explain the effect of laser parameters on the ablation and its products. Their studies illustrated that the rate of ablation will be different for different pulse energies, and the productivity of nanomaterial fabrication followed a linear relationship with the pulse energy. At higher energies, the volume of melt reservoir will be high and, consequently, a higher number of particles can be fabricated. But in the sense of NS fabrication, up to a certain energy corrugation was observed to be good, beyond which corrugation became random structuring. The main reason behind this is the occurrence of many other dynamics such as fragmentation, phase explosion, boiling, and vaporization (Ivanov and Zhigilei 2003, Upadhyay et al. 2008, Zhigilei et al. 2009). Similarly, laser spot diameter at the focus also plays a crucial role in determination of the products of the ablation. If the beam waist lies exactly on the surface of the target, microstructures were observed, whereas the nanoripples with a periodicity of the order 400 nm were observed when the target is placed beyond or before the focal plane. At the central part of the beam, waist fragmentation mechanism prevails, while at the edges, thermal mechanisms prevail. The absorbed energy density per unit time and unit volume by the material at the focal plane is related to the gradient of the Poynting vector. If the laser beam is very tightly focused on the target surface, it dumps the entire energy and may increase the productivity of NPs but simultaneously loses its control on the fabrication of structures with sharp boundaries since tight focusing leads to a gradient of temperature and pressure and, consequently, an inhomogeneous gradient of the Poynting vector. In the case of a moderately focused laser beam, the gradient of the pressure and temperature is smoother, leading to fine ablation of the target surface with a better performance in fabrication of both NPs and NSs. A combined effect of beam waist at the focus and energy per pulse determines the laser fluence, and this combined effect ablates the material in a different way to produce NPs and NSs. Elsayed and coworkers (2013) also investigated the fabrication of gold NPs in water for different laser fluences using 10 ns pulses and discussed increment and decrement about the certain fluence and sizes of the gold NPs. Along with this, ablation of the gold surface exactly at focus and just above/below the focal plane was also investigated. Kabashin and coworkers (2003) investigated the effect of 110 fs pulses in the fabrication of gold NPs in water. They observed that the yield of Au NPs was good at higher laser fluences (1000 J/cm²) than at lower fluences (60 J/cm²). Moreover, NP distribution was observed to be better at lower fluences than at higher fluences. Average size was small in the former case than the latter case for Au NPs. At intermediate fluences, both smaller and larger NPs were observed. Barsch and coworkers (2009) reported the fabrication of Ag, Cu, Mg, and ZrO_2 NPs with ps laser pulses. They made an attempt to explain the fabrication of the mentioned metal and ceramic NPs at different fluences and focal positions.

11.5.4 Number of Pulses

Another important parameter is the number of laser pulses per spot, which strongly affects the ablation yield, size distribution, and the structure on the metal surfaces. The threshold fluence is a function of number of pulses impinging on the point of ablation. For multishot ablation, the threshold fluence is different from the single-shot ablation since the incubation effects diminish the ablation threshold. In the single-shot ablation, absorption of the laser energy by the target follows the Beer–Lambert relation. But in the case of multishot ablation, we cannot apply the Beer–Lambert relation since the reflectivity drops during the impulse of the first few pulses. The efficiency of the input energy coupling increases with the surface structures through the surface plasmons and, thereby, induces losses in reflectivity of the input beam. Leitz and coworkers (2011) discussed the effect of the number of laser pulses on the crater formation for µs, ns, ps, and fs pulses, and diameters of the holes formed were increased to a great extent and blurred in the case of micro (1,000 pulses) and ns (250 pulses) ablation and to a smaller extent in the case of ps pulses (500,000) and fs pulses (5,000).

11.6 RESULTS AND DISCUSSION

11.6.1 EXPERIMENTAL DETAILS

The schematic of a typical ablation in the ultrashort pulse regime is demonstrated in Figure 11.2a. Ultrafast experiments ablation were carried out using a chirped pulse-amplified (CPA) Ti:sapphire laser system (LEGEND, Coherent) delivering nearly transform-limited laser pulses of ~2 ps (or ~40 fs) duration, 1 kHz repetition rate, at 800 nm. Initial experiments were performed with a Ag substrate submerged in double-distilled water (DDW) (2–3 mm above the Ag sample) in a Pyrex cell that was placed on a motorized X–Y stage. Plane polarized (s-polarization) laser pulses were allowed to focus vertically onto the Ag substrate through a planoconvex lens of focal length 25 cm. The position of the focus was approximated to lie at the point where plasma was generated. X–Y translation stages, interfaced to Newport ESP 300 motion controller, were utilized to draw periodic structures on the Ag substrate with separations of ~5, ~25, ~50, ~75 μm. Typical energy

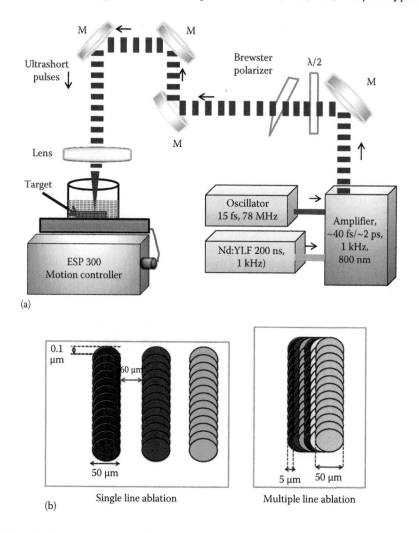

(a)

(b)   Single line ablation   Multiple line ablation

FIGURE 11.2 (a) Schematic of the ps laser ablation experimental setup. (b) Pulses incident on target. (Reproduced with permission from Hamad, S., Podagatlapalli, G.K., Tewari, S.P., and Venugopal Rao, S., Influence of picosecond multiple/single line ablation on copper nanoparticles fabricated for surface enhanced Raman spectroscopy and photonics applications, *J. Phys. D Appl. Phys.*, 46, 485501. © 2013 by the IOP Publishing.)

used was ~150 μJ per pulse. For example, in the case of ~5 μm, a structure was initially drawn on the Ag substrate through the movement of the X-stage followed by vertical movement of ~5 μm (using Y stage), and second-line structure was drawn on the substrate through the movement of the X-stage in the opposite direction. Thus, ablation was carried out for ~25, ~50, and ~75 μm separations. The schematic of multiple/double/single ablation carried out on the Ag substrate is explained in Figure 11.2b. The scanning speeds of the X–Y stages were 0.4 and 0.5 mm/s, respectively. For the case of Bessel beam ablation, the same ~2 ps/~40 fs laser pulses were utilized. Ag targets were purchased from Alfa Aeser (1 mm thick, 99.9% pure) and were immersed in HPLC grade DDW/acetone located in a pyrex cell, and the whole setup was placed on a motorized nanodirect X–Y–Z stage. The typical thickness of the liquid layer on the target surface was ~6 mm. ps/fs laser pulses were focused onto the Ag target immersed in DDW/acetone by an axicon (base angle = 25°). To match the focal plane exactly with the target surface, initially the focus was adjusted on the target in the absence of any liquid. Later, depending on the thickness of liquid layer, focal plane displacement (with respect to the focal plane in air) was estimated and corrected accordingly. Ablation of the Ag target in DDW was performed with ~2 ps laser pulses wherein the target has been moved by X–Y stages (Newport) to draw periodic lines at a separation of ~60 μm (total of 120 lines). The speeds of the X and Y stages were 0.05 and 0.5 m/s, respectively. Accelerations of X and Y stages were 0.05 and 0.5 m/s², respectively. Input pulse energies utilized in the ps ablation case were ~400, ~600, ~800, and ~1000 μJ. Similarly, ablation of the Ag target in acetone was carried out by ~40 fs laser pulses. In this case, the Ag target in the focal plane of the Bessel beam was moved by NTS NanoDirect stage in which the vertical stage (Z) was utilized to adjust the focal point on the target surface and the other two were utilized to draw periodic lines (each line length was 5 mm, total lines ~80) on Ag targets with a separation of 25 μm. In this fs ablation case, pulse energies utilized were of ~600, ~800, and ~1000 μJ. The speeds of the X and Y stages utilized in fs ablation were 100 and 500 μm/s, respectively. Typical duration of each sample ablation was ~40 min. After completion of the ablation, targets were removed and cleaned properly. Similarly, the colloidal solution was taken in air-tightened vessels to prevent it from oxidation. Focusing of Gaussian pulses using an axicon and a transverse profile at the focus is shown in Figure 11.3a. Ag colloids prepared in DDW with ~2 ps laser pulses had different degrees of gray coloration, whereas in acetone (with ~40 fs laser pulses), they exhibited diverse contrasts of golden-yellow coloration. Cu NPs and NSs have been prepared by ablation of a Cu target in liquid media with ~2 ps pulses. Pure Cu targets were washed with acetone after sonication to eliminate any residual organic impurities from the surface. The laser pulses were allowed to focus onto the Cu substrate using a planoconvex lens (f = 25 cm). The beam diameter ($2\omega_0$) estimated at the focus in air was ~40 μm. The typical level of liquid above the metal surface was ~7 mm. Typical fluence used was ~8 J/cm². The scanning speeds of X–Y stages were 0.1 mm/s in each direction. The motorized stages were scanned in such a way as to draw periodic lines on the surface with the given spacing of (1) ~60 μm and (2) ~5 μm. Based on the scanning configuration, two types of ablation were carried out: (1) single line ablation wherein the average number of pulses incident per spot on the target was ~500 (0.1 μm/ms speed, 1 pulse/ms, assuming a spot diameter of ~50 μm on the target in liquid); and (2) double line ablation in which the target surface was ablated twice (consequently the effective number of pulses per spot was estimated to be ~1000).

11.6.2 Results from Ag Target Ablation in Water

11.6.2.1 Ag Targets Ablated with Different Fluences and Scanning Conditions

The prepared Ag colloids by multiple line, double line, single line ablation are designated as NP-1, NP-2, NP-3, and NP-4 and the corresponding substrates as SS-1, SS-2, SS-3, and SS-4 for separations of 5, 25, 50, and 75 μm, respectively (Podagatlapalli et al. 2012). As per the mechanism shown in Figure 11.2b, SS-1 was influenced by multiple line ablations, whereas SS-2 was influenced by double line ablation, since the estimated beam waist on the Ag substrate was >20 μm.

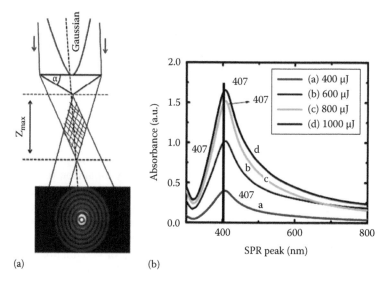

FIGURE 11.3   (a) Geometry of a focused Gaussian pulse using conical lens (Axicon) and formation of Bessel beam pattern. (b) UV–Vis absorption spectra of Ag colloids prepared in double-distilled water. (Reproduced from Podagatlapalli, G.K. et al., *Laser Phys. Lett.*, 12, 036003, 2015. © Astro Ltd. With permission.)

Consequently, complete overwriting (ablation of preablated portion) was carried out for SS-1 and partial overwriting for SS-2. Apart from these, SS-3 and SS-4 were obtained through single ablation. Furthermore, Ag substrates (for a fixed scan separation of 20 μm) were also fabricated at different fluences of ~4, ~8, ~12, ~16, and ~20 J/cm². To achieve superior fabrication rate of NPs, the laser beam should be focused exactly on the surface of the substrate. The theoretical beam waist (ω_0) estimated at the focus (in air) was ~10 μm, while the dimensions of fabricated structures were measured to be ~20 μm in the single ablation case. Subsequent to the completion of the scan, Ag substrates were removed from the liquid and preserved after proper cleaning. Colloidal suspensions were preserved in air-tightened glass vials. The absorption spectra of obtained colloidal solutions and laser exposed substrates were recorded (immediately after ablation) using a UV–Vis spectrometer (Jasco-V-670) equipped with an integrating sphere and working in the spectral range of 250–2500 nm. The energy dispersive X-ray (EDX) spectra and high resolution transmission electron microscope (HRTEM) images of the NP-1, NP-2, NP-3, and NP-4 were recorded to identify the metallic fingerprints and NP size distribution along with their crystallographic information. Morphology of the laser-exposed portions of SS-1, SS-2, SS-3, and SS-4 was characterized by a field emission electron microscope (FESEM) (Ultra 55 from Carl ZEISS) and an AFM. Suspensions of Ag NPs were characterized by transmission electron microscope (TEM) (SEI cecnai G2 S-Twin 200 kV instrument), providing an estimate of the size distribution and morphologies of NPs. Ag colloids were centrifuged on carbon-coated copper grids and analyzed using TEM operated at 200 keV. Details of the colloids and their characterization data have been reported elsewhere (Podagatlapalli et al. 2013). HRTEM data of a single Ag NP exhibited parallel Ag lattice planes having a separation of 2.2 Å corresponding to (111) lattice planes of silver. Selected area electron diffraction (SAED) patterns revealed the polycrystalline nature of the Ag nanospheres for all the four cases. In all of the cases, Ag NPs with polycrystallinity were evident from a concentric ring pattern in SAED images. Measured lattice constants from the SAED pattern agreed well with the reported values of silver and oxidized phases (minimal) of silver. The recorded UV–Vis absorption spectra of NP-1, NP-2, NP-3, and NP-4 exhibited localized surface plasmon resonance (LSPR) peaks in the neighborhood of 412, 417, 407, and 407 nm, respectively.

Depending on the geometry of ablation, the Ag NSs are labeled as SS-1 (line separation of 5 μm, multiple line ablation), SS-2 (line separation of 25 μm, double line ablation), and SS-3/SS-4 (line separation of 50 μm, 75 μm, single line ablation). The surface morphologies of the laser-exposed portions of Ag substrates, characterized by FESEM, are presented in Figure 11.4a through d, depicting NSs on SS-1 and SS-2 with dimensions of ~100 nm. There were no such NSs on SS-3 and SS-4, which was confirmed from FESEM images. Because of multiple/double line ablation, we could not observe distinguishable demarcation between two consecutive line structures in SS-1 and SS-2, whereas for SS-3 and SS-4, we could observe separation between the structures. The morphologies of ablated metallic surfaces suffer from incubation effects (Podagatlapalli et al. 2013). The density and height of the nanosized domes were different for SS-1 and SS-2. In addition to lateral NSs, there were small NP grains on both SS-1 and SS-2 substrates. These structures with Ag NP grains could enhance the Raman signals, and therefore, SS-1 and SS-2 were used for Raman measurements of adsorbed rhodamine 6G (R6G) molecules. The Raman spectra recorded for R6G on the laser-ablated surfaces SS-1 and SS-2 are shown in Figure 11.5a for 532 nm excitation and in Figure 11.5b for 785 nm excitation. Excitation with 785 nm demonstrated higher Raman signature enhancement (compared to 532 nm excitation) and lower fluorescence background from the absorbed analytes. Probably, the dome-like NSs provided

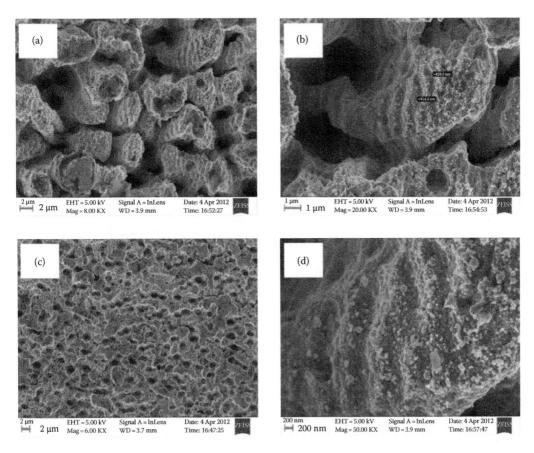

FIGURE 11.4 FESEM images of the laser-exposed portions in Ag substrate in water (a) and (b) dome-like structures formed on the substrate SS-1 because of the over wring, and its closer view (c) and (d) surface morphology of SS-2 shows the dome-like structures. Closer view images show the fabricated Ag NPs grains on the substrate. The scale bar in (a), (b), and (c) is 2 μm, while in (d) it is 200 nm. (Reprinted with permission from Podagatlapalli, G.K., Hamad, S., Sreedhar, S., Prasad, M.D., and Venugopal Rao, S., Silver nano-entities through ultrafast double ablation in aqueous media for surface enhanced Raman scattering and photonics applications, *J. Appl. Phys.*, 113, 073106. © 2013, American Institute of Physics.)

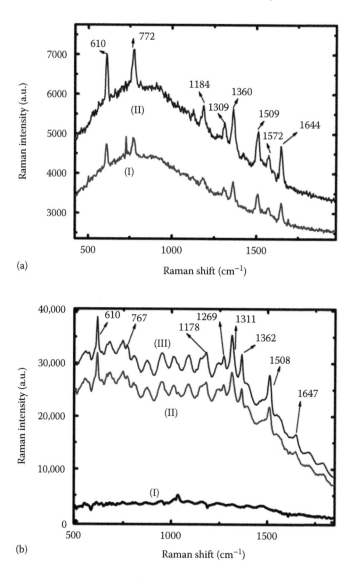

FIGURE 11.5   Raman signals recorded from R6G molecules (12 μM) in methanol (a) with micro-Raman spectrometer (excitation at 532 nm). Red and blue spectra represent the enhanced Raman signatures of R6G from SS-1 and SS-2 and (b) with bulk Raman spectrometer (excitation at 785 nm). Black spectrum was from Ag plain surface. Red and blue spectra represent the enhanced Raman signatures of R6G from SS-1 and SS-2. Signal collection time for both spectrometers was 5 s. (Reprinted with permission from Podagatlapalli, G.K., Hamad, S., Sreedhar, S., Prasad, M.D., and Venugopal Rao, S., Silver nano-entities through ultrafast double ablation in aqueous media for surface enhanced Raman scattering and photonics applications, *J. Appl. Phys.*, 113, 073106. © 2013, American Institute of Physics.)

high local fields that contributed to significant Raman enhancements. Intensity enhancements of ~2439 were obtained for the prominent aromatic C–C stretch mode corresponding to 1362 cm^{-1} of R6G (12 μM) from SS-1 excited with 785 nm. This measurement was carried out by comparing SERS spectra of R6G with reference spectra recorded at a higher concentration from the glass side (nonplasmonic platform).

Figure 11.5 depicts the Raman spectra of R6G overlapped with the fluorescence background. This could be explained on the basis of the distance between the analyte molecule and the nanomaterial.

Estimated intensity enhancements (I_{SERS}/I_{Raman}) for each mode of R6G from the multiply ablated Ag NSs (SS-1) and doubly ablated Ag NSs (SS-2) with the excitation wavelengths of 532 and 785 nm were ~10^3. The performance of SS-1 and SS-2 targets was evaluated by plotting the Raman signal intensity enhancements. It was observed that superior Raman signal intensity enhancements were observed from SS-2 (i.e., doubly ablated Ag NSs) for both the excitation wavelengths of 532 and 785 nm. Our observations demonstrated that the substrate with 25 μm line separations provided optimal Raman enhancements due to double line ablation compared to the substrate obtained with multiple line ablation. We believe that multiple line ablation, probably, caused a partial washout of the nanoroughness compared to double line ablation. Measurements of R6G Raman enhancements revealed that the performance of Ag NSs obtained from double line ablation was superior to NSs fabricated by multiple line ablation. Therefore, we could conclude that scan separation of 25 μm (double line ablation) provided better NSs and, hence, optimal Raman enhancements as a result of double ablation.

Later, the combined effect of double line ablation on Ag targets with diverse input fluences was investigated. For a fixed scan separation (25 μm), Ag substrates were ablated in DDW at 4, 8, 12, 16, and 20 J/cm^2 input fluences. Figure 11.6 illustrates surface morphologies of Ag substrates fabricated at different fluences (for a fixed line separation 25 μm). Along with the formation of NSs, Ag NP grains were formed on these substrates. SERS activity of these Ag substrates was investigated for RDX ($C_3H_6N_6O_6$; 1,3,5-trinitroperhydro-1,3,5-triazine) molecules dissolved in acetonitrile (ACN). Raman spectra recorded for RDX analyte adsorbed on the Ag substrates fabricated with different laser fluences are illustrated in Figure 11.7a and b for 532 and 785 nm excitations, respectively. The spectra were collected in the range of 250–3500 cm^{-1}, which covers most of the Raman bands of RDX. In the recorded SERS spectra of RDX from the Ag nanostructured target (Figure 11.7a), only few peaks (Besner et al. 2007, 2008, Robitaille et al. 2013) were observed compared to the conventional bulk Raman spectra (Figure 11.7b) of RDX. Raman signatures observed in the case of 785 nm excitation were higher in number compared with 532 nm excitation. The modes corresponding to 383 and 924 cm^{-1} (Figure 11.7a) represent the signatures of ACN, and the remaining were from RDX. Intensity enhancements for the observed Raman modes of RDX were estimated, and typical numbers are provided as follows. In the case of excitation of 532 nm, modes corresponding to 2947, 2255, and 3005 cm^{-1} (C–H stretch) exhibited ~582, ~652, and ~86 times of the enhancement, respectively, compared to the same modes observed from higher concentrations of RDX molecules adsorbed on the glass slide. Some of the other modes observed were 1167 and 1381 cm^{-1} (CH$_2$ twisting). However, fluorescence prevailed in the spectra obtained from the substrates prepared at fluences of ~16 and ~20 J/cm^2 (Podagatlapalli et al. 2013). With 785 nm excitation (Figure 11.7b), the dominant modes were 856 cm^{-1} (N–N stretch + NO$_2$ axial scissoring), 924 cm^{-1} (CH$_2$ rocking or combination), 1240 cm^{-1} (N–N and symmetric + NO$_2$ stretch and may be with CH$_2$ twist), 1314 cm^{-1} (N–N stretching + CH$_2$ stretch), 1393 cm^{-1} (CH$_2$ twisting), 1457 cm^{-1} (CH$_2$ scissoring), and 1560 cm^{-1} (ONO equatorial stretching). The observed Raman modes of RDX are in good agreement with earlier reports (Sylvestre et al. 2005). The SERS activity of Ag targets fabricated at different fluences for double line ablation was evaluated by plotting intensity enhancements of RDX Raman modes with respect to the fluence at which NSs were prepared. In the case of 532 nm excitation, strong Raman signal enhancements were observed from the Ag substrates fabricated at 12 J/cm^2. Our data analysis suggested that most of the RDX Raman modes were enhanced strongly from the Ag NSs fabricated at 16 J/cm^2. From the analysis of Raman signal enhancements of RDX with 532 nm and 785 nm excitations illustrated, optimal fluences were ~12 and ~16 J/cm^2, respectively (Podagatlapalli et al. 2013).

11.6.2.2   Ag Targets Ablated with Different Angles of Incidence

Morphologies of laser-exposed portions of Ag substrates fabricated with different angles of incidence were investigated through FESEM and AFM imaging techniques. Figure 11.8a and c

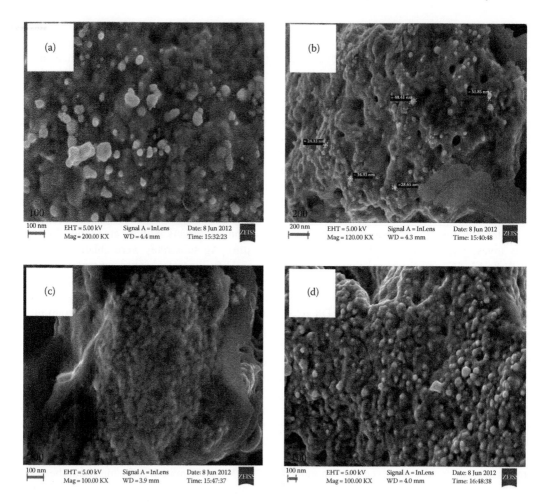

FIGURE 11.6 FESEM images (closer view of the structure) of the laser exposed portions in Ag substrate in water prepared for different laser fluences: (a) 4 J/cm², (b) 8 J/cm², (c) 12 J/cm², and (d) 16 J/cm². The images depict fabricated grains of Ag NPs on the substrate. The scale bar in (a) and (d) is 100 nm, and for (b) and (c), it is 200 nm. (Reprinted with permission from Podagatlapalli, G.K., Hamad, S., Sreedhar, S., Prasad, M.D., and Venugopal Rao, S., Silver nano-entities through ultrafast double ablation in aqueous media for surface enhanced Raman scattering and photonics applications, *J. Appl. Phys.*, 113, 073106. © 2013, American Institute of Physics.)

illustrates the FESEM images of laser-exposed portions of Ag substrates corresponding to NS-5 and NS-15. Figure 11.8b and d represents the AFM images of NS-5 and NS-15, respectively. AFM images exhibited lateral NSs of dimensions <200 nm. Similarly, Figure 11.9 depicts the surface morphologies of NS-30 and NS-45. Figure 11.9a and c illustrates the FESEM images of laser-exposed portions of the Ag substrate corresponding to NS-30 and NS-45, respectively. Figure 11.9b and d illustrates the AFM images of NS-30 and NS-45, respectively. AFM images confirmed the presence of lateral NSs on the surface of NS-30. Ablation of targets was carried out in a controlled manner without changing the ablation conditions except the angle of incidence. The surface topology of laser-exposed portions of the substrates demonstrated different morphologies for NS-30, which is evident from Figure 11.9b.

Nanostructured Ag targets fabricated at oblique laser incidence in DDW are labeled as NS-5 (Ag target ablated at 5° of laser incidence), NS-15 (Ag target ablated at 15° of laser incidence), NS-30 (Ag target ablated at 30° of laser incidence), and NS-45 (Ag target ablated at 45° of laser incidence).

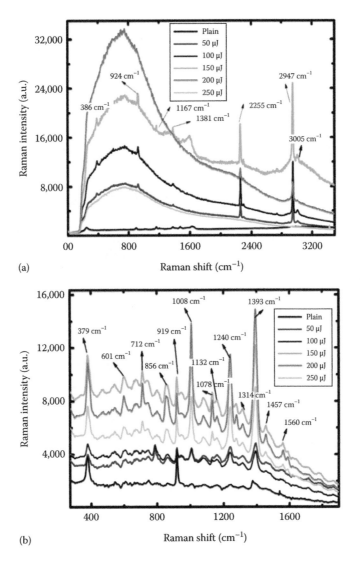

(a)

(b)

FIGURE 11.7 (a) Micro-Raman spectra of adsorbed RDX molecules (excitation at 532 nm) in acetonitrile. The maximum enhancement of Raman signal was observed at pulse energy 150 μJ (fluence of 12 J/cm²). (b) Bulk Raman spectra of the adsorbed RDX (excitation at 785 nm) in acetonitrile. The maximum enhancement of Raman signal was observed at pulse energy of 200 μJ (fluence of 16 J/cm²). More signatures were observed with 785 nm excitation compared to 532 nm excitation. For both the cases, the time of integration was 5 s. (Reprinted with permission from Podagatlapalli, G.K., Hamad, S., Sreedhar, S., Prasad, M.D., and Venugopal Rao, S., Silver nano-entities through ultrafast double ablation in aqueous media for surface enhanced Raman scattering and photonics applications, *J. Appl. Phys.*, 113, 073106. © 2013, American Institute of Physics.)

The Raman spectra of R6G/ANTA adsorbed on NS-5, NS-15, NS-30, and NS-45 were recorded with micro-Raman (WiTech ALPHA 300 instrument) using a continuous wave (cw) Nd:YAG laser at 532 nm and bulk Raman spectrometers (Ocean Optics) with a cw Ar⁺ laser at 785 nm. In the micro-Raman spectrometer, the laser beam was focused on the substrate with an objective lens (100×) whose beam waist estimated was ~700 nm. Raman signals were collected in the backscattering geometry. In bulk Raman spectrometer (excitation at 785 nm), the laser beam (diameter of ~1 mm) was directed toward the sample without any focusing lenses, and Raman signals were

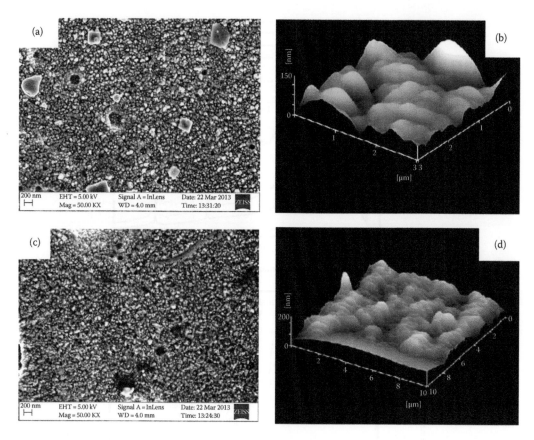

FIGURE 11.8 (a) FESEM imaging of laser-exposed Ag substrates NS-5. (b) AFM imaging of laser-exposed Ag substrates NS-5. (c) FESEM imaging of laser-exposed Ag substrates NS-15. (d) AFM imaging of laser-exposed Ag substrates NS-15. (Reprinted from *Appl. Surf. Sci.*, 303, Podagatlapalli, G.K., Hamad, S., Mohiddon, M.A., and Venugopal Rao, S., Effect of oblique incidence on silver nanomaterials fabricated in water via ultrafast laser ablation for photonics and explosives detection, 217–232. © 2014, with permission from Elsevier.)

collected in the backscattering geometry. All the spectra were calibrated to Raman peak of silicon wafer at 520 cm⁻¹. The Raman spectra of R6G (10 μL) were recorded from the structured Ag substrates. First, the performance of NS-5, NS-15, NS-30, and NS-45 was investigated by recording the Raman spectra of R6G (25 μM) with excitation wavelengths of 785 and 532 nm. Figure 11.10a through d depicts the Raman spectra of R6G (~10 μL drop) placed on the laser-exposed portion to form a monolayer of analyte (excitation wavelength of 785 nm). Multilayered analyte adsorption on NSs generally inhibits the activity of Raman signal enhancements since the multilayers screen the effect of evanescent fields from NSs. Furthermore, the SERS activity of the substrate depends on the distance between NSs and analyte molecule (typically should be few nm). To compare the Raman spectra (excitation wavelength of 785 nm) of the analyte from the laser-ablated Ag surface, the Raman spectra of R6G from the plain Ag surface (blue) and Si substrate (black) were also recorded. The estimated Raman mode intensity enhancements of R6G (25 μM) at 1360 cm⁻¹ were ~13, ~12, ~25, and ~14 from NS-5, NS-15, NS-30, and NS-45, respectively. Similarly, the Raman spectra of R6G (25 nM) were also recorded from four substrates after proper cleaning, and comparison is shown in Figure 11.11.

FIGURE 11.9 (a) FESEM imaging of laser-exposed Ag substrates NS-30. (b) AFM imaging of laser-exposed Ag substrates NS-30. (c) FESEM imaging of laser-exposed Ag substrates NS-45. (d) AFM imaging of laser-exposed Ag substrates NS-45. (Reprinted from *Appl. Surf. Sci.*, 303, Podagatlapalli, G.K., Hamad, S., Mohiddon, M.A., and Venugopal Rao, S., Effect of oblique incidence on silver nanomaterials fabricated in water via ultrafast laser ablation for photonics and explosives detection, 217–232. © 2014, with permission from Elsevier.)

The performance of each substrate was evaluated using the enhancement factors (EFs) given by (Podagatlapalli et al. 2014):

$$EF = \frac{I_{SERS}}{I_{Raman}} \frac{N_{Raman}}{N_{SERS}}$$

where

I_{SERS} is the integrated intensity of R6G Raman band under consideration from the nanostructured Ag substrate

I_{Raman} is the integrated intensity of the same Raman band from higher molar R6G (0.25 M) on the silicon substrate

Similarly, N_{SERS} is the number of analyte molecules constituting the first monolayer on the nanostructured substrate surface under the laser spot area and N_{Raman} is the number of molecules covered by the beam waist in the bulk R6G solution on Si target. N_{SERS} and N_{Raman} were estimated to be

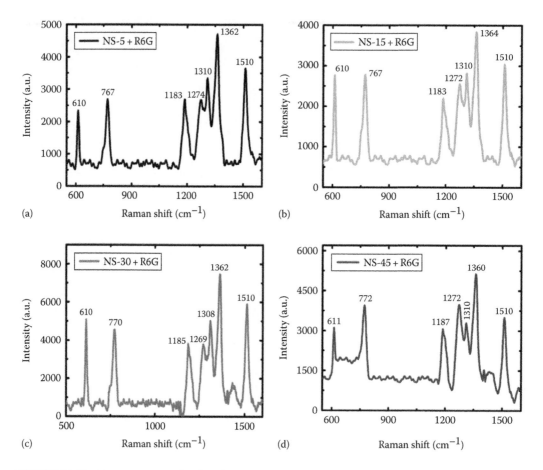

FIGURE 11.10   Raman spectra of R6G (25 μM) recorded from the (a) NS-5, (b) NS-15, (c) NS-30, and (d) NS-45 with an excitation wavelength 785 nm. Time of integration for each spectrum was 10 s. (Reprinted from *Appl. Surf. Sci.*, 303, Podagatlapalli, G.K., Hamad, S., Mohiddon, M.A., and Venugopal Rao, S., Effect of oblique incidence on silver nanomaterials fabricated in water via ultrafast laser ablation for photonics and explosives detection, 217–232. © 2014, with permission from Elsevier.)

~120 and ~5 × 10^9, respectively. In this estimation, the adsorption factor was considered to be 0.3. We arrived at this estimate by considering the beam waist at the focus to be ~0.65 μm for 532 nm excitation and the size of the R6G molecule as 2 × 10^{-18} m^2. For aromatic C–C stretch mode corresponding to 1362 cm^{-1}, estimated EFs were ~2.5 × 10^8, ~2.6 × 10^8, ~5 × 10^9, and ~3.4 × 10^8 from the substrates NS-5, NS-15, NS-30, and NS-45, respectively. Estimated intensity enhancements and EFs revealed that a considerably superior performance of NS-30 (prepared at 30°) compared to others for both μM and nM concentrations of R6G excited with 785 and 532 nm wavelengths.

After cleaning (several times with acetone) and sonication, four Ag NSs were utilized again to record Raman spectra of an explosive molecule 5-amino, 3-nitro, -1H-1,2,4-nitrozole (ANTA) dissolved in ACN (5 mM, 5 μM). The spectra are shown in Figure 11.12a and b for mM and μM concentrations, respectively. As shown in Figure 11.12, dominant modes were 470, 488 cm^{-1} (both are NO$_2$ deformation), 724 cm^{-1} (ring deformation), 843 cm^{-1} (NO$_2$ deformation + ring deformation), 1026 cm^{-1} (N1–N2–C3 bend), 1125 cm^{-1} (NN-symmetric stretch), 1340 cm^{-1} (C–NO$_2$ symmetric stretch), 1530 cm^{-1} (C–NH$_2$ asymmetric stretch + NH$_2$ bend), and 1588 cm^{-1} (C–NH$_2$ symmetric stretch + NH$_2$ bend). For C–NO$_2$ symmetric stretch mode corresponding to 1340 cm^{-1}, estimated values of N_{SERS} and N_{Raman} were ~1.6 × 10^9 and ~4 × 10^{11}, respectively. Estimated EFs

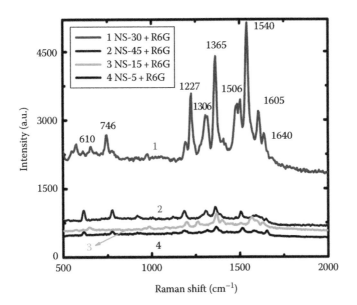

FIGURE 11.11 Raman spectra of R6G (25 nM) recorded from the NS-5, NS-15, NS-30, and NS-45 with an excitation wavelength 532 nm and an integration time 5 s. (Reprinted from *Appl. Surf. Sci.*, 303, Podagatlapalli, G.K., Hamad, S., Mohiddon, M.A., and Venugopal Rao, S., Effect of oblique incidence on silver nanomaterials fabricated in water via ultrafast laser ablation for photonics and explosives detection, 217–232. © 2014, with permission from Elsevier.)

for the 1340 cm^{-1} mode were ~4.1 × 10^3, ~5.5 × 10^3, ~9.7 × 10^4, and ~4 × 10^3 from Ag substrates NS-5, NS-15, NS-30, and NS-45, respectively (using adsorption factor [η]—0.6). Similarly, estimated values of N_{SERS} and N_{Raman} were ~0.8 × 10^6 and ~1.6 × 10^{11}, respectively, for 5 µM concentration of ANTA (C–NO$_2$ symmetric stretch mode corresponding to 1340 cm^{-1}). Estimated EFs were ~8.5 × 10^4, ~4.2 × 10^5, ~1.1 × 10^6, and ~4 × 10^5 for the Ag substrates NS-5, NS-15, NS-30, and NS-45, respectively (adsorption factor [η]—0.4). As per our observations, estimated EFs for R6G and ANTA (both mM and µM) from NS-30 demonstrated stronger enhancements (9.7 × 10^4, 1.1 × 10^6) compared to the other three substrates (Podagatlapalli et al. 2014).

The most important reason behind the obtained enhancement of the Raman signatures from the corrugated surface achieved by laser ablation is that it probably acted like a random grating, thus supporting the excitation of both the localized surface plasmons (presence of Ag NPs grains) and surface plasmon polaritons (presence of random grating). Coupling of incident photons to surface plasmons of nanomaterials determines the magnitude of evanescent field in the vicinity of nanomaterials. The essential condition for coupling in case of surface plasmon polaritons is the spacing of the grating on a metallic target. Similarly, the size and shape of nanomaterials (NSs and NPs) and wavelength of excitation are the key parameters to couple incident light to the localized surface plasmons. Effective coupling of incident photons with surface plasmons of nanomaterials enables nanotips to act as periodically oscillating dipole antennas with an oscillation frequency of incident radiation. As a result, an analyte molecule in the vicinity of nanotip experiences collective electric field of incident photons and field provided by a radiating dipole. Consequently, enhancement of the Raman signals from both plasmons (localized and propagating) plays a critical role. Further, we believe that the Raman signal enhancement of analytes from the laser-exposed portions of Ag substrates depends on the degree of roughness and topology of the surfaces. Ablation of Ag target at 30° incident angle enabled the surface to contain more NSs and random gratings compared to NS-5, NS-15, and NS-45. In the case of NS-30, coupling of incident photons to the Ag surface at 30° incidence angle could have resulted in efficient ablation and formation of NSs. Thus, NS-30 was

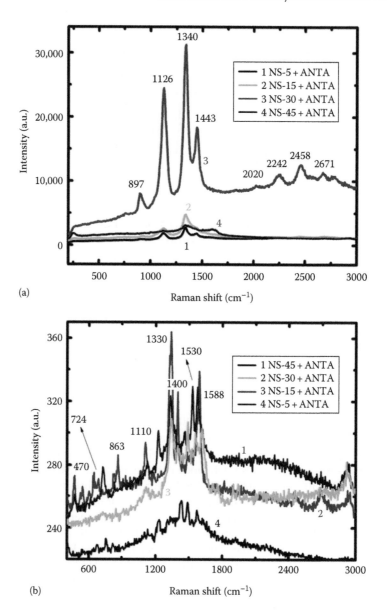

FIGURE 11.12 Raman spectra of ANTA recorded from NS-5, NS-15, NS-30, and NS-45 with an excitation wavelength 532 nm for (a) 5 mM concentration and (b) 5 μM concentration. Time of integration for each spectrum was 5 s for data presented in (a) and 0.5 s for data presented in (b). (Reprinted from *Appl. Surf. Sci.*, 303, Podagatlapalli, G.K., Hamad, S., Mohiddon, M.A., and Venugopal Rao, S., Effect of oblique incidence on silver nanomaterials fabricated in water via ultrafast laser ablation for photonics and explosives detection, 217–232. © 2014, with permission from Elsevier.)

consistent to provide large EFs and efficient trace detection. Furthermore, we succeeded in utilizing the Ag NSs for the identification of two different analytes separately. For both analytes, R6G and ANTA targets fabricated at 30° demonstrated a very significant Raman signal elevation.

11.6.2.3 Ag Targets Ablated with ps/fs Bessel Beams

Laser ablation dynamics of metals in liquids simultaneously depend on laser parameters such as wavelength, repetition rate, pulse duration, energy per pulse, and focusing geometry. Most of the experiments in this field were carried out by laser pulses with focused Gaussian pulses with conventional convex lenses. In conventional Gaussian beam focusing experiments, efficient nanomaterial generation is possible when the spot size (beam waist) on the target is very small. However, the least beam waist minimizes the depth of focus (DOF) (Z_{max}), which, in fact, introduces artifacts and creates a loss of control on the position of beam waist for placing the sample exactly at focus. The focusing of Gaussian pulses through an axicon (shown in Figure 11.3) eliminates the aforesaid problems by offering a considerable DOF. Additionally, it maintains comparatively similar spot size without any translational spread due to diffraction effects. Moreover, these nondiffracting Bessel beams exhibit nearly a constant intensity profile along their propagation. Salient features of Bessel beams, such as self-reconstruction and stability under nonlinear propagation, make them extensively useful in extremely localized and controlled energy deposition in transparent materials. The interference of conical wave fronts produced by an axicon is the primary reason to produce the Bessel beam that has the field and intensity distribution given by $E(\rho) = E_0 \exp(-\rho^2/\omega^2)$, where E_0 is the on-axis field amplitude, ρ is the radial distance from the propagation axis Z, and ω is the radius of Gaussian beam. We performed ablation experiments to exploit the (1) uniqueness of nondiffracting Bessel beams over conventional Gaussian beam and (2) compared the performance of Ag NPs fabricated by picosecond (ps)/femtosecond (fs) laser pulses. The LSPR peak position for the Ag colloids prepared using Bessel beam and input energies of ~400, ~600, ~800, and ~1000 µJ was positioned at 407 nm, which is contrary to the reports on ablation till date obtained using Gaussian laser pulses (Podagatlapalli et al. 2015). Although the absorbance of colloids (measure of yield) changed in accordance with the input energy, LSPR peak position did not change. LSPR peak position is the combined effect of all neighboring NPs covered under the diameter of the light beam path. Other morphological characterizations such as TEM and UV–Vis absorption provide global information on average sizes of NPs. LSPR peak possibly shifts due to the sizes of neighboring NPs. But in the present case, it was different, demonstrating the possibility of fabricating NPs with similar average sizes even at four different input energies.

Four Ag nanostructured targets were fabricated in DDW with ~2 ps pulses at ~400, ~600, ~800, and ~1000 µJ. Similarly, three more Ag nanostructured targets were fabricated in acetone with ~40 fs pulses at ~600, ~800, and ~1000 µJ. These two sets of Ag NSs fabricated with ps and fs Bessel beams were utilized to detect/identify explosive molecules, 2,4,6,8,10,12-hexanitro-2,4,6,8,10,12-hexaazaisowurtzitane (CL-20). NSs were obtained for high-input laser energies such as ~400, ~600, ~800, and ~1000 µJ, only. At lower energies (<400 µJ), we could not observe NSs on laser-exposed portions. Morphologies were investigated by FESEM, and these surfaces were utilized to record the Raman spectra of adsorbed CL-20 (5 µM). Raman spectra were recorded with an excitation wavelength of 532 nm. Figures 11.13 and 11.14 demonstrate the FESEM images of laser-exposed Ag targets ablated with ~2 ps and ~40 fs pulses, respectively. Insets depict the SERS spectra (red) of CL-20 (5 µM) from the laser-exposed portions, and blue curves represent the 0.1 M CL-20 Raman spectra from silicon wafer. Our data suggested that the highly elevated mode in CL-20 was 1330 cm^{-1} (CH bend + NO symmetric stretch) and EFs were estimated by considering I_{SERS} and I_{Raman} of the mode mentioned from NSs as well as the silicon target. Estimated EFs were ~1.1 × 10^6, ~3.4 × 10^6, ~3.3 × 10^5, and ~3.8 × 10^5 for nanostructured targets prepared with ~2 ps pulses at ~400, ~600, ~800, and ~1000 µJ, respectively. Similarly, EFs of the Ag targets prepared with ~40 fs laser pulses were estimated as 1.24 × 10^6, 1.9 × 10^6, and 8.07 × 10^5. From the comparison of EFs, we found that in both regimes, the order of EF was similar, but a slightly lower enhancement was observed from the Ag NSs prepared with ~2 ps pulses. To estimate these factors, beam waist of the excitation wavelength (532 nm) at the focus was considered to be ~650 nm. On comparison of the EFs, we found that in both regimes, the order of magnitude was the same (~10^6) (Podagatlapalli et al. 2013).

11.6.3 Cu Targets Ablated with Different Energies and Scanning Conditions

Cu colloidal NPs have been produced by ps laser ablation in four different liquid media. To avoid the ambiguity, Cu NPs prepared through multiple line (single line) ablation in acetone, dichloromethane

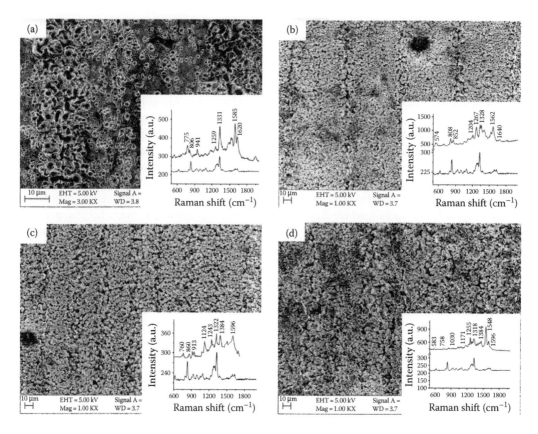

FIGURE 11.13 FESEM images Ag targets with Bessel beams at energies (a) 400 μJ, (b) 600 μJ, (c) 800 μJ, and (d) 1000 μJ. Insets show SERS spectra in red of CL-20 (5 μM) from the laser-exposed portions and blue represent the 0.1 M CL-20 Raman spectra from silicon wafer. Time of integration was 0.5 s for all the cases. (Reproduced from Podagatlapalli, G.K. et al., *Laser Phys. Lett.*, 12, 036003, 2015. © Astro Ltd. With permission.)

(DCM), ACN, and chloroform (CHCl$_3$) were labeled as MCuNP1 (SCuNP1), MCuNP2 (SCuNP2), MCuNP3 (SCuNP3), and MCuNP4 (SCuNP4), respectively. Figure 11.15 summarizes the Raman spectra recorded for (1) MCuNP1 and SCuNP1, (2) MCuNP2 and SCuNP2, (3) MCuNP3 and SCuNP3, (4) MCuNP4 and SCuNP4, respectively. In the case of MCuNP1, the Raman signal intensity for the 531.04 cm^{-1} mode was ~2179, whereas the Raman intensity for pure acetone was ~9, and therefore, the intensity enhancement was estimated to be ~242. Our observation was that Raman signatures in MCuNP1 were elevated better compared to SCuNP1. In the case of MCuNP2 and SCuNP2, the intensity enhancements were ~11 and ~5 for the 283.4 cm^{-1} mode and ~5.5 and ~5 for the 701.6 cm^{-1} mode, respectively. For the case of MCuNP3 and SCuNP3, the intensity enhancements were ~54.8 and ~30.7 for the 379.05 cm^{-1} mode, ~50 and ~29.6 for the 919.53 cm^{-1} mode, and ~16.3 and ~11.6 for the 1374.92 cm^{-1} mode, respectively. Similarly, in the case of MCuNP4 and SCuNP4, the intensity enhancements were ~26.9 and ~5.6 for the 258.5 cm^{-1} mode, ~24.5 and ~5.1 for the 364.5 cm^{-1} mode, ~24.5 and ~5.8 for the 665.9 cm^{-1} mode, and 51.3 and 10.9 for the 754.6 cm^{-1} mode, respectively. Our studies clearly suggest that these NPs have potential for SERS studies of other analyte molecules (Hamad et al. 2013).

Figure 11.16 illustrates the FESEM images of fabricated NSs on the surface of Cu ablated in (1) acetone (MCuNS1), (2) DCM (MCuNS2), (3) ACN (MCuNS3), and (4) chloroform (MCuNS4),

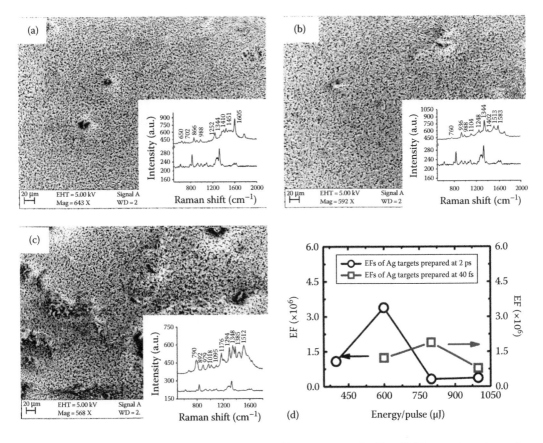

FIGURE 11.14 FESEM images Ag targets with Bessel beams prepared with ~40 fs laser pulses at energies (a) 600 μJ, (b) 600 μJ, and (c) 800 μJ. Insets show SERS spectra in red of CL-20 (5 μM) from the laser-exposed portions and (d) comparison of enhancement factors for the Ag targets prepared with Bessel beams with ~2 ps and ~40 fs pulses. Time of integration was 0.5 s for all the cases. (Reproduced from Podagatlapalli, G.K. et al., *Laser Phys. Lett.*, 12, 036003, 2015. © Astro Ltd. With permission.)

and insets present their respective images with lower resolution (2 μm scale). The surface morphology of multiply ablated Cu substrates MCuNS1 (Figure 11.16a) and MCuNS3 (Figure 11.16c) revealed the formation of weakly distinguishable laser-induced periodic surface structures (LIPSS) with a period of ~400 nm along with cylindrical Cu NP grains (length ~300 nm, diameter ~20 nm) and the formation of LIPSS with a periodicity 250–300 nm, respectively. In the case of MCuNS2 (Figure 11.16b), the surface topography of the Cu substrate consisted of pillar-shaped structures of a few micrometers. Moreover, most of the cubic NPs with the size of 200 nm were formed on top of the pillar. Similarly in the case of MCuNS4 (Figure 11.16d), rough-ended pillars with sizes of a few micrometers were formed on the Cu surface unlike in the case of MCuNS2. Figure 11.17 depicts the FESEM image morphologies of fabricated NSs on the surface of Cu substrates when ablation was carried out in (1) acetone (SCuNS1), (2) DCM (SCuNS2), (3) ACN (SCuNS3), and (4) chloroform (SCuNS4) via single line ablation, and insets of Figure 11.17 depict their respective images with lower resolution (2 μm scale). We estimated that a lesser number of pulses (~500) per spot were incident on the surface compared to the multiple ablation case (~4000). LIPSS with periods of ~300 and ~400 nm for SCuNP1 (Figure 11.17a) and SCuNP3 (Figure 11.17c), respectively, were observed. In addition to this, spherical NP grains (with NP size of ~50 nm) were formed on the periodic structure surfaces in the case of SCuNP1, which was not observed in SCuNP3 case. Similarly, randomly corrugated

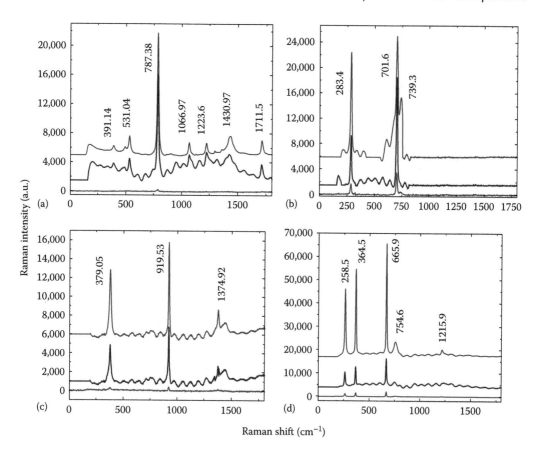

FIGURE 11.15 Raman spectra of (a) MCuNP1 (blue/top), SCuNP1 (wine/middle), and pure acetone (red/bottom); (b) MCuNP2 (blue/top), SCuNP2 (wine/middle), and pure DCM (red/bottom); (c) MCuNP3 (blue/top), SCuNP3 (wine/middle), and pure ACN (red/bottom); and (d) MCuNP4 (blue/top), SCuNP4 (wine/middle), and pure chloroform (red/bottom) recorded with an excitation wavelength of 785 nm and an integration time of 5 s. (Reproduced with permission from Hamad, S., Podagatlapalli, G.K., Tewari, S.P., and Venugopal Rao, S., Influence of picosecond multiple/single line ablation on copper nanoparticles fabricated for surface enhanced Raman spectroscopy and photonics applications, *J. Phys. D Appl. Phys.*, 46, 485501. © 2013 by the IOP Publishing.)

structures along with NP grains and less number of triangular NPs (~250 nm size) were observed for the cases SCuNP2 (Figure 11.17b) and SCuNP4 (Figure 11.17d), respectively (Hamad et al. 2014).

The observed ripples could be the resultant of the interference of the incident electromagnetic wave and surface-scattered electromagnetic wave (SEW) as proposed by earlier groups working on LIPSS. The periodicity of the ripples is related to the wavelength and angle of incidence of the laser beam as in the following equation:

$$\Lambda_\perp = \frac{\lambda}{\lambda / \lambda_s \pm \sin\theta}$$

where

$\Lambda_\perp$ is the period of ripples
λ is the incident wavelength of the laser
λ_s is the wavelength of the surface wave

For a normal incidence, the periodicity of the ripples is nearly equal to the λ of SEW.

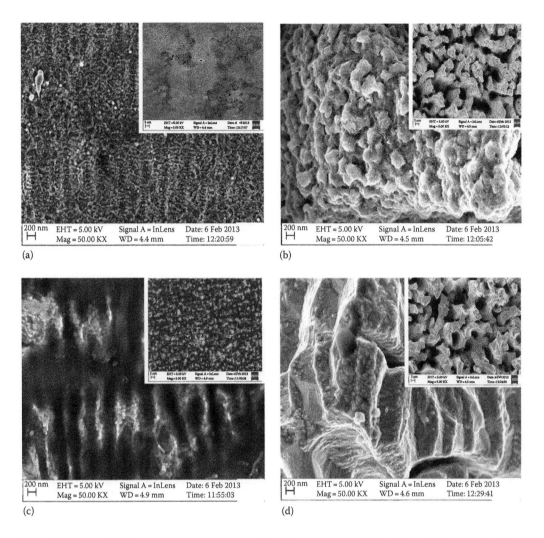

FIGURE 11.16 FESEM images of substrates (a) MCuNS1 in acetone, (b) MCuNS2 in DCM, (c) MCuNS3 in ACN, and (d) MCuNS4 in CHCl₃ fabricated through ultrafast multiple line ablation of Cu. Insets show their corresponding views on a larger scale (2 μm scale). (Reprinted with permission from Hamad, S., Podagatlapalli, G.K., Mohiddon, M.A., and Venugopal Rao, S., Cost effective nanostructured copper substrates prepared with ultrafast laser pulses for explosives detection using surface enhanced Raman scattering, *Appl. Phys. Lett.*, 104, 263104. © 2014, American Institute of Physics.)

Raman spectra of various analytes (ANTA, TNT, and R6G of concentrations 10^{-4} and 10^{-6} M) were recorded using a micro-Raman spectrometer using an excitation source at 532 nm and a 100× microscope objective. Typically, a tiny drop (volume of ~10 μL) of ANTA was placed on all of the 8 Cu NS substrates to achieve a monolayer of the analyte molecules, and they are Cu LIPSS + cylindrical NPs (MCuNS1), pillar-shaped structures + cubical NPs (MCuNS2), LIPSS with 250 nm periodicity (MCuNS3), rough-ended pillars (MCuNS4), LIPSS + spherical NP grains (SCuNS1), randomly corrugated structures + spherical NP grains (SCuNS2), LIPSS with 400 nm periodicity (SCuNS3), and small number of triangular NPs with ~250 nm size (SCuNS4). Raman spectra were typically recorded for four times at various positions on the target, and their average spectra were considered. Later on, utilized Cu substrates were cleaned and sonicated with ACN, and again the same substrates were utilized for *the second time* to record the Raman spectra of R6G. Subsequently, proper cleaning procedures were again followed, and the same substrates

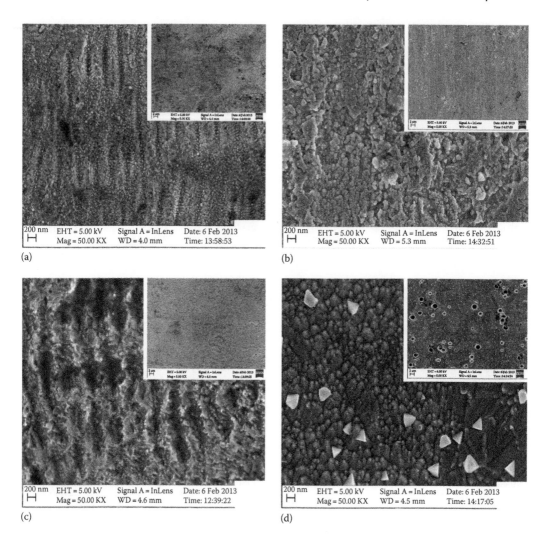

FIGURE 11.17 FESEM images of substrates (a) SCuNS1 in acetone, (b) SCuNS2 in DCM, (c) SCuNS3 in ACN, and (d) SCuNS4 in chloroform fabricated through ultrafast single line ablation of Cu. Insets show corresponding views on a larger scale (2 μm scale). (Reprinted with permission from Hamad, S., Podagatlapalli, G.K., Mohiddon, M.A., and Venugopal Rao, S., Cost effective nanostructured copper substrates prepared with ultrafast laser pulses for explosives detection using surface enhanced Raman scattering, *Appl. Phys. Lett.*, 104, 263104. © 2014, American Institute of Physics.)

were once more utilized, for the *third time*, to record the Raman spectra of TNT. The time separation between these three trials was approximately few weeks. Raman spectra of ANTA from Cu NSs fabricated via ps multiple and single line ablation are shown in Figure 11.18a and b. Significantly observed modes from the data were C–NO$_2$ symmetric stretching mode (1341 cm^{-1}), N–N symmetric stretching mode (1130 cm^{-1}), NO$_2$ deformation + ring deformation (840 cm^{-1}), ring torsion + NO$_2$ deformation (589 cm^{-1}), C–N symmetric stretch mode (1480 cm^{-1}), N4–C5–N1 bending mode (960 cm^{-1}), ring deformation mode (1070 cm^{-1}), and ring deformation + N–H bend (1303 cm^{-1}). The Raman spectra of ANTA from Cu substrates were compared with the normal Raman spectra of ANTA obtained using a Si target to estimate the EFs for 1341 cm^{-1} mode. The intensity for 1341 cm^{-1} mode for MCuNP1 (LIPSS and cylindrical NPs) was ~5301 (arbitrary units). However, the normal Raman intensity of ANTA from Si target for the same mode recorded

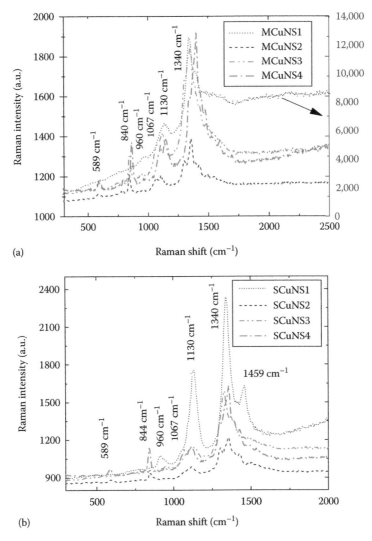

FIGURE 11.18 SERS spectra recorded from ANTA molecule adsorbed on different (a) multiple line and (b) single line ablated Cu NSs. MCuNS1 (SCuNS1)—red/short dotted line, MCuNS2 (SCuNS2)—blue/short dashed line, MCuNS3 (SCuNS3)—orange/short dashed line, MCuNS4 (SCuNS4)—olive/dashed dotted dotted line. The integration time was 5 s (2 μm scale). (Reprinted with permission from Hamad, S., Podagatlapalli, G.K., Mohiddon, M.A., and Venugopal Rao, S., Cost effective nanostructured copper substrates prepared with ultrafast laser pulses for explosives detection using surface enhanced Raman scattering, *Appl. Phys. Lett.*, 104, 263104. © 2014, American Institute of Physics.)

with the same experimental conditions was ~10 (arbitrary units). The intensity enhancement (I_{SERS}/I_{Raman}), therefore, was estimated to be ~530. We estimated that ~50% of analyte molecules were adsorbed on the Cu substrates (the adsorption coefficient (η) used was 0.5). EFs calculated for the 1340 cm^{-1} mode of ANTA molecule were 1.2×10^6, 3.1×10^4, 7.3×10^4, 7.1×10^4, 1.4×10^5, 3.6×10^4, 7.1×10^4, and 7×10^4 for MCuNS1, MCuNS2, MCuNS3, MCuNS4, SCuNS1, SCuNS2, SCuNS3, and SCuNS4, respectively. In the case of ANTA, large Raman signal enhancements were observed from MCuNS1 and SCuNS1. In particular, in the case of MCuNS1 (SCuNS1), nanocylindrical grains (NP grains) were observed on the top of LIPSS. Consequently, these substrates could have provided high local fields, which resulted in the observed strong enhancements (Hamad et al. 2014). Similarly, the Raman spectra of R6G (data presented in Figure 11.19) from

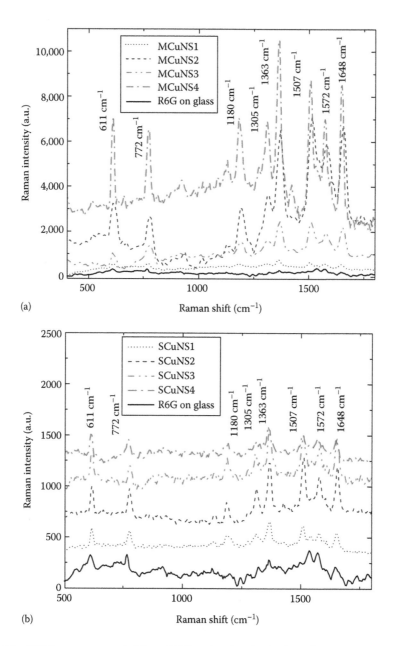

FIGURE 11.19 SERS spectra recorded from R6G molecule adsorbed on different (a) multiple line and (b) single line ablated Cu NSs. MCuNS1 (SCuNS1)—red/short dotted line, MCuNS2 (SCuNS2)—blue/short dashed line, MCuNS3 (SCuNS3)—orange/dashed dotted dotted line, MCuNS4 (SCuNS4)—olive/dashed dotted line, and R6G on glass—black solid lines. The integration time was 5 s. (Reprinted with permission from Hamad, S., Podagatlapalli, G.K., Mohiddon, M.A., and Venugopal Rao, S., Cost effective nanostructured copper substrates prepared with ultrafast laser pulses for explosives detection using surface enhanced Raman scattering, *Appl. Phys. Lett.*, 104, 263104. © 2014, American Institute of Physics.)

the cleaned multiply and singly ablated Cu substrates recorded for the *second time* were analyzed. The prominent enhanced modes of R6G were the aromatic C–C stretch, C–H out-plane bend, and C–C–C ring in-plane bend at 1363, 772, and 610 cm^{-1}, respectively. Evaluation of EFs was carried out by comparing reference spectra of R6G (0.25 M concentration) recorded from a cover slip. At low concentrations (~5 μM) of R6G, it is believed that 43% of molecules (η ~ 0.43) were adsorbed on Cu NS substrates. Estimated EFs of R6G for the 1363 cm^{-1} mode compared with the same mode in normal Raman spectra were 6.2×10^5, 9.1×10^6, 2.6×10^6, 1.8×10^7, 4.6×10^5, 1.1×10^6, 9.1×10^5, and 3.8×10^5 for MCuNS1, MCuNS2, MCuNS3, MCuNS4, SCuNS1, SCuNS2, SCuNS3, and SCuNS4, respectively. Among the substrates investigated, MCuNS2, whose surface topography comprised of μm sized pillars, and MCuNS4, containing roughened pillars along with NP grains, demonstrated stronger enhancements (Hamad et al. 2014).

Figure 11.20 illustrates the Raman spectra of the *third* testing sample TNT, recorded from 8 Cu targets subsequent to appropriate cleaning. The recorded Raman spectra revealed that NO$_2$ symmetrical stretch mode corresponding to 1362 cm^{-1} was predominantly elevated. Additionally, two more peaks were observed: one at 1616 cm^{-1}, which corresponds to C–C aromatic stretching vibration, and another mode at 790 cm^{-1} (C–H out of plane bend). Corresponding EFs calculated for the 1362 cm^{-1} mode were $\sim 4.2 \times 10^4$, $\sim 7.7 \times 10^4$, $\sim 2.2 \times 10^5$, $\sim 1.6 \times 10^4$, $\sim 2.5 \times 10^5$, $\sim 5.8 \times 10^4$, $\sim 1.9 \times 10^5$, and $\sim 5.2 \times 10^4$ for MCuNS1, MCuNS2, MCuNS3, MCuNS4, SCuNS1, SCuNS2, SCuNS3, and SCuNS4, respectively. Raman spectra collected from three analytes using the same Cu targets demonstrated a small reduction in enhancements from analyte 1 to analyte 3 (Hamad et al. 2014). However, the enhancements were reasonably large even in the third trial. Our experimental data demonstrated that nanostructured Cu targets possess potential for elevating the Raman signatures of diverse analytes. We believe that three reasons could possibly contribute to the Raman signal enhancement: (1) The coupling of incident photons to localized and propagating surface plasmons leads to the elevation of evanescent fields (these evanescent fields influence the individual vibrating dipoles to reradiate more number of photons, thus enhancing the Raman signal), (2) the aforementioned coupling of incident photons to LSPR might have fulfilled the condition that the dimensions and shape of the nanomaterials allowed for perpendicular surface plasmon vibrations with respect to the planar surface of the target, (3) whenever the separation between the nanodimensional materials (NPs or nanotips) is optimized, then combined electron resonances of individual structures enhance the probability of coupling, and hence enormous amounts of evanescent fields are possible and sometimes specifically compatible to certain modes of interest. Consequently, some substrates demonstrated better Raman signature enhancement for a specific molecule. The estimated EFs in our experiments were found to be superior compared to EFs obtained with Cu targets prepared by other methods.

11.6.4 Cu Targets Ablated in Chloroform

In these experiments, the Cu metal target was placed at the bottom of a Pyrex cell filled with liquid of ~5 mm thickness above the surface of metal target. The spot size on the sample surface was estimated to be ~90 μm. Typical fluence of ~2.5 J/cm^2 and an ablation time of 30 min were used. The liquid-filled Pyrex cell with Cu target, placed on the X–Y stage (Newport), was translated to draw periodic lines on Cu target at a separation of ~60 μm. Morphologies of the fabricated Cu NSs in CHCl$_3$ were characterized by FESEM, and the data are presented in Figure 11.21a. The FESEM data demonstrated formation of micrometer-sized structures with sharp edges and magnified image depicted random NSs with roughness of ~700 nm. The surface activity of Cu NS was investigated by recording the Raman spectra of three adsorbed molecules (R6G [1×10^{-9} and 5×10^{-7} M], ANTA [1×10^{-6} and 2×10^{-4} M], FOX-7 [1×10^{-6} and 2×10^{-4} M]) on the Cu NS. The Raman spectra

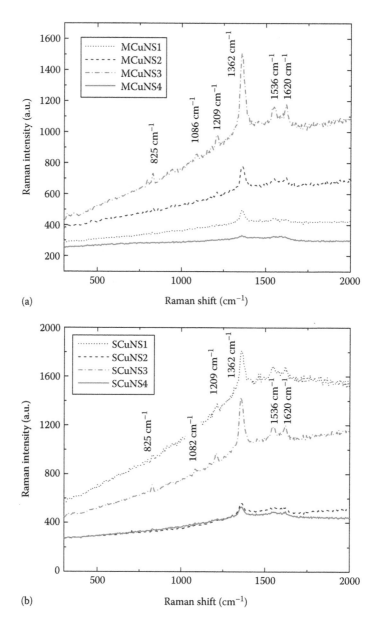

FIGURE 11.20 SERS spectra recorded from TNT molecule adsorbed on different (a) multiple line and (b) single line ablated Cu NSs. MCuNS1 (SCuNS1)—red/short dotted line, MCuNS2 (SCuNS2)—blue/short dashed line, MCuNS3 (SCuNS3)—orange/short dashed dotted line, MCuNS4 (SCuNS4)—olive/solid line. The integration time was 5 s. (Reprinted with permission from Hamad, S., Podagatlapalli, G.K., Mohiddon, M.A., and Venugopal Rao, S., Cost effective nanostructured copper substrates prepared with ultrafast laser pulses for explosives detection using surface enhanced Raman scattering, *Appl. Phys. Lett.*, 104, 263104. © 2014, American Institute of Physics.)

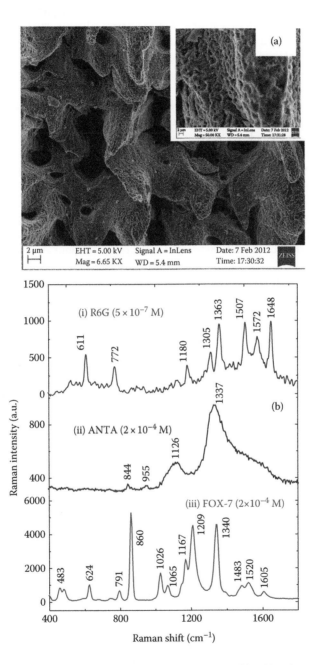

FIGURE 11.21 (a) FESEM image of NSs on the Cu substrate prepared in chloroform and (b) SERS spectra recorded from (i) R6G (blue/top—5×10^{-7} M), (ii) ANTA (wine/middle—2×10^{-4} M), and (iii) FOX-7 (red/bottom—2×10^{-4} M) adsorbed on Cu NSs. (Reprinted from *Chem. Phys. Lett.*, 621, Hamad, S., Podagatlapalli, G.K., Mohiddon, M.A., and Venugopal Rao, S., Surface enhanced fluorescence from corroles and SERS studies of explosives using copper nanostructures, 171–176. © 2015, with permission from Elsevier.)

were recorded at 532 nm with a 0.9 NA objective. SERS spectra of R6G (1×10^{-9} and 5×10^{-7} M) were recorded with 10–20 µL deposited on Cu NS to form a monolayer. An average of 10 spectra collected is illustrated in Figure 11.21b. EFs were evaluated by comparing the SERS spectra with the normal Raman spectra of R6G that were deposited on a silicon surface for the most prominent Raman peak of 1363 cm^{-1}. In case of R6G, at 1×10^{-9} and 5×10^{-7} M concentrations, the absorbed molecules were estimated to be ~35% and ~43%. The estimated EFs for 1363 cm^{-1} mode were ~6×10^{8} and ~1.8×10^{7} for 1×10^{-9} and 5×10^{-7} M concentrations, respectively. Large enhancement in the case of R6G could be due to the rough edges arising from micrometer-sized structures on the Cu surface. Later, the substrate was cleaned by sonication in acetone for half an hour and dried at room temperature. After 2 weeks, the same Cu substrate was utilized again by recording the SERS spectra of an explosive molecule (ANTA) at 1×10^{-6} and 2×10^{-4} M concentrations. SERS spectra of the ANTA molecule adsorbed on the Cu NSs substrate are shown in Figure 11.21b. The main characteristic mode was detected at 1337 cm^{-1} with lower intensity along with the 1577 cm^{-1} mode assigned to C–NO$_2$ symmetrical stretch and C–NH$_2$ symmetrical stretch + NH$_2$ bend, respectively. For 2×10^{-4} M concentration, the characteristic mode at 1337 cm^{-1} was elevated with good signal to noise ratio, and some more modes were detected at 844, 955, and 1126 cm^{-1}, which were not observed in SERS spectra at 1×10^{-6} M concentration. We estimated that ~47% and ~50% ANTA molecules were adsorbed on the NS substrate for the concentrations of 1×10^{-6} and 2×10^{-4} M, respectively. The evaluated EFs for the 1337 cm^{-1} mode were ~1.4×10^{7} and ~2.8×10^{5} at 1×10^{-6} and 2×10^{-4} M concentrations, respectively. One day later, the SERS spectra of another explosive molecule (FOX-7) were recorded for the third time after appropriate cleaning of the substrate. SERS spectra of FOX-7 (Figure 11.21b) demonstrated the characteristic Raman bands corresponding to symmetric C–NO$_2$ stretching and NH wagging at 1340 cm^{-1} with lower intensity at a concentration of 1×10^{-6} M, whereas at 2×10^{-4} M concentration, the mode intensity was clearly enhanced. Moreover, the Raman mode at 1483 cm^{-1} was observed in both the cases. The calculated EFs for the 1340 cm^{-1} mode were ~2.5×10^{5} and ~4.2×10^{4} for 1×10^{-6} and 2×10^{-4} M concentrations, respectively (Hamad et al. 2015).

11.6.5 Cu Targets Ablated with Corroles and Surface-Enhanced Fluorescence/SERS Studies

In this study, Cu targets were ablated in chloroform and in diluted triphenyl corrole (TPC, 2.5×10^{-5} M) and tri tolyl corrole (TTC, 2.5×10^{-5} M) solutions to fabricate chlorinated Cu NPs and suspensions of chlorinated Cu NPs in corroles, respectively. Briefly, the Cu target was placed at the bottom of the Pyrex cell filled with liquid. Typical height of the liquid above the surface of metal target was ~5 mm. The spot size on the sample surface was estimated to be ~90 µm. Typical fluence of ~2.5 J/cm^2 and an ablation time of 30 min were used. The liquid-filled Pyrex cell with Cu target was placed on a motorized X–Y stage, which was controlled by an ESP 300 motion controller. The stages (Newport) were translated to draw periodic lines on Cu target at a separation of ~60 µm. To avoid ambiguity, the fabricated Cu colloids in TTC, TPC, and pure chloroform are labeled as NP-1, NP-2, and NP-3, respectively.

Figure 11.22a and c illustrates the TEM images of NP-1 and NP-2, and insets show the size distribution. It was observed that the morphologies of fabricated NPs were spherical in nature. Following a Gaussian fit of the size distribution histogram average sizes of NPs were found to be ~9.5 and ~38 nm in NP-1 and NP-2, respectively. Figure 11.22b and d illustrates the SAED spectrum of NP-1 and NP-2, respectively. The shape of the SAED pattern revealed that the particles were nanocrystalline in structure and the measured interplanar separations (d) for NP-1 were 3.2 and 2.6 Å, in agreement with those of copper chloride (CuCl) crystal planes [(200) and (211)] and 2.2, 1.82,

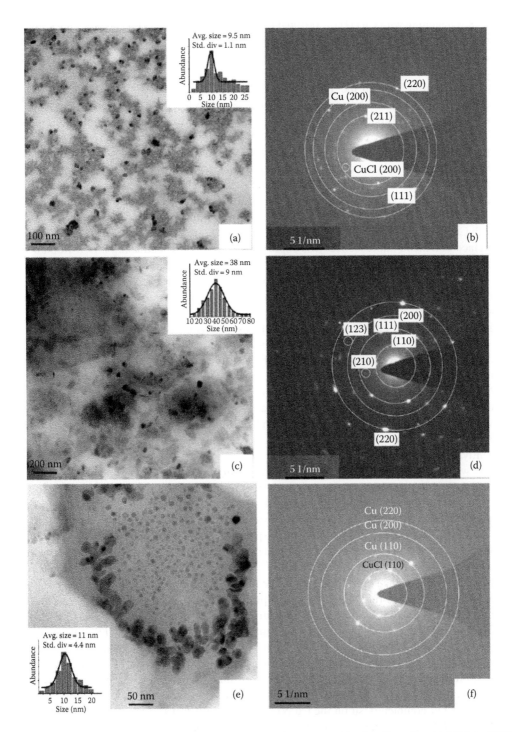

FIGURE 11.22 TEM micrographs of corrole/Cu colloids prepared in liquid media: (a) NP1 in TTC, (c) NP2 in TPC, and (e) NP3 in chloroform. Insets of (a), (c), and (e) illustrate the size distribution (b), (d), and (f) SAED patterns of NP1 in TTC, NP2 in TPC, and NP3 in chloroform. (Reprinted from *Chem. Phys. Lett.*, 621, Hamad, S., Podagatlapalli, G.K., Mohiddon, M.A., and Venugopal Rao, S., Surface enhanced fluorescence from corroles and SERS studies of explosives using copper nanostructures, 171–176. © 2015, with permission from Elsevier.)

and 1.28 Å matching with Cu (111), (200), and (220) planes. In the case of NP-2, "d" values were 4.6, 2.8, and 1.7 Å in good agreement with CuCl planes of (110), (210), and (123) and the atomic spacing values of 2.19, 1.81, and 1.36 Å in good agreement with Cu crystal planes of (111), (200), and (220). SAED results revealed that Cu NPs and CuCl NPs were formed in the colloidal solutions (Hamad et al. 2015). Similarly, from the TEM images of NP-3 presented in Figure 11.22e, it is evident that a majority of produced NPs were smaller in size with a few NPs with large dimensions. The mean diameter of small particles, evaluated by Gaussian fit, was ~11 nm (inset of Figure 11.22e). Similarly, the estimated average size of NPs with larger dimensions was ~25 nm. Figure 11.22f illustrates the SAED pattern of NP-3 and it is evident that NP-3 phase was nanocrystalline. The ring pattern in SAED was indexed with cubic phase of CuCl crystal planes, which were confirmed from the measured interplanar spacing (estimated from the diameter of rings in SAED image). The prevalence of CuCl (110) with "d" spacing of 4.5 Å and Cu (111), (200), and (220), with corresponding "d" spacing values of 2.18, 1.81, and 1.29 Å, respectively, confirmed the formation of cubic phase for both Cu and CuCl NPs.

Surface-enhanced fluorescence (SEF) spectra have been recorded using micro-Raman spectrometer with a 532 nm excitation source, and spectra were collected through a charge couple device (CCD) camera. The laser beam was focused on analytes placed on Cu substrates using microscopic objective (100×, NA = 0.9), and backscattered signal was collected through the same objective. Two types of SEF measurements have been performed. A tiny drop (~10 μL) of (1) manually mixed corroles solution (1 nM) and chlorinated Cu NPs and (2) colloidal corrole Cu NPs were deposited on the cover slips. Three SEF spectra were recorded at different positions on the substrate, and an average of the three spectra obtained was considered for analysis. Typical acquisition time used was 5 s. Figure 11.23 demonstrates the SEF spectra from (1) NP-1 and (2) NP-2. The enhancements were estimated by comparing with characteristic bands obtained in normal fluorescence spectra (NFS) of TTC (100 μM) and TPC (100 μM) at higher concentrations. Comparison of SEF and NFS of both TTC and TPC revealed that the intensity of fluorescence band in SEF spectra was enhanced by 13 and 333 times, respectively. Evidently, similar bands appeared in SEF (622 nm for TTC and 646 nm for TPC) spectrum and NFS as well. We did not observe any visible peak shift in SEF spectra compared to the peaks in NFS. A tiny drop (~10 μL) of manually mixed solution of corroles (1 × 10⁻⁹ M) and Cu/CuCl NPs in chloroform was placed on the glass slide and dried. Recorded SEF spectra of these dried samples are shown in Figure 11.23c and d. The spectra illustrated very high intensity at 626 and 645 nm for TTC and TPC, respectively. SEF spectra of lower concentration (10⁻⁹ M) analytes (TTC and TPC) were compared with the higher concentration (1 × 10⁻⁴ M) spectra to estimate the EFs and were found to be ~2.7 × 10⁵ and ~7.1 × 10⁵ for TTC and TPC, respectively (Hamad et al. 2015). Fluorescence enhancements in the present case, possibly, occurred due to (1) electromagnetic mechanism that motivates the amplification of incident field through the excitation of surface plasmons and (2) increase in quantum yield and radiative decay rates. In the case of TTC–Cu–Cl and TPC–Cu–Cl colloidal solutions, fluorophores (TTC and TPC) were strongly adsorbed on metal surfaces inciting quenching of fluorescence (Hamad et al. 2015). Moreover, there could be reabsorption of the fluorescence emission band through the attached Cu NPs whose SPR band falls in the range of fluorescence band of fluorophore. Due to the reabsorption, fluorescence might have quenched significantly. In the second case, fluorophores (TTC and TPC) might not completely sit on the Cu NPs surface since it was a passive mixing of Cu NPs to TTC and TPC. In this case, due to the local field effect of Cu NPs that were in the proximity of fluorophores, significant fluorescence enhancement was observed with very minimal quenching of fluorescence. Furthermore, reabsorption of fluorescence emission band by the Cu NPs in the neighborhood was possibly absent since their surface was passivated by the CuCl layer (Hamad et al. 2015).

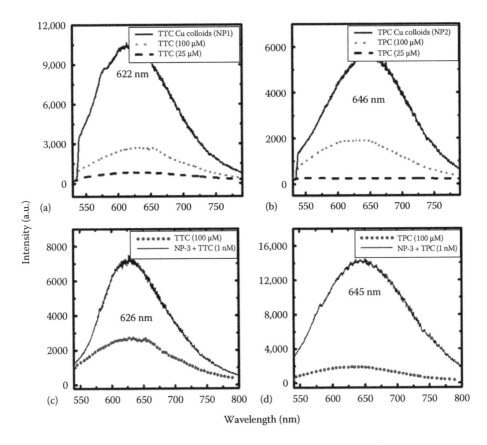

FIGURE 11.23 SEF spectra recorded from (a) TTC Cu colloids (NP1) and (b) TPC Cu colloids (NP2). TTC (100 μM) and TPC (25 μM) spectra are represented by orange/dotted line and black/dashed line, respectively. The spectrum of NP1 (NP2) is represented by blue/solid line. SEF spectra recorded from (c) TTC and (d) TPC corroles adsorbed on NP3 individually. TTC (TPC) spectra are represented by red/dotted lines and TTC/TPC + NP3 by blue/solid lines. The integration time used was 5 s. (Reprinted from *Chem. Phys. Lett.*, 621, Hamad, S., Podagatlapalli, G.K., Mohiddon, M.A., and Venugopal Rao, S., Surface enhanced fluorescence from corroles and SERS studies of explosives using copper nanostructures, 171–176. © 2015, with permission from Elsevier.)

11.7 NONLINEAR OPTICAL STUDIES OF Cu, Ag, AND Al COLLOIDS

11.7.1 NLO PROPERTIES OF Cu COLLOIDS

The third-order NLO properties of the Cu colloids (Cu NPs in acetone, CuCl NPs in DCM, CuO NPs in ACN, and $CuCl_2$ NPs in chloroform) were investigated at a wavelength of 800 nm with ~2 ps pulses using the standard Z-scan technique. These Cu colloids were generated by multiple line (single line) ablation of Cu in acetone, DCM, ACN, and chloroform, which were designated as MCuNP1 (SCuNP1), MCuNP2 (SCuNP2), MCuNP3 (SCuNP3), and MCuNP4 (SCuNP4), respectively. Open-aperture Z-scan data provide information about the nonlinear absorption (related to the imaginary part of third-order NLO susceptibility, $\chi^{(3)}$) properties, while the closed-aperture (CA) Z-scan data provide information on sign and magnitude of the nonlinear refractive index n_2 (related to the real part of $\chi^{(3)}$). The linear transmittance (LT) was typically >95% for all NPs except MCuNP1, MCuNP3 (>80%).

Figure 11.24 depicts the complete open-aperture data obtained for all investigated 8 NPs at both lower (open squares—80 GW/cm²) and higher (open stars—125 GW/cm²) peak intensities. Three distinct behaviors were observed: (1) reverse saturable absorption (RSA) type of behavior that could be attributed to two-photon absorption (2PA; β), (2) an effective three-photon absorption (3PA; γ_{eff}) type of behavior, and (3) a switching mechanism from saturable absorption (SA) to RSA. The open-aperture data of MCuNP1 (Figure 11.24a) and MCuNP3 (Figure 11.24e) illustrate the behavior of switching from SA to RSA at lower peak (open squares—80 GW/cm²) intensity and pure RSA at

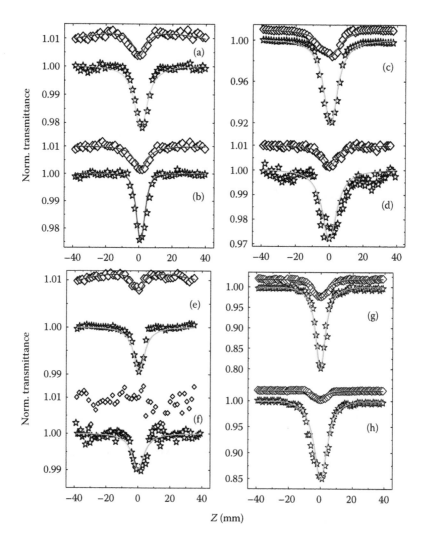

FIGURE 11.24 Open-aperture Z-scan curves obtained for (a) MCuNP1, (b) SCuNP1 with varying input intensities I_{00} = 80 GW/cm² (open squares), I_{00} = 125 GW/cm² (open stars). Open-aperture Z-scan curves obtained for (c) MCuNP2, (d) SCuNP2 with varying input intensities I_{00} = 80 GW/cm² (open squares), I_{00} = 125 GW/cm² (open stars). Open-aperture Z-scan curves obtained for (e) MCuNP3, (f) SCuNP3 with varying input intensities I_{00} = 80 GW/cm² (open squares), I_{00} = 125 GW/cm² (open stars) and (g) MCuNP4, (h) SCuNP4 with varying input intensities I_{00} = 80 GW/cm² (open squares), I_{00} = 125 GW/cm² (open stars). Solid lines are the theoretical fits. (Reproduced with permission from Hamad, S., Podagatlapalli, G.K., Tewari, S.P., and Venugopal Rao, S., Influence of picosecond multiple/single line ablation on copper nanoparticles fabricated for surface enhanced Raman spectroscopy and photonics applications, *J. Phys. D: Appl. Phys.*, 46, 485501. © 2013 by the IOP Publishing.)

higher peak intensity (open stars—125 GW/cm²). The data obtained at higher peak intensities were fitted efficiently using only β. At lower peak intensity, SCuNP1 (Figure 11.24b) data were fitted to β, while SCuNP3 (Figure 11.24f) did not show any NLO behavior. At higher peak intensity, both SCuNP1 and SCuNP3 samples exhibited RSA type of behavior, and the data were fitted successfully using γ_{eff} (2PA + ESA) and β, respectively. The resonances of SCuNP1 (565 and 310 nm) and SCuNP3 (588 nm) might have led to this observation. The open-aperture data of MCuNP2 (Figure 11.24c), SCuNP2 (Figure 11.24d), MCuNP4 (Figure 11.24g), and SCuNP4 (Figure 11.24h) revealed that 2PA was dominant at both peak intensities with 800 nm excitation, which can be accredited to the SPR band of nanocolloids formed at nearly 400 nm (Hamad et al. 2015).

Nonlinear absorption in plasmonic metal NPs is governed by the transitions of conduction/valence band electrons and is explained as follows: (1) completely occupied "d" band to unoccupied conduction band and/or (2) within the conduction band (ground states to excited states) due to plasmon resonances and (3) free carrier absorption from excited conduction band to high-lying states. Figure 11.25 depicts a generic energy band diagram portraying various mechanisms of (1) SA: in the cases of MCuNP1 and McuNP3, we observed residual absorption at 800 nm and, therefore, expected saturation at lower peak intensities with probable transition from "d" band to the "p" (conduction) band and is depicted by **Ia** in Figure 11.25. However, at higher peak intensities, one expects 2PA as depicted by **Ib** with transitions from ground state conduction band (GSCB) to excited state conduction band (ESCB). Due to two-photon resonance (800 + 800 nm) between GSCB and ESCB, 2PA dominates with negligible contribution from SA at higher peak intensities. (2) Instantaneous 2PA with possible mechanisms is depicted in the second box marked as **IIa** (between "d" band and GSCB) and **IIb** (between GSCB and ESCB) and (3) 2-step 3PA (2PA + ESA) at very high peak intensities as depicted by processes in boxes **IIIa** and **IIIb** (excited state absorption induced from ESCB in this case). The figure is generic in the sense that we have assumed a fixed gap between the states and different photon energies, whereas in reality, the band gaps could be different (including SPR positions) for these eight different NPs with single energy photon excitation (1.55 eV for 800 nm ps photon). The experimental data were fitted with appropriate equations to estimate nonlinear absorption

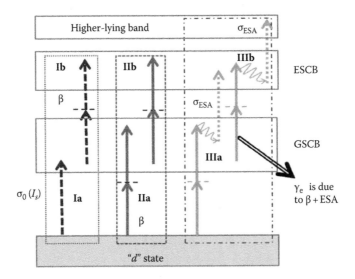

FIGURE 11.25 A generalized energy band diagram of the observed CuNPs (pure Cu NPs, CuO NPs, CuCl NPs, CuCl₂ NPs) explaining various nonlinear absorption phenomena. ESCB refers to excited state conduction band and GSCB refers to ground state conduction band. (Reproduced with permission from Hamad, S., Podagatlapalli, G.K., Tewari, S.P., and Venugopal Rao, S., Influence of picosecond multiple/single line ablation on copper nanoparticles fabricated for surface enhanced Raman spectroscopy and photonics applications, *J. Phys. D Appl. Phys.*, 46, 485501. © 2013 by the IOP Publishing.)

coefficients. Among all, MCuNP4 exhibited largest β, n_2, and $\chi^{(3)}$ values (Hamad et al. 2013). The values of nonlinear refractive index (n_2) were obtained for all colloidal solutions at an intensity of 33 GW/cm² using CA Z-scan method, and the data are shown in Figure 11.26. Figure 11.26a illustrates the CA data of MCuNP1 (triangles) and SCuNP1 (circles), which exhibit positive nonlinearity (like solvent) with $n_2 = 6 \times 10^{-16}$ cm²/W, smaller than the solvent magnitude ($n_2 = 19 \times 10^{-16}$ cm²/W), and the observed sign of n_2 for MCuNP1 and SCuNP1 was negative. Figure 11.26b through d represent CA data of MCuNP2 (SCuNP2), MCuNP3 (SCuNP3), and MCuNP4 (SCuNP4). The observed sign of the nonlinearity was negative for MCuNP2 (SCuNP2) and positive for MCuNP3 (MCuNP3). The data were fitted using standard equation to estimate the magnitudes of n_2. The contribution from solvents (DCM, ACN, and chloroform) (positive nonlinearity) was also identified. The order of magnitude of n_2 was found to be ~10^{-15} cm²/W for all the colloids investigated, and the order of magnitude for β was ~10^{-11} cm/W (Hamad et al. 2013).

11.7.2 NLO PROPERTIES OF Ag COLLOIDS

Picosecond multiple/double line ablation of Ag targets in DDW was carried out to obtain Ag colloids. Ag colloids prepared in the multiple line ablation are labeled as Ag NP-1, whereas the colloids prepared via double line ablation are labeled as Ag NP-2. Average sizes obtained for Ag NP-1 and Ag NP-2 were ~13 and ~17 nm, respectively (Podagatlapalli et al. 2013). In both multiple line/double line ablations, a broad plasmon band was observed, and even the absorbance of both colloids was similar. When compared to single line ablation, multiple line/double line ablation provided large plasmon band (which was evident from the recorded UV–Vis absorption spectra) widths and average sizes. These Ag NPs were observed to have dark gray coloration compared to the other Ag NPs prepared

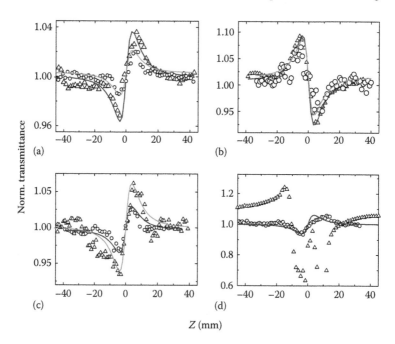

FIGURE 11.26 Closed-aperture Z-scan curves obtained for (a) MCuNP1 (open triangles), SCuNP1 (open circles), (b) MCuNP2 (open triangles), SCuNP2 (open circles), (c) MCuNP3 (open triangles), SCuNP3 (open circles), and (d) MCuNP4 (open triangles), SCuNP4 (open circles). Closed-aperture studies were performed at a peak intensity of 33 GW/cm². Solid lines are the theoretical fits. (Reproduced with permission from Hamad, S., Podagatlapalli, G.K., Tewari, S.P., and Venugopal Rao, S., Influence of picosecond multiple/single line ablation on copper nanoparticles fabricated for surface enhanced Raman spectroscopy and photonics applications, *J. Phys. D: Appl. Phys.*, 46, 485501. © 2013 by the IOP Publishing.)

by double line/single line ablation. Open-aperture Z-scan was performed at peak intensities ~83 and ~138 GW/cm², while the CA study was carried out at ~28 GW/cm². Open-aperture data of NP-1 (LT ~90%) presented in Figure 11.27a demonstrated a switching behavior (from SA to RSA) with a saturation intensity of ~6.5 × 10⁷ W/cm² at lower peak intensity (83 GW/cm², β = 2 × 10⁻¹² cm/W) represented by open circles in the plot. RSA was observed at higher peak intensities (138 GW/cm²) with a strong 2PA coefficient (β = 4.5 × 10⁻¹² cm/W) whose data points are indicated by open stars. The experimental data were fitted with standard equations. The CA data were fitted with theoretical equations to extract n_2. The CA study was carried out at low peak intensity (~28 GW/cm²) to get the information on nonlinear refraction (NLR) of the sample. The CA data are shown in Figure 11.27b and the measured n_2 was ~3.4 × 10⁻¹⁶ cm²/W. Figure 11.27c depicts the open-aperture data of Ag NPs prepared by double line ablation (NP-2) (LT was ~92%), which signifies pure RSA both at lower (β = 6.2 × 10⁻¹² cm/W) and higher (β = 1.78 × 10⁻¹¹ cm/W) peak intensities. The CA data for NP-2 are shown in Figure 11.27d and n_2 retrieved was ~4 × 10⁻¹⁵ cm²/W from the fits of experimental data. The n_2 value of the pure water was higher than the n_2 of NP-2, indicating that the sign of n_2 of NP-2 was nega-tive. The measured $\chi^{(3)}$ values for NP-1 and NP-2 were 2.3 × 10⁻¹⁴ and 2.7 × 10⁻¹³ e.s.u., respectively (Podagatlapalli et al. 2013). The colloids obtained with different angles of incidence were studied

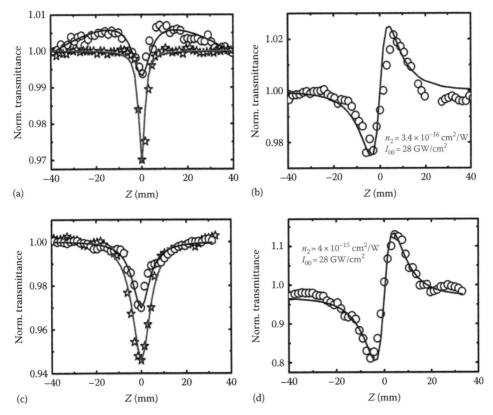

FIGURE 11.27 Open-aperture Z-scan curves obtained for (a) Ag colloids NP-1 with varying input intensi-ties I_{00} = 83 GW/cm² (open circles), I_{00} = 138 GW/cm² (stars). Blue and red colors represent low intensities and high intensities, respectively, (b) closed-aperture Z-scan curves obtained for Ag colloids NP-1 at peak inten-sity 28 GW/cm², (c) open-aperture Z-scan curves obtained for Ag colloids NP-2 with varying input intensities I_{00} = 83 GW/cm² (open circles), I_{00} = 138 GW/cm² (stars), and (d) closed-aperture Z-scan curves obtained for Ag colloids NP-2 at peak intensity 28 GW/cm². Solid lines are the theoretical fits. (Reprinted with permission from Podagatlapalli, G.K., Hamad, S., Sreedhar, S., Prasad, M.D., and Venugopal Rao, S., Silver nano-entities through ultrafast double ablation in aqueous media for surface enhanced Raman scattering and photonics applications, *J. Appl. Phys.*, 113, 073106. © 2013, American Institute of Physics.)

using the fs DFWM technique (at 800 nm), and the magnitude of third-order NLO susceptibility was found to be 10^{-14} e.s.u. (Podagatlapalli et al. 2014).

11.7.3 NLO PROPERTIES OF Al COLLOIDS

Al NPs were fabricated in oxygen-free polar and nonpolar liquids such as carbon tetrachloride (CCl_4) and chloroform ($CHCl_3$) using ~40 fs laser pulses. Laser ablation of bulk Al target in aqueous media provided Al colloid NPs in CCl_4 and $CHCl_3$. Average sizes of Al NPs in CCl_4 and $CHCl_3$ (estimated using image-J software) were ~33 ± 4.5 and 15 ± 5 nm, respectively. The prepared Al colloids were utilized to investigate NLO behavior using ~2 ps laser pulses with the Z-scan technique. The nonlinear absorption of the colloidal NPs was investigated using open-aperture Z-scan method (Podagatlapalli et al. 2012). We could explain how the presence of Al NPs in solvents influences the NLO properties at different input energies. This investigation was performed by ~2 ps laser at a repetition rate of 1 kHz and central wavelength at 800 nm as an excitation source. The colloidal solution was placed in a quartz cuvette of ~1 mm path length in a sample holder located on a motorized stage (Newport ILS 250 PP). Z-scan studies were performed by focusing the 4 mm diameter input beam using an achromatic doublet ($f = 20$ cm). The beam waist at the focus was ~25 μm. Laser beam was allowed to pass through the colloidal solution, and the transmitted light was observed by a photodiode (SM1PD2A) along with lock-in amplifier (7265 DSP from Signal Recovery). The translational stage and photodiode were interfaced to a personal computer.

Figure 11.28a through c depicts the open-aperture Z-scan data of Al colloids in CCl_4 recorded at three input peak intensities of 96, 130, and 190 GW/cm², respectively. The optical properties of

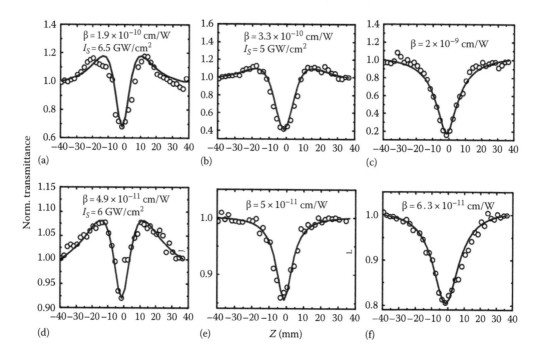

FIGURE 11.28 Open-aperture Z-scan curves obtained for Al colloidal-CCl_4 NPs with varying input intensities (a) $I_{00} = 96$ GW/cm², (b) $I_{00} = 130$ GW/cm², and (c) $I_{00} = 190$ GW/cm². Open-aperture Z-scan curves obtained for Al colloidal-CCl_4 NPs with varying input intensities (d) $I_{00} = 69$ GW/cm², (e) $I_{00} = 110$ GW/cm², and (f) $I_{00} = 140$ GW/cm². Solid lines are theoretical fits. (Reprinted from *Chem. Phys. Lett.*, 530, Podagatlapalli, G.K., Hamad, S., Sreedhar, S., Tewari, S.P., and Venugopal Rao, S., Fabrication and characterization of aluminum nanostructures and nanoparticles obtained using femtosecond ablation technique, 93–97. © 2012, with permission from Elsevier.)

metal NPs can be described using transitions of conduction electrons between the discrete energy states in the quantum wells provided by the LSPRs of NPs. Large enhancements of the local field over the surface of NPs can be obtained at the plasmon resonance frequency. At lower peak intensities, the behavior was switching from SA to RSA (Figure 11.28a), and at higher peak intensities, the behavior switched completely to pure RSA (Figure 11.28c). Pure CCl_4 did not show any nonlinear absorption (NLA), but the presence of NPs (40%–50% LT) introduced NLA. The summary of obtained NLO coefficients for Al colloids in CCl_4 is presented elsewhere (Podagatlapalli et al. 2012). Similarly, Figure 11.28d through f demonstrates the open-aperture data of $CHCl_3$ colloids (65% LT) recorded at the same experimental conditions as mentioned for Al colloids in CCl_4, and input peak intensities utilized were ~69, ~110, and ~140 GW/cm^2, respectively. At lower peak intensities, the behavior was switching type (SA to RSA) (Figure 11.28d), and at higher intensities (Figure 11.28f), the behavior switched completely to RSA (pure RSA) with strong 2PA. The order of magnitude of β was ~10^{-9} cm/W for all the colloids investigated (Podagatlapalli et al. 2012). Our future investigations will comprise evaluating the nonlinearity in these NPs using fs/ps pulses in the entire visible and near-IR spectral regions. Furthermore, we will investigate the pure electronic nonlinearities using low repetition rate ultrashort laser pulses.

11.8 ANTIBACTERIAL STUDIES OF Cu, Ag COLLOIDS

Three solvents, namely, acetone, DCM, and ACN were used for NPs fabrication. The effect of CuNP1 (acetone), CuNP2 (DCM), and CuNP3 (ACN) on bacterial growth was assessed against the gram-negative bacteria *Escherichia coli* using agar well diffusion and viability assay methods. Investigation of the antibacterial activity of different Cu NPs against *E. coli*, gram-negative bacteria, by agar well diffusion and viability assay methods was accomplished (Syed et al. 2012). Bacterial suspension was applied uniformly on the surface of Muller–Hinton agar plates. Subsequently, agar wells were filled with 200 μL of three different Cu NPs and their solvents. The plates were incubated at 37°C for 24 h, and later the diameter of inhibition zone was measured. In viability assay method, *E. coli* culture was grown overnight in Luria–Bertani (LB) broth. Aliquots of 100 μL of culture were subcultured in fresh LB media (2 mL) and treated with NPs and pure solvents that were incubated for 3 h. Inoculums without NPs served as negative control in these measurements. Hundred microliters from each test was taken out for plating it on LB agar plates. The bacterial viability was observed visually, and bacterial colonies were scrutinized after 12 h of incubation at 37°C.

Figure 11.29a and b illustrate the agar well diffusion method, exhibiting the zone of inhibition of Cu NPs fabricated in different liquids and positive control, against *E. coli* (gram-negative bacteria). Figure 11.29a demonstrates that CuNP1 and CuNP2 inhibit the growth of *E. coli*, gram-negative bacteria, and the zones of inhibition were 25 and 28 mm, respectively. The effect of CuNP2 on the inhibition zone was greater than CuNP1, which is consistent with the earlier published results (Syed et al. 2012). Figure 11.29b shows that the zone of inhibition was maximum (35 mm) when oxidized NPs (CuNP3) were used. Observed solvent effect on the bacterial inhibition was minimal. Positive Cu ions released from NPs interact with negative charges of microbes during the interaction of NPs with bacterial microorganisms leading to cellular distortions and causing bacteria to lose its viability. The results of viability assay method displayed a drastic reduction in a number of bacterial colonies after NPs treatment. The corresponding images of these agar plates are shown in Figure 11.29c through f. Figure 11.29c represents the untreated agar plate, d through f illustrates the treated *E. coli* (gram-negative bacteria) with CuNP3, CuNP2, and CuNP1, respectively. The assay method was sensitive than the agar well diffusion technique, even detecting the background activity of solvents. The background activity of three solvents (ACN, DCM, acetone) was less (by at least 75%) than the activity exhibited by those solvents that were potentiated with NPs evident from the bacterial growth on LB plates. The Ag NPs dissolved in ACN also showed remarkable antibacterial activity (26 mm). In the *viability assay*, *E. coli* culture was grown overnight in LB. Aliquots of

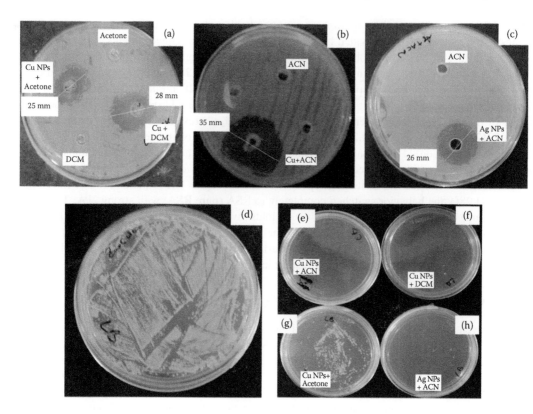

FIGURE 11.29 (a–c) Agar well diffusion method: zone of growth inhibition: (a) *E. coli* with Cu NPs in acetone and DCM (dichloromethane), (b) *E. coli* with Cu NPs in acetonitrile (ACN), (c) *E. coli* with Ag NPs in ACN (Hamad et al. 2012), (d–h) viability assay method: (d) agar growth plate showing *E. coli* after incubation and agar plate of captured *E. coli* with (e) Cu NPs in ACN, (f) Cu NPs in DCM, (g) Cu NPs in acetone, and (h) Ag NPs in ACN.

100 μL of the culture were subcultured in fresh LB media (2 mL) and treated with Ag NPs and pure solvents, and only inoculums without NPs served as negative control. Cultures were incubated for 3 h. Hundred microliters from each test was taken and plated on LB agar plates and then incubated at 37°C overnight. Bacterial growth was investigated and colonies were counted. The data for Ag NPs are presented in Figure 11.29h. Our studies clearly demonstrated that the antibacterial capability of Cu and Ag NPs achieved ULAL.

11.9 CONCLUSIONS

We were successful in fabricating several metallic (Al, Ag, Cu, etc.) NPs and NSs utilizing the technique of ULAL. Several parameters (solvent, angle of laser pulses incidence, Bessel beam, etc.) were investigated for their effect on the ablation products. The achieved NPs and NSs were demonstrated to possess potential for several applications including SERS studies of explosive molecules, and their photonic and antibacterial properties were studied. Some of the highlights of our studies are summarized as follows:

1. Multiple/double/single line ablation differs in the effective pulse number per spot and degree of over writing. During the process of single/double/multiple line ablation, roughness offered by the Ag target varied as a function of the degree of overwriting. The differences in roughness and number of pulses per spot demonstrated different yields of NPs and morphologies of NSs.

2. Double line ablated Ag NSs fabricated at different fluences demonstrated higher Raman intensity enhancements for the explosive molecules RDX (~mM). Fluence of 16 J/cm^2 used in the fabrication of Ag NSs showed superior SERS performance (Podagatlapalli et al. 2013).

3. Ablation with nondiffracting Bessel beams rooted out the ambiguities on short DOF in conventional Gaussian beam focusing through spherical lenses. First-order non-diffracting Bessel beams were generated by focusing Gaussian beam using an axicon (or conical lens). Ablation of Ag in DDW/acetone with Bessel beams revealed the fabrication of Ag NPs with the same dimensions even for different energies (both in ps/fs regimes). Ag colloids prepared in DDW (ps ablation) at different input pulse energies exhibited the SPR peak at the same wavelength. A similar phenomenon was observed in fs case (Ag in acetone). Outcome of Bessel beam ablation might be from the unusual dynamics of cavitation bubbles produced at the time of ablation. Identical average sizes were tentatively assigned to the simultaneous fabrication/fragmentation of Ag NPs by Bessel beam transverse profile. Identification/detection of CL-20 (5 μM) was achieved through Ag NSs prepared by Bessel beam ablation (both ps and fs cases) and the estimated EFs were >10^6.

4. Ablation at 30° angle demonstrated the possibility of higher coupling efficiency of incident photons to the surface plasmons on Ag target enhancing the effectiveness of ablation. Consequently, higher yields of Ag NPs and well-structured nanosurfaces were achieved. SERS studies of R6G/ANTA illustrated that Ag NSs fabricated at 30° angle of incidence provided higher EFs compared to others (Podagatlapalli et al. 2014).

5. Our recent SERS studies on NPs and NSs of Au–Ag alloys prepared using fs ablation clearly demonstrated the capability of detecting secondary explosives of CL-20 and FOX-7 (Podagatlapalli et al. 2015). Other combinations (Au–Cu, Ag–Cu) of alloy NPs and NSs are being investigated.

6. Cu substrates fabricated using ps multiple/single line technique were utilized exultantly to investigate the surface plasmon (localized and propagating)–mediated enhancements of different analytes using SERS studies. The reproducibility of the Cu targets for the trace-level detection of explosive molecules was dealt with a simple cleaning procedure. Trace-level detection of explosive compounds such as ANTA and TNT was performed along with R6G individually after subjecting surfaces to appropriate cleaning procedures and the obtained EFs were >10^5.

7. TTC–Cu–Cl and TPC–Cu–Cl colloidal solutions in TTC, TPC, and Cu/CuCl nanocomposites in chloroform were prepared by means of ps laser ablation of bulk Cu target. Significant enhancements (~10^5) in the fluorescence were obtained for TTC/TPC placed on Cu/CuCl nanocomposites compared to the case wherein fluorophores were directly adsorbed (EF ~333) on the substrates (TTC–Cu–Cl, TPC–Cu–Cl colloids). The possible reasons behind the enhancement were deliberated in detail. The enhancement in fluorescence was attributed to the generated localized electromagnetic field of the Cu NPs (Hamad et al. 2015).

8. Photonic and antibacterial properties of Cu and Ag NPs were successfully demonstrated, and the NLO coefficients obtained for these NPs are on par or better than some of the recently reported successful NLO moieties.

11.10 FUTURE SCOPE

Metallic NPs and NSs achieved through laser ablation (with ns, ps, or fs pulses) find further applications in a variety of fields. This chapter could not cover most of these applications. For example, Guo and coworkers extensively worked on laser structuring of metallic targets (Vorobyev and Guo 2005, 2008, 2013) and tailored their reflectivity properties. Vorobyev and Guo (2005) demonstrated a significant absorption enhancement due to nanostructuring of gold. They believed that the physical mechanism of the enhanced absorption was due to collective effects of nanostructural, microstructural, and macrostructural surface modifications induced by ablation incurred by fs laser pulses. Vorobyev and coworkers (2009) applied the fs laser blackening technique directly to a tungsten incandescent lamp filament and radically brightened the tungsten lamp and enhanced its emission efficiency to approach 100%. Vorobyev and Guo (2013) again studied the origin and formation of random surface NSs produced on metals using the fs laser ablation technique. Tan and coworkers (2013) in their recent review article clearly demonstrated the possibility of preparing functional nanomaterials using ULAL. They also provided the avenues for various applications of these functional nanomaterials. Several new studies (De Bonis et al. 2014, Jiang and Pinchuk 2015, Li et al. 2014, Linic et al. 2015, Maximova et al. 2015, Nguyen et al. 2015, Sportelli et al., 2015, Tan et al. 2013a,b, 2015, Weng et al. 2014, Zhang et al. 2015) reported in the last couple of years as well provide a glimpse of potential of the fs laser pulses for creating metallic and nonmetallic NPs and NSs for various applications. For example, Linic and coworkers (2015) reported the significance of metal NPs in photochemical transformations. Sportelli and group (2015) used laser ablation–synthesized NPs for food packaging applications. Finally, Zhang and coworkers (2015) reported intense THz emission from random metallic NSs with an fs pulse irradiation.

ACKNOWLEDGMENTS

All the authors acknowledge continuous financial support from DRDO, India, through a grants-in-aid scheme and fellowships. The authors also acknowledge the support of Dr. Md. A. Mohiddon during the Raman measurements presented in this work. The authors also wish to acknowledge the partial financial support from UPE-I and UPE-II of University of Hyderabad, India.

REFERENCES

Alnassar, S.I., Akman, E., Oztoprak, B.G., Kacar, E., Gundogdu, O., Khaleel, A., Demir, A. 2013. Study of the fragmentation phenomena of TiO_2 nanoparticles produced by femtosecond laser ablation in aqueous media. *Opt. Laser Technol.* 51: 17–23.

Amendola, V., Meneghetti, M. 2007. Controlled size manipulation of free gold nanoparticles by laser irradiation and their facile bioconjugation. *J. Mater. Chem.* 17: 4705–4710.

Amendola, V., Dini, D., Polizzi, S., Shen, J., Kadish, K.M., Calvete, M.J.F., Hanack, M., Meneghetti, M. 2009a. Self-healing of gold nanoparticles in the presence of zinc phthalocyanines and their very efficient nonlinear absorption performances. *J. Phys. Chem. C* 113: 8688–8695.

Amendola, V., Meneghetti, M. 2009b. Laser ablation synthesis in solution and size manipulation of noble metal nanoparticles. *Phys. Chem. Chem. Phys.* 11: 3805–3821.

Amendola, V., Meneghetti, M. 2013. What controls the composition and the structure of nanomaterials generated by laser ablation in liquid solution? *Phys. Chem. Chem. Phys.* 15: 3027–3046.

Amendola, V., Polizzi, S., Meneghetti, M. 2006. Laser ablation synthesis of gold nanoparticles in organic solvents. *J. Phys. Chem. B* 110: 7232–7237.

Amendola, V., Polizzi, S., Meneghetti, M. 2012. Laser ablation synthesis of silver nanoparticles embedded in graphitic carbon matrix. *Sci. Adv. Mater.* 4: 497–500.

Amendola, V., Riello, P., Meneghetti, M. 2011a. Magnetic nanoparticles of iron carbide, iron oxide, iron@iron oxide, and metal iron synthesized by laser ablation in organic solvents. *J. Phys. Chem. C* 115: 5140–5146.

Amendola, V., Riello, P., Polizzi, S., Fiameni, S., Innocenti, C., Sangregorio, C., Meneghetti, M. 2011b. Magnetic iron oxide nanoparticles with tunable size and free surface obtained via a "green" approach based on laser irradiation in water. *J. Mater. Chem.* 21: 18665–18673.

Amendola, V., Scaramuzza, S., Litti, L., Meneghetti, M., Zuccolotto, G., Rosato, A., Nicolato, E. et al. 2014. Magneto-plasmonic Au–Fe alloy nanoparticles designed for multimodal SERS-MRI-CT imaging. *Small* 10: 2476–2486.

Amoruso, S., Ausanio, G., Barone, A.C., Bruzzese, R., Gragnaniello, L., Vitiello, M., Wang, X. 2005a. Ultrashort laser ablation of solid matter in vacuum: A comparison between the picosecond and femtosecond regimes. *J. Phys. B: Atom. Mol. Opt. Phys.* 38: L329–L338.

Amoruso, S., Ausanio, G., Bruzzese, R., Vitello, M., Wang, X. 2005b. Femtosecond laser pulse irradiation of solid targets as a general route to nanoparticle formation in a vacuum. *Phys. Rev. B* 71: 033406.

Ancona, A., Sportelli, M.C., Trapani, A., Picca, R.A., Palazzo, C., Bonerba, E., Mezzapesa, F.P., Tantillo, G., Trapani, G., Cioffi, N. 2014. Synthesis and characterization of hybrid copper-chitosan nano-antimicrobials by femtosecond laser-ablation in liquids. *Mater. Lett.* 136: 397–400.

Arboleda, D.M., Santillán, J.M.J., Herrera, L.J.M., Van Raap, M.B.F., Zélis, P.M., Muraca, D., Schinca, D.C., Scaffardi, L.B. 2015. Synthesis of Ni nanoparticles by femtosecond laser ablation in liquids: Structure and sizing. *J. Phys. Chem. C* 119: 13184–13193.

Arruebo, M., Pacheco, R.F., Ibarra, M.R., Santamaria, J. 2007. Magnetic nanoparticles for drug delivery. *NanoToday* 2: 22–32.

Asahi, T., Mafuné, F., Rehbock, C., Barcikowski, S. 2015. Strategies to harvest the unique properties of laser-generated nanomaterials in biomedical and energy applications. *Appl. Surf. Sci.* 348: 1–3.

Bagga, K., Barchanski, A., Intartaglia, R., Dante, S., Marotta, R., Diaspro, A., Sajti, C.L., Brandi, F. 2013. Laser-assisted synthesis of *Staphylococcus aureus* protein-capped silicon quantum dots as bio-functional nanoprobes. *Laser Phys. Lett.* 10: 065603.

Balling, P., Schou, J. 2006. Femtosecond-laser ablation dynamics of dielectrics: Basics and applications for thin films. *Rep. Prog. Phys.* 76: 036502.

Barchanski, A., Funk, D., Wittich, O., Tegenkamp, C., Chichkov, B.N., Sajti, C.L. 2015. Picosecond laser fabrication of functional gold-antibody nanoconjugates for biomedical applications. *J. Phys. Chem. C* 119: 9524–9533.

Barcikowski, S., Hahn, A., Guggenheim, M., Reimers, K., Ostendorf, A. 2010. Biocompatibility of nano-actuators: Stem cell growth on laser-generated nickel–titanium shape memory alloy nanoparticles. *J. Nanopart. Res.* 12: 1733–1742.

Barcikowski, S., Mafune, F. 2011. Trends and current topics in the field of laser ablation and nanoparticle generation in liquids. *J. Phys. Chem. C* 115: 4985.

Barcikowski, S., Meńdez-Manjón, A., Chichkov, B., Brikas, M., Račiukaitis, G. 2007. Generation of nanoparticle colloids by picosecond and femtosecond laser ablations in liquid flow. *Appl. Phys. Lett.* 91: 083113.

Barmina, E.B., Stratakis, E., Fotakis, C., Shafeev, G.A. 2010. Generation of nanostructures on metals by laser ablation in liquids: New results. *Quant. Electron.* 40: 1012–1020.

Barmina, E.V., Barberoglu, M., Zorba, V., Simakin, A.V., Stratakis, E., Fotakis, C., Shafeev, G.A. 2009. Surface nanotexturing of tantalum by laser ablation in water. *Quant. Electron.* 39: 89–93.

Bärsch, N., Jakobi, J., Weiler, S., Barcikowski, S. 2009. Pure colloidal metal and ceramic nanoparticles from high-power picosecond laser ablation in water and acetone. *Nanotechnology* 20: 445603.

Batani, D. 2010. Short-pulse laser ablation of materials at high intensities: Influence of plasma effects. *Laser Part. Beams* 28: 235–244.

Batani, D., Vinci, T., Bleiner, D. 2014. Laser-ablation and induced nanoparticle synthesis. *Part. Beams* 32: 1–7.

Bertorelle, F., Ceccarello, M., Pinto, M., Fracasso, G., Badocco, D., Amendola, V., Pastore, P., Colombatti, M., Meneghetti, M. 2014. Efficient AuFeOx nanoclusters of laser-ablated nanoparticles in water for cells guiding and surface-enhanced resonance Raman scattering imaging. *J. Phys. Chem. C* 118: 14534–14541.

Besner, S., Kabashin, A.V., Meunier, M. 2007. Two-step femtosecond laser ablation-based method for the synthesis of stable and ultra-pure gold nanoparticles in water. *Appl. Phys. A Mater. Sci. Process.* 88: 269–272.

Besner, S., Kabashin, A.V., Meunier, M., Winnik, F.M. 2005. Fabrication of functionalized gold manoparticles by femtosecond laser ablation in aqueous solutions of biopolymers. *Proc. SPIE* 5969: 59690B.

Besner, S., Kabashin, A.V., Winnik, F.M., Meunier, M. 2008. Ultrafast laser based "green" synthesis of non-toxic nanoparticles in aqueous solutions. *Appl. Phys. A Mater. Sci. Process.* 93: 955–959.

Besner, S., Kabashin, A.V., Winnik, F.M., Meunier, M. 2009. Synthesis of size-tunable polymer-protected gold nanoparticles by femtosecond laser-based ablation and seed growth. *J. Phys. Chem. C* 113: 9526–9531.

Besner, S., Meunier, M. 2010. Laser synthesis of nanomaterials. In *Laser Precision Microfabrication*, Sugioka, K., Meunier, M., and Piqué, A. (Eds.). Springer Series in Materials Science, Springer-Verlag, Berlin, Germany, Vol. 135, pp. 163–187.

Boulais, E., Lachaine, R., Hatef, A., Meunier, M. 2013. Plasmonics for pulsed-laser cell nanosurgery: Fundamentals and applications. *J. Photochem. Photobiol. C Photochem. Rev.* 17: 26–49.

Chelnokov, E., Rivoal, M., Colignon, Y., Gachet, D., Bekere, L., Thibaudau, F., Giorgio, S., Khodorkovsky, V., Marine, W. 2012. Band gap tuning of ZnO nanoparticles via Mg doping by femtosecond laser ablation in liquid environment. *Appl. Surf. Sci.* 258: 9408–9411.

Chen, J.K., Beraun, J.E. 2001. Numerical study of ultrashort laser pulses interaction with metal films. *Numer. Heat Trans. A Appl.* 40: 1–20.

Chen, Y.-H., Yeh, C.-S. 2001. A new approach for the formation of alloy nanoparticles: Laser synthesis of gold–silver alloy from gold–silver colloidal mixtures. *Chem. Commun.* 4: 371–372.

Chichkov, B.N., Momma, C., Nolte, S., von Alvensleben, F., Tunnermann, A. 1996. Femtosecond, picosecond and nanosecond laser ablation of solids. *Appl. Phys. A Mater. Sci. Process.* 63: 109–115.

Compagnini, G., Messina, E., Puglisi, O., Cataliotti, R.S., Nicolosi, V. 2008. Spectroscopic evidence of a core–shell structure in the earlier formation stages of Au–Ag nanoparticles by pulsed laser ablation in water. *Chem. Phys. Lett.* 457: 386–390.

Compagnini, G., Scalisi, A.A., Puglisi, O. 2003. Production of gold nanoparticles by laser ablation in liquid alkanes. *J. Appl. Phys.* 94: 7874–7877.

Compagnini, G., Sinatra, M.G., Messina, G.C., Patan, G., Scalese, S., Puglisi, O. 2012. Monitoring the formation of inorganic fullerene-like MoS_2 nanostructures by laser ablation in liquid environments. *Appl. Surf. Sci.* 258: 5672–5676.

Dallaire, A.-M., Rioux, D., Rachkov, A., Patskovsky, S., Meunier, M. 2012. Laser-generated Au–Ag nanoparticles for plasmonic nucleic acid sensing. *J. Phys. Chem. C* 116: 11370–11377.

De Bonis, A., Sansone, M., Alessio, L.D., Galasso, A., Santagata, A., Teghil, R. 2013. Dynamics of laser-induced bubble and nanoparticles generation during ultra-short laser ablation of Pd in liquid. *J. Phys. D Appl. Phys.* 46: 445301.

De Bonis, A., Sansone, M., Galasso, A., Santagata, A., Teghil, R. 2014. The role of the solvent in the ultrashort laser ablation of palladium target in liquid. *Appl. Phys. A Mater. Sci. Process.* 117: 211–216.

De Giacomo, A., Dell'Aglio, M., Santagata, A., Gaudiuso, R., De Pascale, O., Wagener, P., Messina, G.C., Compagnini, G., Barcikowski, S. 2013. Cavitation dynamics of laser ablation of bulk and wire-shaped metals in water during nanoparticles production. *Phys. Chem. Chem. Phys.* 15: 3083–3092.

Del Fatti, N., Vallée, F. 2001. Ultrafast optical nonlinear properties of metal nanoparticles. *Appl. Phys. B Lasers Opt.* 73: 383–390.

Delfour, L., Itina, T.E. 2015. Mechanisms of ultrashort laser-induced fragmentation of metal nanoparticles in liquids: Numerical insights. *J. Phys. Chem. C* 119: 13893–13900.

Dell'Aglio, M., Gaudiuso, R., De Pascale, O., De Giacomo, A. 2015. Mechanisms and processes of pulsed laser ablation in liquids during nanoparticle production. *Appl. Surf. Sci.* 348: 4–9.

Dolgaev, S.I., Simakin, A.V., Voronov, V.V., Shafeev, G.A., Bozon-Verduraz, F. 2002. Nanoparticles produced by laser ablation of solids in liquid environment. *Appl. Surf. Sci.* 186: 546–551.

Elsayed, K.A., Imam, H., Ahmed, M.A., Ramadan, R. 2013. Effect of focusing conditions and laser parameters on the fabrication of gold nanoparticles via laser ablation in liquid. *Opt. Laser Technol.* 45: 495–502.

Eroshova, O.I., Perminov, P.A., Zabotnov, S.V., Gongal'skii, M.B., Ezhov, A.A., Golovan', L.A., Kashkarov, P.K. 2012. Structural properties of silicon nanoparticles formed by pulsed laser ablation in liquid media. *Cryst. Rep.* 57: 831–835.

Fan, G., Ren, S., Qu, S., Wang, Q., Gao, R. 2014. Stability and nonlinear optical properties of Cu nanoparticles prepared by femtosecond laser ablation of Cu target in alcohol and water. *Opt. Commun.* 330: 122–130.

Fan, M., Lai, F., Chou, H., Lu, W., Hwang, B., Brolo, A.G. 2013. Surface-enhanced Raman scattering (SERS) from Au:Ag bimetallic nanoparticles: The effect of the molecular probe. *Chem. Sci.* 4: 509–515.

Fazio, E., Neri, F. 2013. Nonlinear optical effects from Au nanoparticles prepared by laser plasmas in water. *Appl. Surf. Sci.* 272: 88–93.

Gamaly, E.G., Rode, A.V., Luther-Davies, B., Tichonchuk, V.T. 2002. Ablation of solids by femtosecond lasers: Ablation mechanism and ablation thresholds for metals and dielectrics. *Phys. Plasmas* 9: 949–957.

Ganeev, R.A., Baba, M., Ryasnyansky, A.I., Suzuki, M., Kuroda, H. 2004. Characterization of optical and nonlinear optical properties of silver nanoparticles prepared by laser ablation in various liquids. *Opt. Commun.* 240: 437–448.

Ganeev, R.A., Baba, M., Ryasnyansky, A.I., Suzuki, M., Kuroda, H. 2005. Laser ablation of GaAs in liquids: Structural, optical, and nonlinear optical characteristics of colloidal solutions. *Appl. Phys. B Lasers Opt.* 80: 595–601.

Gökce, B., Van't Zand, D.D., Menéndez-Manjón, A., Barcikowski, S. 2015. Ripening kinetics of laser-generated plasmonic nanoparticles in different solvents. *Chem. Phys. Lett.* 626: 96–101.

Guisbiers, G., Wang, Q., Khachatryan, E., Arellano-Jimenez, M.J., Webster, T.J., Larese-Casanova, P., Nash, K.L. 2015. Anti-bacterial selenium nanoparticles produced by UV/VIS/NIR pulsed nanosecond laser ablation in liquids. *Laser Phys. Lett.* 12: 016003.

Guo, S., Wang, E. 2011. Noble metal nanomaterials: Controllable synthesis and application in fuel cells and analytical sensors. *NanoToday* 6: 240–264.

Hajiesmaeilbaigi, F., Motamedi, A. 2007. Synthesis of Au/Ag alloy nanoparticles by Nd:YAG laser irradiation. *Laser Phys. Lett.* 4: 133–137.

Hamad, S. 2014. Ultrafast laser fabricated nanoparticles and nanostructures: Characterization, spectroscopy and applications. PhD thesis, University of Hyderabad, Hyderabad, India.

Hamad, S., Podagatlapalli, G.K., Mohiddon, M.A., Venugopal Rao, S. 2014b. Cost effective nanostructured copper substrates prepared with ultrafast laser pulses for explosives detection using surface enhanced Raman scattering. *Appl. Phys. Lett.* 104: 263104.

Hamad, S., Podagatlapalli, G.K., Mohiddon, M.A., Venugopal Rao, S. 2015. Surface enhanced fluorescence from corroles and SERS studies of explosives using copper nanostructures. *Chem. Phys. Lett.* 621: 171–176.

Hamad, S., Podagatlapalli, G.K., Nageswara Rao, S.V.S., Pathak, A.P., Venugopal Rao, S. 2014a. Excited state dynamics of silicon nanocrystals fabricated using ultrafast laser ablation in liquids. In *12th International Conference on Fiber Optics and Photonics*, OSA Technical Digest (online) (Optical Society of America, 2014), paper T3A.48.

Hamad, S., Podagatlapalli, G.K., Sreedhar, S., Tewari, S.P., Venugopal Rao, S. 2012. Femtosecond and picosecond ablation of aluminum for synthesis of nanoparticles and nanostructures and their optical characterization. *Proc. SPIE* 8245: 82450L.

Hamad, S., Podagatlapalli, G.K., Tewari, S.P., Venugopal Rao, S. 2013. Influence of picosecond multiple/single line ablation on copper nanoparticles fabricated for surface enhanced Raman spectroscopy and photonics applications. *J. Phys. D Appl. Phys.* 46: 485501.

Hamad, S., Podagatlapalli, G.K., Vendamani, V.S., Nageswara Rao, S.V.S., Pathak, A.P., Tewari, S.P., Venugopal Rao, S. 2014c. Femtosecond ablation of silicon in acetone: Tunable photoluminescence from generated nanoparticles and fabrication of surface nanostructures. *J. Phys. Chem. C* 118: 7139–7151.

Harilal, S.S., Freeman, J.R., Diwakar, P.K., Hassanein, A. 2014. *Laser-Induced Breakdown Spectroscopy.* Springer Series in Optical Sciences, Springer, Berlin, Germany, Vol. 182, pp. 143–166, Chapter 6.

Hatanaka, K., Itoh, T., Asahi, T., Ichinose, N., Kawanishi, S., Sasuga, T., Fukumura, H., Masuhara, H. 1999. Femtosecond laser ablation of liquid toluene: Molecular mechanism studied by time-resolved absorption spectroscopy. *J. Phys. Chem. A* 103: 11257–11263.

Hatef, A., Meunier, M. 2015. Plasma-mediated photothermal effects in ultrafast laser irradiation of gold nanoparticle dimers in water. *Opt. Exp.* 23: 1967–1980.

Hu, H., Liu, T., Zhai, H. 2015. Comparison of femtosecond laser ablation of aluminum in water and in air by time-resolved optical diagnosis. *Opt. Exp.* 23: 628–635.

Hu, X., Gong, H., Xu, H., Wei, H., Cao, B., Liu, G., Zeng, H., Cai, W. 2011. Influences of target and liquid media on morphologies and optical properties of ZnO nanoparticles prepared by laser ablation in solution. *J. Am. Ceramic Soc.* 94: 4305–4309.

Ibrahimkutty, S., Wagener, P., Menzel, A., Plech, A., Barcikowski, S. 2012. Nanoparticle formation in a cavitation bubble after pulsed laser ablation in liquid studied with high time resolution small angle X-ray scattering. *Appl. Phys. Lett.* 101: 103104.

Intartaglia, R., Bagga, K., Brandi, F. 2014. Study on the productivity of silicon nanoparticles by picosecond laser ablation in water: Towards gram per hour yield. *Opt. Exp.* 22: 3117–3127.

Intartaglia, R., Bagga, K., Brandi, F., Das, G., Genovese, A., Di Fabrizio, E., Diaspro, A. 2011. Optical properties of femtosecond laser-synthesized silicon nanoparticles in deionized water. *J. Phys. Chem. C* 115: 5102–5107.

Intartaglia, R., Bagga, K., Genovese, A., Athanassiou, A., Cingolani, R., Diaspro, A., Brandi, F. 2012a. Influence of organic solvent on optical and structural properties of ultra-small silicon dots synthesized by UV laser ablation in liquid. *Phys. Chem. Chem. Phys.* 14: 15406–15411.

Intartaglia, R., Bagga, K., Scotto, M., Diaspro, A., Brandi, F. 2012b. Luminescent silicon nanoparticles prepared by ultrashort pulsed laser ablation in liquid for imaging applications. *Opt. Mater. Exp.* 2: 510–518.

Intartaglia, R., Barchanski, A., Bagga, K., Genovese, A., Das, G., Wagener, P., Di Fabrizio, E., Diaspro, A., Brandi, F., Barcikowski, S. 2012c. Bioconjugated silicon quantum dots from one-step green synthesis. *Nanoscale* 4: 1271–1274.

Intartaglia, R., Das, G., Bagga, K., Gopalakrishnan, A., Genovese, A., Povia, M., Di Fabrizio, E., Cingolani, R., Diaspro, A., Brandi, F. 2013. Laser synthesis of ligand free bimetallic nanoparticles for plasmonic applications. *Phys. Chem. Chem. Phys.* 15: 3075–3082.

Ishikawa, Y., Kawaguchi, K., Shimizu, Y., Sasaki, T., Koshizaki, N. 2006. Preparation of Fe–Pt alloy particles by pulsed laser ablation in liquid medium. *Chem. Phys. Lett.* 428: 426–429.

Ismail, R.A., Sulaiman, G.M., Abdulrahman, S.A., Marzoog, T.R. 2015. Antibacterial activity of magnetic iron oxide nanoparticles synthesized by laser ablation in liquid. *Mater. Sci. Eng. C* 53: 286–297.

Itina, T.E. 2011. On nanoparticles formation by laser ablation in liquids. *J. Phys. Chem. C* 115: 5044–5048.

Ivanov, D.S., Zhigilei, L.V. 2003. Combined atomistic-continuum modeling of short-pulse laser melting and disintegration of metal films. *Phys. Rev. B* 68: 064114.

Jain, P.K., Huang, X., El-Sayed, I.H., El-Sayed, M. 2008. Noble metals on the nanoscale: Optical and photo-thermal properties and some applications in imaging, sensing, biology, and medicine. *Acc. Chem. Res.* 41: 1578–1586.

Jiang, K., Pinchuk, A.O. 2015. Noble metal nanomaterials: Synthetic routes, fundamental properties, and promising applications. *Solid State Phys.*, 66: 131–211.

Jiang, Y., Liu, P., Liang, Y., Li, H.B., Yang, G.W. 2011. Promoting the yield of nanoparticles from laser ablation in liquid. *Appl. Phys. A: Mater. Sci. Process.* 105: 903–907.

Jin, Z., Gu, W., Shi, X., Wang, Z., Jiang, Z., Liao, L. 2014. A novel route to surface-enhanced Raman scattering: Ag nanoparticles embedded in the nano-gaps of a Ag substrate. *Adv. Opt. Mater.* 2: 588–596.

Kabashin, A.V., Meunier, M. 2003. Synthesis of colloidal nanoparticles during femtosecond laser ablation of gold in water. *J. Appl. Phys.* 94: 7941–7943.

Kabashin, A.V., Meunier, M. 2006. Laser ablation-based synthesis of functionalized colloidal nanomaterials in biocompatible solutions. *J. Photochem. Photobiol. A Chem.* 182: 330–334.

Kabashin, A.V., Meunier, M., Kingston, C., Luong, J.H.T. 2003. Fabrication and characterization of gold nanoparticles by femtosecond laser ablation in an aqueous solution of cyclodextrins. *J. Phys. Chem. B* 107: 4527–4531.

Kamat, P.V. 2002. Photophysical, photochemical and photocatalytic aspects of metal nanoparticles. *J. Phys. Chem. B* 106: 7729–7744.

Karavanskii, V.A., Simakin, A.V., Krasovskii, V.I., Ivanchenko, P.V. 2004. Nonlinear optical properties of colloidal silver nanoparticles produced by laser ablation in liquids. *Quant. Electron.* 34: 644–648.

Kazakevich, P.V., Simakin, A.V., Voronov, V.V., Shafeev, G.A. 2006. Laser induced synthesis of nanoparticles in liquids. *Appl. Surf. Sci.* 252: 4373–4380.

Kazakevich, P.V., Voronov, V.V., Simakin, A.V., Shafeev, G.A. 2004. Production of copper and brass nanoparticles upon laser ablation in liquids. *Quant. Electron.* 34: 951–956.

Kelly, K.L., Coronado, E., Zhao, L.L., Schatz, G.C. 2003. The optical properties of metal nanoparticles: The influence of size, shape, and dielectric environment. *J. Phys. Chem. B* 107: 668–677.

Kenth, S., Sylvestre, J.-P., Fuhrmann, K., Meunier, M., Leroux, J.-C. 2011. Fabrication of paclitaxel nanocrystals by femtosecond laser ablation and fragmentation. *J. Pharma. Sci.* 100: 1022–1030.

König, J., Nolte, S., Tünnermann, A. 2005. Plasma evolution during metal ablation with ultrashort laser pulses. *Opt. Exp.* 13: 10597–10607.

Kuladeep, R., Jyothi, L., Ali, S.S., Deepak, K.L.N., Narayana Rao, D. 2012. Laser-assisted synthesis of Au–Ag alloy nanoparticles with tunable surface plasmon resonance frequency. *Opt. Mater. Exp.* 2: 161–172.

Kuzmin, P.G., Shafeev, G.A., Bukin, V.V., Garnov, S.V., Farcau, C., Carles, R., Warot-Fontrose, B., Guieu, V., Viau, G. 2010. Silicon nanoparticles produced by femtosecond laser ablation in ethanol: Size control, structural characterization, and optical properties. *J. Phys. Chem. C* 114: 15266–15273.

Lachaine, R., Boulais, E., Meunier, M. 2014. From thermo- to plasma-mediated ultrafast laser induced plasmonic nanobubbles. *ACS Photon.* 1: 331–336.

Lalayan, A.A. 2005. Formation of colloidal GaAs and CdS quantum dots by laser ablation in liquid media. *Appl. Surf. Sci.* 248: 209–212.

Lee, I., Han, S.W., Kim, K. 2001. Production of Au–Ag alloy nanoparticles by laser ablation of bulk alloys. *Chem. Commun.* 18: 1782–1783.

Leitz, K.-H., Redlingshöer, B., Reg, Y., Otto, A., Schmidt, M. 2011. Metal ablation with short and ultrashort laser pulses. *Phys. Proc.* 12(Part 2): 230–238.

Li, Y.-J., Chiu, W.-J., Unnikrishnan, B., Huang, C.-C. 2014. Monitoring thrombin generation and screening anticoagulants through pulse laser-induced fragmentation of biofunctional nanogold on cellulose membranes. *ACS Appl. Mater. Int.* 6: 15253–15261.

Liang, Y., Liu, P., Xiao, J., Li, H.B., Wang, C.X., Yang, G.W. 2014. A general strategy for one-step fabrication of one-dimensional magnetic nanoparticle chains based on laser ablation in liquid. *Laser Phys. Lett.* 11: 056001.

Linic, S., Aslam, U., Boerigter, C., Morabito, M. 2015. Photochemical transformations on plasmonic metal nanoparticles. *Nat. Mater.* 14: 567–576.

Link, S., Mostafa El-Syed, A. 2003. Optical properties and ultrafast dynamics of metallic nanocrystals. *Annu. Rev. Phys. Chem.* 54: 331–366.

Liu, B., Hu, Z., Che, Y., Chen, Y., Pan, X. 2007. Nanoparticle generation in ultrafast pulsed laser ablation of nickel. *Appl. Phys. Lett.* 90: 044103.

Liu, P., Cai, W., Fang, M., Li, Z., Zeng, H., Hu, J., Luo, X., Jing, W. 2009. Room temperature synthesized rutile TiO₂ nanoparticles induced by laser ablation in liquid and their photocatalytic activity. *Nanotechnology* 20: 285707.

Liu, P., Cui, H., Wang, C.X., Yang, G.W. 2010. From nanocrystal synthesis to functional nanostructure fabrication: Laser ablation in liquid. *Phys. Chem. Chem. Phys.* 12: 3942–3952.

Lorazo, P., Lewis, L.J., Meunier, M. 2003. Short-pulse laser ablation of solids: From phase explosion to fragmentation. *Phys. Rev. Lett.* 91: 225502/1–225502/4.

Lorazo, P., Lewis, L.J., Meunier, M. 2006. Thermodynamic pathways to melting, ablation, and solidification in absorbing solids under pulsed laser irradiation. *Phys. Rev. B* 73: 134108.

Makarov, G.N. 2013. Laser applications in nanotechnology: Nanofabrication using laser ablation and laser nanolithography. *Phys. Usp.* 56: 643–682.

Margetic, V. 2002. Femtosecond laser ablation. PhD thesis, Institute of Spectrochemistry and Applied Spectroscopy, Dortmund, Germany.

Maximova, K., Aristov, A., Sentis, M., Kabashin, A.V. 2015. Size-controllable synthesis of bare gold nanoparticles by femtosecond laser fragmentation in water. *Nanotechnology* 26: 065601.

Mendivil, M.I., Krishnan, B., Sanchez, F.A., Martinez, S., Aguilar-Martinez, J.A., Castillo, G.A., Garcia-Gutierrez, D.I., Shaji, S. 2013. Synthesis of silver nanoparticles and antimony oxide nanocrystals by pulsed laser ablation in liquid media. *Appl. Phys. A Mater. Sci. Process.* 110: 809–816.

Menéndez-Manjon, A., Barcikowski, S., Shafeev, G.A., Mazhukin, V.I., Chichkov, B.N. 2010. Influence of beam intensity profile on the aerodynamic particle size distributions generated by femtosecond laser ablation. *Laser Part. Beams* 28: 45–52.

Menéndez-Manjón, A., Schwenke, A., Steinke, T., Meyer, M., Giese, U., Wagener, P., Barcikowski, S. 2013. Ligand-free gold-silver nanoparticle alloy polymer composites generated by picosecond laser ablation in liquid monomer. *Appl. Phys. A Mater. Sci. Process.* 110: 343–350.

Menéndez-Manjón, A., Wagener, P., Barcikowski, S. 2011. Transfer-matrix method for efficient ablation by pulsed laser ablation and nanoparticle generation in liquids. *J. Phys. Chem. C* 115: 5108–5114.

Mortazavi, S.Z., Parvin, P., Reyhani, A., Golikand, A.N., Mirershadi, S. 2011. Effect of laser wavelength at IR (1064 nm) and UV (193 nm) on the structural formation of palladium nanoparticles in deionized water. *J. Phys. Chem. C* 115: 5049–5057.

Murphy, C.J., Sau, T.K., Gole, A.M., Orendorff, C.J., Gao, J., Gou, L., Hunyadi, S.E., Li, T. 2005. Anisotropic metal nanoparticles: Synthesis, assembly, and optical applications. *J. Phys. Chem. B* 109: 13857–13870.

Narayanan, K.B., Sakthivel, N. 2010. Biological synthesis of metal nanoparticles by microbes. *Adv. Colloid Int. Sci.* 156: 1–13.

Nath, A., Das, A., Rangan, L., Khare, A. 2012. Bacterial inhibition by Cu/Cu₂O nanocomposites prepared via laser ablation in liquids. *Sci. Adv. Mater.* 4: 106–109.

Nedderson, J., Chumanov, G., Cotton, T.M. 1993. Laser ablation of metals: A new method for preparing SERS active colloids. *Appl. Spectrosc.* 47: 1959–1964.

Nguyen, V., Yan, L., Si, J., Hou, X. 2015. Femtosecond laser-induced size reduction of carbon nanodots in solution: Effect of laser fluence, spot size, and irradiation time. *J. Appl. Phys.* 117: 084304.

Nichols, W.T., Sasaki, T., Koshizaki, N. 2006a. Laser ablation of a platinum target in water. I. Ablation mechanisms. *J. Appl. Phys.* 100: 114911.

Nichols, W.T., Sasaki, T., Koshizaki, N. 2006b. Laser ablation of a platinum target in water. II. Ablation rate and nanoparticle size distributions. *J. Appl. Phys.* 100: 114912.

Nichols, W.T., Sasaki, T., Koshizaki, N. 2006c. Laser ablation of a platinum target in water. III. Laser-induced reactions. *J. Appl. Phys.* 100: 114913.

Nie, S., Emory, S.R. 1997. Probing single molecules and single nanoparticles by surface-enhanced Raman scattering. *Science* 275: 1102–1106.

Niu, K.Y., Yang, J., Kulinich, S.A., Sun, J., Li, H., Du, X.W. 2010. Morphology control of nanostructures via surface reaction of metal nanodroplets. *J. Am. Chem. Soc.* 132: 9814–9819.

Noël, S., Hermann, J., Itina, T. 2007. Investigation of nanoparticle generation during femtosecond laser ablation of metals. *Appl. Surf. Sci.* 253: 6310–6315.

Nolte, S., Chichkov, B.N., Welling, H., Shani, Y., Liebermann, K., Terkel, H. 1999. Nanostructuring with spatially localized femtosecond laser pulses. *Opt. Lett.* 24: 914–916.

Olea-Mejía, O., Fernández-Mondragón, M., Rodríguez-de la Concha, G., Camacho-López, M. 2015. SERS-active Ag, Au and Ag–Au alloy nanoparticles obtained by laser ablation in liquids for sensing methylene blue. *Appl. Surf. Sci.* 348: 66–70.

O'Malley, S.M., Amin, M., Borchert, J., Jimenez, R., Steiner, M., Fitz-Gerald, J.M., Bubb, D.M. 2014. Formation of rubrene nanocrystals by laser ablation in liquids utilizing MAPLE deposited thin films. *Chem. Phys. Lett.* 595–596: 171–174.

Onodera, Y., Nunokawa, T., Odawara, O., Wada, H. 2013. Upconversion properties of Y_2O_3:Er,Yb nanoparticles prepared by laser ablation in water. *J. Lumin.* 137: 220–224.

Ou, Y., Wang, L., Zhu, L., Wan, L., Xu, Z. 2014. In-situ immobilization of silver nanoparticles on self-assembled honeycomb-patterned films enables surface-enhanced Raman scattering (SERS) substrates. *J. Phys. Chem. C* 118: 11478–11484.

Papagiannouli, I., Aloukos, P., Rioux, D., Meunier, M., Couris, S. 2015. Effect of the composition on the nonlinear optical response of Au_xAg_{1-x} nano-alloys. *J. Phys. Chem. C* 119: 6861–6872.

Pelton, M., Aizpuru, J., Bryant, G. 2008. Metal-nanoparticle plasmonics. *Laser Photon. Rev.* 2: 136–159.

Pereira, A., Cros, A., Delaporte, P., Georgiou, S., Manousaki, A., Marine, W., Sentis, M. 2004. Surface nano-structuring of metals by laser irradiation: Effects of pulse duration, wavelength and gas atmosphere. *Appl. Phys. A Mater. Sci. Process.* 79: 1433–1437.

Perez, D., Beland, L.K., Deryng, D., Lewis, L.J., Meunier, M. 2008. Numerical study of the thermal ablation of wet solids by ultrashort laser pulses. *Phys. Rev. B* 77: 014108.

Perez, D., Lewis, L.J. 2003. Molecular-dynamics study of ablation of solids under femtosecond laser pulses. *Phys. Rev. B* 67: 184102.

Petersen, S., Jakobi, J., Barcikowski, S. 2009. In situ bioconjugation—Novel laser based approach to pure nanoparticle-conjugates. *Appl. Surf. Sci.* 255: 5435–5438.

Podagtlapalli, G.K. 2014. Ultrafast laser ablation of metals in liquids for explosives detection using SERS and development of CARS experiments. PhD thesis, University of Hyderabad, Hyderabad, India.

Podagtlapalli, G.K., Hamad, S., Mohiddon, M.A., Venugopal Rao, S. 2014a. Effect of oblique incidence on silver nanomaterials fabricated in water via ultrafast laser ablation for photonics and explosives detection. *Appl. Surf. Sci.* 303: 217–232.

Podagtlapalli, G.K., Hamad, S., Mohiddon, M.A., Venugopal Rao, S. 2015a. Fabrication of nanoparticles and nanostructures using ultrafast laser ablation of silver with Bessel beams. *Laser Phys. Lett.* 12: 036003.

Podagtlapalli, G.K., Hamad, S., Sreedhar, S., Prasad, M.D., Venugopal Rao, S. 2013. Silver nano-entities through ultrafast double ablation in aqueous media for surface enhanced Raman scattering and photonics applications. *J. Appl. Phys.* 113: 073106.

Podagtlapalli, G.K., Hamad, S., Sreedhar, S., Tewari, S.P., Venugopal Rao, S. 2012. Fabrication and characterization of aluminum nanostructures and nanoparticles obtained using femtosecond ablation technique. *Chem. Phys. Lett.* 530: 93–97.

Podagtlapalli, G.K., Hamad, S., Venugopal Rao, S. 2014b. Fabrication of hybrid Ag–Au nanomaterials for explosives detection. In *12th International Conference on Fiber Optics and Photonics*, OSA Technical Digest (online) (Optical Society of America, 2014), paper S5A.51.

Podagtlapalli, G.K., Hamad, S., Venugopal Rao, S. 2014c. Silver nanomaterials in aqueous media fabricated with non-diffracting picosecond Bessel beam and applications. In *Light, Energy and the Environment*, OSA Technical Digest (online) (Optical Society of America, 2014), paper JW6A.13.

Podagtlapalli, G.K., Hamad, S., Venugopal Rao, S. 2015b. Trace-level detection of secondary explosives using hybrid silver-gold nanoparticles and nanostructures achieved with femtosecond laser ablation. *J. Phys. Chem. C* 119: 16972–16983.

Povarnitsyn, M.E., Itina, T.E., Levashov, P.R., Khishchenko, K.V. 2013. Mechanisms of nanoparticle formation by ultra-short laser ablation of metals in liquid environment. *Phys. Chem. Chem. Phys.* 15: 3108–3114.

Povarnitsyn, M.E., Itina, T.E., Sentis, M., Khishchenko, K.V., Levashov, P.R. 2007. Material decomposition mechanisms in femtosecond laser interactions with metals. *Phys. Rev. B* 75: 235414.

Pradeep, T., Anshup. 2009. Noble metal nanoparticles for water purification: A critical review. *Thin Solid Films* 517: 6441–6478.

Qua, L., Song, Q., Li, Y., Peng, M., Li, D., Chen, L., Fossey, J.S., Long, Y. 2013. Fabrication of bimetallic microfluidic surface-enhanced Raman scattering sensors on paper by screen printing. *Anal. Chim. Acta* 792: 86–92.

Radziemski, J., Creamers, D.A. 1989. *Laser Induced Plasma and Applications.* Marcel Dekker, New York.

Rehbock, C., Jakobi, J., Gamrad, L., van der Meer, S., Tiedemann, D., Taylor, U., Kues, W., Rath, D., Barcikowski, S. 2014. Current state of laser synthesis of metal and alloy nanoparticles as ligand-free reference materials for nano-toxicological assays. *Beilstein J. Nanotechnol.* 5: 1523–1541.

Riabinina, D., Chaker, M., Margot, J. 2012. Dependence of gold nanoparticle production on pulse duration by laser ablation in liquid media. *Nanotechnology* 23: 135603.

Rioux, D., Laferrière, M., Douplik, A., Shah, D., Lilge, L., Kabashin, A.V., Meunier, M.M. 2009. Silicon nanoparticles produced by femtosecond laser ablation in water as novel contamination-free photosensitizers. *J. Biomed. Opt.* 14: 021010.

Robitaille, A., Boulais, E., Meunier, M. 2013. Mechanisms of plasmon-enhanced femtosecond laser nanoablation of silicon. *Opt. Exp.* 21: 9703–9710.

Saikiran, V., Vendamani, V.S., Hamad, S., Nageswara Rao, S.V.S., Venugopal Rao, S., Pathak, A.P. 2014. 150 MeV Au ion induced modification of Si nanoparticles prepared by laser ablation. *Nucl. Instrum. Methods Phys. Res. B* 333: 99–105.

Sajti, C.L., Sattari, R., Chichkov, B.N., Barcikowski, S. 2010. Gram scale synthesis of pure ceramic nanoparticles by laser ablation in liquid. *J. Phys. Chem. C* 114: 2421–2427.

Sakamoto, M., Fujistuka, M., Majima, T. 2009. Light as a construction tool of metal nanoparticles: Synthesis and mechanism. *J. Photochem. Photobiol. C Photochem. Rev.* 10: 33–56.

Salminen, T., Dahl, J., Tuominen, M., Laukkanen, P., Arola, E., Niemi, T. 2012. Single-step fabrication of luminescent GaAs nanocrystals by pulsed laser ablation in liquids. *Opt. Mater. Exp.* 2: 799–813.

Santagata, A., Guarnaccio, A., Pietrangeli, D., Szegedi, Á., Valyon, J., De Stefanis, A., De Bonis, A. et al. 2015. Production of silver–silica core–shell nanocomposites using ultra-short pulsed laser ablation in nanoporous aqueous silica colloidal solutions. *J. Phys. D Appl. Phys.* 48: 205304.

Sasaki, T., Shimizu, Y., Koshizaki, N. 2006. Preparation of metal oxide-based nanomaterials using nanosecond pulsed laser ablation in liquids. *Photochem. Photobiol. A Chem.* 182: 335–341.

Sau, T.K., Rogach, A.L. 2010. Nonspherical noble metal nanoparticles: Colloid-chemical synthesis and morphology control. *Adv. Mater.* 22: 1781–1804.

Schwenke, A., Dalüge, H., Kiyan, R., Sajti, C.L., Chichkov, B.N. 2013. Non-agglomerated gold-PMMA nanocomposites by in situ-stabilized laser ablation in liquid monomer for optical applications. *Appl. Phys. A: Mater. Sci. Process.* 111: 451–457.

Schwenke, A., Wagener, P., Nolte, S., Barcikowski, S. 2011. Influence of processing time on nanoparticle generation during picosecond-pulsed fundamental and second harmonic laser ablation of metals in tetrahydrofuran. *Appl. Phys. A Mater. Sci. Process.* 104: 77–82.

Semaltianos, N.G. 2010. Nanoparticles by laser ablation. *Crit. Rev. Solid State Mater. Sci.* 35: 105–124.

Semaltianos, N.G., Logothetidis, S., Perrie, W., Romani, S., Potter, R.J., Edwardson, S.P., French, P., Sharp, M., Dearden, G., Watkins, K.G. 2010. Silicon nanoparticles generated by femtosecond laser ablation in a liquid environment. *J. Nanopart. Res.* 12: 573–580.

Serkov, A.A., Barmina, E.V., Simakin, A.V., Kuzmin, P.G., Voronov, V.V., Shafeev, G.A. 2015. Generation of core–shell nanoparticles Al@Ti by laser ablation in liquid for hydrogen storage. *Appl. Surf. Sci.* 348: 71–74.

Shafeev, G.A. 2008. Formation of nanoparticles under laser ablation of solids in liquids. In *Nanoparticles: New Research*, Lombardi, S.L. (Ed.), Nova Science Publishers, Inc. pp. 1–37.

Shah, A., Rahman, L.-U., Qureshi, R., Rehman, Z.-U. 2012. Synthesis, characterization and applications of bimetallic (Au–Ag, Au–Pt, Au–Ru) alloy nanoparticles. *Rev. Adv. Mater. Sci.* 30: 133–149.

Shaheen, M.E., Gagnon, J.E., Fryer, B.J. 2013. Femtosecond laser ablation of brass in air and liquid media. *J. Appl. Phys.* 113: 213106.

Simakin, A.V., Voronov, V.V., Kricihenko, N.A., Shafeev, G.A. 2004. Nanoparticles produced by laser ablation of solids in liquid environment. *Appl. Phys. A Mater. Sci. Process.* 79: 1127–1132.

Simakin, A.V., Voronov, V.V., Shafeev, G.A., Brayner, R., Bozon-Verduraz, F. 2001. Nanodisks of Au and Ag produced by laser ablation in liquid environment. *Chem. Phys. Lett.* 348: 182–186.

Siuzdak, K., Sawczak, M., Klein, M., Nowaczyk, G., Jurga, S., Cenian, A. 2014. Preparation of platinum modified titanium dioxide nanoparticles with the use of laser ablation in water. *Phys. Chem. Chem. Phys.* 16: 15199–15206.

Sportelli, M.C., Ancona, A., Picca, R.A., Trapani, A., Volpe, A., Trapani, G., Cioffi, N. 2015. Laser ablation synthesis in solution of nanoantimicrobials for food packaging applications. *Mater. Res. Soc. Symp. Proc.* 1804: 37–42.

Spyropoulos, G.D., Stylianakis, M.M., Stratakis, E., Kymakis, E. 2012. Organic bulk heterojunction photovoltaic devices with surfactant-free Au nanoparticles embedded in the active layer. *Appl. Phys. Lett.* 100: 213904.

Stratakis, E. 2012. Nanomaterials by ultrafast laser processing of surfaces. *Sci. Adv. Mater.* 4: 407–431.

Stratakis, E., Barberoglou, M., Fotakis, C., Viau, G., Garcia, C., Shafeev, G.A. 2009a. Generation of Al nanoparticles via ablation of bulk al in liquids with short laser pulses. *Opt. Exp.* 17(15): 12650–12659.

Stratakis, E., Zorba, V., Barberoglou, M., Fotakis, C., Shafeev, G.A. 2009b. Femtosecond laser writing of nanostructures on bulk Al via its ablation in air and liquids. *Appl. Surf. Sci.* 255(10): 5346–5350.

Stratakis, E., Zorba, V., Barberoglou, M., Fotakis, C., Shafeev, G.A. 2009c. Laser writing of nanostructures on bulk Al via its ablation in liquids. *Nanotechnology* 20(10): 105303.

Syed, H., Podagatlapalli, G.K., Hussian, A., Ahmed, N., Sreedhar, S., Tewari, S.P., Rao, S.V. 2012. Fabrication of metal nano-entities using ultrafast ablation for SERS, photonics, and biomedical applications. In *International Conference on Fibre Optics and Photonics*, OSA Technical Digest (online) (Optical Society of America, 2012), paper MPo.31.

Sylvestre, J.-P., Kabashin, A.V., Sacher, E., Meunier, M. 2005. Femtosecond laser ablation of gold in water: Influence of the laser-produced plasma on the nanoparticle size distribution. *Appl. Phys. A Mater. Sci. Process.* 80: 753–758.

Sylvestre, J.-P., Kabashin, A.V., Sacher, E., Meunier, M., Luong, J.H.T. 2004a. Stabilization and size control of gold nanoparticles during laser ablation in aqueous cyclodextrins. *J. Am. Chem. Soc.* 126: 7176–7177.

Sylvestre, J.-P., Poulin, S., Kabashin, A.V., Sacher, E., Meunier, M., Luong, J.H.T. 2004b. Surface chemistry of gold nanoparticles produced by laser ablation in aqueous media. *J. Phys. Chem. B* 108: 16864–16869.

Tagami, T., Imao, Y., Ito, S., Nakada, A., Ozeki, T. 2014. Simple and effective preparation of nano-pulverized curcumin by femtosecond laser ablation and the cytotoxic effect on C6 rat glioma cells in vitro. *Int. J. Pharma.* 468: 91–96.

Tan, D., Lin, G., Liu, Y., Teng, Y., Zhuang, Y., Zhu, B., Zhao, Q., Qiu, J. 2011. Synthesis of nanocrystalline cubic zirconia using femtosecond laser ablation. *J. Nanopart. Res.* 13: 1183–1190.

Tan, D., Liu, X., Dai, Y., Ma, G., Meunier, M., Qiu, J.A. 2015. Universal photochemical approach to ultra-small, well-dispersed nanoparticle/reduced graphene oxide hybrids with enhanced nonlinear optical properties. *Adv. Opt. Mater.* 3: 836–841.

Tan, D., Yamada, Y., Zhou, S., Shimotsuma, Y., Miura, K., Qiu, J. 2013a. Photoinduced luminescent carbon nanostructures with ultra-broadly tailored size ranges. *Nanoscale* 5: 12092–12097.

Tan, D., Zhou, S., Qiu, J., Khusro, N. 2013b. Preparation of functional nanomaterials with femtosecond laser ablation in solution. *J. Photochem. Photobiol. C Photochem. Rev.* 17: 50–68.

Taylor, U., Barchanski, A., Garrels, W., Klein, S., Kues, W., Barcikowski, S., Rath, D. 2011. Toxicity of gold nanoparticles on somatic and reproductive cells. *Adv. Exp. Med. Biol.* 733: 125–133.

Tian, X., Guan, X., Luo, N., Yang, F., Chen, D., Peng, Y., Zhu, J., He, F., Li, L., Chen, X. 2014. In vivo immunotoxicity evaluation of Gd_2O_3 nanoprobes prepared by laser ablation in liquid for MRI preclinical applications. *J. Nanopart. Res.* 16: 2594.

Tiedemann, D., Taylor, U., Rehbock, C., Jakobi, J., Klein, S., Kues, W.A., Barcikowski, S., Rath, D. 2014. Reprotoxicity of gold, silver, and gold-silver alloy nanoparticles on mammalian gametes. *Analyst* 139: 931–942.

Tilaki, R.M., Zad, A.I., Mahdavi, S.M. 2007. Size, composition and optical properties of copper nanoparticles prepared by laser ablation in liquids. *Appl. Phys. A Mater. Sci. Process.* 88: 415–419.

Truong, S.L., Levi, G., Bozon-Verduraz, F., Petrovskaya, A.V., Simakin, A.V., Shafeev, G.A. 2007a. Generation of Ag nanospikes via laser ablation in liquid environment and their activity in SERS of organic molecules. *Appl. Phys. A Mater. Sci. Process.* 89: 373–376.

Truong, S.L., Levi, G., Bozon-Verduraz, F., Petrovskaya, A.V., Simakin, A.V., Shafeev, G.A. 2007b. Generation of nanospikes via laser ablation of metals in liquid environment and their activity in surface-enhanced Raman scattering of organic molecules. *Appl. Surf. Sci.* 254: 1236–1239.

Tsuji, T., Kakita, T., Tsuji, M. 2003. Preparation of nano-size particles of silver with femtosecond laser ablation in water. *Appl. Surf. Sci.* 206: 314–320.

Tsuji, T., Mizuki, T., Yasutomo, M., Tsuji, M., Kawasaki, H., Yonezawa, T., Mafuné, F. 2010. Efficient fabrication of substrates for surface-assisted laser desorption/ionization mass spectrometry using laser ablation in liquids. *Appl. Surf. Sci.* 257: 2046–2050.

Upadhyay, A.K., Inogamov, N.A., Rethfeld, B., Urbassek, H.M. 2008. Ablation by ultrashort laser pulses: Atomistic and thermodynamic analysis of the processes at the ablation threshold. *Phys. Rev. B* 78: 045437.

Usui, H., Shimizu, Y., Sasaki, T., Koshizaki, N. 2005. Photoluminescence of ZnO nanoparticles prepared by laser ablation in different surfactant solutions. *J. Phys. Chem. B* 109: 120–124.

Vadavalli, S., Valligatla, S., Neelamraju, B., Dar, M.H., Chiasera, A., Ferrari, M., Desai, N.R. 2014. Optical properties of germanium nanoparticles synthesized by pulsed laser ablation in acetone. *Front. Phys.* 2: 57.

Vendamani, V.S., Hamad, S., Saikiran, V., Pathak, A.P., Venugopal Rao, S., Ravi Kanth Kumar, V.V., Nageswara Rao, S.V.S. 2015. Synthesis of ultra-small silicon nanoparticles by femtosecond laser ablation of porous silicon. *J. Mater. Sci.* 50: 1666–1672.

Venugopal Rao, S., Podagatlapalli, G.K., Hamad, S. 2014. Ultrafast laser ablation in liquids for nanomaterials and applications. *J. Nanosci. Nanotechnol.* 14: 1364–1388.

Von Der Linde, D., Sokolowski-Tinten, K. 2000. Physical mechanisms of short-pulse laser ablation. *Appl. Surf. Sci.* 154: 1–10.

Vorobyev, A.Y., Guo, C. 2005. Enhanced absorptance of gold following multi-pulse femtosecond laser ablation. *Phys. Rev. B* 72: 195422.

Vorobyev, A.Y., Guo, C. 2008. Colorizing metals with femtosecond laser pulses. *Appl. Phys. Lett.* 92: 041914.

Vorobyev, A.Y., Guo, C. 2013. Direct femtosecond laser surface nano/microstructuring and its applications. *Laser Photon. Rev.* 7: 385–407.

Vorobyev, A.Y., Makin, V.S., Guo, C. 2009. Brighter light sources from black metal: Significant increase in emission efficiency of incandescent light sources. *Phys. Rev. Lett.* 102: 234301.

Wang, C.X., Liu, P., Cui, H., Yang, G.W. 2005. Nucleation and growth kinetics of nanocrystals formed upon pulsed-laser ablation in liquid. *Appl. Phys. Lett.* 87: 201913.

Wang, T., Hu, F., Ikhile, E., Liao, F., Li, Y., Shao, M. 2015. Two-step-route to Ag–Au nanoparticles grafted on Ge wafer for extra-uniform SERS substrates. *J. Mater. Chem.* 3: 559–563.

Weng, C.-I., Cang, J.-S., Chang, J.-Y., Hsiung, T.-M., Unnikrishnan, B., Hung, Y.-L., Tseng, Y.-T., Li, Y.-J., Shen, Y.-W., Huang, C.-C. 2014. Detection of arsenic(III) through pulsed laser-induced desorption/ionization of gold nanoparticles on cellulose membranes. *Anal. Chem.* 86: 3167–3173.

Xia, Y., Rogers, J.A., Paul, K.E., Whitesides, G.M. 1999. Unconventional methods for fabricating and patterning nanostructures. *Chem. Rev.* 99: 1823–1848.

Yan, Z., Chrisey, D.B. 2012. Pulsed laser ablation in liquid for micro-/nanostructure generation. *J. Photochem. Photobiol. C Photochem. Rev.* 13: 204–223.

Yang, G.W. 2007. Laser ablation in liquids: Applications in the synthesis of nanocrystals. *Prog. Mater. Sci.* 52: 648–698.

Yang, G.W. 2011. *Laser Ablation in Liquids: Principles and Applications in the Preparation of Nanomaterials.* Pan Stanford Publishing, Singapore.

Zakharko, Y., Rioux, D., Patskovsky, S., Lysenko, V., Marty, O., Bluet, J.-M., Meunier, M. 2011. Direct synthesis of luminescent SiC quantum dots in water by laser ablation. *Phys. Stat. Solidi Rapid Res. Lett.* 5: 292–294.

Zavedeev, E.V., Petrovskaya, A.V., Simakin, A.V., Shafeev, G.A. 2006. Formation of nanostructures upon laser ablation of silver in liquids. *Quant. Electron.* 36: 978–980.

Zeng, H., Du, X.-W., Singh, S.C., Kulinich, S.A., Yang, S., He, J., Cai, W. 2012. Nanomaterials via laser ablation/irradiation in liquid: A review. *Adv. Funct. Mater.* 22: 1333–1353.

Zhakhovskii, V.V., Inogamo, N.A., Petrov, Y.V., Ashitkov, S.I., Nishihara, K. 2009. Molecular dynamics simulation of femtosecond ablation and spallation with different interatomic potentials. *Appl. Surf. Sci.* 255: 9592–9596.

Zhang, L., Mu, K., Zhao, J., Wu, T., Wang, H., Zhang, C., Zhang, X.-C. 2015. Intense thermal terahertz-to-infrared emission from random metallic nanostructures under femtosecond laser irradiation. *Opt. Exp.* 23: 14211–14218.

Zheng, Z., Shan, G., Li, J., Chen, Y., Liu, Y. 2014. Au/Ag nano alloy shells as near infrared SERS nano-probe for the detection of protein. *Mater. Res. Exp.* 1: 045408.

Zhigilei, L.V., Lin, Z., Ivanov, D.S. 2009. Atomistic modeling of short pulse laser ablation of metals: Connections between melting, spallation, and phase explosion. *J. Phys. Chem. C* 113: 11892–11906.

Zimbone, M., Buccheri, M.A., Cacciato, G., Sanz, R., Rappazzo, G., Boninelli, S., Reitano, R., Romano, L., Privitera, V., Grimaldi, M.G. 2015. Photocatalytical and antibacterial activity of TiO_2 nanoparticles obtained by laser ablation in water. *Appl. Catal. B Environ.* 165: 487–494.

12 Exploring the LaAlO₃/SrTiO₃ Two-Dimensional Electron Gas

From Fundamental to Technical Applications

Ngai Yui Chan, Fan Zhang, Kit Au, Wing Chong Lo, Helen La Wa Chan, and Jiyan Dai

CONTENTS

12.1 INTRODUCTION

"Often, it may be said that the interface is the device," as stated by Prof. Herbert Kroemer in his Nobel lecture in 2000 [1]. A rich set of fundamental physical phenomena can be explored in the complex oxide interface. At the interface, surprising novel properties were discovered, which are not present in their bulk counterparts. The interface between lanthanum aluminate (LaAlO₃) (LAO) and strontium titanate (SrTiO₃) (STO) is a remarkable system that was discovered in 2004 [2];

despite both materials being insulators, a quasi-two dimensional electron gas (2DEG) can be found at their interface. The discovery of 2DEG at the interface between the LAO and STO heterostructure with intriguing properties such as the ability of nanopatterning [3], magnetism [4], superconductivity [5], resistance switching behavior [6], and photoconductivity [7] has been studied extensively in recent years. The study of such properties of the LAO/STO interface is one of emerging areas in the field of condensed matter physics.

Intense research has been conducted to explain the observed conducting mechanism, including the electronic reconstruction (polar catastrophe model) [8], the presence of oxygen vacancies [9–11], and the cation intermixing [12,13] at the interface. However, the exact mechanism to explain these phenomena is still under debate, and researchers continue to investigate this system.

For natural analogue of conventional semiconductor heterostructure, from the view of electron mobility, the LAO/STO is not a feasible candidate for high-speed electronic device applications, with mobility around 5–10 cm^2 V^{-2} s^{-1} at room temperature and up to 10^3 cm^2 V^{-2} s^{-1} at lower temperatures [2]. For comparison, graphene, a well-known two-dimensional material, has an electron mobility higher than 15,000 cm^2 V^{-2} s^{-1} at room temperature [14], high-mobility semiconductor 2DEG systems, such as the $Ga_xAl_{1-x}As/GaAs$ heterostructure, have electron mobility 3 orders of magnitude higher than that of LAO/STO at low temperatures [15], and the molecular beam epitaxy–grown $Mg_xZn_{1-x}O/ZnO$ oxide 2DEG system has a mobility exceeding 700,000 cm^2 V^{-2} s^{-1} [16]. However, the LAO/STO heterostructure, which exhibits novel properties like sharp switching from insulating/conducting state [17] and optical transparency [18], has the ability for combination with other oxides that have outstanding properties.

The significant results observed in this system pave way for potential applications in multifunctional oxide-based electronic devices. Even though the discovery of LAO/STO 2DEG was included in the *Science Magazine* Top 10 Breakthroughs of the Year 2007 [19], a relatively weak area in this field of research remains—the device application for such an oxide interface 2DEG. The main difficulty in applying this 2DEG system in electronic devices is its relatively low response to external stimulus such as light and gas (since gas cannot react and be absorbed onto the heterostructure). In this chapter, a method to modify the surface of the LAO/STO heterostructure to significantly enhance its response to ultraviolet light and gases is introduced. This method has brought this fascinating 2DEG system closer to real-world applications in sensing devices.

The works presented in this chapter are separated into four main sections, concerning the fabrication of the LAO/STO heterostructure and the fabrication of the polar liquid sensor, ultraviolet sensor, and gas sensor. The study of the LAO/STO heterostructure provides an unexplored regime to understand its potential for device application.

12.2 LAO/STO PREPARATION AND PROPERTIES

To fabricate the atomically aligned heterostructure or interface, an atomically flat substrate is required. TiO_2-site terminated $SrTiO_3$ substrate was prepared with conventional $SrTiO_3$ (001) substrate, with miscut angle of the substrate smaller than 0.5°. The preparation procedure of the $LaAlO_3$/$SrTiO_3$ heterostructure is summarized in Figure 12.1. First, the as-received $SrTiO_3$ substrates were treated with buffered hydrofluoric acid (BHF) by NH_4F:HF = 7:1, followed by a heat treatment process to obtain an atomically flat substrate [20–22]. Then, a $LaAlO_3$ thin film was grown on the treated $SrTiO_3$ substrate using pulsed laser deposition (the deposition parameters are summarized in Table 12.1); during the deposition, the growth of the thin film was monitored by reflection high energy electron diffraction (RHEED). Finally, electrical wires were bonded at the corners (by the van der Pauw geometry) for subsequent electrical properties measurement.

The roughness of the top layer increases with thin film deposition. A decrease in the intensity of the RHEED beam is observed. The intensity of the beam reaches a minimum when half of the monolayer of the $LaAlO_3$ thin film is deposited on top of the $SrTiO_3$ substrate. Afterward, the intensity of the RHEED beam increases until a full monolayer of the $LaAlO_3$ is deposited on the $SrTiO_3$

FIGURE 12.1 Fabrication process of the LaAlO₃/SrTiO₃ heterostructure. Surface treatment process was performed on the as-received SrTiO₃ (001) substrate. LaAlO₃ thin film was deposited on top of the TiO₂-site-terminated substrate. After the deposition, electrical wires were bonded with the van der Pauw geometry of the LaAlO₃/SrTiO₃ heterostructure for subsequent electrical measurement.

TABLE 12.1

Experimental Parameters Used for Thin Film Deposition

Target to substrate distance	60 mm
Laser energy	250–280 mJ
Repetition rate of the laser	1 Hz
Deposition temperature	750°C
Base vacuum	$<10^{-5}$ Pa

surface. From the RHEED measurement shown in Figure 12.2a, it is observed that each oscillation corresponds to one monolayer deposition (with thickness ~0.4 nm), so the surface crystallinity of the sample can be determined. In addition, as shown in Figure 12.2b and c, the electron diffraction pattern of the bare SrTiO₃ substrate and after deposition (LaAlO₃/SrTiO₃ heterostructure) can be obtained.

X-ray diffraction (XRD) and transmission electron microscopy (TEM) were used to investigate the quality of the LaAlO₃ thin film and its epitaxial relationship with the SrTiO₃ substrate after deposition. XRD analysis is one of the nondestructive techniques for analyzing a wide range of materials. Figure 12.3 shows a θ–2θ scan of the 20 unit cells (uc) (~8 nm)-thick LaAlO₃ thin film on the TiO₂-site-terminated SrTiO₃ substrate in the range of 2θ = 18°–28°, where only the (001) reflection of the LaAlO₃ can be observed with the (001) reflection of SrTiO₃ substrate, indicating a c-axis-oriented growth. In the XRD diffractogram, Kiessig fringes were obtained, resulting from the constructive and destructive interference between the film and the film–substrate interface, which indicates an abrupt and highly ordered crystalline structure between the LaAlO₃/SrTiO₃ interfaces.

The LaAlO₃/SrTiO₃ heterostructure was characterized by cross-sectional TEM. The cross-sectional TEM sample was prepared by mounting the LAO/STO sample face to face and mechanically grinding the cross section to 20 μm and then finally ion-milling to a thickness below 100 nm. As shown in Figure 12.4, the LaAlO₃–SrTiO₃ interface is atomically sharp and abrupt, implying that the surface of the STO is atomically flat. The bright dots representing LaAlO₃ unit cells can be clearly identified, from which the number of layers can be counted accurately with an uncertainty of one unit cell. The LAO thin film, as shown in the figure, has a thickness of around 20 uc.

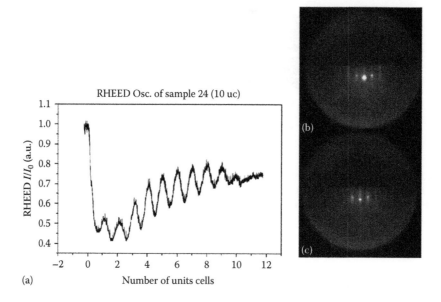

FIGURE 12.2 (a) RHEED oscillations during the growth of $LaAlO_3$ thin film on top of the $SrTiO_3$ substrate with 10 unit cell (uc) thickness in $LaAlO_3$. (b) RHEED diffraction along the (001) plane of the $SrTiO_3$ before the deposition. (c) RHEED diffraction of the $LaAlO_3$ plane after the deposition, which confirms the crystallinity of the film.

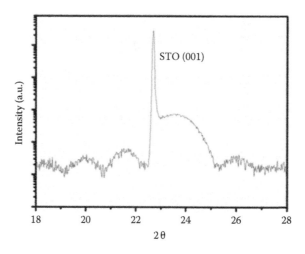

FIGURE 12.3 XRD pattern of the $LaAlO_3/SrTiO_3$ heterostructure.

A variety of electrical property characterizations can be conducted by employing the Physical Properties Measurement System. Sheet resistance versus temperature can be measured while the temperature of the system varies during measurement. The resistance measurement of the sample was constructed in the van der Pauw geometry [23] for a sample with size of 5×5 mm², where four aluminum wires were ultrasonically bonded on the sample corners and ohmic behavior was observed from the linear current–voltage characteristics.

To clarify that the as-grown LAO/STO sample at the interface is a standard metallic conductor, its sheet resistance and mobility as a function of temperature was measured as shown in Figure 12.5. To reduce the effect of photoexcitation of excess charge carrier, the experiment was

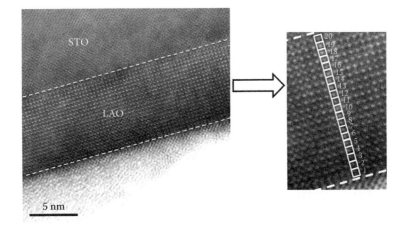

FIGURE 12.4   High-resolution TEM image of the LAO/STO interface. The number of unit cells (around 20 uc) can be counted.

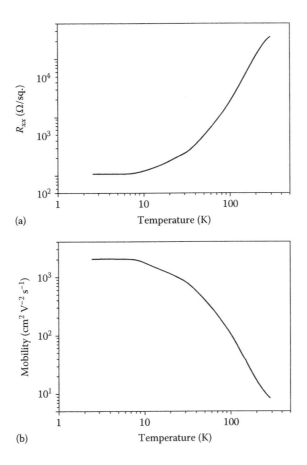

FIGURE 12.5   (a) Sheet resistance and (b) the mobility of the LAO/STO sample as a function of temperature.

performed by shielding the sample from any light for 1 day before the experiment. It should be noted that, although the structures of $LaAlO_3$/SrO-site-terminated $SrTiO_3$ and of $LaAlO_3$/TiO_2-site-terminated $SrTiO_3$ heterostructure are very similar, different electronic properties can be measured. The former gives insulating properties while the latter shows metallic properties. In this thesis, the $LaAlO_3$/TiO_2-site-terminated $SrTiO_3$ heterostructure is investigated; the n-type sheet carrier charge density is measured to be $n_s = 2 \times 10^{13}$ cm^{-2} by a Hall effect measurement. As shown in the figure, sheet resistance decreases with temperature, and the mobility of the interface increases up to $\mu = 10^3$ cm^2 V^{-1} s^{-1} at 2.5 K.

12.3 LAO/STO DEVICE APPLICATIONS

12.3.1 POLAR LIQUID MOLECULE–INDUCED TRANSPORT PROPERTY MODULATION AT THE LAO/STO HETEROINTERFACE

The conducting mechanism behind the LAO/STO interface was reconciled by the intrinsic polar catastrophe model [2,17]. In this model, a proposed charge transfer from the LAO layer (including surface) to the interface is associated with a buildup of the internal electric field in the LAO layer, which originates from the polar discontinuity between LAO (alternating charged layers) and STO (charge-neutral layers). Such an internal electric field, also predicted from first principles calculations for STO/LAO/vacuum stacks [24,25], was evidenced by various experiments [26–28]. For example, an experimental observation of an expansion of the LAO c-axis in ultrathin layers indicated strong electrostrictive effect produced by the dielectric LAO [29]. Recent experimental reports also clearly revealed the importance of the correlation between the interface and the surface charge states, making it possible to manipulate the interfacial conduction using a conducting atomic force microscopy technique [6,30] by handling with the surface charges and adsorbates [31–33], or a capping layer of STO [34]. The fact that the interface conducting states can be highly sensitive to surface adsorbates leads to a strong possibility of LAO/STO interfaces functioning as polar molecule sensors.

Recently, Xie et al. [35] reported that surface adsorption of polar liquids, such as water, induced a huge change (by factor of 3) in the conductivity at LAO/STO interfaces. This conductivity change has been attributed to an increase in sheet carrier density of the 2DEG by more than 2×10^{13} cm^{-2}, suggesting that the adsorbent has a great influence on charge transfer from the film surface to the interface. These results also suggested that sensor application is possible utilizing the surface interface coupling-induced conductivity modulation. In this chapter, the room temperature current–voltage (I–V) relationships of the LAO/STO interface using an in-plane field effect transistor device structure have been characterized, where the polar molecules act as a gate voltage to affect the source/drain current across the interface channel. The local conductivity of the buried interface changes dramatically from metallic to semiconducting when a droplet of the polar liquid such as water is put onto the film surface. These observations provide a direct evidence of the built-in electric field in LAO layer and demonstrate that such a fascinating interface system can be used for high-performance polar molecule sensors.

12.3.1.1 Polar Liquid Sensing

I–V curves of the interface were measured before and after placing a droplet of liquid at the center of the sensor. A variety of liquids, including water, acetone, gasoline, hexane, etc., have been tested. Figure 12.6a shows the I–V curves with and without a water droplet on top of the LAO surface. It can be seen that for the sample tested in air, the I–V characteristic is typically linear with perfect Ohmic behavior, illustrating a metallic conduction at the LAO/STO interface. When a drop of deionized (DI) water is added to the exposed LAO surface, the I–V curves change to a typical field-effect transistor (FET) source–drain type current with a strong saturation at high voltages.

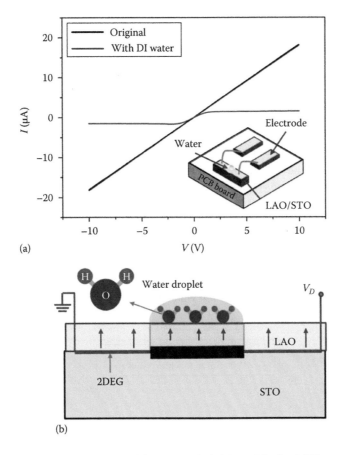

FIGURE 12.6 (a) *I–V* curves of the LAO/STO interface with/without deionized (DI) water droplet on top of LAO surface. The inset is a schematic diagram showing the sensor structure. (b) Schematic diagram illustrating water molecule alignment along the electrostatic field direction on the sample surface. The larger arrow indicates relatively larger polarization field, and the smaller arrow represents smaller polarization field in the LAO layer.

This metal/semiconductor/metal junction-like behavior indicates that there is a strong field effect induced by water molecules that changes the local interfacial conduction from metallic to semiconducting. Since LAO is a polarized ionic oxide with the polarization pointing from the substrate to the surface [35], water molecules, which are polar, have a tendency to align their dipole moments with the electrostatic field to reach the most stable state (Figure 12.1b). As also illustrated in Reference 35, the water molecules' alignment weakens the local electric potential in LAO and thus modifies the band structure of LAO. For samples with water droplet on top of the LAO surface, the weakening of the local electric field in the LAO layer underlying the water droplet results in a transition of the interfacial transport property from metallic 2DEG to n-type semiconducting dominated, as shown in Figure 12.6b. Therefore, a metal/n-type semiconductor/metal junction, analogous to a narrow channel field-effect transistor, can be formed, whose *I–V* characteristic far deviates from linear relation. The liquid was kept on top of the sample surface while the measurements were performed. The blowing off of all visible liquid immediately after placing the droplet recovers the *I–V* curve to its perfect Ohmic behavior with a smaller resistance. After the sample was heated to above 373 K, the linear *I–V* characteristics resumed. It was also noted that nonpolar liquids such as gasoline and hexane have no obvious influence on the *I–V* curve (not shown since they are exactly the same as that without water droplet).

12.3.1.2 Schottky-Diode-Like Junction

It is interesting to see that the 2DEG is sensitive to the position of water droplet on the LAO surface. When the water droplet is placed off from the device center (for this experiment, there is no insulating parylene coating), the I–V curve becomes asymmetric. Figure 12.7a shows a rectified I–V characteristic when the water droplet is close to one side of the electrodes. A Schottky-diode-like junction is realized (Figure 12.7a) in this polar molecule–modulated oxide interface. A schematic diagram is shown in Figure 12.7b to illustrate the mechanism of forming the Schottky junction.

12.3.1.3 Mechanism Discussion

Figure 12.8 illustrates the band diagram qualitatively demonstrating the proposed conducting mechanism across the junctions of 2DEG and n-type semiconducting LAO/STO interface under the water droplet. Figure 12.8a shows the state when water droplet is placed at the center of the sensor without applying a bias voltage V_D, where two Schottky junctions are formed; while under a V_D, the band structure changes, and the current is restricted by the thickness of the depletion layer through a tunneling mechanism (Figure 12.8b). In the case when water droplet is placed at one side closing to the electrode, where one Schottky junction is formed between the metallic and n-type semiconducting LAO/STO interface, the current is rectified. Figure 12.8c and d demonstrates the band structure of this junction. No saturation was observed when positive V_D was applied. This indicates that, for

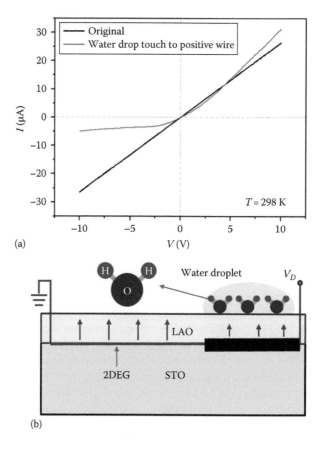

(a)

(b)

FIGURE 12.7 (a) I–V curves of the LAO/STO interface with water on top of LAO surface but close to one end of electrode. (b) Schematic diagram illustrating water molecule alignment along the electrostatic field direction on the sample surface.

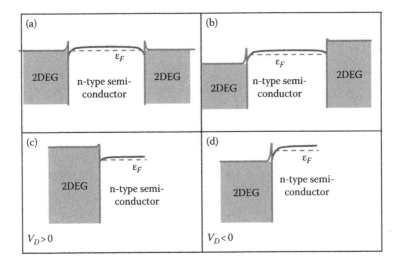

FIGURE 12.8 Schematic band diagrams of Schottky junctions formed along the LAO/STO interface of the sample with water droplet at the center of the sample for (a) zero bias voltage and (b) nonzero bias voltage. (c, d) Schematic band diagrams when water droplet is close to one side of the electrode: (c) when $V_D > 0$ and (d) when $V_D < 0$.

the 2DEG, the Fermi level of the metallic region is higher than that of the n-type semiconducting region, which is different from a typical Schottky diode. The typical charge carrier density of 2DEG at the LAO/STO interface, usually with a value of ~10^{13} cm^{-2}, is much lower than that of normal metals. As a result, a nonnegligible local depletion layer in metallic 2DEG region can be formed near the junction. In this scenario, when $V_D > 0$, as shown in Figure 12.8c (2DEG is negatively biased), the depletion layer in the metallic region is suppressed. Therefore, a quasi-Ohmic behavior can be observed. While for $V_D < 0$, as shown in Figure 12.8d (2DEG is positively biased), the current is likely to be limited due to the locally formed barrier in metallic region. As a consequence, a "rectifier" is realized.

In order to study the tuning effect to the transport property of the 2DEG by water molecules when they change from liquid to solid (ice), the water temperature was cooled below its ice point by adding liquid nitrogen (LN₂) around the sample. As can be seen in Figure 12.9, after the water droplet is frozen to LN₂ temperature, the linear *I–V* characteristic of the interface resumed, but with a larger slope (lower resistance) due to the decreased temperature. By comparing the result to a control sample, where there is no water droplet on the surface, one can see that, at LN₂ temperature, ice has no effect to the *I–V* characteristics. Figure 12.10 shows the curve for the measured current when we freeze the water droplet to ice at LN₂ temperature and then melt ice back to water, where the ice and melting points are indicated. During the current measurement, a fixed bias voltage of 10 V at V_D is applied.

In order to more clearly understand the mechanism of the 2DEG tuning by water molecules at difference temperatures, water temperature is adjusted below its ice point with a cryostat placed around the sample. Figure 12.11 shows temperature-dependent *I–V* curves of the LAO/STO interface with water droplet/ice on top. One can see that when water transforms to ice at a temperature of 260 K, which is lower than ice point, the *I–V* curve shows a hysteresis characteristic, when it is linear when the voltage is less than 20 V, then decreasing when voltage increases. This phenomenon may be due to the melting of the most interfacial layer of water molecules due to Joule heating induced by current; while at a temperature much lower than the ice point, there is negligible hysteresis. Other unknown reasons may also be responsible for this interesting phenomenon, and further

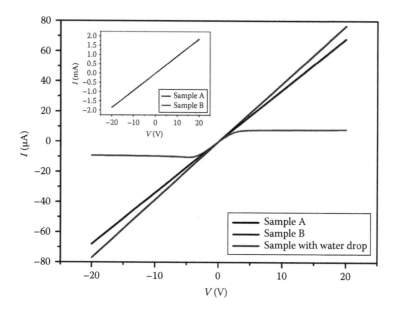

FIGURE 12.9 *I–V* curves of the LAO/STO interface for samples A (control) and B with water on top of the LAO surface. The inset is *I–V* curves of samples A and B at LN_2 temperature, where the water droplet in sample B changes to ice. The two curves are overlapped and indistinguishable.

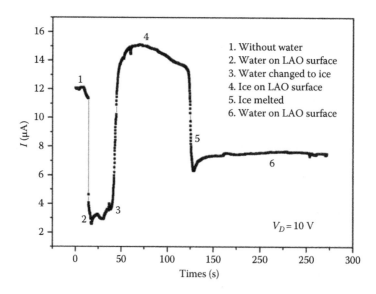

FIGURE 12.10 Current–time curve of the LAO/STO interface with water/ice on top of the sample surface when a fixed voltage is added on V_D.

study is needed. Theoretically, from the first principle calculations, the adsorption of a thin layer of polar water molecules on the surface of $LaAlO_3$ can remarkably enhance the carrier density of the interfacial 2DEG by at least 50%, which is qualitatively consistent with reported experimental results [36].

These results suggest that this type of sensor may be extended to applications in polar gas molecule sensors if a catalytic layer can be coated on the sensor surface. Another application as a biosensor, such as a DNA sensor, can be expected due the fact that many types of DNA carry negative charges.

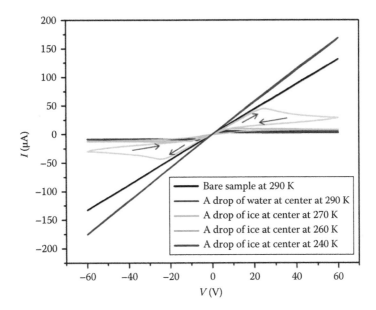

FIGURE 12.11 Temperature-dependent *I–V* curves of the LAO/STO interface with water or ice on top of the LAO surface.

12.3.2 Palladium Nanoparticle Enhanced Giant Photoconductivity at LaAlO₃/SrTiO₃ 2DEG Heterostructure

12.3.2.1 Pd/LAO Film Growth

A LAO film was deposited on a TiO_2-terminated $SrTiO_3$ (001) substrate with a size of 5 mm × 5 mm by laser molecular beam epitaxy (laser-MBE) using a krypton flouride (KrF) Excimer laser with wavelength 248 nm, at a repetition rate of 1 Hz, and a single crystalline $LaAlO_3$ target [37]. During deposition, the substrate temperature was maintained at 750°C with base vacuum lower than 2×10^{-5} Pa; the growth was monitored by reflection high energy electron diffraction (RHEED). In order to reduce oxygen vacancies in the film, after deposition, the samples were annealed *in situ* at a reduced temperature of 550°C at 1000 Pa O_2 for 1 h. The sample was then cooled down to room temperature in the same ambient. After deposition, Hall and sheet resistance measurements were made to characterize the 2DEG nature of the interface. Palladium nanoparticles (NPs) were deposited on LAO/STO surface using DC magnetron sputtering under a power of 15 W at room temperature with Ar gas (99.995%) of 10 sccm flow rate and a base pressure of 100 mTorr. The deposition time was optimized to make sure that the Pd NPs did not form a complete conducting path on the LAO/STO surface. A TEM grid with carbon film was placed close to the sample during the deposition of palladium for subsequent TEM structural analysis (since it is room temperature growth, the Pd NPs on the carbon film can roughly represent the structure of the Pd NPs on LAO surface). High-resolution TEM was used for the nanoparticles analysis. The size of the Pd NPs may affect the experimental result. Figure 12.12 shows the TEM images for the Pd NPs deposited by our magnetron sputtering system for 6 and 15 s. As can be seen in Figure 12.12a and b, the Pd NPs deposited for 6 s gave the best uniformity with sizes close to ~2 nm. Increasing the deposition time of Pd NPs up to 15 s, as shown in Figure 12.12c and d, increases the size of the Pd NPs but is not uniformly distributed (ranging from 2 to 10 nm). Therefore, to minimize the effect of Pd size distribution to the electrical performance, we choose to use Pd NPs with a mean size of ~2 nm. Increasing the time of Pd deposition results in a Pd thin film, which can conduct electricity.

The samples were ultrasonically wire-bounded with Al wires and the two-probe electrical characteristics were measured by the electrometer under a DC bias of 10 V in a stainless steel vacuum

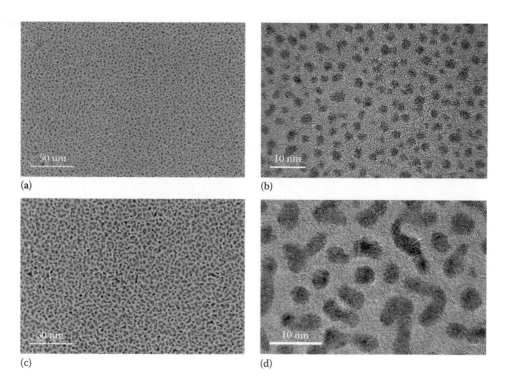

FIGURE 12.12 (a) Low-magnification TEM images of the Pd NPs deposited by magnetron sputtering for 6 s, (b) high-resolution images showing the uniformly distributed Pd NPs with size around ~2 nm, (c) low magnification TEM images of the Pd NPs deposited for 15 s, and (d) the corresponding high-resolution images showing Pd particles with various sizes.

chamber with oxygen, dry air, and argon, respectively. Oxygen and argon gases are of 99.9995% purity. Ultraviolet light source was induced into the chamber for photoresistance measurement. Lights with different wavelengths were generated with a standard system equipped with a mono-chromator and a dual-channel power meter.

LaAlO$_3$ thin films with 5, 10, 20, and 40 uc thicknesses were deposited on TiO$_2$-terminated STO (001) substrates, where the growth conditions were the same as those reported elsewhere [38]. After the deposition, Pd NPs were deposited on top of the surface of the LAO/STO hetero-structure by means of DC magnetron sputtering at room temperature. This room-temperature-grown Pd NPs layer has very weak adhesion and can be removed easily by surface cleaning using cotton tips with methanol, and thus should not lead to an epitaxial growth on the LAO layer. Control samples (LAO/STO) without Pd NPs deposition were also fabricated as a reference. Pd NPs were also deposited on TiO$_2$-site-terminated STO substrate (without LAO) showing that Pd NPs were well below the conduction percolation threshold and no photoconductivity effect was observed.

12.3.2.2 Surface Morphology

Figure 12.13a is an AFM image showing the surface morphology of the Pd NPs–coated LAO/STO sample. The disappearance of surface terrace suggests the coverage of LAO surface by Pd NPs. Figure 12.13b is a HRTEM image of Pd NPs on carbon film, where it can be seen that the mean particle size of the Pd NPs is ~2 nm and particles are uniformly distributed. In order to confirm the existence of Pd NPs on LAO surface and to know their structural characteristics, a TEM carbon film is used together with LAO/STO when coating the Pd NPs. The high-resolution images of the Pd NPs also show clear lattice fringes of Pd, and the inset selected area electron diffraction (SAED)

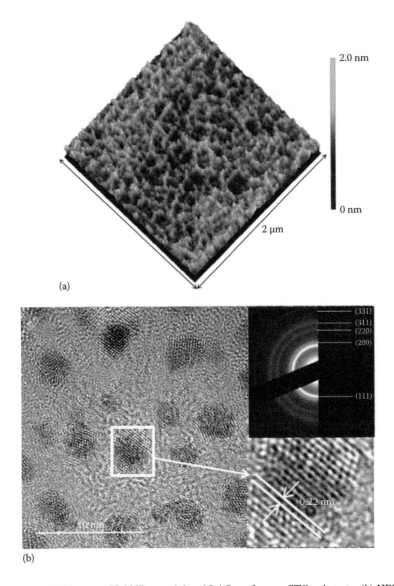

FIGURE 12.13 (a) AFM image of Pd NPs–modulated LAO surface on STO substrate. (b) HRTEM images showing uniformly distributed Pd NPs with an average size of around 2 nm. The lattice spacing of the particles is identified to be around 0.22 nm corresponding to Pd (111) atomic plane. The inset is a SAED pattern showing the polycrystalline structure of the Pd NPs.

pattern can be indexed as the nanocrystalline fcc structure of Pd. (Figure 12.14 shows the surface morphology of the bare TiO₂-terminated STO substrate and LAO/STO surface by AFM where clear terrace can be observed.) Figure 12.15a and c shows the morphology of the Pd NPs on the carbon film, and the energy dispersive x-ray analysis result from Pd NPs is shown in Figure 12.15b and d.

It is noted that, the Pd NPs on top of the LAO/STO surface can be removed easily using cotton tips. Figure 12.15c and d shows the TEM analysis and EDS results for the Pd NPs after removing the Pd NPs by cotton tips. The Pd NPs show the same morphology as what we observed in Figure 12.15a, and the EDX result is same as shown in Figure 12.15b. The actual surface morphology has been studied by preparing cross-sectional TEM samples for the Pd/LAO/STO structure. However, the picture obtained is not very meaningful since during sample preparation, where rinsing and heating the substrates were inevitable, Pd NPs were rinsed away or aggregate during the treatment.

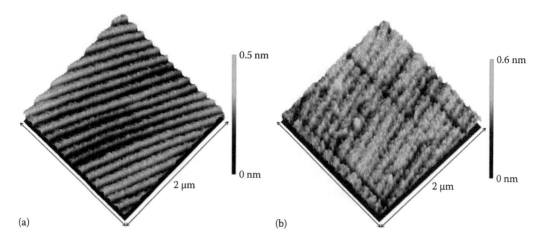

FIGURE 12.14 (a) Bare TiO_2-site-terminated STO (001) substrate. (b) Surface morphology of the surface after 5 uc LAO thin film deposition; clear terraces can still be observed.

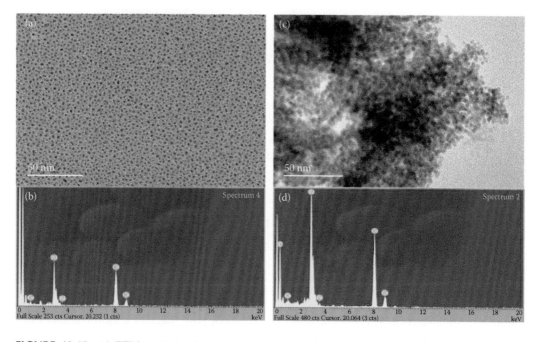

FIGURE 12.15 (a) TEM analysis of the Pd NPs on the carbon grid showing uniform distribution. (b) Elemental analysis was performed with energy dispersive system (EDX); the two significant peaks at low energy correspond to C-Kalpha and O-Kalpha. The existence of palladium atoms in the nanoparticles can be observed in the peak listed in 2.4 eV. (c) TEM analysis of Pd NPs removed with cotton tips from the sample's surface. (d) The corresponding EDX analysis.

12.3.2.3 Photoresponse under UV Light and Room Light

With $LaAlO_3$ surface modification by Pd NPs, $LaAlO_3/SrTiO_3$ (LAO/STO) interfacial 2DEG presents giant optical switching effect with photoconductivity on/off ratio as high as 750% under UV light irradiation for wavelength shorter than 400 nm. This giant optical switching behavior can be explained by the Pd nanoparticle's catalytic effect and surface/interface charge coupling, which will be discussed in the following section.

Photoresponses of bare (i.e., without Pd NPs coating) and Pd NPs–coated LAO/STO samples under a repeatable on/off UV light (365 nm) are shown in Figure 12.16. It can be seen that when the UV light is turned on, both samples change from a high-resistance state to low-resistance state; when the light is off, the resistance of the interface increases as a function of time. The resistance of the interface decreases within 0.1 s after the UV light is irradiated on both samples, which is then followed by a slow photorecovery process. The increase in resistance of the Pd NP–coated sample at each time interval is much higher than the bare sample. Within 2 min after the light is switched off in each cycle, the resistance is yet to be saturated; this can be attributed to the persistent photocon-ductivity effect [39–41]. This result reveals that after the surface modification by Pd NPs, the LAO/STO interfacial 2DEG becomes more sensitive to UV light, making it possible for its application as UV light sensor.

The interface of the Pd NPs–coated LAO/STO sample is also very sensitive to UV radiation from tube lamps, and Figure 12.17 shows the typical photoemission of the conventional tube lamp, where UV radiation with wavelength lower than 365 nm can be observed. Figure 12.18 shows the response of the Pd NPs–coated LAO/STO sample under room light, and around 30% change in resistance can be observed under on/off room light conditions. Overexposure to fluorescent tube lamp (FT) was found to be harmful to humans [42,43]. The phosphor coating inside a typical fluorescent tube lamb absorbs UV radiation and emits a broad range of visible light; however, not all UV radiation can be absorbed by this phosphor coating.

12.3.2.4 Wavelength-Dependent Photoconductivity

To understand the mechanism of such Pd NPs–induced enhancement effect, wavelength and time dependences of the photoconductivity for the two samples were characterized. Figure 12.19 illus-trates the wavelength-dependent photoconductivity of the LAO/STO samples (with and without Pd NPs on surface) at room temperature. The photoconductivity of the heterostructure was measured at 10 V bias, and the samples were irradiated by light with wavelength ranging from 700 to 300 nm (from low photon energy to high photon energy). The samples were put in the dark environment before measurement and the measurement started after the resistance reached a stable state. It is noted that, in the dark environment, the resistance of the Pd NPs surface–modulated LAO/STO heterostructure (~3 MΩ) is much higher than that of the bare LAO/STO 2DEG, which is ~0.25 MΩ. The change in the photoconductivity is defined by $(I_{photo}-I_{dark}/I_{dark}) \times 100\%$, where I_{photo} corresponds

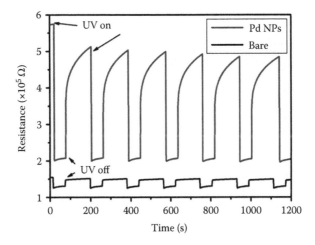

FIGURE 12.16 Photoresponse characteristics of the Pd NPs–coated LAO/STO (red) and LAO/STO (black) heterostructure showing the reversible switching behavior under periodic illumination of a 365 nm UV light with incident power densities 10 mW/cm^2.

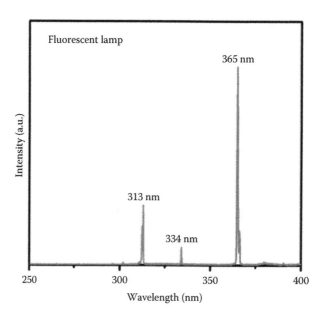

FIGURE 12.17 Photoemission spectra of the conventional tube lamp, where UV emission with wavelength lower than 365 nm can be observed.

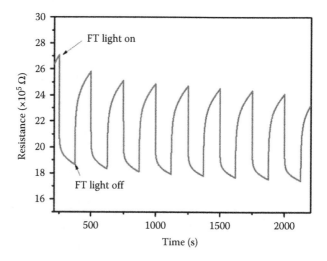

FIGURE 12.18 Photoresponse characteristics of the Pd NPs–coated LAO/STO heterostructure showing the reversible switching behavior under periodic illumination of a fluorescent (FT) light.

to the photocurrent under light illumination and I_{dark} corresponds to the dark current. Strong photoconductivity is observed at ~380 nm, while the increase in photoconductivity for both samples in the range of wavelength 420–700 nm is not obvious. The change in photocurrent for the bare sample at 380 nm is only 18% (a small increase in 420–700 nm is due to the surface trap state), but for the Pd NPs–coated LAO/STO sample, the change is 750%, which is a giant photoconductivity. It is worth noting that the 380 nm UV light is around the band-gap energy of STO substrate (3.3 eV), suggesting that the increase of photocurrent is due to the absorption of photon energy by STO and the generation of electron–hole pairs; while the electrons diffuse to the interface as a contribution to

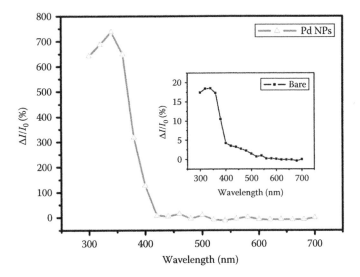

FIGURE 12.19 Wavelength-dependent photoresponse, that is, change of photocurrent for the Pd NPs–coated and bare (inset) LAO/STO samples under 10 V bias.

the carriers. By contrast, for the photons with energy below the bandgap energy, only very limited number of charge carriers can be generated corresponding to the mid-gap defect states of the STO substrate from oxygen vacancies during the deposition of the thin film.

12.3.2.5 Time-Dependent Resistance of the LAO/STO Interface

In order to investigate the effect to photoconductivity in the UV range, time-dependent resistance of the interface for the bare and Pd NPs–coated LAO/STO samples were measured at different ambient conditions (oxygen, dry air, and argon). To circumvent the scattering of the data from sample to sample and rule out the effects of the unwanted experimental factors, the Pd NPs–coated LAO/STO heterostructure was prepared from the same piece of sample and the Pd NPs were grown under the same condition for all measurements. As shown in Figure 12.20a, for the recovery process of the LAO/STO sample, the resistance of the 2DEG increases less than 30% in an hour after switching off the UV light in the oxygen and argon environments, suggesting that the change of gas does not affect the transport properties of the interface. However, very different electrical responses for the Pd NPs–coated LAO/STO sample under oxygen, dry air, and argon were observed, as shown in Figure 12.20b. With argon gas in the chamber, resistance of the Pd NP/LAO/STO sample recovers very slowly, with a recovery speed similar to that in the bare sample, as shown in Figure 12.20a. When the gas ambient is changed from argon to dry air, the speed of the recovery process increases as the result of oxygen adsorption where Pd NPs are treated as the available surface adsorption sites for oxygen molecules. Further studies show that recovery of resistance is even faster under pure oxygen gas because more oxygen molecules are available for chemisorption on the Pd NPs surface. Figure 12.20a and b shows the significant difference of electrical response between the Pd NPs–modulated and the bare LAO/STO samples, suggesting enhanced photoconductivity upon UV light illumination, that is, the Pd NPs on the LAO/STO heterostructure significantly enhance the photoconductivity of the interface.

The curve under a different environment has been fitted using the stretched exponential relaxation law as listed in Equation 12.1.

$$I_{photo}(t) = I_{photo}(0)\exp-\left(\frac{t}{\tau}\right)^{\beta} \qquad (12.1)$$

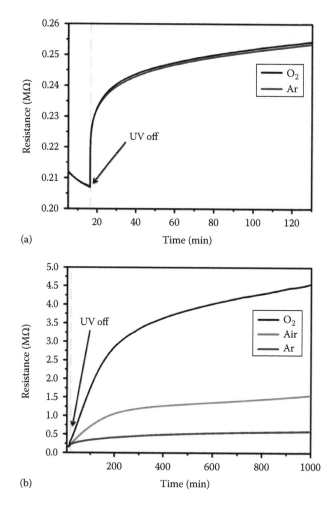

FIGURE 12.20 Photorecovery process in oxygen, dry air, and argon in bare (a) and Pd NPs-coated (b) LAO/STO samples at room temperature.

The results are shown in Figure 12.21, and the fitting parameters are listed in Table 12.2, where $I_{photo}(t)$ is the photocurrent of the interface at specific time t, τ is the characteristic relaxation time, and β is the stretching parameter with value between 0 to 1. For β close to a value of 1, it shows a Debye-like behavior, a model used to describe relaxation behavior. The rate of change of the resistance is initially rapid, but it becomes continually slower as time progresses.

12.3.2.6 Current Voltage Characteristics under Different Ambient

The I–V characteristics for the bare and Pd NPs–coated LAO/STO samples were measured under the same condition. Before the measurement, both samples were put in the dark for 24 h. The I–V curves for bare LAO/STO sample in the dark and under UV light are shown in Figure 12.22a, where ohmic conducting behavior of the interface can be seen, and the resistances were found to be 26 kΩ in the dark and 20 kΩ under UV light. The I–V curve of the bare sample measured under normal room light illumination is almost the same as shown in Figure 12.22a, showing less change in resistance. Figure 12.22b shows the I–V curves for the Pd NPs–coated sample measured in the dark, room light, and UV irradiation. It is apparent that the I–V curves will exhibit very large changes in slope under different illuminations, indicating a significant change in resistance at the interface.

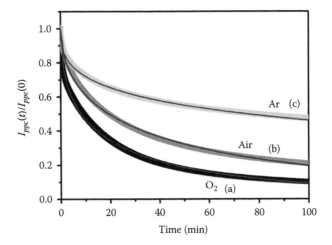

FIGURE 12.21 Photorecovery process in (a) oxygen, (b) dry air, and (c) argon. I_{ppc} refers to the persistent photocurrent.

TABLE 12.2
Characteristic Relaxation Time τ and Stretching Parameter β Fitted by $I_{photo}(t) = I_{photo}(0)\exp{-(t/\tau)\beta}$ in Different Ambients

	τ (min)	β
In oxygen	17.47	0.498
In dry air	39.57	0.520
In argon	201.94	0.371

12.3.2.7 Mechanism Discussion

To explain this enhanced photoconductivity effect of the Pd NPs–coated sample, we propose the following model (as shown in Figure 12.23). It is believed that the presence of Pd NPs and the resultant charge coupling and exchange with the oxygen-deficient LAO film facilitate the interfacial redox reaction at room temperature [10,44–46]. From the view point of electronic band structure, the work function of the Pd NPs relative to the electron affinity of STO determines the transfer of electrons from the interface to the Pd NPs. Metals such as Pd with work function (5.6 eV) larger than the electron affinity of STO (4.0 eV) result in a strong suppression of 2DEG carrier densities at the interface due to the reduction of an internal built-in electric field in the LAO layer; a similar effect has been observed in a recent first principle calculation [47–49]. In addition to this Pd NPs electron affinity–induced degradation of the 2DEG, the Pd NPs catalytic electrochemical reaction with oxygen molecules is believed to play a more important role in reducing the sheet carrier density (n_s) of the LAO/STO 2DEG. The measured n_s for bare sample is ~6.5 × 10^{13} cm^{-2} under UV irradiation and ~1.2 × 10^{13} cm^{-2} in the dark; while for the Pd NPs–coated LAO/STO, n_s under UV irradiation is measured to be ~3.9 × 10^{13} cm^{-2} and in the dark it is ~3.4 × 10^{12} cm^{-2}.

To compare the conducting behavior between the bare LAO/STO and Pd NPs–coated LAO/STO samples, the resistance–temperature ratios of the samples were measured from room temperature down to 15 K in the "dark" and "under UV irradiation" environments; the UV light with wavelength 365 nm was irradiated on the sample during cooling, and the results are shown in Figure 12.24.

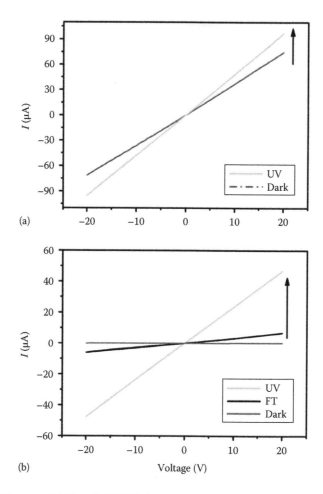

FIGURE 12.22 *I–V* curves of (a) bare LAO/STO interface in the dark and under UV light and (b) Pd NPs–coated LAO/STO in the dark, under fluorescent tube lamp (FT), and UV light.

One can see from Figure 12.24a and b that the bare LAO/STO sample shows metallic behavior under UV irradiation and the dark environment. For the Pd NPs–coated sample, the interface shows metallic behavior under UV irradiation as shown in Figure 12.24c, while in the dark (as shown in Figure 12.24d), the conducting behavior of the interface changes to insulating. This result suggests that the presence of metallic elements on the LAO layer affects the electrical properties of the interface. It is well known that Pd NPs are very sensitive to the surrounding environment due to the large surface-to-volume ratio and the Pd NPs surface usually absorbs oxygen molecules when exposed to air [50,51]. The Pd NPs catalytically activate the dissociation of molecular oxygen, and due to the catalytic effect, the interaction between oxygen molecules from the ambient and Pd NPs results in the dissociation of O_2, that is, $O_2 + 2e^- \rightarrow 2O^-$. A similar phenomenon has been observed in Pd NPs–coated ZnO systems [51,52].

It was proposed that when oxygen molecules from the ambient were absorbed on the exposed surface of Pd NPs, electrons were extracted from the ZnO conduction band, and depletion layers were formed at the surface of ZnO, causing a decrease in carrier concentration. In the case of the LAO/STO system, we believe that oxygen molecules dissociate and generate chemisorbed oxygen species due to the enhanced interaction by Pd NPs on the surface, and the oxygen molecules tend to capture free electrons from the Pd NPs and cause interfacial 2DEG electrons to diffuse out to the LAO film or the Pd NPs. Therefore, with the absorption of oxygen on the Pd NPs surface, O^- tends

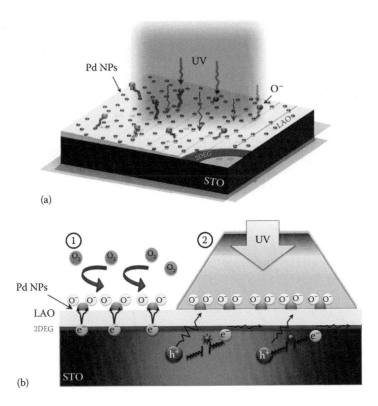

FIGURE 12.23 (a) Schematic diagram of the Pd NPs–surface modulated LAO/STO with UV light irradiation. (b) Schematic depiction of the processes: *Process 1*—oxygen molecules absorbed on Pd NPs attract electrons from 2DEG, making it negatively charged; *Process 2*—UV light irradiates on the surface and excites electron–hole pairs inside STO, the photoexcited electrons become the excess charge carriers to the 2DEG interface, and photoexcited holes diffuse through LAO and combine with the negatively charged oxygen on the surface of LAO.

to "spillover" the surface of the LAO [53], resulting in more negative charge carriers being trapped on the LAO surface, and, therefore, the carrier density of the interfacial 2DEG decreases and the resistance of the interface increases.

On the other hand, when UV light irradiates on the surface, more electrons are induced in the 2DEG interface due to the generation of photon-excited electrons and holes in the STO substrate (suggested by the fact that the photocurrent starts to increase when the photon energy exceeds the band-gap energy of STO). The persistent increase of the photoresistivity when light is off is due to the slow recombination of electrons and holes [39]; the slow recombination process is believed to be due to the internal electric field buildup from the polar discontinuity in the LAO layer. This field tends to separate electrons and holes (electrons move toward the interface and holes move toward the surface), and, therefore, the recombination is generally prohibited; but the interdiffusion of electrons and holes still results in slow recombination.

Electrical measurement were conducted on samples with Pd NPs coated on the 10 and 40 uc LAO/STO, and the electrical performance of the sample varies with testing condition as shown in Figure 12.25. Compared to the 5 uc Pd NPs–coated LAO/STO sample. It is apparent that the 10 and 40 uc thick samples show a slower and smaller photoresponse, as per the data given in Table 12.3. It is believed that the thicker LAO film hinders the diffusion of electrons to the Pd NPs due to the increased diffusion length and so the ratio of resistance recovery is not as high as in the case of the 5 uc.

In summary, the Pd NPs modification of the LAO/STO heterostructure surface exhibits a large oxygen-sensitive photoconductivity at room temperature. This interesting phenomenon is attributed

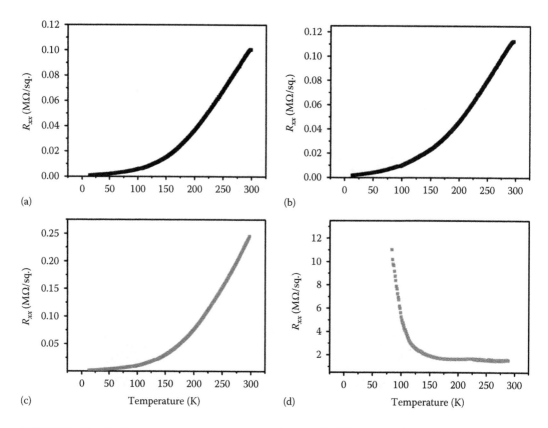

FIGURE 12.24   Resistance-temperature curve of the bare LAO/STO sample (a) under UV and (b) in the dark, for Pd NPs–coated LAO/STO samples (c) under UV and (d) in the dark.

to Pd NPs' catalytic effect and surface/interface charge coupling. These results are interesting in physics and can probably be used for sensor applications such as UV light sensing and gas sensing. The persistent photocurrent can be attributed to the existence of internal field in the LAO layer. These also provide evidence for the polar catastrophe model for the explanation of conducting inter-face, as discussed widely in the literature.

12.3.3   Highly Sensitive Gas Sensor by the $LaAlO_3/SrTiO_3$ Heterostructure with Pd NPs Surface Modulation

12.3.3.1   Gas Sensing Characteristics of Pd NPs–Coated LAO/STO

Pd NPs with size around 2 nm were deposited by DC magnetron sputtering in pure argon ambient on the surface of LAO/STO heterostructure at room temperature [54]. Figure 12.26a schematically shows the device structure, and the actual picture of the sample is shown in Figure 12.26b, where the aluminum wires are ultrasonically bonded at the four corners for electrical characterization. It is apparent that the Pd NPs, which appear as dark spots, are distributed uniformly and densely on the surface of the LAO layer. The Pd NPs are crystallized and the density does not lead to a complete conduction percolation path. The high-resolution image shown in Figure 12.26c and the ring-shaped selected area electron pattern (SAED) shown in Figure 12.26d indicate the nanocrystalline fcc structure of the Pd NP. The small size and high density distribution of the nanoparticles should enhance the responsivity to the targeted gases [55,56].

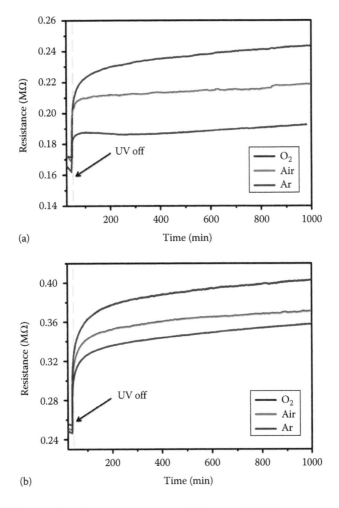

FIGURE 12.25 Thickness-dependent response to gases (a) 10 uc and (b) 40 uc.

TABLE 12.3
Change of Resistance in Different Gas Environments

	5 uc (%)	10 uc (%)	40 uc (%)
Bare LAO/STO	42	13	27
In argon	130	18	45
In dry air	633	29	48
In oxygen	2071	41	57

12.3.3.1.1   Effect of Gas Concentration

Figures 12.27 through 12.29 show the hydrogen gas sensing characteristics for the Pd NPs–coated LAO/STO heterostructure with thickness of 5 uc at room temperature and elevated (80°C) temperature, where it is of particular interest that the conductance of the sample is extremely sensitive to H$_2$ gas. It is noticed that our pristine LAO/STO heterostructure does not show any response to H$_2$ gas as reflected from its current–voltage characteristics, and there is no experimental report either to show the gas sensing ability of such heterostructure. The current–voltage characteristics for the

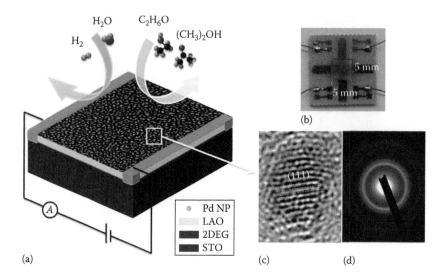

FIGURE 12.26 (a) Schematic diagram of the Pd NPs–coated LAO/STO heterostructure (the surface is a low-magnification TEM image representing the Pd NPs), (b) actual picture of the device, (c) high-resolution TEM image of a single Pd NP, and (d) SAED pattern of the Pd NP.

bare LAO/STO and Pd NPs–coated LAO/STO sample at 20 ppm H_2 and pure oxygen are shown in Figure 12.30. Figure 12.30a shows the I–V response of the interface, where typical ohmic behavior with same resistance can be seen in the O_2 and H_2 ambient, suggesting that the change of gas ambient does not affect the electrical properties of the bare LAO/STO interface. However, as shown in Figure 12.30b, the electrical response of the Pd NPs–coated LAO/STO interface changes after the exposure to H_2 and O_2, and the I–V response shows a decrease in resistance for the 5 uc sample after exposure to 20 ppm H_2 gas. This suggests that the response to H_2 gas is mainly attributed to the presence of Pd NPs on the surface of LAO/STO.

It is believed that the reason H_2 gas sensing in LAO/STO 2DEG has not been observed before is due to the low reactivity of the LAO surface to any gas. The Pd NPs on the LAO surface, therefore, play a crucial role in the functionalization of H_2 gas sensing. It is also worth mentioning that the response to a gas is an interfacial effect, since the experiment with surface electrode contact without connecting to the interface does not show any measurable conductance and response upon exposure to hydrogen.

For a typical gas sensor, three physical parameters are usually studied for the sensing performance, including sensitivity $S(\%)$, response time τ_R, and recovery time τ_s. Among them, $S(\%)$ is defined as the relative variation of the current at the interface and can be calculated as $S = \Delta I/I_0$ (%), where $\Delta I = I_g - I_0$ (I_0 is the current of the interface before exposing to hydrogen and I_g is the peak current in presence of hydrogen). The parameters of τ_R and τ_s are defined respectively as the time reach 90% of the saturated current of full response and recovery.

The transient response of the gas sensor to hydrogen at room temperature is shown in Figure 12.27a, illustrating the variations in current in the interface corresponding to On (for 5 min) and Off of H_2 gas flow for two cycles with 20 ppm H_2 gas concentration. The results show that the interfacial current quickly increases when the device is exposed to H_2 gas, with $\tau_R \sim 7.3$ min, followed by a slow approach to saturation. After the current in the interface saturated, the hydrogen gas inside the chamber was pumped out and synthetic air was flushed to the sample for the recovery process; the current of the interface decreased and essentially resumed the original state with $\tau_s \sim 36.7$ min. The responses of the Pd NPa–coated LAO/STO heterostructure measured at room temperature was also recorded in the environment with hydrogen gas at different concentrations

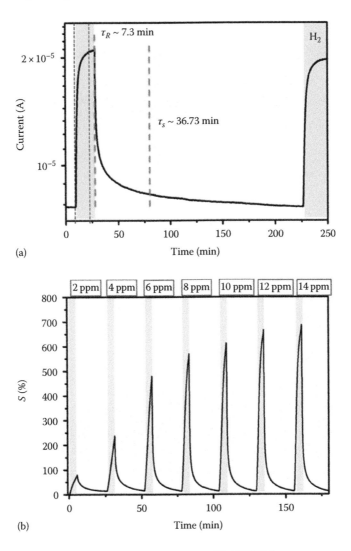

FIGURE 12.27 (a) Electrical response for the Pd NPs–coated LAO/STO heterostructure exposed to 20 ppm H$_2$ gas loading and de-loading cycle at room temperature. (b) Real-time response of sensitivity under exposure to H$_2$ gas with various concentrations (2–14 ppm) at room temperature.

(from 2 to 14 ppm). The results shown in Figure 12.27b are plotted as the relationship between sensitivity and time. A detectable change in current of ~80% was achieved for the sample with hydrogen concentration as low as ~2 ppm. It is apparent that exposing the sensor to a higher concentration of hydrogen gas induces faster increase of the current to a higher peak value. These sawtooth responses are superimposed on a monotonically increasing background; this fact can be attributed to the charge accumulation and slow recovery rate of the sensor operated at room temperature. Nevertheless, it shows that the Pd NPs–modulated LAO/STO heterostructure is very sensitive to H$_2$ gas.

Temperature is an important factor that greatly influences the hydrogen sensing response based on the catalytic effect. Usually, higher temperature would lead to higher sensing performance due to the lowering of activation energy for gas adsorption and desorption of devices. Figure 12.28a shows the cyclic performance of the device with 20 ppm H$_2$ operated at 80°C, where the sensor shows both shorter, τ_R ~ 0.8 min, and longer, τ_s ~ 0.95 min, response times, which are much faster compared to

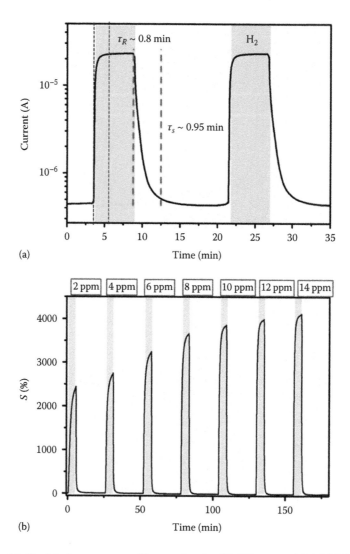

FIGURE 12.28 (a) Electrical response to H_2 gas for the Pd NPs–coated LAO/STO heterostructure. (b) Sensitivity response at 80°C.

the sensor operated at room temperature. The shorter response time can be explained by faster diffusion and dissociation of hydrogen molecules at higher temperature, where the surface has higher reactivity and more active sites for H_2 molecules adsorption to Pd NP. Figure 12.28b shows the sensor's response to H_2 gas with different concentrations, where it can be seen that at higher temperature, the sensitivity is much higher. A response with sensitivity up to 2400% has been recorded for a device exposed to 2 ppm H_2 gas.

As shown in Figure 12.29a, the response time τ_R of the Pd NPs–coated LAO/STO heterostructure at 80°C generally tends to be shortened (for $[H_2] = 2$ ppm, $\tau = 3.2$ min and for $[H_2] = 20$ ppm, $\tau = 0.8$ min). Upon exposure to dry air for the recovery process, dissolved hydrogen on Pd NPs reacts with the oxygen in air and forms H_2O by the reaction of $2H_2 + O_2 = 2H_2O$. At relatively low working temperatures, the response time is longer due to slow desorption of the water molecules formed on the surface. The sensor was able to detect H_2 gas with low concentration at high temperature, while it was relatively less sensitive to low H_2 concentrations at room temperature. Figure 12.29b plots the

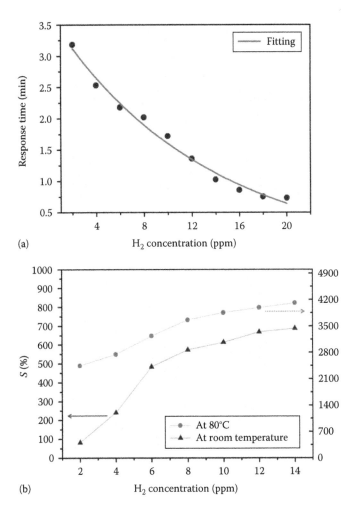

(a)

(b)

FIGURE 12.29 (a) The relationship between response time and hydrogen concentration for a sensor operated at 80°C. (b) Combined sensor sensitivity at room temperature and 80°C.

sensitivities for the sensor to various concentrations of H$_2$ gas at room temperature and at 80°C. It is apparent that at low H$_2$ concentration, the sensitivity increases with the increase of H$_2$ concentration; while at high concentration, it begins to saturate probably due to the lack of adsorption sites on the Pd NPs surface.

12.3.3.1.2   Effect of Oxygen Concentration in Hydrogen Gas

Instead of testing the sensor using H$_2$ gas balanced in argon, the sensor was found to be able to detect H$_2$ gas balanced with air as well, where coexistence of oxygen gas (oxidizing gas) during hydrogen gas (reducing gas) detection should play a very important role in the sensing characteristics of the Pd NPs–coated LAO/STO heterostructures. Figure 12.31 shows the time-dependent interfacial current of the Pd NPs–coated LAO/STO sensor exposed to various concentrations of H$_2$ in synthetic air (10, 40, 380, and 3500 ppm) at 80°C, with oxygen gas recovery after each H$_2$/air measurement. With the introduction of the gas mixture to the sensor, the current of the interface increases, while it also increases with the increase in hydrogen concentration, showing a faster increase in current to a higher peak value.

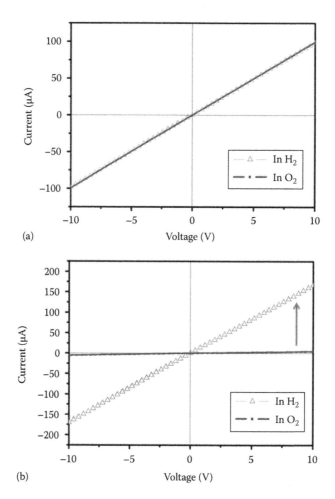

(a)

(b)

FIGURE 12.30 Current–voltage characteristics in H_2 and O_2 ambient of the (a) bare LAO/STO and (b) Pd NPs–coated LAO/STO samples.

12.3.3.1.3 Effect of LAO Thickness

It is of particular interest to show the sensors' response to H_2 gas for the samples with different LAO thicknesses, so a series of samples with different LAO thicknesses (10, 20, and 40 uc) and the same configuration were tested. Figure 12.32 shows the response curves of the sensors to different hydrogen concentrations at room temperature and at 80°C. All the samples respond to hydrogen gas and the responses tend to saturate at a high hydrogen concentration. Compared to the sample of 5 uc LAO thickness, smaller and slower gas responses for higher LAO thicknesses have been observed. It is believed that a thicker LAO film reduces the field effect and hinders the diffusion of charge carrier to the Pd NPs with the increase in diffusion length, resulting in the heterostructure's lower sensitivity to H_2 gas.

12.3.3.2 Response to Other Gases

The working mechanism of the gas sensor lies in the conversion of electrical conductivity due to surface reactions such as oxidation or reduction that is caused by different gas exposure. Oxidizing (reducing) gases can serve as electron-withdrawing (donating) groups and change the channel carrier concentration through charge coupling with the interface; while the adsorbents from the ambient gas molecules interact strongly with the Pd NPs–coated LAO/STO heterostructure. In addition

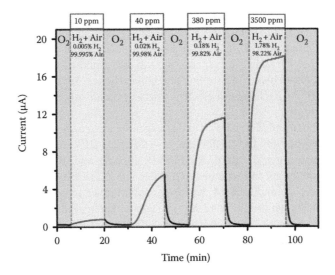

FIGURE 12.31　Time-dependent interfacial current changes of the Pd NPs–coated LAO/STO sensor exposed to various concentrations of H$_2$ gas in synthetic air (10, 40, 380, and 3500 ppm) at 80°C and then back to O$_2$ gas for recovery process. The concentration of the H$_2$/air gas was monitored by using the precalibrated mass flow controllers. Hydrogen gas and synthetic air were introduced to the gas mixer by a two-way valve using two separate mass flow controllers, with the test gases allowed to flow through the test chamber with the sensor installed.

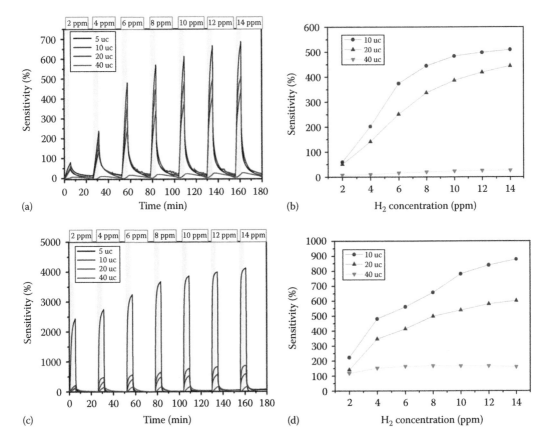

FIGURE 12.32　Real-time sensitivity response of Pd NPs–coated LAO/STO heterostructure with different LAO layer thicknesses at (a, b) room temperature and (c, d) 80°C.

to H_2 gas, other reducing gases such as ethanol, acetone, and water vapor were tested. The sensor was put in the argon ambient to obtain the base line and a fixed amount of the chemical vapor was flowed to the sensor for sensitivity test. As shown in Figure 12.33a and b, sensitivities to acetone and ethanol gases reached ~64% and 48%, respectively, and the response times were $\tau_{R,acetone}$ ~ 25 min and $\tau_{R,ethanol}$ ~ 8 min, respectively, for acetone and ethanol gases. We believe that this gas-sensing characteristic is due to the exchange of electrons between ionosorbed species and nanoparticles, as well as the LAO film.

The reaction of ethanol and acetone with ionic oxygen species can be described by $CH_3CH_2OH + 6O^- \rightarrow 2CO_2 + 3H_2O + 6e^-$ and $CH_3COCH_3 + O^- \rightarrow CH_3CO^+CH_2 + OH^- + e^-$. The reductive gas reduces the molecular oxygen on top of the Pd NPs surface, and, thus, the electron concentration at the interface increases and resumes to a stable state. The behavior of the device to water vapor is similar to acetone and ethanol, however, much higher sensitivity and shorter response time are observed. As shown in Figure 12.33c, the conductance of the Pd NPs–coated LAO/STO heterostructure increases with sensitivity up to ~8000% after exposure to H_2O vapor at room temperature.

The H_2O molecules should remove adsorbed oxygen on the LAO surface. As mentioned by Xie et al. [35], H_2O are known to alter the charge carrier density of the LAO/STO interface. Sensing of the H_2O molecules with reasonable value of sensitivity and slow response and recovery have been reported, and H_2O molecules were proposed as the major polar gases in the air that can influence

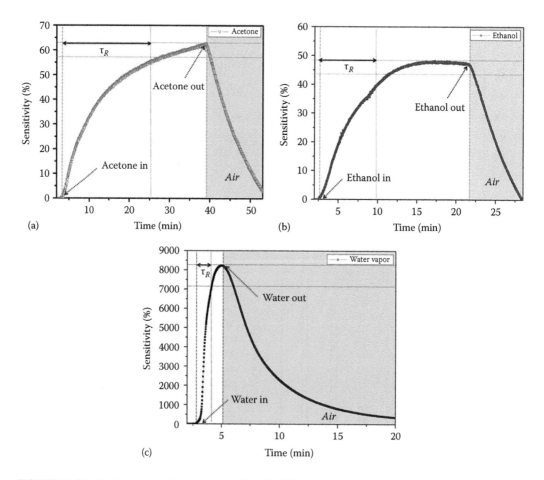

FIGURE 12.33 Real-time sensitivity response of the Pd NPs–coated LAO/STO heterostructure to (a) acetone, (b) ethanol, and (c) water vapor in argon ambient at room temperature.

interface conductance. One probable reaction is that water vapor can be dissociated on the Pd surface giving rise to H$^+$ and OH$^-$ ions, that is, H$_2$O↔H$^+$ + (OH)$^-$. A first principle calculation shows that H$_2$O binds strongly to the AlO$_2$ outer surface [57,58] and modulates the conductivity at the interface. However, as dry air is fed in for recovery process, the recovery rate is much slower than the case in acetone and ethanol. Water molecules adsorbed on the Pd NPs surface with slower evaporation rate lead to lower surface adsorption for chemisorptions of oxygen species, resulting in a slower recovery rate.

12.3.3.3 Mechanism Discussion

A model (as shown in Figure 12.34) has been proposed to explain the enhanced sensitivity to gases for the Pd NPs–coated LAO/STO heterostructure. Palladium has been used in sensing applications due to its high diffusion coefficient, solubility, and selectivity with respect to hydrogen [59]. As shown in the diagram, direct adsorption of the gaseous molecules or the following by-products formed by the reaction on the Pd NPs surface occur on top of the LAO layer. The presence of Pd NPs results in charge coupling and exchange with the oxygen-deficient LAO film, which facilitates the interfacial redox reaction of oxygen [44,46,60], where the sensing mechanism can be explained in terms of oxidizing/reducing gas effect. Therefore, the surface is active and promotes further adsorption of oxygen from the atmosphere due to the presence of Pd.

The surface of the Pd contains a number of oxygen-related physisorption species such as O$_2^-$ and O$^-$, which "spillover" on the surface. On the other hand, when the heterostructure is exposed to hydrogen gas, hydrogen acts as the reducing gas, which decreases the concentration of oxygen species on the surface and eventually increases the concentration of the charge carrier at the interface. Electrically, the work function of the Pd metal relative to the electron affinity of STO determines the transfer direction of electrons between the 2DEG to the Pd NPs. The presence of Pd metal on the surface with work function 5.6 eV is larger than the electron affinity of STO (4.0 eV), leading to a strong suppression of 2DEG carrier densities at the LAO/STO interface due to a reduction in the builtin electric field in the LAO layer; a similar phenomenon has been observed based on the first principle calculation in the LAO/STO system [47,48].

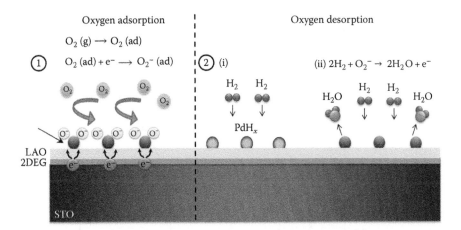

FIGURE 12.34 Proposed mechanism for H$_2$ gas sensing process: *Process 1*: the oxygen molecules are adsorbed on the Pd NPs and attract electrons from the 2DEG, making the Pd NPs negatively charged, which decrease the charge carrier concentration of the interface; *Process 2*: (i) Pd NPs changes to palladium hydride (PdH$_x$) after exposure to hydrogen gas and apparently decrease the work function of Pd, which lowers the barrier height of the interface. (ii) The atomic hydrogen reacts with the oxygen molecules adsorbed on the Pd NPs surface, the hydrogen molecules react with the adsorbed oxygen species, 2H$_2$ + O$_2^-$ (ad) → 2H$_2$O + e$^-$, and the as-released electrons enhance the conductivity of the interfaces.

As reported, the hydrogen-sensing response of semiconductor-based gas sensors with noble metal electrodes is related to Schottky contact between the noble metal electrode and the heterostructure [53,61,62]. The hydrogen molecules may be adsorbed and dissociated into hydrogen atoms on Pd surface and dissolve into the Pd bulk; and consequently the Pd NPs change to palladium hydride (PdH$_x$) that apparently decrease the work function of Pd and lower the barrier height, resulting in less suppression of the carrier density at the interface, and, therefore, fewer electrons from the interface may transfer to (PdH$_x$) [63,64].

In addition to the Pd NPs electron affinity–induced modulation of the interfacial conductivity, the Pd NPs catalytic electrochemical reaction with oxygen molecules and the consequence of charge exchange with the LAO film is believed to play a more important role in reducing the charge carrier density of the LAO/STO 2DEG. The Pd NPs surface usually absorbs dissociated oxygen molecules when exposed to air [50]. It is known that the transport properties of the Pd NPs–coated LAO/STO heterostructure are strongly affected by the gas environment, especially oxygen, and the conductance of the interface decreases when the device is exposed to oxygen ambient [37]. The Pd NPs act as catalysts that activate the dissociation of molecular oxygen [50], and, therefore, the presence of ambient oxygen has a considerable influence on the performance of Pd NPs–coated LAO/STO 2DEG.

When the sensor is exposed to H$_2$ gas, the hydrogen molecules react with the adsorbed oxygen species, $2H_2 + O_2^- \text{ (ad)} \rightarrow 2H_2O + e^-$, and the as-released electrons will enhance the conductivity of the interface. Afterward, when air is introduced into the device, oxygen molecules in air react with the surface adsorbed-hydrogen atoms to form H$_2$O and evaporate due to exothermic reaction. This reaction process is exothermic and the produced H$_2$O molecules are desorbed quickly from the surface. A depletion region at the interface will be rebuilt by the adsorbed oxygen species on the Pd surface and results in O$_2^-$ or O$^-$ formation.

To study the importance of the presence of oxygen species for the sensing response, three carrier gases, synthetic air (~20% oxygen balanced in 80% argon gas), pure oxygen, and argon, were selected to study the behavior of recovery process after the sensor was exposed to hydrogen. As shown in Figure 12.35, the resistance of the LAO/STO 2DEG has been measured for the sensor exposed to different carrier gases.

First, measurement began with the sensor that was exposed to pure oxygen ambient environment to obtain a baseline (0.83 MΩ). Second, the resistance of the sample decreased to ~0.17 MΩ when 20 ppm hydrogen gas was introduced for 5 min. Third, synthetic air was introduced to the sample

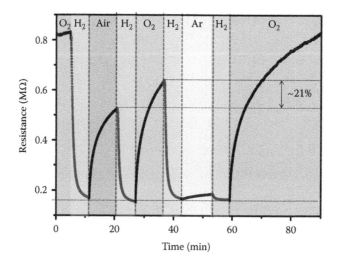

FIGURE 12.35 Variation of the resistance of the Pd NPs–coated LAO/STO sensor exposed to 20 ppm hydrogen with recovery process conducted under synthetic air, pure oxygen, and argon gas.

for 10 min, which enabled the recovery process, and the resistance of the sample reached ~0.53 MΩ. Fourth, hydrogen gas was introduced again to the sensor for 5 min and the resistance of the sample reached ~0.17 MΩ. Then, pure oxygen was introduced to the sample for 10 min, and the resistance of the sample reached ~0.64 MΩ after the recovery process. Compared with the case where the sample recovers in synthetic air, the resistance of the sample differs for ~21%, which implies that the recovery process is extremely sensitive to the presence of oxygen. There was little or no recovery when this experiment was conducted in argon gas. The result demonstrates the key role of the oxygen adsorbed on the surface of the LAO/STO heterostructure.

It should be noted that the type of chemisorbed oxygen species depends strongly on temperature, where a relatively higher temperature favors the redox reaction, and the interface 2DEG will become more sensitive for H_2 sensing. It may be argued that the hydrogen sensitivity for the Pd NPs–coated LAO/STO may be due to the volume expansion of the Pd NPs lattice. For bulk palladium, a lattice expansion as high as 3%–4% has been reported [65], and the hydrogen sensing mechanism has been explained by the tunneling current between the Pd NPs. To rule out this effect, Pd NPs were also deposited on the TiO_2-site-terminated STO substrate (without LAO), and the results show that the Pd NPs are well below the conduction percolation threshold in hydrogen ambient.

Indeed, the proposed mechanism should be supported by experimental data [66]. The use of Kelvin probe force microscopy (KPFM) can determine the work function of the sample surface, and the ambient controlled KPFM is needed to measure the work function of the sample under different gas ambient. With KPFM, the work function of the surface can be determined at atomic scales. In order to have a deeper understanding to the observed phenomenon of the Pd NPs–coated LAO/STO sample, the transport and the KPFM measurement were performed in a glove box, the measurement was made under nitrogen, nitrogen (98%)/hydrogen (2%) and oxygen ambient. In order to suppress the influence of the persistent photocurrent, the samples, including the bare LAO/STO and the Pd NPs–coated LAO/STO, were stored in the dark for at least 1 day.

The conductance (G) and the work function (W) of the Pd NPs–coated LAO/STO heterostructure show significant response upon exposure to hydrogen gas. The conductance of the Pd NPs–coated LAO/STO heterostructure ($G_{Pd/LAO/STO}$) is smaller than that of the bare LAO/STO heterostructure ($G_{LAO/STO}$), suggesting that the presence of Pd NPs influences the sheet carrier density of the LAO/STO interface significantly. According to Xie et al. [35], the sheet carrier density σ is directly related to the uncompensated potential, V_{uncom}. A potential drop across the polar LAO film is given by

$$\Delta\sigma = -\frac{\varepsilon_{LAO}}{ed_{LAO}}\Delta V_{uncom} \qquad (12.2)$$

where
  V_{uncom} is the uncompensated potential
  e is the unit charge
  ε_{LAO} is the dielectric constant of LaAlO₃

Due to the large work function of the Pd element, a Schottky contact is formed at the Pd–LAO interface and hence the potential profile of the LAO thin film changes. Such band bending alters the uncompensated potential, V_{uncom}, and the work function. Assuming that the electron mobility of the LAO/STO heterostructure and the Pd NPs–coated LAO/STO heterostructure is the same, the relation $G_{LAO/STO} - G_{Pd/LAO/STO} = (\sigma_{LAO/STO} - \sigma_{Pd/LAO/STO}) \times e \times \mu$ can be obtained. A large difference of $G_{Pd/LAO/STO}$ and the work function of Pd NPs–coated LAO/STO has been observed in the nitrogen and hydrogen/nitrogen ambient. Two possible processes, that is, the PdH_x formation and the dissociate or adsorption of hydrogen gas on the LAO/STO surface, can be considered. PdH_x has a smaller work function compared with Pd, which subsequently lowers the V_{uncom} and raises the charge carrier concentration σ, according to Equation 12.2. Thus, hydrogen adsorption on the Pd NPs–coated

LAO STO surface can readily explain the notable decrease of the work function of the Pd NPs–coated LAO/STO heterostructure.

The gas-sensing ability has been attributed to charge coupling between the desorbed gas molecules/Pd NPs and the LAO/STO interface through the LAO layer. These results not only promise the potential interest for understanding the oxide interfacial 2DEG, but also its application in all-oxide devices, and, thus, opening a new route to complex oxide physics and ultimately to the design of devices in oxide electronics.

12.4 CONCLUSION AND PERSPECTIVES

In this chapter, the interfacial properties of the oxide heterostructure and its application in sensing device were reviewed. In particular, through surface modification, the LAO/STO 2DEG was demonstrated to have great potential in polar liquid sensor, UV light sensor, and gas sensor. The following conclusions can be made.

The potential application of LAO/STO interfacial 2DEG as polar molecule sensor was demonstrated with an explanation of the sensing mechanism, where a Schottky junction model was proposed. These results present another evidence to the existence of a polarization field inside the LAO layer due to polar catastrophe and charge exchange between the LAO layer and the STO surface.

With $LaAlO_3$ surface modification by Pd NPs, the $LaAlO_3/SrTiO_3$ interfacial 2DEG presents a giant optical switching effect with a photoconductivity on/off ratio as high as 750% under UV light irradiation. Strong photoconductivity has been observed at wavelengths around 380 nm, while the increase of photoconductivity in the wavelength range of 420–700 nm is not obvious, suggesting that the increase in photocurrent is due to the absorption of photon energy by the STO substrate from the generation of electron–hole pairs. For the Pd nanoparticle decorated LAO/STO heterostructure, the recovery process is found to be dependent on oxygen concentration, which can be fitted by the stretched exponential relaxation law. The palladium electron affinity induced the degradation of the interface and the palladium nanoparticle catalytic electrochemical reaction with oxygen molecules is believed to play an important role in the UV sensing mechanism. This result shows that the LAO/STO heterostructure can probably be used for sensor applications in the field of UV light sensing and gas sensing.

A highly sensitive hydrogen (H_2) gas sensing characteristic has been demonstrated on the same system where the Pd NPs are accommodated to the surface of LAO/STO 2DEG heterostructure to trigger its sensitivity and selectivity to gases. The surface-modified LAO/STO heterostructure is demonstrated to be highly sensitive to different gases. Hydrogen, oxygen, ethanol, acetone, and water vapor can be detected via changes in the transport properties of the interface. The sensor is found to be able to detect hydrogen gas with concentration as low as ~2 ppm at room temperature. Exposing the sensor to a higher concentration of hydrogen induces a faster increase in sensitivity. Increasing temperature would lead to higher sensing performance due to the lowering of activation energy for gas adsorption and desorption on the devices, and shorter response/recovery time is observed for sensors operated at 80°C. The gas-sensing ability has been attributed to charge coupling between the desorbed gas molecules/Pd NPs and the LAO/STO interface through the LAO layer.

These results provide further evidence for understanding the mechanism of the oxide interfacial 2DEG. It suggests that there is an energy well at the interface and an internal electric field in the LAO film. Movable electrons induced by oxygen vacancy can flow in and out of the interface driven by the internal field. The persistent photocurrent may be due to the internal field that hinders the recombination of electron and holes generated by UV light irradiation. The results reported in this chapter demonstrate the LAO/STO 2DEG application in all-oxide devices, and thus opens a new route to complex oxide physics and ultimately to the design of devices in oxide electronics.

ACKNOWLEDGMENTS

Ngai Yui Chan thanks the support of the Hong Kong PhD Fellow Scheme from the Research Grants Council of Hong Kong No. RUY3. Jiyan Dai acknowledges the support from the Hong Kong NSFC/RGC under Grant No. N-PolyU517/14 and the Strategic Importance Projects of the Hong Kong Polytechnic University under Grants No. 1-ZE25 and 1-ZVCG.

REFERENCES

1. H. Kroemer. Quasi-electric fields and band offsets: Teaching electrons new tricks. Nobel Lecture, December 8, 2000, http://go.nature.com/5SFA6C. Accessed on April 12, 2016.
2. A. Ohtomo and H. Y. Hwang. A high-mobility electron gas at the LaAlO$_3$/SrTiO$_3$ heterointerface. *Nature*, 427(6973), 423–426, 2004.
3. C. W. Schneider, S. Thiel, G. Hammerl, C. Richter, and J. Mannhart. Microlithography of electron gases formed at interfaces in oxide heterostructures. *Appl. Phys. Lett.*, 89(12), 122101, 2006.
4. A. Brinkman, M. Huijben, M. van Zalk, J. Huijben, U. Zeitler, J. C. Maan, W. G. van der Wiel, G. Rijnders, D. H. A. Blank, and H. Hilgenkamp. Magnetic effects at the interface between non-magnetic oxides. *Nat. Mater.*, 6(7), 493–496, 2007.
5. N. Reyren, S. Thiel, A. D. Caviglia, L. F. Kourkoutis, G. Hammerl, C. Richter, C. W. Schneider et al. Superconducting interfaces between insulating oxides. *Science*, 317(5842), 1196–1199, 2007.
6. C. Cen, S. Thiel, G. Hammerl, C. W. Schneider, K. E. Andersen, C. S. Hellberg, J. Levy, and J. Mannhart. Nanoscale control of an interfacial metal-insulator transition at room temperature. *Nat. Mater.*, 7(4), 298–302, 2008.
7. P. Irvin, Y. Ma, D. F. Bogorin, C. Cen, C. W. Bark, C. M. Folkman, C. B. Eom, and J. Levy. Rewritable nanoscale oxide photodetector. *Nat. Photon.*, 4(12), 849–852, 2010.
8. N. Nakagawa, H. Y. Hwang, and D. A. Muller. Why some interfaces cannot be sharp. *Nat. Mater.*, 5(3), 204–209, 2006.
9. A. Kalabukhov, R. Gunnarsson, J. Börjesson, E. Olsson, T. Claeson, and D. Winkler. Effect of oxygen vacancies in the SrTiO$_3$ substrate on the electrical properties of the LaAlO$_3$/SrTiO$_3$ interface. *Phys. Rev. B*, 75, 121404R, 2007.
10. Z. Q. Liu, C. J. Li, W. M. Lü, X. H. Huang, Z. Huang, S. W. Zeng, X. P. Qiu et al. Origin of the two-dimensional electron gas at LaAlO$_3$/SrTiO$_3$ interfaces: The role of oxygen vacancies and electronic reconstruction. *Phys. Rev. X*, 3(2), 021010, 2013.
11. M. Basletic, J. L. Maurice, C. Carrétéro, G. Herranz, O. Copie, M. Bibes, E. Jacquet, K. Bouzehouane, S. Fusil, and A. Barthélémy. Mapping the spatial distribution of charge carriers in LaAlO$_3$/SrTiO$_3$ heterostructures. *Nat. Mater.*, 7(8), 621–625, 2008.
12. L. Qiao, T. C. Droubay, T. C. Kaspar, P. V. Sushko, and S. A. Chambers. Cation mixing, band offsets and electric fields at LaAlO$_3$/SrTiO$_3$(001) heterojunctions with variable La:Al atom ratio. *Surf. Sci.*, 605(15–16), 1381–1387, 2011.
13. S. A. Chambers, M. H. Engelhard, V. Shutthanandan, Z. Zhu, T. C. Droubay, L. Qiao, P. V. Sushko et al. Instability, intermixing and electronic structure at the epitaxial LaAlO$_3$/SrTiO$_3$(001) heterojunction. *Surf. Sci. Rep.*, 65(10–12), 317–352, 2010.
14. J. E. Drut and T. A. Lähde. Is graphene in vacuum an insulator? *Phys. Rev. Lett.*, 102, 023802, 2009.
15. K. Hirakawa, H. Sakaki, and J. Yoshino. Mobility modulation of the two-dimensional electron gas via controlled deformation of the electron wave function in selectively doped AlGaAs–GaAs heterojunctions. *Phys. Rev. Lett.*, 54, 1279, 1985.
16. J. Falson, D. Maryenko, Y. Kozuka, A. Tsukazaki, and M. Kawasaki. Magnesium doping controlled density and mobility of two-dimensional electron gas in Mg$_x$Zn$_{1-x}$O/ZnO heterostructures. *Appl. Phys. Exp.*, 4, 091101, 2011.
17. S. Thiel, G. Hammerl, A. Schmehl, C. W. Schneider, and J. Mannhart. Tunable quasi-two-dimensional electron gases in oxide heterostructures. *Science*, 313(5795), 1942–1945, 2006.
18. J. Mannhart and D. G. Schlom. Semiconductor physics: The value of seeing nothing. *Science*, 430, 620–621, 2004.
19. Breakthrough of the Year, The runners-up. *Science*, 318(5858), 1844–1849, December 21, 2007.
20. M. Kawasaki, K. Takahashi, and T. Maeda. Atomic control of the SrTiO$_3$ crystal surface. *Science*, 266(5190), 1540–1542, 1994.

21. M. Kawasaki, A. Ohtomo, and T. Arakane. Atomic control of $SrTiO_3$ surface for perfect epitaxy of perovskite oxides. *Appl. Surf. Sci.*, 107, 102–106, 1996.

22. G. Koster, B. L. Kropman, G. J. H. M. Rijnders, D. H. A. Blank, and H. Rogalla. Quasi-ideal strontium titanate crystal surfaces through formation of strontium hydroxide. *Appl. Phys. Lett.*, 73(20), 2920, 1998.

23. L. J. van der Pauw. A method of measuring specific resistivity and Hall effect of discs of aribtrary shape. *Philips Res. Reports*, 13, 1–9, 1958.

24. Z. Popović, S. Satpathy, and R. Martin. Origin of the two-dimensional electron gas carrier density at the $LaAlO_3$ on $SrTiO_3$ interface. *Phys. Rev. Lett.*, 101, 256801, 2008.

25. J. Lee and A. A. Demkov. Charge origin and localization at the n-type $SrTiO_3/LaAlO_3$ interface. *Phys. Rev. B*, 78, 193104, 2008.

26. G. Singh-Bhalla, C. Bell, J. Ravichandran, W. Siemons, Y. Hikita, S. Salahuddin, A. F. Hebard, H. Y. Hwang, and R. Ramesh. Built-in and induced polarization across $LaAlO_3/SrTiO_3$ heterojunctions. *Nat. Phys.*, 7, 80–86, 2011.

27. C. W. Bark, D. A. Felker, Y. Wang, Y. Zhang, H. W. Jang, C. M. Folkman, J. W. Park et al. Tailoring a two-dimensional electron gas at the $LaAlO_3/SrTiO_3$ (001) interface by epitaxial strain. *Proc. Natl. Acad. Sci.*, 108(12), 4720–4724, 2011.

28. S. Pauli, S. Leake, B. Delley, M. Björck, C. Schneider, C. Schlepütz, D. Martoccia, S. Paetel, J. Mannhart, and P. Willmott. Evolution of the interfacial structure of $LaAlO_3$ on $SrTiO_3$. *Phys. Rev. Lett.*, 106(3), 036101, 2011.

29. C. Cancellieri, D. Fontaine, S. Gariglio, N. Reyren, A. Caviglia, A. Fête, S. Leake et al. Electrostriction at the $LaAlO_3/SrTiO_3$ interface. *Phys. Rev. Lett.*, 107, 056102, 2011.

30. C. Cen, S. Thiel, J. Mannhart, and J. Levy. Oxide nanoelectronics on demand. *Science*, 323(5917), 1026–1030, 2009.

31. F. Bi, D. F. Bogorin, C. Cen, C. W. Bark, J. W. Park, C. B. Eom, and J. Levy. Water-cycle' mechanism for writing and erasing nanostructures at the $LaAlO_3/SrTiO_3$ interface. *Appl. Phys. Lett.*, 97(17), 173110, 2010.

32. Y. Xie, C. Bell, T. Yajima, Y. Hikita, and H. Y. Hwang. Charge writing at the $LaAlO_3/SrTiO_3$ surface. *Nano Lett.*, 10(7), 2588–2591, 2010.

33. Y. Xie, C. Bell, Y. Hikita, and H. Y. Hwang. Tuning the electron gas at an oxide heterointerface via free surface charges. *Adv. Mater.*, 23(15), 1744–1747, 2011.

34. R. Pentcheva, M. Huijben, K. Otte, W. E. Pickett, J. E. Kleibeuker, J. Huijben, H. Boschker et al. Parallel electron–hole bilayer conductivity from electronic interface reconstruction. *Phys. Rev. Lett.*, 104, 166804, 2010.

35. Y. Xie, Y. Hikita, C. Bell, and H. Y. Hwang. Control of electronic conduction at an oxide heterointerface using surface polar adsorbates. *Nat. Commun.*, 2, 494, 2011.

36. W. C. Lo, K. Au, N. Y. Chan H. Huang, C.-H. Lam, and J. Y. Dai. First principles study of transport properties of $LaAlO_3/SrTiO_3$ heterostructure with water adsorbates. *Solid State Commun.*, 169, 46–49, 2013.

37. N. Y. Chan, M. Zhao, N. Wang, K. Au, J. Wang, L. W. H. Chan, and J. Dai. Palladium nanoparticle enhanced giant photoresponse at $LaAlO_3/SrTiO_3$ two-dimensional electron gas heterostructures. *ACS Nano*, 7, 8673–8679, 2013.

38. K. Au, D. F. Li, N. Y. Chan, and J. Y. Dai. Polar liquid molecule induced transport property modulation at $LaAlO_3/SrTiO_3$ heterointerface. *Adv. Mater.*, 24(19), 2598–2602, 2012.

39. A. Rastogi and R. C. Budhani. Solar blind photoconductivity in three-terminal devices of $LaAlO_3/SrTiO_3$ heterostructures. *Opt. Lett.*, 37(3), 317–319, 2012.

40. A. Tebano, E. Fabbri, D. Pergolesi, G. Balestrino, and E. Traversa. Room-temperature giant persistent photoconductivity in $SrTiO_3/LaAlO_3$ heterostructures. *ACS Nano*, 6, 1278–1283, 2012.

41. E. Di Gennaro, U. Scotti, C. Aruta, C. Cantoni, A. Gadaleta, A. R. Lupini, D. Maccariello, D. Marré et al. Persistent Photoconductivity in 2D Electron Gases at Different Oxide Interfaces, *Adv. Optical Mater.*, 1, 834–843, 2013.

42. A NEMA Lighting Systems Division Document, Ultraviolet radiation from fluorescent lamps, Lamp Section, National Electrical Manufacturers Association, Rosslyn, VA, May 4, 1999.

43. Health and Consumers Scientific Committees, Health effects of artificial light, http://ec.europa.eu/health/scientific_committees/opinions_layman/artificial-light/en/about-artificial-light.htm. Accessed on April 12, 2016.

44. S. W. Lee, Y. Liu, J. Heo, and R. G. Gordon. Creation and control of two-dimensional electron gas using Al-based amorphous oxides/$SrTiO_3$ heterostructures grown by atomic layer deposition. *Nano Lett.*, 12(9), 4775–4783, 2012.

45. F. Trier, D. V. Christensen, Y. Z. Chen, A. Smith, M. I. Andersen, and N. Pryds. Degradation of the interfacial conductivity in LaAlO$_3$/SrTiO$_3$ heterostructures during storage at controlled environments. *Solid State Ion.*, 230, 12–15, 2012.

46. Y. Chen, N. Pryds, J. E. Kleibeuker, G. Koster, J. Sun, E. Stamate, B. Shen, G. Rijnders, and S. Linderoth. Metallic and insulating interfaces of amorphous SrTiO$_3$-based oxide heterostructures. *Nano Lett.*, 11(9), 3774–3778, 2011.

47. R. Arras, V. G. Ruiz, W. E. Pickett, and R. Pentcheva. Tuning the two-dimensional electron gas at the LaAlO$_3$/SrTiO$_3$(001) interface by metallic contacts. *Phys. Rev. B*, 85, 125404, 2012.

48. A. Janotti, L. Bjaalie, L. Gordon, and C. G. Van de Walle. Controlling the density of the two-dimensional electron gas at the SrTiO$_3$/LaAlO$_3$ interface. *Phys. Rev. B*, 86, 241108, 2012.

49. J. Lee, C. Lin, and A. Demkov. Metal-induced charge transfer, structural distortion, and orbital order in SrTiO$_3$ thin films. *Phys. Rev. B*, 87, 165103, 2013.

50. I. Meusel, J. Hoffmann, J. Hartmann, M. Heemeier, M. Bäumer, J. Libuda, and H. Freund. The interaction of oxygen with alumina-supported palladium particles. *Catal. Lett.*, 71(1), 5–13, 2001.

51. A. Bera and D. Basak. Pd-nanoparticle-decorated ZnO nanowires: Ultraviolet photosensitivity and photoluminescence properties. *Nanotechnology*, 22, 265501, 2011.

52. Y. Chang, J. Xu, Y. Zhang, S. Ma, L. Xin, L. Zhu, and C. Xu. Optical properties and photocatalytic performances of Pd modified ZnO samples. *J. Phys. Chem. C*, 113(43), 18761–18767, 2009.

53. A. Kolmakov, D. O. Klenov, Y. Lilach, S. Stemmer, and M. Moskovits. Enhanced gas sensing by individual SnO$_2$ nanowires and nanobelts functionalized with Pd catalyst particles. *Nano Lett.*, 5(4), 667–673, 2005.

54. N. Y. Chan, M. Zhao, J. Huang, K. Au, M. H. Wong, H. M. Yao, W. Lu et al. Highly sensitive gas sensor by the LaAlO$_3$/SrTiO$_3$ heterostructure with Pd nanoparticle surface modulation. *Adv. Mater.*, 26(34), 5962–5968, 2014.

55. F. Yang, S. Kung, M. Cheng, J. C. Hemminger, and R. M. Penner. Smaller is faster and more sensitive: The effect of wire size on the detection of nanowires. *ACS Nano*, 4(9), 5233–5244, 2010.

56. M. A. Lim, H. Kim, C. Park, Y. W. Lee, S. W. Han, and Z. Li. A new route toward ultrasensitive, flexible chemical sensors: Metal nanotubes by wet-chemical synthesis along sacrificial nanowire templates. *ACS Nano*, 6(1), 598–608, 2012.

57. F. Li, M. Liang, W. Du, M. Wang, Y. Feng, Z. Hu, L. Zhang, and E. G. Wang. Writing charge into the n-type LaAlO$_3$/SrTiO$_3$ interface: A theoretical study of the H$_2$O kinetics on the top AlO$_2$ surface. *Appl. Phys. Lett.*, 101(25), 251605, 2012.

58. Y. Li and J. Yu. Modulation of electron carrier density at the n-type LaAlO$_3$/SrTiO$_3$ interface by water adsorption. *J. Phys. Condens. Matter*, 25(26), 265004, 2013.

59. S. Semancik and T. B. Fryberger. Model studies of SnO$_2$ based gas sensors: Vacancy defects and Pd additive effects. *Sens. Actuat. B Chem.*, 1, 97–102, 1990.

60. F. Trier, D. V. Christensen, Y. Z. Chen, A. Smith, M. I. Andersen, and N. Pryds. Degradation of the interfacial conductivity in LaAlO$_3$/SrTiO$_3$ heterostructures during storage at controlled environments. *Solid State Ion.*, 230, 12–15, 2013.

61. H. Hasegawa and M. Akazawa. Hydrogen sensing characteristics and mechanism of Pd/AlGaN/GaN Schottky diodes subjected to oxygen gettering. *J. Vac. Sci. Technol. B Microelectron. Nanom. Struct.*, 25(4), 1495, 2007.

62. K. Skucha, Z. Fan, K. Jeon, A. Javey, and B. Boser. Palladium/silicon nanowire Schottky barrier-based hydrogen sensors. *Sens. Actuat. B Chem.*, 145(1), 232–238, 2010.

63. T. C. Lin and B. R. Huang. Palladium nanoparticles modified carbon nanotube/nickel composite rods (Pd/CNT/Ni) for hydrogen sensing. *Sens. Actuat. B Chem.*, 162(1), 108–113, 2012.

64. M. G. Chung, D. H. Kim, D. K. Seo, T. Kim, H. U. Im, H. M. Lee, J. B. Yoo, S. H. Hong, T. J. Kang, and Y. H. Kim. Flexible hydrogen sensors using graphene with palladium nanoparticle decoration. *Sens. Actuat. B Chem.*, 169(13382), 387–392, 2012.

65. F. Favier, E. C. Walter, M. P. Zach, T. Benter, and R. M. Penner. Hydrogen sensors and switches from electrodeposited palladium mesowire arrays. *Science*, 293(5538), 2227–2231, 2001.

66. H. Kim, N. Y. Chan, J. Dai, and D. W. Kim. Enhanced surface-and-interface coupling in Pd-nanoparticle-coated LaAlO$_3$/SrTiO$_3$ heterostructures: Strong gas- and photo-induced conductance modulation. *Sci. Rep.*, 5, 8531, 2014.

13 Novel Nanoelectronic Device Applications of Nanocrystals and Nanoparticles

Z. Liu

CONTENTS

13.1 INTRODUCTION

Nanotechnology builds on advances in microelectronics for almost half a century. The miniaturization of electrical components greatly increased the utility and portability of computers, imaging equipment, microphones, and other electronics. Scaling down of electronic device sizes is the fundamental strategy for improving the performance of integrated circuits. The downscaling of sizes of metal-oxide-semiconductor field-effect transistors (MOSFETs), yielding higher speeds and larger packing densities at a lower cost for each generation of manufacturing technology, has been the basis of the development of semiconductor industry for the past several decades [1,2]. However, in the early years of the twenty-first century, the scaling of MOSFETs entered the deep sub-50 nm regime. In this deep nanoscaled regime, fundamental limits of MOSFETs and technological challenges with regard to the scaling of MOSFETs are encountered. In order to extend the prodigious progress of IC performance, it is essential to explore new design, architectures, and physical mechanisms in the search for the next device breakthrough. Next-generation nanoelectronic devices may have an operation principle effective to utilize quantum-mechanical effects for smaller dimensions and thus provide a new functionality beyond that is attainable with MOSFETs.

Semiconductor and metal nanoparticles are of a unique electronic nature due to the quantum-mechanical rules governing their tiny nanoscale sizes. Based on nanocrystals and nanoparticles, which are synthesized as standalone films or dispersed in certain dielectric matrices, many new electronic devices have been developed to further exploit the performance and efficiency of nanoelectronics. In this chapter, several kinds of novel nanoelectronic devices based on nanocrystals and nanoparticles will be introduced, including single-electron devices, memristors, self-learning devices, and flexible or printing thin-film transistors (TFTs).

13.2 SINGLE-ELECTRON DEVICES

Single-electron devices are promising as new nanoscale devices because they retain their scalability even at the atomic scale; moreover, they can control the motion of even a single electron [3,4]. Therefore, if single-electron devices are used as ultra large-scale integration (ULSI) elements, then such ULSI circuits will have the attributes of extremely high integration and extremely low power consumption. In this respect, scalability means that the performance of electronic devices increases with the reduction of device dimensions. Power consumption is roughly proportional to the number of the electrons transferred from the voltage source to the ground in logic operations. Therefore, the utilization of single-electron devices in ULSI circuits is expected to reduce the power consumption of the ULSI circuits. The operation and operation principle of single-electron transistors (SETs) is briefly explained in this section. Next, the analytical device model of a SET, which is a typical functional single-electron device, for circuit simulation is derived and the methodology of designing logic circuits with SETs is discussed. Finally, some applications of single-electron transistors are introduced.

13.2.1 SINGLE-ELECTRON BOX

Let a small conductor (traditionally called an *island*) be electrically neutral; that is, the material has exactly as many electrons as it has protons in its crystal lattice. In this state, the island does not generate any appreciable electric field beyond its borders, and a weak external force may bring in an additional electron from outside. In most single-electron devices, this injection is carried out by electron tunneling through an energy barrier created by a thin insulating layer. Now the net charge of the island becomes $-e$ instead of being neutral, and an electric field E is produced, which repulses the following electrons to be added into the *island*. Though the fundamental charge $e = -1.6 \times 10^{-19}$ C is very small, the field E, which is inversely proportional to the square of the *island* size, may become rather strong for nanoscale-sized structures. For example, the field is as large as about 140 kV/cm on the surface of a 10 nm sphere in vacuum [5].

Figure 13.1 shows the conceptually simplest device, the single-electron box [6]. The device consists of just one small island separated from a larger electrode (as the electron source) by a tunnel barrier. An external electric field may be applied to the island using another electrode (i.e., the gate electrode) separated from the island by a thicker insulator that prohibits electron tunneling.

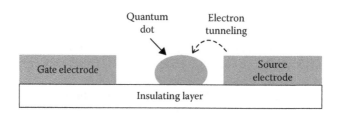

FIGURE 13.1 Schematic of a single-electron box. The single-electron box consists of a quantum dot, an electrode close to the dot allowing electron tunneling, and a gate electrode coupled to the dot, which is able to modulate the energy levels in the dot.

The applied electric field on the gate electrode changes the electrochemical potential of the island and thus determines the conditions of electron tunneling. Elementary electrostatics shows that the free (Gibbs) energy of the system may be described as [5]

$$W = \frac{Q^2}{2C_\Sigma} + \left(\frac{C_0}{C_\Sigma}\right)QU + \text{const.} \tag{13.1}$$

where
  $Q = -ne$ is the island charge (n is the number of uncompensated electrons)
  C_0 is the island gate capacitance
  C_Σ is the total capacitance of the island (including C_0)

Usually, this expression can be rewritten as [5]

$$W = \frac{(ne - Q_e)^2}{2C_\Sigma} + \text{const.} \tag{13.2}$$

where parameter Q_e, defined as [5]

$$Q_e \equiv UC_0, \tag{13.3}$$

which is usually called the "external charge." An elementary calculation using these equations shows that Q is a step-like function of Q_e, that is, of the gate voltage, with a fixed distance between the neighboring steps [5]:

$$\Delta Q_e = e, \tag{13.4}$$

$$\Delta U = \frac{e}{C_0} = \text{const.} \tag{13.5}$$

13.2.2 COULOMB BLOCKADE EFFECTS

The free energy $F(n)$ of a system having n electrons in the island is expressed as

$$F(n) = W_C(n) - A(n), \tag{13.6}$$

where
  $W_C(n)$ is the charging energy
  $A(n)$ is the work done by the gate voltage in order to make the electron number increase from
    0 to n [7]

When tunneling events do not occur, the tunneling junction behaves like a normal capacitor, and the polarization charge on the capacitors does not have to be associated with a discrete number of electrons, n. This polarization charge is essentially due to a rearrangement of the electron gas with respect to the positive background of ions. Therefore, the polarization charge takes a continuous range of values, although the number of electrons in the quantum dot takes a discrete number of electrons, n. The polarization charges on the tunneling junction and gate capacitor are obtained from the following relationship [7]:

$$Q_t - Q_g = -ne, \tag{13.7}$$

$$\frac{Q_t}{C_t} + \frac{Q_g}{C_g} = V_g, \tag{13.8}$$

where Q_t and Q_g are the polarization charges on the tunneling junction and the gate capacitor, respectively. By using Q_t and Q_g, the charging energy $W_C(n)$ of the quantum dot is expressed as [7]

$$W_C(n) = \frac{e^2 n^2}{2C_\Sigma} + \frac{1}{2} \frac{C_t C_g V_g^2}{C_\Sigma}, \tag{13.9}$$

where $C_\Sigma = C_t + C_g$. In addition, the work $A(n)$ done by the gate voltage source in order to make the electron number of the quantum dot change from 0 to n is expressed as [7]

$$A(n) = \int I(t) \cdot V_g dt = Q_g V_g = en \frac{C_g}{C_\Sigma} V_g + \frac{C_t C_g V_g^2}{C_\Sigma}. \tag{13.10}$$

In order to maintain the electron number in the quantum dot, the following condition is required: $F(n) < F(n \pm 1)$.

Free energy change $\Delta F(n, n + 1)$ that accompanies a transition of the electron number from n to $n + 1$ is also simply expressed with critical charges Q_c as [7]

$$\Delta F(n, n+1) = F(n+1) - F(n) = \frac{e}{C_t}(Q_t - Q_c). \tag{13.11}$$

Experimentally, a typical single-electron tunneling and Coulomb blockade effect was reported by Chan et al. [8] in a nanostructure incorporated with gold nanocrystals (Au NCs). Figure 13.2 shows the HfAlO/Au NCs/HfAlO trilayer structure on a p-Si substrate. The Au NCs are uniformly distributed between the HfAlO control layer and the tunnel layer. The thicknesses of the tunnel layer and the control layer are about 15 and 20 nm, respectively. The interface between the HfAlO tunnel layer and the silicon substrate is mainly due to the oxygen diffusion into the sample during postdeposition annealing. The inset shows the plane-view TEM image of the trilayer structure, where we can see that the self-organized and uniformly distributed Au NCs are spherical in shape, where the range of diameter of Au NCs is from 3 to 4 nm, with a density of about 1.6×10^{12} cm^{-2}.

Figure 13.3 shows the I–V curves of the capacitor structure (with and without Au NCs) measured with positive bias at different temperatures [8]. Coulomb blockade effect can be observed in the trilayer structure in the figure, indicated by the periodically modulated current as the bias voltage increases. As apparent from Figure 13.3, the curve of the control layer (without Au NCs) does not show any steps in the I–V curve, suggesting that the Au NCs are responsible for the Coulomb blockade effect. From the I–V curve measured at 20 K, an average Coulomb voltage can be determined as $\Delta V = 31$ mV, which corresponds to the Coulomb capacitance $C = 5.1$ aF based on the relationship $\Delta V = e/C$. It is interesting to notice that peaks instead of flat steps are observed in the I–V curves as shown in Figure 13.3. It is believed that the current peaks are due to the interplay of resonant tunneling and Coulomb blockade effect [8].

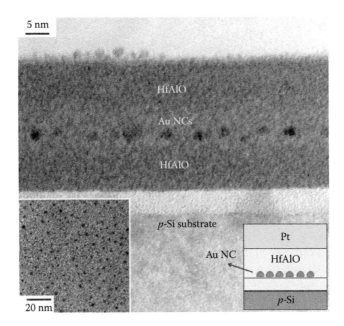

FIGURE 13.2 Cross-sectional HRTEM image of the trilayer floating gate memory structure, where Au NCs are evenly distributed in HfAlO matrix. The inset shows the plane-view HRTEM image of the Au NCs. (From Chan, K.C. et al., *Appl. Phys. Lett.*, 92, 143117, 2008.)

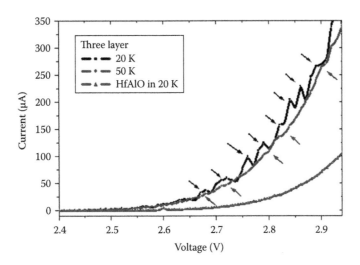

FIGURE 13.3 *I–V* curves at 20 and 50 K, showing the steps and the peaks in current. For the control layer with only HfAlO, there are no steps or peaks that can be observed. (From Chan, K.C. et al., *Appl. Phys. Lett.*, 92, 143117, 2008.)

13.2.3 WORKING PRINCIPLES OF SETs

SETs are three-terminal switching devices, which can transfer electrons one by one from source to drain. The schematic structure of SETs is shown in Figure 13.4. As shown in the figure, the structure of SETs is almost the same as that of MOSFETs. However, SETs have tunneling junctions in place of *pn*-junctions of the MOSFETs and a quantum dot in place of the channel region of the MOSFETs.

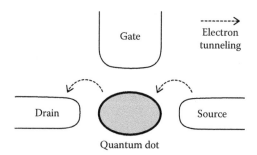

FIGURE 13.4 Schematic structure of a single-electron transistor.

The electrical potential of the quantum dot (island) can be tuned by the gate, capacitively coupled to the island. As shown in Figure 13.5a, in the blocking state, no accessible energy levels are within tunneling range of the electron on the source contact. All energy levels on the island electrode with lower energies are occupied; therefore, the electron on the source side is not able to tunnel through the isolating junction (dashed arrow, meaning tunneling is prohibited). However, when a positive voltage is applied to the gate electrode, the energy levels of the island electrode are lowered. The electron can now tunnel onto the island (solid arrow), occupying a previously vacant energy level, as illustrated in Figure 13.5b. From there it can tunnel onto the drain electrode where it inelastically scatters and reaches the drain electrode Fermi level.

The energy levels of the island electrode are evenly spaced with a separation of ΔE. ΔE is the energy needed to transfer each subsequent electron to the island, which acts as a self-capacitance C. The lower the C, the bigger the ΔE. It is crucial for ΔE to be larger than the energy of thermal fluctuations $k_B T$; otherwise, an electron from the source electrode can always be thermally excited onto an unoccupied level of the island electrode, and no blocking can be observed.

The source-to-drain current of SETs can be calculated by using the tunneling rate of an electron through the tunneling junction. The rate $\Gamma(n, n + 1)$ of an electron tunneling through a tunneling junction, which accompanies a transition of the electron number in the dot from n to $n + 1$, is given by [4]

$$\Gamma(n, n+1) = \frac{1}{e^2 R_t} \frac{\Delta F(n, n+1)}{1 - \exp\left[-\Delta F(n, n+1)/k_B T\right]}, \quad (13.12)$$

where

$\Delta F(n, n + 1)$ is the free energy change that accompanies the tunneling

R_t is the tunneling resistance of the junction

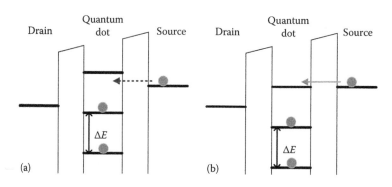

FIGURE 13.5 Energy levels of source, island (i.e., the quantum dot) and drain in a SET for both the blocking state (a) and the transmitting state (b).

In the following, the tunneling rate through the junction of the source is denoted by $\Gamma_s(n, n+1)$ and the tunneling rate through the junction of the drain is denoted by $\Gamma_d(n, n+1)$. The probability p_n of finding n electrons in the dot may change by leaving this state or by coming into this state from the states $n-1$ or $n+1$ [4]

$$\frac{dp_n}{dt} = \Gamma_{tot}(n+1,n)p_{n+1} + \Gamma_{tot}(n-1,n)p_{n-1} - [\Gamma_{tot}(n,n+1) + \Gamma_{tot}(n,n-1)]p_n, \tag{13.13}$$

where

$$\Gamma_{tot}(n,n+1) = \Gamma_s(n,n+1) + \Gamma_d(n,n+1).$$

A normalization condition exists:

$$\sum_{n=-\infty}^{+\infty} p_n = 1. \tag{13.14}$$

The current (I) of a SET is obtained by [4]

$$I = e\Sigma p_n \left[\Gamma_s(n,n+1) - \Gamma_d(n,n+1) \right]. \tag{13.15}$$

The characteristics of single-electron devices are usually calculated with numerical simulators [9,10].

13.2.4　FABRICATION OF Si-BASED SETs

Since single-electron phenomena can be observed in any conductive substances, single-electron devices are fabricated using a variety of materials such as aluminum, GaAs heterostructures, and silicon [11,12]. However, in order to utilize single-electron devices as elemental devices of ULSI circuits, silicon SETs are essential. This can be achieved if fabrication techniques of nanometer-scaled silicon quantum dots are established. Besides, fabricating silicon SETs by using MOS processes is advantageous because we can use highly advanced fabrication tools developed for complementary metal oxide semiconductor (CMOS) ULSI circuits.

Generally, silicon quantum dots formation is possible through two approaches: (1) patterned by fine-lithography techniques and (2) grown by deposition processes. Using the former approach, it is possible to accurately define the structures and positions of quantum dots. Takahashi et al. proposed a novel silicon quantum dot fabrication called pattern-dependent oxidation (PADOX) [13]. In 2000, Ono et al. developed the vertical version of PADOX (V-PADOX) [14]. In PADOX, laterally broad 2-D regions are essential for tunneling barrier formation. The process is also known as Si width modulation. On the other hand, in V-PADOX, vertically broad, namely, thick, 2-D regions are utilized for tunneling barrier formation, known as thickness modulation. The advantage of the V-PADOX is that it makes it possible to form two tiny islands in a small area by utilizing not a lithographic process but the oxidation process, which includes the accumulation of stress in small structures. Thus, two SETs can be fabricated in an extremely small area. Both PADOX and V-PADOX are able to be combined with the conventional CMOS. However, the latter approach is favorable from the viewpoints of throughput and fabricated quantum-dot sizes.

13.2.5　DEVICE APPLICATIONS OF SETs

The implementation of SETs is able to increase the packaging density and to decrease the power consumption of integrated circuits. Therefore, many logic circuits consisting of SETs have been developed. There are generally two methods for implementing logic operations in the circuits of SETs: (1) by representing a bit by a single electron and using field effect to transfer electrons one by one and (2) by representing a bit by more than one electron and using field effect to switch the current on or off. The former method is more attractive from the power consumption standpoint. However, in that case even one erroneous electron caused by noise or thermal agitation will completely alter the logic operation results. Therefore, from the viewpoint of operation reliability, the latter method is preferable. Logic circuits realized by the former method resemble those consisting of charge-coupled devices, whereas logic circuits realized by the latter method resemble those consisting of MOSFETs. Since present logic circuits consist of MOSFETs, the latter method is preferable.

Figure 13.6a shows the realization of a SET based on nanocrystals reported by Klein et al. [15]. CdSe nanocrystals with a diameter of 5.5 nm were bound to two closely spaced gold leads by bifunctional linker molecules. The leads were fabricated on a degenerately doped silicon wafer, which was used as the gate electrode to tune the charge state of the CdSe nanocrystal. Figure 13.6b shows a field-emission scanning electron microscope (SEM) image of a completed device [15]. It can be seen that a number of nanocrystals are distributed in the ~10 nm gap between the two electrodes. Although the gap is so small and even filled with nanocrystals, most devices have immeasurably high impedance ($R > 100$ GΩ). Only about 1 in 20 have a measurable resistance, typically in the range 10 MΩ to 1 GΩ. Devices with relatively small resistance typically behave like transport occurs through a single nanocrystal, even though the number of nanocrystals in the junction region is quite large as can be seen in Figure 13.6b. Since electron tunneling through the linker molecules has an exponential decay length of less than 1 Å [15], only a well-placed nanocrystal, within 2 nm of each lead, can contribute to the current conduction.

Figure 13.7 shows the linear device conductance (G) measured at $T = 4.2$ K as a function of the gate voltage (V_g) applied to the heavily doped Si substrate. As the gate bias was swept from −9 to −5 V, the conductance increased to a peak and then declined back to zero. The insets in Figure 13.7 show the plots of the current (I) flowing through the nanocrystal as a function of the voltage (V) applied between the two leads at two fixed values of V_g. The I–V measured at V_g, which is away from the linear response peak corresponding to V_g of ~0.7 V, shows a suppressed conductance at small V whereas the I–V characteristic taken at the center of the peak has a finite linear conductance for small V. The conductance peak observed in Figure 13.7 is a Coulomb oscillation. The suppressed conductance on each side of the peak is a consequence of the finite energy required to add (remove) an electron to (from) the nanocrystal in its ground state. This energy is analogous to the electron affinity (ionization energy) of a molecule. The peak occurs when the two charge

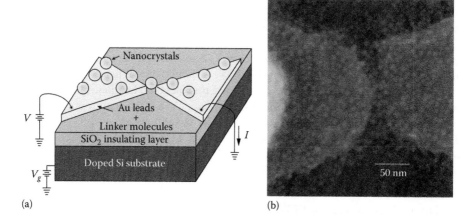

(a)   (b)

FIGURE 13.6 (a) Schematic of the single-electron device based on CdSe nanocrystals sit on top of two closely placed Au leads. The setup of measurement probes applied on the devices is also demonstrated. (b) Field-emission scanning electron microscope (SEM) image of CdSe nanocrystals with 5.5 nm in diameter distributed over and in the 10 nm gap of two 13 nm thick leads. (From Klein, D.L. et al., *Nature*, 389, 699, 1997.)

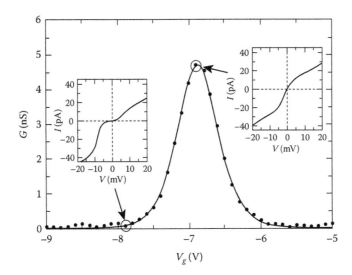

FIGURE 13.7 Conductance (G) versus gate voltage (V_g) for a single nanocrystal transistor measured at $T = 4.2$ K. The dots are the measured values and the solid curve is a fit to the data by using the standard Coulomb blockade model [16] with a temperature $T = 5$ K. The insets show the $I–V$ characteristics measured at the gate voltages, indicated by arrows. (From Klein, D.L. et al., *Nature*, 389, 699, 1997.)

states of the nanocrystal have the same total (including electrostatic) energy and an extra electron can therefore hop on and off the nanocrystal at no energy cost. The maximum size of the gap in the $I–V$ curves away from the peak provides a measure of the addition (removal) energy for electrons.

A map of the differential conductance (dI/dV) of the device as a function of both V and V_g is plotted as a gray scale in Figure 13.8. The evolution of the Coulomb gap with V_g can be seen in Figure 13.8a and b. The Coulomb gap is indicated with light-colored diamond-shaped regions in the figure. It can be observed that the Coulomb gap is zero at Coulomb oscillation; it increases to a maximum value approximately halfway between two oscillations, and then decreases to zero at the next Coulomb oscillation. Figure 13.8c shows a schematic evolution of the Coulomb gap for successive electrons along with the inferred addition energies, which range from ~15 to 30 meV. Additional $I–V$ measurements on eight other samples without a gate have been performed and the typical addition energies are determined to be in the range of 30–60 meV.

SETs have the properties of low power and good scalability; therefore, they are very promising to be used in future ULSI circuits. The working physics and mechanisms of SETs are discussed in this section. Researchers have already developed a number of processes to form the essential quantum dots mainly based on semiconductor nanocrystals. But in order to operate SET circuits at room temperature, the size of the quantum dot must be much smaller than 10 nm. Fabricating a structure smaller than 10 nm is still quite difficult with the present technology. The development of ULSI nanofabrication techniques is desirable to realize such nanoscale devices. Another challenge is operation stability, which involves removing random background charge or, alternatively, providing continuous charge transfer in nanoscale resistors. In future, it is important to develop an operation scheme under which the functionality of SET circuits can surpass that of conventional CMOS circuits.

13.3 MEMRISTORS

A memristor, a contraction of "memory resistor," is a two-terminal electronic device whose resistance can be precisely modulated by the current or flux flowing through it. It was theoretically proposed by Leon Chua in 1971 as the fourth basic circuit element, other than the resistor, capacitor,

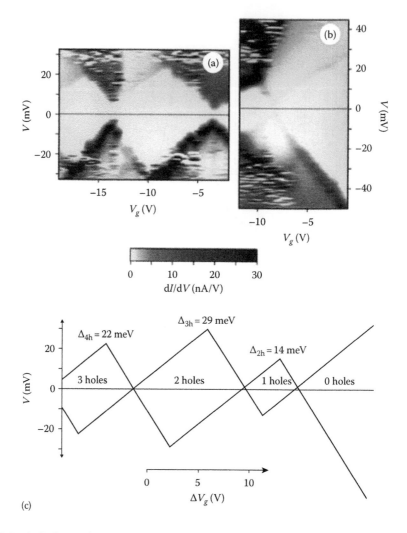

FIGURE 13.8 (a, b) Composite grayscale plots of the differential conductance (dI/dV) of a CdSe nanocrystal plotted as a function of both V_g and V. The white diamond shaped regions correspond to the Coulomb gap. (c) A schematic illustration of the data from (a) and (b), indicating the inferred number of holes on the nanocrystal as a function of V_g and the addition energy for successive holes. (From Klein, D.L. et al., *Nature*, 389, 699, 1997.)

and inductor [17]. Figure 13.9 shows the four fundamental two-terminal circuit elements and illustrates the relationship between different circuit variables [18]. It can be seen that the memristor is defined by memristance, which can be described by a relationship between charge (q) and flux (φ).

Mathematically, memristive systems can be described by the following set of equations [18]:

$$v = R(w,i)i, \tag{13.16}$$

$$\frac{dw}{dt} = f(w,i), \tag{13.17}$$

where
  R is the resistance
  w is the internal state variable
  f is a function describing the change of w as a function of inputs and the current state

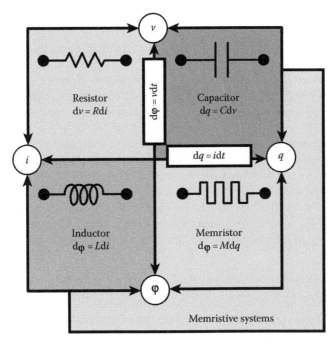

FIGURE 13.9 The four fundamental circuit elements—resistor, inductor, capacitor, and memristor—and the functional relationships of a pair of the four basic circuit variables (i.e., voltage [V], current [I], charge [q], and flux [φ]) defined by the circuit elements (i.e., resistance [R], inductance [L], capacitance [C], and memristance [M]). (From Strukov, D.B. et al., *Nature*, 453, 80, 2008.)

Although the memristor is physically a simple two-terminal device, it can exhibit highly complex behaviors when different terms are incorporated into these equations. As a result, memristor devices existed only in theory and papers for almost 40 years.

In 2008, a research group lead by R.S. Williams from Hewlett Packard demonstrated the first workable memristor in a laboratory, which was composed of nanoscale titanium oxide thin films sandwiched by two platinum electrodes [18]. The TiO_x-based device exhibited a fascinating property, which was the dependence of device resistance on time-varying current by means of a chemical reaction in the interface region between TiO_2 and TiO_x layers. Specifically, the boundary between the high-resistance, pure TiO_2 layer and the low-resistance TiO_x layer can be brought to a nanometer scale upon the application of electric fields between the top and bottom electrodes. Since the finding of the missing piece in the puzzle of circuit elements shown in Figure 13.9, intensive studies on memristors have been carried out in multiple aspects, including materials, device structures, and neural applications.

13.3.1 Memristors Based on Fe_3O_4 Nanoparticle Assemblies

Regarding nanocrystals and nanoparticles, Kim et al. reported a new addition to the memristor family with nanoparticle assemblies consisting of an infinite number of monodispersed single crystalline Fe_3O_4 nanoparticles [19]. A nonhydrolytic chemical method was used to prepare single crystalline Fe_3O_4 nanoparticles, which have diameters of 7, 9, 12, and 15 nm with a well-controlled size monodispersity ($\sigma \approx 5\%$). The surface ligands of synthesized nanoparticles were carefully removed by washing with tetramethylammonium hydroxide solution for a pristine, inorganic surface. The nanoparticle assemblies in the form of compact pellets with dimensions of $0.5 \times 1 \times 4$ mm were then fabricated by coldpressing in a die under 160 Pa for

15 min. There was no heat treatment during the preparation of the pellets, and thus the surface properties of the nanoparticles were not altered.

The voltage–current (V–I) characteristics of the nanoparticle assembly pellets were measured using a conventional four-point probing method, as shown in Figure 13.10a. In the case of nanoparticles with a diameter (D) of 7 nm (see Figure 13.10b), a type of hysteresis is observed at room temperature, as shown in Figure 13.10c. The bistable V–I characteristics illustrated that the switching in measured voltages are directly related to the existence of hysteretic behavior as the current sweeping was applied in sequences of 1–6, corresponding to $0 \to +20$ nA $\to 0 \to -20$ nA $\to 0$. As the current was increased from 0 to +20 nA, the switching from a low-resistance state ($R_{ON} \approx 2 \times 10^7$ Ω) to a high-resistance state ($R_{OFF} \approx 4 \times 10^8$ Ω) occurred at $I \approx +16$ nA, as shown in Figure 13.10c. In contrast, ohmic behavior was seen when the current decreases from +20 nA to 0. The same behavior was observed when the current sweeping direction was reversed. This type of instantaneous switching from the low-resistance (i.e., on) state to the high-resistance (i.e., off) state occurred in a typical R_{OFF}/R_{ON} ratio of 20:1. Stable V–I characteristics were observed when sweeps were carried out repeatedly, suggesting good endurance in the memristors based on Fe_3O_4 nanoparticle assemblies.

13.3.2 MEMRISTORS BASED ON AMORPHOUS Si EMBEDDED WITH Ag NANOPARTICLES

Similar to the TiO_2/TiO_x bilayer structure invented by the HP research group, a memristor based on Ag nanoparticle–embedded Si thin films was reported by Jo et al. [20]. The memristor device consists of a bottom tungsten nanowire electrode, a sputtered silicon layer (2–4 nm), an amorphous silicon

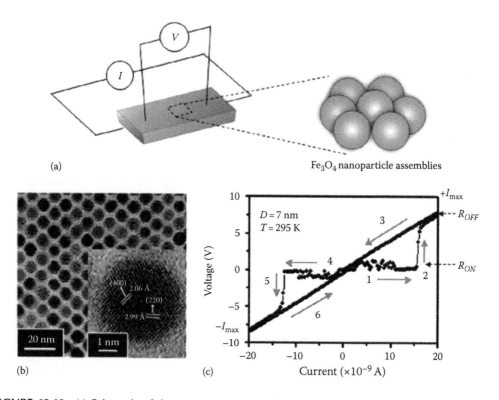

FIGURE 13.10 (a) Schematic of the measurement setup for V–I analysis of the nanoparticle assemblies. (b) Transmission electron microscopy (TEM) and high-resolution TEM (HRTEM) images of 7 nm sized Fe_3O_4 nanoparticles. (c) V–I characteristics measured at room temperature for 7 nm sized Fe_3O_4 nanoparticle assemblies. (From Kim, T.H. et al., *Nano Lett.*, 9, 2229, 2009.)

(a-Si) layer (2.5–4.5 nm) deposited by plasma-enhanced chemical vapor deposition (PECVD), a co-sputtered silver and silicon layer (20–30 nm thick), and a top chrome/platinum nanowire electrode, as schematically illustrated in Figure 13.11. The bottom tungsten nanowire electrode was fabricated by e-beam lithography, mask layer lift-off, and reactive ion etching. After that, a-Si deposition, co-sputtering, and top metal lift-off processes were performed, followed by another reactive ion etching step to remove the co-sputtered layer outside the cross-point regions defined by the perpendicularly crossed top and bottom electrodes.

As schematically shown in Figure 13.11, the device consists of multiple film structure with active layer of sputtered Ag and Si, which has a properly designed Ag/Si mixture ratio gradient. More specifically, a Ag-rich (high-conductivity) region and a Ag-poor (low-conductivity) region were formed, respectively, in the fabrication process, as shown in Figure 13.11. Typically, resistance switching devices require an electroforming process to activate the switching property regardless of switching material. During electroforming, metal ions or particles are injected into the host medium and cause semipermanent structural modifications inside the insulating storage layer. The forming process creates localized conducting paths (filaments), resulting in discrete, abrupt resistance switching characteristics. Through co-sputtering of Ag and Si, nanoscale Ag particles and clusters were incorporated into the Si host medium during device fabrication, and a uniform conduction front between the Ag-rich and Ag-poor regions was formed. As a result, the forming process was not required in the Ag/Si-based devices. In addition, instead of discrete, localized conducting filament formation, the continuous motion of the conduction front under applied bias in the co-sputtered memristor device results in reliable "analog" switching behaviors (i.e., a gradual change in resistance instead of abrupt changes), as shown in Figure 13.12a.

Figure 13.12a shows the measured current (blue lines) through the memristor as a function of the applied voltage across the device for five consecutive positive voltage sweepings followed by five consecutive negative voltage sweepings. Being different from resistive switching devices that show abrupt conductance switching, the conductance in devices based on Ag-embedded Si thin films continuously increased (or decreased) during the positive (or negative) voltage sweeps, and the slope of $I–V$ curve in each subsequent sweep picked up where the last sweep left off.

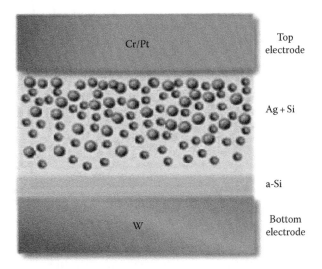

FIGURE 13.11 The device structure of a two-terminal memristor. The active layer is a two-layered structure based on amorphous Si thin films with and without Ag nanoparticles, respectively. (From Jo, S.H. et al., *Nano Lett.*, 10, 1297, 2010.)

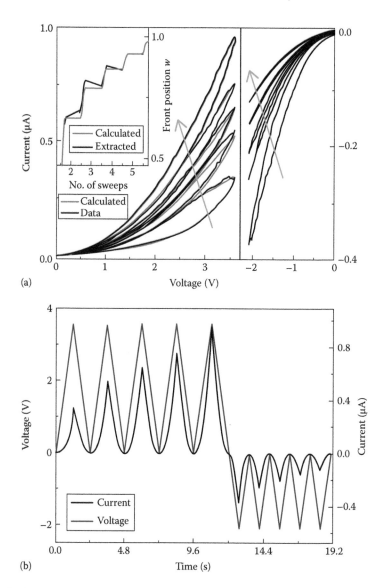

FIGURE 13.12 Memristor characteristics in Ag-embedded a-Si thin-film structures. (a) Measured (blue lines) and calculated (orange lines) *I–V* characteristics of the memristor (device size: 500 nm × 500 nm). The inset shows comparison of calculated and extracted values of the normalized Ag front position *w* during positive DC sweepings. (b) The replotted current and voltage data from (a) as a function of time, highlighting the current change in sequential voltage sweeps. (From Jo, S.H. et al., *Nano Lett.*, 10, 1297, 2010.)

The *I–V* relationship in this device can be well fitted by a memristor circuit model (orange lines in Figure 13.12a) described as [18]

$$i(t) = \frac{1}{R_{ON} w(t) + R_{OFF}(1 - w(t))} v(t), \tag{13.18}$$

where $w(t)$ represents the normalized position of the conduction front between the Ag-rich and Ag-poor regions within the active device layer and its value varies between 0 and 1. When a positive voltage bias is applied on the device, Ag ions driven by the electric field move from the Ag-rich region to the Ag-poor region and increase w, and vice versa. As $w(t)$ approaches 1 (or 0), the device reaches the smaller (highest) resistance state with a resistance of R_{ON} (or R_{OFF}). In this model, the position $w(t)$ can be assumed to be a linear function of the flux-linkage $\varphi(t) = \int v(t)\mathrm{d}t$ through the device. Equation 13.18 can then be rewritten as [20]

$$i(t) = G(\varphi(t))v(t), \tag{13.19}$$

which is the equation for a flux-controlled memristor and $G(\varphi(t))$ is the memductance. However, bias voltages with amplitude greater than a threshold (V_T) of 2.2 V were required to drive the Ag ions inside the a-Si matrix and voltages smaller than V_T have negligible effect on the memristor resistance.

Based on the memristor model discussed, the current values (orange lines) were calculated as a function of the voltage sweepings and plotted in Figure 13.12a as orange lines [20]. The measured current (blue lines) is also shown in Figure 13.12a and a good consistency can be observed between the measured and calculated curves. In addition, the values of $w(t)$ (orange lines) used to calculate the current during five consecutive positive voltage sweepings is also included as an inset of Figure 13.12a, by assuming $w(t)$ to be a linear function of the flux-linkage $\varphi(t)$. In comparison, the values of $w(t)$ can be also extracted from the data using Equation 13.18, which are plotted in the inset of Figure 13.12a, indicated with blue lines. The good agreements between the calculated and the measured values in $i(t)$ and $w(t)$ suggest that the device characteristics above the V_T can be explained by the flux-controlled memristor model using Equations 13.18 and 13.19, where the front position is roughly a linear function of the flux-linkage. Due to the existence of a threshold effect, the device is actually not a true memristor but falls in the more broadly defined memristive device category [21].

The flux-controlled memristor model introduced suggests that the conductance in the device can be incrementally modulated by tuning the duration and sequence of the applied programming voltage. Figure 13.13a shows the memristor response when the device was programmed by 100 consecutive identical positive pulses of 3.2 V, 300 µs followed by 100 identical consecutive negative voltage pulses of −2.8 V, 300 µs [20]. The device current was measured at a small read voltage of 1 V after each programming pulse. As expected from the previous sweeping I–V characteristics of the device shown in Figure 13.12a, the application of positive voltage pulses (i.e., the potentiating signal, P) incrementally increases the memristor current (equivalently the conductance), and the application of negative voltage pulses (i.e., depressing signal, D) incrementally decreases the memristor current or conductance, as can be seen from Figure 13.13a. It is thus noticed that the memristor conductance can be controlled and tuned by the flux-linkage $\varphi(t)$ as a result of the application of programming pulses. This effect was further demonstrated in Figure 13.13b involving the application of mixed bipolar pulses. Mixed positive and negative voltage pulses with fixed pulse height but different duration were applied to the device and the change in memristor conductance (ΔG) was measured after each pulse. A proportional correlation between ΔG and the duration of the applied pulses was observed. More specifically, the application of a longer positive (or negative) pulse caused a larger increase (or decrease) in the memristor conductance, and vice versa.

13.3.3 Electronic Synapses Realized with Si-Based Memristors

A synapse bears striking resemblance to the memristor, introduced in the previous section. Similar to a biological synapse, the conductance of a memristor can be incrementally tuned by carefully controlling the flux-linkage through it. Realization of an electronic device that can emulate the functions of biological synapse is essential to the development of next-generation computing technology. Currently, the computers based on the von Neumann architecture, which fetch, decode, and execute instructions in a sequential process, are ideal to solve structured problems, such as mathematical problems. For solving problems where interactions with the real world are involved and no specific instructions are defined, modern digital computers become very inefficient, and such cases usually require systems with great complexity and huge power consumption. Though current digital computers possess the computing speed and capability to emulate brain functionality of certain animals, the associated complexity and energy dissipation in the system grows exponentially along the hierarchy

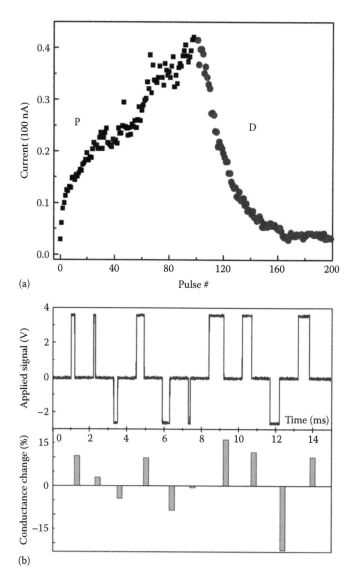

FIGURE 13.13 The evolution of the memristor current as a result of the programming pulses. (a) The incremental increase and decrease of device current read at 1 V as a result of consecutive positive potentiating (P) pulses (3.2 V, 300 μs) followed by negative depressing (D) pulses (−2.8 V, 300 μs). (b) Mixed potentiating and depressing pulses with different pulse widths used for programming (top); and change of the measured device conductance after the application of each pulse. The conductance change was normalized to the maximum memristor conductance. (From Jo, S.H. et al., *Nano Lett.*, 10, 1297, 2010.)

of animal intelligence. For example, to perform certain cortical simulations at the cat scale even at 83 times slower firing rate, the IBM team had to employ Blue Gene/P (BG/P), which is a supercomputer equipped with 147,456 CPUs and 144 TB of main memory [20,22]. On the other hand, being dramatically different from the von Neumann digital architecture, which stores and processes data separately in modern computers, brains of biological creatures are configured in large connectivity between neurons (~10^4 in a mammalian cortex), offering highly parallel processing power and high efficiency. The connectivity strength between two neurons (i.e., the so-called synaptic weight) can be precisely adjusted by the ionic flow through them and it is widely believed that the adaptation of synaptic weights enables the memorization and learning ability of biological systems.

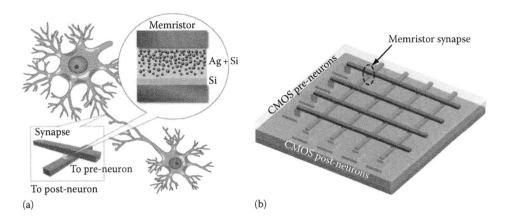

FIGURE 13.14 (a) Schematic illustration of the concept of using the memristor introduced in Figure 13.11 to emulate a synapse between two neurons. (b) Schematic of a neuromorphic system with CMOS neurons and memristor synapses in a cross-point configuration. (From Jo, S.H. et al., *Nano Lett.*, 10, 1297, 2010.)

Jo et al. demonstrated the experimental implementation of biological synapses with nanoscale silicon-based memristors, which can emulate an advanced synaptic function, namely, spiking-time-dependent plasticity (STDP) [20]. STDP is an important synaptic adaption rule for competitive Hebbian learning and it can be achieved in a hybrid synapse/neuron circuit composed of CMOS neurons and nanoscale memristor synapses, as illustrated in Figure 13.14a. These demonstrations provide a direct experimental support for memristor-based neuromorphic systems. Figure 13.14b shows the implementation of a crossbar hardware structure, where a two-terminal memristor synapse is formed at each cross-point and connects CMOS-based pre- and postsynaptic neurons [20]. The crossbar-configured network based on memristors can potentially offer connectivity and function density comparable to those of biological systems. In such a network, data will be processed and stored in a parallel approach, which is analogous to biological systems rather than digital computers. As shown in Figure 13.14b, every CMOS neuron in the "pre-neuron" layer is directly connected to every neuron in the "post-neuron" layer at each cross-point with unique synaptic weights. A high synaptic density of 10^{10} cm^{-2} can be potentially obtained for crossbar networks with 100 nm pitch [20].

Memristor devices based on Ag nanoparticles embedded with a-Si thin films are capable of emulating biological synapses with properly designed CMOS neuron components to provide programming voltage pulses with controlled pulse width and height [20]. The neuron circuit consists of two CMOS-based integrate-and-fire neurons connected by a nanoscale memristor with an active device area of 100 nm × 100 nm. Specifically, being analogous to the synaptic behavior of biological systems, the neuron circuit generates a potentiating or a depressing pulse across the memristor synapse when the presynaptic neuron spikes before or after the postsynaptic neuron, with the pulse width being an exponentially decaying function of the relative neuron spike timing ($\Delta t = t_{pre\text{-}neuron} - t_{post\text{-}neuron}$, where $t_{pre\text{-}neuron}$ and $t_{post\text{-}neuron}$ are the time when the presynaptic neuron and postsynaptic neuron fire spikes, respectively).

Figure 13.15 shows the measured change of the memristor synaptic weight after each neuron spiking event obtained in the hybrid CMOS-neuron/memristor-synapse circuit [20]. When the pre-neuron spikes before (or after) the post-neuron, the synaptic weight of the memristor increases (or decreases). Moreover, as can be seen from Figure 13.15, the change in the synaptic weight versus the spike timing difference Δt can be well fitted with exponential decay functions. It is thus verified that an STDP function analogous to that of biological synaptic systems can be obtained in such memristor synapses.

Figure 13.16 presents the endurance of the device after continuous applications of the potentiation and depression pulses [20]. As can be seen in Figure 13.16b, after 1.5×10^8 times of operations with P/D pulses, the device shows certain degradation but the evolution of device conductance

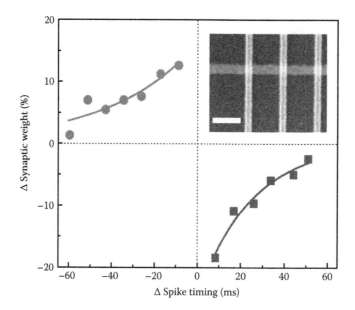

FIGURE 13.15 The measured change of synaptic weight in the memristor as a function of the relative neuron spiking timing (Δt), demonstrating the STDP function. The inset shows an SEM image of the fabricated memristor crossbar array with a scale bar of 300 nm. (From Jo, S.H. et al., *Nano Lett.*, 10, 1297, 2010.)

still follows the P/D modulation. Assuming that biological synapses are communicated at a rate of 1 Hz [23], this endurance provides the capability of continuous synaptic operation for around 5 years. The endurance characteristic together with the large density offered by the simple two-terminal memristor synapses in the crossbar configuration, as shown in Figure 13.14b, make the hybrid CMOS–neuron/memristor–synapse approach very promising for the hardware implementation of biology-inspired neuromorphic systems.

13.3.4 Electrochemical Dynamics of Metallic Nanoparticles in Dielectrics

In device applications, nanocrystals and nanoparticles are usually embedded in certain matrices with insulating or semiconducting properties. The ability to control and manipulate nanoscale inclusions in the solid-state medium will greatly expand the functions of materials and create new devices for electronic and optoelectronic applications. Therefore, it is vital to understand how nanoparticles behave in electrical and chemical aspects. As mentioned above, novel devices like memristors and electronic synapses are mostly based on metallic nanoclusters dispersed in a dielectric matrix. Thus, in this section, the metallic nanoparticles in dielectrics are taken as an example to elaborate the electrochemical dynamics in the nanocomposite materials.

Yang et al. studied the origin of the dynamic growth and migration processes of metal nanoclusters in dielectrics with microscopic techniques [24]. In situ transmission electron microscopy (TEM) studies were carried out on samples consisting of individual metal nanoclusters embedded in a dielectric host film. Detailed analysis and tracking of individual clusters were conducted and the migration inside the bulk of the film was investigated. By tracing the dynamic motion of nanoclusters under electric field in an in situ TEM, fundamental atomic/ionic dynamics were revealed. The experimental strategy is illustrated in Figure 13.17a through d, where Ag nanoclusters were used. As illustrated in Figure 13.17a, Ag nanoclusters were included in the SiO_2 film by annealing a stacked $SiO_2/Ag/SiO_2$ structure with an Ag layer of ~1 nm thickness. Au electrodes were then deposited onto the $SiO_2/Ag/SiO_2$ stack. Subsequently, successive focused ion beam lift-out, patterning, and thinning processes were performed to prepare an array of Au/SiO_2/Ag nanoclusters/SiO_2/Au

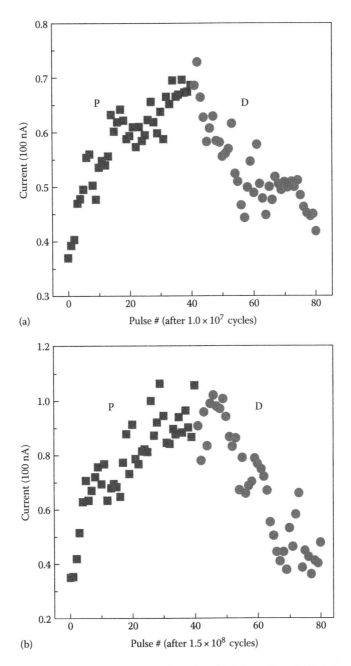

FIGURE 13.16 Endurance of the memristor device after (a) 1.0×10^7 and (b) 1.5×10^8 potentiating/depressing (P/D) cycles. In each test, 3.1 V, 800 μs potentiating pulses, −2.9 V, 800 μs depressing pulses, and 1 V, 2 ms read pulses were used. After each P or D pulse, the device conductance was measured by a read pulse and recorded. (From Jo, S.H. et al., *Nano Lett.*, 10, 1297, 2010.)

devices for cross-sectional TEM studies. As shown in Figure 13.17d, the top Au electrode of a device is contacted by a movable W probe that applies voltages during in situ TEM analysis. Figure 13.17e shows a TEM image for an embedded Ag nanoparticle along its [011] zone axis, and Figure 13.17f displays the corresponding fast Fourier transformation results, which clearly reveal the crystalline face-centered cubic (FCC) structure of the Ag nanocluster.

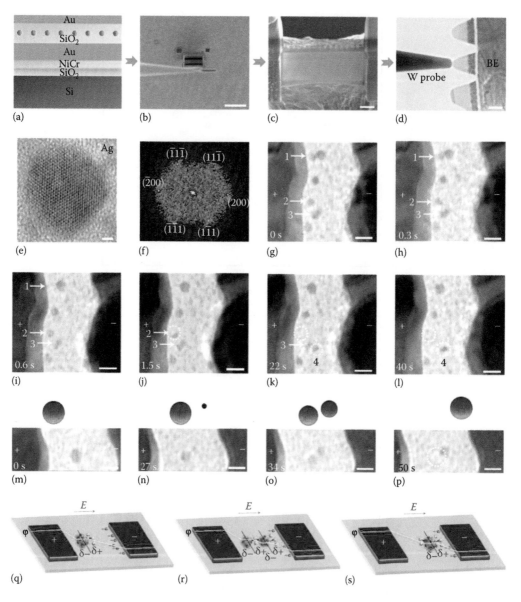

FIGURE 13.17 Dynamic growth and migration of Ag nanoclusters in a SiO$_2$ film. (a) Schematic of the device structure consisting of Ag nanoclusters embedded SiO$_2$ film. (b) SEM image showing the FIB lift-out process. Scale bar: 20 mm. (c) SEM image showing an array of samples prepared by FIB. Scale bar: 1 mm. (d) TEM image showing the in situ experimental setup. Scale bar: 200 nm. (e) High-resolution TEM image and (f) the corresponding fast Fourier transformation results. Scale bar: 1 nm. (g–l) Real-time TEM images revealing the electrical field–induced growth and migration of Ag nanoclusters. Scale bar: 10 nm. (m–p) In situ TEM images together with simulation results showing the dissolution and growth processes between two Ag nanoclusters. Scale bar: 10 nm. (q) Schematic of a polarized Ag nanocluster embedded in SiO$_2$ under electric field, overlaid with the electric potential. (r) New cluster nucleation and formation of another bipolar electrode nearby. (s) The dissolution of the original cluster and the growth of the new cluster lead to an effective movement of a cluster along the electric field. (From Yang, Y. et al., *Nat. Commun.*, 5, 4232, 2014.)

The evolution and migration of embedded Ag nanoclusters can be regularly observed when an electric field (~0.8 MV/cm corresponding to an applied voltage of 3 V) is applied across the Au electrode, as shown in Figure 13.17g through l. The Ag nanoclusters formed in SiO_2 have diameters of ~5 nm. At location 1, dynamic interactions between two Ag nanoclusters were observed. As shown in Figure 13.17g through i, the nanocluster on the left was gradually dissolved and included into another cluster on the right side, resulting in an overall movement of Ag nanoclusters along the direction of the applied electric field. Similarly, clusters 2 and 3 were also gradually dissolved. Cluster 2 disappeared with a void left in the film, as shown in Figure 13.17j and k. As for cluster 3, its dissolution resulted in the creation of a new cluster (namely, cluster 4) nearby, indicated by the lower circle in Figure 13.17k and l. Figure 13.17m through p shows the interactions between another pair of Ag nanoclusters, illustrating the gradual dissolution of the original nanocluster in the upstream and the nucleation and growth of a new nanocluster in the downstream. Effective growth along the field direction was achieved as the original Ag nanocluster was completely dissolved in a location where a void was left behind (indicated by a circle in Figure 13.17p).

The dynamic process of Ag nanoclusters revealed by in situ TEM can be explained by electrochemical redox processes in bipolar electrodes with strong influence from kinetic effects. Under an electric field inside the dielectric, the metal nanoclusters can be treated as bipolar electrodes, with an effective cathode (δ−) facing the field direction and an effective anode (δ+) on the opposite side, as shown in Figure 13.17q. In this model, electrochemical oxidation processes occur on the anode side of the clusters and Ag^+ ions are generated. The Ag^+ ions drift under the applied electric field and then reduced upon reaching a cathode (e.g., the cathode side of another Ag cluster in the downstream). In the Au/SiO_2/Ag nanoclusters/SiO_2/Au devices, there is no Ag supply to the first Ag cluster upstream, so it experiences only net oxidation and thus gradually dissolves. The created Ag^+ ions can be reduced and deposited onto the second nanocluster downstream. If oxidation on the anode side of the second nanocluster is suppressed due to kinetic limitations, the second nanocluster will experience a net growth. As a result, an effective cluster displacement along the electric field is realized by the merging of the first and second nanoclusters downstream, as shown in Figure 13.17q through s. On the other hand, the redox and migration processes of a specific nanocluster are determined by the local structural and or electrochemical environment sensed by the cluster and are affected by film inhomogeneity and electric field distribution, so the dynamic growth and migration are stochastic in nature.

It is found that the formation of the second nanocluster inside the SiO_2 film requires a nucleation process, which is the initial formation of a critical nucleus. This process can be regarded as a chemical process of formation of stable atomic configuration to minimize the excess surface energy and allow further growth. As a result, the nucleation process is spatially selective instead of being homogeneous, as shown in Figure 13.17. If nucleation could not proceed, the Ag^+ ions can reach the Au counter electrode and become reduced there. This appears to be the case for cluster 2 in Figure 13.17g through l. However, a significant geometric change may not be observed in this case due to the much larger volume of the electrode.

These analyses show that kinetic factors of the electrochemical redox reactions and ionic drift strongly influence the dynamics of metal nanoclusters. The evolution of the shape and size of a nanocluster depends on the competition between the oxidation and reduction rates on the cathode and anode sides of the bipolar electrodes. The relationship can be described as [24]

$$\frac{dV^i}{dt} \propto \Gamma^i_{red} - \Gamma^i_{ox}, \tag{13.20}$$

where

V^i is the volume of cluster i

Γ^i_{ox} and Γ^i_{red} is the oxidation and reduction rate, respectively, on the anode and cathode sides of cluster i, respectively

For two neighboring nanoclusters (i and $i-1$) with small distance, the exchange of Ag ions can be approximated as [24]

$$\Gamma_{red}^{i} = \Gamma_{ox}^{i-1}, \tag{13.21}$$

accounting for the conservation of the Ag material.

The experimental results show that the clusters in the upstream merge into their counterparts in the downstream, for example, clusters 1 and 3 in Figure 13.17g through l and the cluster in Figure 13.17m. Figure 13.17m through p shows the consistency between the experimental observations and simulation results using Equations 13.20 and 13.21 by assuming $\Gamma_{ox}^{i} = 0$ for the second nanocluster. Other behaviors such as displacement (movement) of nanoclusters along the electric field toward the counter electrode can also be expected from Equation 13.20, as shown in Figure 13.17s. Since SiO_2 is an amorphous insulator, the conditions for ion migration and nucleation are inhomogeneous and depend strongly on the local microstructure, electric field distribution, electrochemical environment, and the presence of defects, as observed in Figure 13.17g through p. Different metal systems were extensively studied by Yang et al. as well, and it is concluded that the apparent movements of metal nanoclusters are always along the electric field direction. However, different electric fields are required for the migration of metals within the measurement duration.

In situ TEM studies were also conducted on filament effects in actual resistive memory structures [24]. As shown in Figure 13.18a, devices with a $Ag/SiO_2/W$ structure were formed by evaporating SiO_2 film onto a W probe, which then connects with a high-purity Ag wire. Measurement results for a device with a ~40 nm SiO_2 film are shown in Figure 13.18b through e.

The filament growth in the SiO_2 film started with the appearance of several Ag clusters near the Ag electrode after ~3 min (Figure 13.18b). These clusters then behave as bipolar electrodes during subsequent growth. As expected, over time these clusters move closer to the W electrode, following the splitting–merging processes (Figure 13.17m through p). Due to the higher concentration of Ag^+ ions near the Ag electrode and therefore the higher probability to overcome the nucleation barrier, more Ag clusters will be nucleated near the Ag electrode inside SiO_2, and this repeated nucleation and growth leads to a filament shape, as observed in Figure 13.18d. The growth mode results in a conically shaped filament (Figure 13.18d and e) with its base at the Ag/SiO_2 interface and expanding toward the inert W electrode. Such a growth direction can be explained by the generalized framework discussed earlier. The overall growth of the filament can then be characterized by the step-by-step movements of Ag cluster components driven by the electrochemical kinetics, as illustrated by Figure 13.18f through j. The details include (f) Ag ionization at the active Ag electrode; (g) nucleation of Ag nanoclusters near the active electrode, where the nanoclusters act as bipolar electrodes in subsequent filament growth; (h) growth of Ag nanoclusters at new positions that are closer to the W electrode; (i) migration of Ag nanoclusters leads to void formation marked by the dashed line; and (j) continued filament growth by void refilling from the active electrode.

In addition, the formation of a void was observed in the SiO_2 film at the interface region in Figure 13.18d and e after large amounts of Ag atoms passed through. The void formation indicates the plastic deformation of the dielectric film due to increased mechanical stress. Similar voids have also been observed in the cases of discrete nanoclusters (e.g., Figure 13.17j through l and p). Apart from facilitating fast ion diffusion, the voids provide space for continuous movement of Ag clusters in a bootstrapping fashion, as shown in Figure 13.18d and e. As the existing Ag nanoclusters move away toward the W electrode, new nanoclusters can move in and fill the void. The continuous movement of Ag clusters facilitates filament growth toward the inert W electrode. The observation of dynamic void formation and filament growth processes is fundamentally important for the understanding and optimization of memristive device operations.

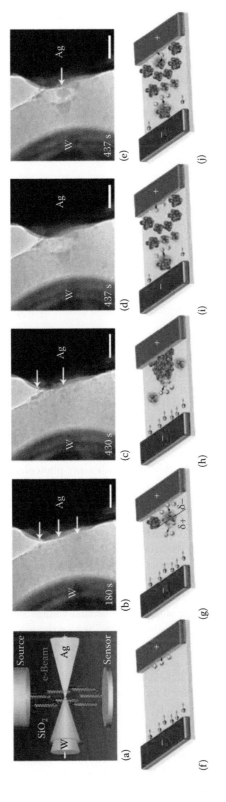

FIGURE 13.18 In situ TEM observation of bootstrapping filament growth and void formation. (a) Schematic of the experimental setup. (b–e) Real-time TEM images showing continuous filament growth in a 40 nm SiO$_2$ film under the application of 8 V. Scale bar: 20 nm. (f–j) Schematic illustration of the bootstrapping filament growth mode. (From Yang, Y. et al., *Nat. Commun.*, 5, 4232, 2014.)

13.4 SELF-LEARNING DEVICES

As one kind of smart materials, nanoparticles/nanoclusters-related materials can be used in smart devices for neural applications. Neural networks have an important characteristic, which is the learning ability, and thus human/living organisms can react according to input signals based on their experiences. Such neural networks have been realized with conventional MOS devices, but complicated software programming with additional devices and circuits are required due to lack of inherent self-learning ability in the conventional devices [25]. As a result, due to hardware complexity and huge energy dissipation, biomimetic computing architectures with self-learning abilities are difficult to be employed in current computational systems, which is similar to the case discussed [22]. Realization of devices with inherent self-learning abilities could be one of the fundamental solutions to overcome this obstacle.

13.4.1 SELF-LEARNING BEHAVIORS IN Ni-RICH NiO THIN FILMS

Liu et al. demonstrate a resistive switching device based on a Ni-rich nickel oxide thin film that has an inherent self-learning ability similar to that of the neural networks in the human brain [26]. Depending on the experience of voltage pulses or sweeping in a way similar to that of the human brain, the device exhibited multiple resistance states corresponding to short-term memory (STM) or long-term memory (LTM). In addition, through the modulation of its resistance according to the number of external electrical pulses, the device was able to mimic neuroplasticity, which is another important function of neural networks.

As shown in Figure 13.19a, device fabrication started with thermal growth of 200 nm SiO_2 thin film on a p-type Si wafer. A 10 nm Ni layer followed by 200 nm Au layer were deposited onto the SiO_2 film with electron-beam evaporation to form the bottom electrode. A Ni-rich nickel oxide thin film with a thickness of ~60 nm was then deposited onto the bottom electrode. The deposition was carried out by radio-frequency (rf) (13.6 MHz) magnetron sputtering of a NiO target (99.99% in purity) with an Ar flow rate of 75 sccm and the rf power of 150 W. A 120 nm Au/10 nm Ni layer was deposited onto the nickel oxide thin film by electron-beam evaporation to form top electrodes. The devices actually had a simple metal-insulator-metal (MIM) structure. An analysis on the chemical states of the synthesized nickel oxide thin film with x-ray photoelectron spectroscopy (XPS) revealed that the Ni:O atomic ratio was 1.5:1, indicating that the as-deposited nickel oxide thin film was Ni rich. Naturally, the elemental Ni component was believed to be dispersed in the NiO matrix as Ni nanoparticles or nanoclusters.

The device shown in Figure 13.19a is basically a typical MIM structure used in the resistive random access memory (RRAM); therefore, it exhibits a resistive switching behavior [26]. Figure 13.19b shows the typical $I-V$ characteristics of the device in voltage-sweeping measurement with a current compliance of 0.1 A. When the voltage stated with 0 V was increased to ~1.8 V, there was an abrupt increase in the current, showing that the device resistance was switched from a high-resistance state to a low-resistance state. In the second voltage sweeping, a sudden drop of current appeared when the voltage reached ~0.6 V, showing that the resistance was switched back to the high-resistance state. This is a typical resistive switching behavior, which can be controlled by voltage sweeping applied to the electrodes of the device. The formation and rupture of conductive filaments in the material system are believed to be responsible for the resistive switching [26]. The self-learning ability was realized based on the resistive switching characteristics of the NiO thin film–based device, and achieving STM, LTM, and a transition from STM to LTM was possible with this device.

The memorization was realized in the device as a function of the number of the electrical pulses applied across the top and bottom electrodes [26]. Pulses of 0.5 V, 0.1 s apart were applied to the device in a low-resistance state, and the evolution of device resistance was recorded. As shown in Figure 13.20a, three memory stages can be observed, which are the unmemorized stage, the STM

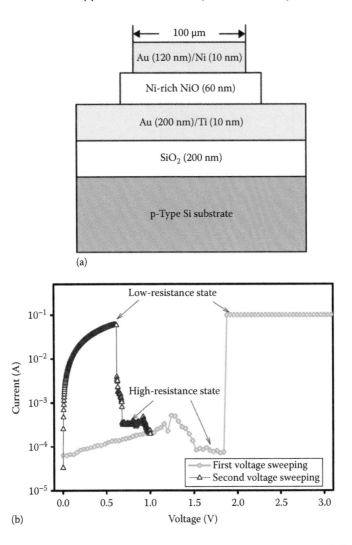

FIGURE 13.19 (a) Schematic of the self-learning device structure and (b) *I–V* characteristics of the device in the voltage sweeping measurement. (From Liu, Y. et al., *Appl. Phys. A*, 105, 855, 2011.)

stage, and the LTM stage. In the unmemorized stage of the first 16 pulses, the device resistance was ~5 Ω and there was no obvious change. This stage imitates an unmemorized state in human neural networks, meaning the information is not stored in the brain (i.e., no neuronal connections are established) although we experience some external impression. In the next 17th–36th pulses corresponding to the STM stage, the device resistance increased gradually with the number of applied pulses. This is an intermediate state between the unmemorized stage and the LTM stage since the resistance is not stable. These slight increases in device resistance in the intermediate states originated from the rupture of localized filaments in the material system [27]. In this stage, the memorization is formed and gradually enhanced with external impression. However, it is still not stable, and the memorization can be easily lost. Experimental results in Figure 13.20b show that the STM state can switch back to the initial state after a waiting time of ~1500 s, exhibiting a memory loss behavior. In the LTM stage, the resistance was stabilized in a saturation state. The saturated resistance can be maintained for a considerably long time if the conductive filaments in the material system are fully ruptured due to Joule heating process [28,29]. The stable memorization is analogous to the LTM in human brain.

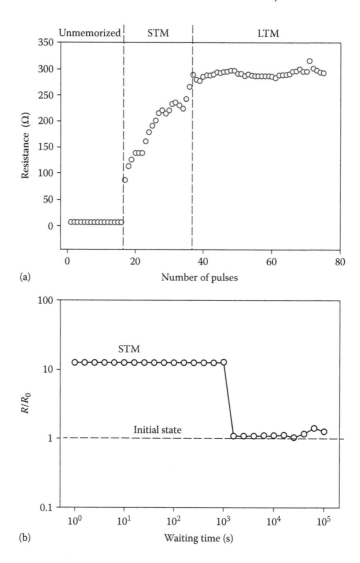

FIGURE 13.20 (a) The evolution of device resistance as a function of the number of pulses and (b) the change of normalized resistance of the STM state with waiting time. (From Liu, Y. et al., *Appl. Phys. A*, 105, 855, 2011.)

As can be seen from Figure 13.20a, for the STM to occur, 16 pulses of 0.5 V, 0.1 s are required. If the voltage magnitude or the duration of applied pulses changes, a different number of pulses will be needed to trigger the STM formation. Figure 13.21 shows the dependence of minimum number of pulses on the voltage magnitude and pulse width for the STM to occur in the device. The STM was considered to be formed when the resistance was increased by an order. For each pulse voltage/width, 10 devices located at different places on the wafer were measured to obtain the minimum number of pulses required statistically. Small pulse voltages (i.e., below 0.4 V) were not able to form STM regardless of the pulse width. This is due to the fact that localized Joule heating and or electric fields at small voltages are not sufficient to cause the rupture of localized filaments, which are formed in the low-resistance states [27]. For the pulse with a width of 0.005 s, an average of 43 pulses with a voltage magnitude of 0.5 V were required to form the STM; however, the average number was reduced to 20 if 0.6 V pulses were used. On the other hand, with the voltage magnitude

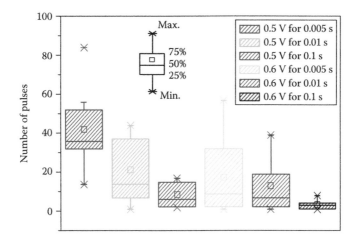

FIGURE 13.21 Minimum numbers of pulses required for the STM to occur for different combinations of voltage magnitude and pulse width. (From Liu, Y. et al., *Appl. Phys. A*, 105, 855, 2011.)

fixed at 0.5 or 0.6 V, the average number of pulses decreased as well when the pulse width was increased, as can be seen from Figure 13.21. In addition, the resistance change is also a function of the interval between applied electrical pulses. For a longer interval between pulses, more pulses would be needed to switch the device. The dependence of minimum number of pulses for the STM to occur on the voltage magnitude and pulse width is related to the heat accumulation process during the rupture of the conductive filaments. Pulses with a larger voltage magnitude, a larger pulse width, a larger duty cycle (i.e., a short interval in between pulses), or a combination of these are able to accumulate a high density of heat energy in a localized region and, thus, easily switch the device from the unmemorized state to the STM state through the melting of conductive filaments.

Instead of voltage pulses, repeated forward and backward voltage sweepings can also be used to realize memorization transition from STM to LTM [26]. Figure 13.22a shows the dual-sweeping *I–V* characteristics for 10 cycles. A quasi-linear *I–V* characteristic was observed for each sweeping, suggesting that ohmic conduction might be the dominant current conduction process in the device. Device resistance was obtained at +0.05 V in backward sweepings, and the evolution of the resistance with measurement cycles is shown in Figure 13.22b. As shown in Figure 13.22a, the *I–V* curves show a hysteresis in the forward and backward measurement, and the device current continuously decreases with the sweeping cycles. In the first five cycles, very minimal hysteresis can be observed, indicating that the device is maintained in the unmemorized state. The resistance in this state is almost a constant of ~4 Ω as shown in Figure 13.22b. For the sixth cycle, a large hysteresis occurs, and the resistance is suddenly increased to 25 Ω. The resistance increased again to ~52 Ω for the seventh cycle, indicating that the device was entering the STM state. For the 9th and 10th cycles, the two curves were almost consistent. No significant hysteresis can be observed and the resistance was saturated at ~78 Ω. It is believed that the device was in the LTM state after the ninth dual-sweeping measurement.

Figure 13.23 shows the statistical minimum number of cycles required for the STM to occur as a function of the ending voltage of a sweeping cycle. The results were similar to the dependence of the minimum number of the pulses required for the STM to occur on the pulse voltage described earlier. As can be seen in Figure 13.23, the average minimum number of cycles decreases when the ending voltage is increased. For example, six cycles of voltage sweeping were required to switch the device into STM for an ending voltage of 0.4 V; however, only four cycles were needed for an ending voltage of 0.5 V. This indicated that a larger portion of conductive filaments were ruptured in a cycle with a higher ending voltage.

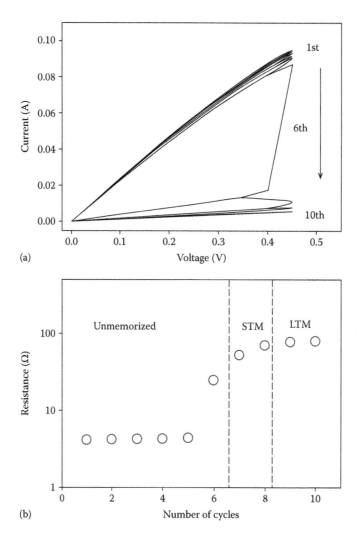

FIGURE 13.22 (a) *I–V* characteristics of the forward and backward voltage sweepings and (b) evolution of the resistance measured at +0.05 V in the backward sweeping with sweeping cycles. (From Liu, Y. et al., *Appl. Phys. A*, 105, 855, 2011.)

Figure 13.24 shows a possible circuit application of the self-learning device. The input voltage pulses applied on the V_{in} have a minimum value of 0 V and a maximum value of V_{im}. The output pulses at the V_{out} terminal should have a voltage magnitude changing with time, as shown in Figure 13.24. It should have a minimum voltage of 0 V and a maximum value of V_{om}, which can be described by

$$V_{om} = V_{im} \left[1 + \frac{R_1}{(R_2 + R_D)} \right]^{-1}, \tag{13.22}$$

where R_D is the resistance of the self-learning device, which changes with the number of pulses. R_1 and R_2 can be carefully chosen according to the range of device resistance to make the circuit sensitive. Since R_D will gradually increase to a saturated value with the number of applied pulses, the output voltage will also increase to a saturated value with the number of pulses. The illustration

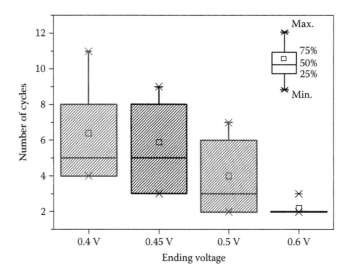

FIGURE 13.23 Minimum number of voltage sweeping cycles required for the STM to occur as a function of the ending voltage of the sweeping cycle. (From Liu, Y. et al., *Appl. Phys. A*, 105, 855, 2011.)

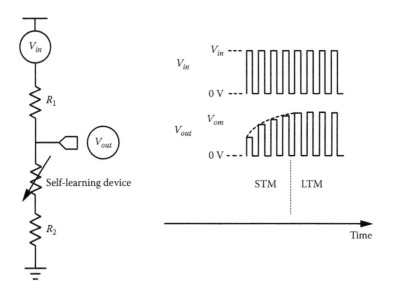

FIGURE 13.24 A possible circuit implementation of the self-learning device. (From Liu, Y. et al., *Appl. Phys. A*, 105, 855, 2011.)

of input and output signals as a function of time is illustrated in Figure 13.24, indicating that the memorization from STM to LTM can be realized with such a circuit implementation based on the self-learning device discussed earlier.

Neuroplasticity is used to describe brain changes in response to experience [30]. It is mainly based on the creation of interconnections between neurons. Similarly, in the resistive switching device, the internal structure changes through formation and rupture of conductive filaments as a result of the application of a series of electrical pulses. The change in internal structure is demonstrated by a sharp change or switching in device resistance. The resistance can be increased by electrical pulses, as shown in Figure 13.20a, and a sharp switching from a low-resistance state to a high-resistance state can be achieved with pulses with suitable voltage and width. On the other hand, sharp switching from a high-resistance state to a low-resistance state can be also achieved with pulses with sufficiently large voltage and pulse width if a high-resistance state already exists. Figure 13.25 shows the response of the resistance of a device in a high-resistance state to the application of electrical pulses of 1.9 V, 0.1 s. The high resistance of ~10^4 Ω was maintained for five pulses. However, the application of the sixth pulse switched the device to a low-resistance state of 19 Ω. The resistance remained unchanged for the application of subsequent pulses. This indicates that the device could be switched from a high-resistance state to a low-resistance state by electrical pulses without experiencing any intermediate states. It is believed that stable filaments were fully formed in the device at the sixth pulse, which is similar to a neuroplasticity behavior, where neuronal connections are created after certain times of experiences.

Figure 13.26 shows the statistical minimum numbers of pulses required to switch a device from a high-resistant state to a low-resistance state at various pulse voltages with the pulse width fixed at 0.1 s [26]. It was found that voltages below 1.6 V were not able to switch a device from the high-resistance state to the low-resistance state. Twenty-five pulses at 1.7 V on average over five measured devices were needed for the switching to occur, and the average number became smaller for pulses with higher voltages.

13.4.2 Transient Memory Loss Behavior

In biological systems, "memory" is an essential building block for learning. Unlike modern semiconductor memories, human memory is by no means eternal. Yet, forgetfulness is not always a disadvantage since it helps to remove redundant information and release memory storage for more important or more frequently accessed information. Moreover, it is thought to be necessary for individuals to adapt to new environments. Eventually, only memories that are of great significance are transformed from STM into LTM through stimulation based on repeated experiencing. No matter what the exact mechanisms are, forgetting is an indisputable fact. Simply speaking, short-term memory retention can be attributed to transient activation or transmission of certain chemicals, which could be ions, transmitters, receptors, etc. In the human brain, memory immediately starts to decay after an impression. For stimulations with a longer interval in between, memory decay will be more significant and more impressions will be needed to achieve the STM and the LTM.

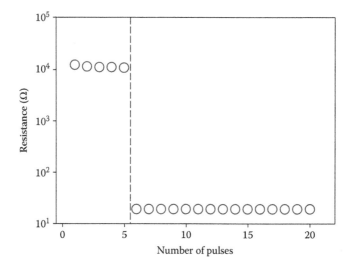

FIGURE 13.25 Device resistance in response to electrical pulses resembling neuroplasticity in human brain. The resistance was measured at +1.9 V. (From Liu, Y. et al., *Appl. Phys. A*, 105, 855, 2011.)

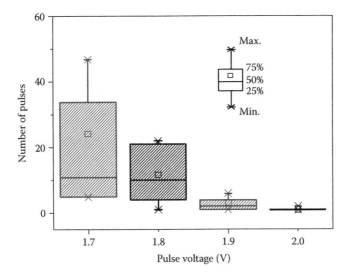

FIGURE 13.26 Minimum numbers of pulses needed to switch a device from the high-resistance state to the low-resistance state at various pulse voltages with the pulse width fixed at 0.1 s. (From Liu, Y. et al., *Appl. Phys. A*, 105, 855, 2011.)

Hu et al. reported transient memory loss effect in resistive switching devices based on Ni-rich NiO thin films [31]. Similar to the learning-behavior introduced in Section 13.4.1, the device based on Ni-rich NiO thin film shows a gradual change in conductance under the application of consecutive voltage pulses [31], resembling the STM to LTM transition in biological systems, as shown in Figure 13.27. The difference is that the device exhibited a conductance increment rather than a resistance increment (conductance reduction), shown in Figure 13.20a. As shown in Figure 13.27,

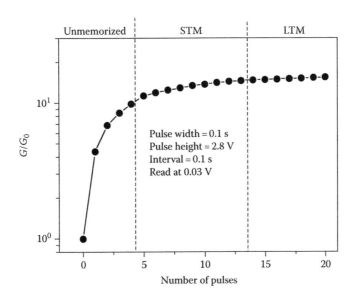

FIGURE 13.27 Device conductance (G) measured at 0.03 V as a function of the number of the applied electrical pulses (2.8 V in voltage magnitude, 0.1 s in width, and 0.1 s in interval between pulses). The initial conductance (G_0) is 387.5 μS. (From Hu, S.G. et al., *Appl. Phys. A*, 109, 349, 2012.)

STM has been achieved after four electrical pulses, which is similar to the situation that our brain can quickly memorize a scene or a matter after several impressions. The memorization is continuously enhanced with the number of voltage pulses until a saturation state that is reached after 13 electrical pulses, and the memory state can be maintained for a long time. The conductance state corresponding to the STM is an unstable intermediate state with a typical retention time of tens of minutes; in contrast, the saturated conductance state corresponding to the LTM has a retention time longer than a few hours. The formation of STM conductance state here is due to the formation of the localized conductive filaments in the material system [27–29]. With more electrical pulses applied, more and or stronger localized filaments are created, and thus the device conductance increases [32]. Once stable and strong filaments are formed, the LTM state with a saturated conductance is formed. On the other hand, trapping of electrons in the oxide film could also affect the conduction. The defects in the oxide, including O vacancies [33] and Ni vacancies [34], can serve as the electron traps that modulate current conduction during electrical stimulations.

Besides the learning ability that yields memorization, memory loss behavior (i.e., the forgetting effect) is another important function of the human brain. The resistive switching device can also be used to mimic memory loss function. Successive voltage pulses with a fixed height of 1.8 V and a width of 0.005 s, but with different intervals (i.e., the off time between two consecutive pulses) of 0.005, 0.05, and 0.5 s, respectively, were applied to the device. The device conductance was measured at a reading voltage of 0.03 V after each pulse. Figure 13.28 shows the evolution of device conductance, which is normalized to the initial conductance of G_0 as a function of the number of voltage pulses for different intervals [31]. As can be seen in this figure, to achieve STM with the conductance increased by one order, 13, 22, and 43 pulses are required for the intervals of 0.005, 0.05, and 0.5 s, respectively. It was believed that voltage-induced Joule heating plays an important role in the process of CF formation in the resistive switching device [28,29,35,36]. Besides, heat diffusion during the interval suppresses the formation of CFs. Therefore, the number of pulses required to form the STM conductance state increases with the interval. This phenomenon is identical to the memory-loss function in the human brain, where a piece of memory could be lost if impressions were not continuously received.

Figure 13.29 shows the dependence of the minimum number of pulses that is required to achieve the STM state on the pulse interval [31]. The occurrence of the STM is defined as when

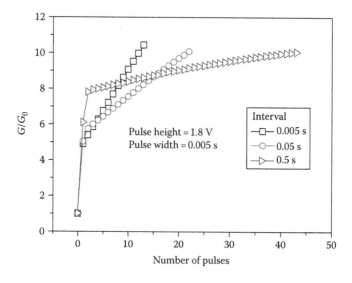

FIGURE 13.28 Change in device conductance with number of pulses for different intervals. The pulse voltage and width are fixed at 1.8 V and 0.005 s, respectively. (From Hu, S.G. et al., *Appl. Phys. A*, 109, 349, 2012.)

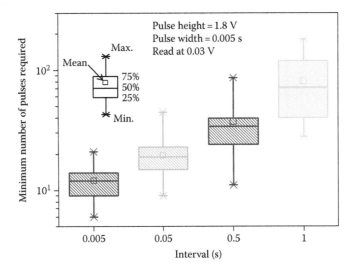

FIGURE 13.29 Statistical minimum number of pulses required to achieve the STM conductance state for different intervals between pulses. (From Hu, S.G. et al., *Appl. Phys. A*, 109, 349, 2012.)

the device conductance is increased by one order. The pulse height and width were fixed at 1.8 V and 0.005 s, respectively. More than 20 devices selected from different locations on the wafer were measured to obtain the statistical average value. The longer the interval between pulses, the larger the number of pulses required on average. For example, 12.2 pulses on average are needed to achieve the STM for an interval of 0.005 s; while 81.9 pulses on average are required for an interval of 1 s. This characteristic in the resistive switching device can simulate the human brain's behavior—that it is difficult to form the STM with longer intervals between events due to memory loss during pulse intervals, and thus more impressions are required to form a solid memorization.

The self-learning ability of a device has been demonstrated with repeated forward and backward voltage sweepings [25,26]. Similarly, the "forgetting" ability can be also realized with voltage sweepings. Figure 13.30 shows typical *I–V* characteristics of the device measured during the repeated forward and backward voltage sweepings between 0 and 2.2 V. The interval between the forward and backward sweeping cycles was set to 10 and 20 s, with results shown in Figure 13.30a and b, respectively. The device current increases gradually after each sweeping cycle, showing that the conductance increases with voltage sweeping. To achieve the STM state with conductance increased by an order, 11 and 17 sweeping cycles are required for the sweeping intervals of 10 and 20 s, respectively, as shown in Figure 13.30.

Figure 13.31 shows the statistical minimum number of sweeping cycles that is required to form the STM state as a function of the interval. As can be seen clearly in this figure, more sweeping cycles are needed for a longer interval. The results shown in Figure 13.30 and Figure 13.31 suggests that the formation of conductive filaments are depressed during the interval period and thus more sweeping cycles are needed to form sufficient conductive filaments for a longer interval. The depression of the conductive filament formation during intervals is analogous to the memory-loss function in the human brain.

13.5 PRINTING AND FLEXIBLE THIN-FILM TRANSISTORS

Moore's law has been the primary driving force for the advancement of ICs, which leads to the number of transistors per microelectronic chip being doubled every 1½ years. However, the cost of a chip per unit of area has remained relatively static for decades. The prevalent Si-based technology

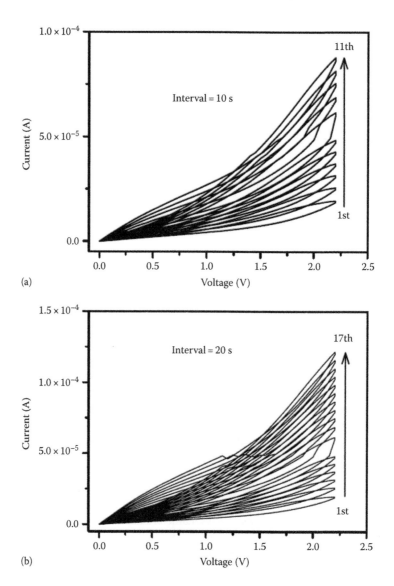

FIGURE 13.30 Measured *I–V* characteristics of the device in the repeated forward and backward voltage sweepings, with the interval between two consecutive voltage sweeping being (a) 10 s and (b) 20 s, respectively. (From Hu, S.G. et al., *Appl. Phys. A*, 109, 349, 2012.)

is not always a cost-effective solution for certain IC products, such as the popular wearable devices nowadays. Hence, there is interest in developing solution-based techniques (e.g., printing and spin-coating) for microelectronics fabrication that are inexpensive, allow fabrication on flexible substrates (e.g., plastics and fiber papers), and have large-area synthesizing ability.

The primary focus has been on organic materials for solution-based printing [37,38]. Organic semiconductors such as poly(3-hexylthiophene) based on solution-based synthesis approaches have demonstrated a field effect mobility of ~0.1 cm^2 V^{-1} s^{-1} [38]. The low mobilities of the solution-processed organic thin-film transistors (TFTs) make them inappropriate for applications in microprocessors, display drivers, and active matrix backplanes for high-resolution displays. Inorganic semiconductors are the basis for almost all high-performance microelectronic devices. They can

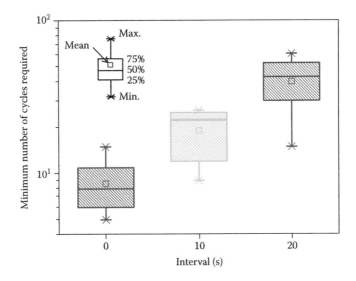

FIGURE 13.31 Statistical minimum number of the sweeping cycles required to achieve the STM conductance state for different intervals. (From Hu, S.G. et al., *Appl. Phys. A*, 109, 349, 2012.)

have an intrinsic mobility as high as 1000 cm^2 V^{-1} s^{-1} and a lifetime more than 50 years. Inorganic microelectronic devices and circuits are generally fabricated with conventional Si-related techniques such as thermal oxidation, evaporation, sputtering, chemical vapor deposition, ion implantation, photolithography, and so on. Unfortunately, the solution-based approach is not a convenient way to synthesize inorganic semiconductors since they are not intrinsically soluble in any convenient solvents and do not offer structural variability as a mechanism to alter their solubility [39].

13.5.1 Printable CdSe Nanocrystal for Inorganic TFTs

Ridley et al. reported a pioneering work on the method of printing inorganic thin films by using stable CdSe nanocrystal solutions as precursors to form polycrystalline thin films [39]. Unlike its bulk crystalline counterpart, the population of surface atoms in a nanocrystal is of nonnegligible percentage. Therefore, nanocrystals have properties somewhere between those of a bulk solid and an atomic species, exhibiting size-dependent physical, electrical, and optical properties [40].

A typical synthesis was carried out in a nitrogen-filled glove box under room temperature. Methanolic solutions of the reagents CdI$_2$ and Na$_2$Se (at 7.4 mM concentrations) were prepared. Equimolar amounts of the reagents were then added to the reaction vessel, which was charged with pyridine. Pyridine generally constituted one-third of the total reaction volume. The reaction proceeds instantaneously, forming a yellow solution that quickly precipitates the nanocrystalline product. The nanocrystals are then isolated by centrifugation and decanting of the supernatant, and the soluble by-product, NaI, is removed by repeatedly dispersing the product in methanol and isolating the particles after centrifugation and decanting. After dissolution of the nanocrystals in pyridine, they are precipitated by addition of hexane and again isolated by centrifugation and decanting. Finally, the nanocrystals are redissolved in pyridine and filtered through a 0.2 mm filter.

After synthesis of CdSe nanocrystals, the fabrication of TFT devices were carried out in the glove box as well [39]. A single drop of the pyridine CdSe nanocluster solution was deposited onto the area between the source and drain electrodes (Cr/Au) of a TFT structure. The CdSe nanocrystals were heated by placing the wafer on a hot plate. Various heat treatments ranging from 150°C to 350°C were used to sinter the nanocrystals in the channel region. Upon cooling, Norland Optical

Adhesive 73, a photocurable polymer adhesive, was deposited and cured to encapsulate the CdSe channel region of the TFT. The fabricated device was a coplanar inverted TFT consisting of n^+-Si gate, 100 nm thermal SiO_2, Cr/Au (10 nm/100 nm) source and drain contacts, and a channel layer with a length of 8 μm and width of 293 μm [39]. The TFTs were measured in air in a dark box at room temperature. It was found that a field effect was not observed from unencapsulated devices and those devices heated to 150°C. However, TFTs processed at 250°C were found to exhibit a field effect, with the mobilities proportional to the treatment temperature.

Figure 13.32a shows the electrical characteristic of the TFT ramped from room temperature at 80°C/min and held at 350°C for an hour [39]. With a gate sweep of $V_{GS} = -40$ to $+40$ V at $V_{DS} = 2.5$ V, the I_D–V_{GS} curve shows an ON/OFF ratio of 3.1×10^4, a linear regime mobility of 1 cm^2 V^{-1} s^{-1}

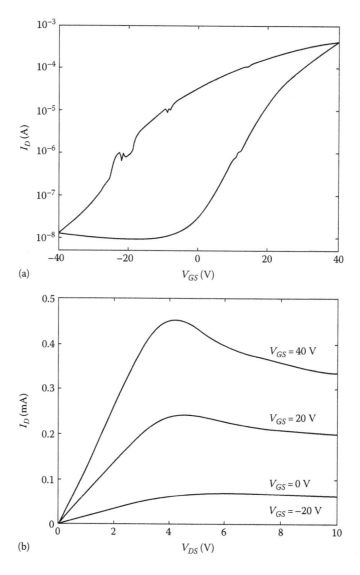

(a)

(b)

FIGURE 13.32 Electrical characteristics of a TFT based on printed CdSe nanocrystals. (a) Logarithmic I_D–V_{GS} curves for a gate sweep of $V_{GS} = -40$ to $+40$ V at $V_{DS} = 2.5$ V, and (b) I_D–V_{DS} curves with a gate sweeping from $V_{GS} = -40$ to $+40$ V in steps of 20 V. (From Ridley, B.A. et al., *Science*, 286, 746, 1999.)

(extracted by equating the slope of an I_D–V_{GS} plot to $(W/L)\mu C_{ox}V_{DS}$ at $V_{DS} = 2.5$ V, where C_{ox} is the capacitance of the gate oxide), a threshold voltage of 6.7 V (extracted from the V_{GS} intercept of an I_D–V_{GS} plot), and a subthreshold slope of ~7 to 10 V/dec. The hysteresis observed in Figure 13.32a may be due to interface states at the semiconductor–insulator interface or simply due to poor encapsulation and oxidation of the TFT. The negative resistance seen in the saturation region of the I_D–V_{DS} characteristics shown in Figure 13.32b is also likely to be caused by hysteresis in the semiconductor film. The mobility of these TFTs is approximately an order of magnitude larger than the mobility of printed organic TFTs. The field-effect mobility can be increased further by improvements in nanocrystal solution purity and processing techniques, which is very promising for practical circuit applications. It is also worth mentioning that the processing temperatures for this device are compatible with high glass transition temperature plastic substrates, such as polyimide.

13.5.2 SPIN-COATED CdSe NANOCRYSTALS FOR FLEXIBLE TFTs

Using the same CdSe nanocrystals (NCs) but with different synthesizing approach, Kim et al. reported flexible, low-voltage, and high-performance TFT devices, and also demonstrated fully functional integrated circuits based on such devices [41]. In order to prepare the solution for spin coating, highly monodispersed CdSe NCs with compact thiocyanate ligands were synthesized in a nitrogen glove box [42,43]. NCs with thiocyanate ligands were then redispersed in dimethylformamide and spincast on the top of flexible substrates to form uniform, crack-free, and randomly close-packed NC thin films as semiconducting channels in TFTs. Unlike organic semiconductors, the NC thin films have a morphology and mobility largely insensitive to surface roughness except with a root-mean-squared value as large as a few nanometers [44]. To fabricate flexible devices, either a 25 or 50 mm thick polyimide, covered with 30 nm of atomic layer deposited (ALD) Al_2O_3 at 250°C, was used as the substrate. The Al_2O_3 encapsulated the polyimide substrate as a buffer layer, which can eliminate severe delamination and cracking of the deposited NC thin films and metal electrodes during subsequent thermal processing. A 20 nm thick Al back gate was deposited onto the buffer layer by thermal evaporation through a shadow mask. The device was exposed to oxygen plasma to form some native Al_2O_3 on the Al gate and to create additional hydroxyl groups. A 30 nm ALD Al_2O_3 was grown as the gate dielectric layer. The measured capacitance of the Al_2O_3 dielectric layer was 0.253 ± 0.019 mF/cm^2, allowing for low-voltage operation. In addition, the gate leakage was characterized at pA levels, indicating the combination of CdSe NC thin film channel and the ALD Al_2O_3 dielectric is suitable for flexible electronic applications [41]. Inside a nitrogen glove box, In/Au (50 nm/40 nm) electrodes were thermally deposited through a shadow mask onto the NC thin film to complete back-gate/top-contact TFTs. Figure 13.33a shows the schematic of the whole device structure, and Figure 13.33b demonstrates the real circuits fabricated on a Kapton substrate based on the CdSe NC TFTs.

Figure 13.33c shows typical output characteristics (I_D–V_{DS}) [41]. An n-type device behavior that is modulated by a small positive voltage as low as 2 V can be observed from this figure. The extracted electron field-effect mobilities from the transfer curves (I_D–V_G) shown in Figure 13.33d are 21.9 ± 4.3 cm^2 V^{-1} s^{-1} in the linear regime ($V_{DS} = 0.1$ V) and are 18.4 ± 3.6 cm^2 V^{-1} s^{-1} in the saturation regime ($V_{DS} = 2$ V). These CdSe NC TFTs built on plastic substrates operate well at low voltages. Good device performances, including high I_{ON}/I_{OFF} ratio (>10^6), low subthreshold swing ($S = 0.28 \pm 0.09$ V/dec), low threshold voltage ($V_T = 0.38 \pm 0.15$ V), and low hysteresis ($\Delta V_T = 0.25 \pm 0.07$ V) at $V_{DS} = 2$ V were achieved in these devices. The low hysteresis was attributed to passivation of the NC surface by indium and the selection of ALD Al_2O_3 as the gate dielectric, which helped to reduce the density of trap states at the NC surface and at the semiconductor–gate dielectric interface [43]. In addition, the small variation in device parameters and large-area uniformity of these NC TFTs enables their integration in flexible TFT–based integrated circuits.

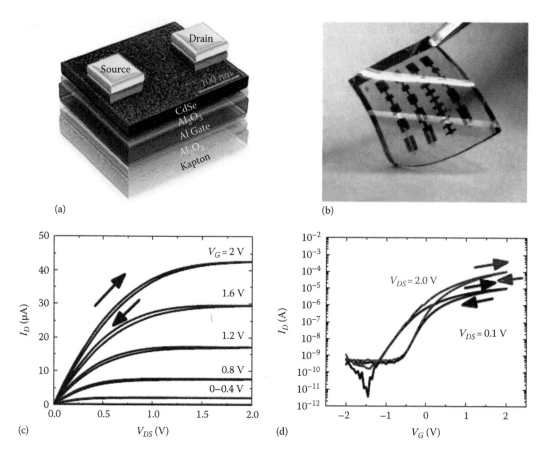

FIGURE 13.33 Flexible TFTs based on CdSe nanocrystals. (a) Schematic and (b) photograph of a flexible TFT based on CdSe NCs atop a Kapton substrate. (c) Output I_D–V_{DS} and (d) transfer I_D–V_G characteristics. (From Kim, D.K. et al., *Nat. Commun.*, 3, 1216, 2012.)

REFERENCES

1. G. Baccarani, M. R. Wordeman, and R. H. Dennard, Generalized scaling theory and its application to a 1/4 micrometer MOSFET design, *IEEE Transactions on Electron Devices*, 31, 452–462, 1984.
2. R. H. Dennard, V. L. Rideout, E. Bassous, and A. R. LeBlanc, Design of ion-implanted MOSFET's with very small physical dimensions, *IEEE Journal of Solid-State Circuits*, 9, 256–268, 1974.
3. K. K. Likharev, Single-electron transistors: Electrostatic analogs of the DC SQUIDS, *IEEE Transactions on Magnetics*, 23, 1142–1145, 1987.
4. H. Grabert and M. H. Devoret, *Single Charge Tunneling: Coulomb Blockade Phenomena in Nanostructures*. New York: Plenum Press, 1992.
5. K. K. Likharev, Single-electron devices and their applications, *Proceedings of the IEEE*, 87, 606–632, 1999.
6. D. V. Averin and K. K. Likharev, Single electronics: A correlated transfer of single electrons and cooper pairs in systems of small tunnel junctions, in: *Mesoscopic Phenomena in Solids*, B. L. Altshuler, P. A. Lee, and W. Richard Webb, eds. Amsterdam, the Netherlands: Elsevier, 1991.
7. K. Uchida, Single-electron devices for logic applications, in: *Nanoelectronics and Information Technology—Advanced Electronic Materials and Novel Devices*, R. Waser, ed., 3rd ed. John Wiley & Sons, Weinheim, Germany, 2012.
8. K. C. Chan, P. F. Lee, and J. Y. Dai, Single electron tunneling and Coulomb blockade effect in HfAlO/Au nanocrystals/HfAlO trilayer nonvolatile memory structure, *Applied Physics Letters*, 92, 143117, 2008.

9. C. Wasshuber, H. Kosina, and S. Selberherr, SIMON-A simulator for single-electron tunnel devices and circuits, *IEEE Transactions on Computer-Aided Design of Integrated Circuits and Systems*, 16, 937–944, 1997.

10. M. Kirihara, K. Nakazato, and M. Wagner, Hybrid circuit simulator including a model for single electron tunneling devices, *Japanese Journal of Applied Physics*, 38, 2028, 1999.

11. T. A. Fulton and G. J. Dolan, Observation of single-electron charging effects in small tunnel junctions, *Physical Review Letters*, 59, 109–112, 1987.

12. U. Meirav, M. A. Kastner, and S. J. Wind, Single-electron charging and periodic conductance resonances in GaAs nanostructures, *Physical Review Letters*, 65, 771–774, 1990.

13. Y. Takahashi, Silicon single-electron devices for logic applications, in: *Proceeding of the 32nd European Solid-State Device Research Conference*, Firenze, Italy, 2002, pp. 61–70.

14. Y. Ono, Y. Takahashi, K. Yamazaki, M. Nagase, H. Namatsu, K. Kurihara et al., Fabrication method for IC-oriented Si single-electron transistors, *IEEE Transactions on Electron Devices*, 47, 147–153, 2000.

15. D. L. Klein, R. Roth, A. K. L. Lim, A. P. Alivisatos, and P. L. McEuen, A single-electron transistor made from a cadmium selenide nanocrystal, *Nature*, 389, 699–701, 1997.

16. L. P. Kouwenhoven and P. L. McEuen, Single electron tunneling through a quantum dot, in: *Nanoscience and Technology*, G. Timp, ed. New York: AIP Press.

17. L. O. Chua, Memristor—The missing circuit element, *IEEE Transactions on Circuit Theory*, 18, 507–519, 1971.

18. D. B. Strukov, G. S. Snider, D. R. Stewart, and R. S. Williams, The missing memristor found, *Nature*, 453, 80–83, 2008.

19. T. H. Kim, E. Y. Jang, N. J. Lee, D. J. Choi, K.-J. Lee, J.-T. Jang et al., Nanoparticle assemblies as memristors, *Nano Letters*, 9, 2229–2233, 2009.

20. S. H. Jo, T. Chang, I. Ebong, B. B. Bhadviya, P. Mazumder, and W. Lu, Nanoscale memristor device as synapse in neuromorphic systems, *Nano Letters*, 10, 1297–1301, 2010.

21. L. O. Chua and K. Sung Mo, Memristive devices and systems, *Proceedings of the IEEE*, 64, 209–223, 1976.

22. R. Ananthanarayanan, S. K. Esser, H. D. Simon, and D. S. Modha, The cat is out of the bag: Cortical simulations with 109 neurons, 1013 synapses, in: *Proceedings of the Conference on High Performance Computing Networking, Storage and Analysis*, Portland, OR, 2009, pp. 1–12.

23. V. Lev-Ram, S. T. Wong, D. R. Storm, and R. Y. Tsien, A new from of cerebellar long-term potentiation is postsynaptic and depends on nitric oxide but not cAMP, *Proceedings of the National Academy of Sciences of the United States of America*, 99, 8389, 2002.

24. Y. Yang, P. Gao, L. Li, X. Pan, S. Tappertzhofen, S. Choi et al., Electrochemical dynamics of nanoscale metallic inclusions in dielectrics, *Nature Communications*, 5, 4232, 2014.

25. T. Hasegawa, T. Ohno, K. Terabe, T. Tsuruoka, T. Nakayama, J. K. Gimzewski et al., Learning abilities achieved by a single solid-state atomic switch, *Advanced Materials*, 22, 1831–1834, 2010.

26. Y. Liu, T. P. Chen, Z. Liu, Y. F. Yu, Q. Yu, P. Li et al., Self-learning ability realized with a resistive switching device based on a Ni-rich nickel oxide thin film, *Applied Physics A*, 105, 855–860, 2011.

27. D. C. Kim, S. Seo, S. E. Ahn, D.-S. Suh, M. J. Lee, B.-H. Park et al., Electrical observations of filamentary conductions for the resistive memory switching in NiO films, *Applied Physics Letters*, 88, 202102, 2006.

28. A. Sawa, Resistive switching in transition metal oxides, *Materials Today*, 11, 28–36, 2008.

29. R. Waser and M. Aono, Nanoionics-based resistive switching memories, *Nature Materials*, 6, 833–840, 2007.

30. R. J. Davidson and A. Lutz, Buddha's brain: Neuroplasticity and meditation [in the spotlight], *IEEE Signal Processing Magazine*, 25, 176–174, 2008.

31. S. G. Hu, Y. Liu, T. P. Chen, Z. Liu, Q. Yu, L. J. Deng et al., Realization of transient memory-loss with NiO-based resistive switching device, *Applied Physics A*, 109, 349–352, 2012.

32. M.-J. Lee, S. Han, S. H. Jeon, B. H. Park, B. S. Kang, S.-E. Ahn et al., Electrical manipulation of nanofilaments in transition-metal oxides for resistance-based memory, *Nano Letters*, 9, 1476–1481, 2009.

33. M. C. Ni, S. M. Guo, H. F. Tian, Y. G. Zhao, and J. Q. Li, Resistive switching effect in $SrTiO_3$–δ/Nb-doped $SrTiO_3$ heterojunction, *Applied Physics Letters*, 91, 183502, 2007.

34. W.-L. Jang, Y.-M. Lu, W.-S. Hwang, T.-L. Hsiung, and H. P. Wang, Point defects in sputtered NiO films, *Applied Physics Letters*, 94, 062103, 2009.

35. S. Seo, M. J. Lee, D. H. Seo, E. J. Jeoung, D.-S. Suh, Y. S. Joung et al., Reproducible resistance switching in polycrystalline NiO films, *Applied Physics Letters*, 85, 5655–5657, 2004.

36. K. Tsunoda, K. Kinoshita, H. Noshiro, Y. Yamazaki, T. Iizuka, Y. Ito et al., Low power and high speed switching of Ti-doped NiO ReRAM under the unipolar voltage source of less than 3 V, in: *2007 IEEE International Electron Devices Meeting—IEDM'07*, Piscataway, NJ, December 10–12, 2007, pp. 767–770.

37. Z. Bao, A. Dodabalapur, and A. J. Lovinger, Soluble and processable regioregular poly(3-hexylthiophene) for thin film field-effect transistor applications with high mobility, *Applied Physics Letters*, 69, 4108–4110, 1996.

38. H. Sirringhaus, N. Tessler, and R. H. Friend, Integrated optoelectronic devices based on conjugated polymers, *Science*, 280, 1741–1744, 1998.

39. B. A. Ridley, B. Nivi, and J. M. Jacobson, All-inorganic field effect transistors fabricated by printing, *Science*, 286, 746–749, 1999.

40. A. P. Alivisatos, Electrical studies of semiconductor-nanocrystal colloids, *MRS Bulletin*, 23, 18–23, 1998.

41. D. K. Kim, Y. Lai, B. T. Diroll, C. B. Murray, and C. R. Kagan, Flexible and low-voltage integrated circuits constructed from high-performance nanocrystal transistors, *Nature Communications*, 3, 1216, 2012.

42. A. T. Fafarman, W.-K. Koh, B. T. Diroll, D. K. Kim, D.-K. Ko, S. J. Oh et al., Thiocyanate-capped nanocrystal colloids: Vibrational reporter of surface chemistry and solution-based route to enhanced coupling in nanocrystal solids, *Journal of the American Chemical Society*, 133, 15753–15761, 2011.

43. J.-H. Choi, A. T. Fafarman, S. J. Oh, D.-K. Ko, D. K. Kim, B. T. Diroll et al., Bandlike transport in strongly coupled and doped quantum dot solids: A route to high-performance thin-film electronics, *Nano Letters*, 12, 2631–2638, 2012.

44. T. Sekitani, U. Zschieschang, H. Klauk, and T. Someya, Flexible organic transistors and circuits with extreme bending stability, *Nature Materials*, 9, 1015–1022, 2010.

Index

A

Ag target ablation in water
  with different angles of incidence
    coupling of incident photons, 387
    EFs, 385–387
    FESEM and AFM imaging techniques, 381–382, 384–385
    Raman spectra of ANTA, 386, 388
    Raman spectra of R6G, 384, 386–387
    Raman spectra of R6G/ANTA, 383–384
  with different fluences and scanning conditions
    ablation mechanism, 376–377
    EDX spectra, 378
    FESEM images, 378–379
    HRTEM images, 378
    micro-Raman spectra of adsorbed RDX molecules, 381, 383
    Raman spectra of R6G molecules, 379–381
    SAED patterns, 378
    surface morphologies, 381–382
  with ps/fs Bessel beams
    FESEM images, 389–391
    Gaussian beam focusing experiments, 389
    laser parameters, 388
    LSPR peak position, 389
    nonlinear propagation, 389
Alloyed semiconducting QDs, 296–297
Amplified spontaneous emission (ASE), 115
Anodized alumina films (AAFs), 347
Antibacterial activity, Cu, Ag colloids, 409–410
Au-functionalized ZnO sensor
  chemical sensor, 79
  depletion layer, 81
  dynamic response and recovery, 79–81
  sensing mechanism, 81
  sensitivity, 78
  sensor measurements, 78
Auger spectroscopy
  advantages, 158
  Ag $M_4N_{45}N_{45}$ Auger spectra and photoelectron spectra, 159–160
  Auger chemical shift, 158
  binding energies, 159–160
  deconvoluted photoelectron spectra, 159, 161–162
  modified Auger parameter, 159
  photoionization process, 158
  thermal annealing, 162
  XPS measurements, 158–159

B

Bulk-structure lasers, 96
Buried InAs/InP quantum dots
  carrier-induced refractive index, 104–105
  carrier scattering processes, 103–104
  complex optical susceptibility, 105
  electronic structure, 101–102
  gain, 104–106
  line-width enhancement factor, 106–107
  macroscopic and microscopic polarization, 105
  phenomenological gain compression factor, 106

C

CdS-sensitized ZnO nanorods
  CdS nanoparticle-decorated ZnO nanorods and P3HT polymer blending, 63
  current density *vs.* voltage characteristics, 64
  energy band diagram, 66
  EQE, 65–66
  FESEM image, 64
  PEDOT:PSS layer, 64
Compound semiconductor nanocrystals
  CdSe nanocrystals
    hydrothermal methods, 3
    MPA-capped CdSe nanocrystals synthesis, 3
    TEM, 3–4
    XRD, 3–5
  CdS nanocrystals
    aqueous-phase synthesis, 4–5
    colloid synthesis technique, 5
    microemulsion by ultrasonic irradiation, 5
    TEM, 5–6
    wet chemical precipitation method, 5
    XRD, 6–7
  chemical methods, 2
  CuCl nanocrystals
    Bridgman method, 12
    chemical methods, 12
    Czochralski method, 12
    hydrothermal methods, 12
    SEM characterization, 12–13
    XRD, 14
  GaAs nanocrystals
    AFM, 8, 10
    electrochemical techniques, 7
    laser ablation, 7–8
    TEM, 8–9
    wet process, 8
    XRD, 8–9
  III–V compound, 7
  InAs nanocrystals
    AFM, 11–12
    chemical synthesis, 9
    deposition, 9–10
    epitaxial method, 10
    SEM, 10–11
    surface passivation, 9
    TEM, 10–11
    XRD, 11, 13
  potential applications, 2
  TOP/TOPO/HPA method, 2–3
  TOP/TOPO method, 2

Printed and bound by CPI Group (UK) Ltd, Croydon, CR0 4YY

01/11/2024

01782604-0012